Practical Electrical Engineering

Sergey N. Makarov • Reinhold Ludwig
Stephen J. Bitar

Practical Electrical Engineering

Second Edition

Springer

Sergey N. Makarov
ECE Department
Worcester Polytechnic Institute
Worcester, WA, USA

Reinhold Ludwig
ECE Department
Worcester Polytechnic Institute
Worcester, MA, USA

Stephen J. Bitar
Worcester Polytechnic Institute
Worcester, MA, USA

ISBN 978-3-319-96691-5 ISBN 978-3-319-96692-2 (eBook)
https://doi.org/10.1007/978-3-319-96692-2

Library of Congress Control Number: 2018952468

This Springer imprint is published by the registered company Springer Nature Switzerland AG
The registered company address is: Gewerbestrasse 11, 6330 Cham, Switzerland

To Antonina, Margot, and Juliette

Contents

Contents

Chapter 1: From Physics to Electric Circuits

Overview

Prerequisites:
- Knowledge of university physics: electricity and magnetism

Objectives of Section 1.1:
- Show that the electric voltage and the electric potential may be treated as two equivalent quantities
- Define the electric voltage—work per unit charge—in the form of a line integral and show its independence on the integration path for conservative fields
- Relate voltage to the potential energy of the electric field
- Introduce three-dimensional potential distributions and realize the guiding function of metal wires
- Formulate and understand major conditions of electrostatics of conductors
- Visualize surface charge distributions in the electrostatic case

Objectives of Section 1.2:
- Introduce electric current density as a function of the applied electric field
- Visualize steady-state current flow in a single conductor along with the associated electric potential/voltage distribution
- Visualize electric and magnetic-field distributions for a two-wire DC transmission line
- Obtain initial exposure to the Poynting vector
- Realize that electric power is transferred via Poynting vector even in DC circuits
- Indicate a path toward circuit problems where the field effects become important

Objectives of Section 1.3:
- Review basic hydraulic (fluid mechanics) analogies for DC circuit elements
- Present major hydraulic analogies for dynamic circuit elements in AC circuits
- Briefly discuss hydraulic analogies for semiconductor components

Application Examples:
- Human body subject to applied voltage
- Human body in an external electric field

S. N. Makarov et al., *Practical Electrical Engineering*, https://doi.org/10.1007/978-3-319-96692-2_1

Keywords:
Electricity, Electric field intensity, Electric field, Electric field magnitude, Lines of force, Electric potential, Electric voltage, Line integral, Contour integral, Conservative field, Potential energy of the electric field, Voltage drop, Voltage difference, Ground reference, Neutral conductor, Common conductor, Voltage versus ground, Equipotential lines, Volumetric charge density, Surface charge density, Gauss' theorem, Equipotential surface, Self-capacitance, Electrostatic discharge, Effect of electrostatic discharge on integrated circuits, Boundary element method, Electric current density, Material conductivity, Transmission line, Direct current (DC), Electric load, Ideal wire, Kirchhoff's voltage law (KVL), Magnetic field, Magnetic-field intensity, Ampere's law, Cross (vector) product, Poynting vector, Poynting theorem, Wireless communications, Wireless power transfer, Fluid mechanics analogy of an electric circuit, Hydraulic analogy of an electric circuit, Voltage source (hydraulic analogy), Resistance (hydraulic analogy), Current source (hydraulic analogy), Capacitance (hydraulic analogy), Inductance (hydraulic analogy), Electric transformer (hydraulic analogy), NMOS transistor (hydraulic analogy), Bipolar junction transistor (hydraulic analogy)

Section 1.1 Electrostatics of Conductors

This introductory chapter is *optional* in the sense that the reader does not need its content as a prerequisite for the subsequent chapters. The aim of this chapter is to illustrate why electric circuits trace their origin to electromagnetic fields. The chapter highlights several field concepts which form the theoretical foundation of electric circuits. At the same time it makes clear why, for the majority of electric circuits, the electric and magnetic fields are often ignored without affecting the final results. When this is the case, the electric circuits and components follow useful and simple hydraulic analogies discussed below.

1.1.1 Charges, Coulomb Force, and Electric Field

Electric Charges

The smallest electric charge is known as the elemental charge of an electron, $q = -1.60218 \times 10^{-19}\text{C}$ (*coulombs*). In electrical engineering, we deal with much larger charges, which, for this reason, are assumed to be infinitely divisible. There are no positive movable charges in metal conductors. Therefore, when we talk about positive charges, it is implied that we have a *lack of electrons* at a certain location, e.g., at the surface. Oppositely, the negative charge is the *excess of electrons* at a certain location.

Definition of the Electric Field

Electrostatics plays a key role in explaining the operations of electric capacitors and all semiconductor devices. The word *electricity* is derived from the Greek word for amber. Probably Thales of Miletus was the first who discovered, about 600 B.C., that amber, when rubbed, attracts light objects. An electrostatic force acting on a charge q is known as the *Coulomb force*. This Coulomb force is a vector; it is measured in *newtons* (or N)

$$\vec{F} = q\vec{E} \ \ [\text{N}] \tag{1.1}$$

Equation (1.1) is the definition of the *electric field intensity* vector, $\vec{E}$, often called the *electric field*. This electric field is created by other (remote or nearby) charges. In the general case, the electric field exists both in free space and within material objects, whether conductors or dielectrics. The electric field $\vec{E}$ is measured in *volts* (V) *per meter* (V/m). The *field magnitude*, $E = \sqrt{E_x^2 + E_y^2 + E_z^2} = |\vec{E}|$, has the same units. From Eq. (1.1),

$$1\text{V} = \frac{1 \ \text{N} \times 1\text{m}}{1 \ \text{C}} = \frac{1 \ \text{J}}{1 \ \text{C}} \tag{1.2}$$

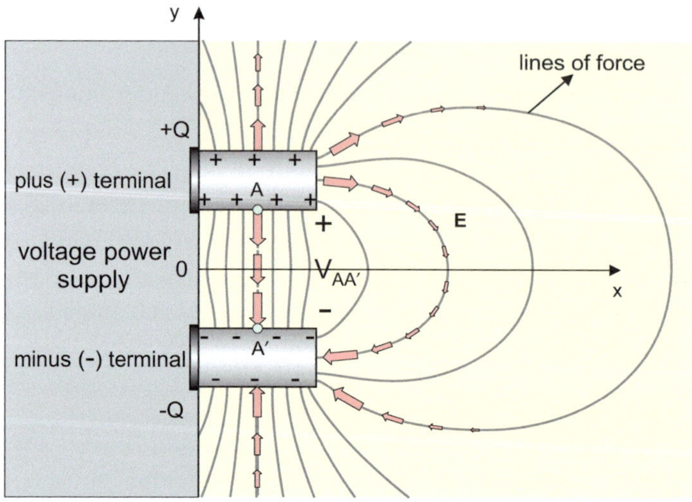

Fig. 1.1. Electric field emanating from a voltage supply with open-circuit output terminals.

Electric Field of a Laboratory Power Source

As an example, we consider an electric field generated by a laboratory voltage source turned on. Figure 1.1 shows the realistic electric field distribution modeled numerically. The power supply has two output terminals shaped as two metal cylinders. The plastic cover is ignored. Nothing is connected to the terminals yet. Still, the power supply already performs a "charge separation": the plus (+) terminal is charged positively (total charge is $+Q$), whereas the minus (−) terminal is charged negatively (total charge is $-Q$). Those charges are schematically shown in Fig. 1.1. As a result, an electric field is created. The electric field in Fig. 1.1 is directed along particular lines, which we call *lines of force*. This electric field surrounds the power supply terminals. Every line of force starts at the positive terminal and ends at the negative terminal. The strength of the electric field everywhere in space is linearly proportional to the *supply voltage* as studied next. However, the field shape always remains exactly the same.

1.1.2 Electric Potential and Electric Voltage

The *electric potential* φ measured in *volts* (V) and *electric voltage* V measured in *volts* (V) are two *identical* quantities once they refer to the same *observation point* A and to the same *reference point* A'. Both terms may be used; the electric potential is frequently denoted by V. The potential is more common in physics. Work is done against the electric forces when a charge is moved in an electric field. The *electric potential* or *electric voltage* $V_{AA'}$ between points A (#1) and A' (#2) is work in joules per coulomb (per unit charge) to bring a positive charge from reference point A' (#2) to observation point A (#1), i.e., *against* the electric field—see Fig. 1.1. The work per unit charge over a short straight vector distance $d\vec{l}$ is

$$-\frac{1}{q}\vec{F}\cdot d\vec{l}=-\vec{E}\cdot d\vec{l}=-Edl\cos\theta \tag{1.3}$$

where θ is the angle between $\vec{E}$ and $d\vec{l}$, $E=\left|\vec{E}\right|$, and $dl=\left|d\vec{l}\right|$. The total work or $V_{AA'}$ is the sum of all such small contributions conventionally written in the form of an integral

$$V_{AA'}=-\int_{A'}^{A}\vec{E}\cdot d\vec{l}=\int_{A}^{A'}\vec{E}\cdot d\vec{l} \tag{1.4}$$

The integral in Eq. (1.4) is a *line integral*, also called a *contour integral*. In the general case, it is evaluated along a curve connecting points A and A'. In the particular case of Fig. 1.1, this curve is just a straight line.

Exercise 1.1: Assume for simplicity that the electric field along the line of force from A to A' in Fig. 1.1 is strictly uniform and has the magnitude of 50 V/m. The line length is 0.02 m. Find voltage (or potential) $V_{AA'}$.

Answer: $V_{AA'}=-\int_{A'}^{A}\vec{E}\cdot d\vec{l}=-\int_{0}^{0.02\text{m}}50\times\cos\pi\times dl=50\frac{\text{V}}{\text{m}}\times 0.02\ \text{m}=1\ \text{V}.$

The electrostatic field (and any slowly varying electric field) is called a *conservative field*. There are two equivalent definitions of a conservative field:

1. Electric voltage or electric potential $V_{AA'}$ is *path independent*; it only depends on the position of A and A', but not on the shape of the curve between A and A'.
2. The line integral in Eq. (1.4) over any closed contour is zero.

The equivalence of these definitions is proved by treating two different integration contours between A and A' as two parts of one closed contour. The independence of the integration path suggests that the voltage is equal to the *potential energy of a unit charge* in the electric field. Strictly speaking, it is the change in the potential energy.

1.1.3 Electric Voltage Versus Ground

The voltage between two arbitrary points $V_{AA'}$ (e.g., between two terminals of a resistor) is a convenient measure when analyzing electric circuits with discrete circuit components. In this case, it is called the *voltage drop* (or *voltage difference*) across a circuit element. At the same time, it is equally convenient to define the "global" or absolute voltage between an arbitrary point in space A and some *fixed* point A', which is assigned the voltage value *zero*. This fixed point (or the set of points) is called the *ground reference*. In general, the ground reference may be chosen arbitrarily. In physics of localized conductors, it is customary to choose the ground at infinity. In electrical engineering, the ground reference is either the physical (earth) ground or some *neutral (common) conductor* assigned to zero volts. Thus, the *absolute voltage versus ground* denoted by $V(\vec{r})$ is still defined by Eq. (1.4) where point A is now characterized by the position vector $\vec{r}$. By definition, it becomes zero when $\vec{r}$ approaches ground, i.e., A'. The equivalent representation of Eq. (1.4) for conservative fields may be shown to be

$$\vec{E}(\vec{r}) = -\mathrm{grad}V(\vec{r}) = -\nabla V(\vec{r}) \tag{1.5}$$

Thus, the electrostatic field is expressed as the gradient of the electric potential or of the (absolute) electric voltage everywhere in space. In other words, it means the electric field does not have closed loops, but starts and ends at the charges. Equation (1.5) is of significant value since it replaces a complicated vector $\vec{E}$ by the single scalar voltage V.

Exercise 1.2: Electric voltage/potential with respect to ground is given in Cartesian coordinates by $V(\vec{r}) = -y$ [V]. Determine the electric field everywhere in space.

Answer: $E_y = 1\mathrm{V/m}$, $E_x = E_z = 0$.

As an example, we choose the x-axis in Fig. 1.1 as the ground reference. The positive supply terminal is chosen to have a voltage equal to +0.5 V versus ground, and the negative terminal is assigned a voltage equal to −0.5 V versus ground. Those values will uniquely determine charges $\pm Q$ in Fig. 1.1. The function $V(\vec{r})$ is now plotted using the lines of equal potential, or *equipotential lines*. The result is shown in Fig. 1.2. It will be proved next that the surface of any metal (or other) conductor in electrostatics is an *equipotential surface*. All points on this surface have the *same* value of the electric potential: +0.5 V for the plus terminal and −0.5 V for the minus terminal in Fig. 1.2. Using Eq. (1.5) it can be verified that the equipotential lines and the lines of force are always *perpendicular* to each other; you can see this in Fig. 1.2.

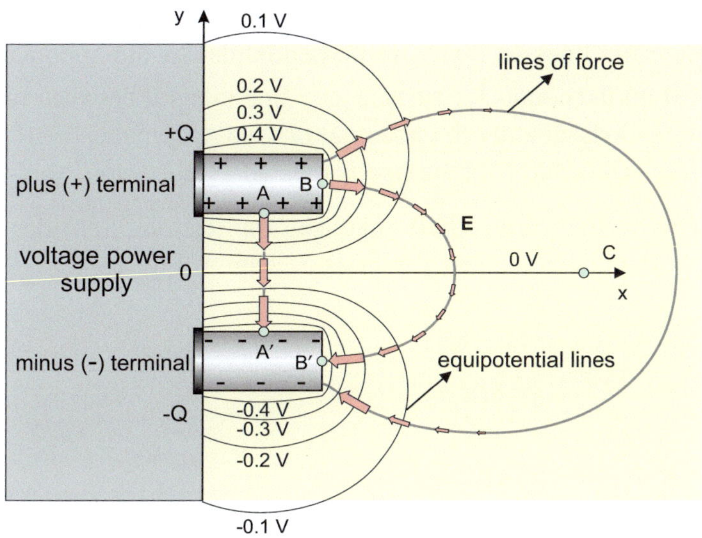

Fig. 1.2. Electric field and electric potential (electric voltage) of a 1-V voltage supply. Equipotential lines are thin solid curves, while the lines of force are the thicker curves.

Exercise 1.3: In Fig. 1.2, determine voltage differences: $V_{BB'}$, $V_{A'A}$, V_{AB}, V_{AC}, $V_{B'C}$.

Answer: $V_{BB'} = 1$ V, $V_{A'A} = -1$ V, $V_{AB} = 0$ V, $V_{AC} = 0.5$ V, $V_{B'C} = -0.5$ V.

1.1.4 Equipotential Conductors

Consider a *conductor*, which is characterized by a sufficient number of free charges. The conductor is subject to an applied electric voltage, or to an applied electric field, or to those effects combined. The *charge density in the conductor* is the *difference* between the concentration of ions (positive charges) and electrons (negative charges) multiplied by the charge of an electron. The charge density can be either *volumetric*, measured in C/m^3, or *surface*, measured in C/m^2. The following is true when dealing with electrostatics:

1. The electric field everywhere *within* the conductor is *zero*, $\vec{E} = \vec{0}$. Otherwise, the Coulomb force given by Eq. (1.1) will act on free charges and cause permanent charge motion (current flow), which is impossible.
2. The volumetric charge density everywhere within the conductor is *zero*. If this were not true, then according to *Gauss' theorem*, there would be a nonzero electric field within the conductor, which is impossible based on statement #1.
3. The surface charge density is *not* zero. In fact, the surface charge is distributed so as to assure that statement #1 is satisfied.
4. For the electric field given by Eq. (1.5), the *tangential component* of the electric field $\vec{E}_t$ is *continuous* across an interface. Since $\vec{E}_t = \vec{0}$ inside the conductor, $\vec{E}_t$

must be zero over the entire surface of the conductor too. This is seen in Figs. 1.1 and 1.2 where the lines of force are perpendicular to the conductor surface.

5. Since $\vec{E}_t = \vec{0}$ on the conductor surface, any line integral between two points on this surface is *zero*. Consequently the potential or voltage remains the *same* for any point on this surface. The conductor surface is thus an *equipotential surface*.

These statements have an immediate practical application. Consider two conductors (wires) attached to the power supply terminals as seen in Fig. 1.3.

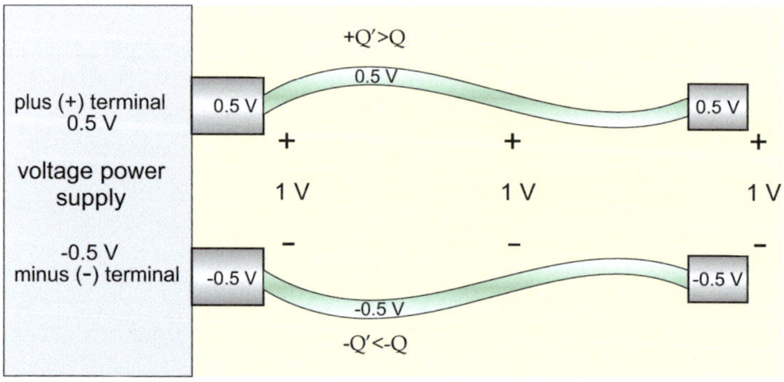

Fig. 1.3. Voltage source from Fig. 1.2 with two wires connected. There is still no current flow.

Everywhere along the upper wire, the voltage is +0.5 V with respect to ground. Furthermore, along the lower wire, the voltage is −0.5 V with respect to ground. Everywhere in space the voltage difference between the two wires is therefore 1 V. The wires may be extremely long. The charge $\pm Q'$ required to maintain the corresponding voltage difference of 1 V increases when the combined area of the metal conductors increases. The conducting wires thus "guide" the electric field to a remote point. Without the attached wires, the field would be spread out in space as seen in Figs. 1.1 and 1.2.

Exercise 1.4: In Fig. 1.3, two wires happen to be very close to each other at a certain location; they are separated by 1 mm. What is approximately the electric field strength at this location?

Answer: On the order of 1000 V/m, the wire isolation has little influence.

Exercise 1.5: In Fig. 1.3, a human body is in contact with the upper wire. What is the body's voltage versus ground set at 0 V?

Answer: Since the human body is still a conductor, its voltage is +0.5 V.

1.1.5 Use of Coulomb's Law to Solve Electrostatic Problems

The theory of electrostatics solely relies upon the distribution of surface charges, since no other charges exist. An important case is a conductor subject to voltage V applied from one terminal of the voltage source; the connection position does not matter. The other terminal is usually located very far away, ideally at infinity; it is customarily assigned a value of 0 V. As a result, the conductor acquires an extra positive charge Q if V is positive. The quantities of interest are the value of Q itself and the resulting surface charge distribution. The ratio Q/V is the *self-capacitance* of the conductor. Consider a complicated conductor—a human body. The idea of the solution is simple and elegant. The entire body surface is divided into many *small* elements with a *constant* charge density each. We assign an unknown charge q_i to every such element with number i $(i = 1, \ldots, N)$ and center position $\vec{r}_i$. Charge q_i is very similar to the *point charge*. It follows from *Coulomb's law* that it generates the electric potential in space given by

$$V(\vec{r}) = \frac{q_i}{4\pi\varepsilon_0|\vec{r} - \vec{r}_i|} \tag{1.6}$$

Here, $\varepsilon_0 = 8.85419 \times 10^{-12}$ F/m is the *electric permittivity* of air. The net voltage of the jth element is the sum of all such voltage contributions, i.e.,

$$V(\vec{r}_j) = \sum_{i=1}^{N} \frac{q_i}{4\pi\varepsilon_0|\vec{r}_j - \vec{r}_i|} = 1 \text{ V}, \quad j = 1, .., N \tag{1.7}$$

Equations (1.7) forms a system of N equations for N unknown charges. The diagonal terms need a special treatment. By solving this algebraic system of equations using linear algebra, we obtain the unknown charges; their sum is the net charge Q. Figure 1.4 shows the surface charge distribution found this way for two human subjects. Emphasize that the charge (and the strongest electric field) concentrates at the *sharpest parts* of the body: elbows, hands, feet, and the head. The total excess body charge Q in both cases is approximately 50×10^{-12} C. This is a very small charge; the same charge is stored in a 50 pF capacitor at 1 V. The method of this example is widely used in electrostatic simulations including modeling *electrostatic discharge* and its effect on *integrated circuits*.

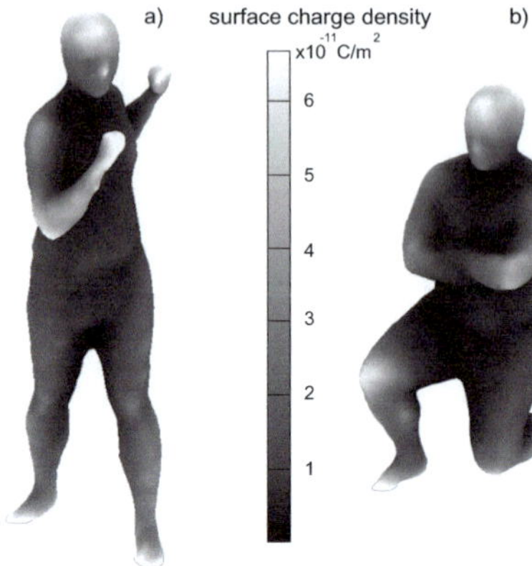

Fig. 1.4. Surface charge distribution over the human body based on an applied voltage of 1 V. Both subjects are ECE graduate students.

A human beneath the power line is subject to an electric field. A similar solution applies given that the body surface is still the equipotential surface. Figure 1.5 shows the surface charge distribution for two human subjects in a vertical electric field of 1 V/m. The negative charges concentrate close to the head, whereas the positive charges concentrate in the lower body. Figure 1.5 also shows the electric potential distribution around the body. Dense equipotential lines mean a high local electric field. The local field may exceed the external field by a factor of 10 or more. All results are linearly scaled with the applied electric field.

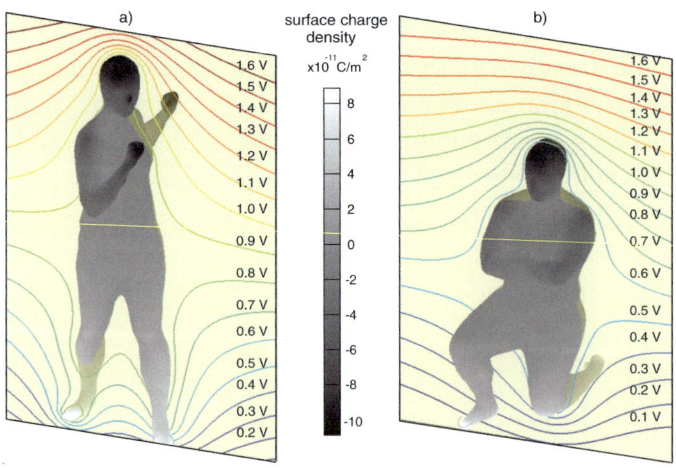

Fig. 1.5. Human body subject to an applied electric field: surface charge and potential distribution.

Section 1.2 Steady-State Current Flow and Magnetostatics

1.2.1 Electric Current

An electric current in a material is the directed motion of *free positive* charges. Consequently, the electrons, which are often the only free charges in the conductor, move in the opposite direction. The *electric current density* $\vec{j}$ everywhere in the material is a vector, and it is measured in amperes per meter square, i.e., A/m^2. The *total current* I through a conductor with the uniform current density is the product of the current density magnitude and the conductor cross section A; I is measured in amperes. Circuits with lumped components deal with the total currents only. The current density (and the total current) is directly proportional to the applied electric field,

$$\vec{j}(\vec{r}) = \sigma \vec{E}(\vec{r}) \tag{1.8a}$$

at any point of interest $\vec{r}$. Here, σ is the *material conductivity* with units of siemens/m, i.e., S/m. Note that 1 S = 1/Ω. If there is no electric field, there is no electric current in the material and vice versa. In metals, the conductivity σ is very high. Therefore, even a *vanishingly small* electric field inside a metal conductor creates a *large* electric current.

Exercise 1.6: An AWG #00 (American Wire Gauge) aluminum wire has the conductivity of 4.0×10^7 S/m and the diameter of 9.266 mm. Determine the total current in the wire when the electric field inside the wire is 0.01 V/m. This value is more than 10,000 times less than the field between the terminals of a 5-V laboratory supply separated by 2 cm.

Answer: 27 A.

For steady-state current flow, an electrostatic potential exists; it is given by Eq. (1.5). Therefore, the current density can be expressed as the gradient of the potential,

$$\vec{j}(\vec{r}) = -\sigma \nabla V(\vec{r}) \tag{1.8b}$$

1.2.2 Difference Between Current Flow Model and Electrostatics

In electrostatics, there are no charges and there is no electric field within conductors; only the surface charges exist. The steady-state current problem is quite different. The volumetric uncompensated charge density still does not exist within the conductor, but the electric field does. Along with this, the surface charge density is always present, similar to electrostatics. Physically, the difference arises from the different boundary conditions. The conductor surface is divided into two distinct parts: the surface of electrodes, S_e, where the voltage (or current) is applied, and the rest of conductor surface, S_c, which is in contact with air. The following is true for the steady-state current flow:

1. Everywhere on the surface S_c, the current density component perpendicular to the surface is zero, i.e., $\vec{j} \cdot \vec{n} = 0$, where $\vec{n}$ is the unit normal vector to the surface, and dot denotes the scalar product of two vectors. In other words, no current can flow from the conductor into air.
2. On the surface of electrodes S_e, the voltage is given: for example, +0.5 V on the left electrode and −0.5 V on the right electrode. Alternatively, the inflowing current, $\vec{j} \cdot \vec{n}$, may be given.

How Does the Conductor "Guide" the Electric Field?

We consider the current flow in a conducting cylinder with two circular electrodes shown in Fig. 1.6b. The electrostatic counterpart of the problem is given by the same coaxial electrode pair in air, see Fig. 1.6a. The electrodes have the radius a; they are separated by $25a$. The electrode voltages are ±0.5 V. The exact value of the cylinder conductivity does not matter; the same results will be obtained. The electrostatic problem and the steady-state current problem are both solved as described in the previous section. Results of both solutions are given in Fig. 1.6 where the equipotential lines and the electric field vectors are plotted. Some general observations from this figure are worthy of note:

1. The current-carrying conductor "guides" the electric field as shown in Fig. 1.6b which, otherwise, would be spread out in space; see Fig. 1.6a.
2. In the long conductor, the electric field and the electric current are both directed along the conductor axis; they are *uniform* across any conductor cross section, which is simultaneously an *equipotential surface*. In other words, current flow in the long conductor is *one dimensional*, like water flow in a pipe. This is also true if the conductor is bent or has a noncircular cross section
3. The voltage decreases linearly along the conductor from the most positive to the most negative value. The voltage drop per unit length is constant; it is only a function of the applied voltage.

It is seen from Fig. 1.6b that within in a current-carrying conductor of length l:

$$E = \frac{V}{l} \tag{1.9}$$

where E is the magnitude of the electric field (its direction is along the conductor axis), and V is the voltage across the conductor (1 V in the present case). Equation (1.9) is a simplified version of Eq. (1.4) for uniform fields. In many textbooks, it is used to derive Ohm's law. Note that the electric field in Fig. 1.6b is not continuous across the conductor-air interface. A component of the electric field perpendicular to the conductor boundary suddenly appears. This component is due to the surface charges on the conductor–air interface (not shown in the figure).

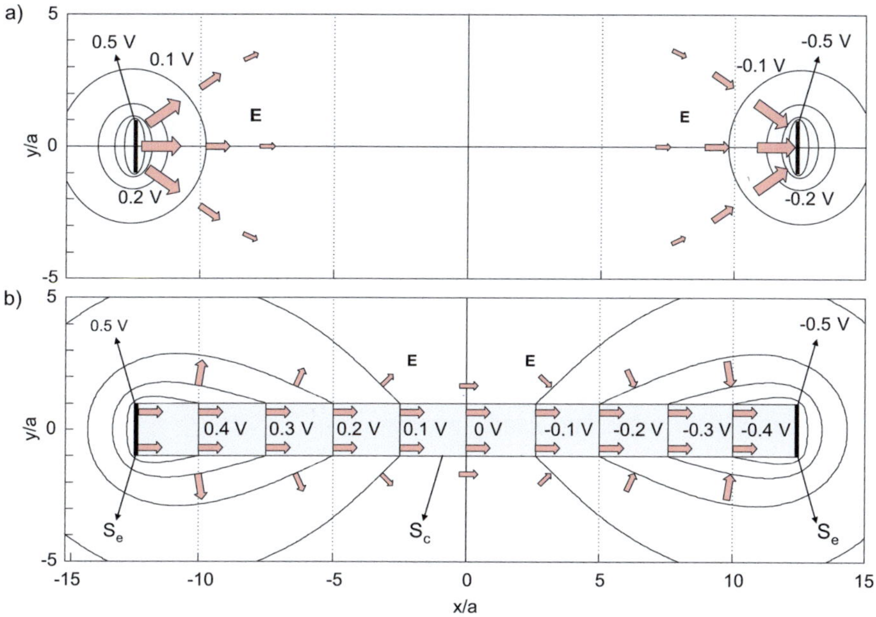

Fig. 1.6. (**a**) Two electrodes subject to ± 0.5 V in free space; (**b**) the same electrodes but with a conducting cylinder between them. Equipotential lines and electric field vectors are shown.

Exercise 1.7: A common AWG #22 copper wire is used in the laboratory to form a coil with the radius of 0.1 m and 100 turns. This coil is subject to an applied voltage of 0.5 V. Determine the total current in the wire if its diameter is 0.64516 mm; the copper conductivity is 5.8×10^7 S/m. *Hint:* find the electric field in the wire first.

Answer: 0.1509 A.

1.2.3 Physical Model of an Electric Circuit

Thus far, we have considered the current flow in only one wire. However, a simple *electric circuit* uses two wires and it includes at least three elements:

1. A voltage power supply, which in steady-state (*direct current* or *DC*) case generates a constant voltage between its terminals.
2. An *electric load* that consumes electric power. The load may be modeled as a resistant material of a much smaller conductivity.
3. *Two wires*, which extend from the source to the load. Those wires form a *transmission line*. In laboratory settings, both wires may be arbitrarily bent.

Figure 1.7 shows a physical model of the simple circuit. We study its electric part first. In Fig. 1.7a, the highest electric field magnitude is observed exactly between the two conducting wires, at the line connecting its centers. At the same time, the electric field in the wires is usually very small. Still, sufficient current flows there due to the high material conductivity; see Eq. (1.8a). However, the electric field in the load cylinder is *not* small. This is indicated by denser equipotential surfaces. In Fig. 1.7a, the equipotential surfaces are all separated by 0.05 V. There is a net voltage drop of 0.3 V along each wire and the voltage drop of 0.4 V across the load, resulting in a total potential drop of 1 V. This equality is KVL (*Kirchhoff's voltage law*). Ideally, when the conductivity of the wires is infinitely high, the *entire* source voltage appears across the load cylinder. The field in the wires becomes vanishingly small, but enough current still flows. The wire of infinite conductivity, or the *ideal wire*, is a useful abstraction.

Exercise 1.8: If the voltage drop along each wire in Fig. 1.7a were 0.01 V, what would be the value of the voltage across the load?

Answer: 0.98 V.

Exercise 1.9: In contrast to Fig. 1.7b, the field within the wire in Fig. 1.6b is not small, regardless of its conductivity. Why is it so?

Answer: There is no load in Fig. 1.6b; the entire voltage drop is purposely forced to occur across the wire.

1.2.4 Magnetostatics and Ampere's Law

We next study the magnetic part of the circuit in Fig. 1.7b. The same electric current I flows in the entire circuit. The current in one of the conductors in Fig. 1.7b creates the *magnetic field* (*magnetic-field intensity*) $\vec{H}$ with units of A/m around the conductor. If the conductors are considered to be sufficiently long, the magnitude (absolute value) of the field $\vec{H}(\vec{r})$ anywhere in space, except within the conductor, is given by

$$H(\vec{r}) = \frac{I}{2\pi|\vec{r}|} \tag{1.10}$$

Equation (1.10) is a particular form of *Ampere's law* for an infinite straight wire of the total current I. The vector $\vec{H}$ forms concentric circles around the wire; its direction follows the right-hand rule. When two wires are present, as in Fig. 1.7b, the resulting combined magnetic field is the vector sum of two solutions given by Eq. (1.10) for two conductors

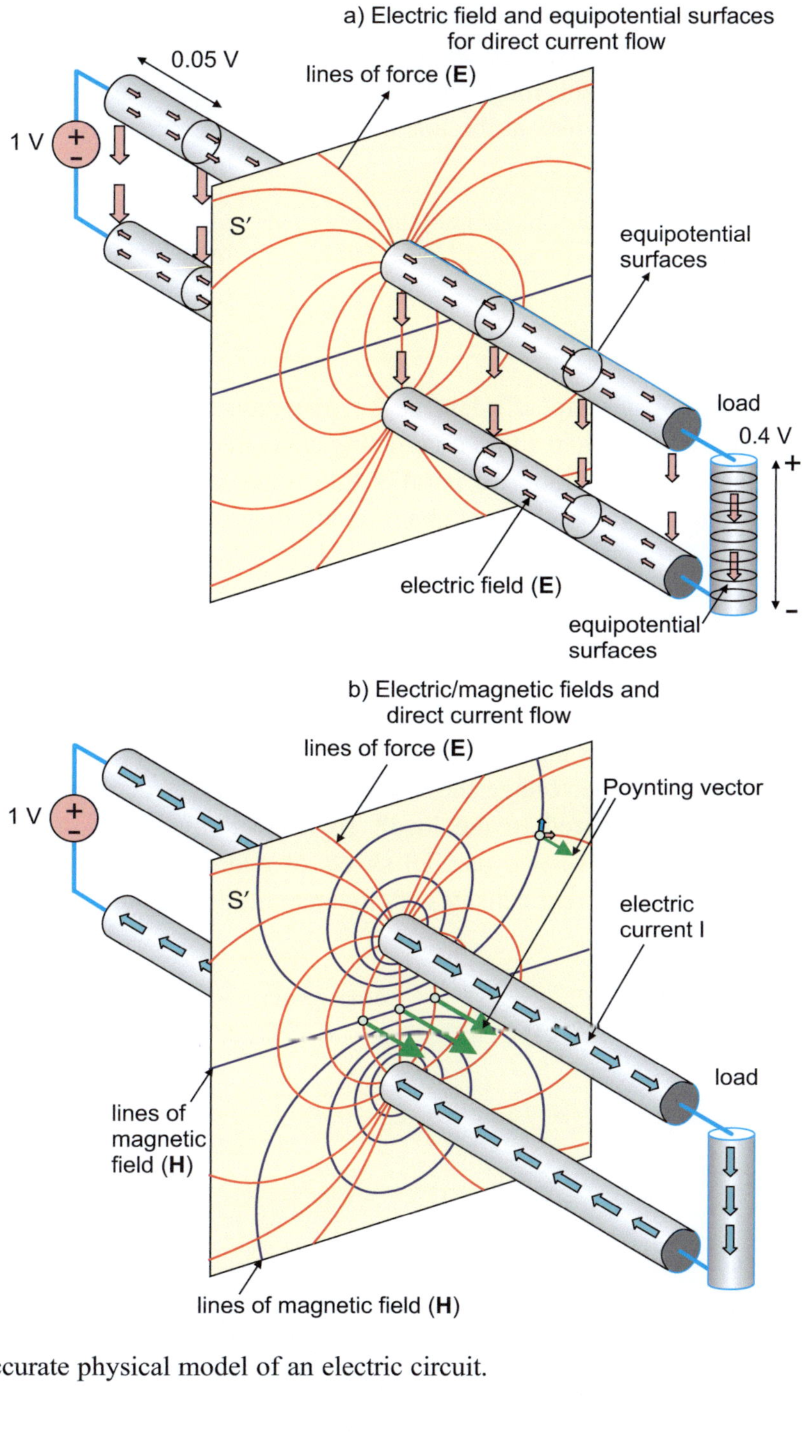

Fig. 1.7. Accurate physical model of an electric circuit.

having the opposite current directions, respectively. The lines of the combined magnetic field are shown in Fig. 1.7b. They are always *perpendicular* to the lines of force for the electric field. The magnetic-field magnitude in Fig. 1.7b also has its maximum exactly between the two conducting wires, at the line connecting its centers.

Exercise 1.10: Determine the magnetic-field magnitude in the middle between two parallel long wires carrying current of ± 1 A each and separated by 2 cm.

Answer: $H(\vec{r}) = 31.8$ A/m.

Another useful form of Ampere's law is the magnetic field of an infinite "current sheet," i.e., when current flows in a thin conducting sheet in one direction. The sheet may be thought of as an infinite number of parallel thin wires carrying the same current. If j [A/m] is the current density per unit of sheet width, then the resulting constant field is

$$H = \frac{1}{2} j = \text{const} \tag{1.11}$$

1.2.5 Origin of Electric Power Transfer

We know from physics classes that electric power P delivered to the load is given by the product $P = VI$ where V is the voltage across the load and I is the current through it. How exactly is this power transferred to the load? To answer this question, we assume a transmission line in the form of two parallel sheets of width W and spacing l shown in Fig. 1.8. Each sheet carries current density with the magnitude j per unit of sheet length.

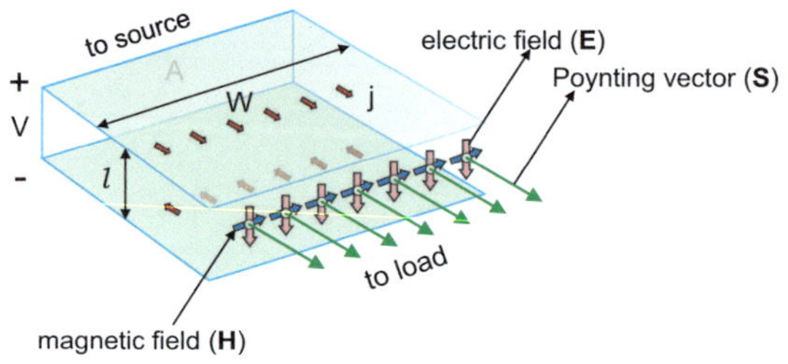

Fig. 1.8. Transmission line in the form of two current sheets.

When the material conductivity is infinite, the voltage between the two sheets is the load voltage V. For $l/W << 1$, Eq. (1.9) yields the electric field within the transmission line, $E = V/l$. The magnetic field is found using Eq. (1.11). The result is $H = j$ since both sheets contribute to the field within the line. Next, we define a vector

$$\vec{S} = \vec{E} \times \vec{H} \tag{1.12}$$

where the symbol $\times$ stands for the *cross product* or *vector product*. The vector $\vec{S}$ is called the *Poynting vector*; it is shown in Fig. 1.8. The *units for the Poynting vector* are given by power per unit area, i.e., $1\mathrm{V/m} \times 1\ \mathrm{A/m} = 1\ \mathrm{W/m^2}$. Thus, the Poynting vector characterizes the *directional energy flux density* in space. Its magnitude in Fig. 1.8 is given by $S = EH = Vj/l = VI/(lW)$ where $I = jW$ is the net current in every conductor. Multiplying the Poynting vector by the area $A = lW$ where the fields are concentrated, we obtain the remarkable result,

$$AS = VI = P \tag{1.13}$$

In other words, the power is transferred by the fields. This result is perhaps less important for wired circuits where the fields are directly linked to charges and currents. However, it is important for the transition from wired to wireless circuits. At a sufficiently high frequency, significant electric and magnetic fields will be radiated by an antenna into *empty space*. These fields will carry power flux density given by Eq. (1.12).

Exercise 1.11: Two conductors extending from the source to the load are parallel 1-cm-wide perfectly conducting sheets separated by 1 mm. The (vertical) electric field between the plates is 100 V/m; the (horizontal) magnetic field is 100 A/m. Determine electric power delivered to the load. Assume no field fringing.

Answer: $P = 0.1\mathrm{W}$.

Section 1.3 Hydraulic and Fluid Mechanics Analogies

1.3.1 Hydraulic Analogies in the DC Steady State

In DC and low-frequency alternating circuits, the electric and magnetic fields outside conductors may be ignored without affecting the final results. When this is the case, the electric circuits follow rather precisely useful *fluid mechanics* (or *hydraulic*) *analogies*. Major hydraulic analogies are depicted in Fig 1.9. Let us consider a water pump connected to a filter in Fig. 1.9a—left. The water pump creates a constant pressure difference p between its terminals, which forces water to move through the filter. Although the constant pressure (torque) pump is less common than a water pump of constant flux, we can use it as an analogy since it exactly corresponds to the *voltage source*, which maintains constant voltage difference V in Fig. 1.9a—right. The filter can be thought of as an electric resistance: it opposes the water flow, and a certain pressure difference, or voltage, is required to overcome this resistance. Electric current corresponds to fluid velocity times the tube cross section which constitutes the total water flux. For the entirely closed pumping system in Fig. 1.9a, the water pressure inside the system can have an arbitrary reference level p_0. This level may be quite different from the ambient atmospheric pressure. Similarly, an isolated electric circuit may have an arbitrary voltage V_0 versus ground, due to static charge accumulation.

The condition $V_0 = 0$ will be achieved by grounding the circuit. Figure 1.9b–d also specifies hydraulic analogies for the separate circuit elements. We consider steady-state flow of incompressible fluid. Emphasize that the DC voltage source in Fig. 1.9c is analogous to a *constant-torque water pump*, which creates a constant pressure difference

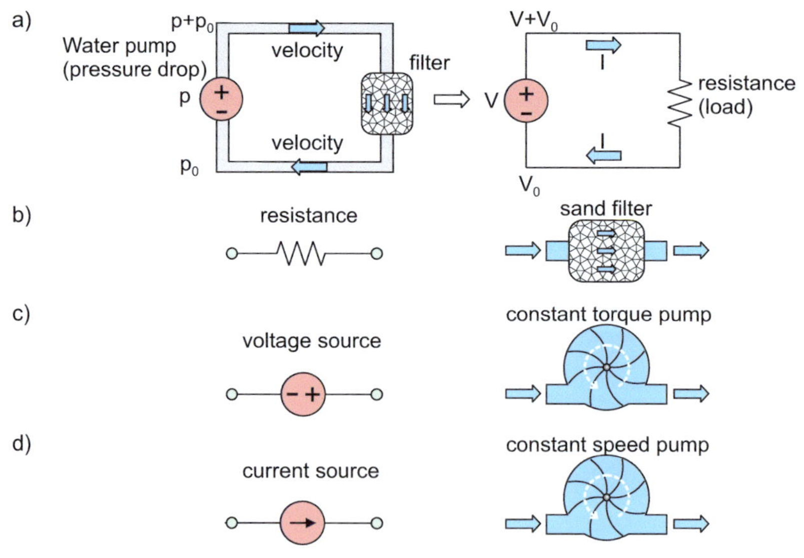

Fig. 1.9. Hydraulic analogies in the DC steady state.

between its terminals, whereas a *DC electric current source* in Fig. 1.9d is similar to a *constant-speed water pump*, which creates constant fluid flux.

1.3.2 Analogies for Alternating-Current (AC) Circuits

This case illustrated in Fig. 1.10 corresponds to alternating flow of incompressible fluid in a piping system. The *AC voltage source* corresponds to a harmonically (sinusoidally) oscillating piston in Fig. 1.10a, with a constant-torque amplitude. A *capacitance* in Fig. 1.10b is represented by a *flexible membrane*. The capacitance value, C, corresponds to the *inverse stiffness*, $1/k$, of the membrane, also called the *compliance*. When $k \to \infty$ and $C \to 0$ (a rigid membrane), the capacitance value tends to zero. The membrane becomes a solid wall, which blocks the alternating fluid flow entirely. In another limiting case ($k \to 0$ or $C \to \infty$), the membrane has no effect on the fluid flow. Intermediate cases correspond to a partial blocking. A massive wheel with a rotational inertia in Fig. 1.10c represents an *inductance*. The inductance value, L, corresponds to the *mechanical mass* m of the wheel.

When $m \to \infty$ or $L \to \infty$, the wheel does not respond to fluid oscillations and blocks the alternating fluid flow entirely. In the opposite case ($m \to 0$ or $L \to 0$), the wheel has no effect on the fluid flow. Intermediate cases correspond to a partial blocking.

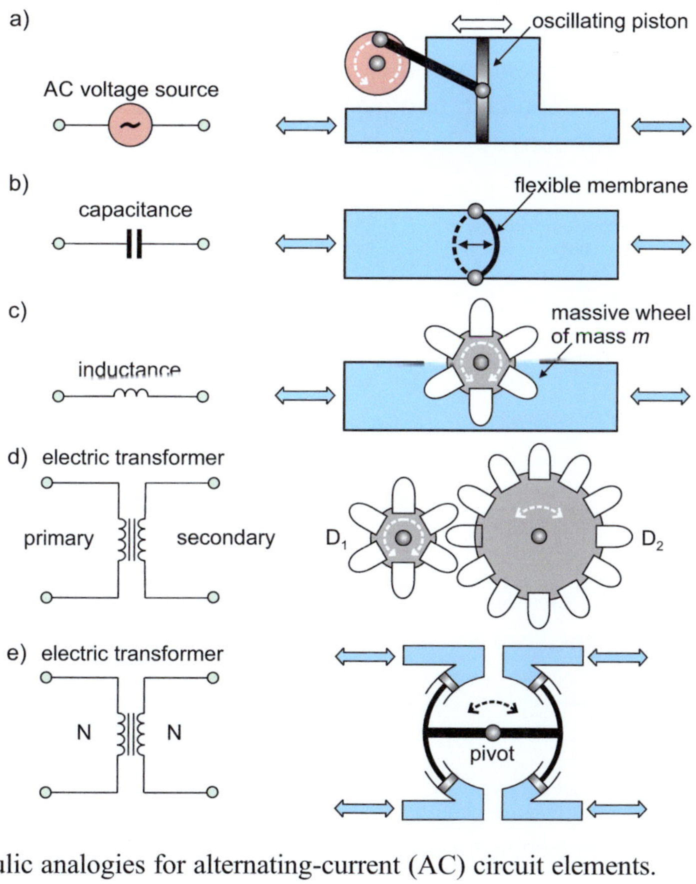

Fig. 1.10. Hydraulic analogies for alternating-current (AC) circuit elements.

An *electric transformer* shown in Fig. 1.10d operates with alternating currents. One mechanical analogy is a gear transmission or gearbox. In terms of angular speed ω [rad/s] and developed torque T $[\mathrm{N}\cdot\mathrm{m}]$, one has $T_2 = (D_2/D_1)T_1$, and $\omega_2 = (D_1/D_2)\omega_1$, where $D_{1,2}$ are pitch diameters of gear wheels. Here, torque is the voltage and speed is the current. $D_{1,2}$ are similar to the number of turns, $N_{1,2}$, of the primary and secondary coils of the transformer, respectively. This analogy ignores the field effect—magnetic coupling between the coils. Therefore, it will fail in the DC case. A more realistic transformer analogy is shown in Fig. 1.10e. The model with four pistons transforms power from one circuit to another in the AC case only. It is drawn for a 1:1 transformer. When a transformer with a turn ratio of 2:1 is required, the area of output pistons is doubled. This doubles the output current, but the output voltage (the force) will be halved.

1.3.3 Analogies for Semiconductor Circuit Components

Semiconductor circuit components are similar to *fluid valves*, which are either externally controlled or are controlled by the fluid-flow pressure itself. Figure 1.11a shows a hydraulic analogy for a *diode*. This picture highlights its major function: a one-way valve. Fig. 1.11b illustrates the operation of an n-channel *metal-oxide-semiconductor field-effect transistor*, or NMOS transistor. This transistor is a valve controlled by a third voltage terminal. For another *bipolar junction transistor* or BJT in Fig. 1.11c, not only the control voltage but also the control current is important. In other words, to keep the valve open, we must also supply a small amount of current (fluid) at the control terminal.

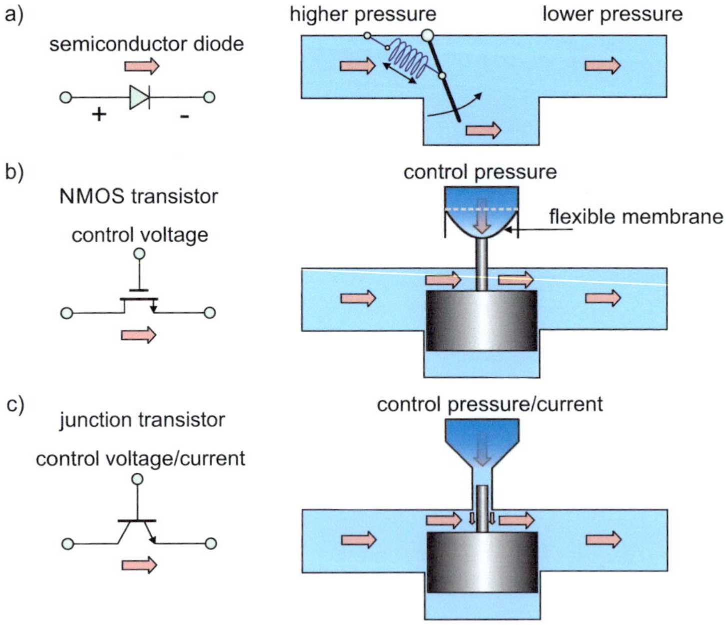

Fig. 1.11. Analogies for semiconductor circuit components.

Summary

<table>
<tr><th colspan="2">Electrostatics</th></tr>
<tr><td>Electric voltage/electric potential
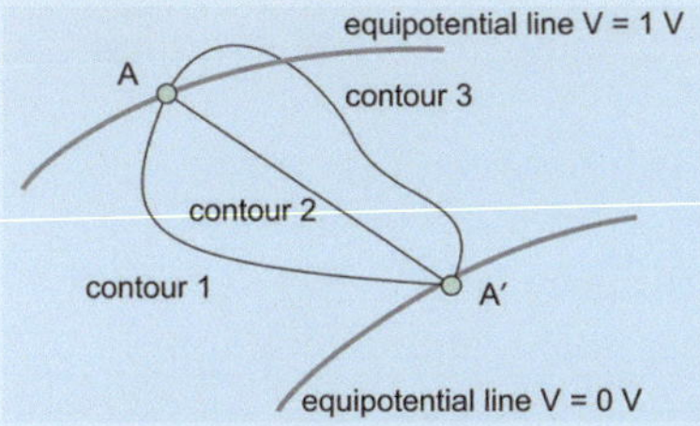</td><td>$V_{AA'} = 1$ V $\Rightarrow$Work of 1 J is necessary to bring the 1 C of charge from point A' to point A against the field;
$V_{AA'} = -\int_{A'}^{A} \vec{E} \cdot d\vec{l}$ for any contour 1, 2, or 3;
$\vec{E}(\vec{r}) = -gradV(\vec{r}) = -\nabla V(\vec{r})$ for potential $V(\vec{r})$ everywhere in space including materials;
$V = lE$ in uniform fields (most important).</td></tr>
<tr><td>Coulomb force on charge q
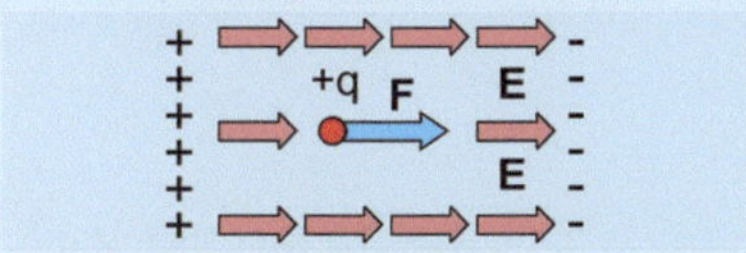</td><td>$\vec{F} = q\vec{E}$ [N]
The force is directed along the field for positive charges and against the field for negative charges</td></tr>
<tr><td>Gauss law
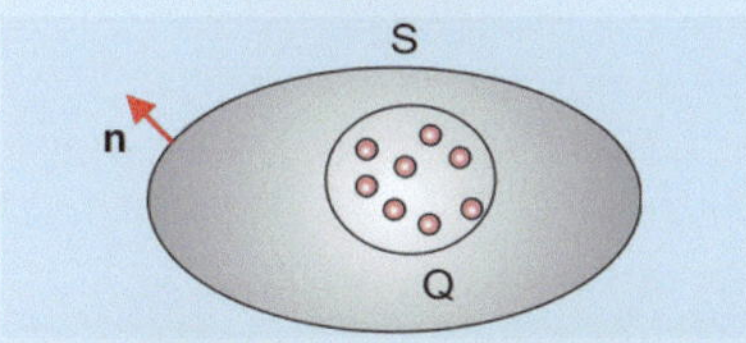</td><td>Total flux of the electric field through closed surface S times the permittivity is the total charge enclosed by S.
$\frac{Q}{\varepsilon_0} = \int_S \vec{E} \cdot \vec{n}\, dS$ (ε_0 =8.854×10^{-12} F/m)</td></tr>
<tr><td>Equipotential conductors
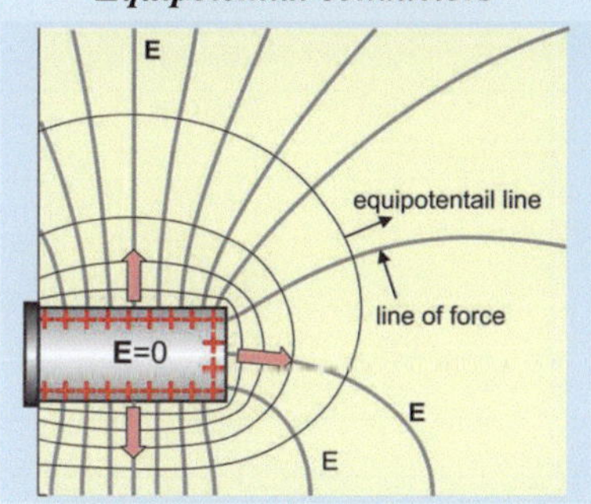</td><td>Within conductor(s) with applied voltage:
– Electric field is exactly zero;
– Volumetric charge density (C/m^3) is exactly zero.
On surface(s) of conductor(s) with applied voltage:
– Every point has the same voltage (conductor surface is equipotential surface);
– Surface charge density (C/m^2) exists;
– Emanating electric field is perpendicular to the surface (tangential field is zero: $\vec{E}_t = 0$)
Outside conductor(s):
– Equipotential lines and lines of force are perpendicular to each other</td></tr>
<tr><td>Voltage between two wires
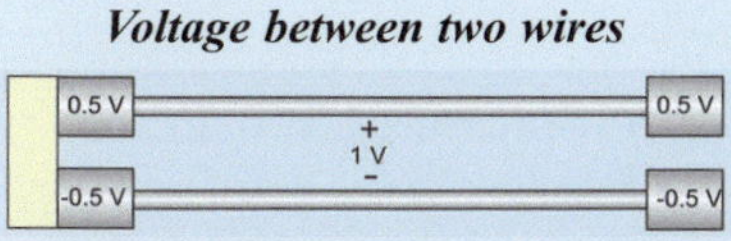</td><td>– Metal wires “guide” the electric field/voltage to a remote point
– Given the same voltage, charges on wires increase when their length increases.</td></tr>
<tr><td>Electrical induction in electrostatics
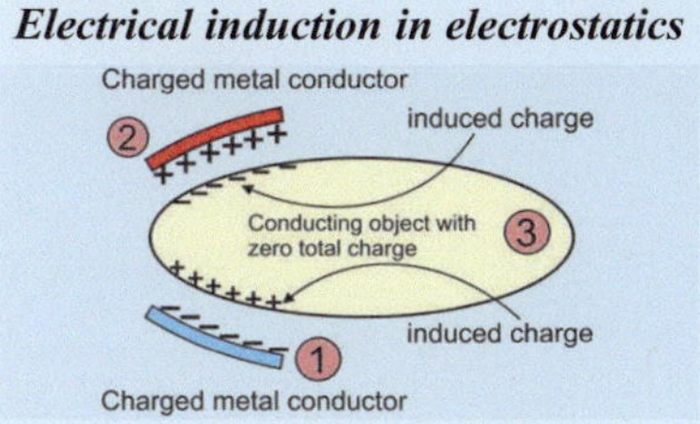</td><td>– Conductors 1 and 2 are subject to applied voltages V_1, V_2;
– Conductor 3 has zero net charge;
– Conductor 3 acquires certain voltage V_3;
– Surface charges in conductor 3 are separated as shown in the figure.</td></tr>
</table>

(continued)

<table>
<tr><th colspan="2">Steady-state current flow</th></tr>
<tr><td>Associated electric field
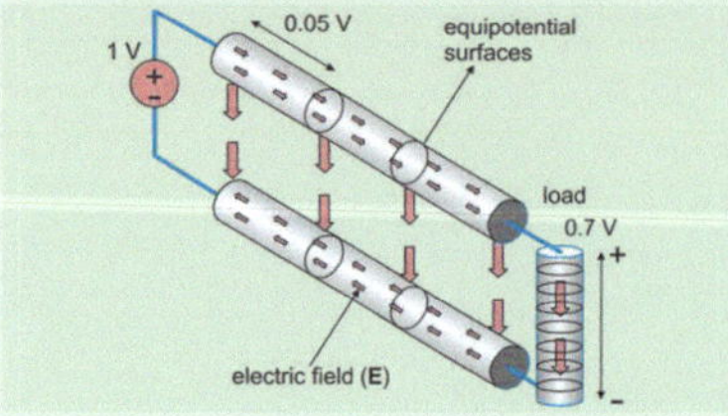
</td><td>– Current density (A/m^2) and total current I exist if and only if there is an electric field within the material;
$$\vec{j}(\vec{r}) = \sigma \vec{E}(\vec{r});$$
– There is little field in the conductor wires of a high conductivity;
– The field is high in air gap between two wires;
– The field is equally high within the load;
– Both wires carry total surface charges $\pm Q$ defined by supply voltage, wire separation, and wire length</td></tr>
<tr><td>Simple formulas for wire conductors
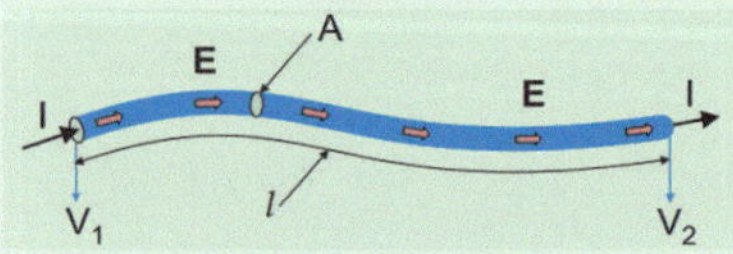
</td><td>$$I = AE\sigma$$
$$I = \pi r^2 E\sigma$$
$$E = (V_1 - V_2)/l$$</td></tr>
<tr><td>Non-uniform current flow
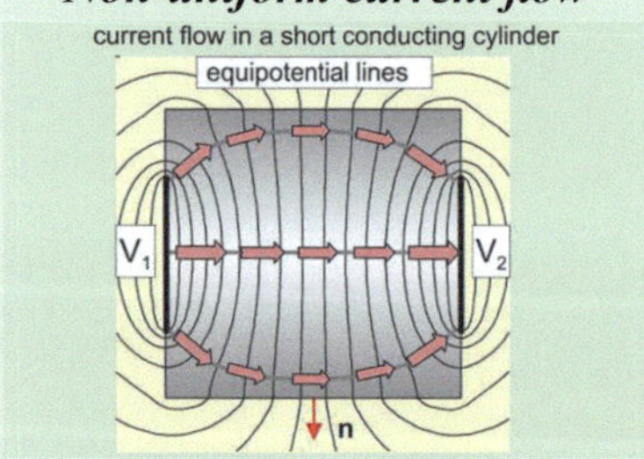
</td><td>Compare:
In the steady-state current-flow model:
– Electric field within the conductor is not zero;
– The conductor surface is not the equipotential surface
In the electrostatic model:
– Electric field within the conductor is zero;
– The conductor surface is the equipotential surface.</td></tr>
<tr><td>Current conservation law
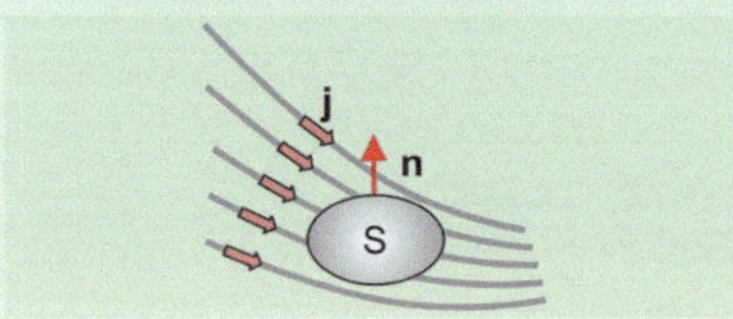
</td><td>$$\int_S \vec{j}\cdot\vec{n}\,dS = 0$$
There are no sources and sinks of electric current within a conductor except the surface electrodes</td></tr>
<tr><th colspan="2">Magnetostatics</th></tr>
<tr><td>Ampere's law in a general form
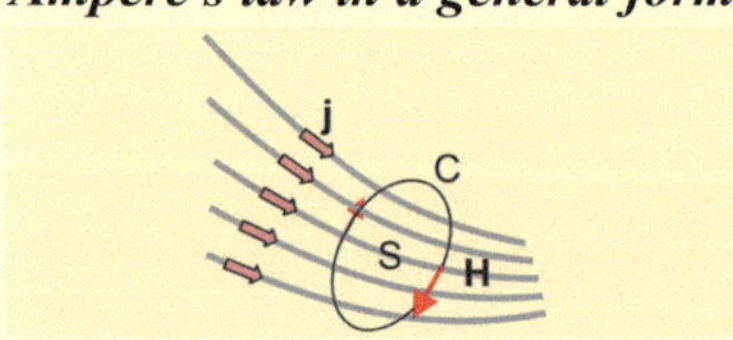
</td><td>Line integral of the magnetic field over a closed contour is the total current enclosed by this contour
$$\int_S \vec{H}\cdot d\vec{l} = \int_S \vec{j}\cdot\vec{n}\,dS = I_{\text{enc}}, j \text{ is measured in A/m}^2$$</td></tr>
<tr><td>Particular forms of Ampere's law
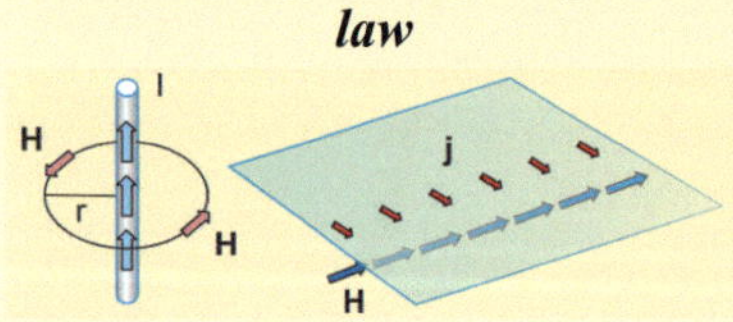
</td><td>Wire: $H(\vec{r}) = \frac{I}{2\pi r}, \quad r = |\vec{r}|$ (particular form)
Sheet of current: $H = \dfrac{j}{2} = const, j$ is in A/m</td></tr>
</table>

Problems

1.1 Electrostatics of Conductors

1.1.2 Electric Potential and Electric Voltage

1.1.3 Electric Voltage Versus Ground

1.1.4 Equipotential Conductors

Problem 1.1. Determine voltage (or potential) V_{AB} and V_{BA} (show units) given that the electric field between points A and B is uniform and has the value of 5 V/m. Point A has coordinates (0, 0); point B has coordinates (1, 1).

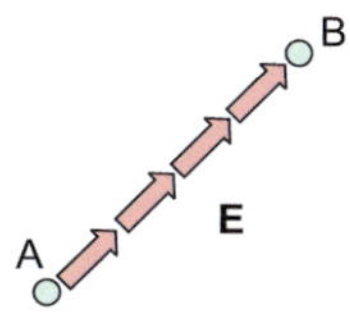

Problem 1.2. Determine voltages V_{AB}, V_{BD}, and V_{BC} given that the electric field shown in the figure that follows is uniform and has the value of (A) 10 V/m, (B) 50 V/m, and (C) 500 V/m.

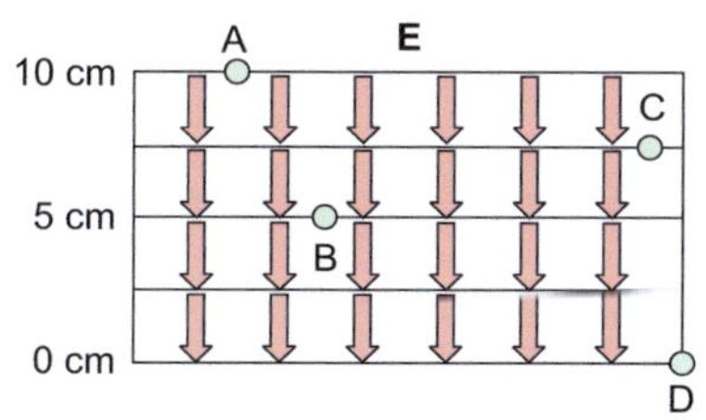

Problem 1.3. Assume that the electric field along a line of force AA' has the value $1 \times l$ V/m where $0 \le l \le 1$ m is the distance along the line. Find voltage (or potential) $V_{AA'}$.

Problem 1.4. The electric potential versus ground is given in Cartesian coordinates by $V(\vec{r}) = -x + y - z$ [V]. Determine the corresponding electric field everywhere in space.

Problem 1.5. The figure below shows the electric potential distribution across the semiconductor pn-junction of a Si diode. What kinetic energy should the positive charge (a hole) have in order to climb the potential hill from anode to cathode given that the hill "height" (or the built-in voltage of the pn-junction) is $V_{bi} = 0.7$ V? The hole charge is the opposite of the electron charge. Express your result in joules.

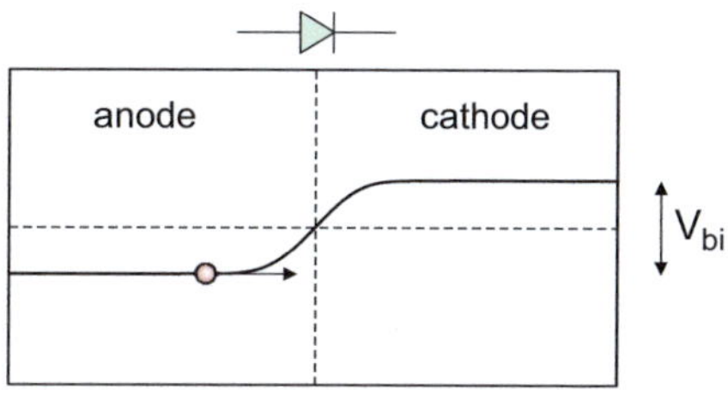

Problem 1.6. Is the electric field shown in the figure that follows conservative? Justify your answer.

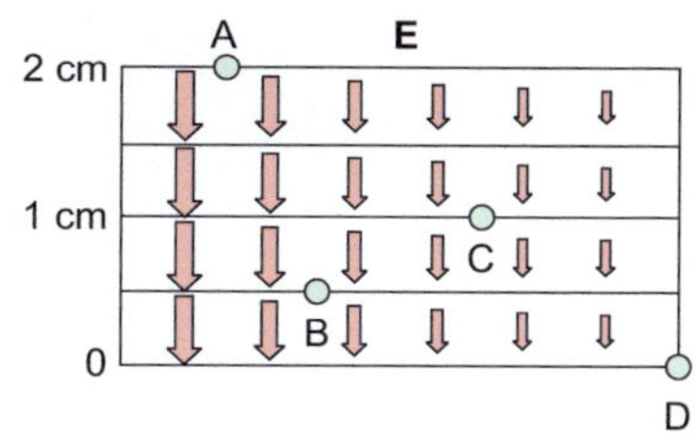

Problem 1.7. List all conditions for voltage and electric field used in electrostatic problems.

Problem 1.8. The figure below shows a 345 kV power tower used in MA, USA—front view. It also depicts electric potential/voltage and electric field distributions in space:

A. Determine which figure corresponds to the electric potential and which to the magnitude of the electric field.
B. Provide a detailed justification of your answer.

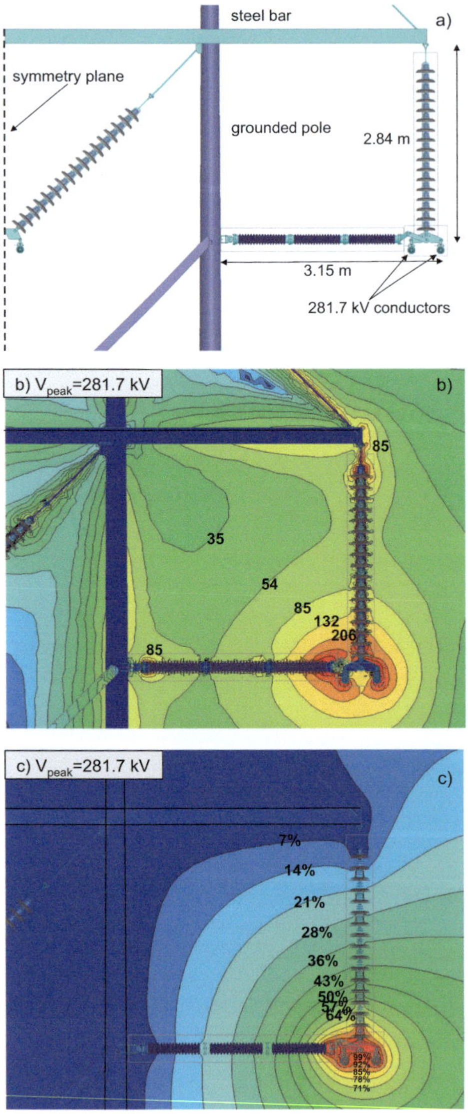

Problem 1.9. Figure that follows shows some isolated conductors. Determine the following voltage differences: V_{AB}, V_{AC}, V_{CB}.

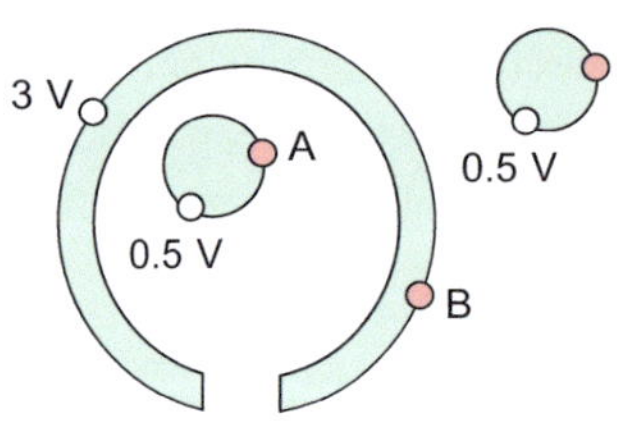

Problem 1.10. In your circuit, two wires connected to a 10-V voltage supply happen to be very close to each other at a certain location; they are separated by 2 mm.

A. What is the voltage between the wires at this location?

B. What is approximately the electric field strength at this location?

Problem 1.11. Is the figure that follows correct? Black curves indicate metal conductors. Why yes or why not?

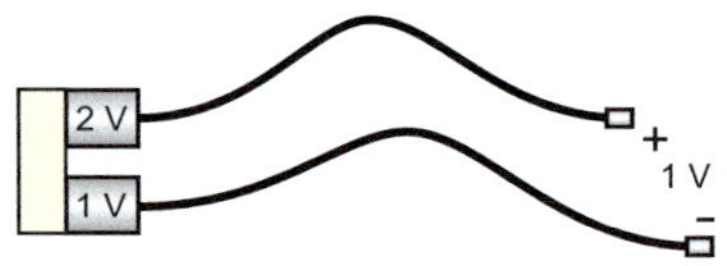

1.2 Steady-state Current Flow and Magnetostatics

1.2.1 Electric Current

1.2.2 Difference Between Current Flow Model and Electrostatics

1.2.3 Physical Model of an Electric Circuit

Problem 1.12. List the conditions for voltage and electric field used in the steady-state electric current problems.

Problem 1.13. Figure that follows shows the lines of force and equipotential lines for a DC current flow in a conductor due to two electrodes. List all mistakes of this drawing.

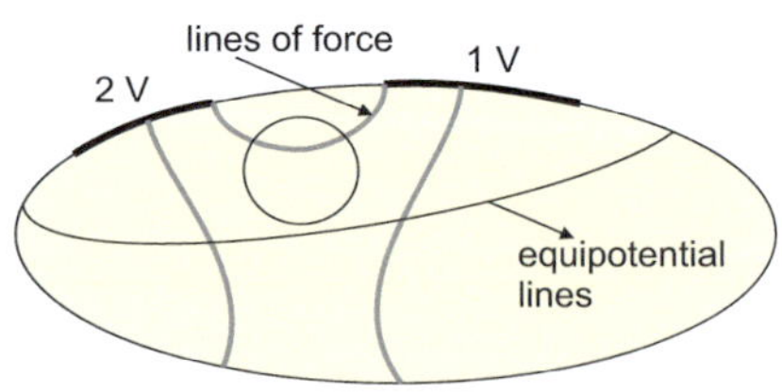

Problem 1.14. An AWG #10 (American Wire Gauge) aluminum wire has the conductivity of 4.0×10^7 S/m and the diameter of 2.58826 mm.

Determine the total current in the wire when the electric field inside the wire is (A) 0.001 V/m; (B) 0.005 V/m.

Problem 1.15. A copper wire (AWG #24) in the form of a coil with the radius of 0.1 m and 1000 turns is subject to applied voltage of 5 V. Determine the total current in the wire if its diameter is 0.51054 mm; the copper conductivity is 5.8×10^7 S/m.

Problem 1.16. The figure below shows a conducting cylinder of radius $R = 1$ cm, length $L = 5$ cm, and conductivity $\sigma_1 = 1.0$ S/m in air. Two electrodes are attached on both cylinder sides; the electrode radius is exactly the cylinder radius. Electrode voltages are exactly ± 1 V:

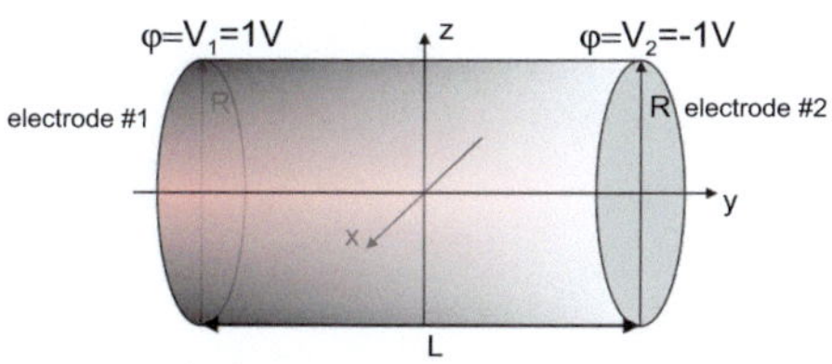

A. Determine and sketch to scale the electric potential everywhere inside the cylinder and on its surface.
B. Determine and sketch to scale the electric field everywhere inside the cylinder.
C. Attempt to sketch the electric potential distribution outside the cylinder.
D. Repeat tasks A and B when the voltage electrodes are replaced by current electrodes with the applied electric current density of ± 1 A/m^2.

1.2.4 Magnetostatics/Ampere's Law

1.2.5 Origin of Electric Power Transfer

Problem 1.17. A DC magnetic field of 1000 A/m is measured between two parallel wires of an electric circuit separated by 0.5 m. What is the circuit current?

Problem 1.18. Two conductors running from the source to the load are two parallel 0.5-cm-wide thin plates of infinite conductivity. The (vertical) electric field between the plates is 100 V/m; the (horizontal) magnetic field between the plates is 100 A/m. The load power is 0.1 W. Assuming no field fringing, determine (A) plate separation, (B) load voltage, and (C) load current.

Problem 1.19. Repeat the previous problem when the electric field between the plate electrodes increases by a factor of two, but the magnetic field decreases by a factor of two.

1.3 Hydraulic and fluid mechanics analogies

Problem 1.20. For the hydraulic setup shown in the figure, draw its electrical counterpart (an electric circuit) using the circuit symbols.

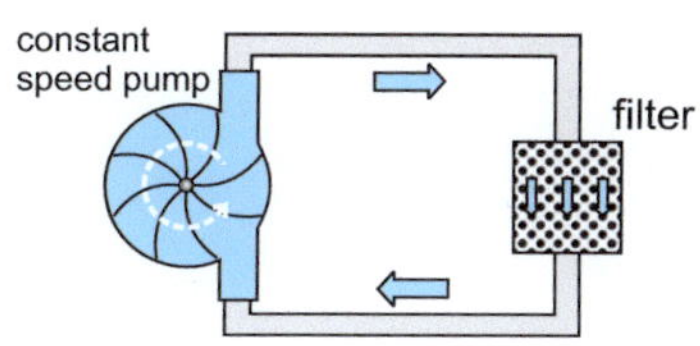

Problem 1.21. For the hydraulic setup shown in the figure, present its electrical counterpart (an electric circuit). Note a connection to a large reservoir with atmospheric pressure.

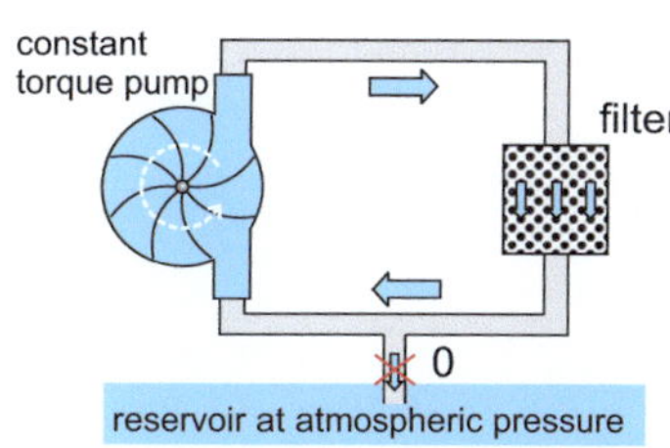

Part I

DC Circuits: General Circuit Theory—Operational Amplifier

Chapter 2: Major Circuit Elements

Overview

Prerequisites:

- Knowledge of university physics: electricity and magnetism (optional)
- Knowledge of vector calculus (optional)

Objectives of Section 2.1:

- Realize the difference between circuit elements and circuit components
- Review (derive) Ohm's law
- Become familiar with the v-i characteristic of the resistance including limiting cases
- Realize the importance of ohmic losses in long cables
- Become familiar with discrete fixed resistors and with resistive sensing elements

Objectives of Section 2.2:

- Realize the meaning of a passive nonlinear circuit element and its v-i characteristic
- Define two resistance types (static and dynamic) for a nonlinear passive circuit element
- Present two examples of nonlinear elements: ideal diode and a threshold switch

Objectives of Section 2.3:

- Introduce the concept of independent voltage and current sources and become familiar with their v-i characteristics
- Introduce the concept of practical voltage/current sources including their v-i characteristics
- Obtain initial exposure to the operation principles of voltage sources including specific examples

Objectives of Section 2.4:

- Become familiar with the concept of a dependent source
- Become familiar with four major types of dependent sources
- Obtain initial exposure to transfer characteristics of dependent sources
- Become familiar with ideal time-varying and AC sources

S. N. Makarov et al., *Practical Electrical Engineering*, https://doi.org/10.1007/978-3-319-96692-2_2

Objectives of Section 2.5:
- Formalize the meaning of voltmeter and ammeter from the viewpoint of open and short circuits
- Obtain a clear understanding of the circuit ground and its role in the circuit
- Review different ground types

Application Examples:
- Power loss in transmission lines and cables
- Resistive sensing elements
- DC voltage generator with permanent magnets
- Chemical battery

Keywords:
Circuit elements, Circuit components, *v*-*i* characteristic, Resistance, Polarity, Voltage difference, Voltage drop, Voltage polarity, Passive reference configuration, Ohm's law, Linear passive circuit element, Conductance, Siemens, mho, Short circuit, Open circuit, Ohmic conductor, Mobility of charge carriers, Material conductivity, Material resistivity, Electric circuit, American Wire Gauge (AWG), Resistor, Fixed resistors, Surface Mound Devices (SMD) (resistance value, geometry size), Variable resistor, Potentiometer, Resistive sensors, Photoresistor, Photocell, Negative temperature coefficient (NTC), Thermistor equation, Thermistor constant, Thermocouple, Peltier-Seebeck effect, Strain gauge, Strain sensitivity, Gauge factor, Strain gauge equation, Potentiometric position sensor, Nonlinear passive circuit elements, Non-ohmic circuit elements, Radiation resistance, Ideal diode, Shockley equation, Static resistance, Dynamic resistance, Small-signal resistance, Differential resistance, Incremental resistance, (DC) Operating point, Quiescent point, Electronic switch, Solid-state switch, Switch threshold voltage, Two-terminal switch, Three-terminal switch, Unidirectional switch, Bidirectional switch, Independent ideal voltage source, Active reference configuration, Nonlinear passive circuit elements, Non-ohmic circuit elements, Radiation resistance, Static resistance, Ideal diode, Shockley equation, Dynamic resistance, Small-signal resistance, Differential resistance, Incremental resistance, (DC) Operating point, Quiescent point, Electronic switch, Solid-state switch, Switch threshold voltage, Two-terminal switch, Three-terminal switch, Unidirectional switch, Bidirectional switch, Independent ideal voltage source, Active reference configuration, Practical voltage source, Maximum available source current, Maximum available source power, Open-circuit source voltage, Short-circuit circuit current, Internal source resistance, Independent ideal current source, Practical current source, Charge separation principle, Faraday's law of induction, Lorentz force, Instantaneous generator voltage, Average generator voltage, Battery voltage, Battery capacity, Battery energy storage, Dependent sources, Voltage-controlled voltage source, Current-controlled voltage source, Voltage-controlled current source, Current-controlled current source, Open-circuit voltage gain, Transresistance, Transconductance, Short-circuit current gain, Voltage amplifier, Current amplifier, Transresistance amplifier, Transconductance amplifier, Transfer characteristic, AC voltage source, Ideal voltmeter, Ideal ammeter, Earth ground, Chassis ground, Common (neutral) terminal (ground), Forward current, Return current, Absolute voltages in a circuit

Section 2.1 Resistance: Linear Passive Circuit Element

2.1.1 Circuit Elements Versus Circuit Components

Circuit Elements

Similar to mechanical mass, spring, and damper used in analytical dynamics, *circuit elements* are simple hypothetic ideal models. Every circuit element is characterized by its *unique voltage/current dependence* called the *υ-i characteristic*. Most of the *υ-i* characteristics reflect general physical laws. A list of the circuit elements includes:

1. Resistance
2. Capacitance
3. Inductance
4. Ideal electric transformer
5. Voltage source (independent and dependent)
6. Current source (independent and dependent)
7. Ideal switch
8. Ideal (Shockley) diode
9. Logic gates (NOT, AND, OR).

Circuit elements may be *linear* (resistance) or *nonlinear* (ideal diode), *passive* (resistance) or *active* (voltage source), *static* (resistance) or *dynamic* (capacitance/inductance), or both. Although all circuit elements studied here are static ones, the extension to the case of time-varying voltage and current is often trivial.

Circuit Components

Circuit components are numerous hardware counterparts of the circuit elements. Examples of the circuit components include resistor, capacitor, inductor, battery, etc. The circuit components may be modeled as combinations of the ideal circuit elements with one dominant desired element (e.g., resistance) and several parasitic ones (e.g., parasitic inductance and capacitance of a physical resistor). Another example is a battery, which is modeled as an ideal voltage source in series with a (small) resistance. In practice, we attempt to model any existing or newly discovered circuit component as a *combination* of the well-known circuit elements. The same is valid for more complicated structures targeted by electrical, mechanical, and biomedical engineers. An example is a human body, the response of which is modeled as a combination of resistance and capacitance.

2.1.2 Resistance

Symbols and Terminals

Figure 2.1 shows the circuit symbol for *resistance* with current direction and voltage polarity: positive voltage applied to the left terminal and a negative voltage applied to the right terminal cause a current to flow from left to right, as depicted in Fig. 2.1b. As a circuit element, the resistance is fully symmetric: terminals 1 and 2 in Fig. 2.1 may be interchanged without affecting its operation. Thus, the resistance does not have *a polarity*.

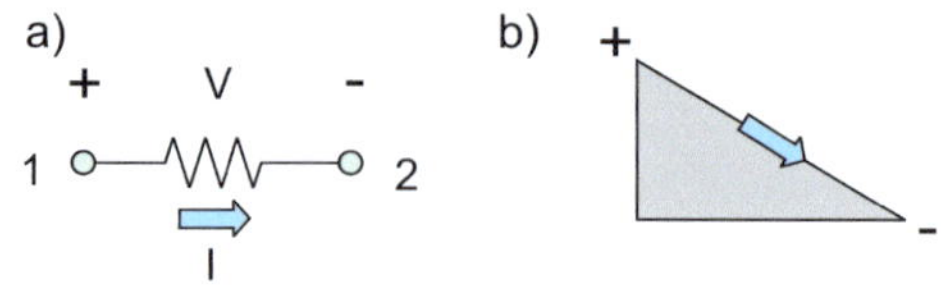

Fig. 2.1. Resistance symbol along with the voltage and current.

Voltage Across the Resistance

The *voltage difference* (or *voltage drop* or simply *voltage*) across the resistance, *V*, in Fig. 2.1 is a *signed quantity*. The voltage is measured in volts (V), named in honor of Italian physicist Alessandro Volta (1745–1827), who invented the first battery. Plus and minus signs across the resistance indicate the *voltage polarity*. Specifically, a plus sign denotes a (presumably) higher *absolute voltage level* versus *ground* than the minus sign; see Fig. 2.1b. For example, let us assume that the value *V* in Fig. 2.1a is positive and equal to 1 V. This means that the electric field spends positive work equal to

$$1\mathrm{V} \times 1\mathrm{C} = 1\mathrm{J} \tag{2.1}$$

when moving the charge of 1 C through the resistance from left to right in Fig. 2.1a. Similarly, the positive work of one joule is to be spent by an external force to move one coulomb of charge across a potential difference of one volt *against* the electric field. In power electronics, quantities of 1 kV (1000 V), even 1 MV (10^6 V) for voltage, are customary. In sensors and cellular phone circuits, for example, voltages are usually much lower. Values of 1 mV (10^{-3} V) or even 1 μV (10^{-6}V) may be recorded. Voltage applied to the resistance causes electric current flow.

Current Through Resistance: Passive Reference Configuration

The net current *I* flowing through the resistance is shown in Fig. 2.1a by an arrow. The current is measured in amperes (A), named in honor of French physicist and mathematician André-Marie Ampère (1775–1836). For example, the value of *I* in Fig. 2.1a is 1 A. This means that one coulomb of charges passes through the resistance in one second:

$$1\mathrm{A} \times 1\mathrm{s} = 1\mathrm{C} \tag{2.2}$$

The electric current flow *through* the resistance (and any other circuit element) is a *directed quantity*; the arrow shows its direction. A useful fluid mechanics analogy for the resistance is water (electric current) that flows down the "voltage" hill in Fig. 2.1b. The relation between voltage polarity and current direction depicted in Fig. 2.1b is known as the *passive reference configuration*. It is commonly used for all passive circuit elements such as resistances, diodes, capacitances, and inductances. Physically, the passive reference configuration means that the resistance consumes electric power, but does not create it. In power electronics, currents of several A, even kA (1000 A), are customary.

In digital and communication circuits, however, currents are usually low; therefore, units of 1 μA (10^{-6}A) or 1 mA (10^{-3}A) are commonly used.

Ohm's Law: Resistance and Conductance

According to *Ohm's law*, the voltage V across the resistance and the current I through the resistance are related by a simple linear expression

$$V = RI \tag{2.3}$$

with the proportionality constant R known as the *resistance*. This expression was first established by German mathematician and physicist Georg Simon Ohm (1789–1854) in 1827 but was coldly received by the scientific community at that time. It took nearly 14 years before the Royal Society of London finally recognized his work and his discovery is now known as Ohm's law. The unit of resistance R carries his name ohm and the Greek symbol Ω. The unit follows from Eq. (2.3) as volt over ampere:

$$1\ \Omega = \frac{1\ \text{V}}{1\ \text{A}} \tag{2.4}$$

The resistance is the *linear passive circuit element*. Resistance values vary typically between 1 Ω and 100 MΩ. The reciprocal of the resistance is the conductance, G:

$$G = \frac{1}{R} \tag{2.5}$$

The unit of conductance, Ω^{-1}, is called *siemens* (S) in honor of Ernst Werner von Siemens (1816–1892), a German inventor and the founder of what is today Europe's largest electrical engineering company (Siemens AG). An older American equivalent of that unit is *mho* (℧) or ohm spelled backwards! Conductance is useful in the circuit analysis.

Exercise 2.1: A voltage of 20 V is applied to a 1-MΩ resistance. Determine the current through the resistance.

Answer: 20 μA.

Exercise 2.2: A voltage of 20 V is applied to a 1-mS conductance. Determine current through the conductance.

Answer: 20 mA.

2.1.3 υ-*i* Characteristic of the Resistance: Open and Short Circuits

Figure 2.2 plots the linear dependence given by Eq. (2.3) for two distinct resistances. The corresponding plot is known as the *υ-i characteristic* (or the *υ-i dependence*). We use small letters *υ-i* to maintain consistency with the following study of time-varying circuits. The *υ-i* characteristic is the "business card" of the circuit element—every circuit element has its own *υ-i* characteristic. Once the *υ-i* characteristic is known, the circuit element is characterized completely. The following is true with reference to Fig. 2.2a:

1. The slope of the *υ-i* dependence for the resistance is equal to $1/R$ or G.
2. Smaller resistance leads to a stepper *υ-i* dependence (large currents).
3. Larger resistance leads to a flatter *υ-i* dependence (small currents).
4. The negative part of the *υ-i* dependence simply means simultaneous switching voltage polarity and current direction, respectively, in Fig. 2.1.

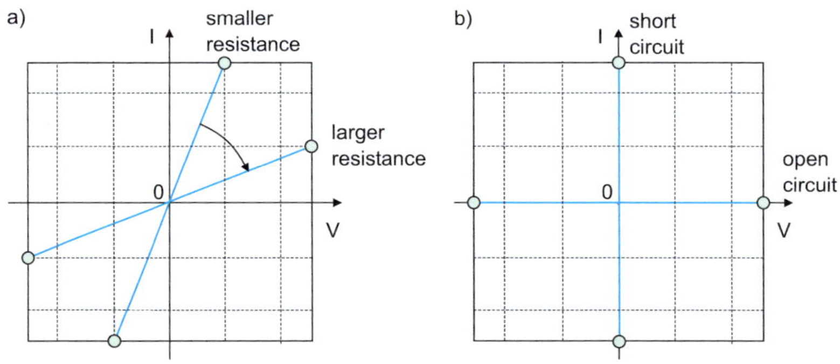

Fig. 2.2. *υ-i* Characteristics for resistances and for the short and open circuits, respectively.

Open and Short Circuits

Two limiting cases of the resistance *υ-i* characteristics are the *short circuit* and the *open circuit*, as seen in Fig. 2.3.

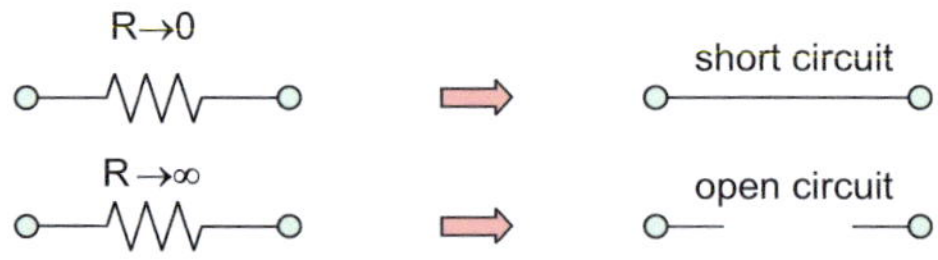

Fig. 2.3. Transformation of resistance to a short and an open circuit, respectively.

When $R \to 0$, the resistance becomes a short circuit, or an ideal wire. There is no voltage drop V across the wire, but any current I can flow through it. Therefore, the *υ-i* characteristic of the short circuit is the straight vertical line in Fig. 2.2b. When $R \to \infty$, the resistance becomes an open circuit, or an ideal vacuum gap. There is no current I through the gap at any value of the applied voltage V. Therefore, the *υ-i* characteristic of the open circuit is the straight horizontal line in Fig. 2.2b.

Exercise 2.3: Every vertical division in Fig. 2.2a is 0.1 A; every horizontal division is 1 V. Find resistances for two *υ-i* dependencies in the figure.

Answer: $R = 4\ \Omega$ and $R = 25\ \Omega$, respectively.

Exercise 2.4: An *ideal switch* is open when $V < 0$ and is closed when $V \geq 0$. Plot the *υ-i* characteristic given that only a positive current $I > 0$ can flow.

Answer: Horizontal line $I = 0$ when $V < 0$ and vertical line $V = 0$ when $I > 0$.

2.1.4 Power Delivered to the Resistance

Voltage V across the resistance is work in joules necessary to pass 1 C of charge through the resistance. Since there are exactly I coulombs passing through the resistance in one second, the power P delivered to the resistance must be the product of work per unit charge and the number of charges passing through the element in one second: $P = VI$. The power P has indeed the units of watts $(1\ \text{V} \times 1\ \text{A} = 1\ \text{J}/1\ \text{s} = 1\ \text{W})$. When the *υ-i* characteristic of the resistance is examined, the power is equal to the area of the shaded rectangles in Fig. 2.4. Using Ohm's law, Eq. (2.3) gives us three equivalent definitions of the absorbed power by a resistance:

$$P = VI \quad \text{Basic definition, valid for any passive circuit element} \tag{2.6}$$

$$P = \frac{V^2}{R} \quad \text{Power for resistance in terms of voltage} \tag{2.7}$$

$$P = RI^2 \quad \text{Power for resistance in terms of current} \tag{2.8}$$

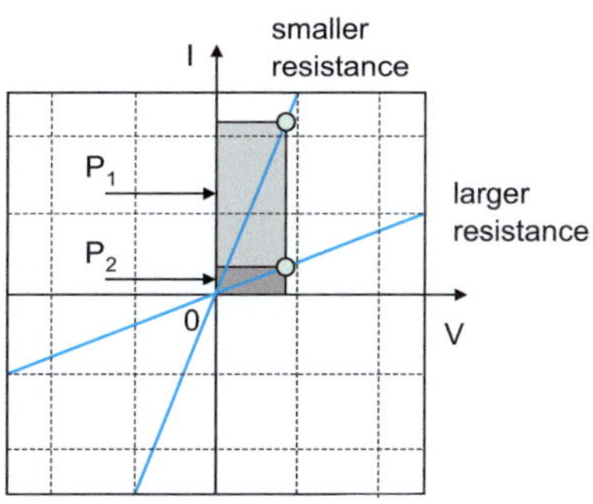

Fig. 2.4. *υ-i* Characteristics for the resistances and power rectangles. $P_{1,2}$ are absorbed powers.

Despite their obvious nature, all three equations are useful in practice. In particular, Eq. (2.7) indicates that a small resistance absorbs more power than the large resistance *at the same applied voltage*; this is seen in Fig. 2.4. Imagine for a moment that we know the voltage across the resistance, but do not know the current. This happens if a number of circuit elements are connected in parallel to a known voltage source. Then Eq. (2.7) is used to find the power. However, if the current is known, but the voltage is not (a number of elements connected in series to a current source), then Eq. (2.8) is employed.

Example 2.1: A voltage of 10 V is applied to a 2.5-Ω resistance. Determine the absorbed electric power in three possible ways as stated by Eqs. (2.6) through (2.8).

Solution: To apply two of the three equations, current through the resistance is needed. From Ohm's law, $I = V/R = 4$ A. The electric power delivered to the resistance can thus be determined in three ways:

$$P = VI = 40\,\text{W} \quad \text{Basic power definition, passive reference configuration}$$

$$P = \frac{V^2}{R} = 40\,\text{W} \quad \text{Power for resistance in terms of voltage}$$

$$P = RI^2 = 40\,\text{W} \quad \text{Power for resistance in terms of current}$$

2.1.5 Finding Resistance of Ohmic Conductors

An *ohmic conductor* satisfies Ohm's law given by Eq. (2.3). Finding its resistance is equivalent to the derivation of Ohm's law under certain assumptions. Let us consider a conducting circular cylinder subject to an applied voltage V in Fig. 2.5. The cylinder has length l and a cross-sectional area A.

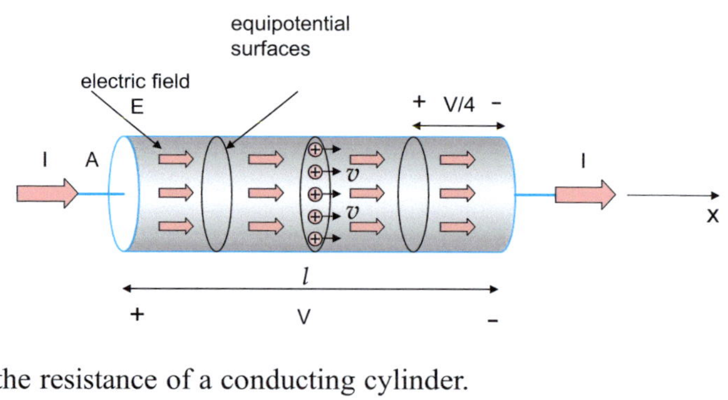

Fig. 2.5. Finding the resistance of a conducting cylinder.

Finding Total Current

The net electric current in a metal or other conductor is defined as the net flux of *positive* charge carriers directed along the conductor axis x:

$$I = Aqnv \tag{2.9}$$

Here, qn is the *volumetric charge density* of *free charges* q with concentration n in coulombs per cubic meter, C/m^3, and v is the magnitude of the average charge velocity in m/s. In the one-dimensional model of the current flow, the average velocity vector is directed along the x-axis seen in Fig. 2.5. Since the electrons have been historically assigned a negative charge, the electric current direction is *opposite* to the direction of electron motion in a conductor. The electron carries an elemental charge of $-q = -1.60218 \times 10^{-19}$ C. Because A and qn are constants for a given conductor, the electric current is simply associated with the charge's mean velocity v.

Finding Average Carrier Velocity

In order to find v, we use the following method. The total voltage drop V applied to a sufficiently long, conducting cylinder is *uniformly* distributed along its length following the equally spaced equipotential surfaces; this is schematically shown in Fig. 2.5. This fact has been proved in Chapter 1. Such a voltage distribution corresponds to a *constant uniform* electric field within the cylinder, which is also directed along the cylinder axis. The magnitude of the field, E, with the units of V/m, is given by

$$E = V/l \tag{2.10}$$

The electric field creates a Coulomb force acting on an individual positive charge q. The Coulomb force is directed along the field; its magnitude F is given by

$$F = qE \tag{2.11}$$

The key is a *linear* relation between the charge velocity v and force F or, which is the same, a linear relation between the charge velocity v and the applied electric field E, i.e.,

$$v = \mu E \tag{2.12}$$

where μ is the so-called mobility of charge carriers, with the units of m^2/(V·s). Carrier mobility plays an important role in semiconductor physics. With the help of Eqs. (2.10) and (2.12), the expression for the total current Eq. (2.9) is transformed to

$$V = \left(\frac{l}{Aqn\mu}\right)I = \left(\frac{l}{A\sigma}\right)I = RI, \quad \sigma = qn\mu, \quad R = \frac{l}{A\sigma} \tag{2.13}$$

This is the expression for the resistance of a cylindrical conductor. *Material conductivity* σ is measured in S/m. Its reciprocal is the *material resistivity* $\rho = 1/\sigma$ measured in Ω·m.

Example 2.2: Estimate resistance R of a small doped Si disk with the length l of 5 μm, cross section of $A = 10^{-4}$ cm^2, uniform electron, concentration (carrier concentration) of $n = 10^{17}$ cm^{-3}, and carrier mobility of $\mu_n = 1450$ cm^2/(V·s).

Solution: Resistance calculations are usually simple when the one-dimensional model of a conducting cylinder or a disk is used. However, one must be careful with the units. Units of cm are customary in semiconductor physics. Therefore, one should first convert all different units of length to meters (or to centimeters). After that, we use the definition of the resistance given by Eq. (2.13) and obtain (units of meters are used):

$$R = \frac{l}{Aqn\mu_n} = \frac{5 \times 10^{-6}}{10^{-8} \times 1.602 \times 10^{-19} \times 10^{23} \times 0.145} = 0.215\ \Omega \tag{2.14}$$

Table 2.1 lists conductivities of common materials. What is the major factor that determines the conductivity of a particular conducting material? According to Eq. (2.13), there are two such parameters: charge concentration and charge mobility.

Table 2.1. DC conductivities of conductors, semiconductors, and insulators (25 °C, multiple sources).

Material	Class	σ (S/m)	Material	Class	σ (S/m)
Silver	Conductor	6.1×10^7	Seawater	Semiconductor	4
Copper	Conductor	5.8×10^7	Human/animal tissues	Semiconductor	0.1–2.0
Gold	Conductor	4.1×10^7	Germanium	Semiconductor	2
Aluminum	Conductor	4.0×10^7	Fresh water	Semiconductor	0.01
Brass	Conductor	2.6×10^7	Wet soil	Semiconductor	0.01–0.001
Tungsten	Conductor	1.8×10^7	Dry soil	Semiconductor	0.001–0.0001
Zinc	Conductor	1.7×10^7	Intrinsic silicon (Si)	Semiconductor	4.4×10^{-4}
Nickel	Conductor	1.5×10^7	Gallium arsenide (GaAs)	Semiconductor	10^{-6}
Iron	Conductor	1.0×10^7	Glass	Insulator	10^{-12}
Tin	Conductor	0.9×10^7	Porcelain	Insulator	10^{-14}
Lead	Conductor	0.5×10^7	Hard rubber	Insulator	10^{-15}
Graphite	Conductor	0.003×10^7	Fused quartz	Insulator	10^{-17}
Carbon	Conductor	0.003×10^7	Teflon	Insulator	10^{-23}
Magnetite	Conductor	0.002×10^7			

It is mostly the different concentration of free charge carriers n that makes the resistance of two materials quite different. For example, $n = 8.46 \times 10^{28}\mathrm{m}^{-3}$ in copper (a good conductor), whereas it may be $n = 10^{16}\mathrm{m}^{-3}$ in a moderately doped silicon crystal (doped semiconductor). However, it is also the difference in mobility μ that represents the "friction" experienced by the "gas" of free charges with density n that is moving through the solid or liquid conductor under the applied voltage (electric field).

Exercise 2.5: Using Table 2.1, determine the total resistance of an aluminum wire having a length of 100 m and a cross-sectional area of 1 mm^2.

Answer: 2.5 Ω.

2.1.6 Application Example: Power Loss in Transmission Wires and Cables

All metal wires and cables are ohmic conductors. Electric power absorbed by an ohmic conductor is transformed into heat. This is known as electric power loss. We can apply Eqs. (2.6)–(2.8) and Eq. (2.13) in order to determine the loss of electric power in transmission lines and/or cables. This question has significant practical importance. Figure 2.6 outlines the corresponding *electric circuit*. The electric circuit is a closed path for electric current. Resistance R_L characterizes the load. Only Ohm's law is used to analyze this circuit, along with the current continuity. No other circuit laws are necessary. We also consider a voltage source in Fig. 2.6. The voltage sources will be studied next.

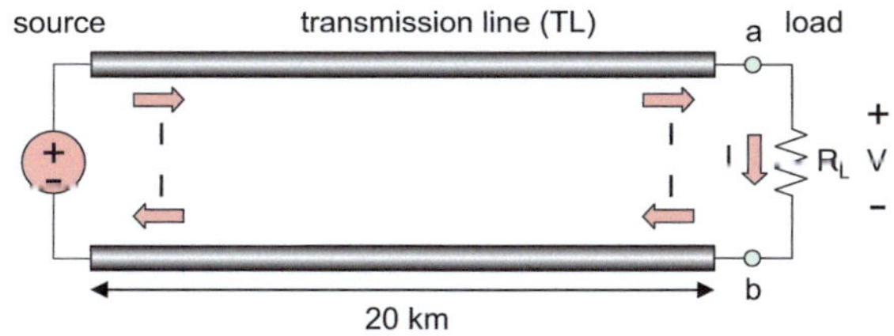

Fig. 2.6. A long transmission power line carrying a steady-state current I to the load resistance.

According to Eq. (2.13), the wire resistance is inversely proportional to its diameter. In the USA, the *American Wire Gauge* (AWG) system was developed to classify the wire diameters of conductors. You probably have heard an electrician refer to a gauge 12 household wiring. This implies a wire diameter of about 2 mm, or 0.08″. Table 2.2 reports common AWG numbers and maximum current strengths.

Table 2.2. American Wire Gauge (AWG) wire parameters. The maximum current is given for solid copper (Source: Handbook of Electronic Tables and Formulas for American Wire Gauge).

AWG #	Diameter (inches)	Diameter (mm)	Resistance per 1000 ft or 304.8 m (Ω)	Maximum current in (A) for power transmission
24	0.0201	0.51054	25.67	0.577
22	0.0254	0.64516	16.14	0.92
20	0.0320	0.81280	10.15	1.50
18	0.0403	1.02362	6.385	2.30
16	0.0508	1.29032	4.016	3.70
14	0.0640	1.62814	2.525	5.90
12	0.0808	2.05232	1.588	9.30
10	0.1019	2.58826	0.999	15.0
Gauges 10 through 1 are not shown				
0 (1 aught)	0.3249	8.252	0.09827	150
00 (2 aught)	0.3648	9.266	0.07793	190
000 (3 aught)	0.4096	10.404	0.06180	239
0000 (4 aught)	0.4600	11.684	0.04901	302

Example 2.3: An AWG 0 aluminum transmission grid cable schematically shown in Fig. 2.6 has a wire diameter of 8.25 mm and a cross-sectional area of 53.5 mm^2. The conductivity of aluminum is 4.0×10^7 S/m. The total cable length (two cables must run to a load) is 40 km. The system delivers 1 MW of DC power to the load. Determine the power loss in the cable when *load* voltage V and *load* current I are given by:

1. $V = 40$ kV and $I = 25$ A
2. $V = 20$ kV and $I = 50$ A
3. $V = 10$ kV and $I = 100$ A

Why is high-voltage power transmission important in power electronics?

Solution: We find the total cable resistance from Eq. (2.13):

$$R = \rho\frac{L}{A} = \frac{L}{\sigma A} = \frac{40 \times 10^3}{4.0 \times 10^7 \times 53.5 \times 10^{-6}} = 18.7\Omega \tag{2.15}$$

The same load current I flows through the load modeled by a resistor R_L and through the cables in Fig. 2.6. Therefore, power loss in the cables may be found using Eq. (2.8). Knowing the load voltage (or the voltage across the cable) is not necessary. The power loss in the cables is thus given by $P = RI^2$. For the three different cases corresponding to the same load power, we obtain

Example 2.3 (cont.):

$$\begin{aligned}&1.\ P = RI^2 = 11.7\,\text{kW or } 1.17\ \%\text{ of the load power}\\&2.\ P = RI^2 = 46.8\,\text{kW or } 4.68\ \%\text{ of the load power}\\&3.\ P = RI^2 = 187\,\text{kW or } 18.7\ \%\text{ of the load power}\end{aligned} \tag{2.16}$$

Clearly, the high-voltage power transmission allows us to reduce the power loss in long cables very significantly while transmitting the same power to the load. Therefore, the high-voltage transmission lines passing through the country have typical voltages between 100 kV and 800 kV.

In circuit analysis in the laboratory, we usually consider *ideal* or perfectly conducting wires whose resistance is zero. This is justified since the wire lengths for most practical circuit applications are so short that the voltage drop is negligibly small.

2.1.7 Physical Component: Resistor

Fixed Resistors

Resistance is constructed intentionally, as a separate circuit component. This component is called the *resistor*. A common axial leaded carbon film 0.25-W resistor deployed on a solderless protoboard is seen in Fig. 2.7a. Those carbon film resistors are typically manufactured by coating a homogeneous layer of pure carbon on high-grade ceramic rods. After a helical groove is cut into the resistive layer, tinned connecting leads of electrolytic copper are welded to the end-caps. The resistors are then coated with layers of tan-colored lacquer. The common *Surface Mount Device* (SMD) thin-film resistor is shown in Fig. 2.7b. Manufacturing process variations result in deviations from the normal resistor values; they are known as tolerances and reported to the end user through an *extra color ring* (for leaded axial resistors) or an *extra digit* (for SMD resistors). Typical power ratings for the axial resistors are 1/6 W, ¼ W, ½ W, 1 W, 2 W, and 3 W. When the power delivered to the resistor considerably exceeds the particular rating, the resistor may burn out, releasing a prominent "carbon" smell. The axial resistors have color codes shown in Fig. 2.7c. To find the value of the resistor depicted in Fig. 2.7a, we first encounter the tolerance code, which will typically be gold, implying a 5 % tolerance value. Starting from the *opposite* end, we identify the first band, and write down the number associated with that color; in Fig. 2.7a it is 9 (white). Then, we read the next band (brown) and record that number; it is 1. After this we read the multiplier black, which is 0. The resistor value in Fig. 2.7a is consequently $R = 91 \times 10^0 = 91\ \Omega$.

The surface mount resistors, also known as SMD resistors, do not have color codes. The SMD resistors are labeled numerically as **102** $= 10 \times 10^2 = 1\ \text{k}\Omega$, **271** $= 27 \times 10^1 = 270\ \Omega$, etc. Along with this, the SMD resistors, similar to other SMD components, do have codes corresponding to their geometry size. Each size is described as a four-digit number. The first two digits indicate length, and the last two digits indicate

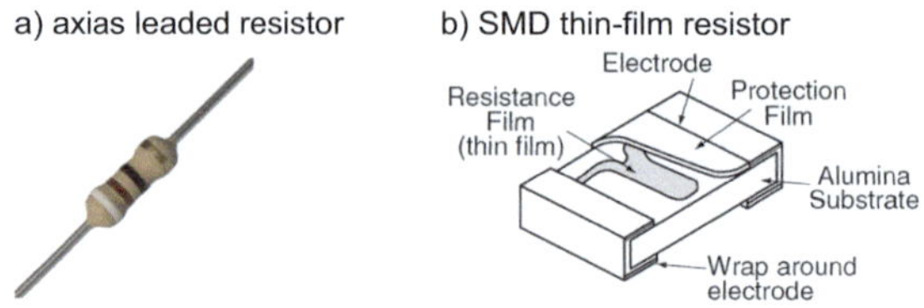

c) Color codes for axial resistors

Black	Brown	Red	Orange	Yellow	Green	Blue	Violet	Gray	White
0	1	2	3	4	5	6	7	8	9

Fig. 2.7. (**a**) A leaded axial resistor, (**b**) a thin-film resistor, and (**c**) color codes for leaded resistors.

width (in 0.01″, or 10 mils units). Some popular SMD resistor sizes are 0603 (0.06″ × 0.03″, or 60 × 30 mils, or 1.6 × 0.8 mm), 0805 (0.08″ × 0.05″), and 1206 (0.12″ × 0.06″).

Variable Resistors (Potentiometers)

The simplest *variable resistor* is a *potentiometer*. A picture is shown in Fig. 2.8a along with its equivalent electric schematic in Fig. 2.8b. The potentiometer is used either as a voltage divider, discussed later in the text, or as a variable resistor. By rotating the potentiometer shaft, it is possible to obtain any resistance value up to the maximum potentiometer value. The adjustable resistance is obtained between terminals 1 and 2 or 2 and 3 of the potentiometer, respectively. You should remember that the potentiometer is a nonpolar device. This means that it can be placed into the circuit in any orientation. With this knowledge the joking engineer telling you that "all resistors in your circuit are backwards" should not cause any fear.

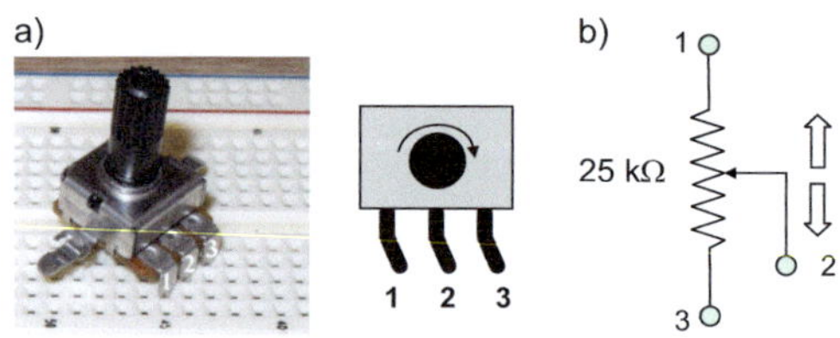

Fig. 2.8. A rotary 25-kΩ potentiometer rated at 0.25 W and its equivalent circuit schematic.

2.1.8 Application Example: Resistive Sensors

There are a variety of sensor types—*resistive sensors*—which use electric resistance variation to measure a mechanical or a thermal quantity. Some of them are shown in Fig. 2.9. As a first example, we consider a temperature sensor based on a *thermistor* (a resistor), with a resistance that varies when ambient temperature changes; see Fig. 2.9a.

The word thermistor is a contraction of the words "thermal" and "resistor." As a second example of a resistance subject to ambient conditions, we will consider a *photoresistor* (*photocell*) shown in Fig. 2.9b. The final example is a *strain gauge* shown in Fig. 2.9c.

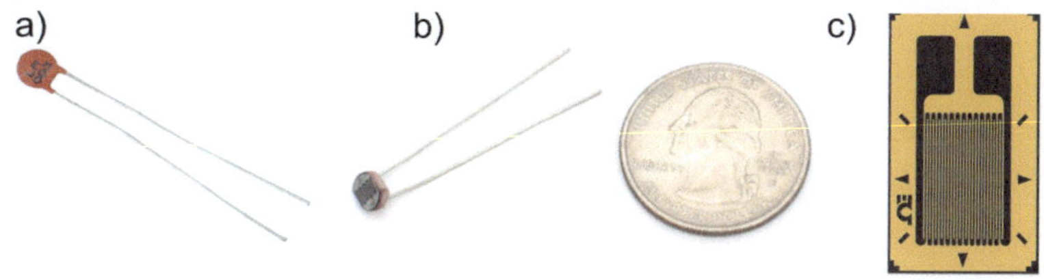

Fig. 2.9. Sensing elements which change their resistances when ambient conditions change.

Thermistor

The thermistor changes its resistance as temperature increases or decreases. General-purpose thermistors are made out of *metal oxides* or other *semiconductors*. Successful semiconductor thermistors were developed almost simultaneously with the first transistors (1950s). For a metal-oxide thermistor, its resistance *decreases* with increasing temperature. Increasing the temperature increases the number of free carriers (electrons) and thus increases the sample conductivity (decreases its resistance). Shown in Fig. 2.9a is a very inexpensive NTC—*negative temperature coefficient*—leaded thermistor. According to the manufacturer's datasheet, it reduces its resistance from approximately 50 kΩ at room temperature (about 25 °C) by 4.7 % for every degree Celsius (or Kelvin) and reaches about 30 kΩ at body temperature according to the *thermistor equation*:

$$R_1 = R_2 \exp\left[B\left(\frac{1}{T_1} - \frac{1}{T_2}\right)\right] \tag{2.17}$$

where T_1, T_2 are two *absolute* temperatures always given in degrees K. Temperature T_2 corresponds to a room temperature of 25 °C so that $R_2 = R_{25°\mathrm{C}}$, temperature T_1 is the observation temperature, and B is the *thermistor constant*, which is equal to 4200 K in the present case. Equation (2.17) is a nontrivial result of the solid-state physics theory. We emphasize that Eq. (2.17) is more accurate than the temperature coefficient of the thermistor—the above referenced value of −4.7 %. Typical applications include temperature measurement, control, compensation, power supply fan control, and printed circuit board (PCB) temperature monitoring. Inexpensive thermistors operate from −30 °C to approximately +130 °C. At higher temperatures, thermocouples should be used.

Thermocouple

Figure 2.9 does not show one more important temperature sensor—the *thermocouple*—which is used to measure large temperatures and large temperature differences. It operates based on a completely different principle. The thermocouple does not significantly change its resistance when temperature changes. Instead, it generates an electric current and the associated voltage, when the junction of the two metals is heated or cooled, known as the *Peltier-Seebeck effect*; this voltage can be correlated to temperature. Therefore, the thermocouple, strictly speaking, is not a resistive sensor.

Photoresistor (Photocell)

An idea similar to the thermistor design applies. Quanta of light incident upon the photocell body create new free charge carriers—new electron-hole pairs in a semiconductor. If the concentration of free charges increases, the resistance of the sample decreases according to Eq. (2.13). The resistance is inversely proportional to the concentration. The photocell in Fig. 2.9b is characterized by very large nonlinear variations of the resistance in response to ambient light.

Strain Gauge

The *strain gauge* measures mechanical strain. The operation is based on Eq. (2.13), which defines the resistance through material conductivity σ, the length of the resistor l, and its cross section A. When the resistor, which may be a trace on the base of a metal alloy, is stretched, its length l increases and its cross section A decreases. Hence, the resistance R increases due to *both* of these effects simultaneously; changes in the resistance may be made visible for small strains. Shown in Fig. 2.9c is an inexpensive uniaxial strain gauge with a nominal resistance of 350 Ω; typical resistances are 120, 350, 600, 700, and 1000 Ω. The gauge changes its resistance R in proportion to the *strain sensitivity* S_G of the wire's resistance, also called the *gauge factor* (GF). For a strain gauge, the relative resistance variation, $\Delta R/R$, is estimated based on known values of the strain sensitivity, S_G, and strain, ε. The *strain gauge equation* has the form

$$\Delta R/R = S_G\varepsilon \tag{2.18}$$

The dimensionless strain sensitivity S_G varies around 2. The strain (a relative elongation) is a dimensionless quantity. It is measured in micro-strains, $\mu\varepsilon$, where one $\mu\varepsilon$ is 10^{-6}. Typical strain values under study are on the order of 1000 $\mu\varepsilon$. Using Eq. (2.18) this yields a relative resistance variation as small as 0.2 %. Because of this, the circuits for the strain measurements should be designed and built with great care. Temperature compensation efforts are also required. Since the relative resistance changes are very small, the strain gauge is a *linear* device: the strain is directly proportional to resistance variations.

Potentiometric Position Sensor

Another general resistive sensor is a *potentiometric* (or *potentiometer*) *position sensor*. Its operation becomes apparent when we rotate the potentiometer dial in Fig. 2.8a. A change in the resistance is directly proportional to the rotation angle. The resistance variation can be converted to voltage variation and then measured. Similar potentiometer sensors for measuring linear motion also exist.

Sensitivity of Resistive Sensors

One major difference between different resistive sensing elements is a very different degree of the relative resistance variations. For the photocell, the resistance variation is up to 100 times. For the thermistor, the resistance variation can be as much as 50 %. For the strain gauge, the resistance variations do not exceed 0.5 %.

Circuit Symbols

There are several similar but not identical standards for *circuit symbols* related to resistance: International standard IEC 60617, American ANSI standard Y32 (IEEE Std 315), etc. Figure 2.10 shows popular circuit symbols for variable resistances.

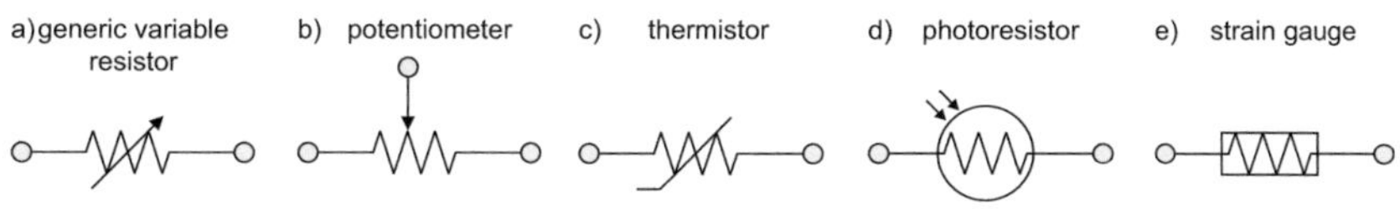

Fig. 2.10. Circuit symbols for variable resistances: (**a**) generic variable resistance, (**b**) potentiometer, (**c**) thermistor, (**d**) photoresistor, and (**e**) strain gauge.

Section 2.2 Nonlinear Passive Circuit Elements

2.2.1 Resistance as a Model for the Load

It might appear at first glance that the resistance causes mainly power loss and heating. However, the concept of heating elements in household appliances or power losses in long cables covers only a small subset of applications. Important resistance applications are related to the resistive sensors studied previously. Resistances are widely used in the circuits to provide different voltage values, i.e., *bias* circuit components such as diodes and transistors. Last but not least, resistances model an arbitrary power-absorbing device, *the load*, which can be mechanical, acoustical, microwave, optical, etc. From a circuit point of view it *does not matter* how the electric power supplied to a load is eventually transformed. The circuit delivers certain power to the load, but cares little about whether the power is converted into heat to warm a heating pan, light to illuminate our house, mechanical power to drive a motor, or electromagnetic radiation generated by a cell phone. Circuit analysis is concerned with the power delivered to a power-absorbing device, *the load*, leaving its utilization and conversion to other engineering disciplines. Therefore, many practical loads can be replaced by a simple *load resistance* R_L. Such a resistance is often called the *equivalent resistance*, $R_L = R_{eq}$. It is also valid for AC circuits. For AC circuits it is convenient to use *rms* (root mean square) voltages, which are equivalent to DC voltages and thus provide us with the same power value delivered to the load. Figure 2.11 shows two examples of the load replacement with the equivalent resistance: a light source radiating visible light and an antenna radiating microwaves.

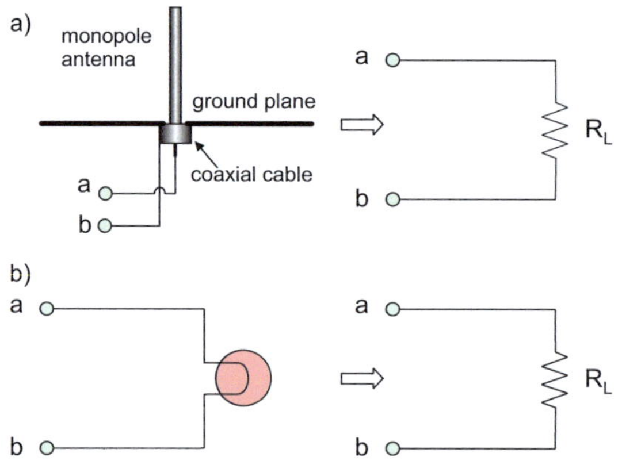

Fig. 2.11. (**a**) Radiating monopole antenna is modeled as a resistance. (**b**) Light source is approximately modeled as a resistance.

Example 2.4: A small commercial monopole antenna shown in Fig. 2.11a is rated at $R_{eq} = 50\ \Omega$ in the ISM band of 902–928 MHz. When an rms voltage of 10 V is applied to the antenna, what is the total amount of power radiated by the antenna?

Example 2.4 (cont.):

Solution: We use Eq. (2.7) and obtain $P = V_{\mathrm{rms}}^2/R_{\mathrm{eq}} = 2$ W. All of this power is radiated in the form of an outgoing electromagnetic wave. There is no heat loss. The resistance R_{eq} is called the *radiation resistance* in such a case. The above analysis is valid only in a certain frequency range.

A load that exactly follows Ohm's law Eq. (2.3) is called the *linear load*. While the transmitting antenna in Fig. 2.11a is a linear load, an incandescent light bulb in Fig. 2.11 is not. Most of the loads deviate from the linear Ohm's law.

2.2.2 Nonlinear Passive Circuit Elements

Nonlinear passive circuit elements do not satisfy Ohm's law with the constant resistance R over a wide range of voltages. Therefore, they are also called *non-ohmic circuit elements*. The non-ohmic elements may be described by a similar expression:

$$V = R(V)I \Leftrightarrow R(V) \equiv \frac{V}{I(V)} \tag{2.19}$$

but with a *variable resistance* $R(V)$. For passive elements, $R(V) > 0$. The resistance $R(V)$ is known as the *static* or *DC resistance*. Figure 2.12 depicts the *υ-i* characteristics for three distinct passive circuit elements.

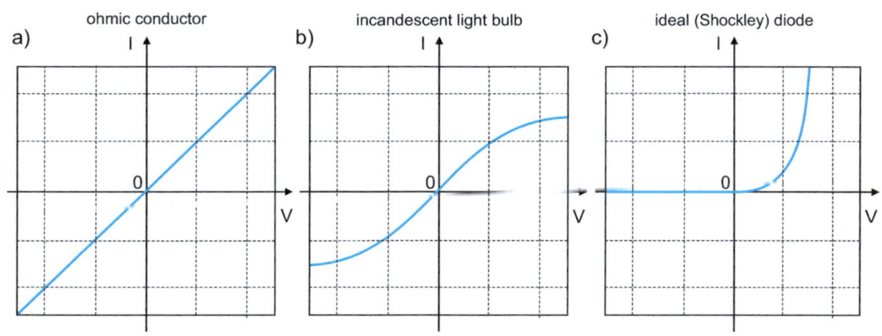

Fig. 2.12. Three *υ-i* characteristics: (**a**) linear—resistance; (**b**) nonlinear—incandescent light bulb; and (**c**) nonlinear—ideal or Shockley diode.

The first element is an ohmic element (*ohmic conductor*) with a constant resistance R. The corresponding *υ-i* characteristic is a straight line—the circuit element is linear. The second element corresponds to an incandescent light bulb. Its resistance R increases when the applied voltage V increases (the conductivity of the radiating filament of wire decreases with increasing absorbed power and wire temperature). Hence, the *υ-i* characteristic bends down and deviates from the straight line—see Fig. 2.12b. This element only approximately follows Ohm's law. It is therefore the nonlinear circuit element. The third element in Fig. 2.12 corresponds to an *ideal (Shockley) diode*. The diode does not conduct at negative applied voltages. At positive voltages, its v-i characteristic is very sharp (exponential). The diode is also the nonlinear circuit element. Strictly speaking, the v-i characteristic of the incandescent light bulb does not belong to the list of circuit elements due to its limited applicability. However, the ideal diode is an important nonlinear circuit element. The nonlinear elements are generally *polar* (non-symmetric) as Fig. 2.12c shows.

2.2.3 Static Resistance of a Nonlinear Element

Once the v-i characteristic is known, we can find the static resistance $R(V)$ of the nonlinear circuit element at any given voltage V_0. For example, the v-i characteristic of the ideal diode shown in Fig. 2.12c is described by the exponential *Shockley equation*:

$$I = I_S\left[\exp\left(\frac{V}{V_T}\right) - 1\right] \tag{2.20}$$

In Eq. (2.20), the constant I_S [A] is the diode *saturation current*. The saturation current is very small. The constant V_T [V] is called the *thermal voltage*.

Example 2.5: Give a general expression for the diode resistance $R(V)$ using Eq. (2.20) and find its terminal values at $V \to 0$ and $V \to \infty$, respectively. Then, calculate static diode resistance R_0 and diode current I_0 when the voltage across the diode is $V_0 = 0.55$ V. Assume that $I_S = 1 \times 10^{-12}$ A and $V_T = 25.7$ mV.

Solution: Using Eq. (2.20) we obtain

$$R(V) = \frac{V}{I_S\left[\exp\left(\frac{V}{V_T}\right) - 1\right]} \tag{2.21}$$

When $V \to 0$, we can use a Taylor series expansion for the exponent. Keeping only the first nontrivial term, one has $\exp(V/V_T) \approx 1 + V/V_T$. Therefore,

$$R(V) \to \frac{V_T}{I_S} \text{ when } V \to 0 \ \text{ (or } V/V_T << 1) \tag{2.22}$$

This value is very large, in excess of 1 GΩ. The diode is thus the open circuit with a good degree of accuracy.

On the other hand, at large V, the exponential factor in Eq. (2.21) greatly increases. Therefore,

$$R(V) \to 0 \text{ when } V \to \infty \tag{2.23}$$

Example 2.5 (cont.):

The diode becomes virtually a short circuit as indicated by the almost vertical slope in Fig. 2.12c. Finally, we obtain the particular values for $V_0 = 0.55\,\text{V}$:

$$R_0 = 275\ \Omega,\quad I_0 = 2.00\ \text{mA} \tag{2.24}$$

2.2.4 Dynamic (Small-Signal) Resistance of a Nonlinear Element

Equally, and perhaps even more important, is the concept of a *dynamic* (or *small-signal*) resistance, r, of the nonlinear circuit element. Other equivalent names include *differential resistance* or *incremental resistance*. Quite often the voltage across the element and the current through it are given by

$$V = V_0 + v,\quad I = I_0 + i \tag{2.25}$$

where V_0 and I_0 are the DC (constant-value) voltage and current related to each other through the static resistance, $V_0 = R_0 I_0$. These values are set with the help of an external DC circuit. On the other hand, quite small components v and i describe a very weak time-varying (AC or pulse) signal. Though weak, this signal contains the major information to be processed. A receiver circuit in your cell phone is an example. Now, how are v and i related to each other? The answer is still given by Ohm's law but written in terms of the *dynamic resistance*, i.e.,

$$v = ri,\quad r \equiv \left.\frac{dV}{dI}\right|_{V=V_0,\ I=I_0} \tag{2.26}$$

This derivative is to be evaluated at the *operating point* V_0, I_0 (also called the *quiescent point* or *Q-point*). Figure 2.13 provides a graphical proof of Eq. (2.26) using the example of an ideal diode. The zoomed-in version of Fig. 2.12c has been used. The *dynamic (small-signal) resistance* is thus the *inverse slope* of the v-i characteristic at the operating point. The exact mathematical proof is performed using a Taylor series expansion.

Exercise 2.6: Determine the small-signal resistance r for the ohmic circuit element with $V = RI,\ R = \text{const}$.

Answer: $r = R$ for any operating point.

The dynamic diode resistance plays a decisive role in the design of amplifiers based on junction transistors. The ideal-diode circuit element becomes a part of the transistor circuit model. The dynamic resistance is also critical for amplifiers which use other transistor types. From the mathematical point of view, finding the static and dynamic resistance is simply finding the function and its first derivative at the operating point.

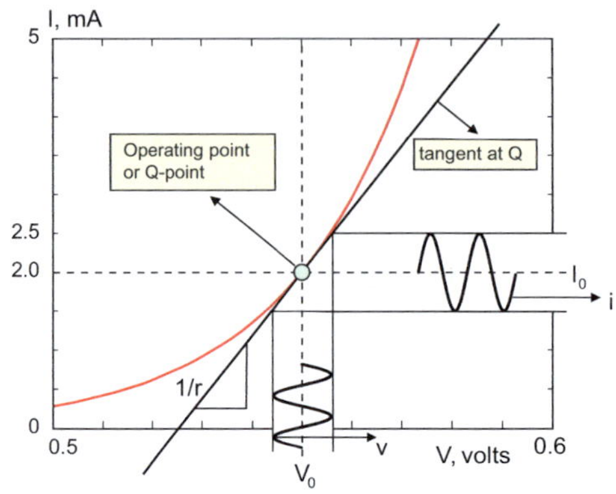

Fig. 2.13. Finding dynamic resistance for an ideal diode.

Example 2.6: Give a general expression for the dynamic diode resistance r using Eq. (2.20) at an arbitrary operating point with current I_0. Then, calculate the dynamic diode resistance r when the voltage across the diode is V_0 =0.55 V. Assume that $I_S = 1 \times 10^{-12}$ A and $V_T = 25.7$ mV.

Solution: Using Eq. (2.20) we obtain

$$V = V_T \ln\left[1 + \frac{I}{I_S}\right] \Rightarrow r(I) = \frac{V_T}{I + I_S} \Rightarrow r \approx \frac{V_T}{I_0} \tag{2.27}$$

since the saturation current I_S may be neglected. At $V_0 = 0.55$ V we obtain $I_0 = 2.00$ mA—see Eq. (2.24). Therefore,

$$r = 12.8\ \Omega \text{ when } V_0 = 0.55 \text{ V}. \tag{2.28}$$

2.2.5 Electronic Switch

In an *electronic switch* (or a *solid-state switch*), an electric quantity (voltage or current) acts as a stimulus—it opens and closes the switch. The electronic switch is inherently a nonlinear device. Figure 2.14 shows a *two-terminal switch* and its (idealized) v-i characteristic. The voltage across the switch V is generated by the main circuit. When this voltage reaches a certain *switch threshold voltage* V_{Th}, the switch becomes closed, it can conduct any current. Further, the voltage across the switch does not change. Switches of this type usually involve pn-junction diodes. For example, the diode v-i characteristic from Fig. 2.12c may approximate the step function in Fig. 2.14b. The *unidirectional switch* shown in Fig. 2.14 can pass the current only in one direction. *Bidirectional switches* can pass current in both directions.

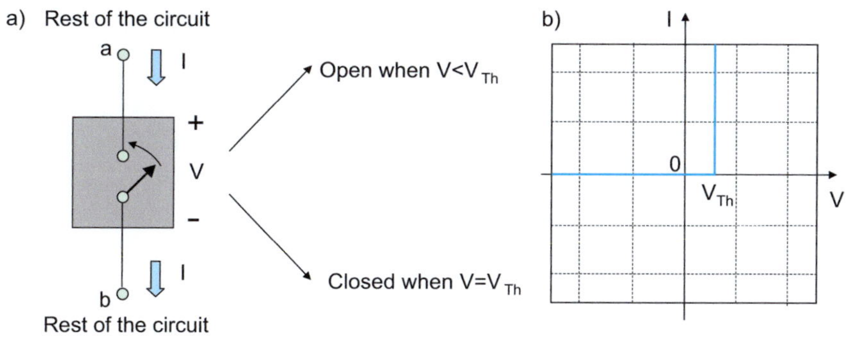

Fig. 2.14. (**a**) Two-terminal unidirectional threshold switch and (**b**) its ideal υ-i characteristic.

Figure 2.15 shows another, *three-terminal electronic switch*. The voltage V controlling the switch operation is now generated by a separate (control) circuit. It is still referenced to the common circuit ground. When the control voltage reaches a certain *switch threshold voltage* V_{Th} *or exceeds* it, the switch closes. The switches of this type involve transistors, either junction or field effect. A distinct feature of the switch in Fig. 2.15 is that the control voltage may have arbitrary values, including $V > V_{\mathrm{Th}}$. Therefore, its υ-i characteristic involves all states to the right of the vertical line in Fig. 2.14.

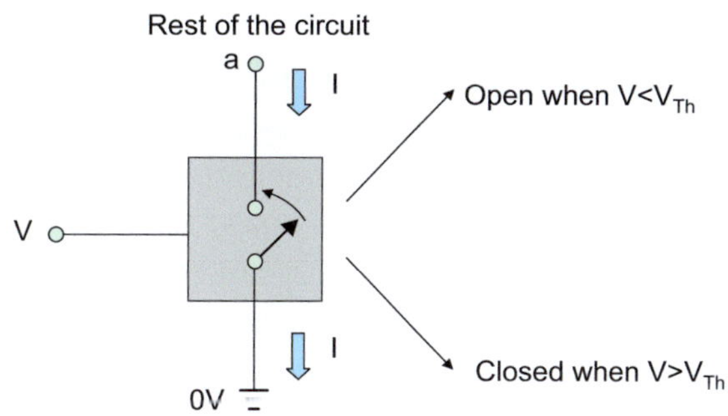

Fig. 2.15. Three-terminal unidirectional threshold switch (pull-down switch).

The switch shown in Fig. 2.15 and its pull-up counterpart are the "heart" of any digital circuit, which is essentially a nonlinear switching circuit. An introduction to digital switching circuits is provided elsewhere.

Exercise 2.7: Based on conditions of example 2.5, determine when a diode switch closes. This condition approximately corresponds to the diode current of 10 mA.

Answer: The diode voltage should be equal to 0.594 V or approximately 0.6 V.

Section 2.3 Independent Sources

2.3.1 Independent Ideal Voltage Source

An *independent ideal voltage source* is an important circuit element. Figure 2.16a shows the corresponding circuit symbol for the DC (steady-state) source. As a circuit element, the voltage source is not symmetric: terminals 1 and 2 (commonly labeled as plus and minus or red and black) may not be interchanged without affecting its operation. In other words, the voltage source is a *polar device*. The voltage source generates a positive *voltage difference* (or *voltage drop* or simply *voltage*) across its terminals, $V_S > 0$, the polarity of which is shown in Fig. 2.16. The term *independent* means that voltage V_S does not vary because of different parameters of an electric circuit (not shown in the figure), which is *implied to be connected* to the voltage source.

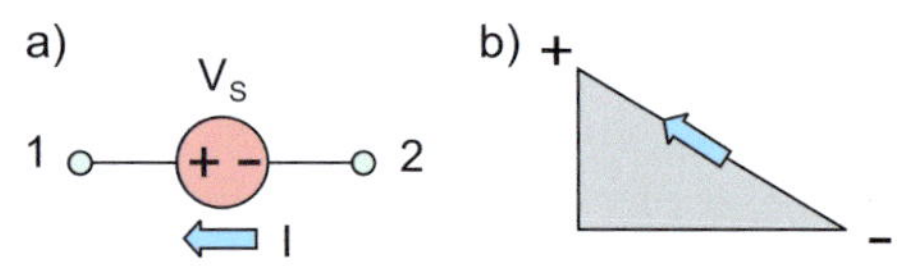

Fig. 2.16. Symbol for an ideal voltage source along with the voltage and current behavior.

Current Through the Voltage Source: Active Reference Configuration

The current *I* flowing through the voltage source is shown in Fig. 2.16 by an arrow. The relation between voltage polarity and current direction depicted in Fig. 2.16 is known as the *active reference configuration*. It is commonly used for all active circuit elements such as voltage and current sources, either dependent or independent. A useful fluid mechanics analogy for the voltage source is water (electric current) that is pushed up the "voltage" hill in Fig. 2.16b by external means. Alternatively, one may think of a water pump that is characterized by a constant pressure drop. The active reference configuration means that the voltage source supplies electric power to the circuit. This configuration is the opposite of the passive reference configuration for the resistance.

υ-i Characteristic of the Voltage Source

Figure 2.17a plots the *υ-i characteristic* of the ideal voltage source. The term *ideal* literally means that the υ-*i* characteristic is a straight vertical line: the ideal voltage source is capable of supplying *any* current to *any* circuit while keeping the *same* voltage V_S across its terminals. In reality, it is not the case since the high currents mean high powers. Therefore, a laboratory power supply—the physical counterpart of the ideal voltage source—will be eventually overloaded as shown in Fig. 2.17b. Figure 2.17c shows a common way of drawing the υ-i characteristic for the voltage source with the axes *interchanged*. For the purposes of consistency, the *x*-axis will always be used as the voltage axis throughout the text.

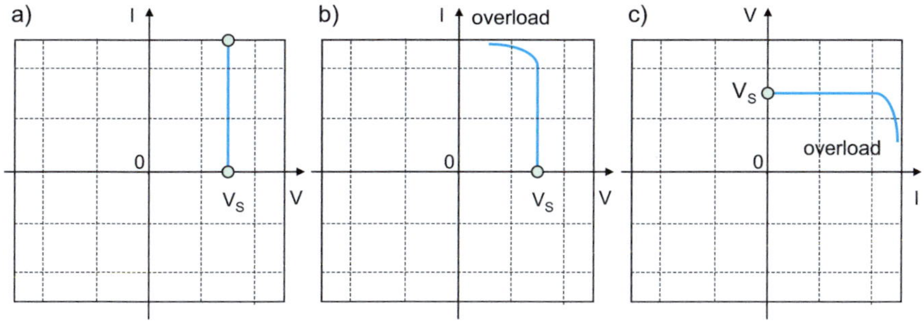

Fig. 2.17. υ-i Characteristics for (**a**) ideal voltage source used in the circuit analysis and (**b**) its physical counterpart—a regulated laboratory power supply. (**c**) Typical way of drawing the υ-i characteristic for the voltage source with the axes interchanged.

Symbols for Independent Voltage Source

Multiple symbols may be used in a circuit diagram to designate the independent ideal voltage source—see Fig. 2.18. All these symbols are equivalent from the circuit point of view, as long as we imply the ideal source. However, their physical counterparts are quite different. The general symbol in Fig. 2.18a implies either an AC to DC converter (the laboratory power source) or a battery. The symbol in Fig. 2.18b relates to a battery and Fig. 2.18c–d depicts battery banks.

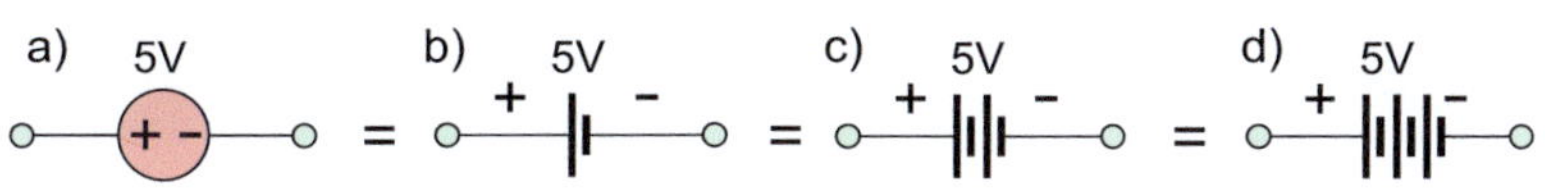

Fig. 2.18. (**a**) Generic DC voltage source, (**b**) single battery, and (**c**) and (**d**) battery banks. All symbols in the circuit diagram are equivalent.

Example 2.7: Solve an electric circuit shown in Fig. 2.19—determine circuit current I and voltage across the resistance V.

Solution: We use the graphical solution—plot the υ-i characteristic of the 2 kΩ resistance and the υ-i characteristic of the voltage source on the same graph to scale—see Fig. 2.19b. The intersection point is the desired solution: $V = 3$ V, $I = 1.5$ mA. Indeed, this simple solution implicitly uses circuit laws (KVL and KCL) studied next.

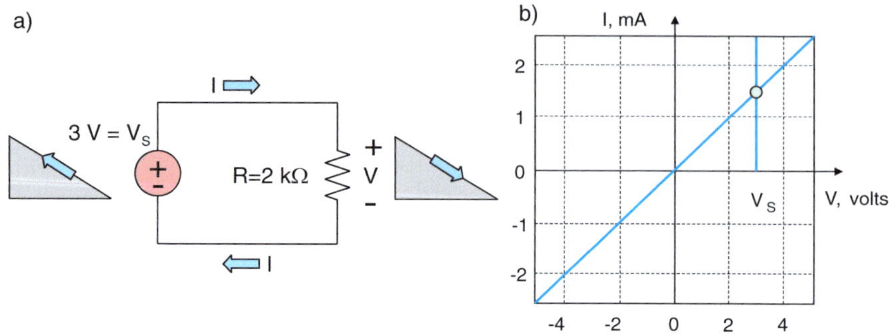

Fig. 2.19. Electric circuit solution in graphical form.

2.3.2 Circuit Model of a Practical Voltage Source

Any *practical voltage source* is modeled as a *combination* of an ideal voltage source V_S and an ideal resistance R *in series*—see Fig. 2.20a. The resistance R reflects the non-ideality of the practical source: it limits the *maximum available source current* and the *maximum available source power* by (similar to the current-limiting resistor):

$$I_{\max} = V_S/R, \quad P_{\max} = V_S I_{\max} = V_S^2/R \tag{2.29}$$

Voltage V_S is called the *open-circuit voltage* of the source for an obvious reason. Similarly, current $I_{\max}$ is called the *short-circuit current* of the source. Once both quantities are measured in laboratory, resistance R (called the *internal source resistance*) may be found using Eq. (2.29).

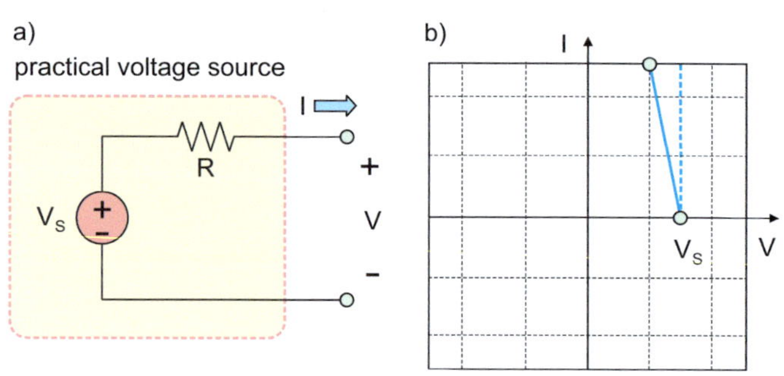

Fig. 2.20. Circuit model of a practical voltage source and its v-i characteristic.

Exercise 2.8: The open-circuit voltage of a voltage source is 9 V; the short-circuit current is 2 A. Determine the internal source resistance.

Answer: 4.5 Ω.

The υ-i characteristic of the practical voltage source is the plot of source current I versus voltage V available from the source in Fig. 2.20a. This voltage is generally less than V_S since any nonzero current I causes a voltage drop of RI across resistance R. One has

$$V = V_S - RI \Rightarrow I = \frac{V_S - V}{R} \tag{2.30}$$

This υ-i characteristic is plotted in Fig. 2.20b by a solid line. The deviation from the straight vertical line characterizes the degree of non-ideality. Emphasize that any laboratory power supply is indeed a practical voltage source. However, using a special circuit, its input is *regulated* so that the output voltage does depend on the output current, at least over a reasonable range of currents. Therefore, instead of Fig. 2.20b we arrive at a more reliable voltage source from Fig. 2.17b.

Exercise 2.9: Determine internal source resistance for the source illustrated in Fig. 2.20b given that every horizontal division is 3 V and every vertical division is 1 A.

Answer: 0.6 Ω.

2.3.3 Independent Ideal Current Source

An *independent ideal current source* is dual of the ideal voltage source. Figure 2.21a shows the corresponding circuit symbol for the DC (constant-current) source. As a circuit element, the constant-current source is *directional*: terminals 1 and 2 indicate the direction of the current flow. The current source generates a positive constant current, $I_S > 0$, which flows from terminal 2 to terminal 1 in Fig. 2.21. The term *independent* means that current I_S does not vary because of different parameters of an electric circuit (not shown in the figure), which is *implied to be connected* to the current source.

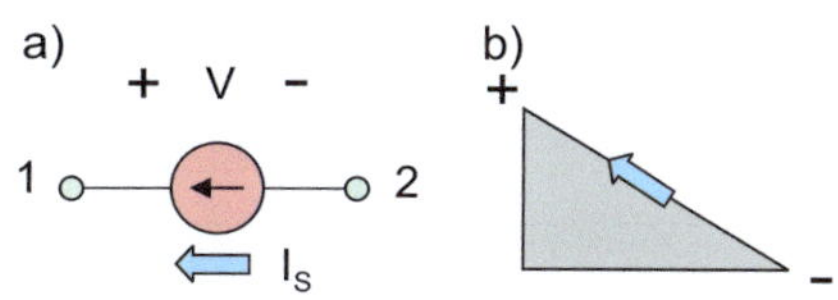

Fig. 2.21. Symbol for the ideal current source along with voltage and current designations.

Voltage Across the Current Source: Active Reference Configuration

Once the current source is connected to a circuit, a voltage V will be created across it. The voltage polarity is indicated in Fig. 2.21a. The relation between voltage polarity and current direction shown in Fig. 2.21 is again the *active reference configuration*, similar to the voltage source. A useful fluid mechanics analogy for the electric current source is a water pump that creates a constant water supply (e.g., 0.5 ft^3/s). Indeed, this water pump will be characterized by a certain pressure difference across its terminals, which is the analogy of voltage V in Fig. 2.21.

υ-i Characteristic of the Current Source

Figure 2.22a plots the *υ-i characteristic* of an ideal current source. Compared to the ideal voltage source, the graph is rotated by 90 degrees. The term *ideal* again means that the *υ-i* characteristic is the straight horizontal line: the ideal current source is capable of creating *any* voltage across its terminals while keeping the *same* current I_S flowing into the circuit. In reality, it is not the case since high voltages mean high powers. Therefore, a laboratory current power supply—the physical counterpart of the ideal current source—will eventually be overloaded as shown in Fig. 2.17b. The current laboratory supplies are rarely used (one common use relates to transistor testing); they are less common than the voltages supplies. However, the current sources are widely used in transistor circuits, both integrated and discrete. There, the current sources are created using dedicated transistors. Furthermore, photovoltaic sources are essentially current sources.

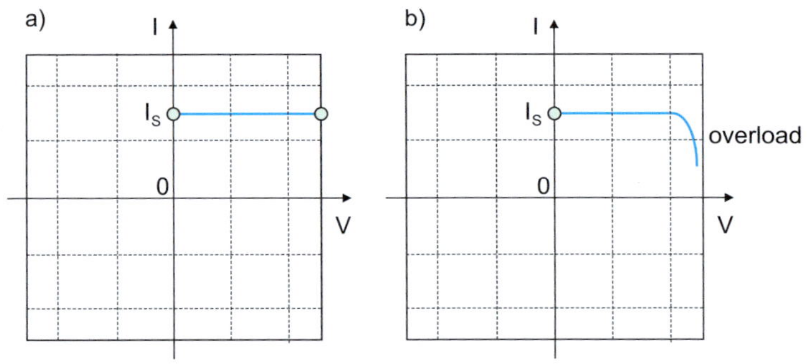

Fig. 2.22. *υ*-i Characteristics of (**a**) an ideal current source and (**b**) its physical counterpart.

Symbols for Independent Current Source

A few equivalent symbols may be used in a circuit diagram to designate the independent ideal current source; see Fig. 2.23. All these symbols are equivalent as long as we imply the ideal source. The symbol in Fig. 2.23a is used in North America, the symbol in Fig. 2.23b is European, and the symbol in Fig. 2.23c may be also found in older texts.

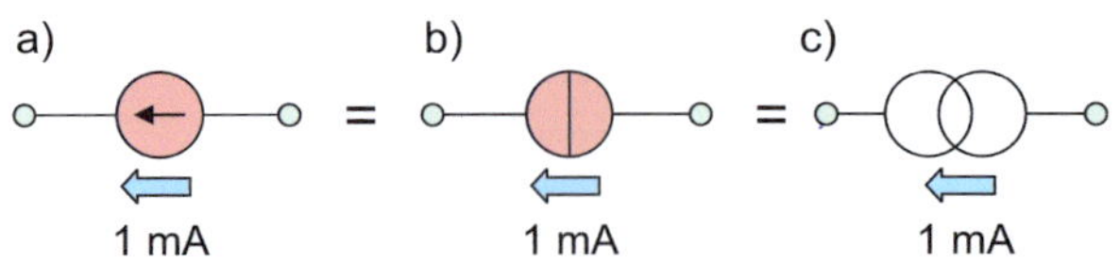

Fig. 2.23. Equivalent symbols of the current source in the circuit diagram.

Example 2.8: Solve an electric circuit shown in Fig. 2.24—determine voltage V across the source and across the resistance.

Solution: Similar to the voltage source, we use a graphical solution—plot the *υ-i* characteristic of the 2-kΩ resistance and the *υ-i* characteristic of the current source on the same graph to scale; see Fig. 2.24b. The intersection point gives us the desired solution: $V = 3$ V. Note that the solutions for this example and the solution for Example 2.7 coincide. This means that, under certain conditions, we can interchange both sources without affecting the circuit performance. Indeed, the graphical solution implicitly uses the circuit laws (KVL and KCL) studied in detail next.

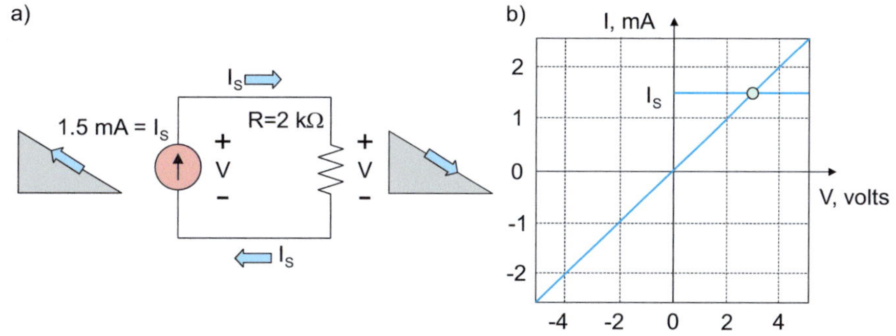

Fig. 2.24. Electric circuit solution in the graphical form.

2.3.4 Circuit Model of a Practical Current Source

Any *practical current source* is modeled as a *combination* of the ideal current source I_S and the ideal resistance R *in parallel*—see Fig. 2.25a. The resistance R reflects the non-ideality of the practical source: it limits the *maximum available source voltage* and the *maximum available source power* by

$$V_{\max} = RI_S, \quad P_{\max} = V_{\max} I_S = RI_S^2 \tag{2.31}$$

Voltage $V_{\max}$ is again called the *open-circuit voltage* of the source. Similarly, current I_S is called the *short-circuit current* of the source. Once both the quantities are measured, resistance R (called the *internal source resistance*) may be found using Eq. (2.31).

Exercise 2.10: The open-circuit voltage of a current source is 9 V; the short-circuit current is 2 A. Determine the internal source resistance.

Answer: 4.5 Ω.

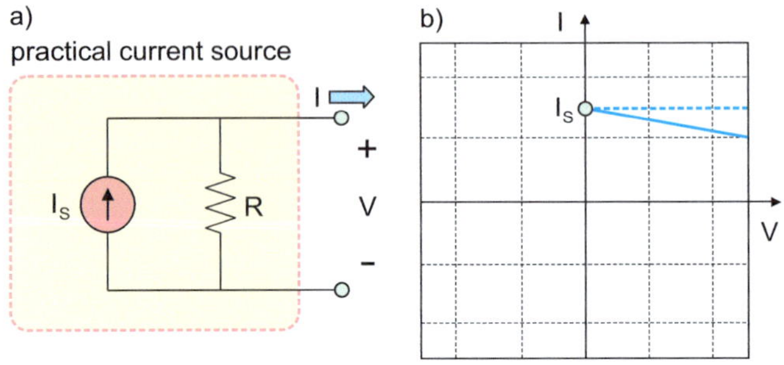

Fig. 2.25. Circuit model of a practical current source and its v-i characteristic.

The v-i characteristic of the practical current source is the plot of current I available from the source versus voltage V across the source in Fig. 2.25b. This current is generally less than I_S since a portion of I_S flows through the internal resistance R, i.e.,

$$I = I_S - \frac{V}{R} \tag{2.32}$$

This v-i characteristic is plotted in Fig. 2.25b by a solid line. The deviation from the straight horizontal line characterizes the degree of non-ideality.

Exercise 2.11: Determine the internal source resistance for the source illustrated in Fig. 2.25b given that every horizontal division is 3 V and every vertical division is 1 A.

Answer: 15 Ω.

2.3.5 Operation of the Voltage Source

Operation of a voltage power supply of any kind (an electric generator, a chemical battery, a photovoltaic cell, etc.) might be illustrated based on the *charge separation principle* very schematically depicted in Fig. 2.26. We need to deliver electric power to a load modeled by an equivalent resistance R_L. First, we consider in Fig. 2.26 a charged capacitor with a charge Q connected to a load resistor R_L at an initial time moment. The capacitor voltage V is related to charge by $V = Q/C$ where C is the (constant) capacitance. The capacitor starts to discharge and generates a certain load current I_L. At small observation times, the change in Q is small, so is the change in V. Therefore, the capacitor initially operates as a voltage power supply with voltage V. However, when time progresses, the capacitor discharges and the voltage V eventually decreases.

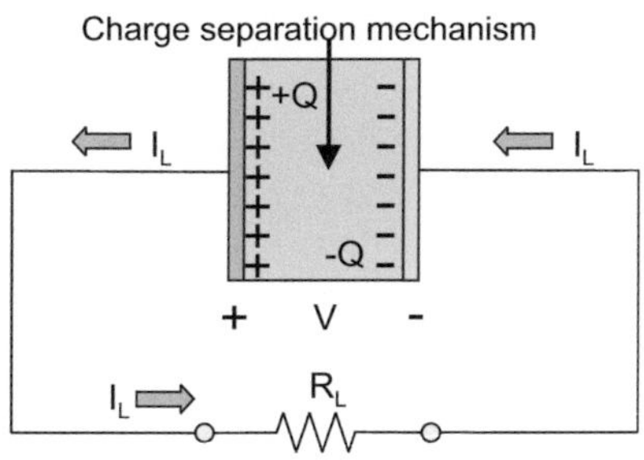

Fig. 2.26. Power source schematically represented as a capacitor continuously charged by a charge separation mechanism—the *charge pump.*

How could we keep V constant, i.e., continuously charge the capacitor? A charge separation mechanism should be introduced between the hypothetic capacitor plates to continuously compensate for the charge leakage. That mechanism may have the forms:

1. For an electromechanical generator, this is the *Lorentz force* that acts on individual electrons in a conductor and pushes them to one conductor terminal while creating the opposite charge density on the opposite conductor terminal. The macroscopic effect of the Lorentz force is the *Faraday's law of induction.*
2. For a battery, these are chemical reactions at the electrodes which cause a charge separation, i.e., positive metal ions dissolve in the electrolyte and leave excess electrons in the metal electrode on the left in Fig. 2.26.
3. For the photovoltaic cell, this is a built-in potential of the semiconductor pn-junction that separates light-generated negative carriers (electrons) and positive carriers (holes) as shown in Fig. 2.26.

Indeed, the capacitor analogy in Fig. 2.26 is only an illustrative approach, especially for electromechanical power generation. Below, we will consider a few specific examples.

2.3.6 Application Example: DC Voltage Generator with Permanent Magnets

A realistic electromechanical voltage source—a basic *DC generator* with permanent magnets—is shown in Fig. 2.27. This generator setup makes use of the *Lorentz force*:

$$\vec{f} \equiv q(\vec{v} \times \vec{B}) \tag{2.33}$$

The Lorentz force acts on charge q moving with a velocity $\vec{v}$ in an external magnetic field with the vector flux $\vec{B}$ measured in tesla (T). The force itself is measured in newtons. The cross symbol in Eq. (2.33) denotes the vector product of two vectors evaluated according to the right-hand rule. Shown in Fig. 2.27a are two permanent magnets (stator of the generator) responsible for creating the magnetic flux $\vec{B}$ emanating from the north pole

(N) and terminating at the south pole (S). The armature (rotor) rotates clockwise in Fig. 2.27b with the armature velocity $\vec{v}$.

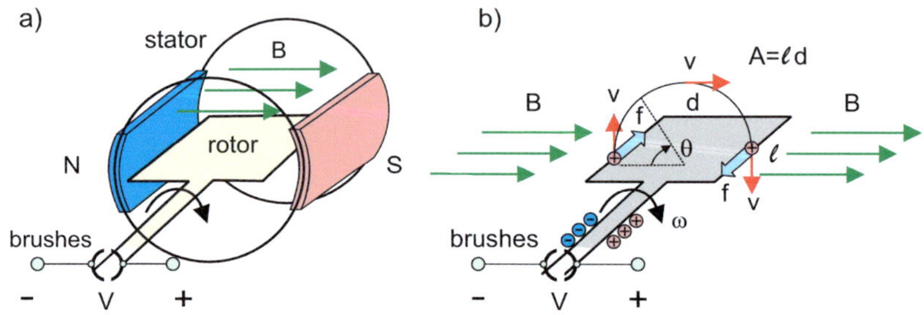

Fig. 2.27. Charge separation in a DC electromechanical generator.

When the flux density $\vec{B}$ is applied, every positive charge $+q$ in the armature segment l will experience a Lorentz force with the magnitude $f = +qvB$ which will move this charge toward the right terminal of the armature in Fig. 2.27b. Similarly, every negative charge $-q$ in the armature would experience the equal but oppositely directed Lorentz force $f = -qvB$ which will move this charge toward the left terminal. Hence, a charge separation occurs along the armature which will give rise to an induced voltage V. Total work W of the Lorentz force on a charge q along the entire armature path in Fig. 2.27b is given by $W = 2lf$. This work divided by the amount of charge determines the equivalent voltage that will be developed on the generator terminals, i.e., the *instantaneous generator voltage* $V = W/q = 2lvB$. If the armature rotates at an angular speed ω (rad/s), the charge velocity perpendicular to the field is given by $v = d/2\omega\cos\theta$ (m/s). Plugging in this expression and averaging over angles θ from 0 to $\pi/2$, we obtain the *average generator voltage* in the form

$$V = ldB\langle\cos\theta\rangle = (2/\pi)AB\omega \quad [\mathrm{V}] \tag{2.34}$$

where A is the armature area. If the rotor has N turns, the result is multiplied by N. The same expression for the voltage is obtained using the *Faraday's law of induction*. A *regulator circuit* is necessary to obtain a flat DC voltage without ripples. Any brushed DC motor operates as a generator when its shaft is rotated with a certain speed. The generated open-circuit voltage may be observed in laboratory with the oscilloscope.

Exercise 2.12: Determine average open-circuit generator voltage in Fig. 2.27 given $A = 0.1\ \mathrm{m}^2$, $B = 0.2\ \mathrm{T}$, $\omega = 20$ rad/s (191 rpm), and the armature with 20 turns.

Answer: 5.1 V.

2.3.7 Application Example: Chemical Battery

A chemical reaction in a battery induces a continuous charge separation. The "charge pump" so constructed, once connected to a load, is able to create a continuous electric current into a load and a voltage difference across it. You are probably aware of the quest to improve the venerable battery. Extensive coverage in the media and in technical journals frequently reports on new chemical compounds and control circuits. They target smaller, more powerful rechargeable batteries for such diverse devices as portable computers, cell phones, sensors, and automobiles. Specifically, it is the automotive sector which implements hybrid vehicle technology where powerful electro motors in conjunction with high-performance batteries are supplementing, even completely replacing, conventional combustion engines. For a standard *chemical battery*, the two important parameters are *battery voltage* and *battery capacity*. The capacity, Q, is a new quantity that is needed because of a battery's inability to provide constant current and power for an infinite time duration. How a battery behaves over time is illustrated in Fig. 2.28 where we monitor the power and current as a function of time. A key time constant is the so-called discharge time, which is critically dependent on the attached load.

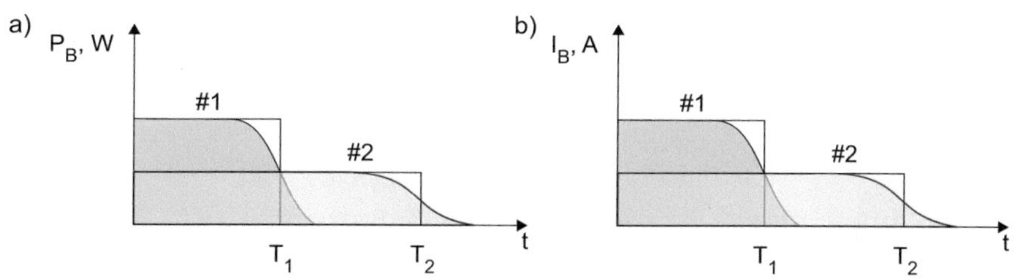

Fig. 2.28. Generic plots of delivered power, P_B, and electric current, I_B, for two different loads labeled #1 and #2. The discharge times T_1 and T_2 correspond to the loads #1 and #2, respectively.

When a load of resistance R_L is connected to the battery, and the battery's internal resistance R is negligibly small compared to that resistance, the circuit current, I_B, and the power delivered by the battery, P_B, are determined based on Ohm's law:

$$I_B = \frac{V_B}{R}, \quad P_B = V_B I_B \tag{2.35}$$

The *total energy*, E_B, stored in the battery and then delivered to the circuit is a *fixed constant*. Its value depends on the battery type and size. The total energy in joules is given by the time integral of delivered power over time, i.e.,

$$E_B = \int_0^\infty P_B(t')dt' \tag{2.36}$$

We can assume that the total energy is a finite constant; it follows from Eq. (2.36) that the delivered power must drop to zero at a finite time T. This is schematically shown in

Fig. 2.28 for two different load resistances, #1 and #2, which require two different circuit currents. Even though the two power curves in Fig. 2.28a are different, the area under those curves, denoting the total energy stored in the battery, remains *the same* to a sufficient degree of accuracy. The battery's terminal voltage V_B also remains *approximately constant* over the entire operation cycle and even afterwards. It is the battery's current I_B that finally sharply decreases with time and causes a drop in power, as seen in Fig. 2.28b. Let us consider the simplest case where the current is a constant for $t < T$ and at $t = T$ drops to zero and stays zero for $t > T$. From Eq. (2.36), it follows that

$$E_B = \int_0^\infty P_B(t')dt' = \int_0^T V_B I_B dt' = [TI_B]\, V_B \tag{2.37}$$

The expression in the square brackets is the definition of the *battery capacity*, Q:

$$Q \equiv TI_B = \frac{E_B}{V_B} \tag{2.38}$$

Since the battery terminal voltage is always known, its capacity determines the total energy stored in the battery. The capacity is measured in A·h (Ah) or for small batteries in mA·h (mAh). The capacity rating that manufacturers print on a battery is based on the product of 20 h multiplied by the maximum constant current that a fresh battery can supply for 20 h at 20°C while keeping the required terminal voltage. The physical size of batteries in the USA is regulated by the American National Standards Institute (ANSI) and the International Electrotechnical Commission (IEC). Table 2.3 lists the corresponding parameters of some common batteries.

Example 2.9: A 12-V battery rated at a capacity of $Q = 100$ A·h may deliver 5 A over a 20-h period, 2.5 A over a 40-h period, or 10 A over a 10-h period. Find the total energy delivered by the battery provided that its internal resistance is negligibly small.

Solution: The total energy delivered by the battery is equal to

$$E_B = V_B Q = 12 \cdot 100 \ \ \text{V} \cdot \text{A} \cdot \text{h} = 1200 \ \ \text{W} \cdot \text{h} = 4.32\,\text{MJ} \tag{2.39}$$

It remains constant for each case. This example shows that the electric energy can be measured either in joules or in Wh or more often in kWh. Clearly, 1 Wh = 3600 J.

Circuit Model of a Battery

As a practical voltage source, a battery always has a small, but finite *internal* resistance, R. Battery's equivalent circuit therefore includes the ideal voltage source and the internal resistance in series—see Fig. 2.29. Even though the values of R are small, the internal resistance has critical implications affecting both the battery's efficiency and its ability to provide a high instantaneous power output. In general, it is difficult to directly measure

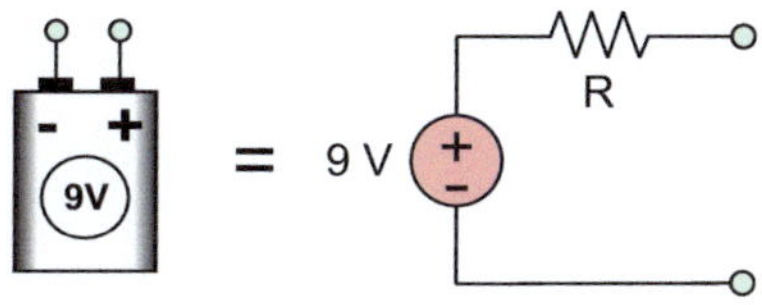

Fig. 2.29. Circuit model of a battery: the ideal voltage source in series with an internal resistor.

the internal resistance of batteries, you would need a calibrated load resistor and sophisticated measurement equipment to precisely measure voltages and currents.

Exercise 2.13: A 12-V battery has an internal resistance of 10 Ω. What are the maximum current and the maximum power that the battery can output?

Answer: $I_{\max} = 1.2\ \text{A},\ P_{\max} = 14.4\ \text{W}$

Many battery types have been developed for a wide range of applications. They differ both in *battery energy storage* per kg of weight, or unit volume, and in power delivery per kg of weight, or per unit volume. In particular, modern heavy-duty, deep-cycle batteries may sport the following properties:

$$\text{Energy storage}: \quad 150\,\text{W}\cdot\text{h/l}, \tag{2.40}$$

$$\text{Power density}: \quad 2\ \text{kW/l}. \tag{2.41}$$

You can compare Eqs. (2.40) and (2.41) with the last row of Table 2.3 and establish the approximate density of the battery device.

Table 2.3. Characteristics of batteries (from multiple datasheets).

Battery size/type	Rechargeable	Voltage (cell)	Capacity (A·h)	Resistance (R)
AAA	No	1.5	1.3	100–300mΩ for alkaline battery per cell
AA	No	1.5	2.9	
C	No	1.5	8.4	
D	No	1.5	20.5	
9 V	No	9.0	0.6	~400 mΩ
Lithium batteries	Yes	3.6–3.7	0.7–1.5	~300 mΩ
Lead acid starter battery (automotive, deep cycle)	Yes	12.6	~600A for 30 s at 32 °F before voltage drops to 7.20 V	<100 mΩ
Deep-cycle marine, electric vehicles	Yes	Variable: ~ 30 W·h per kg of weight, or ~ 108 kJ per kg of weight		~200 mΩ

Section 2.4 Dependent Sources and Time-Varying Sources

2.4.1 Dependent Versus Independent Sources

If the strength of the source (voltage or current) does not vary because of variation of the circuit parameters, the source is an *independent source*. The voltage and current sources considered previously are the independent sources. However, if the strength of the source is controlled by some dedicated circuit parameters, the sources are called *dependent sources*. Figure 2.30 shows circuit symbols (diamonds) for the dependent voltage and current sources. The ideal dependent sources are the important circuit elements, along with the independent sources. We explain the notations in Fig. 2.30 as follows.

1. The ideal independent and dependent sources may generate not only the steady-state (DC) voltages and currents, but also *arbitrary time-varying* voltages and currents. To underscore this fact, we will use the *lowercase* notations for voltages and currents, respectively.
2. The dependent sources generate (or *output*) voltage or current in response to some *input* voltage or current—the stimulus. To underscore this fact, we will use the subscript $_{\mathrm{OUT}}$ for the generated voltage or current strengths. This is in contrast to the subscript $_{\mathrm{S}}$, which always denotes the independent sources.
3. The stimulus voltage and the stimulus current (not yet shown in Fig. 2.30) will be denoted by v_{in} and i_{in}, respectively.

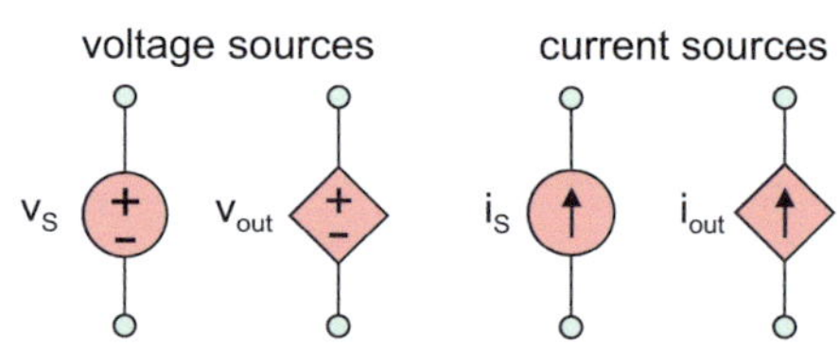

Fig. 2.30. Circuit symbols for ideal independent and dependent sources, respectively. Lowercase notations for voltages and currents are used to underscore possible time variations.

2.4.2 Definition of Dependent Sources

The stimulus voltage v_{in} is the voltage across a certain resistance. Likewise, the stimulus current i_{in} is the current through a certain resistance. Figure 2.31 shows *four major types of dependent sources* where the stimulus and the response are combined into one block—the shaded rectangle. Such a combination reflects the physical reality since this block usually corresponds to a single circuit component—a transistor or an amplifier. Emphasize that the resistances in Fig. 2.31 may be reduced to an *open circuit* for dependent voltage sources or to a *short circuit* for dependent current sources, respectively, if required. Also note that another circuit element may be present in place of the resistance, for example, the ideal diode. The bottom (ground) nodes in every circuit in Fig. 2.31 may be interconnected to emphasize the same voltage reference.

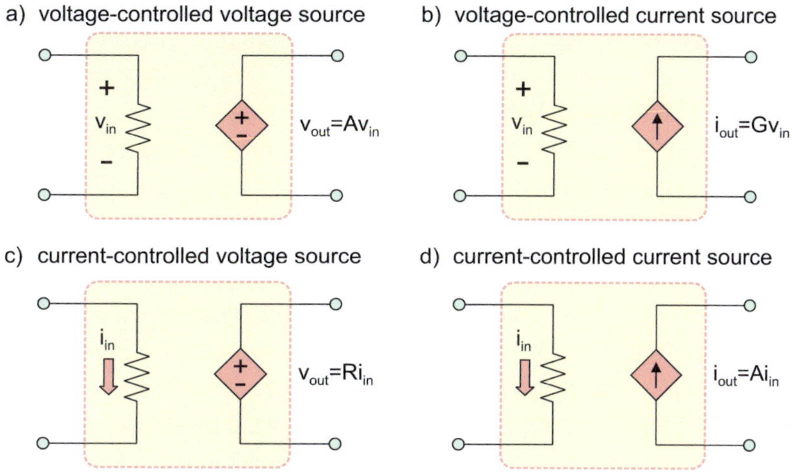

Fig. 2.31. Four major types of dependent sources.

Voltage-Controlled Voltage Source

This dependent source is shown in Fig. 2.31a. The source voltage or the output voltage follows the input voltage according to a linear law

$$v_{\mathrm{out}} = Av_{\mathrm{in}} \tag{2.42}$$

where the dimensionless constant A is called the *open-circuit voltage gain* of the dependent source. However, units of V/V or V/mV are often used. For example, the expressions $A = 5\ \mathrm{V/mV}$ and $A = 5000$ are equivalent. Equation (2.42) is valid *irrespective* of the circuits connected to the dependent source to the right and to the left in Fig. 2.31a. In this sense, the voltage-controlled voltage source is the ideal circuit element. Such a source is a *voltage amplifier.*

Voltage-Controlled Current Source

This dependent source is shown in Fig. 2.31b. The source current or the output current follows the input voltage according to a linear law:

$$i_{\mathrm{out}} = Gv_{\mathrm{in}} \tag{2.43}$$

where the constant G with units of $\mathrm{A/V} = \Omega^{-1} = \mathrm{S}$ is called the *transconductance* of the dependent source, similar to the name conductance. For example, the expressions

$G = 0.5\ \ \mathrm{A/mV}$ and $G = 500\ \ \mathrm{S}$ are equivalent. Emphasize that the transconductance has nothing in common with the conductance (inverse resistance) of a passive resistor. Equation (2.43) is also valid *irrespective* of the circuits connected to the dependent source to the right and to the left in Fig. 2.31b. In this sense, the voltage-controlled current source is again the ideal circuit element. Such a source is a *transconductance amplifier.*

Current-Controlled Voltage Source

This dependent source is shown in Fig. 2.31c. The source voltage or the output voltage follows the input current through the resistance in Fig. 2.31c according to a linear law:

$$v_{\mathrm{out}} = R i_{\mathrm{in}} \tag{2.44}$$

where the constant R with units of $\mathrm{V/A} = \Omega$ is called the *transresistance* of the dependent source, similar to the name resistance. For example, the expressions $R = 5\mathrm{V/mA}$ and $R = 5000\ \ \Omega$ are equivalent. Emphasize that the transresistance has nothing in common with the resistance of a passive resistor. Equation (2.44) is again valid *irrespective* of the circuits connected to the dependent source to the right and to the left in Fig. 2.31c. In this sense, the current-controlled voltage source is also an ideal circuit element. Such a source is a *transresistance amplifier.*

Current-Controlled Current Source

The last dependent source is shown in Fig. 2.31d. The source current or the output current follows the input current according to a linear law:

$$i_{\mathrm{out}} = A\, i_{\mathrm{in}} \tag{2.45}$$

where the dimensionless constant A is called the *short-circuit current gain* of the dependent source. However, units of A/A or A/mA are often used. For example, the expressions $A = 0.5\ \ \mathrm{A/mA}$ and $A = 500$ are equivalent. We repeat that Eq. (2.45) is valid *irrespective* of the circuits connected to the dependent source to the right and to the left in Fig. 2.31d—the voltage-controlled voltage source is the ideal circuit element. Such a source is a *current amplifier.*

2.4.3 Transfer Characteristics

The dependent sources do not possess the v-i characteristic. Instead, a *transfer characteristic* of the source is used, which relates the output voltage or current to the input voltage or current. For example, the transfer characteristic of the voltage-controlled voltage source follows Eq. (2.42). It is a straight line in the v_{in}, v_{out} plane (the *xy*-plane), with the slope equal to A. Other linear transfer characteristics are obtained similarly.

Example 2.10: Solve a circuit shown in Fig. 2.32a—determine current i through the 1-kΩ resistance. The independent voltage source is given by $v_S = 0.5 + 2\cos 2t$ [V]; the open-circuit voltage gain of the dependent voltage source is 5 V/V.

Solution: The input voltage is simply the independent-source voltage, $v_{in} = v_S$. The output voltage is $v_{out} = 5v_{in} = 5v_S$. The output current follows Ohm's law:

$$i = \frac{v_{OUT}}{1\ \mathrm{k}\Omega} = 2.5 + 10\cos 2t \quad [\mathrm{mA}] \tag{2.46}$$

Note that all circuit parameters now become time dependent. However, this does not change the solution compared to the steady-state case.

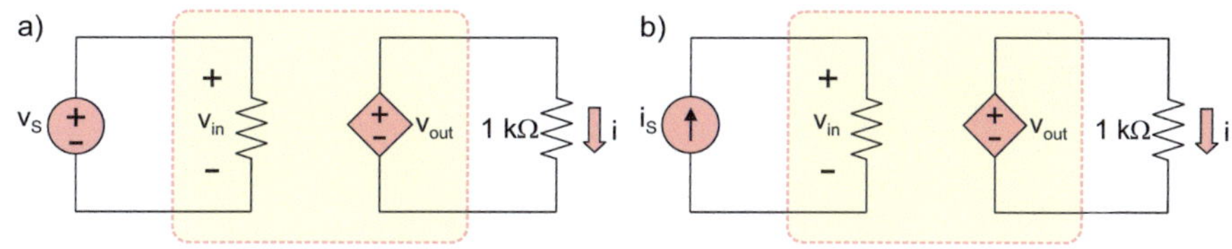

Fig. 2.32. Two circuits with the dependent voltage-controlled voltage source.

Exercise 2.14: Solve an electric circuit shown in Fig. 2.32b—determine current i through the 1-kΩ resistance. The independent current source is given by $i_S = 0.5 + 2\cos 2t$ [mA]; the open-circuit voltage gain of the dependent source is 5 V/V. The leftmost resistance in Fig. 2.32b (often called the *input resistance*) is 1 kΩ.

Answer: $i = 2.5 + 10\cos 2t$ [mA] (the same answer as in Example 2.10).

2.4.4 Time-Varying Sources

Figure 2.33 shows a number of commonly used symbols for the voltage source—the ideal circuit element—which differentiate its time-related behavior. Figure 2.33a shows the steady-state ideal DC voltage source. Figure 2.33b indicates an arbitrary (either steady-state or variable) ideal voltage source. Figure 2.33c–d indicates a time-harmonic ideal *AC* (alternating current) *voltage source* described by a cosine function in the form

$$v_S(t) = V_m \cos(\omega t + \varphi) \quad [\mathrm{V}] \tag{2.47}$$

where V_m is the AC *source amplitude* with the units of volts, ω is the AC *source angular frequency*, and φ is the phase in degrees or radians. The AC current sources do not have special symbols—the symbols from Fig. 2.30 are used. The same is valid for the dependent AC sources, both voltage and current.

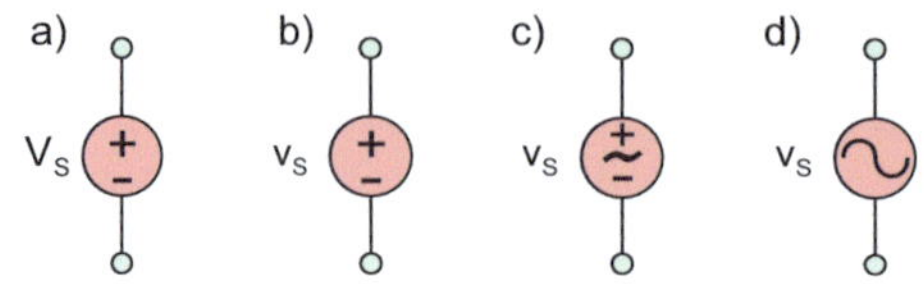

Fig. 2.33. Symbols for independent voltage source which imply (**a**) the DC source (capital V_S), (**b**) an arbitrary source (lowercase v_S), and (**c**) and (**d**) AC time-harmonic sources (lowercase v_S).

AC Source Polarity

Since the voltage in Eq. (2.47) is alternating, the polarity of the AC voltage source is also variable. This circumstance is reflected in Fig. 2.33d where the source polarity is not shown at all. However, for reference purposes, and when the multiple sources of the same frequency are present in the circuit, it is always useful to designate the source polarity. Reversing the AC source polarity means changing the phase in Eq. (2.47) by $\pm 180°$.

Section 2.5 Ideal Voltmeter and Ammeter: Circuit Ground

2.5.1 Ideal Voltmeter and Ammeter

The ubiquitous voltmeter and ammeter are devices designed to measure voltages and currents. Both devices are usually assembled in one unit known as a digital multimeter (DMM). From the circuit point of view, the *ideal voltmeter* is an *open circuit* which conducts zero current as shown in Fig. 2.34. An *ideal ammeter* is a *short circuit* which conducts any current with zero resistance—see the same figure. In reality, the voltmeter will conduct a small leakage current, and the ammeter will exhibit a small resistance.

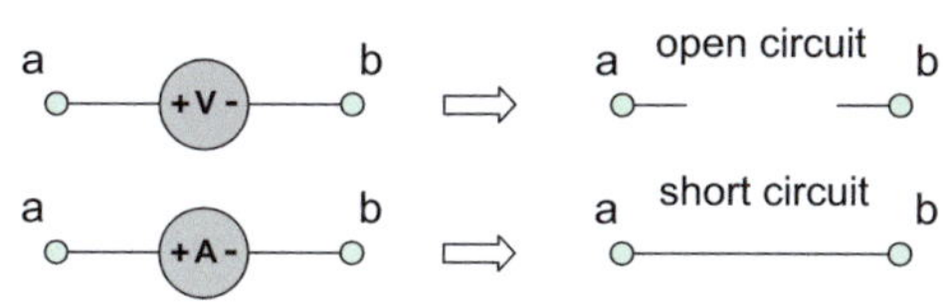

Fig. 2.34. Circuit equivalencies for ideal voltmeter and ammeter.

These features guarantee that the connection of the measurement device will not change the circuit operation. Figure 2.35 shows the proper connection of the voltmeter and ammeter to measure current through circuit element *A* and voltage across this element.

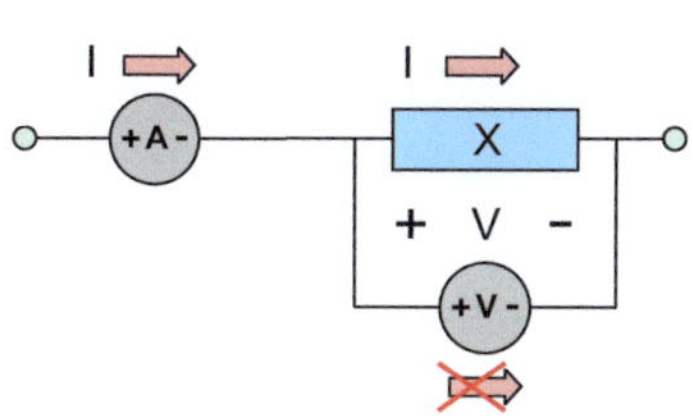

Fig. 2.35. Correct connection of voltmeter and ammeter for voltage and current measurements.

The ammeter is always connected *in series* with element *X*. In other words, to connect the ammeter we must *break* the circuit either before or after element *X*. Since the ammeter has no resistance, it acts just like an ideal wire and thus does not perturb the electric circuit. On the other hand, the voltmeter is always connected *in parallel* with element *X*. The circuit current *I* in Fig. 2.35 cannot flow through the voltmeter, which acts as an open circuit. As required, it will flow through element *X*. We conclude that an ideal voltmeter does not perturb the circuit either. Generally, voltage measurements are simpler to perform than current measurements.

Wrong Connections of Ammeter and Voltmeter

The ammeter connected in parallel will *short out* the element *A*: the current will flow through the ammeter. If the element *A* were a load, there would no longer be a load resistance in the circuit. And with no attached load, the power supply will deliver the largest possible current, which will likely burn out the ammeter fuses or destroy other circuit elements. The voltmeter is an open circuit. Connecting it in series is equivalent to physically breaking the circuit. The circuit will no longer properly function.

2.5.2 Circuit Ground: Fluid Mechanics Analogy

Consider first a fluid mechanics analogy of an ungrounded electric circuit shown in Fig. 2.36a.

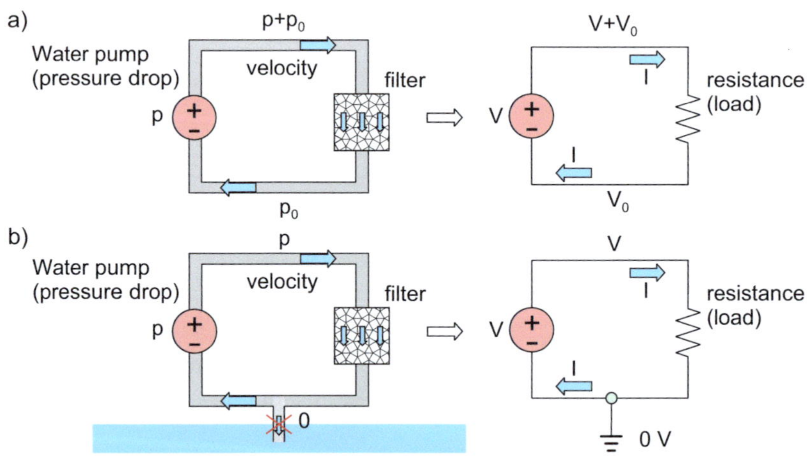

Fig. 2.36. A large reservoir at atmospheric pressure attached to a pumping system serves as an analogy to the ground connection in an electric circuit.

A water pump creates a constant pressure difference p between its terminals, which forces water to move through the filter. The pressure water pump is less common than a water pump of a constant flux; however, it exactly corresponds to the voltage power supply of the electric circuit. For entirely closed (*isolated*) pumping systems, such as those shown in Fig. 2.36a, the water pressure inside the system can in principle have an arbitrary pressure deviation p_0 from the ambient atmospheric pressure. A large p_0 is in practice undesirable since if the system breaks, then a large pressure difference with regard to atmospheric pressure will cause high-speed water leakage. Similarly, an isolated electric circuit may have an arbitrary voltage V_0 versus ground voltage, due to static charge accumulation. We could make the reference level equal to atmospheric pressure (make p_0 equal to zero) if we connect tubing to a large water reservoir at atmospheric pressure as shown in Fig. 2.36b. There is indeed no water flow through such a connection; but the pressure level is normalized. A similar situation takes place for the electric ground shown in Fig. 2.36b. By connecting a point in the circuit to a ground, we normalize the circuit voltage to the earth's voltage level, which we define to be 0 V, and eliminate any static charges. There is *no current flow* through the ground connection, except, maybe, for the first time moment. Therefore, this connection is only a *voltage reference point*. A similar analogy holds for a current source (pump of a constant flux).

2.5.3 Types of Electric Ground

Figure 2.37 shows three different types of electric ground connections. The first one is the *earth ground*. A true earth ground, as defined by the National Electrical Code (USA), physically consists of a conductive pipe or rod driven into the earth to a minimum depth of 8 feet. Obviously, it is not always possible to physically connect the circuit directly to the earth. Some examples include a cell phone, an automobile, or an airplane.

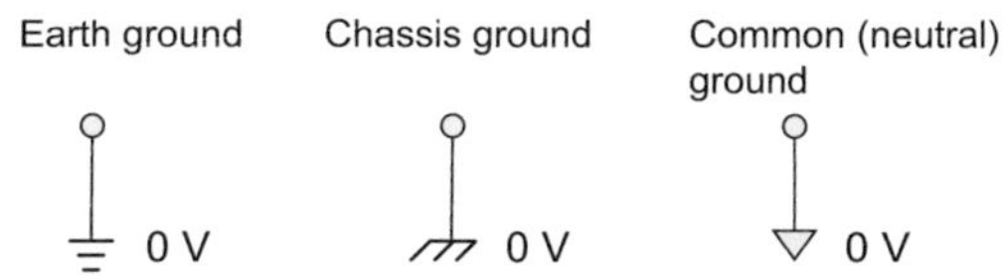

Fig. 2.37. Different ground types: earth ground, chassis ground, and common (neutral) ground.

The second ground type is the *chassis ground*. It is the physical *metal* frame or structure of an automobile, an airplane, a desktop computer, a cell phone, or other electrical devices; the term *case* is very similar in meaning. The chassis ground primarily involves a connection to the metal case. It is implied that the case should eventually discharge due to contact with other objects or with earth. The term *ground plane* for planar printed circuits, which is usually the copper bottom of a printed circuit board, is equivalent to chassis ground. The third ground type in Fig. 2.37 is the *common terminal* or *common ground*. The word *common* is typical for many circuits including the amplifier circuits considered next, when a dual-polarity power supply is used. Here two identical batteries are connected in series, plus to minus. The common terminal of the dual power supply so designed serves as the reference ground; even a metal case is not necessarily required. The AC analog of the common ground is the *neutral terminal* of your wall plug. Frequently, different ground types may be interconnected. For example, the neutral terminal of the wall plug should be connected to earth ground at a certain location. The chassis ground of a large truck may be connected to the physical ground by a little flexible strip nearly touching the asphalt.

2.5.4 Ground and Return Current

We have already seen that electric current in a circuit always flows in closed loops. This is a simple and yet a very critical property of an electric circuit. The steady-state current that flows to a load is sometimes called *forward current*, whereas the current that returns to the power supply is the *return current*. Can the chassis ground itself be used as a part of this loop for the return current? The answer is yes, and Fig. 2.38a depicts this situation as an example. Here a 9 V battery is powering an incandescent light bulb. For the chassis ground in Fig. 2.38a, the circuit is correctly drawn, but putting two wires into the soil, as shown in Fig. 2.38b, will fail owing to the high resistance of the earth. The use of the ground to establish a path for the return current is quite common for the chassis ground (automotive electronics) and also for the common ground. However, it should not be attempted for the earth ground connection. Emphasize that, in many circuit diagrams, the difference between the chassis ground, the common ground, and the true earth ground is often *ignored*. Namely, the symbol of the earth ground used in the circuit often implies either the chassis ground or the common ground, i.e., the (physically grounded or not) current return path.

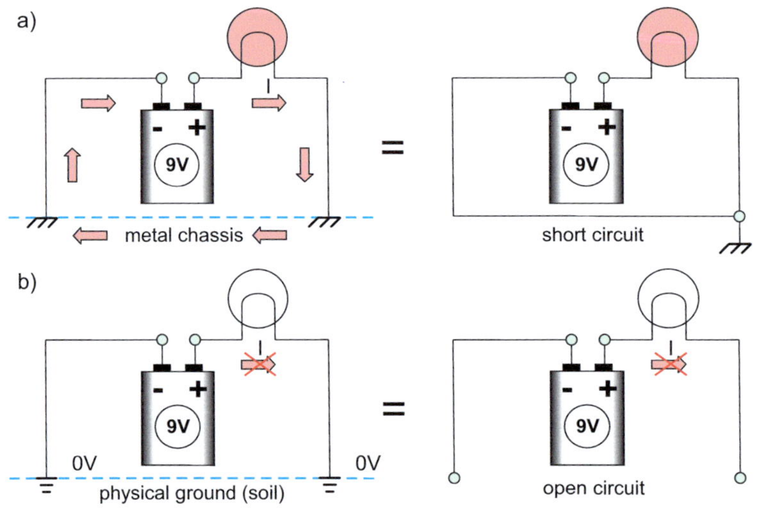

Fig. 2.38. (**a**) The return current path for the chassis ground is metal; it can be replaced by a wire. (**b**) There is no current return path, since soil (dry or wet) is a very poor conductor. The circuit is therefore open and not functioning.

2.5.5 Absolute Voltage and Voltage Drop Across a Circuit Element

The electric ground serves as a voltage reference point in a circuit. It allows us to use *two* types of voltages in the circuit:

1. The *absolute voltage* at a certain circuit node
2. The *voltage drop* or simply the *voltage across a circuit element*

Figure 2.39 shows the concept. Voltages $V_{a,b,c,d}$ are absolute voltages measured versus ground at nodes *a, b, c, d* in Fig. 2.39. Voltages $V_{A,B,C}$ give the voltage drop across the circuit elements A, B, and C. Indeed, the ideal wires remain the equipotential surfaces (have the same absolute voltage). Taking into account the polarity of the voltages $V_{A,B,C}$ shown in Fig. 2.39, one has for the node voltages

$$V_a = 0\ \text{V},\ V_b = V_a + V_A = 10\ \text{V},\ V_c = V_b - V_B = 5\ \text{V},\ V_d = V_c - V_C = 0\ \text{V} \quad (2.48)$$

Note that both voltage types—absolute voltage and voltage across a circuit element—are often denoted by the same letter V (in the DC case) and may be easily misplaced. Both of them are widely used in electric circuit analyses. The hint is that the voltage across a circuit element always has the polarity labeled with $\pm$ sign, whereas the absolute voltage often has not.

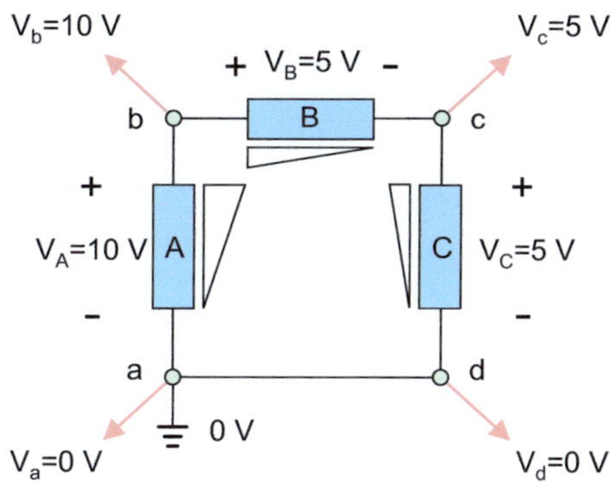

Fig. 2.39. Absolute voltages measured versus ground in a grounded electric circuit and voltages across individual circuit elements. Note that there is no voltage drop across *ideal* wires.

Exercise 2.14: Determine the absolute voltages at nodes 1 through 6 in the circuit shown in Fig. 2.40.

Answer: Clearly, $V_1 = 0\text{V}$ since node 1 is directly connected to ground. Then,

$$V_2 = V_1 + V_{\mathrm{A}} = 6\ \text{V},\quad V_3 = V_2 + V_{\mathrm{B}} = 12\ \text{V},\quad V_4 = V_3 - V_{\mathrm{C}} = 9\ \text{V},$$
$$V_5 = V_4 - V_{\mathrm{D}} = 3\ \text{V}, V_6 = V_5 - V_{\mathrm{E}} = 0\ \text{V}$$

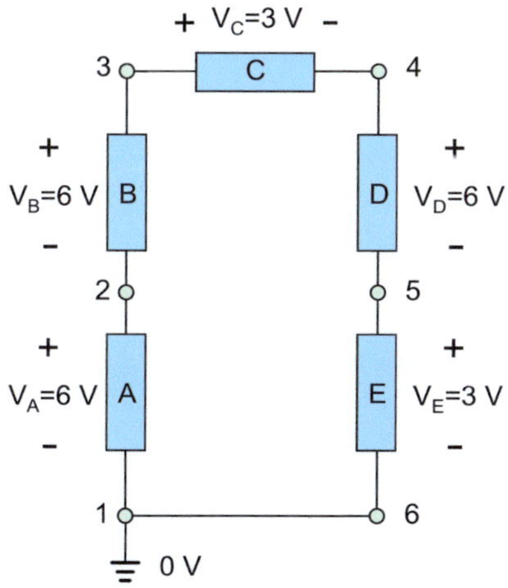

Fig. 2.40. Voltages across circuit elements in a grounded electric circuit.

Summary

<table>
<tr><th colspan="3">Passive circuit elements</th></tr>
<tr><th>Name and symbol</th><th>v–i Characteristic</th><th>Physical counterpart (component)</th></tr>
<tr><td>Resistance
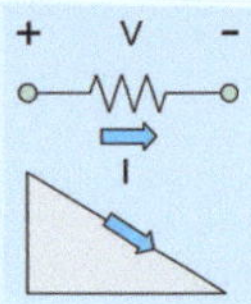</td><td>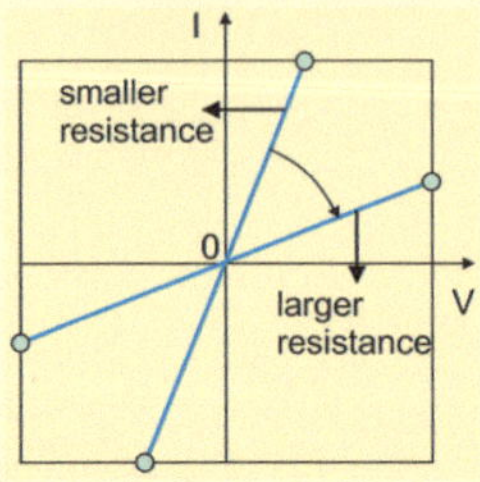</td><td>Resistor
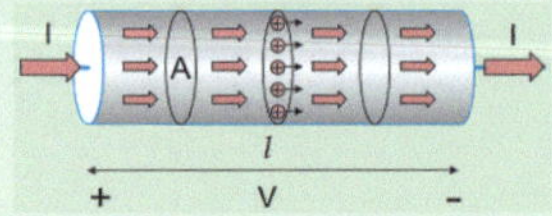
$V = RI, \quad \sigma = qn\mu, \quad R = \frac{l}{A\sigma}$
$P = VI = RI^2 = V^2/R$</td></tr>
<tr><td>Short circuit:
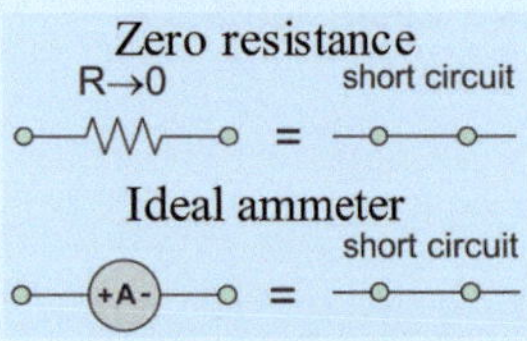</td><td>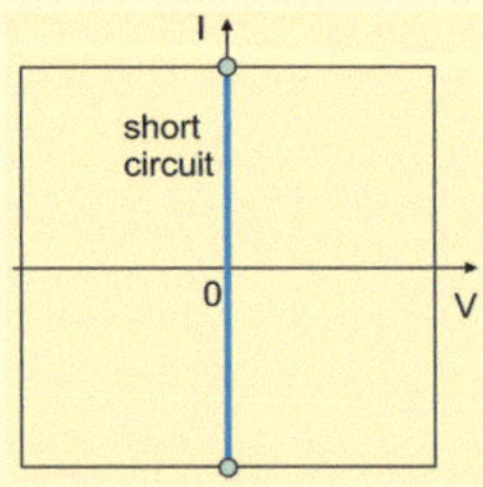</td><td>Short wire of almost zero resistance
Ammeter</td></tr>
<tr><td>Open circuit:
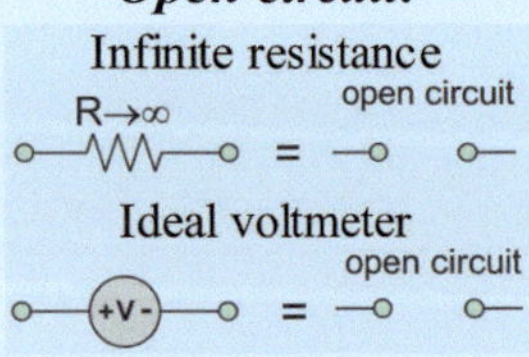</td><td>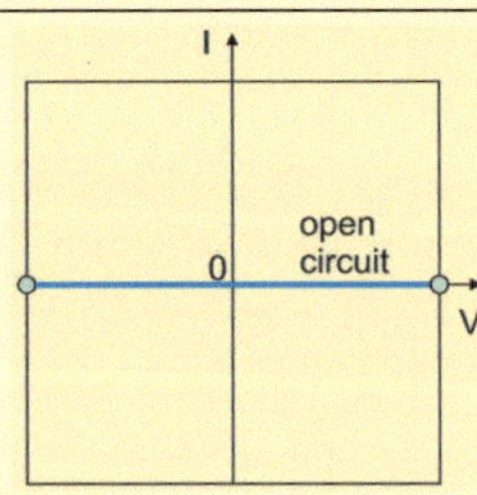</td><td>Air gap of almost infinite resistance
Voltmeter</td></tr>
<tr><td>Ideal diode
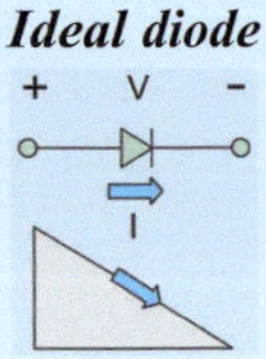</td><td>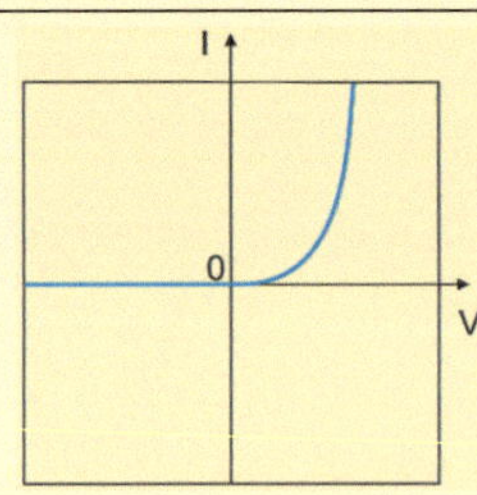</td><td>Electronic diode
Transistor junctions
Solar cell
$I = I_S\left[\exp\left(\frac{V}{V_T}\right) - 1\right]$
Static resistance: $R_0 \equiv V_0/I(V_0)$
Dynamic resist.: $r \equiv dV/dI|_{V_0,\, I_0}$</td></tr>
<tr><td>Threshold switch
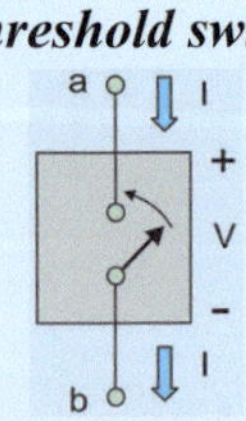</td><td>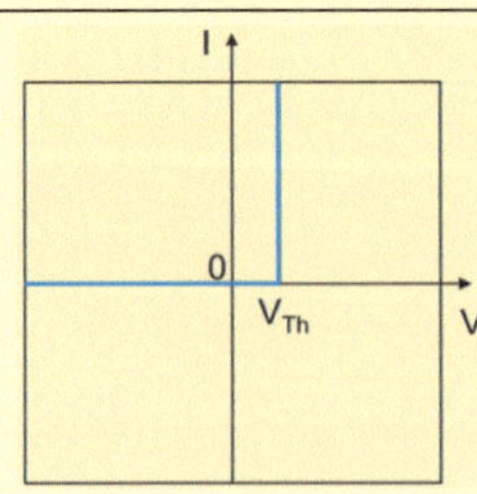</td><td>Diode
Transistor
Open circuit when $V < V_{Th}$
Short circuit when $V = V_{Th}$</td></tr>
</table>

(continued)

Active circuit elements		
Name and symbol	v–i Characteristic	Physical counterpart (component)
Independent voltage source 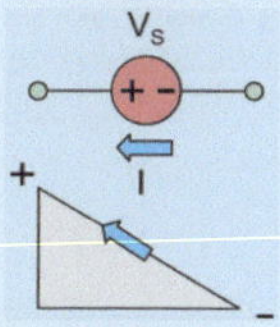	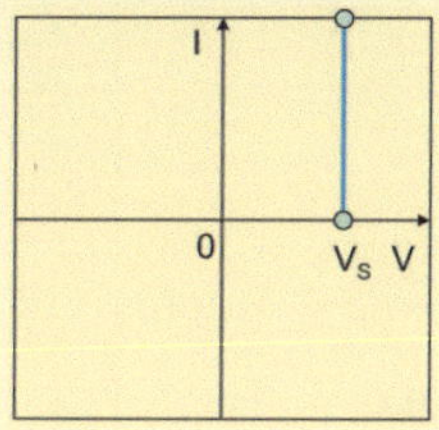	***Practical voltage source*** 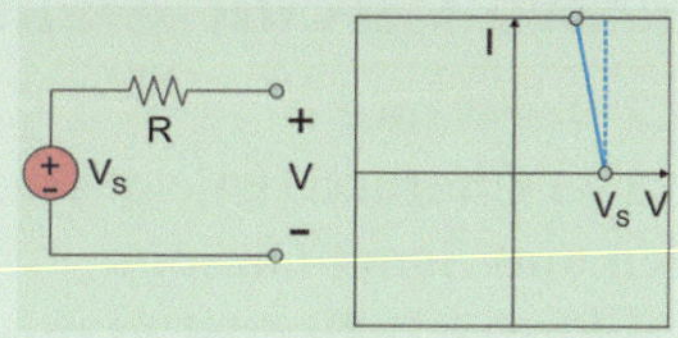
Independent current source 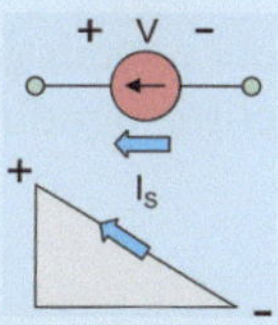	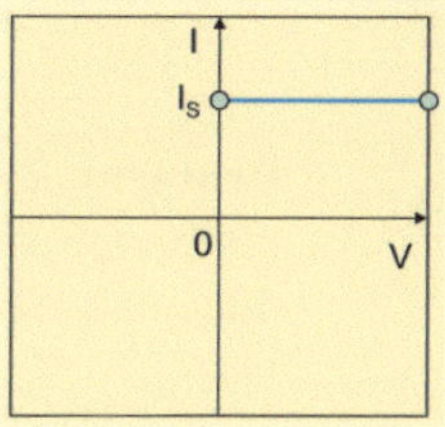	***Practical current source*** 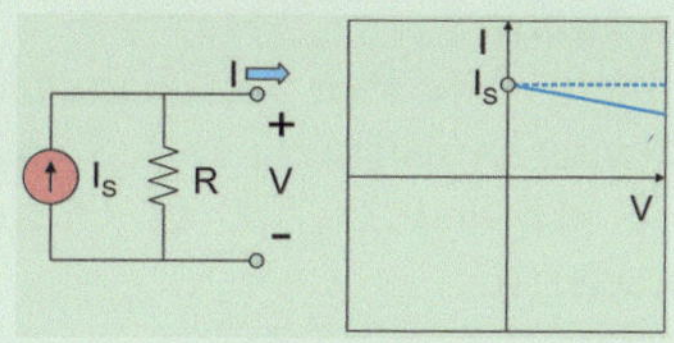
Voltage-controlled voltage source 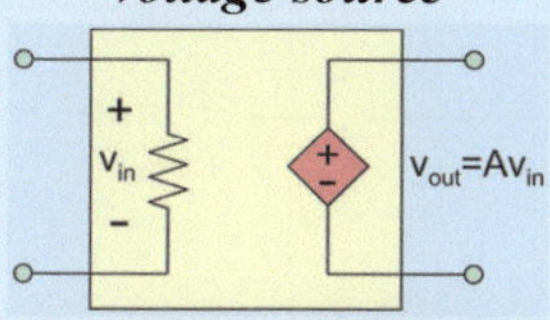	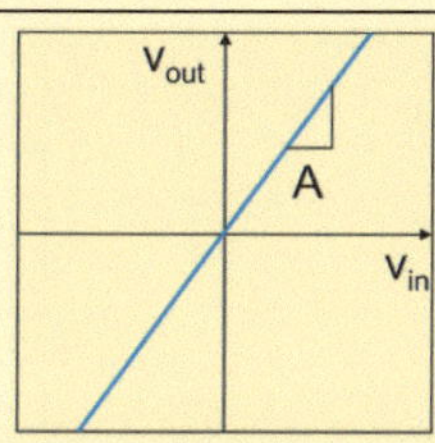	***Transistor Amplifier*** $v_{out} = Av_{in}$ A—open-circuit *voltage* gain [V/V, V/mV] (dimensionless)
Current-controlled voltage source 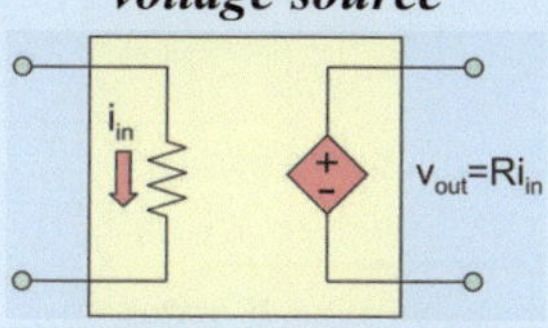	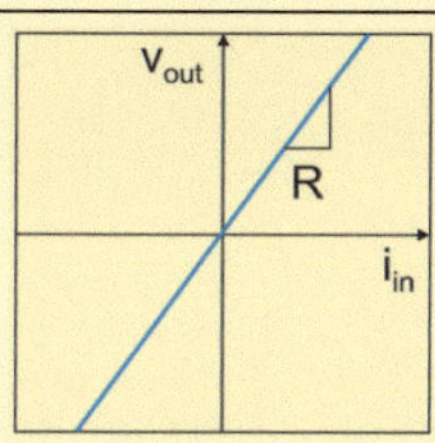	***Transistor Amplifier*** $v_{out} = Ri_{in}$ R—transresistance [V/A, V/mA] (units of resistance, Ω)
Voltage-controlled current source 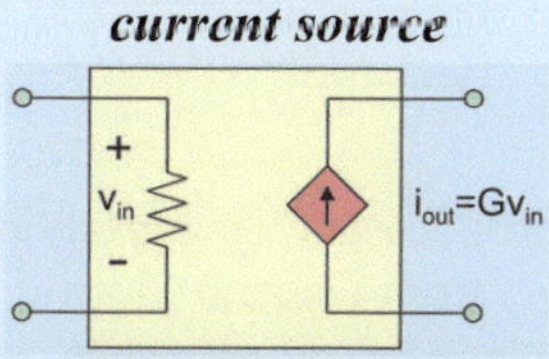	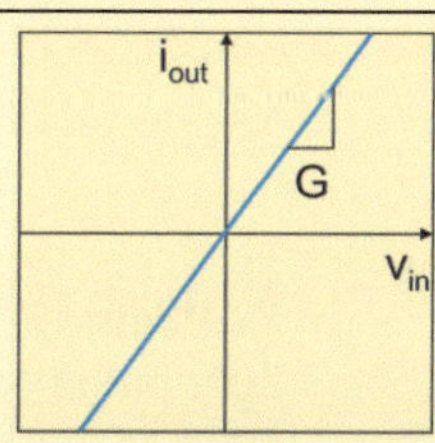	***Transistor Amplifier*** $i_{out} = Gv_{in}$ G—transconductance [A/V] (units of conductance, Ω^{-1})
Current-controlled current source 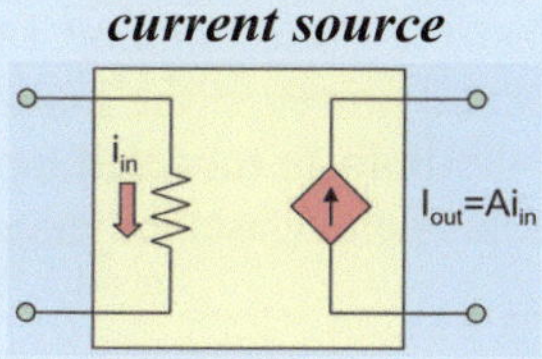	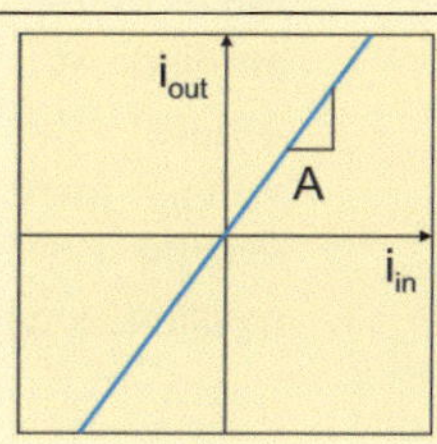	***Transistor Amplifier*** $i_{out} = Ai_{in}$ A—short-circuit *current* gain [A/A, A/mA] (dimensionless)

Problems

2.1 Resistance: Linear Passive Circuit Element

2.1.2 Resistance

2.1.3 *υ-i* Characteristic of the Resistance: Open and Short Circuits

2.1.4 Power Delivered to the Resistance

2.1.5 Finding Resistance of Ohmic Conductors

Problem 2.1. Plot *υ-i* characteristics of the following resistances: (A) 8 kΩ, (B) 2 kΩ, (C) 1 kΩ, and (D) 500 Ω. Clearly label each characteristic.

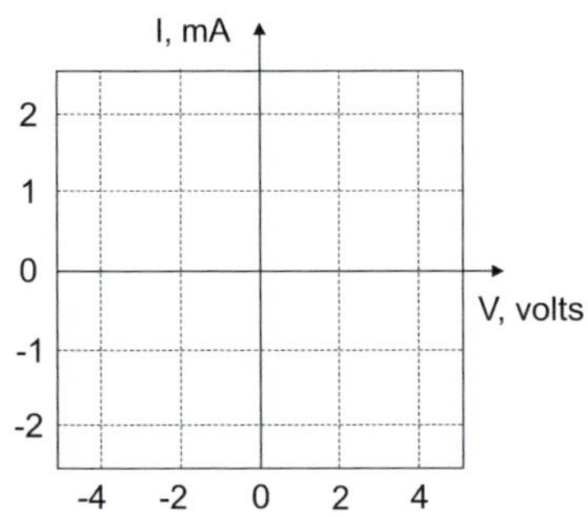

Problem 2.2. Plot *υ-i* characteristics of the following resistances: (A) 1.667 Ω. (B) Open circuit on the same graph.

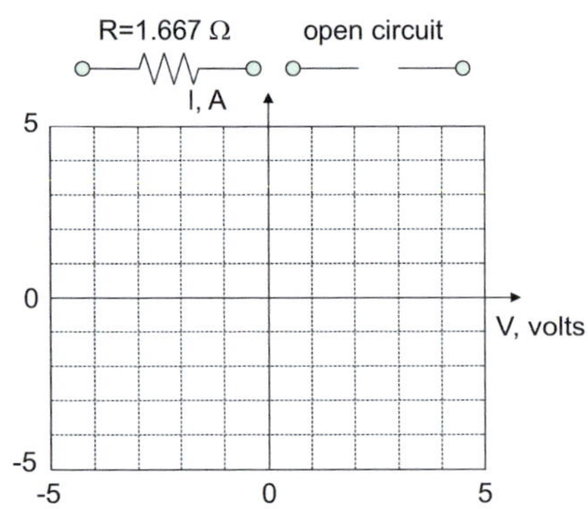

Problem 2.3. Plot *υ-i* characteristics of the following resistances: (A) 2.5 Ω. (B) Short circuit on the same graph.

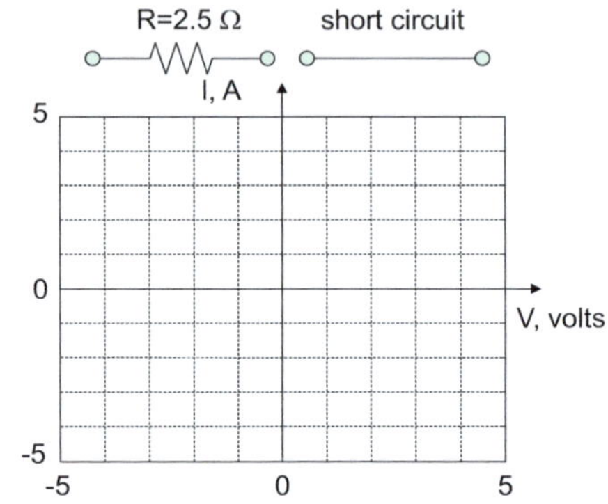

Problem 2.4. Given *υ-i* characteristics of a resistance determine the corresponding conductance. Show units.

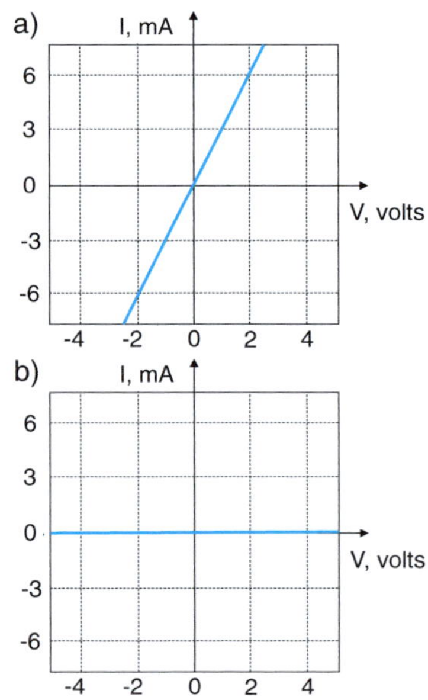

Problem 2.5. An incandescent energy-saving light bulb ("soft white") from General Electric is rated to have the wattage of 57 W when the applied AC voltage is 120 V *rms* (root mean square). This means that the corresponding DC voltage providing the *same* power to the load is exactly 120 V. When the bulb is modeled as a resistance, what is the equivalent resistance value?

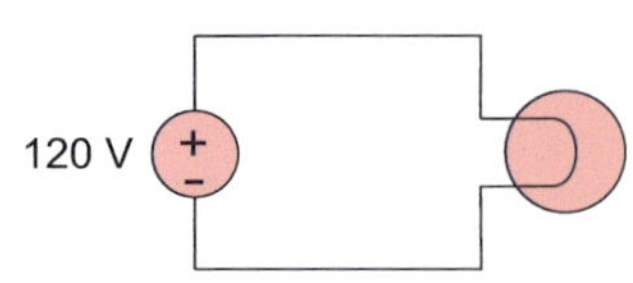

Problem 2.6. The power absorbed by a resistor from the ECE laboratory kit is 0.2 W. Plot the v-i characteristics of the corresponding resistance to scale given that the DC voltage across the resistor was 10 V.

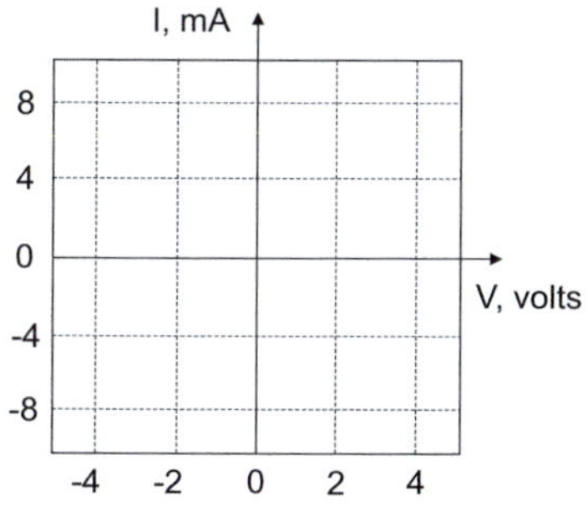

Problem 2.7. The number of free electrons in copper per unit volume is $n = 8.46 \times 10^{28} \frac{1}{\mathrm{m}^3}$. The charge of the electron is -1.60218×10^{-19} C. A copper wire of cross section 0.25 mm^2 is used to conduct 1A of electric current.

A. Sketch the wire, the current direction, and the direction of electron motion.
B. How many coulombs per one second is transported through the conductor?
C. How fast do the electrons really move? In other words, what is the average electron velocity?

Problem 2.8. Repeat the above problem when the conductor's cross section is increased to 5 mm^2.

Problem 2.9. A copper wire having a length of 1000 ft and a diameter of 2.58826 mm is used to conduct an electric current of 5 A.

A. What is wire's total resistance? Compare your answer to the corresponding result of Table 2.2.
B. What is the power loss in the wire?

Problem 2.10

A. A copper wire having a length of 100 m and a cross section of 0.5 mm^2 is used to conduct an electric current of 5 A. What is the power loss in the wire? Into what is this power loss transformed?
B. Solve task A when the wire cross section is increased to 2.5 mm^2.

Problem 2.11. Determine the total resistance of the following conductors:

A. A cylindrical silver rod of radius 0.1 mm, length 100 mm, and conductivity 6.1×10^7 S/m.
B. A square graphite bar with the side of 1 mm, length 100 mm, and conductivity 3.0×10^4 S/m.
C. A semiconductor doped Si wafer with the thickness of 525 μm. Carrier mobility is $\mu = 0.15$ m^2/(V·s). Carrier concentration is $n = 10^{23}$ m^{-3}. Carrier charge is 1.6×10^{-19} C. The resistance is measured between two circular electrodes with the radius of 1 mm each, which are attached on the opposite sides of the wafer. Assume uniform current flow between the electrodes.

a)

b)

c)

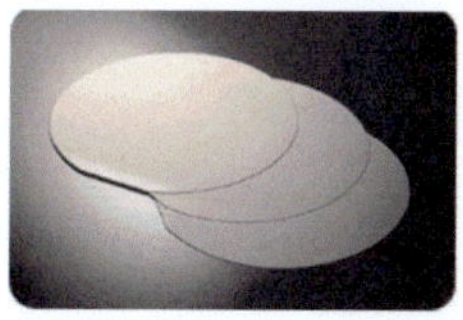

Problem 2.12. A setup prepared for a basic wireless power-transfer experiment utilizes a square multi-turn loop schematically shown in the figure, but *with 40 full turns*. A #22 gauge copper wire with the diameter of 0.645 mm is used. Total loop resistance, R, is needed. Please assist in finding the loop resistance (show units).

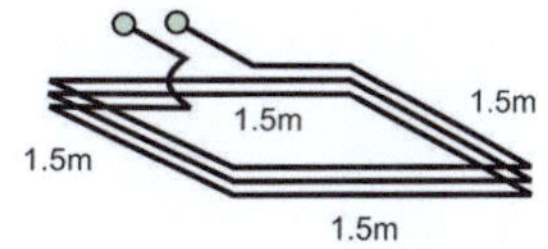

Problem 2.13. Estimate resistance, R_n (show units), of the n-side of a Si pn-junction diode in

the figure that follows. We model the n-side by a Si bar having the following parameters:

1. Length of $L = 0.0005$ cm $= 5$ μm.
2. Cross section of $A = 0.01$ cm $\times$ 0.01 cm $= 1 \times 10^{-4}$ cm^2.
3. *Uniform* electron concentration (carrier concentration) of $n = 10^{17}$ cm^{-3}. This value is typical for a Si diode pn-junction. Carrier mobility is $\mu_n = 1450$ cm^2/(V·s).

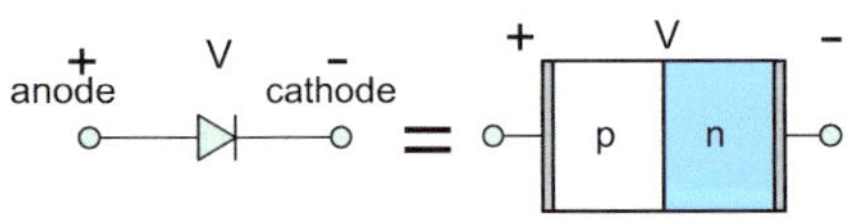

Problem 2.14. Estimate resistance, R_{p} (show units), of the p-side of a Si pn-junction diode in the figure that follows. We model the p-side by a Si bar having the following parameters:

1. Length of $L = 0.0005$ cm $= 5$ μm.
2. Cross section of $A = 0.01$ cm $\times$ 0.01 cm $= 1 \times 10^{-4}$ cm^2.
3. *Uniform* hole concentration (carrier concentration) of $n = 10^{17}$ cm^{-3}. This value is typical for a Si diode pn-junction. Carrier mobility is $\mu_n = 500$ cm^2/(V·s).

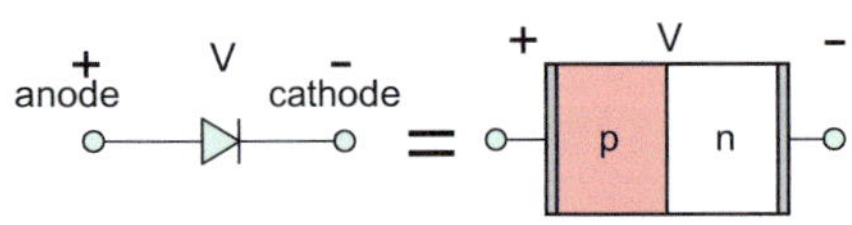

Problem 2.15. A cross section of the most popular NMOS transistor is shown in the following figure.

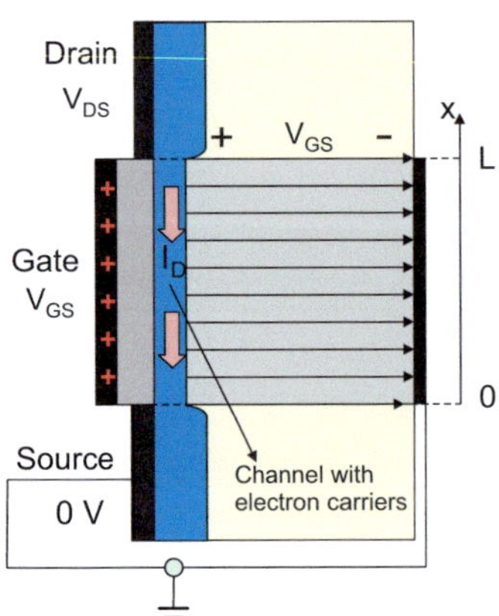

The transistor has three terminals (metal contacts): drain (with voltage $V_{\mathrm{DS}} > 0$ vs. source), gate (with voltage $V_{\mathrm{GS}} > 0$ vs. source), and source itself (grounded). The source is also connected to a metal conductor on the other side of the semiconductor body. Accordingly, there are *two* types of the electric field within the semiconductor body: the *horizontal field* created by V_{GS}, and the *vertical field* created by V_{DS}. The horizontal field fills a *conducting channel* between the drain and the source with charge carriers, but has *no effect* on the vertical charge motion. The resulting carrier concentration in the channel is given by $n = N(V_{\mathrm{GS}} - V_{\mathrm{Th}}) > 0$, $N = \mathrm{const}$, $V_{\mathrm{Th}} = \mathrm{const}$. The individual carrier charge is q. Given the channel cross section A, the carrier mobility μ, and the channel length L, determine transistor current I_{D} and transistor resistance (drain-to-source resistance) R_{DS}. Express both results in terms of quantities listed above *including* V_{GS} and V_{Th}.

2.1.6 Application Example: Power Loss in Transmission Wires and Cables

Problem 2.16. An AWG 0000 aluminum transmission grid cable has the wire diameter of 11.68 mm and the area of 107 mm^2. The conductivity of aluminum is 4.0×10^7 S/m. The total cable length (two cables must run to a load) is 100 km. The system delivers 10 MW of DC power to a load. Determine the power loss in the cable (show units) when load voltage and current are given by:

1. $V = 200$ kV and $I = 50$ A
2. $V = 100$ kV and $I = 100$ A
3. $V = 50$ kV and $I = 200$ A

Why do you think the high-voltage power transmission is important in power electronics?

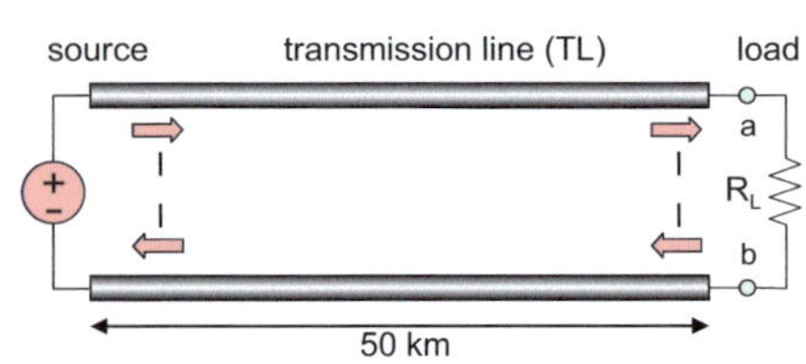

Problem 2.17. Solve the previous problem when the total cable length (two cables must run to a load) is increased to 200 km.

Problem 2.18. An AWG 00 aluminum transmission grid cable has the wire diameter of 9.266 mm. The conductivity of aluminum is 4.0×10^7 S/m. A power transmission system that uses this cable is shown in the figure that follows. The load power is 1 MW. Determine the minimum necessary load voltage V that guarantees us a 1 % relative power loss in the cables.

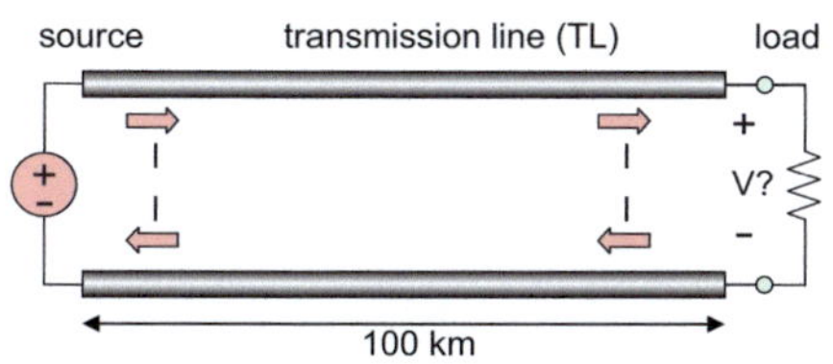

Problem 2.19. An AC-direct micro-hydropower system is illustrated in the figure that follows.

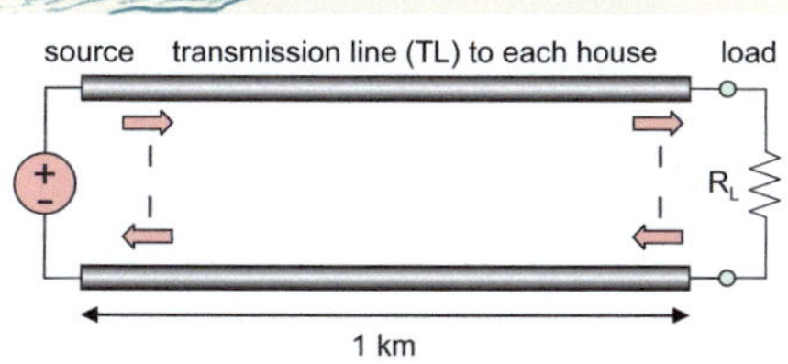

Reprinted from *Micro-Hydropower Systems* Canada 2004, ISBN 0-662-35880-5.

The system uses a single phase induction generator with the rms voltage (equivalent DC voltage) of 240 V. The system serves four small houses, each connected to the generator via a *separate* transmission line with the *same* length of 1000 m. Each line uses AWG#10 aluminum wire with the diameter of 2.59 mm. The conductivity of aluminum is 4.0×10^7 S/m. The house load is an electric range with the resistance of 20 Ω. Determine total power delivered by the generator, P_{total}, total power loss in the transmission lines, P_{loss}, and total useful power, P_{useful} (show units).

2.1.7 Physical Component: Resistor

Problem 2.20. A leaded resistor has color bands in the following sequence: brown, black, red, gold. What is the resistor value?

Problem 2.21. Potentiometer operation may be schematically explained as moving sliding contact #2 in the following figure along a uniform conducting rod with the total resistance of 20 kΩ. Determine resistance between terminals 1 and 2 as well as between terminals 2 and 3 of the potentiometer, when the sliding contact is at one fifth of the rod length.

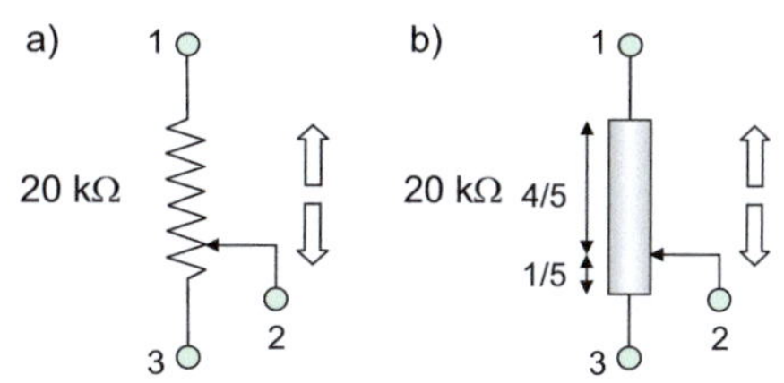

2.2 Nonlinear Passive Circuit Elcments

2.2.2 Nonlinear Passive Circuit Elements

2.2.3 Static Resistance

2.2.4 Dynamic (Small-Signal) Resistance

2.2.5 Electronic Switch

Problem 2.22. A nonlinear passive circuit element—the ideal diode—is characterized by the v-i characteristic in the form $I = I_S\left[\exp\left(\frac{V}{V_T}\right) - 1\right]$ with $I_S = 1 \times 10^{-13}$ A and $V_T = 25.7$ mV. Find the static diode resistance R_0 and the diode current I_0 when (A) $V_0 = 0.40$ V, (B) $V_0 = 0.50$ V, (C) $V_0 = 0.55$ V, and (D) $V_0 = 0.60$ V.

Problem 2.23. Find the dynamic (small-signal) resistance r of a nonlinear passive circuit element—the ideal diode—when the operating DC point V_0, I_0 is given by the solutions to the previous problem. Consider all four cases.

Problem 2.24. A nonlinear passive circuit element is characterized by the v-i characteristic in the form $I = I_S \frac{V/V_S}{\sqrt{1+(V/V_S)^2}}$ with $I_S = 1$ A and $V_S = 1$ V. Plot the v-i characteristic to scale. Next, find the static element resistance R_0, the element current I_0, and the corresponding dynamic element resistance r when (A) $V_0 = 0.1$ V, (B) $V_0 = 1.0$ V, and (C) $V_0 = 5.0$ V.

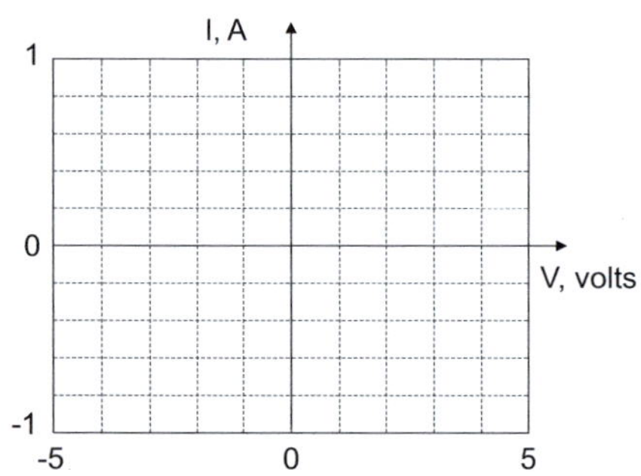

Problem 2.25. Repeat the previous problem when $I_S = 0.5$ A and $V_S = 0.5$ V. All other parameters remain the same. Consider the following DC operating points: (A) $V_0 = 0.05$ V, (B) $V_0 = 0.50$ V, and (C) $V_0 = 2.50$ V.

Problem 2.26. A DC circuit shown in the following figure includes two interconnected passive elements: an ideal diode and a resistance. One possible circuit solution is given by an intersection of two v-i characteristics marked by a circle in the same figure. This solution predicts a non-zero circuit current and a positive voltage across both circuit elements.

A. Is this solution an artifact (a mistake has been made somewhere)?

B. Is this solution true (the circuit so constructed might function)?

Justify your answer.

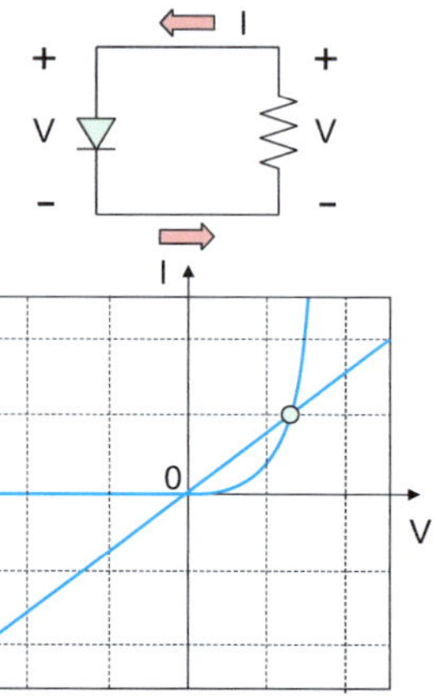

2.3 Independent Sources

2.3.1 Independent Ideal Voltage Source

2.3.2 Circuit Model of a Practical Voltage Source

Problem 2.27. In the following figure, determine if the element is a resistance or a voltage source. Find the power delivered to element A or taken from element A in every case.

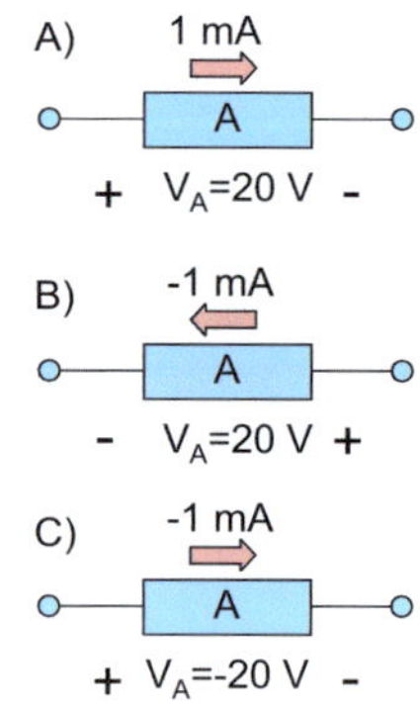

Problem 2.28. Based on voltage and current measurements, determine if the circuit element is a resistance or a voltage source. Readings of the ammeter and voltmeter are shown in the following figure.

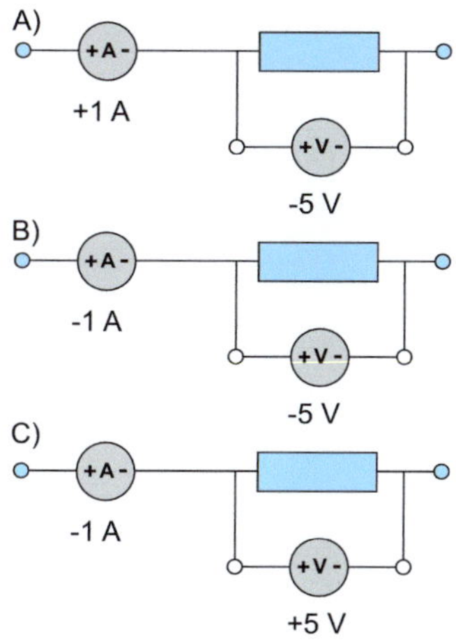

Problem 2.29. The figure that follows shows a circuit with a passive nonlinear circuit element shown by a rectangle.

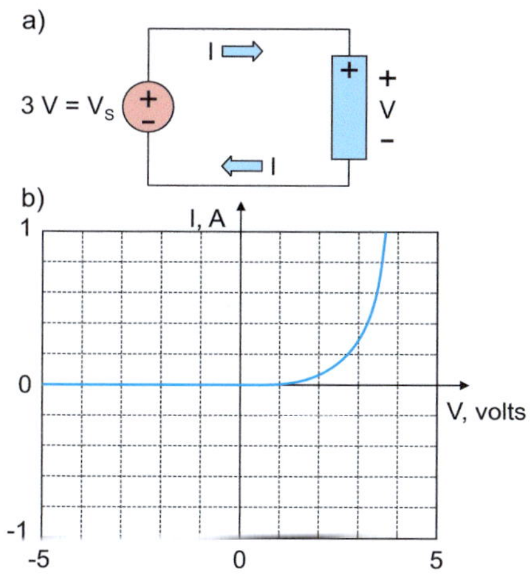

The polarity (direction of current inflow for passive reference configuration) of the element is labeled by a sign plus. The *υ*-*i* characteristic of the element is also shown in the figure. Determine current *I* and voltage *V*.

Problem 2.30. Repeat the previous problem for the circuit shown in the figure that follows.

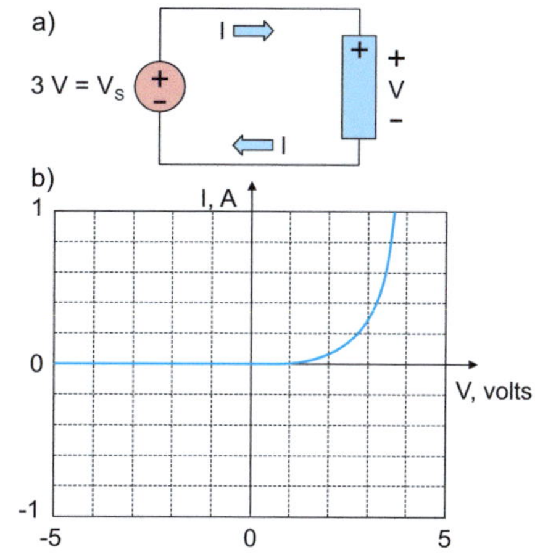

Problem 2.31. Plot to scale the *υ*-*i* characteristic of the practical voltage source shown in the following figure.

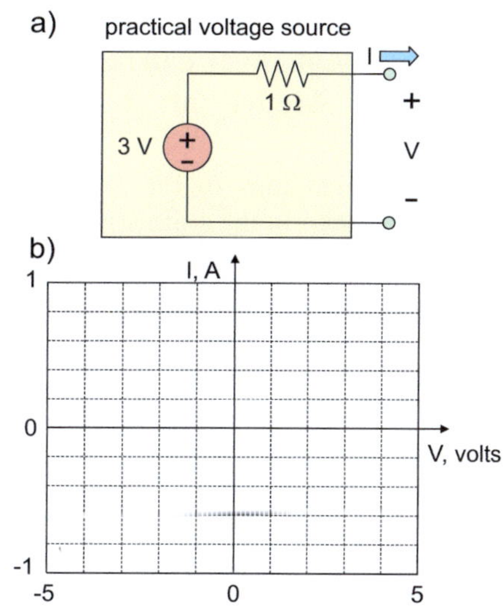

Problem 2.32. The following figure shows a circuit with a passive nonlinear circuit element labeled by a rectangle. Element's polarity (direction of current inflow for passive reference configuration) of the element is indicated by a sign plus. The *υ*-*i* characteristic of the

element is also shown in the figure. Determine circuit current I.

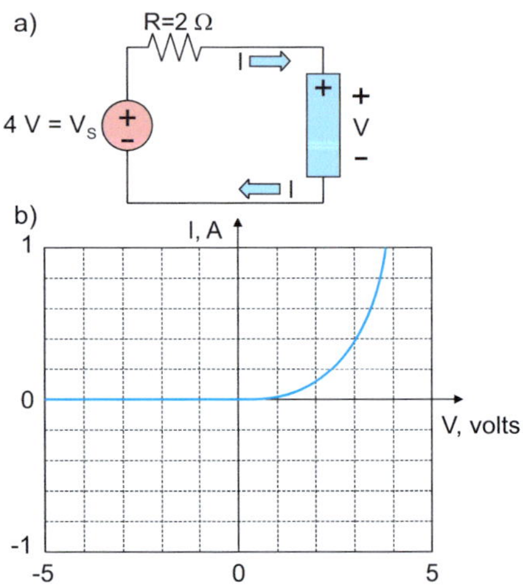

2.3.3 Independent Ideal Current Source

2.3.4 Circuit Model of a Practical Current Source

Problem 2.33. Readings of the ammeter and voltmeter are shown in the following figure.

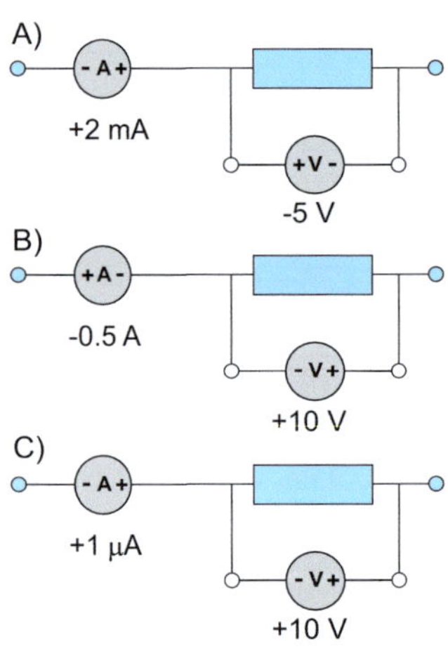

Based on voltage and current measurements, determine if the element is a resistance or a current source. Then, find the power delivered to the circuit element or taken from it in every case.

Problem 2.34. The following figure shows a circuit with a passive nonlinear circuit element shown by a rectangle. Element's polarity (direction of current inflow for passive reference configuration) of the element is labeled by a sign plus. The v-i characteristic of the element is also shown in the figure. Determine current I and voltage V.

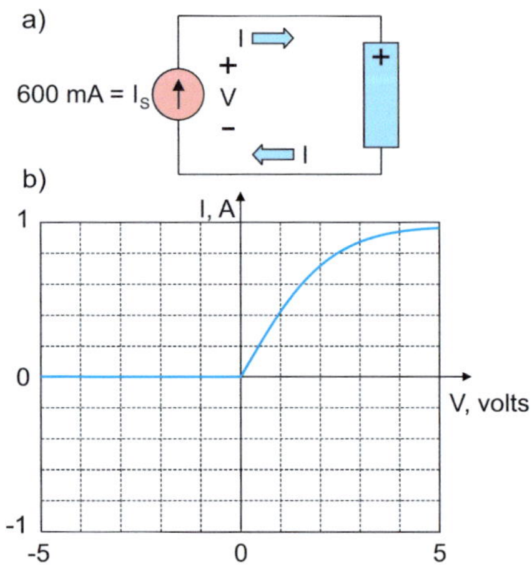

Problem 2.35. Repeat the previous problem for the circuit shown below.

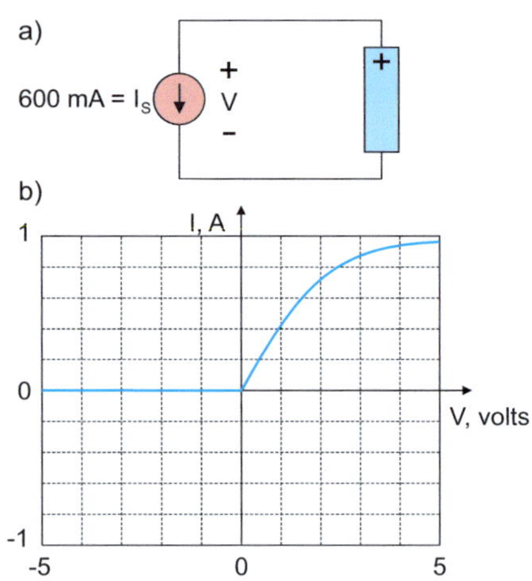

Problem 2.36. Plot to scale the υ-i characteristic of the practical current source shown in the following figure.

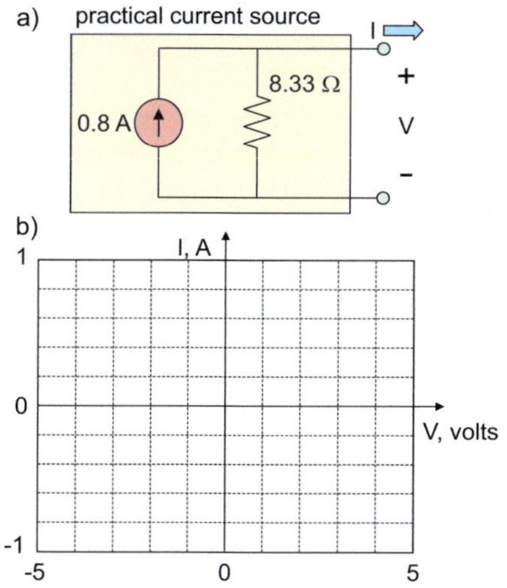

Review Problems

Problem 2.37. For every circuit element shown in the following figure, plot its υ-i characteristic on the same graph.

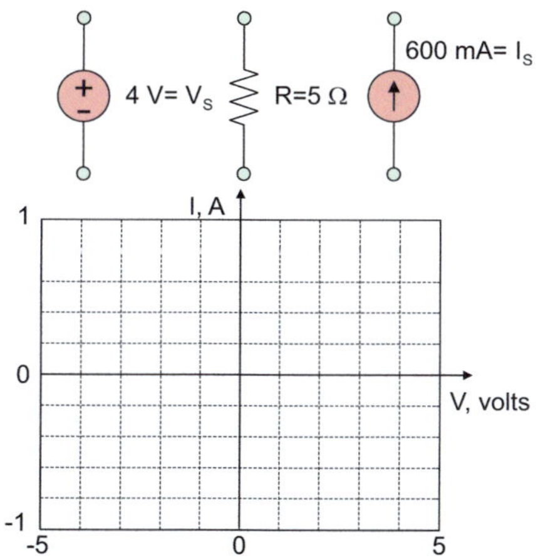

Problem 2.38. Repeat the previous problem: for every circuit element shown in the figure below, plot its υ-i characteristic on the same graph.

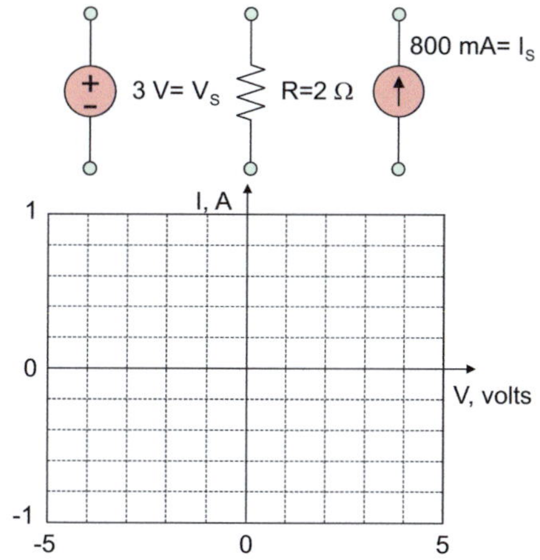

2.3.7 Application Example: Chemical Battery

Problem 2.39. The electronics aboard a certain sailboat consume 96 W when operated from a 24 V source.

A. If a certain fully charged deep-cycle marine battery is rated for 24 V and 100 A h, for how many hours can the electronics be operated from the battery without recharging? (The ampere-hour rating of the battery is the battery capacity—the operating time to discharge the battery multiplied by the current).
B. How much energy in kilowatt hours is initially stored in the battery?

Problem 2.40. A motor of a small, unmanned electric vehicle consumes 120 W and operates from a 24-V battery source. The source is rated for 200 Ah.

A. For how many hours can the motor be operated from the source (a battery bank) without recharging?
B. How much energy in kilowatt hours is initially stored in the battery source?

Problem 2.41. A certain sensing device operates from a 6-V source and consumes 0.375 W of power over a 20-h time period. The source is a combination of four fully charged AAA batteries, 1.5 V each, assembled in series. The batteries discharge by the end of the 20-h period.

A. What is the expected capacity of a typical AAA battery used, in mAh?
B. How much energy in Joules was stored in each AAA battery?

Problem 2.42. How many Joules are in 1 kWh and how many N·m does this correspond to?

2.4 Dependent Sources and Time-Varying Sources

2.4.1 Dependent Versus Independent Sources

2.4.2 Definition of Dependent sources

2.4.3 Transfer Characteristics

2.4.4 Time-Varying Sources

Problem 2.43. Draw circuit diagrams for four major types of dependent sources, label stimulus voltage/current and output voltage/current. Describe operation of each dependent source.

Problem 2.44. Solve an electric circuit shown in the following figure—determine current i through the 2-kΩ resistance. The independent current source is given by $i_S = 0.2 - 1.5\cos 5t$ [mA]; the open-circuit voltage gain of the dependent source is 12 V/V.

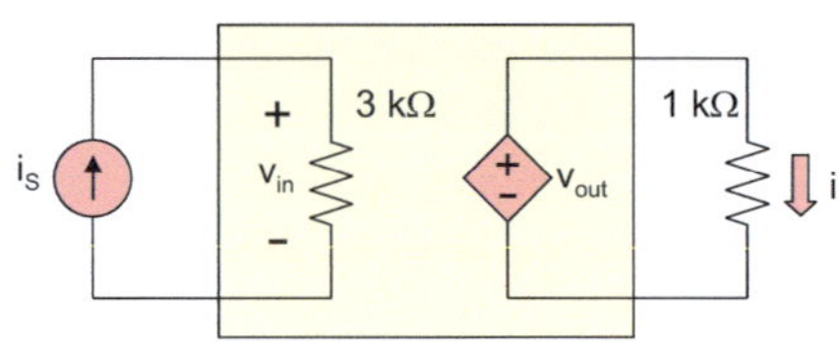

Problem 2.45. Solve an electric circuit shown in the following figure—determine current i through the 2-kΩ resistance. The independent voltage source is given by $v_S = -0.3 + 0.7\cos 6t$ [V]; the transresistance of the dependent source is 250 V/A.

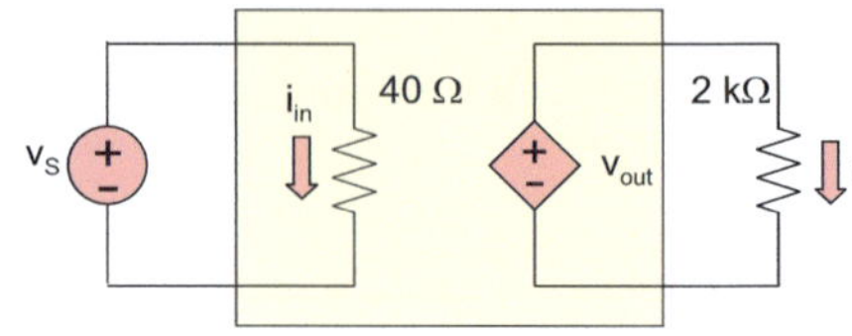

Problem 2.46. Solve an electric circuit shown in the following figure—determine voltage v through the 30-Ω resistance. The independent current source is given by $i_S = -0.05 - 0.2\cos 2t$ [mA]; the transconductance of the dependent source is 10 A/V.

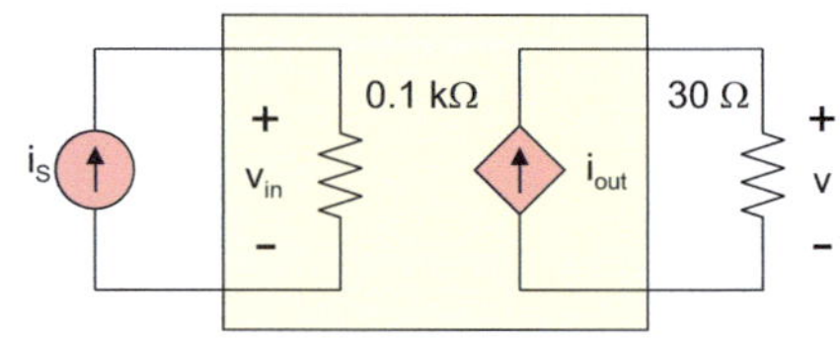

Problem 2.47. Solve an electric circuit shown in the following figure—determine voltage v through the 1-kΩ resistance. The independent voltage source is given by $v_S = 0.05 + 0.1\cos 4t$ [V]; the short-circuit current gain of the dependent source is 10 A/A.

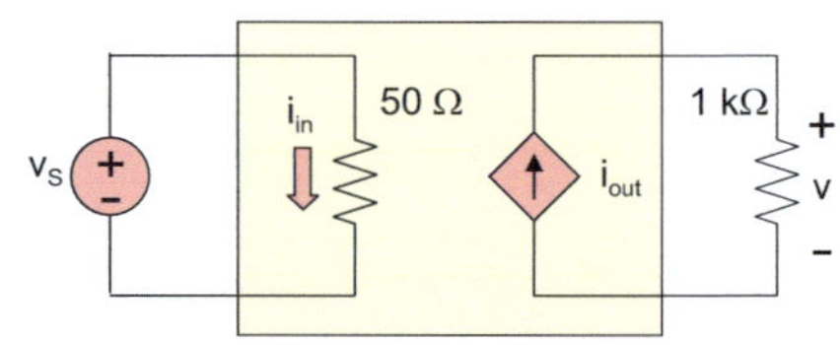

Problem 2.48. Solve an electric circuit shown in the following figure—determine current i through a nonlinear passive circuit element shown by a rectangle. Element's polarity (direction of current inflow for passive reference configuration) is labeled by a sign plus. The v-i characteristic of the element is also shown in the figure. The independent current source is given by $i_S = 0.1$ A; the open-circuit voltage gain of the dependent source is 5 V/V.

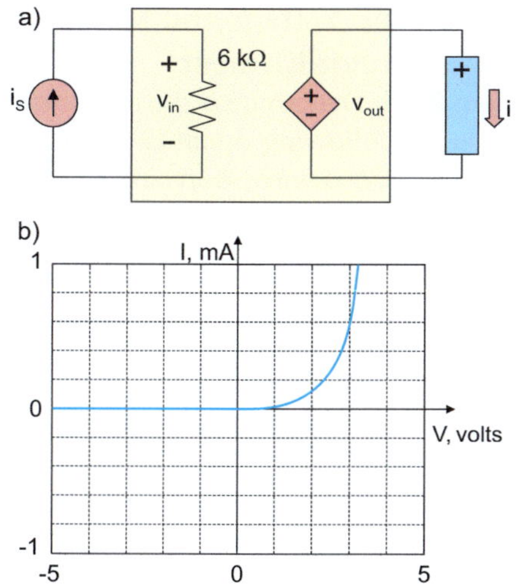

Problem 2.49. Solve the electric circuit shown in the following figure—determine voltage v across the 1-kΩ resistance.

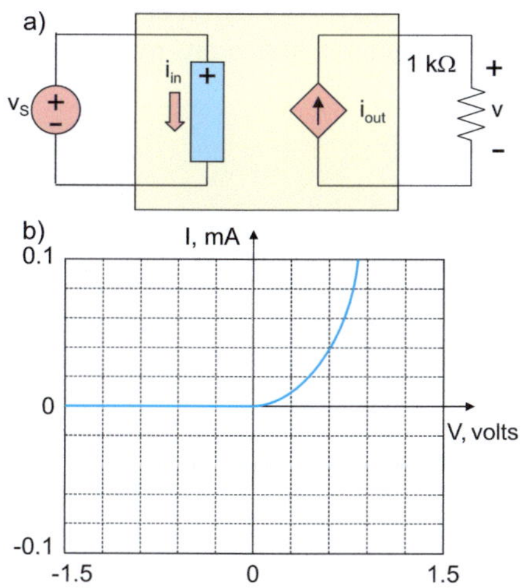

A nonlinear passive circuit element is shown by a rectangle. Element's polarity (direction of current inflow for passive reference configuration) of the element is labeled by a sign plus. The v-i characteristic of the element is also shown in the figure. The independent voltage source is given by $v_S = 0.6$ V; the short-circuit current gain of the dependent source is 100 A/A.

2.5 Ideal Voltmeter and Ammeter: Circuit Ground

2.5.1 Ideal Voltmeter and Ammeter

2.5.3 Types of Electric Ground

2.5.4 Ground and Return Current

Problem 2.50. You attempt to measure electric current through a resistance as part of a circuit. Is the following figure appropriate? What is the current across the 51-Ω resistor?

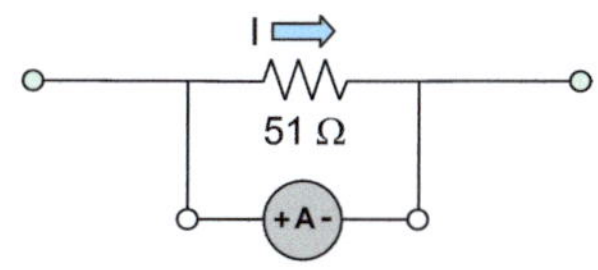

Problem 2.51. You attempt to measure voltage across a resistance in the circuit. Is the following figure correct? What is the voltmeter's reading, assuming an ideal instrument?

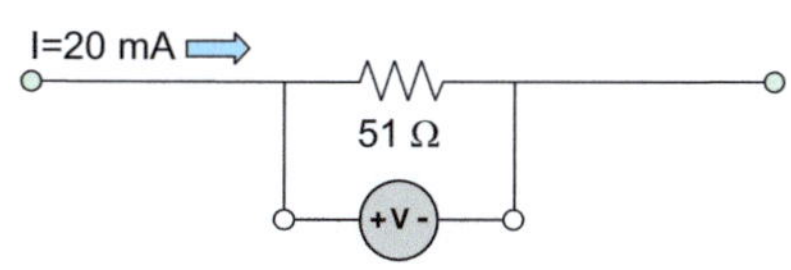

Problem 2.52. Two circuits with an incandescent light bulb are shown in the following figure. Will they function? Explain.

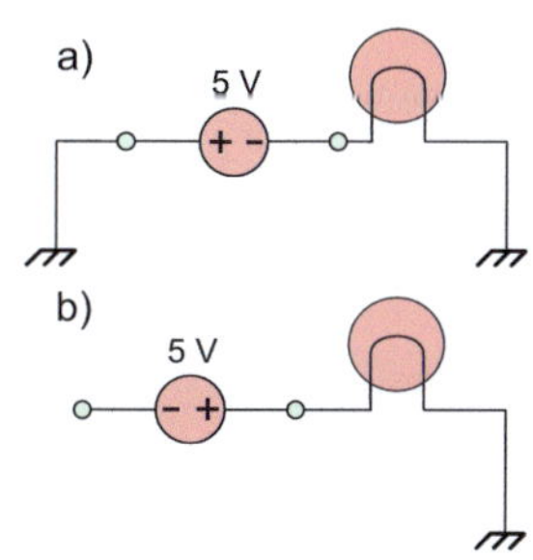

Problem 2.53. A 9-V battery is connected to a 7.5-kΩ resistor shown in the following figure.

Find current I in every case. (a) The negative terminal is left disconnected; (b) the negative terminal is connected to the positive terminal through the resistor; (c) both terminals are connected to chassis ground.

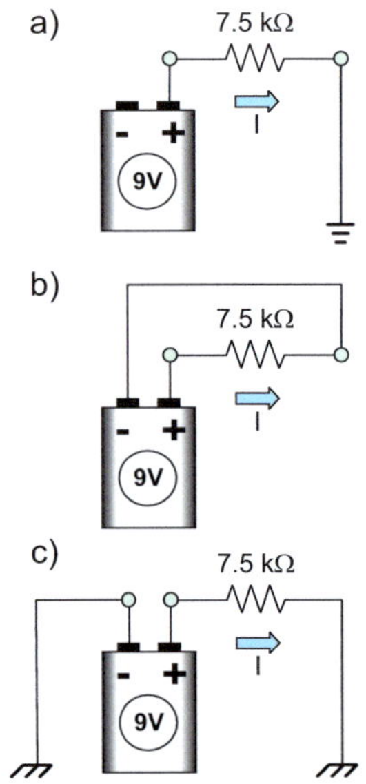

Problem 2.54. What is the voltmeters' (ammeter's) reading in the figure below?

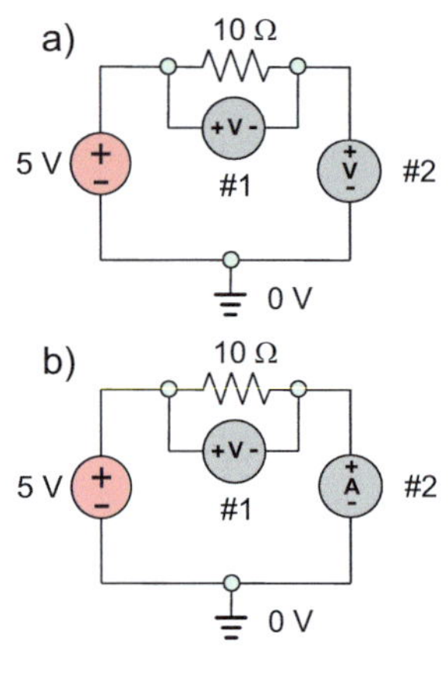

2.5.5 Absolute Voltage and Voltage Drop Across a Circuit Element

Problem 2.55. Determine if the circuit element shown in the following figure is a resistance, a voltage source, or a wire (short circuit). Absolute voltages at points a and b are measured versus ground.

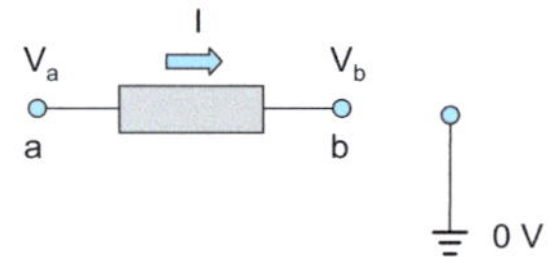

1. $V_a = 3\,\text{V}, \quad V_b = 3\text{V}, \quad I = 1\,\text{A}$
2. $V_a = 3\,\text{V}, \quad V_b = 1\text{V}, \quad I = -1\,\text{A}$
3. $V_a = -2\text{V}, \quad V_b = -5\,\text{V}, \quad I = 2\,\text{A}.$

Problem 2.56. Determine if the circuit element shown in the following figure is a resistance, a voltage source, or a wire (short circuit). Absolute voltages at points a and b are measured versus ground.

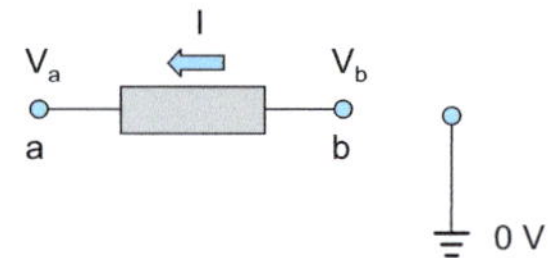

1. $V_a = 6\,\text{V}, \quad V_b = 3\,\text{V}, \quad I = 1\,\text{A}$
2. $V_a = 1\,\text{V}, \quad V_b = 1\,\text{V}, \quad I = -1\,\text{A}$
3. $V_a = -7\text{V}, \quad V_b = -5\text{V}, \quad I = -2\text{A}.$

Problem 2.57. Determine absolute voltages at nodes 1 through 6 in the circuit shown in the following figure.

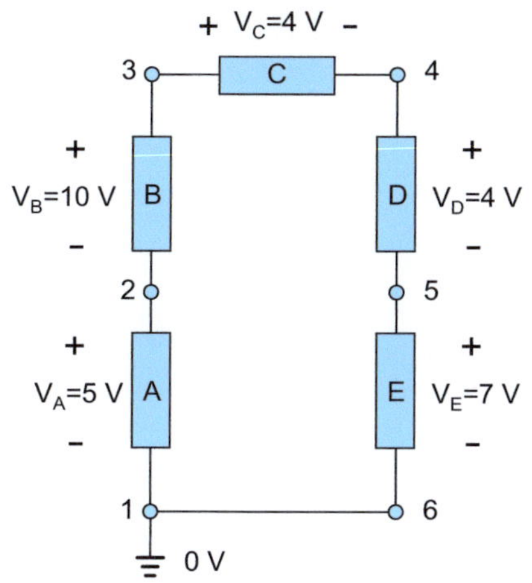

Problem 2.58. Determine voltages across circuit elements A, B, C, D, and E in the circuit shown in the following figure.

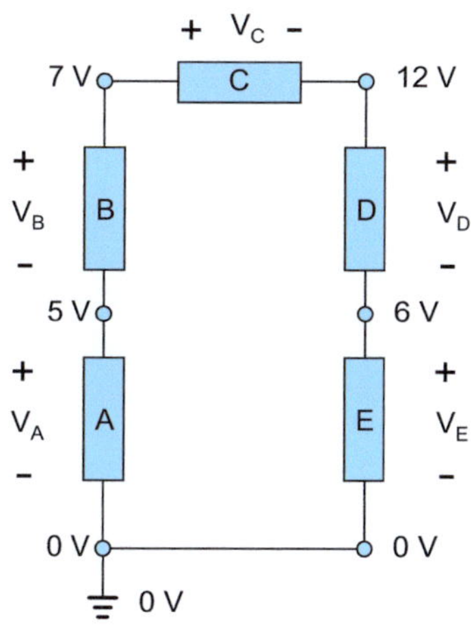

Chapter 3: Circuit Laws and Networking Theorems

Overview

Prerequisites:
- Knowledge of circuit elements, their *v*-*i* characteristics, and Ohm's law (Chap. 2)

Objectives of Sect. 3.1:
- Understand the meaning of an electric network and its topology (nodes, branches, loops, meshes)
- Review the Kirchhoff's current law, its use, and value
- Review the Kirchhoff's voltage law, its use, and value
- Become familiar with the Tellegen's theorem and Maxwell's minimum heat theorem

Objectives of Sect. 3.2:
- Be able to combine sources and resistances in series and parallel
- Practice in the reduction of resistive networks using series/parallel equivalents
- Realize the function and applications of the voltage divider circuit
- Realize the function of the current divider circuit
- Understand the function and applications of the Wheatstone bridge

Objectives of Sect. 3.3:
- Understand the role and place of linear circuit analysis
- Learn the superposition theorem
- Understand the decisive value of superposition theorem for linear circuit analysis
- Learn about immediate applications of the superposition theorem
- Obtain the initial exposure to Y and Δ networks and to T and Π networks

Application examples:
- Voltage divider as a sensor circuit
- Voltage divider as an actuator circuit
- Superposition theorem for a cellphone

S. N. Makarov et al., *Practical Electrical Engineering*, https://doi.org/10.1007/978-3-319-96692-2_3

Keywords:
Electric network, Branches of electric network, Nodes of electric network, Loops of electric network, Meshes of electric network, Essential mesh, Branch currents, Branch voltages, Series connection, Parallel connection, Shunt connection, Kirchhoff's current law, Kirchhoff's voltage law, Maxwell's minimum heat theorem, Tellegen's theorem, Power conservation law for electric networks, One-port network, Equivalent electric networks, Equivalent electric circuits, Series battery bank, Battery pack, Dual-polarity power supply, Common ground of the dual-polarity power supply, Virtual ground of the dual-polarity power supply, Parallel battery bank, Series and parallel combinations (of resistances, of conductances),, Equivalent resistance, Equivalent circuit element, Reduction of resistive networks, Voltage divider circuit, Voltage division rule, Sensor circuit sensitivity, Maximum sensitivity of the voltage divider circuit, Current limiter, Current-limiting resistor, Current divider circuit, Current division rule, Wheatstone bridge (definition of, difference signal, difference voltage, balanced), Linear circuit (definition of, homogeneity, additivity, superposition), Nonlinear circuit (definition of, linearization, dynamic or small-signal resistance), Superposition theorem, Superposition principle, Y network, Δ network, Two-terminal networks, Three-terminal networks, Conversion between Y and Δ networks, Replacing a node by a loop, Δ to Y transformation, Y to Δ transformation, Balanced Y network, Balanced Δ network, Star to delta transformation, T network, T pad, Π network, Π pad, Two-terminal network (definition of, input port, output port)

Section 3.1 Circuit Laws: Networking Theorems

Electric components have to be interconnected to obtain functional circuits which perform specific tasks like driving a motor or monitoring a power plant. Interconnected circuit components form an *electric network*. In turn, any electric network can be evaluated using two simple yet very general laws: Kirchhoff's current law (KCL) and Kirchhoff's voltage law (KVL). Series, parallel, and other combinations of circuit elements, whether linear or not, can be explained and solved using these laws.[1] They were established by Gustav Kirchhoff (1824–1887), a German physicist and mathematician, in 1845, while Kirchhoff was a 21-year-old student at the University of Königsberg in East Prussia. The analyses of all electric circuits are based on KCL and KVL.

3.1.1 Electric Network and Its Topology

An *electric network* can be studied from a general mathematical point of view. If the specific electrical properties are abstracted, there remains a geometrical circuit, characterized by sets of *nodes*, *branches*, and *loops*. These three items form the *topology of an electric network,* and the interconnection of its elements can be represented as a graph. The study of electric network topology makes it possible to:

A. Identify identical circuit blocks in electric circuits which can be drawn in a variety of ways. Examples include series/parallel connections, as well as wye (Y or T) and delta (Δ or Π) blocks as considered in this chapter.
B. Analyze general properties of very large electric circuits such as electric power grids. For example, there are important relationships between the power grid topology and risk identification and mitigation.
C. Relate electric circuits to other disciplines. For example, electric networks have a remarkable ability to model the dynamical behavior of complicated biological systems.

Nodes, Branches, Loops, and Meshes

We consider an electric network with four arbitrary circuit elements A to D as shown in Fig. 3.1a. A *branch* is a two-terminal circuit element. All four two-terminal elements in Fig. 3.1a are therefore branches.

[1] The KCL and KVL concepts are so powerful that they even find applications in equivalent form in magnet systems such as transformers and motors. For example, the magnetic flux in a yoke with air gaps can be model according to KCL and KVL.

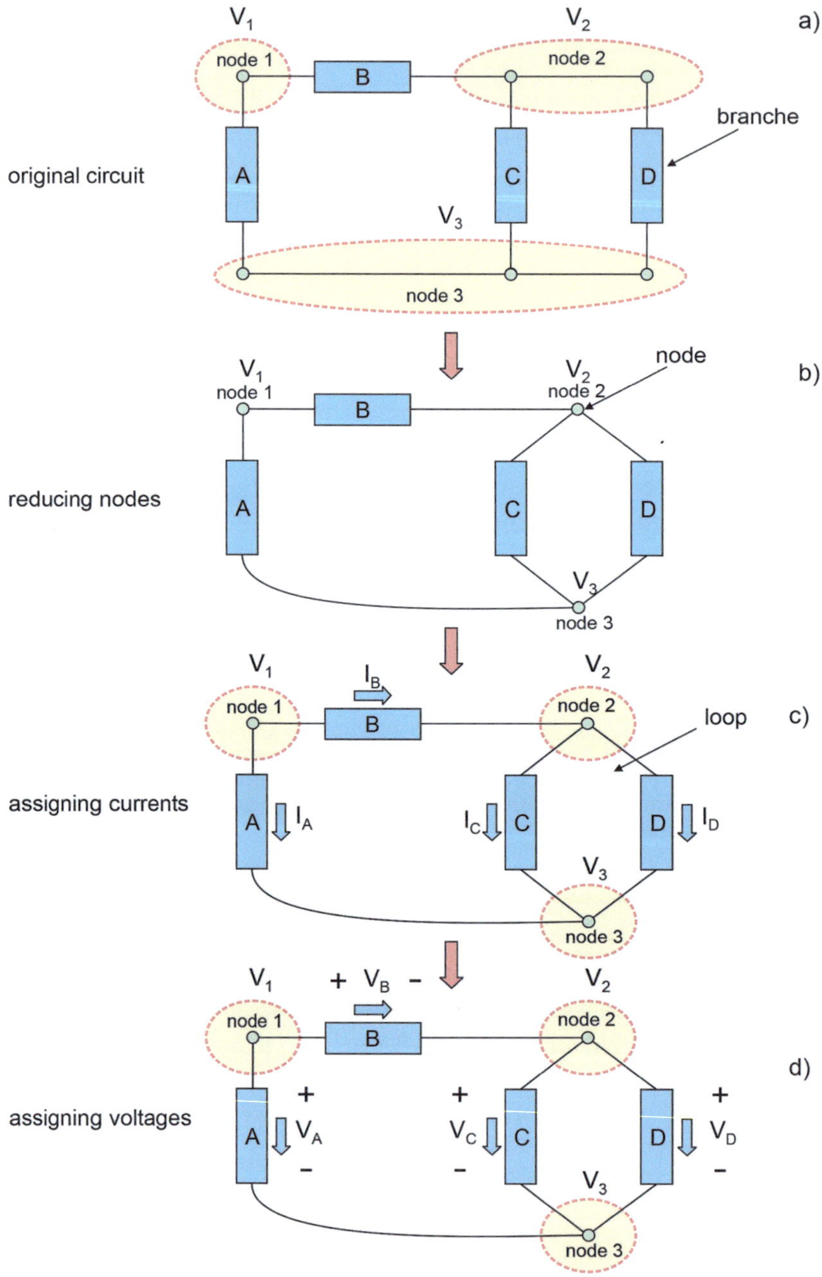

Fig. 3.1. An electric network and assigning branch voltages/currents.

A *node* is a point of interconnection of *two or more* branches. All small six circles in Fig. 3.1a are formally identified as nodes. Every node i can be assigned a certain voltage V_i with respect to circuit ground. If a short circuit (a connecting wire) connects two or more nodes, these nodes constitute a *single node* since they have the same voltage. The circuit may be redrawn to *reduce* the number of nodes and keep only the meaningful nodes (single nodes) with the *distinct voltages* as shown in Fig. 3.1b. The circuits in Fig. 3.1a and in Fig. 3.1b are identical. A *loop* is any closed path between two or more branches. There are *three* loops in Fig. 3.1b. A *mesh* (or *essential mesh*) is a loop that does not contain any other loops within it. There are *two* meshes in Fig. 3.1b. A planar (or two-dimensional) electric network with b branches, n nodes, and m meshes satisfies, after keeping only nodes with distinct voltages (single nodes), the equality

$$b = n + m - 1 \tag{3.1}$$

which is sometimes called the fundamental theorem of network topology. It is proved by considering the electric network as a polygonal graph in two dimensions, where each edge is a branch and each single node is a vertex.

Branch Currents and Voltages

The *branch currents* and their directions may be assigned *arbitrarily*, see Fig. 3.1c. The physical currents either coincide with them or are directed in opposite directions. This can easily be found by checking the sign of the current value once the analysis is complete. If the *branch voltage* polarities have to be assigned *afterward*, they should satisfy *the same* reference configuration for all branches. Let us say the passive reference configuration with regard to the previously assigned current directions is seen in Fig. 3.1d. The branch voltages (voltage drops) in Fig 3.1d are expressed through the node voltages according to:

$$V_A = V_1 - V_3, \quad V_B = V_1 - V_2, \quad V_C = V_D = V_2 - V_3 \tag{3.2}$$

Conversely, the branch voltages may be assigned *arbitrarily* at first. If the directions of the branch currents have to be assigned *afterward*, they should again satisfy the same reference configuration.

Exercise 3.1: Establish whether or not the networks in Fig. 3.1a and in Fig. 3.1b satisfy Eq. (3.1).

Answer: The answer is yes for both figures if we consider single nodes. However, if we consider every small circle in Fig. 3.1a as a node, Eq. (3.1) will not be satisfied.

Series and Parallel Connections

Two or more branches (circuit element) are *in series* if they *exclusively* share a common node. Elements A and B in Fig. 3.1d are in series. Elements B and C are *not* since node 2 is also shared by element D. Two or more branches (circuit elements) are *in parallel* if they are connected to the *same* two nodes. Elements C and D in Fig. 3.1d are in parallel. The parallel connection is also called the *shunt connection*; the parallel element may be called *shunt* (or *shunting*) *elements*.

3.1.2 Kirchhoff's Current Law (KCL)

Let us begin our investigation with KCL. It specifically applies to the *nodes* in an electric network. *Kirchhoff's current law* simply states that *the net current entering the node is zero*. In other words, the sum of inflowing currents is equal to the sum of outflowing currents. This statement is also known as the current conservation law, which is the electrical counterpart of the mass conservation law in fluid mechanics. For N currents entering a node, KCL can be cast in the form

$$\sum_{i=1}^{N} I_i = 0 \tag{3.3}$$

where N denotes the total number of nodal currents. The current directions may be assigned *arbitrarily*; the same results will eventually be obtained. The nodal current is taken with a plus sign if it is entering the node, i.e., the current arrow is directed toward the node. It carries a minus sign if it is leaving the node, i.e., the current arrow points in the opposite direction. The current value itself (positive or negative) is substituted afterward. Interestingly, if this law did not hold, an uncompensated charge could accumulate in a node over time. This uncompensated charge and its associated Coulomb force would eventually destroy the operation of the underlying electric circuit. To illustrate the use of KCL, we consider Fig. 3.2 with four different node types for a collection of branches A, B, C, D which could represent arbitrary circuit elements.

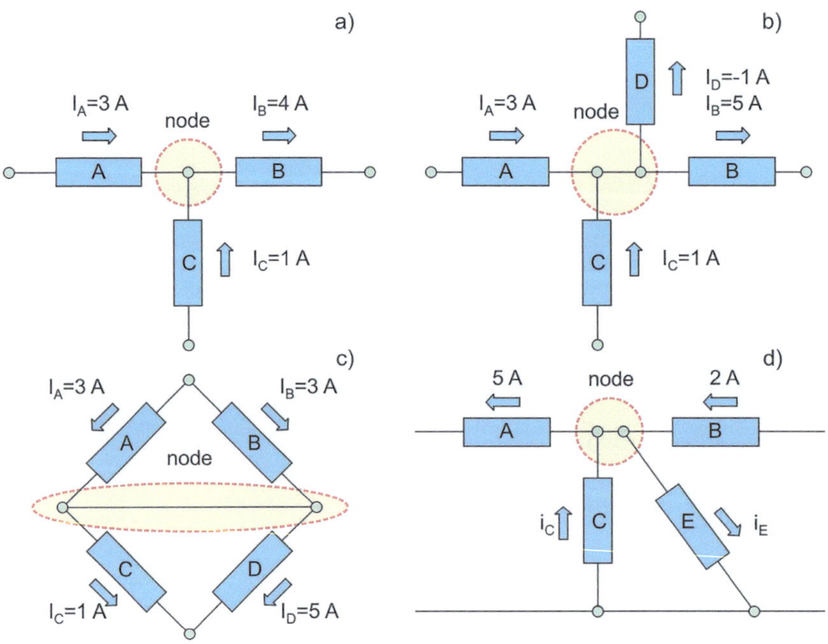

Fig. 3.2. Different types of the nodes in an electric network subject to KCL.

While the node in Fig. 3.2a is simple, more complicated node configurations may be observed in a circuit (see the following figures). You should note that in a node we may move individual joints to one common joint without affecting the circuit's operation. A node transformation to a single joint is a convenient tool used when working with more complicated nodes.

Exercise 3.2: Write KLC for the nodes shown in Fig. 3.2.

Answer: In the case of Fig. 3.2a, $I_A + I_C = I_B$ for any values of I_A, I_B, I_C. In the case of Fig. 3.2b, $I_A + I_C = I_B + I_D$ for any values of I_A, I_B, I_C, I_D. In the case of Fig. 3.2c, $I_A + I_B = I_C + I_D$ for any values of I_A, I_B, I_C, I_D. In the case of Fig. 3.2d, 2A + $I_C = 5A + I_E$ for any values of I_C, I_E.

Example 3.1: In some cases, the use of KCL may be sufficient to determine all currents in an electric network. Solve for the unknown currents I_A, I_C, I_D in a network shown in Fig. 3.3.

Solution: First, we note that the wire connection on the left states that the current along the wire is preserved. This implies $I_A = -1$ A. KCL for node 1 gives

$$1\text{A} + I_C = 5\text{ A} \Rightarrow I_C = 4\text{ A} \tag{3.4a}$$

Next, KCL applied to node 2 yields

$$5\text{A} = I_D + 2A \Rightarrow I_D = 3\text{ A} \tag{3.4b}$$

Thus, the circuit is solved. Node 3 has not been used; it can be employed to check the correctness of the solution: $I_A + I_D + 2A = I_C$ or 4 A = 4 A.

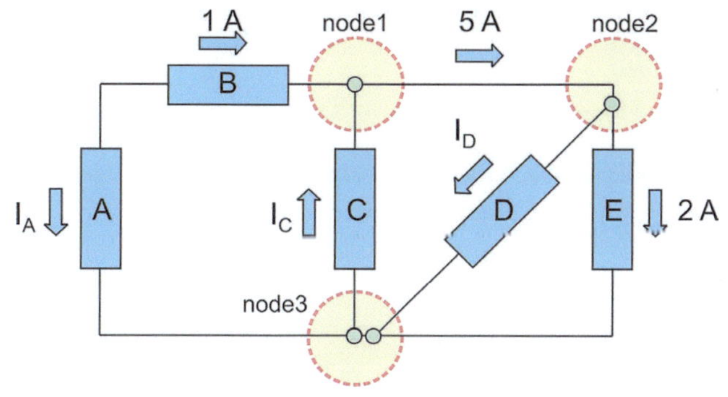

Fig. 3.3. Solving in a network using KCL.

3.1.3 Kirchhoff's Voltage Law (KVL)

This law specifically applies to the *loops* in an electric network. The *Kirchhoff's voltage law* states that *the sum of voltages for any closed loop is zero.* In other words, the total amount of work needed to move a unit electric charge one loop turn is zero.[2] If this law were not applicable, a

[2] The physical counterpart of KVL is Faraday's law of induction in the static case. When there is a variable magnetic field penetrating a wire loop, KVL is no longer valid.

charge moving in a closed loop would constantly accelerate and eventually escape the circuit or constantly decelerate and eventually stop moving. In order to formulate KVL for a closed loop, we need to identify the loop direction. It is usually chosen to be clockwise (see the dotted arrows in Figs. 3.4 and 3.5). KVL in its general form states

$$\sum_{i=1}^{N} V_i = 0 \qquad (3.5)$$

where N is the total number of circuit elements in a loop. The voltage V_i is taken with a plus sign if the loop arrow is entering the positive voltage polarity and with a minus sign otherwise. The polarities of voltages V_i can be assigned *arbitrarily*; the same result is obtained. To demonstrate this fact, in Fig. 3.4 we first consider a simpler network with different voltage polarities. One network branch is purposely designated as a voltage source, while the rest are arbitrary circuit elements.

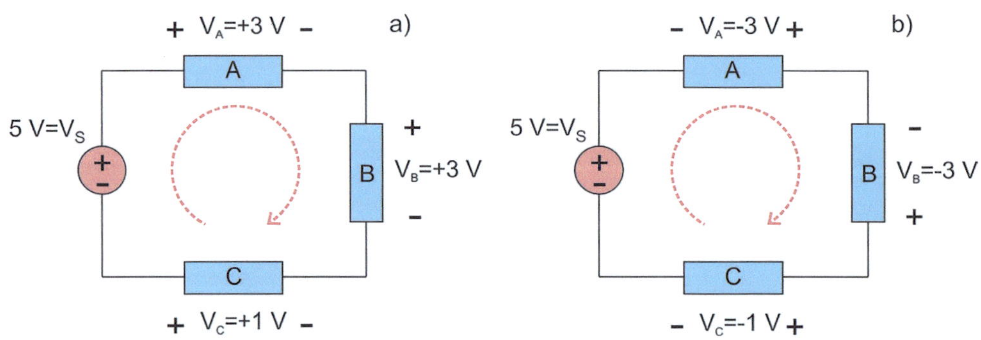

Fig. 3.4. KVL applied to a closed loop with one voltage source and three passive circuit elements. The dotted arrow indicates current flow in clockwise direction.

Example 3.2: Write KVL for the circuit shown in Fig. 3.4 that includes an ideal voltage source and three other circuit elements.

Solution: In the case of Fig. 3.4a, we start with the source and strictly apply our convention of positive and negative polarity based on the prescribed loop arrow direction:

$$-5\text{V} + V_A + V_B - V_C = 0 \qquad (3.6a)$$

In the case of Fig. 3.4b, we have to change signs of V_A, V_B, V_C and thus obtain

$$-5\text{V} - V_A - V_B + V_C = 0 \qquad (3.6b)$$

Note that these two cases only differ by voltage polarities.

Example 3.2 (cont.):

Since reversing voltage polarities in Fig. 3.4b was taken into account by changing the sign of the voltage, both figures yield the identical result after substituting values:

$$-5\ \text{V} + 3\ \text{V} + 3\ \text{V} - 1\ \text{V} = 0 \tag{3.6c}$$

This observation highlights the fact that the voltage polarities for circuit elements may initially be assigned arbitrarily: applying KVL will ultimately lead to the correct signs of the voltage values.

Figure 3.5 shows another electric network with all branches now consistently labeled; the voltages in red denote the actual values with respect to the initially assigned directions.

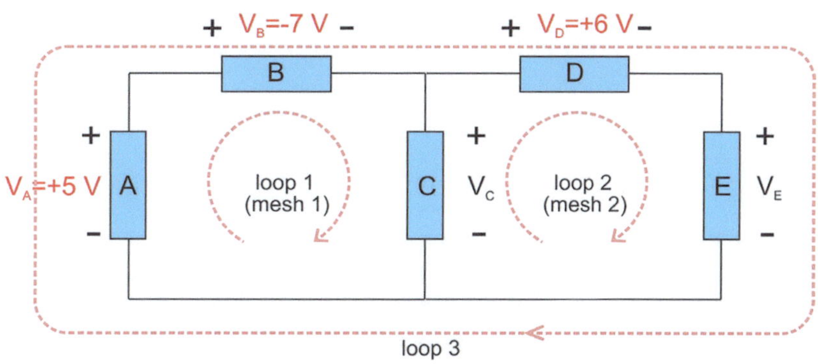

Fig. 3.5. KVL applied to a network with three loops, two meshes, and five circuit elements.

Exercise 3.3: Determine the number of branches, nodes, loops, and meshes for the network shown in Fig. 3.5.

Answer: There are five branches, four single nodes, three loops, and two meshes.

Example 3.3: In some cases, the use of KVL may be sufficient to determine all voltages in an electric network. Solve for the unknown voltages in the network shown in Fig. 3.5.

Solution: The application of KVL for loop 1 (mesh 1) has the form

$$-5\text{V} - 7\text{V} + V_C = 0 \Rightarrow V_C = 12\text{V} \tag{3.6d}$$

Moreover, KVL for loop 2 (mesh 2) results in (V_C is already known)

$$-12\text{V} + 6\text{V} + V_E = 0 \Rightarrow V_E = 6\text{V} \tag{3.6e}$$

Example 3.3 (cont.):

Finally, applying KVL for loop 3 is the solution check:

$$-5\ \text{V} - 7\ \text{V} + 6\ \text{V} + 6\ \text{V} = 0 \tag{3.6f}$$

If at least one of the voltages V_A, V_B, V_D in Fig. 3.5 is unknown, then a unique solution does not exist. Through the use of KCL more information must be acquired.

3.1.4 Power-Related Networking Theorems

This section would be incomplete without an introductory discussion of two power-related networking theorems. The first one is *Maxwell's minimum heat theorem* formulated by James Clerk Maxwell in 1891. It states that, for a *linear* electric network of resistive circuit elements and voltage/current sources, the currents distribute themselves in such a way that the total dissipated power (generated heat) in the resistances is a *minimum*. If there are N_R resistive circuit elements, we can state

$$\sum_{i=1}^{N_R} V_i I_i = \sum_{i=1}^{N_R} R_i I_i^2 = \sum_{i=1}^{N_R} \frac{V_i^2}{R_i} = \min \tag{3.7a}$$

where voltages V_i and currents I_i must all satisfy the passive reference configuration. Indeed, Eq. (3.7a) is equivalent to the condition where the net power generated by all sources is also minimized. Maxwell's minimum heat theorem can be proved using KCL, KVL, and Ohm's law. The second theorem is *Tellegen's theorem* formulated by Bernard D.H. Tellegen in 1952. It postulates that, for an *arbitrary* electric network with a total of N circuit elements of arbitrary (linear or nonlinear, passive or active) nature, the equality

$$\sum_{i=1}^{N} V_i I_i = 0 \tag{3.7b}$$

must hold if all voltages V_i and currents I_i satisfy one (let's say the passive) reference configuration. Tellegen's theorem serves as a *power (or energy) conservation law for all electric networks*. It can be proved based on KCL and KVL *only*. Tellegen's theorem has other important implications and generalizations. To illustrate both theorems, we consider two simple examples.

Example 3.4: Use Maxwell's minimum heat theorem and determine the unknown current x through resistance R_1 in Fig. 3.6a.

Solution: Equation (3.7a) yields

$$R_1 x^2 + R_2 (x - I_S)^2 = \min \tag{3.8a}$$

This function is minimized when its derivative with respect to x is zero. Therefore,

Example 3.4 (cont.):

$$x = \frac{R_2}{R_1 + R_2} I_S \tag{3.8b}$$

which is the current division principle studied in the next section. The voltage division principle may be established similarly.

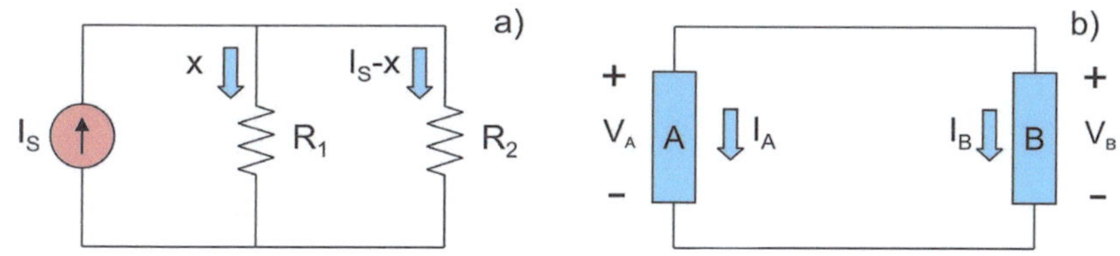

Fig. 3.6. Examples for Maxwell's minimum heat theorem and Tellegen's theorem.

Example 3.5: Prove Tellegen's theorem for the network shown in Fig. 3.6b.

Solution: Both elements follow the passive reference configuration. By KCL we conclude $I_B = -I_A$. By KVL, $V_B = V_A$. Therefore, $V_A I_A + V_B I_B = 0$, which is the simple poof. In practice, element A may be a voltage source, and element B may be a resistance. We can also use the active reference configuration for the source but need to define its power to be *negative* in such a case.

3.1.5 Port of a Network: Network Equivalence

So far we have studied only closed electric networks. A network can have a *port*, through which it is interconnected to another network as shown in Fig. 3.7. The network may be active or passive. This specific network has two (input or output) terminals a and b, which form a *single port*. Thus, the network in Fig. 3.7 is known as a *one-port network*. All series/parallel combinations of sources and resistances are typically one-port networks.

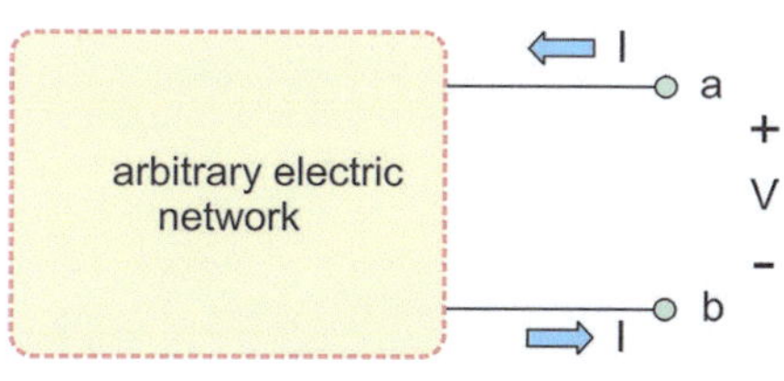

Fig. 3.7. Generic system for establishing a network equivalence.

Two arbitrary one-port networks in the form of Fig. 3.7 are said to be *equivalent networks* (or *equivalent electric circuits*) when their v-i characteristics at terminals a and b *coincide*. This means that for any given voltage, there is the same current entering the network or leaving it and vice versa. Therefore, these networks are indistinguishable.

Section 3.2 Series and Parallel Network/Circuit Blocks

3.2.1 Sources in Series and in Parallel

Series-Connected Battery Bank

The simultaneous use of KCL and KVL allows us to analyze the behavior of combinations of active circuit elements and establish their equivalence. The physical counterparts are various *battery banks*, which are interconnections of the *identical* batteries. Figure 3.8 shows a *series battery bank*, also called a *battery pack*, with two or more batteries connected in series. The battery symbol implies an ideal voltage source.

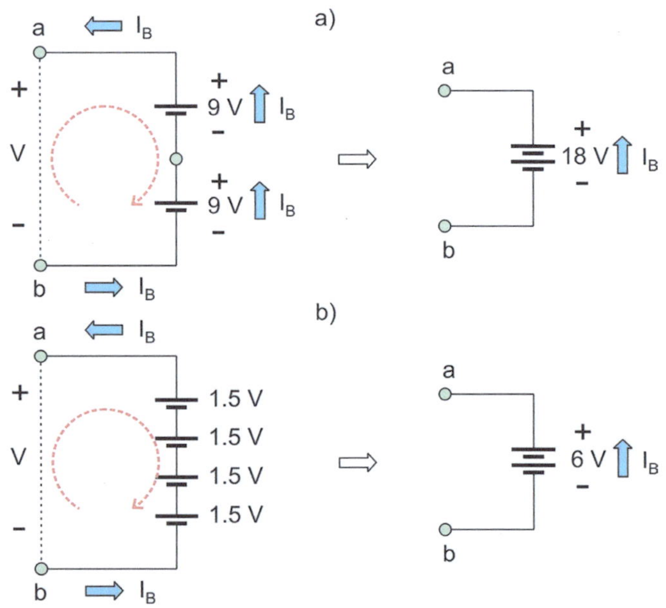

Fig. 3.8. Series combinations of battery cells and their equivalent representations.

We intend to find the resulting voltage and current of this combination. To determine the equivalent voltage, we close the circuit loop shown in Fig. 3.8a by introducing a *virtual* circuit element with terminals a and b, with an unknown voltage V between these terminals. This element simulates the rest of the circuit, which closes the current path. Applying KVL for the loop shown in Fig. 3.8a results in

$$-V+9\ \mathrm{V}+9\ \mathrm{V}=0\Rightarrow V=18\ \mathrm{V} \tag{3.9}$$

Moreover, using KCL, we obtain the same current flows throughout the left-handed circuit of Fig. 3.8a; this current is exactly equal to I_B.[3] Therefore, the series combination of two batteries in Fig. 3.8a is *equivalent* to one battery bank that provides double the voltage, or 18 V, compared to the unit cell. However, it delivers a current of a single cell. The same method can now be applied to multiple battery cells connected in series; one such battery bank is shown in Fig. 3.8b.

Dual-Polarity Voltage Power Supply

Two batteries, or other voltage sources, connected in series can be used as a *dual-polarity power supply* as shown in Fig. 3.9. The middle terminal provides a *virtual ground* for the circuit or the *common ground*. Both negative and positive voltages with respect to the common port can now be created in the circuit. Such a source is of particular importance for operational amplifier circuits and for transistor circuits. Every multichannel laboratory power supply may operate as a dual power supply. The common terminal may (but does not have to) additionally be connected to the earth ground.

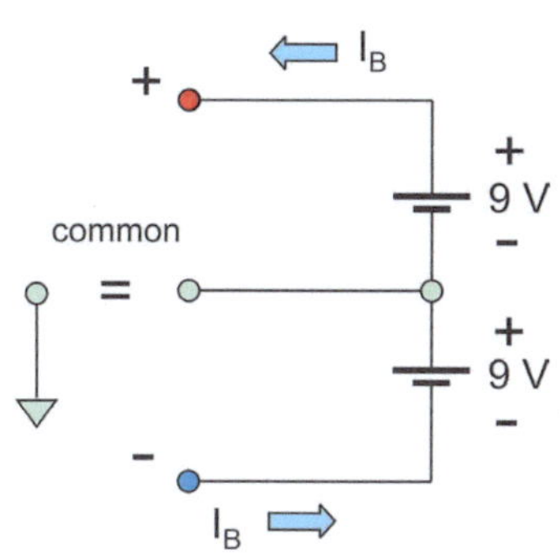

Fig. 3.9. A dual-polarity power supply constructed with two battery cells.

Parallel-Connected Battery Bank

As an alternative to the series connected battery bank, we can investigate the *parallel battery bank* shown in Fig. 3.10. To determine the equivalent voltage of the combination, we again close the circuit loop by introducing a virtual circuit element with terminals a and b and with an unknown voltage V between these terminals. The use of KVL gives

$$V=9\ \mathrm{V} \tag{3.10}$$

Thus, the voltage of the battery bank in Fig. 3.10 is still equal to the unit cell voltage. However, applying KCL to both rightmost nodes in Fig. 3.10 indicates that the current doubles. Therefore, the

[3] If realistic battery cells are capable of delivering different currents when short circuited, then the lowest cell current will flow under short circuit condition.

parallel combination of two batteries is *equivalent* to one battery bank that provides the same voltage of 9 V as one unit cell but at twice the current strength.

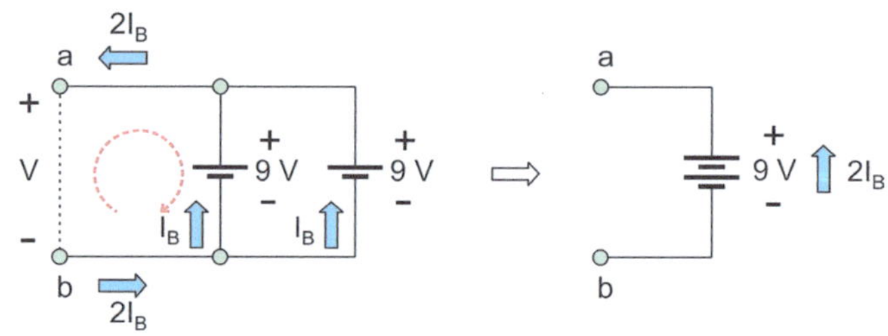

Fig. 3.10. A parallel combination of battery cells and its equivalent single battery representation.

Series Versus Parallel Connection

What is the difference between series and parallel combinations of two 9 V batteries? First, let us find the power delivered to the circuit. We assume $I_B = 1$ A in both cases, even though the specific value of the current is not important. For the series combination in Fig. 3.8a, the delivered power is 18 V × 1 A = 18 W. For the parallel combination in Fig. 3.10, the delivered power is again 9 V × 2 A = 18 W. Thus, as far as the power rating is concerned, there is no difference. You should, however, remember that we always deliver power to a *load*. For the series combination, the implied load resistance becomes 18 V/1 A = 18 Ω. For the parallel combination, the anticipated load resistance will be 9 V/2 A = 4.5 Ω. Thus, it is the load resistance that determines which combination should be used. This question is of great practical importance.

Combinations of Current Sources

Combinations of current sources are studied similarly. They are important for photovoltaic and thermoelectric semiconductor devices.

3.2.2 Resistances in Series and in Parallel

Series Connection

After the sources have been analyzed, we turn our attention to *series and parallel combinations of resistances* (or *conductances*). Their physical counterparts are various circuit loads, for example, individual households connected to the same power grid or individual motors driven by the same source. Figure 3.11 depicts the series combination of two resistances R_1 and R_2.

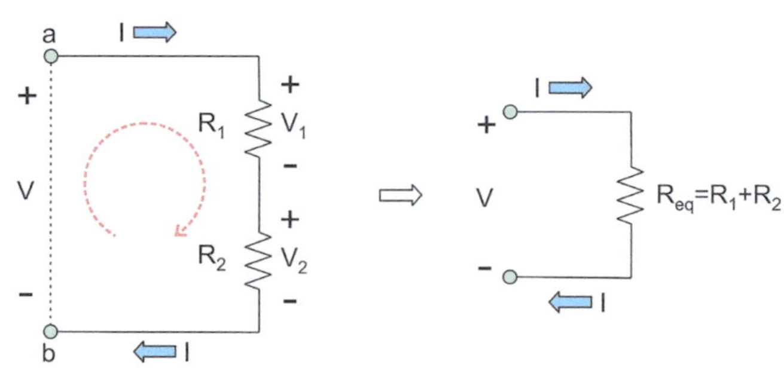

Fig. 3.11. Two resistances in series and their equivalent single resistance representation.

Note that the current direction for both resistances corresponds to a passive reference configuration: current flows "down the voltage hill." Again, we close the circuit loop by introducing a virtual circuit element with terminals a and b and with an unknown voltage V between those terminals. This virtual circuit element, which simulates the rest of the circuit, allows us to close the current path. KVL for the loop shown in Fig. 3.11 results in

$$V = V_1 + V_2 = IR_1 + IR_2 = I(R_1 + R_2) = IR_{eq} \Rightarrow R_{eq} = R_1 + R_2 \tag{3.11a}$$

Thus, two resistances can be replaced by one *equivalent resistance*, which is the sum of the individual resistances. It follows from Eq. (3.11a) that the equivalent resistance gives us the *same* circuit current (and the *same* power into the load) as the original resistance combination does, for any applied voltage. This is the formal description of the equivalent resistance to be generalized later. Equation (3.11a) can easily be extended to any arbitrary number of resistances connected in series. Equation (3.11a) may be also formulated in terms of conductances, the reciprocals of resistances,

$$\frac{1}{G_{eq}} = \frac{1}{G_1} + \frac{1}{G_2} \tag{3.11b}$$

Parallel Connection

Next, we consider the parallel combination of resistances shown in Fig. 3.12. KVL applied to the loop shown in the figure and for another loop between two resistances indicates that the voltages across every resistance are equal to V. KCL applied to either rightmost node results in

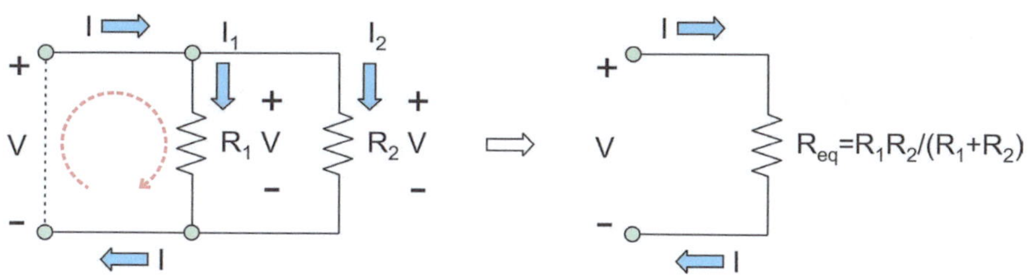

Fig. 3.12. Two resistances connected in parallel and the equivalent resistance.

$$I = I_1 + I_2 = \frac{V}{R_1} + \frac{V}{R_2} = \frac{V}{R_{eq}} \Rightarrow \frac{1}{R_{eq}} = \frac{1}{R_1} + \frac{1}{R_2} \Rightarrow G_{eq} = G_1 + G_2 \tag{3.12}$$

Therefore, the parallel combination of two resistances is equivalent to one resistance, which has a value equal to the reciprocal of the sum of the reciprocal values of both. The equivalent resistance again gives us the same circuit current as the original resistance combination does, for any applied voltage. We emphasize that the equivalent resistance *is always smaller in value than each of the resistances to be combined in parallel.* Note that the conductances simply add up for the parallel combinations. Equation (3.12) can again easily be extended to any arbitrary number of resistances connected in parallel.

Meaning of Equivalent Circuit Element

In summary, the study of series/parallel active and passive circuit elements leads us to the following simple definition of an equivalent circuit element, either passive or active. The *equivalent circuit element* possesses the same υ-i characteristic as the υ-i characteristic of the original circuit, for which voltage and current are acquired at its terminals a and b.

3.2.3 Reduction of Resistive Networks

The reduction of a network of many resistances to a single equivalent resistance is a topic of practical importance. There are many examples of distributed resistive networks. A contemporary example is a rear window defroster, which is a distributed resistive heater. The corresponding solution is usually based on:

(i) Step-by-step use of series/parallel equivalents
(ii) Moving, splitting, or reducing modes
(iii) Reliance on fluid mechanics analogies, which may be helpful for resistive networks

Although a unique solution *always* exists, its practical realization may be quite difficult. The following examples outline the procedure for the *reduction of resistive networks*.

Example 3.6: Find the equivalent resistance between terminals a and b for the resistive network shown in Fig. 3.13a.

Solution: We should not start with terminals a and b but with the *opposite* side of the circuit. First, the three resistances furthest to the right are combined in series in Fig. 3.13a. The next step is the parallel combination of the resulting resistance and the 1.5 kΩ resistance in Fig. 3.13b. The final step is another series combination; this results in the final equivalent resistance value of 1875 Ω. We need to point out again that it is impossible to:

(i) Combine in series two resistances separated by a *node*.
(ii) Move the resistance *through* a node or move the node *through* a resistance.

These restrictions hold for other circuit elements as well, beside resistors.

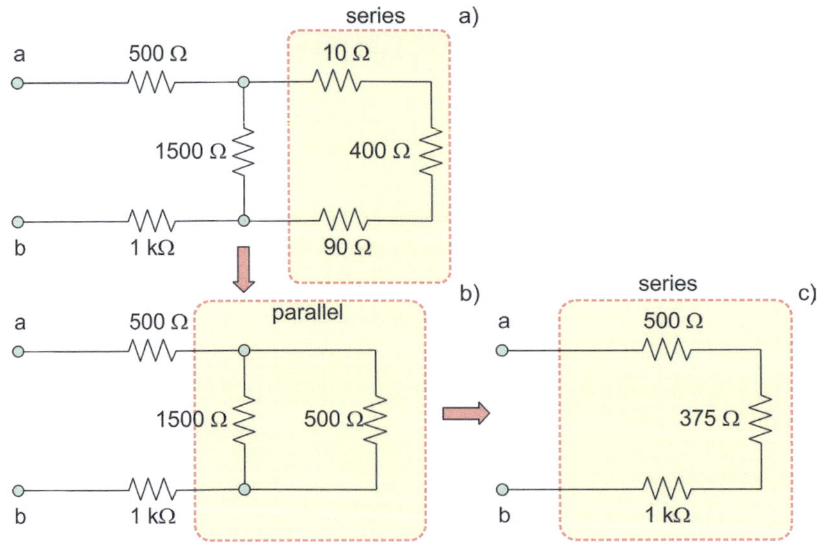

Fig. 3.13. Step-by-step circuit reduction via series and parallel resistance combinations.

Example 3.7: Find the equivalent resistance between terminals *a* and *b* for the resistive network shown in Fig. 3.14a.

Solution: First, we can split and move a node along the wire, leading to a circuit shown in Fig. 3.14b. Next, we combine two pairs of resistances in parallel. The next step, in Fig. 3.14c, is the series combination of three resistances. The last step is the solution of a parallel circuit, leading to the equivalent resistance of 66.67 Ω.

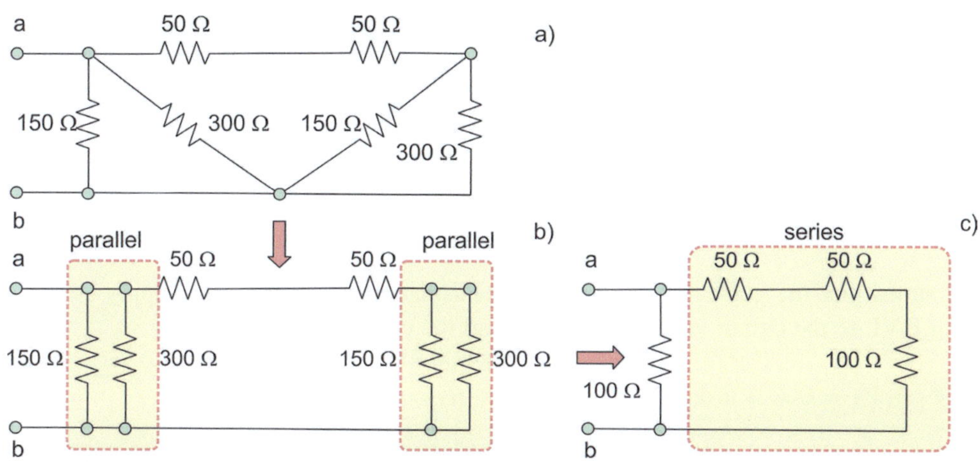

Fig. 3.14. Step-by-step circuit reduction to a single equivalent resistance.

Exercise 3.4: Using a fluid mechanics analogy of identical water flow in two symmetric channels, find the equivalent resistance of the network shown in Fig. 3.15.

Answer: 2.5 kΩ

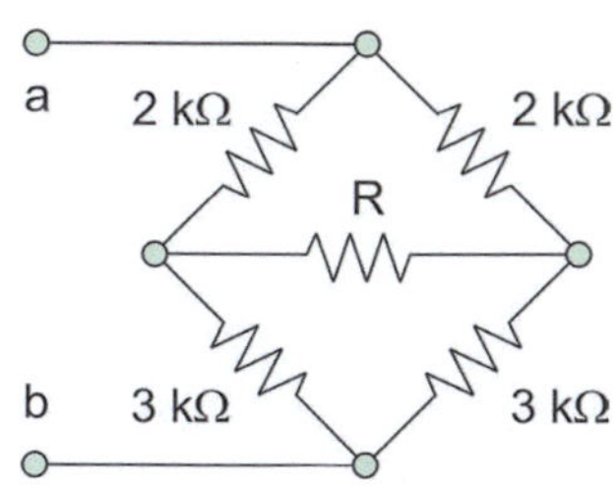

Fig. 3.15. The resistive network discussed in Exercise 3.4.

3.2.4 Voltage Divider Circuit

The purpose of the *voltage divider circuit* is to provide a voltage different from the supply voltage. The voltage divider circuit is associated with *resistances in series*. It is perhaps the most important basic electric circuit or block within another circuit. The voltage division principle is used in sensor circuits, actuator circuits, and bias circuits. Furthermore, any input and output port of a (transistor) amplifier is essentially a voltage divider. We consider a particular voltage divider form, shown in Fig. 3.16, connected to a DC voltage supply. This circuit, as with *any* other electric circuit, can be analyzed by using KCL and KVL *simultaneously*. We prefer using this method although the combination of two resistances in series will provide an equivalent solution.

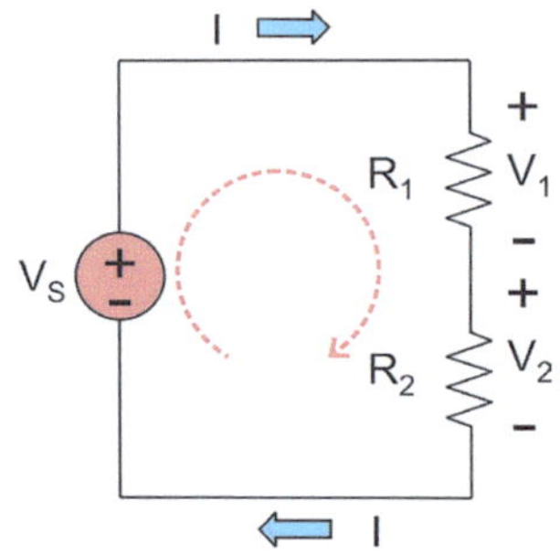

Fig. 3.16. A conventional voltage divider consists of two resistances and an ideal voltage source.

KCL states that the current I through both resistance and the voltage supply is the same. Applying KVL to the circuit loop allows us to find the circuit current:

$$-V_S + V_1 + V_2 = 0 \Rightarrow V_S = V_1 + V_2 = I(R_1 + R_2) \Rightarrow I = \frac{V_S}{R_1 + R_2} \tag{3.13}$$

Once the circuit current I is known, then Ohm's law can be used. This yields *the voltage divider rule*

$$V_1 = \frac{R_1}{R_1 + R_2} V_S, \quad V_2 = \frac{R_2}{R_1 + R_2} V_S \tag{3.14}$$

Equation (3.14) states that the major function of the voltage divider is to divide the voltage of the power source between two resistances in *direct* proportion so that:

(i) The larger resistance always acquires a higher voltage, and the smaller resistance acquires a smaller voltage.
(ii) The individual voltages always add up to the supply voltage, i.e., $V_1 + V_2 = V_S$.

Exercise 3.5: A voltage divider circuit uses a 10 V DC source and two resistances: $R_1 = 5\ \Omega$ and $R_2 = 100\ \Omega$. What are the voltages V_1, V_2 across the resistances?

Answer: $V_1 = 0.48$ V, $\mathrm{V}_2 = 9.52$ V, $V_1 + V_2 = 10$ V.

The voltage divider with multiple resistances $R_1, R_2, \ldots, R_N$ is solved in the form

$$V_S = V_1 + V_2 + \ldots + V_N \Rightarrow I = \frac{V_S}{R_1 + R_2 + \ldots + R_N} \quad \text{and} \quad V_i = \frac{R_i}{R_1 + R_2 + \ldots + R_N} V_S \tag{3.15}$$

3.2.5 Application Example: Voltage Divider as a Sensor Circuit

Consider a resistive sensing element (thermistor, strain gauge, photoresistor, etc.) as denoted by $R_2(x)$ in Fig. 3.17. The element changes its resistance $R_2(x)$ when an external parameter x changes. Parameter x could be temperature, pressure, humidity, solar radiation, or any other physical parameter that undergoes an environmental change. A simple sensor configuration is a direct connection to a voltage source and to the DMM for voltage measurements (see Fig. 3.17a). However, no matter how the sensor resistance changes, the sensor will always output the source voltage. A solution to the this problem is a voltage divider circuit shown in Fig. 3.17b. The extra resistance R_1 is fixed. According to Eq. (3.14), voltage V_2 in Fig. 3.17b varies depending on the influence of $R_2(x)$:

$$V_2 = V_2(x) = \frac{R_2(x)}{R_1 + R_2(x)} V_S \tag{3.16}$$

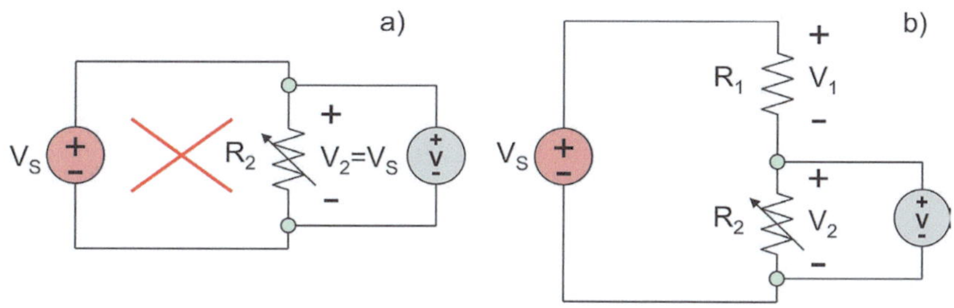

Fig. 3.17. (**a**) Incorrect sensor circuit; (**b**) a sensor circuit on the basis of a resistive voltage divider where $R_2(x)$ changes its resistance depending on the process parameter x.

The variable voltage $V_2(x)$ is measured by the voltmeter. The dependence of V_2 on R_2 is clearly nonlinear in Eq. (3.16). Although Eq. (3.16) can be linearized by choosing a sufficiently large R_1 to make the denominator nearly constant, we will show later that such an operation greatly decreases device *sensitivity*. Let us assume that the external parameter x in Eq. (3.16) changes from a lower limit x_1 to an upper limit x_2, i.e., $x_1 \le x \le x_2$. As a result, the sensing resistance changes monotonically, but not necessarily linearly, from $R' = R_2(x_1)$ to $R'' = R_2(x_2)$. We also assume that if $x_1 \le x \le x_2$, then $R' > R''$. The *sensor circuit's sensitivity*, S, is given by

$$S = \frac{V_2(x_1) - V_2(x_2)}{x_2 - x_1} \quad \left[\frac{\mathrm{V}}{\text{units of } x}\right] \tag{3.17}$$

The sensitivity is expressed in terms of voltage variation per one unit of x. A higher sensitivity implies a larger voltage variation and thus provides a better sensor resolution and improved robustness against noise.

Design of the Sensor Circuit for Maximum Sensitivity

Let us pose the following question: what value should the fixed resistance R_1 assume in order to achieve the *highest sensitivity of the voltage divider sensor*? It is clear that R_1 cannot be very small (otherwise the voltage reading will always be V_S and the sensitivity will be zero) and that R_1 cannot be very large (otherwise the voltage reading will be always 0 V and the sensitivity will be zero). The sensitivity is thus a positive function that is zero at $R_1 = 0$ and at $R_1 = \infty$. According to the extreme value theorem, a global maximum should exist between these two values. We denote the unknown resistance R_1 with variable t, substitute V_2 from Eq. (3.14), and rewrite Eq. (3.17) in the form

$$S = \frac{V_S}{x_2 - x_1}\left(\frac{R'}{t+R'} - \frac{R''}{t+R''}\right) \tag{3.18a}$$

It is convenient to transform this result into a simpler expression $S = V_S S_0 f(t)$, where a constant S_0 is called the *intrinsic* sensitivity of the resistive sensing element and $f(t)$ is the sole function of the first resistance, i.e.,

$$S_0 = \left[\frac{R' - R''}{x_2 - x_1}\right], \quad f(t) = \frac{t}{(R'+t)(R''+t)} \tag{3.18b}$$

This function $f(t)$ is to be maximized. At the function's maximum, the derivative of $f(t)$ versus t should be zero. Using the quotient rule for the differentiation of a fraction, it follows from Eq. (3.18b) that

$$f'(t) = \frac{R'R'' - t^2}{(R'+t)^2(R''+t)^2} \tag{3.18c}$$

The final result following from the condition $f'(t) = 0$ is surprisingly simple

$$t = R_1 = \sqrt{R'R''} \tag{3.18d}$$

In other words, the fixed resistance of the voltage divider circuit should be equal to the *geometric mean* of two extreme resistances of the sensing element itself.

Example 3.8: For the NTC-503 thermistor sensing element, $x_1 = 25^\circ\text{C}$ (room temperature), $x_2 = 37^\circ\text{C}^\circ$ C (approximate temperature of a human body), $R' = 50\ \text{k}\Omega$, and $R'' = 30\ \text{k}\Omega$. What is the sensitivity of the voltage divider sensor if $V_S = 9$ V and (A) $R_1 = 5\ \text{k}\Omega$, (B) $R_1 = 500\ \text{k}\Omega$, and (C) $R_1 = \sqrt{30 \cdot 50} \approx 39\ \text{k}\Omega$?

Solution: We substitute the numbers in Eq. (3.18a) and find the sensitivity. The corresponding sensitivity plot as a function of R_1 is given in Fig. 3.18a. The particular sensitivity values are (A) $S = 39$ mV/°C; (B) $S = 26$ mV/°C; and (C) $S = 95$ mV/°C.

Example 3.9: For the SGT-1/350-TY11 strain gauge, the nominal resistance is 350 Ω. The resistance variation of ±0.1% for the tensile strain ε is observed, and the intrinsic device sensitivity is $S_0 = 700\ \Omega/\varepsilon$. What is the sensitivity of the voltage divider sensor if $V_S = 4.5$ V and (A) $R_1 = 50\ \Omega$, (B) $R_1 = 5\ \text{k}\Omega$, and (C) $R_1 = 350\ \Omega$? For positive sensitivity numbers, interchange $x_{1,2}$ in Eq. (3.18a).

Solution: In this example, $R' = 350.35\ \Omega$ and $R'' = 349.65\ \Omega$. We use Eq. (3.18b) and plot the sensitivity as a function of R_1; this is seen in Fig. 3.18b. The particular values are ($\mu\varepsilon$ are micro-strain units) (A) $S = 0.24$ mV/1000$\mu\varepsilon$; (B) $S = 0.55$ mV/1000$\mu\varepsilon$; and (C) $S = 2.25$ mV/1000$\mu\varepsilon$.

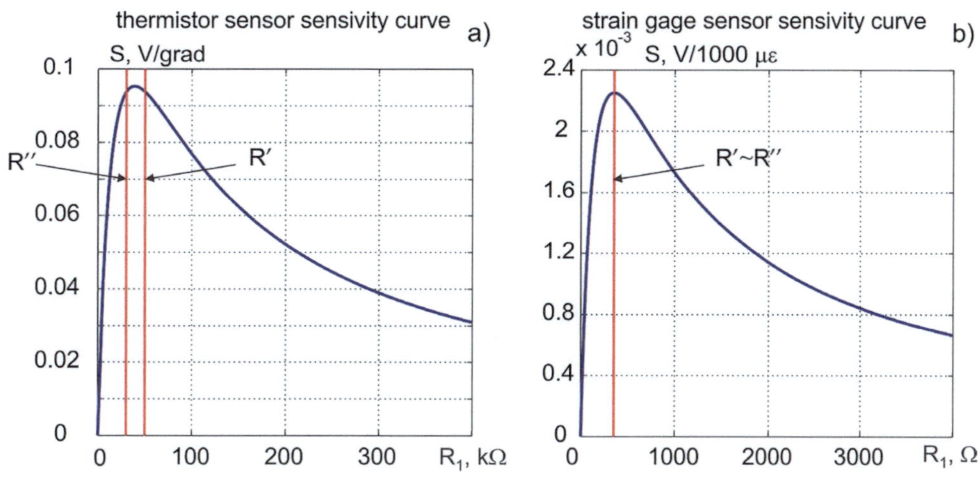

Fig. 3.18. Sensitivity curves for the divider sensor circuits with a thermistor and a strain gage.

3.2.6 Application Example: Voltage Divider as an Actuator Circuit

The circuit shown in Fig. 3.19 contains a *voltage divider circuit block* with one resistive sensing element, i.e., resistance R_1. The block is connected to a three-terminal *electronic switch* (a transistor). A variable voltage V controlling the switch operation is created by the voltage divider. When the control voltage V reaches a certain threshold voltage V_{Th} or exceeds it, the switch closes. A DC motor is now connected to the source, and the rotor starts to spin. Switches of this type involve field-effect transistors. We emphasize that there is no current into the control switch terminal, only the control voltage counts. If R_1 is a thermistor and the DC motor is a fan motor, the entire circuit may operate as a basic temperature controller in an enclosure or in a room. The control voltage versus circuit ground, the reference point, is given by (cf. Eq. (3.16))

$$V(x) \equiv V_2(x) = \frac{R_2}{R_1(x) + R_2} V_S \tag{3.19}$$

where variable x corresponds to the ambient temperature.

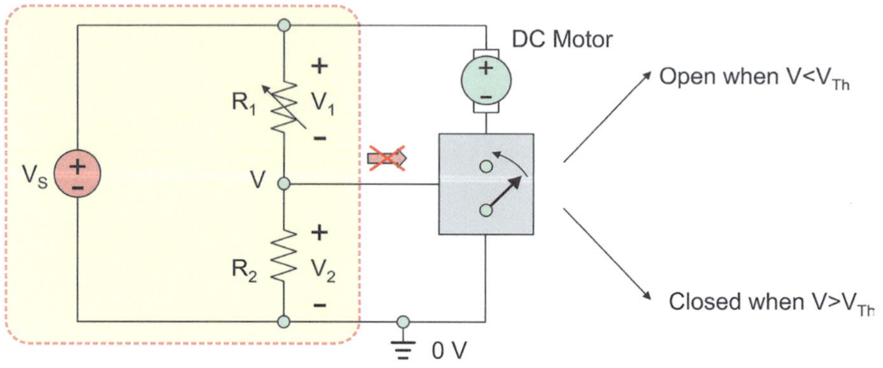

Fig. 3.19. An actuator circuit based on the voltage divider principle.

Example 3.10: The circuit in Fig. 3.19 uses an NTC-503 thermistor sensing element with $R_1 = 50$ kΩ at 25 °C (room temperature) and $R_1 = 30$ kΩ at 37 °C. The fixed resistance (assumed temperature independent) is $R_2 = 12$ kΩ. The threshold voltage of the switch is $V_{\mathrm{Th}} = 2.3$ V. The supply voltage is 9 V. Determine the circuit behavior at 25° C and at 37° C, respectively.

Solution: According to Eq. (3.19), the control voltage at 25° C is equal to 1.74 V. This value is below the threshold voltage. Consequently, the switch is open and the motor is not connected to the source. However, the control voltage at 37° C is 2.57 V. This value is above the threshold voltage and the switch is now closed: the motor is connected to the source and is spinning. Figure 3.20 shows the corresponding laboratory setup. Note that in reality the threshold voltage of the transistor switch is not exactly a constant; it depends on the temperature.

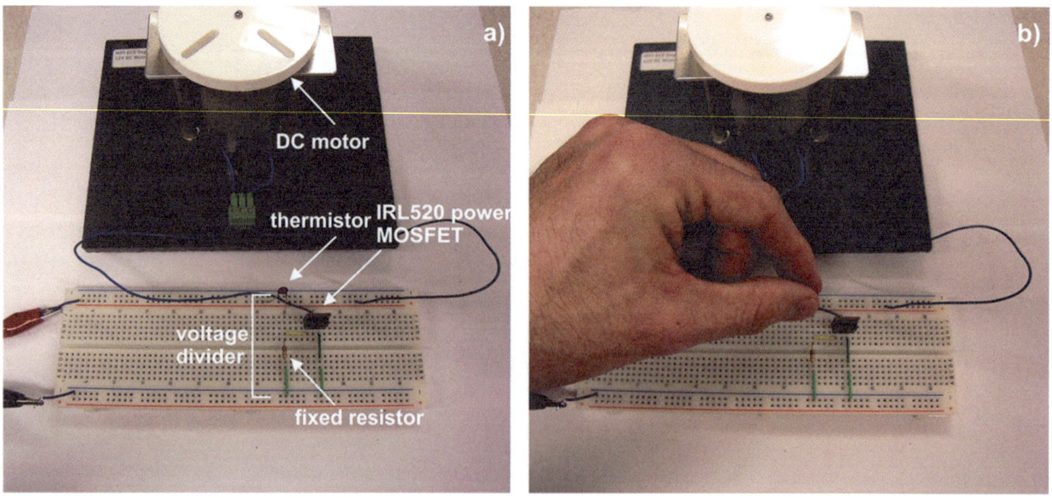

Fig. 3.20. Laboratory realization of the circuit shown in Fig. 3.19.

3.2.7 Current Limiter

The *current limiter* (or simply the *current-limiting resistor*) is a particular case of the voltage divider circuit and is of a significant practical importance. Its concept is shown in Fig. 3.21. A load resistance R_L is connected to the ideal voltage source in series with another (smaller) resistance R, which physically acts as a *current-limiting resistor*.

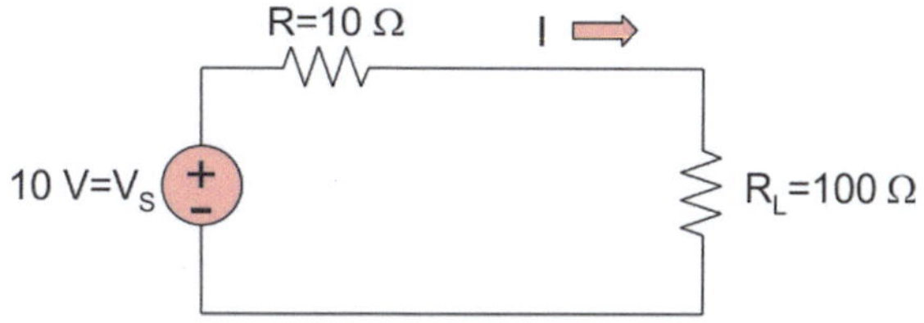

Fig. 3.21. Voltage divider circuit in the current-limiting configuration. The 10 Ω resistance is used to limit the circuit current.

If the load resistance is fixed at a rather high value, the circuit in Fig. 3.21 does not pose any problem, and the current-limiting resistance of 10 Ω becomes insignificant. For example, in Fig. 3.21, the circuit current is 100 mA in the absence of the first resistance R. The power delivered to the load resistance is $P = R_L I^2 = 1$ W. However, the load resistance may be variable, and it may attain really small values. When this happens, the circuit current increases. In Fig. 3.21, it becomes equal to 10 A when the load resistance decreases to 1 Ω and the current-limiting resistor is missing. The power delivered to the load resistance also increases to $P = 100$ W. This large power may overheat and eventually destroy the small-scale load (a thermistor is one example). The role of resistance R is to *limit* the total current when the load resistance is either variable or constant but small. For example, in the circuit of Fig. 3.21, the maximum possible circuit current is

$$I = \frac{V_S}{R + R_L} < \frac{V_S}{R} = 1\text{A} \tag{3.20}$$

irrespective of the value of the load resistance. Therefore, the power delivered to a 1 Ω-load becomes always less than 1 W instead of the initial value of 100 W.

3.2.8 Current Divider Circuit

The *current divider circuit* shown in Fig. 3.22 is associated with *resistances connected in parallel*. It is *dual* to the voltage divider circuit in the sense that the roles of voltage and current are interchanged. Figure 3.22 shows the concept.

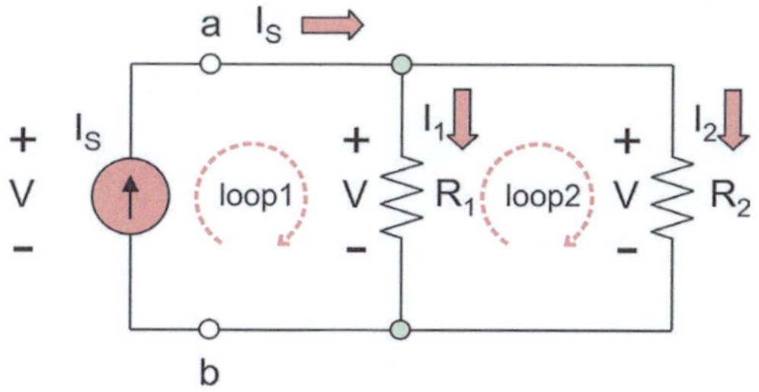

Fig. 3.22. A current divider circuit with an ideal current source.

To solve the circuit, we again prefer to use KCL and KVL *simultaneously*. KCL gives

$$I_S = I_1 + I_2 \tag{3.21}$$

Based on KVL for both loop 1 and loop 2, the voltage across the current source (voltage between its terminals a and b) is equal to the voltage across either resistor and is equal to V. Application of Ohm's law gives the expression for this voltage,

$$I_S = \frac{V}{R_1} + \frac{V}{R_2} \Rightarrow V = \frac{I_S}{\frac{1}{R_1} + \frac{1}{R_2}} = \frac{R_1 R_2}{R_1 + R_2} I_S \tag{3.22}$$

Therefore, we obtain the *current division rule* in the form

$$I_1 = \frac{V}{R_1} = \frac{R_2}{R_1 + R_2} I_S, \quad I_2 = \frac{V}{R_2} = \frac{R_1}{R_1 + R_2} I_S \tag{3.23}$$

Equation (3.23) teaches us that the major function of the current divider is to divide the current of the power source between two resistances in an *inverse* proportion so that:

(i) The larger resistance always acquires the smaller current, and the smaller resistance acquires the larger current.
(ii) The individual currents add up to the source current, i.e., $I_1 + I_2 = I_S$.

In other words, the electric current always chooses a path of least resistance. If one resistance is replaced by a wire the entire source current will flow through the wire; the second resistance will be *shorted out* by the wire.

Exercise 3.6: A current divider uses a 3 mA current source and two resistances: $R_1 = 200\ \Omega$ and $R_2 = 600\ \Omega$. What are the currents I_1, I_2 through the resistances?

Answer: $I_1 = 2.25$ mA, $I_2 = 0.75$ mA. The smaller resistance acquires the larger current.

Example 3.11: The current divider circuit can be assembled with a voltage source as shown in Fig. 3.23. Find currents I_1, I_2 and the total circuit current I.

Solution: The circuit in Fig. 3.23 can be analyzed in a number of ways. Perhaps the simplest way is to recognize that, according to KVL, the voltages across all three elements in Fig. 3.23 are equal to each other and equal to 10 V. Therefore,

$$I_1 = \frac{10\text{V}}{R_1}, \quad I_2 = \frac{10\text{V}}{R_2}, \quad I = I_1 + I_2 = \frac{10\text{V}}{R_{\text{eq}}}, \quad R_{\text{eq}} = \frac{R_1 R_2}{R_1 + R_2} \tag{3.24}$$

Another way to solve the same circuit is to combine resistances in parallel, find the circuit current I, and then apply the current division principle.

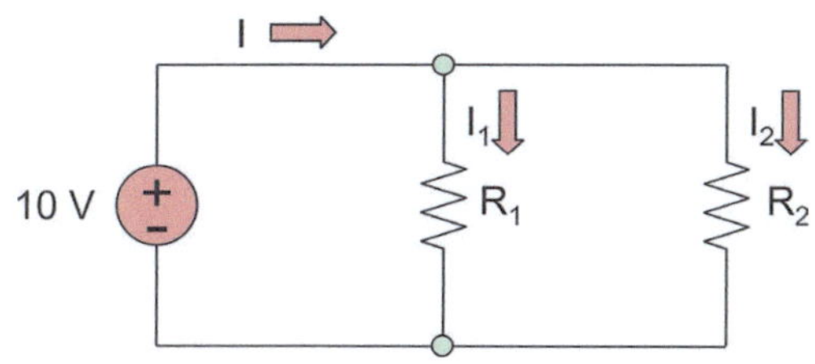

Fig. 3.23. A current divider circuit with a voltage source.

3.2.9 Wheatstone Bridge

The *Wheatstone bridge* was invented by the British scientist and mathematician, Samuel Christie (1784–1865), and first used for resistance measurements by Sir Charles Wheatstone in 1843. It is shown in Fig. 3.24. From Fig. 3.24, we can recognize that the Wheatstone bridge is, in fact, a combination of two *independent* voltage divider blocks: one with two fixed resistances R_1, R_2 and another with one fixed resistance R_3 and some other resistance R_4 denoted here by $R_4 = R(x)$. When the resistance measurements are implied, R_4 is the fixed unknown resistance. However, common modern applications use the Wheatstone bridge as a part of the sensor circuit. In this case, $R(x)$ is a variable resistance (a sensing element), where x is a physical quantity to be measured. The second voltage divider is the voltage divider sensor circuit; the first voltage divider is fixed. Circuit ground (absolute voltage reference) may be introduced as shown in Fig. 3.24.

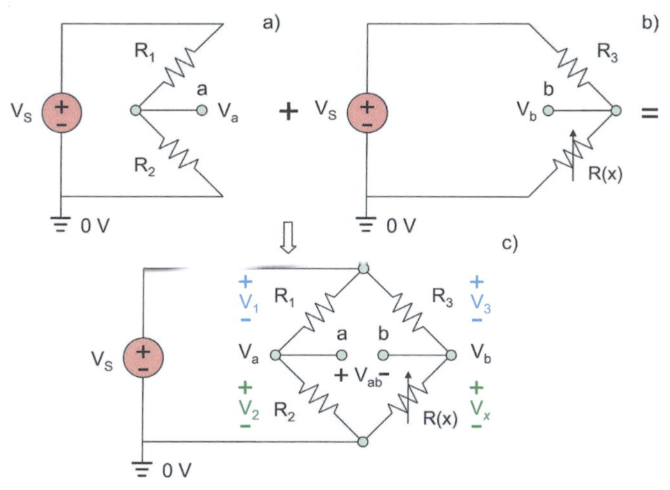

Fig. 3.24. The Wheatstone bridge is a combination of two independent voltage dividers connected to the same voltage source. The second voltage divider is a sensor circuit.

It has been shown that the voltage divider circuit is a basic sensor circuit. Now, why do we need two voltage dividers? The answer to this question will be based on the fact that, with the help of the fixed divider, we can eliminate a *DC voltage offset* in the sensor voltage reading of the "master" voltage divider and thus enable the use of a *difference signal* (and a *difference* amplifier) to amplify the likely very weak sensor voltages. The key point is that the output voltage of the Wheatstone bridge is not V_a or V_b but the *differential voltage* between terminals a and b in Fig. 3.24, $V_{ab} = V_a - V_b$.

Example 3.12:

A. A simple voltage divider circuit with $R_3 = 350\ \Omega$ and $R(x) = 350\ \Omega \pm 0.1\%$ (the strain gauge) in Fig. 3.24b is used for mechanical strain measurements with a 4.5 V voltage power supply. What is the output voltage V_b of the sensor circuit?

B. The same voltage divider circuit is augmented with another fixed voltage divider having $R_1 = R_2 = 350\ \Omega$ to form the Wheatstone bridge shown in Fig. 3.24c. What is the output voltage V_{ab} of the sensor circuit now?

Solution: In case A, we use the voltage division rule and obtain

$$V_b = 2.25\ \text{V} \pm 1.125\ \text{mV} \tag{3.25a}$$

It is difficult to process such voltages since we cannot really amplify them. Amplification of 2.251125 V by a factor of 100 gives 225.1125 V; such a large voltage simply cannot be obtained with the common amplifier circuits. In case B, however, the value of V_b should be subtracted from $V_a = 2.25$ V, which yields the sensor voltage in the form

$$V_{ab} = \mp 1.125\ \text{mV} \tag{3.25b}$$

If we now amplify 1.125 mV by a factor of 100, a conventional value of 0.1125 V would be obtained. Along with this fact, the differential sensor voltage has another significant advantage, which is its immunity against *circuit noise*.

General Model of Wheatstone Bridge

Using the voltage division rule twice, the differential voltage V_{ab} of the Wheatstone bridge in Fig. 3.24c becomes ($R_4 = R(x)$)

$$V_{ab} = \left(\frac{R_2}{R_1 + R_2} - \frac{R_4}{R_3 + R_4}\right) V_S \tag{3.26}$$

The Wheatstone bridge is *balanced* when $V_{ab} = 0$. From Eq. (3.26) one obtains the necessary and sufficient condition for the *balanced Wheatstone bridge*,

$$\frac{R_1}{R_2} = \frac{R_3}{R_4} \tag{3.27}$$

Section 3.3 Superposition Theorem and Its Use

3.3.1 Linear and Nonlinear Circuits

Linear Circuit

The superposition theorem studied in this section is only valid for linear circuits. A *linear circuit* is a circuit that includes only the *linear* circuit elements (elements with a linear or straight υ-i characteristic):

1. Resistance
2. Capacitance (υ-i relationship is time-dependent but still linear)
3. Inductance (υ-i relationship is time-dependent but still linear)
4. Voltage source (independent and linearly dependent)
5. Current source (independent and linearly dependent)

Every linear circuit satisfies both the *homogeneity* and *additivity* properties. To explain those properties, we consider a linear circuit with an input parameter x (input voltage or current) and an output parameter $f(x)$ (output voltage or current). A function f is the characteristic of the circuit itself; it must be a linear function. Namely, when an input parameter is a *linear superposition* $ax_1 + bx_2$ of two individual stimuli, x_1, x_2, the output is also a *linear superposition* of two individual responses, i.e.,

$$f(ax_1 + bx_2) = af(x_1) + bf(x_2) \tag{3.28}$$

For example, if we double all source strengths in a linear circuit, voltages across every passive circuit element and currents through every circuit element will also double.

Nonlinear Circuit and Circuit Linearization

A *nonlinear circuit* will include nonlinear circuit elements, elements with a nonlinear υ-i characteristic. Any circuit with semiconductor components (such as diodes, transistors, solar cells) is a nonlinear circuit. Since the vast majority of electronic circuits include semiconductor components, a legitimate question to ask is what value do the linear circuits have in this case? One answer is given by a *linearization procedure*, which makes it possible to reduce the nonlinear circuit to a linear one, in a certain domain of operating parameters. Mathematically, circuit linearization means that a nonlinear relationship $V(I)$ is expanded into a Taylor series,

$$V(I) = V_0 + \left.\frac{dV}{dI}\right|_{V=V_0,\ I=I_0}(I - I_0) + \ldots \tag{3.29}$$

about a certain operating point V_0, I_0, where only the constant and the *linear* terms are retained. The derivative in Eq. (3.29), the so-called dynamic or small-signal resistance r can now be used in place of the familiar resistance R for the linear ohmic circuit elements.

3.3.2 Superposition Theorem or Superposition Principle

The *superposition theorem*, often called the *superposition principle*, applies to circuits with more than one voltage and/or source. It states that the complete circuit solution is obtained as a *linear superposition* of particular solutions, for every power source separately. In other words, we are zeroing (or *turning off*) all the power sources except for *one*, find the solution, and then add up all such solutions. The following rules apply:

1. To turn off a voltage source, we replace it by a *short circuit* or an ideal wire (see Fig. 3.25a). The voltage across the ideal wire is exactly 0 V.
2. To turn off a current source, we replace it by an *open circuit* or an air gap (see Fig. 3.25b). The current through the gap is exactly 0 A.
3. The dependent sources do not need to be zeroed. They remain the *same* for every particular solution and may affect every particular solution.

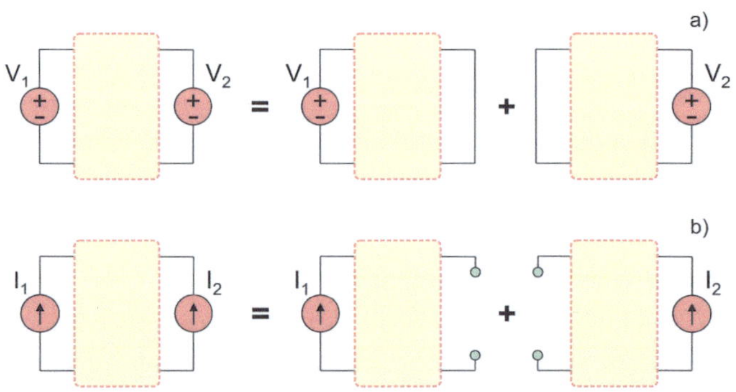

Fig. 3.25. Superposition theorem for two independent sources (either voltage or current).

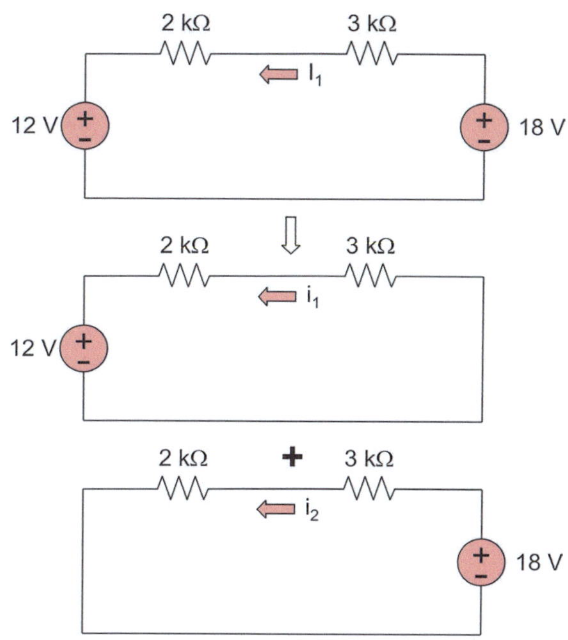

Fig. 3.26. Application of the superposition principle to a circuit with voltage sources.

Example 3.13: Find current I_1 in the circuit shown in Fig. 3.26.

Solution: The first step is shown in the figure; we apply the superposition theorem and obtain two simpler circuits. Each of those circuits is solved using series equivalents. For the circuit with the 12 V power source,

$$i_1 = -12\ \mathrm{V}/5\ \mathrm{k\Omega} = -2.4\ \mathrm{mA} \tag{3.30}$$

The minus sign is due to the fact that the actual circuit current flows in the opposite direction. For the circuit with the 18 V source,

$$i_2 = 18\ \mathrm{V}/5\ \mathrm{k\Omega} = 3.6\ \mathrm{mA} \tag{3.31}$$

The total circuit current I_1 is therefore given by $I_1 = i_1 + i_2 = 1.2$ mA. Circuit voltage across each resistance can be found as well.

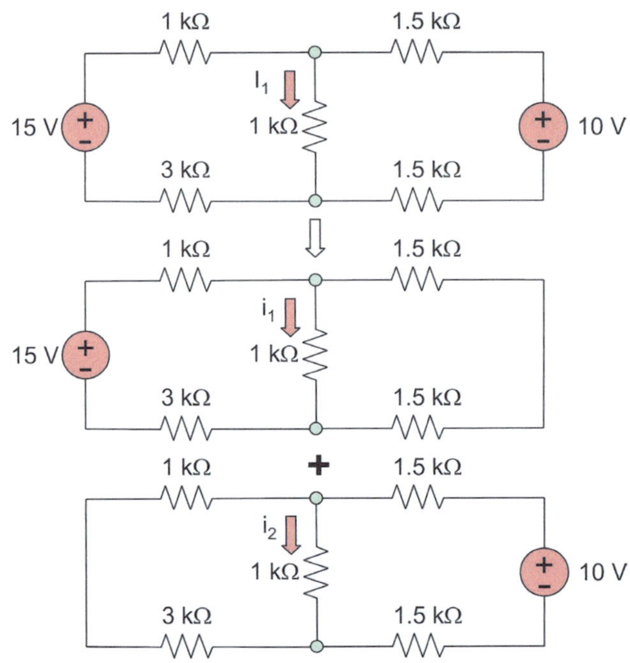

Fig. 3.27. Application of the superposition principle to voltage sources.

Example 3.14: Find current I_1 in the circuit shown in Fig. 3.27.

Solution: The first step is again shown in the figure; we apply the superposition theorem and obtain two simpler circuits. Each of those circuits is solved using series/parallel equivalents. For the circuit with the 15 V power source,

$$R_{eq} = 1\ \text{k}\Omega + 3\ \text{k}\Omega + 0.75\ \text{k}\Omega = 4.75\ \text{k}\Omega \tag{3.32}$$

The circuit current is 3.1579 mA; i_1 is 75% of this value (from current division). For the circuit with the 10 V source, $R_{eq} = 3.8\ \text{k}\Omega$ and i_2 is 2.1053 mA. Thus, current I_1 is given by

$$I_1 = i_1 + i_2 = 4.4737\ \text{mA} \tag{3.33}$$

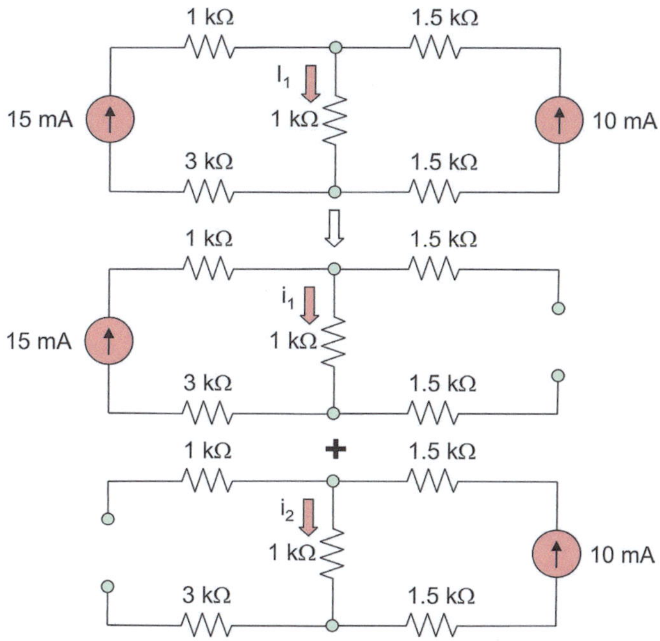

Fig. 3.28. Application of the superposition principle to current sources.

Example 3.15: Find current I_1 in the circuit shown in Fig. 3.28.

Solution: The first step is shown in the figure; we apply the superposition theorem and obtain two simpler circuits by disconnecting current sources. The first circuit predicts $i_1 = 15$ mA, while the second circuit predicts $i_2 = 10$ mA. Therefore,

$$I_1 = i_1 + i_2 = 25 \text{ mA} \tag{3.34}$$

In other words, each of the currents i_1, i_2 flows in its own loop. The solution *does not* depend on any particular resistor value.

The superposition theorem is the direct consequence of circuit linearity. Interestingly, the superposition theorem is applicable not only to DC circuits but also to AC and transient circuits. The superposition theorem *does not* hold for electric power though, since power is the product of voltage and current. The circuit power and the power delivered to individual elements may be correctly obtained only from the *final* solution.

Application Example: Superposition Theorem for a Cellphone

Why do we ultimately need to solve circuits with multiple sources? Look at your cellphone. There is a circuit inside, which receives a very weak radio-frequency input voltage signal from the antenna. This is the first source. The signal is processed and amplified by transistors powered by the cellphone battery. This is the second source (or sources). The signal is then demodulated through interaction with an internal high-frequency signal generator. This is the third source.

Moreover, every component in that circuit is in fact "noisy"; it creates a small thermal noise voltage, which is modeled by its own tiny voltage source. A careful solution of the entire circuit with multiple voltage and current sources including noise sources is possible by superposition. This solution allows electrical engineers to extract the weak input signal from otherwise overwhelming noise and properly design the cellphone circuitry.

3.3.3 Y (Wye) and Δ (Delta) Networks: Use of Superposition

The series and parallel resistance configurations are not the only meaningful network blocks. Situations often arise when the resistances are neither in parallel nor in series. In Fig. 3.29a, b, we introduce two new networking blocks, which are in importance second to the series/parallel equivalents.

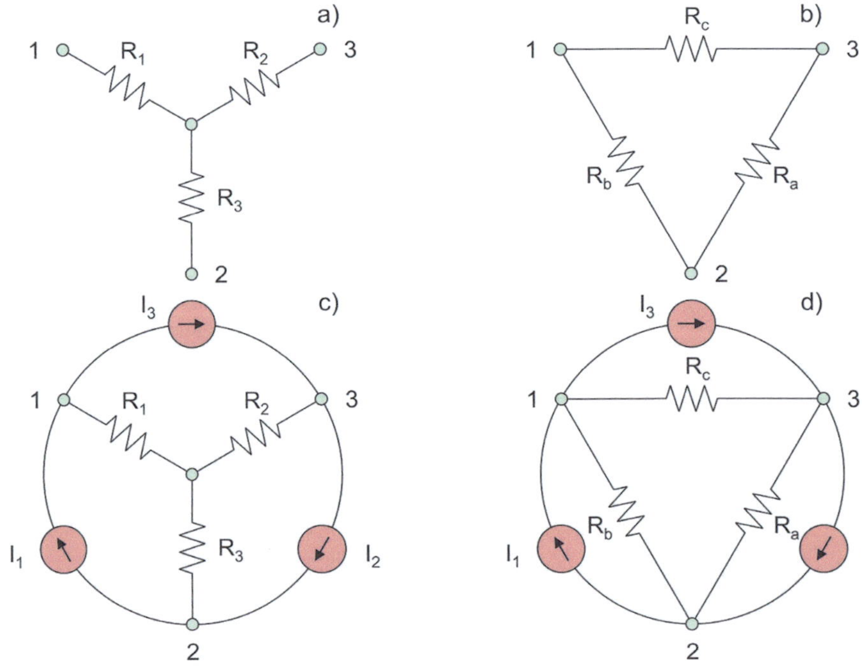

Fig. 3.29. (**a**) Y (wye) and (**b**) Δ (delta) networks; (**c**, **d**) applying the superposition theorem to establish network equivalence.

The first block is known as the *Y (wye) network*. It represents a *nodal* connection of three arbitrary resistances. The second block is known as the Δ *(delta) network*. It represents a *loop* connection of three arbitrary resistances. Both Y and Δ networks have three terminals 1, 2, and 3; they are therefore known as *three-terminal networks*. This is in contrast to series/parallel resistance circuits, which are usually *two-terminal networks*.

The Y and Δ networks occur either independently or as part of a larger network. Important applications include three-phase power electronics circuits, filter circuits, and impedance-matching networks in high-frequency circuits. The theory that follows holds for AC circuits too, when the resistances become general *impedances*.

Conversion Between Y and Δ Networks

A problem of significant practical importance is the *conversion between Y and Δ networks* in Fig. 3.29a, b. This conversion is equivalent to *replacing a node by a loop* (strictly speaking, by a *mesh*) and vice versa in a more complicated circuit or network. Such a replacement may significantly simplify the overall circuit analysis. The conversion is established based on the superposition theorem. Two arbitrary networks are equivalent if their v-i characteristics are the same. In other words, by connecting three arbitrary sources to terminals 1, 2, and 3 of the Y network, we must obtain terminal voltages and currents identical to those of the Δ network with the same sources. We select three current sources I_1, I_2, I_3 in Fig. 3.29c, d. The solution with three sources is obtained as a superposition of three partial solutions, with two sources open-circuited at a time. Let us keep the source I_1 and replace I_2, I_3 by open circuits first. Voltages V_{12} for both networks will be the same when the equivalent resistances R_{12} between terminals 1 and 2 are the same. A similar treatment holds for terminals 1 and 3 (source I_3 is kept), and terminals 2 and 3 (source I_2 is kept), respectively. Therefore, with reference to Fig. 3.29c, d, we have

$$R_{12} = R_1 + R_3 = R_b \| (R_a + R_c) = \frac{R_b(R_a + R_c)}{R_a + R_b + R_c} \tag{3.35a}$$

$$R_{13} = R_1 + R_2 = R_c \| (R_a + R_b) = \frac{R_c(R_a + R_b)}{R_a + R_b + R_c} \tag{3.35b}$$

$$R_{23} = R_2 + R_3 = R_a \| (R_b + R_c) = \frac{R_a(R_b + R_c)}{R_a + R_b + R_c} \tag{3.35c}$$

Next, we add Eq. (3.35a) and Eq. (3.35b) and subtract from this result Eq. (3.35c). This gives us an expression for R_1. To obtain R_2, we add Eq. (3.35b) and Eq. (3.35c) and subtract Eq. (3.35a). R_3 is obtained by adding Eq. (3.35a) and Eq. (3.35c) and subtracting Eq. (3.35b). The result has the form of a *Δ to Y transformation*:

$$R_1 = \frac{R_b R_c}{R_a + R_b + R_c}, \quad R_2 = \frac{R_a R_c}{R_a + R_b + R_c}, \quad R_3 = \frac{R_a R_b}{R_a + R_b + R_c} \tag{3.36}$$

The inverse transformation, a *Y to Δ transformation*, follows

$$R_a = \frac{R_1R_2 + R_1R_3 + R_2R_3}{R_1}, \quad R_b = \frac{R_1R_2 + R_1R_3 + R_2R_3}{R_2}, \quad R_c = \frac{R_1R_2 + R_1R_3 + R_2R_3}{R_3} \tag{3.37}$$

Balanced Y and Δ Networks

When all resistances of the Y network are equal to R_Y, the Y network is said to be *balanced*. When all resistances of the Δ network are equal to R_Δ, the Δ network is also *balanced*. For balanced networks, two previous equations reduce to

$$R_Y = \frac{R_\Delta}{3}, \quad R_\Delta = 3R_Y \tag{3.38}$$

Example 3.16: Find equivalent resistance of a network between terminals *a* and *b* in Fig. 3.30a.

Solution: To simplify the network, we use the Δ to Y transformation and obtain the network shown in Fig. 3.30b. Since all resistances of the bottom Δ network are equal, we can use the simplified Eq. (3.38) to find the new resistance values. The remaining circuit is solved using series/parallel combinations, which gives us $R_{\text{eq}} = 2.2\ \text{k}\Omega$.

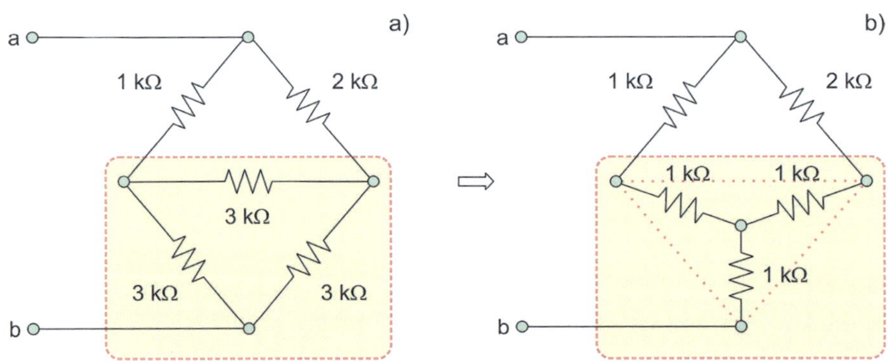

Fig. 3.30. Network simplification using Δ to Y transformations.

The conversions between Y and Δ networks date back to Arthur E. Kennelly (1861–1939), an Indian American engineer who established them in 1899. Note that the transformation between Y and Δ networks is also called the *star to delta transformation*.

3.3.4 T and Π Networks: Two-Port Networks

The Y and Δ networks are equivalent to *T* and Π *networks*, respectively, which are shown in Fig. 3.31. The T and Π networks are predominantly used as *two-port networks* or two-port networking blocks. Every port should have two terminals. To create the two ports with two terminals each, we simply split terminal 2 in Fig. 3.31a, c into two terminals: 2 and 4 in Fig. 3.31b, d, respectively. Port #1 is the *input port* of the network. Port #2 is the *output port* of the network. Multiple two-port networks may be connected in chains. Equations (3.36), (3.37), and (3.38) allow us to establish the connections between the T and Π networks, which are the same as the connections between the Y and Δ networks. Hence, any two-port T network may be replaced by a two-port Π network and vice versa. The T network is sometimes called the *T pad* and the Π network the Π *pad*. Both networks are used as attenuators, filters, and antenna tuners. In the last two cases, capacitances and inductances occur in place of resistances.

Exercise 3.7: A two-port T network in Fig. 3.31b is characterized by $R_1 = 3\ \text{k}\Omega$ and $R_2 = R_3 = 5\ \text{k}\Omega$. Establish resistance values for the equivalent Π network in Fig. 3.31d.

Answer: $R_a = 18.33\ \text{k}\Omega$, and $R_b = R_c = 11\ \text{k}\Omega$.

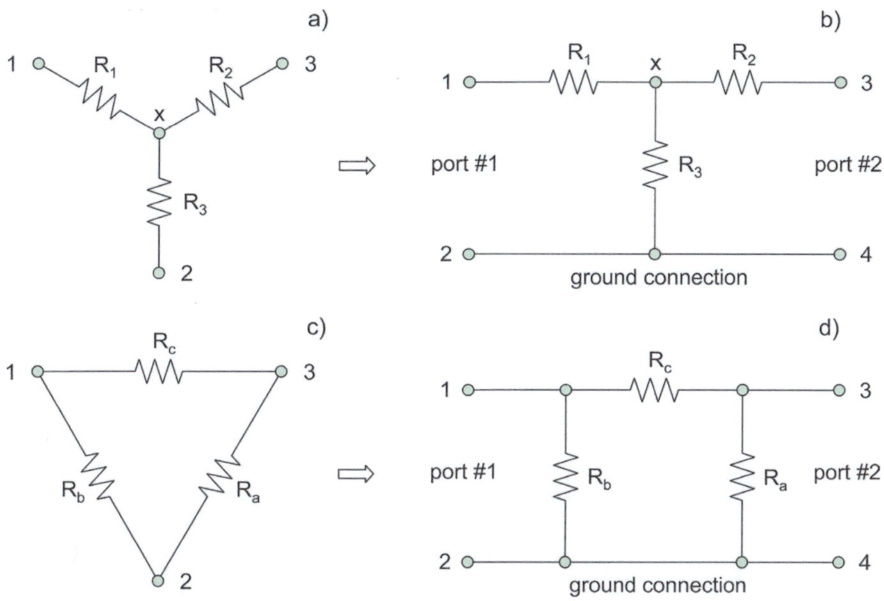

Fig. 3.31. Conversion of Y and Δ three-terminal networks to equivalent T and Π two-port, four-terminal networks. (**a**) Y network. (**b**) Two-port T network. (**c**) Δ network. (**d**) Two-port Π network.

3.3.5 General Character of Superposition Theorem

The superposition theorem makes it possible to analyze not only the two-port electric networks but also various networks with multiple ports such as sensor arrays, antenna arrays, multiple-input and multiple-output (MIMO) communication systems, arrays of magnetic resonance imaging (MRI) coils, etc. It is so widely used that, quite often, we do not even mention its name and consider the corresponding result as "obvious."

Summary

Name	Circuit	Meaning
Network topology	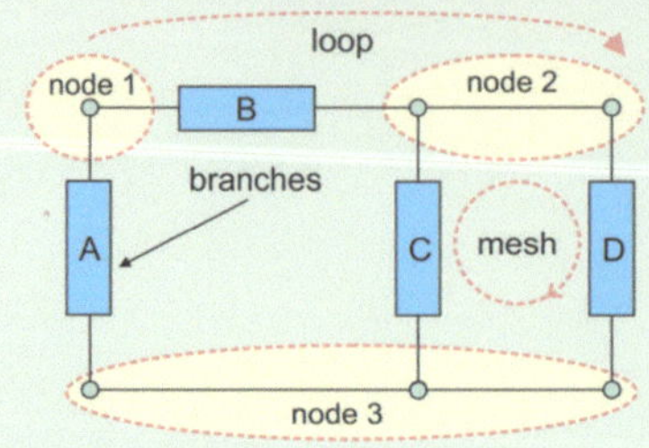	Any electric network consists of b branches, n single nodes (with distinct voltages), l loops, and m (essential) meshes (loops that do not contain any other loops)
Network topology theorem	$b = n + m - 1$	Valid for two-dimensional (planar) electric networks
Kirchhoff's current law (KCL)	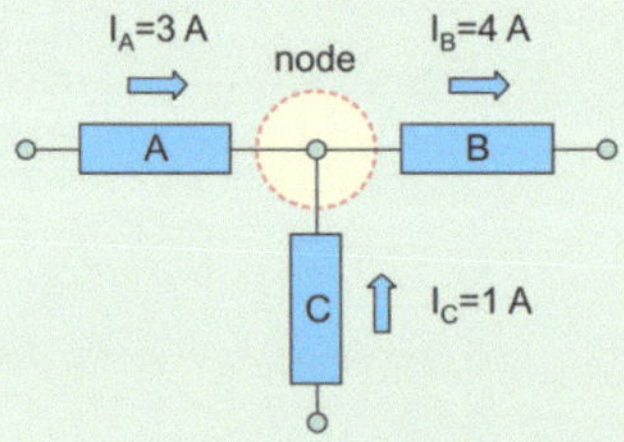	All currents flowing into a circuit node add to zero: $$\sum_{i=1}^{N} I_i = 0$$ **Valid for all circuits: linear and nonlinear, passive and active**
Kirchhoff's voltage law (KVL)	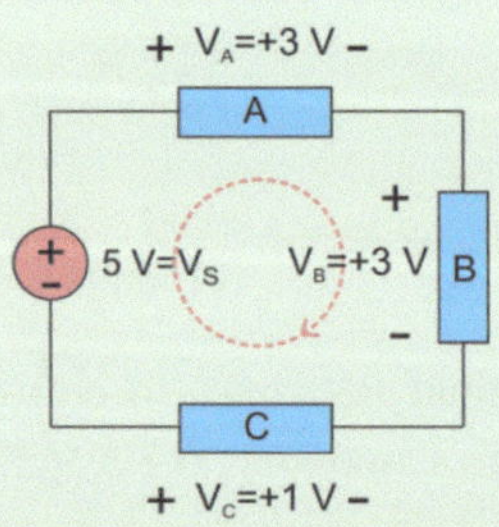	All voltage drops over the elements in a loop add to zero: $$\sum_{i=1}^{N} V_i = 0$$ **Valid for all circuits: linear and nonlinear, passive and active**
Maxwell's minimum heat theorem Tellegen's theorem	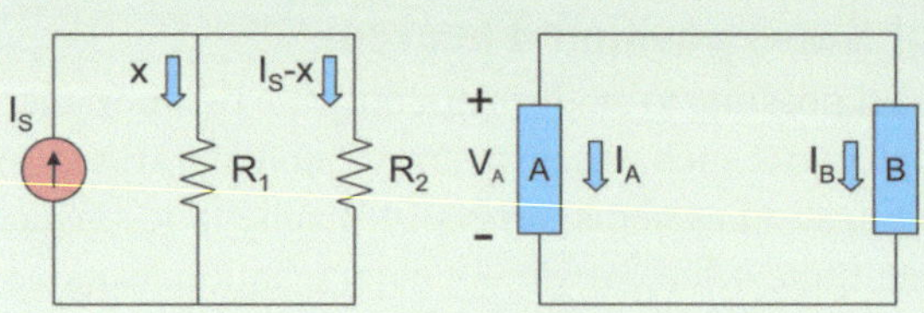	$$\sum_{i=1}^{N_R} R_i I_i^2 = \sum_{i=1}^{N_R} V_i^2 / R_i = \min$$ **Valid only for lin. circuits** $$\sum_{i=1}^{N} V_i I_i = 0$$ **Valid for all circuits**
One-port network and network equivalence	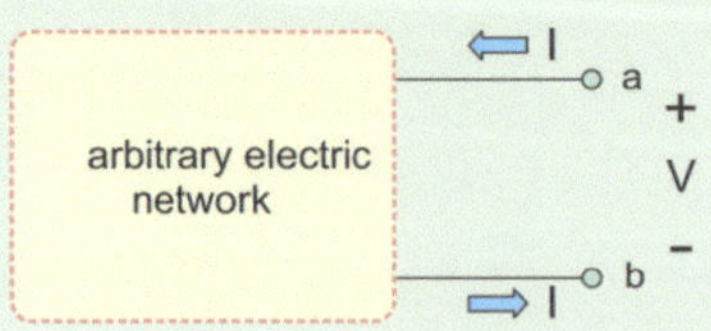	Two electric one-port networks are equivalent when their complete υ-i characteristics coincide **Valid for all circuits: linear and nonlinear, passive and active**

(continued)

Series and parallel source combinations (one-port networks)	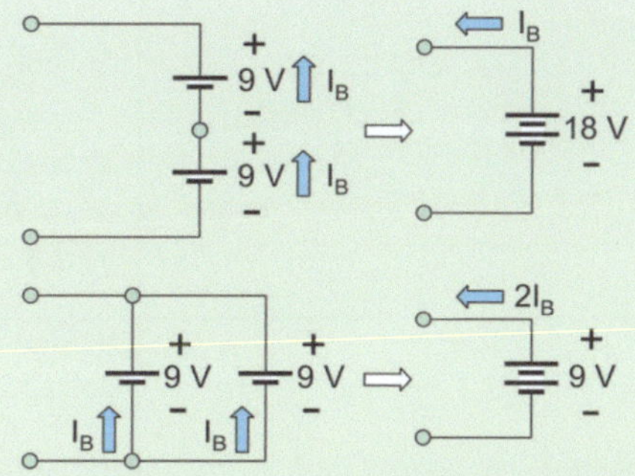	– Voltages are added for series combinations; – Currents are added for parallel combinations; – Current sources are combined similarly **Valid only for ideal voltage and current sources** – circuit elements
Series and parallel resistance or conductance combinations (one-port networks)	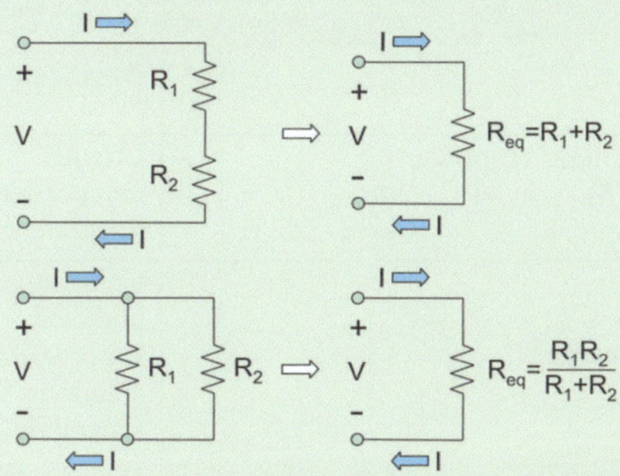	Series combinations: $R_{eq} = R_1 + R_2 + ...R_N$ $\frac{1}{G_{eq}} = \frac{1}{G_1} + \frac{1}{G_2} + ... + \frac{1}{G_N}$ Parallel combinations: $\frac{1}{R_{eq}} = \frac{1}{R_1} + \frac{1}{R_2} + ... + \frac{1}{R_N}$ $G_{eq} = G_1 + G_2 + ...G_N$
Voltage divider circuit	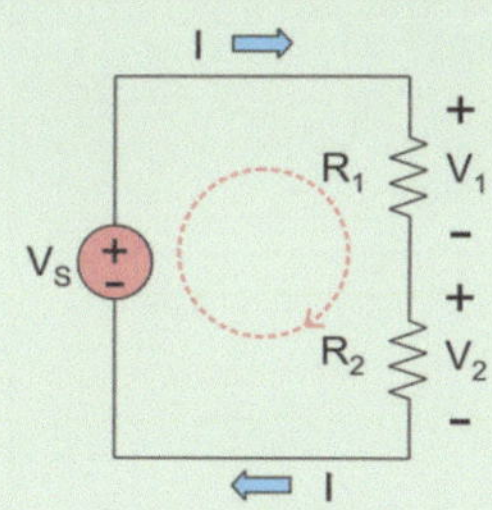	$V_1 = \frac{R_1}{R_1 + R_2} V_S$ $V_2 = \frac{R_2}{R_1 + R_2} V_S$ $V_1 + V_2 = V_S$ Can also be used with arbitrary circuit elements
Voltage divider circuit as a sensor	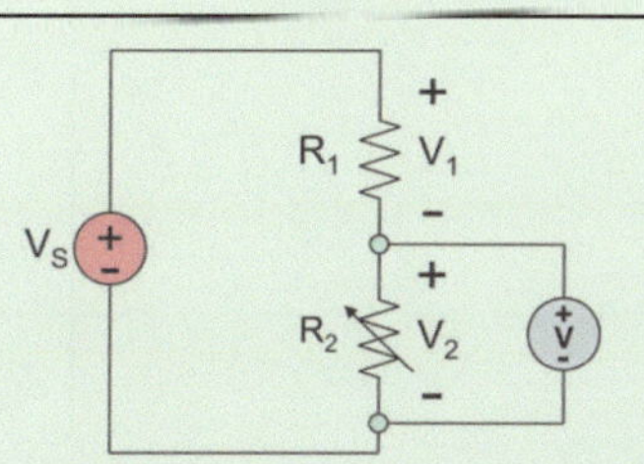	Variable sensor voltage: $V_2 = V_2(x) = \frac{R_2(x)}{R_1 + R_2(x)} V_S$ Maximum sensitivity: $R_1 = \sqrt{R'R''},\ R' < R_2 < R''$
Current-limiting resistor	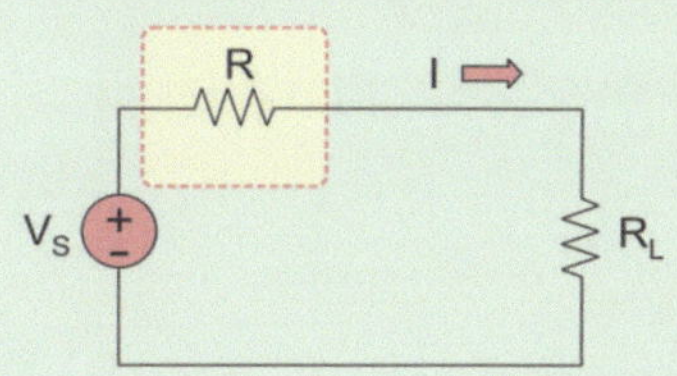	Maximum circuit current at arbitrary load resistance: $I < \frac{V_S}{R}$ **Valid for all circuits**

(continued)

Current divider circuit	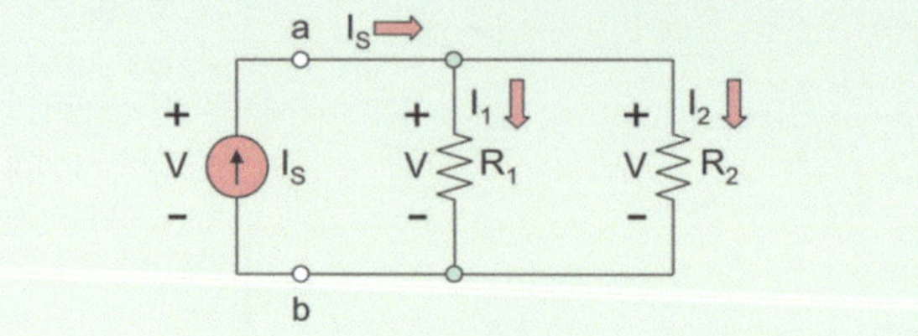	$I_1 = \frac{V}{R_1} = \frac{R_2}{R_1 + R_2} I_S$ $I_2 = \frac{V}{R_2} = \frac{R_1}{R_1 + R_2} I_S$ $I_1 + I_2 = I_S$
Wheatstone bridge	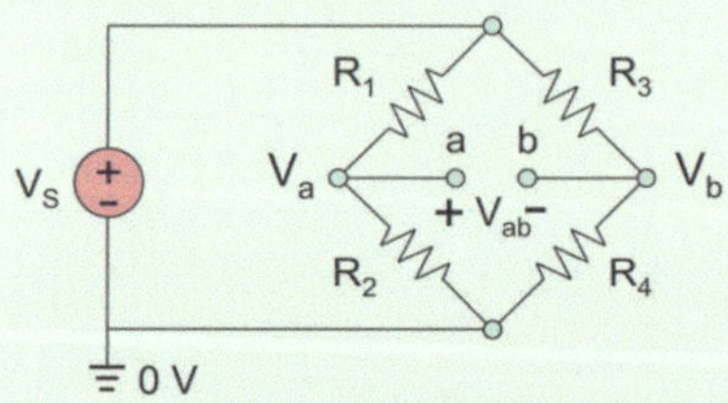	Bridge equation (from two voltage dividers): $V_{ab} = \left(\frac{R_2}{R_1 + R_2} - \frac{R_4}{R_3 + R_4} \right) V_S$ Balanced: $\frac{R_1}{R_2} = \frac{R_3}{R_4}$
Linear circuits	Linear response: $f(ax_1 + bx_2) = af(x_1) + bf(x_2)$	Circuits with resistances, independent sources, and linear dependent sources
Superposition theorem (central theorem in linear circuit analysis)	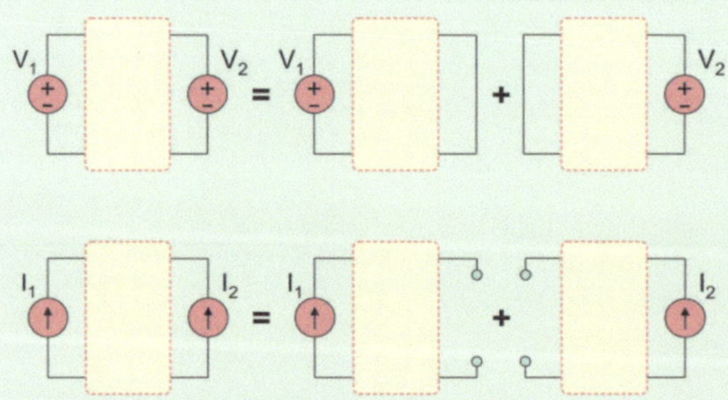	Complete circuit solution is obtained as a *linear superposition* of particular solutions, for every source separately. We turn off all the sources except for one, find the solution, and then add up all such solutions **Valid only for lin. circuits**
Y (wye or star) and Δ (delta) three-terminal network equivalence	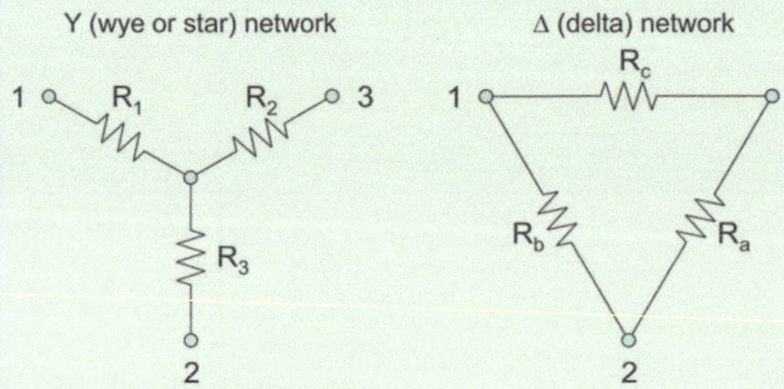	$R_1 = \frac{R_b R_c}{\Delta}$, $R_2 = \frac{R_a R_c}{\Delta}$, $R_3 = \frac{R_a R_b}{\Delta}$, $\Delta = R_a + R_b + R_c$ $R_a = \frac{\Delta}{R_1}$, $R_b = \frac{\Delta}{R_2}$, $R_c = \frac{\Delta}{R_3}$ $\Delta = R_1 R_2 + R_1 R_3 + R_2 R_3$ Node to loop transformation
T and Π two-port networks – derivatives of Y and Δ networking blocks	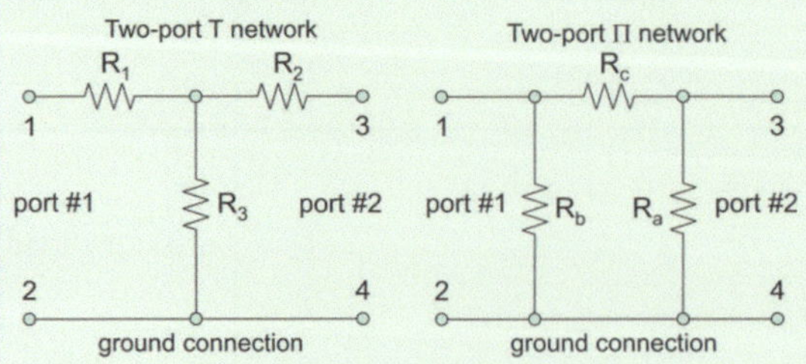	$R_1 = \frac{R_b R_c}{\Delta}$, $R_2 = \frac{R_a R_c}{\Delta}$, $R_3 = \frac{R_a R_b}{\Delta}$, $\Delta = R_a + R_b + R_c$ $R_a = \frac{\Delta}{R_1}$, $R_b = \frac{\Delta}{R_2}$, $R_c = \frac{\Delta}{R_3}$ $\Delta = R_1 R_2 + R_1 R_3 + R_2 R_3$ Conversion is the same as for Y to Δ networks

Problems

3.1 Circuit Laws: Networking Theorems

3.1.1 Electric Network and Its Topology

Problem 3.1. In the network graph shown in the figure below:

A. Find the number of branches, single nodes (nodes with distinct voltages), and meshes.
B. Prove the equality $b = n + m - 1$ for the number of branches b, meshes m, and single nodes n.

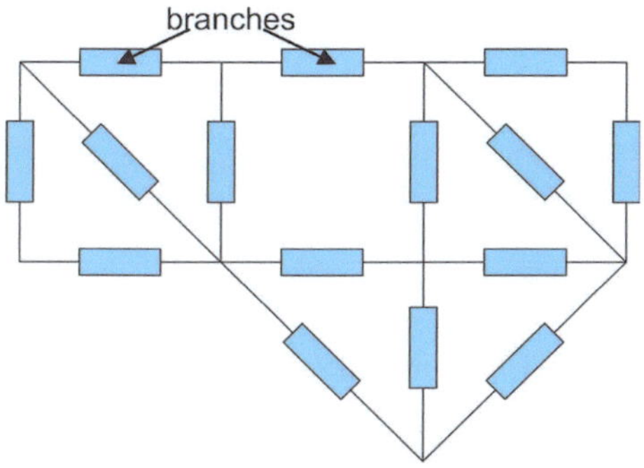

Problem 3.2. Repeat problem 3.1 for the circuits shown in the following figures.

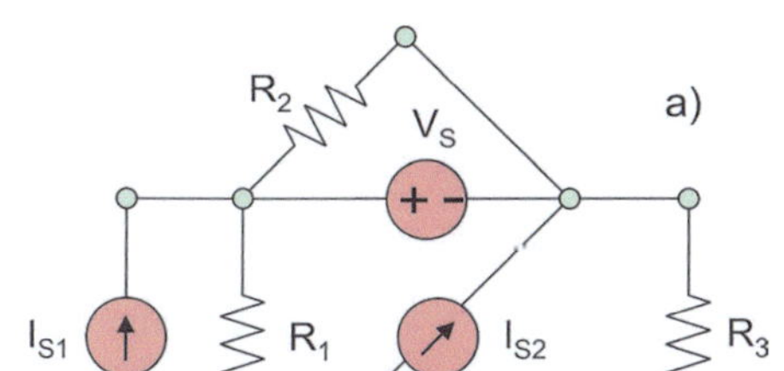

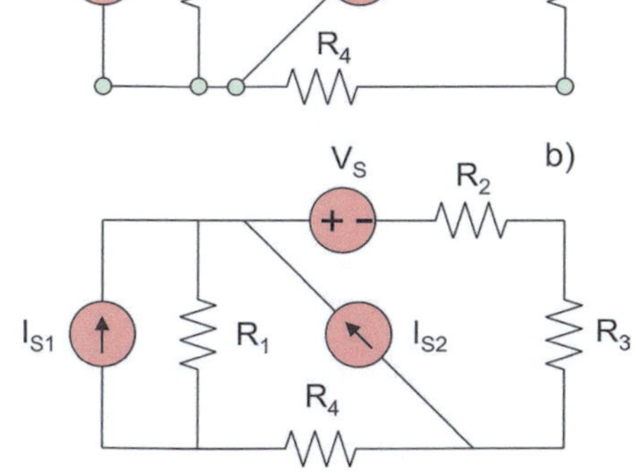

Problem 3.3. Repeat problem 3.1 for the circuit shown in the figure that follows. Each straight segment is now a branch.

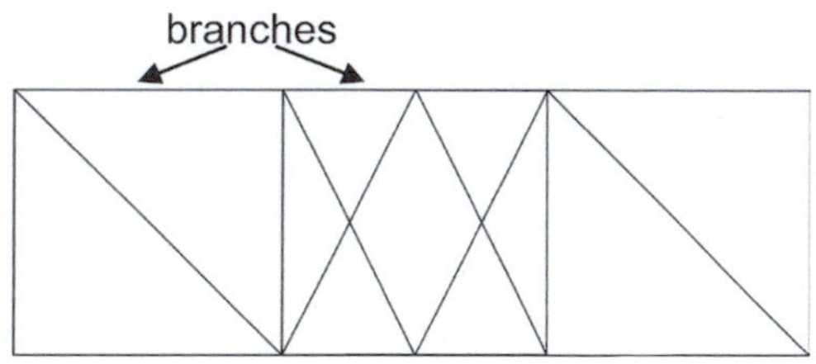

3.1.2 Kirchhoff's Current Law (KCL)

Problem 3.4. Find current I_B for the node shown in the figure that follows.

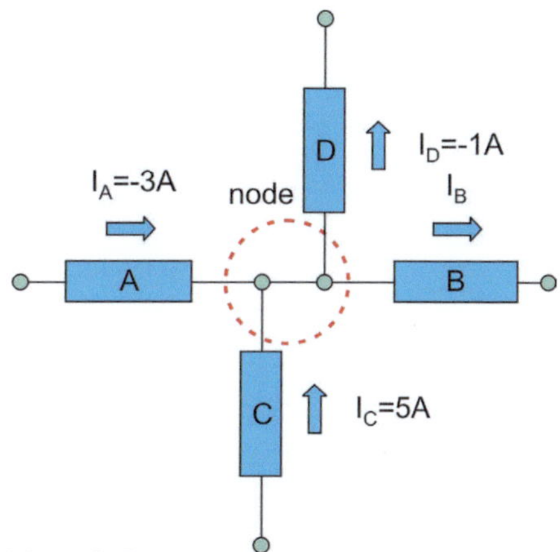

Problem 3.5.

A. Find current I_D for the node shown in the figure below.
B. Redraw this node (and the circuit between terminals a and b) in an equivalent form eliminating the horizontal wire.

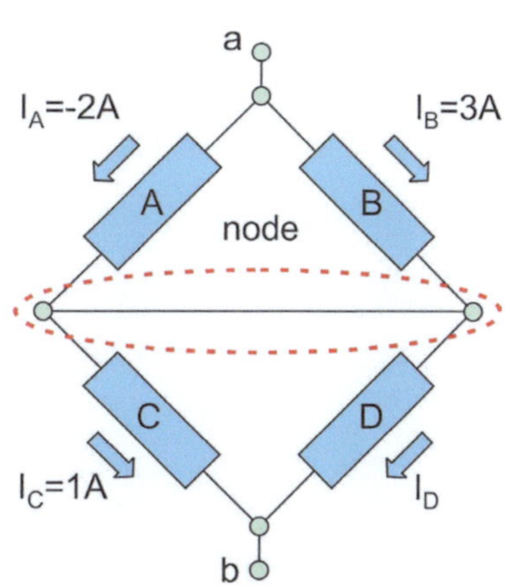

Problem 3.6. Determine current i_C for the circuit shown in the figure that follows.

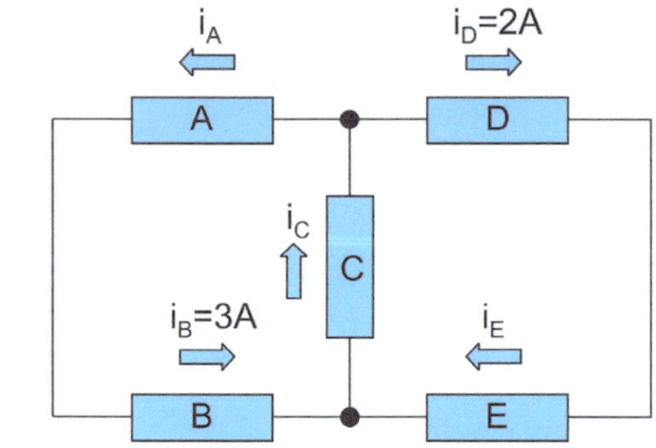

Problem 3.7. Find a relation between currents i_C, i_E for the circuit shown in the figure that follows. Does the problem have a unique solution?

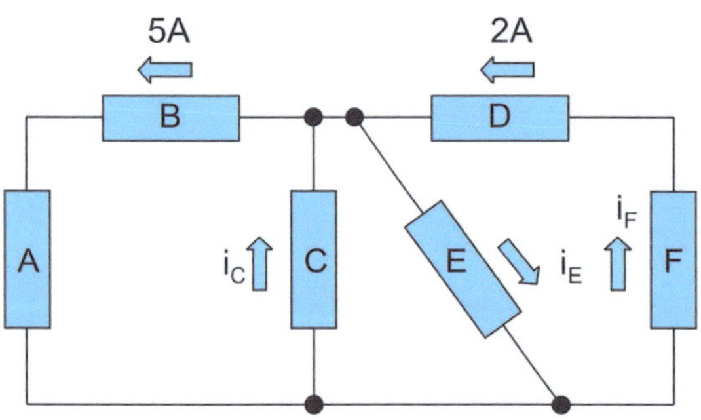

Problem 3.8. Find currents i_A, i_B, i_D, i_H for the circuit shown below.

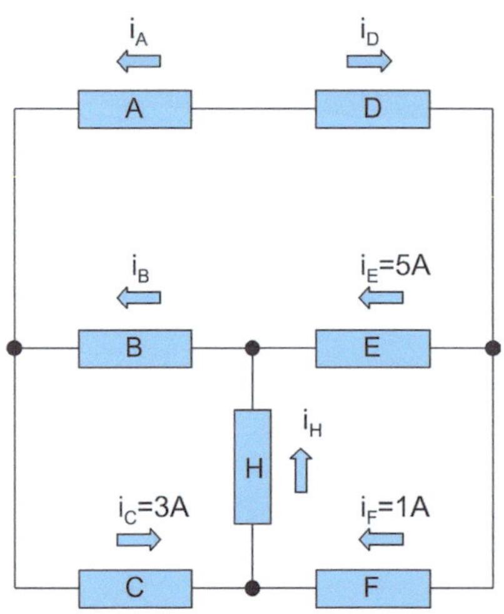

Problem 3.9. Determine currents i_C, i_F, i_H for the circuit shown in the figure that follows.

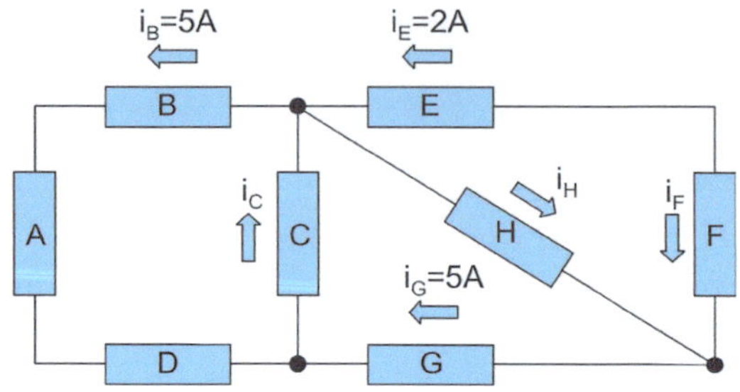

Problem 3.10. Determine currents i_C, i_F, i_H in the figure that follows.

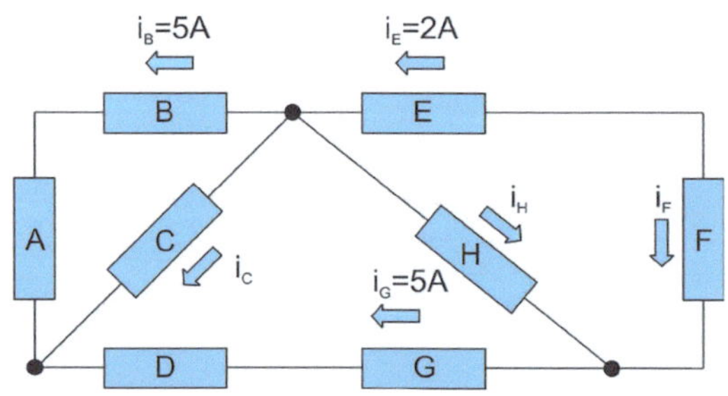

Problem 3.11. Find all unknown currents for the circuit shown in the figure below.

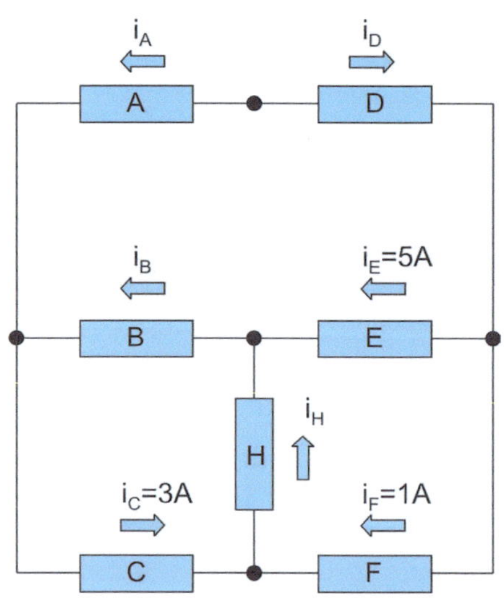

3.1.3 Kirchhoff's Voltage Law (KVL)

Problem 3.12.

A. Find voltage V_C for the circuit shown in the figure that follows;
B. How would the solution change if V_D were equal to 0 V?

C. Could the value $V_E = 0$ V be used in this problem?

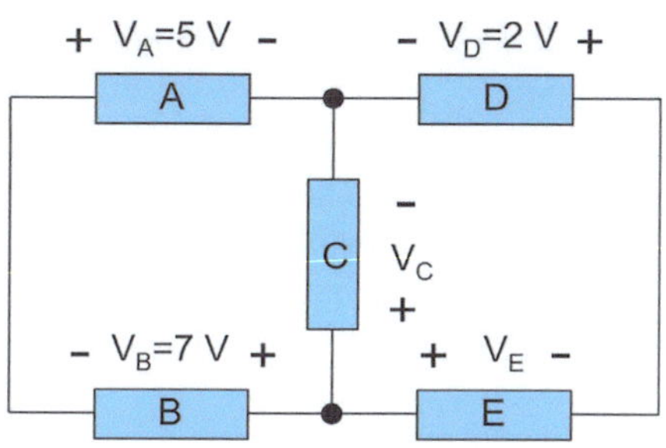

Problem 3.13. Determine voltage V_F in the circuit shown in the figure below.

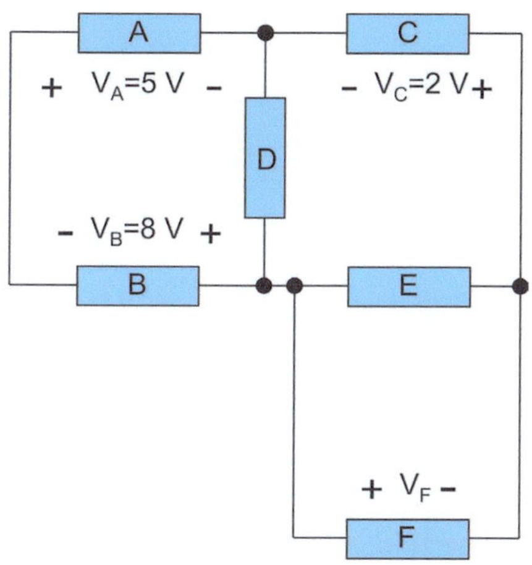

Problem 3.14. Find the unknown voltages V_A, V_B, V_G, V_H for the circuit shown in the figure below.

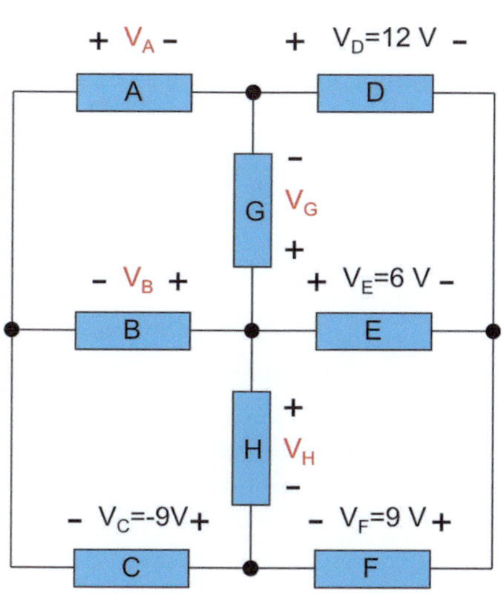

Problem 3.15. Determine voltage V_E for the circuit shown below.

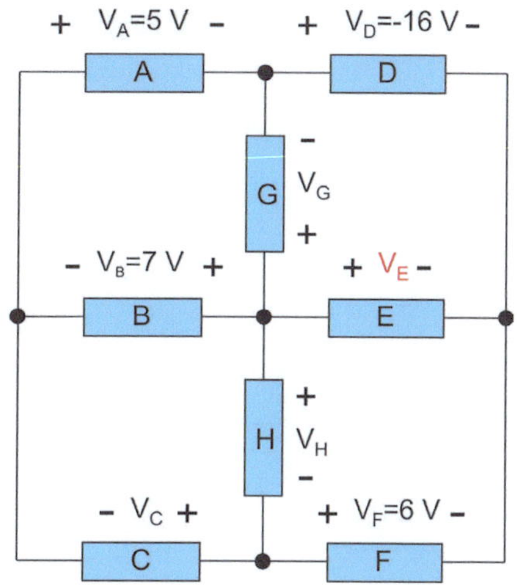

Problem 3.16. Equipotential lines for a human body subject to a vertical electric field with the strength of 1 V/m are shown in the figure that follows.

A. Determine voltages V_{AB}, V_{BC}, V_{CA}.

B. Establish the KVL loop and formulate KVL for these three voltages.

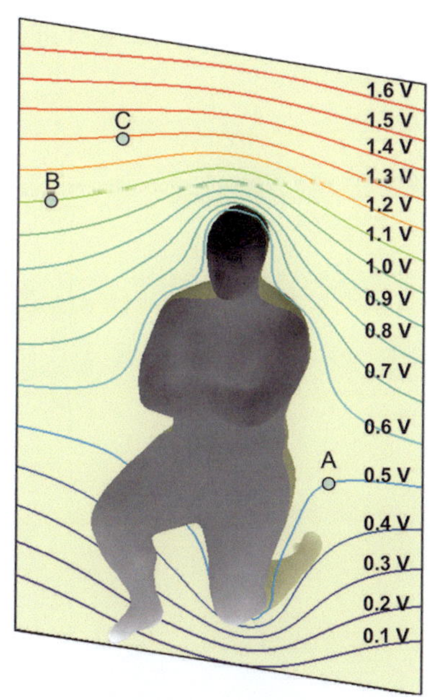

3.1.4 Power-Related Theorems

Problem 3.17. For the circuit shown in the figure that follows:

(i) Determine which circuit elements are resistances and which are sources.
(ii) Find the power (delivered to the circuit or taken from the circuit) for every circuit element.
(iii) Assuming that the powers of the sources are negative, find the sum of all powers in the circuit.

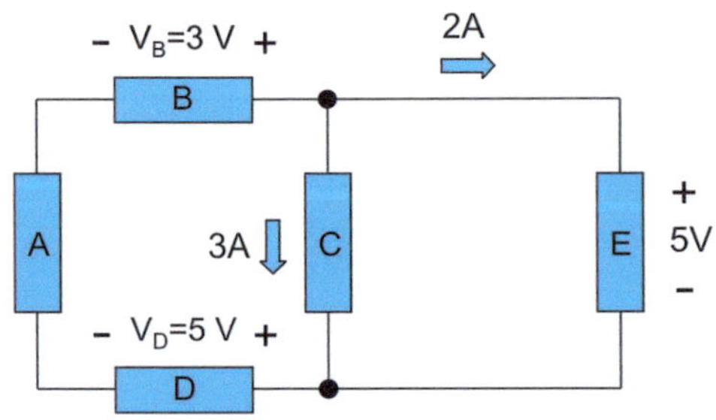

Problem 3.18. For the circuit shown in the figure:

1. Use KVL and KCL to solve for unknown currents and voltages.
2. Find the power (delivered to the circuit or taken from the circuit) for every circuit element.
3. Assuming that the powers of the sources are negative, find the algebraic sum of all powers in the circuit.

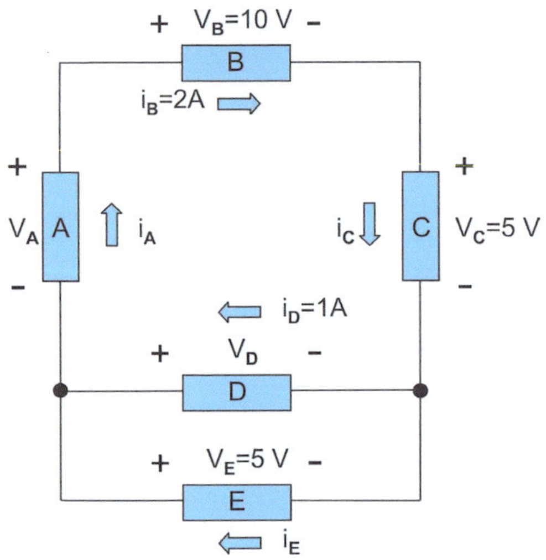

Problem 3.19. For the circuit shown in the figure:

1. Use KVL to solve for the unknown voltages.
2. Use KCL to solve for the unknown currents.
3. For each of six circuit elements, determine if the element is a resistance or a source.
4. Assuming that the source powers are negative, find the algebraic sum of all powers in the circuit.

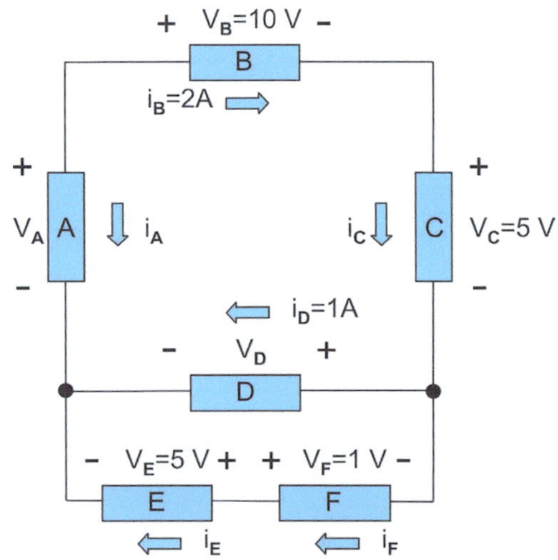

3.2 Series and Parallel Network/Circuit Blocks

3.2.1 Sources in Series and in Parallel

Problem 3.20. The electronic circuits onboard an 18-foot long Parker motor boat consume 96 W when operated from a 24-V source. The source is a combination of two fully charged deep-cycle batteries, each of which is rated for 12 V and 100 ampere hours.

1. Should the batteries be connected in series or in parallel?
2. For how many hours can the electronics be operated from the battery bank without recharging?
3. How much energy in kilowatt hours is initially stored in each battery?

Problem 3.21. A certain sensing device consumes 0.375 W of power over a 20 h period and operates from a 6 V source. The source is a combination of four fully charged AAA batteries, 1.5 V each. The batteries discharge by the end of the 20 h period.

1. Should the batteries be connected in series or in parallel?
2. What is a typical capacity of the AAA battery?
3. How much energy in watt hours is initially stored in each battery?

Problem 3.22. A load has the resistance of 1.5 Ω and requires the applied voltage of 3 V. A number of battery cells are given, each of which is rated for 1.5 V. Each cell may deliver no more than 1 A of current.

1. Construct and draw a battery bank that could be used to drive the load.
2. Is the solution to the problem unique?

Problem 3.23. Repeat the previous problem when the required load voltage is changed to 6 V.

3.2.2 Resistances in Series and in Parallel

3.2.3 Reduction of Resistive Networks

Problem 3.24. Determine the equivalent resistance between terminals *a* and *b*.

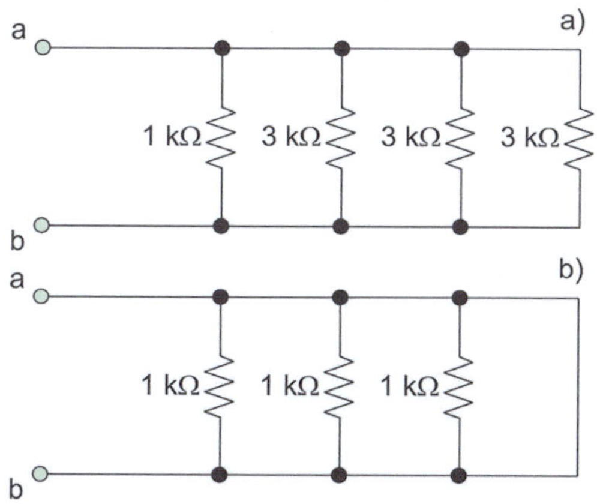

Problem 3.25. The equivalent electric circuit for a car rear window defroster is shown in the figure below. All resistances are equal: $R_1 = \ldots = R_{15} = 10$ Ω. Determine the heat power (power delivered to the defroster).

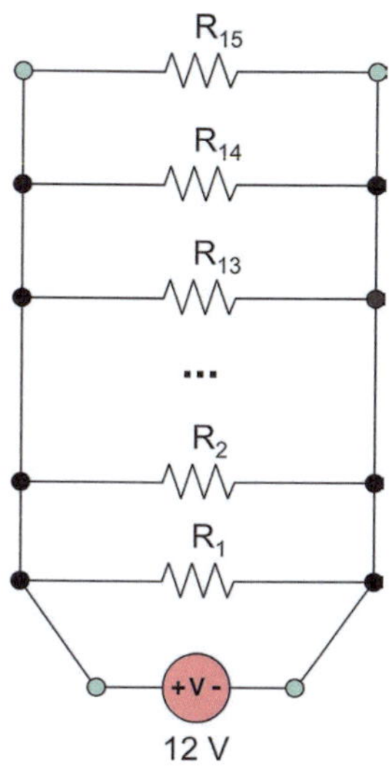

Problem 3.26. Find the equivalent resistance between terminals *a* and *b*.

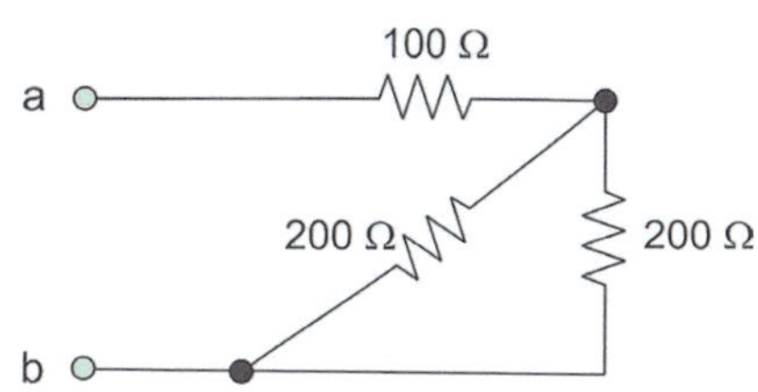

Problem 3.27. Find the equivalent resistance between terminals *a* and *b*.

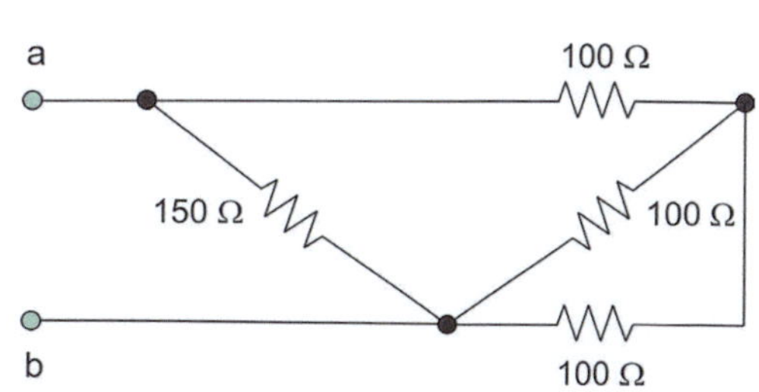

Problem 3.28. Find the equivalent resistance between terminals *a* and *b*.

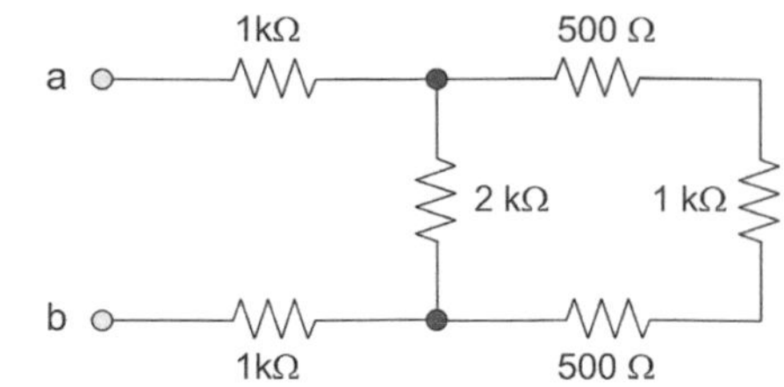

Problem 3.29. Determine the equivalent resistance between terminals *a* and *b*.

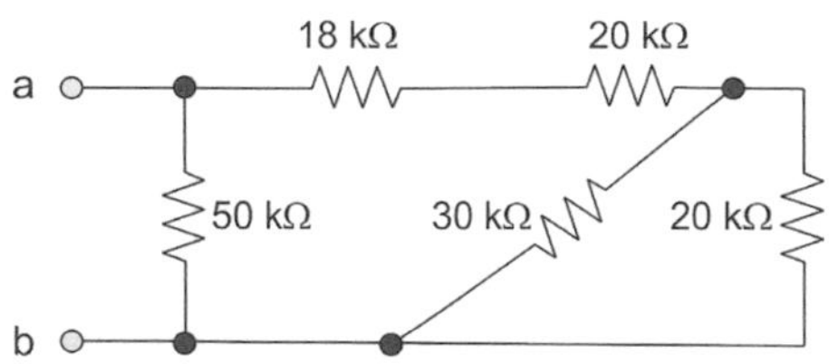

Problem 3.30. Find the equivalent resistance between terminals *a* and *b*.

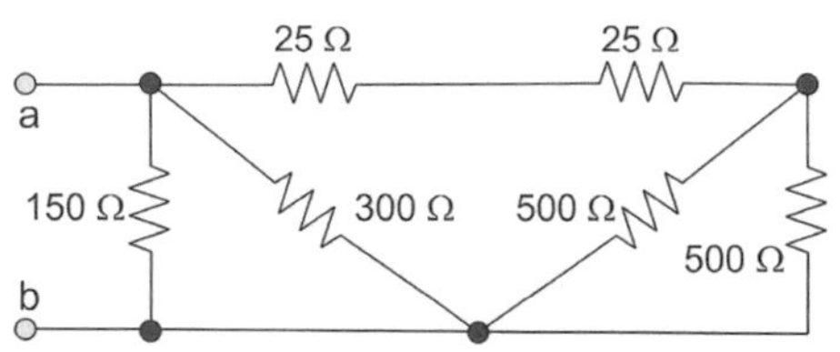

Problem 3.31. Determine the equivalent resistance (resistance between ports a and b) of the network shown in the figure that follows.

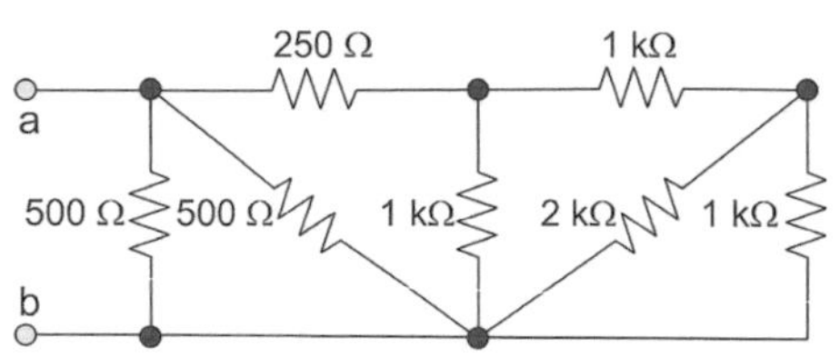

Problem 3.32. Determine the equivalent resistance between terminals *a* and *b*.

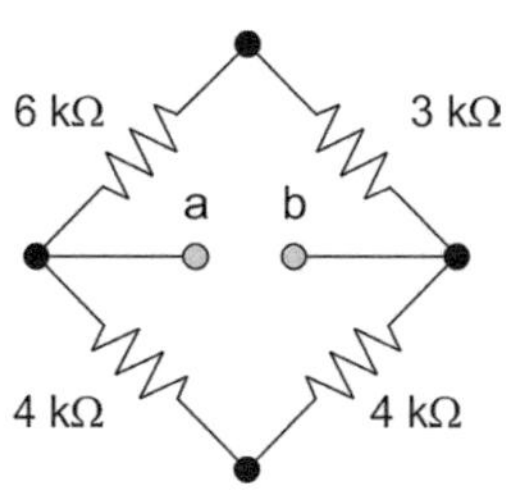

Problem 3.33. Find the equivalent resistance between terminals *a* and *b*.

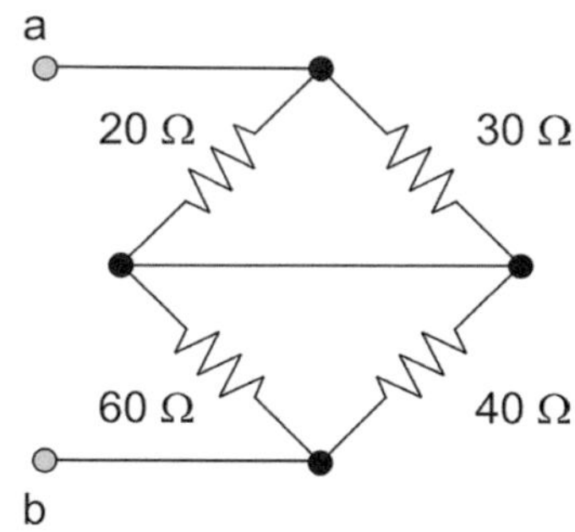

Problem 3.34. Determine the equivalent resistance of the network (resistance between terminals *a* and *b*) shown in the figure below.

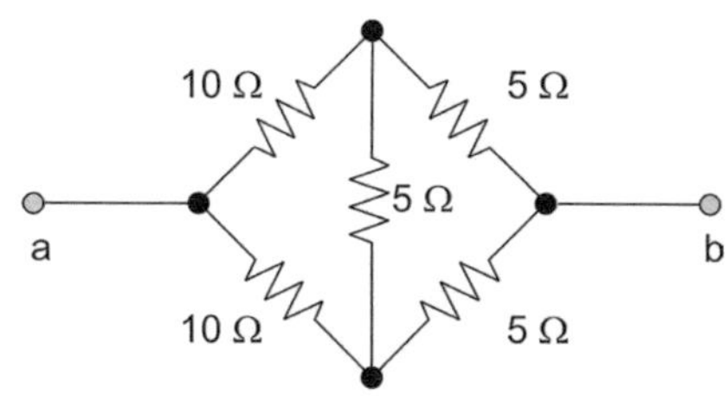

Problem 3.35. Determine the equivalent resistance between terminals *a* and *b* (show units).

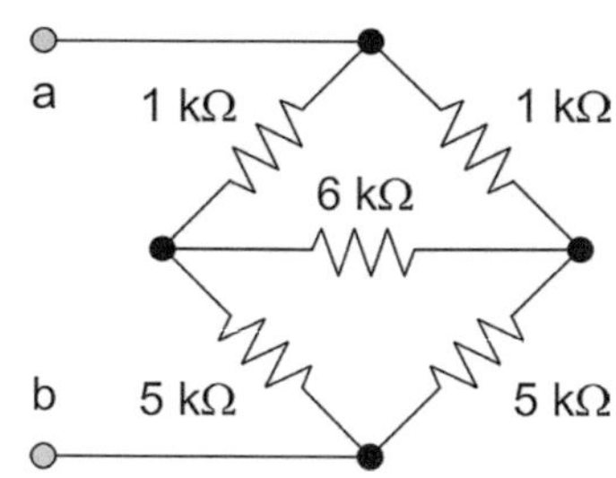

Problem 3.36. A network shown in the figure below is known as a *ladder*. The ladder network includes one particular section with a series ($R_1 = 1\ \Omega$) and a shunt ($R_2 = 1\ \Omega$) resistance, known as the *L*-section. This section is then repeated an infinite number of times to the right. Determine the equivalent resistance between terminals *a* and *b* of the ladder network.

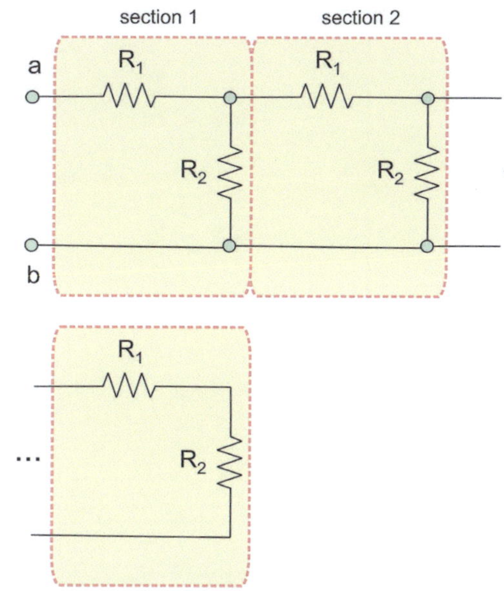

Hint: For the semi-infinite ladder network in the previous figure, the equivalent resistance, R_{eq}, will not change after adding a new section up front as shown in the figure that follows.

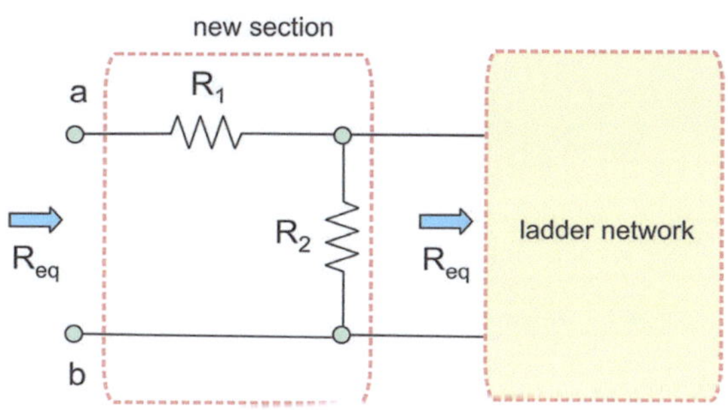

Problem 3.37.

A. Repeat the previous problem for the semi-infinite ladder network circuit shown in the same figure when $R_1 = R_3 = 10\ \Omega$ and $R_2 = 25\ \Omega$.

B. How different is your result from the equivalent resistance of the finite ladder network with only four sections?

Problem 3.38. Another important ladder network type (with the T-section) is shown in the figure.

A. Determine the equivalent resistance between terminals a and b of the ladder network when $R_1 = 2R_3 = 10\ \Omega$ and $R_2 = 10\ \Omega$.

B. How different is your result from the equivalent resistance of the finite ladder network with only four sections?

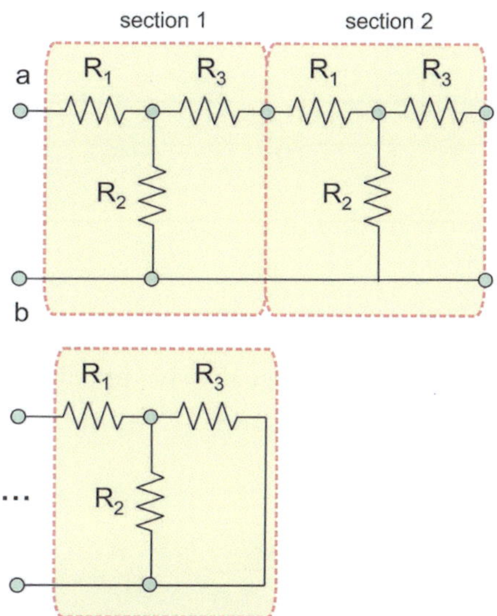

3.2.4 Voltage Divider Circuit

Problem 3.39. Redraw the circuit shown in the figure that follows and plot the distribution of the circuit voltage versus ground reference point to scale between two circuit points a and b as a function of distance x from point a.

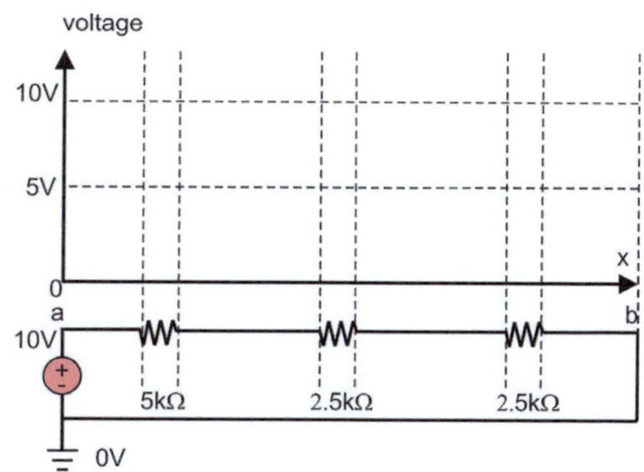

Problem 3.40. Repeat the task of the previous problem for the disconnected circuit shown in the figure below.

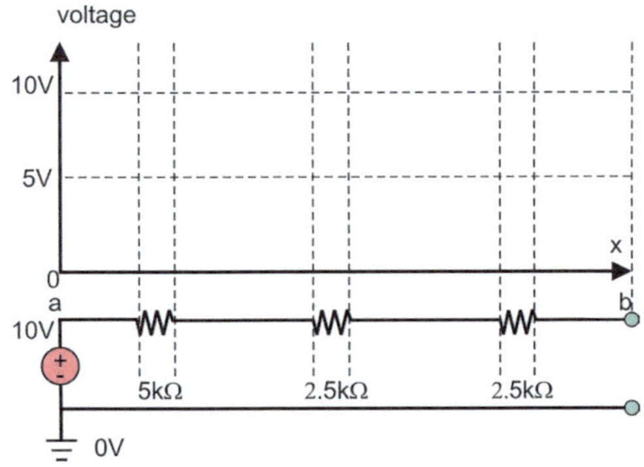

Problem 3.41. For the circuit shown in the figure that follows, use the voltage division principle to calculate V_1, V_2, V_3.

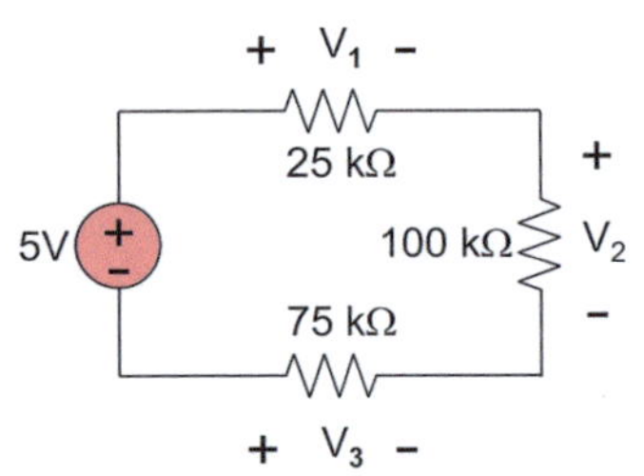

Problem 3.42. For the circuit shown in the following figure:

A. Calculate V_1, V_2, V_3.
B. Find the voltage across the current source.

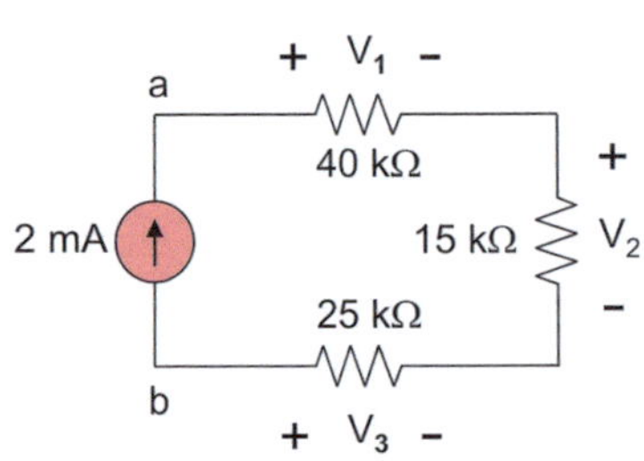

Problem 3.43. For the circuits shown in the figure that follows, determine voltages V_1, V_2, absolute voltage V_a at node a versus ground (show units), and circuit current I.

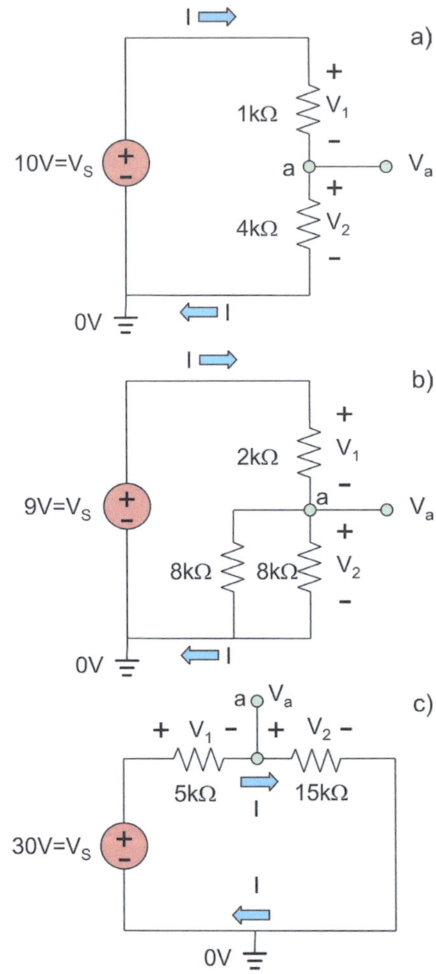

3.2.5 Application Example: Voltage Divider as a Sensor Circuit

3.2.6 Application Example: Voltage Divider as an Actuator Circuit

Problem 3.44. An NTC thermistor-based temperature sensor should operate between 25° C and 65° C from a 6 V DC power supply. The thermistor's resistance changes from $R' = 50$ kΩ to $R'' = 20$ kΩ in this temperature range.

A. Present a circuit diagram for the simple temperature sensor.
B. Determine the value of the unknown resistance for the maximum circuit sensitivity.

C. Determine the maximum circuit sensitivity.

D. What is the circuit sensitivity when the unknown resistance is set to 1 kΩ?

Problem 3.45. In the previous problem, using software of your choice (MATLAB is recommended), plot the circuit sensitivity to scale as a function of the value of the unknown resistance in the range from 1 kΩ to 100 kΩ.

Problem 3.46. A strain gauge with nominal resistance of 120 Ω is used in conjunction with a 2.5 V DC voltage source. Its nominal resistance changes by ±0.2% when the gauge operates in the permissible strain range, which is $\pm 1000\mu\varepsilon$.

A. Present a circuit diagram for the simple strain gauge sensor.

B. Determine the value of the unknown resistance for the maximum circuit sensitivity.

C. Determine the maximum circuit sensitivity.

D. What is the circuit sensitivity when the unknown resistance is set to1 kΩ?

Problem 3.47. In the previous problem, using software of your choice (MATLAB is recommended), plot the circuit sensitivity to scale as a function of the value of the unknown resistance in the range from 10 Ω to 1000 Ω.

Problem 3.48. A voltage divider circuit with $R_1 = 700\ \Omega$ (the fixed resistance) and $R_2 = 700\ \Omega \pm 0.1\%$ (the strain gauge) is used. The voltage power supply is rated at 4.5 V.

A. Show that voltage across the strain gauge varies in the range 2.25 V ± 1.125 mV.

B. Can you derive an analytical formula that gives the voltage variation across $R_2 = R_1 \pm \Delta$ as a *linear* function of an arbitrary (but very small) resistance variation Δ? [*Hint:* use your calculus background—the Maclaurin series versus a small parameter.]

3.2.7 Current Limiter

Problem 3.49. A thermistor is connected to an ideal voltage power source of 9 V. Determine the value of the current-limiting resistor R based on the requirement that the power delivered to the thermistor should be always less than 0.1 W. The lowest possible value of R should be chosen. Consider two cases:

1. Thermistor resistance is exactly 100 Ω.
2. Thermistor resistance changes from 200 Ω to 100 Ω.

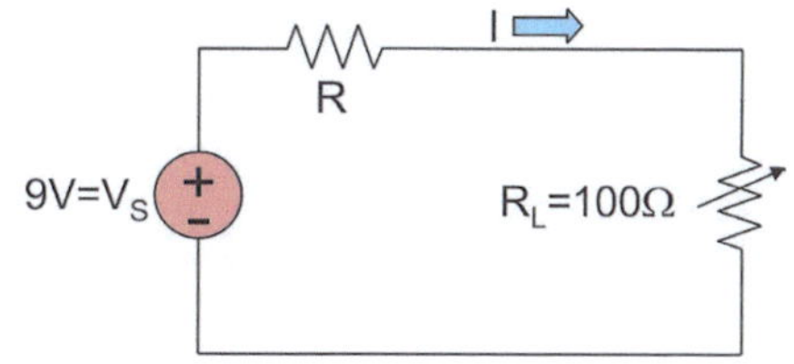

3.2.8 Current Divider Circuit

Problem 3.50. For the circuit shown in the figure that follows:

A. Calculate voltage between terminals *a* and *b*. Show its polarity on the figure.

B. Use the current division principle to calculate branch currents i_1, i_2.

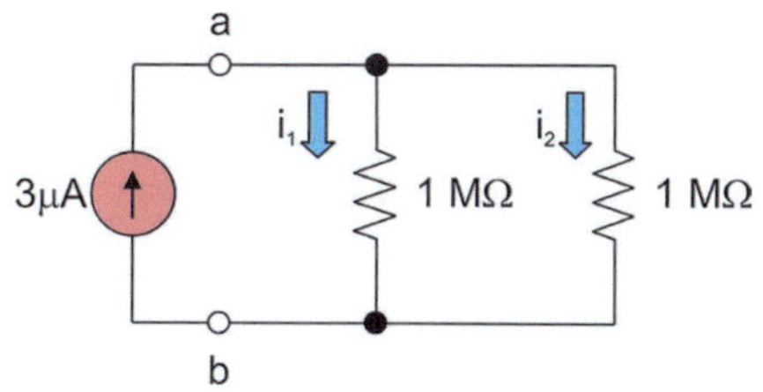

Problem 3.51. Find branch currents i_1, i_2 for the circuit depicted in the figure below.

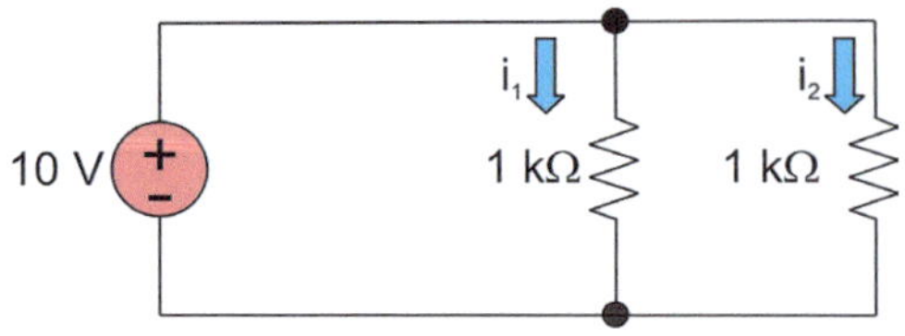

Problem 3.52. Find branch currents i_1, i_2 for the circuit shown in the figure that follows.

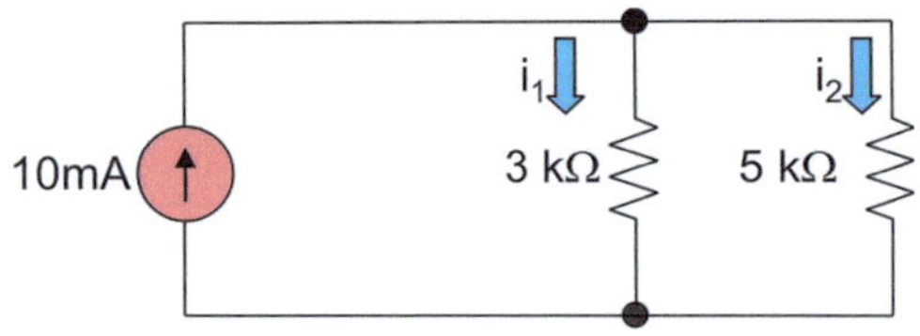

Problem 3.53. Find the voltage between terminals *a* and *b* (voltage across the current power source) for the circuit shown in the figure that follows.

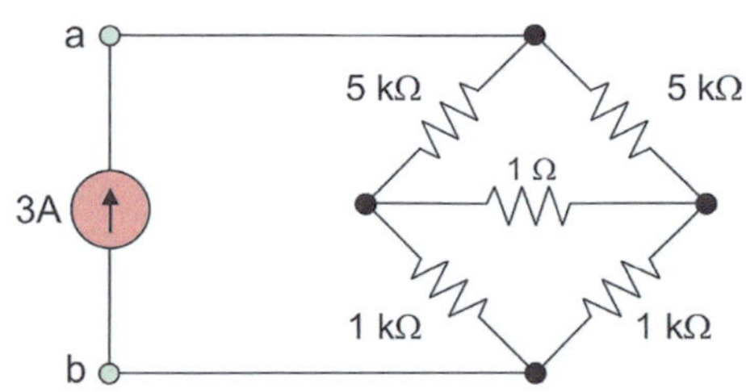

Problem 3.54. The voltage source in the circuit is delivering 0.2 W of electric power. Find *R*.

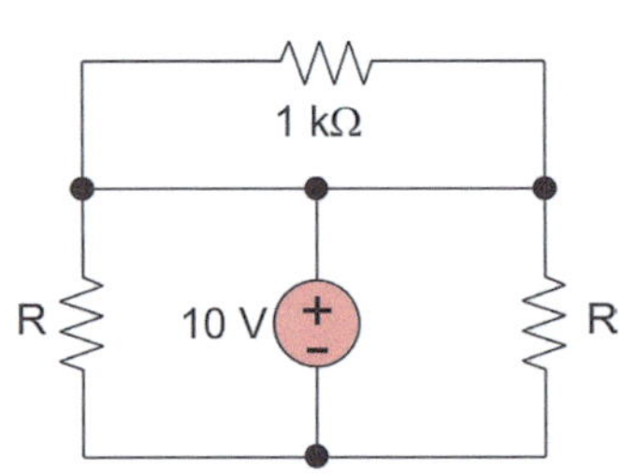

3.2.9 Wheatstone Bridge

Problem 3.55. Describe in your own words the function and the scope of the Wheatstone bridge.

Problem 3.56. Find the voltage between terminals *a* and *b* for the circuit shown in the figure that follows.

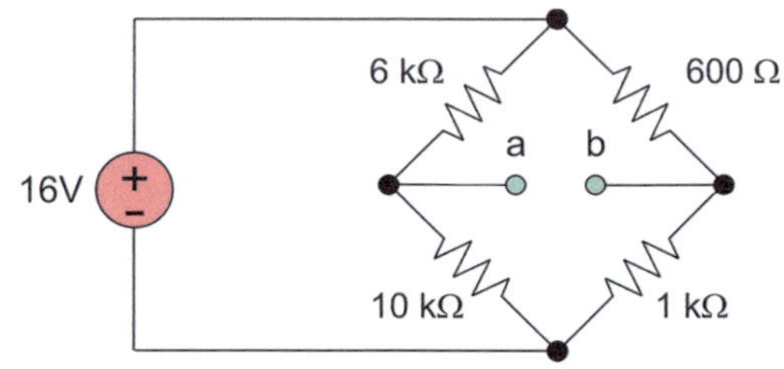

Problem 3.57. Find the voltage between terminals *a* and *b* for the circuits shown in the figure that follows.

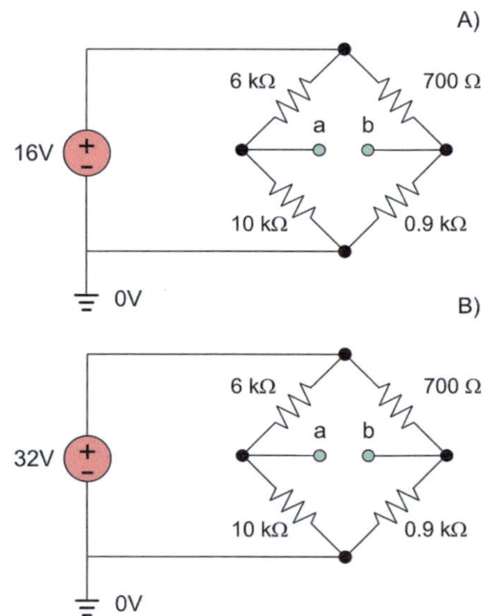

Problem 3.58. You are given:

(i) A photoresistor that changes its resistance from 20 kΩ for brightness to 500 kΩ for darkness
(ii) A 9 V battery
(iii) A voltmeter (DMM)
(iv) Any other necessary precise resistors

Construct (present the circuit diagram) a Wheatstone bridge sensor circuit that:

(i) Has zero voltage reading for brightness

(ii) Has maximum possible voltage reading for darkness (has maximum sensitivity)

Problem 3.59. You are given:

1. A SGT-1/350-TY11 uniaxial strain gauge with the nominal resistance of 350 Ω (no strain)
2. A 4.5 V voltage source
3. Any number of precise resistors, of any value

When tensile strain is applied, a resistance variation up to +0.1% is observed. Present the circuit diagram for the Wheatstone bridge sensor circuit that:

(i) Has zero voltage reading at no strain
(ii) Has maximum possible voltage response when strain is present (has maximum sensitivity)
(iii) Outputs positive voltages when the resistance of the strain gauge increases

Problem 3.60. Resistance of a strain gauge increases when its length increases (a bended surface under test becomes convex) and decreases when its length decreases (a bended surface under test becomes concave). The corresponding strains are the tensile strain and the compressive strain. You are given two strain gauges (#1 and #2), which are attached to opposite sides of a thin bended surface under test, at the same position. The gauge resistance at normal conditions (no bending) is $R = 100\ \Omega$. You are also given any number of fixed resistors, of any value.

1. Suggest and sketch a sensor circuit which will convert changes in the resistance into measurable voltage changes. This circuit should possibly have:
   (a) Zero sensor output voltage at normal sensor conditions (no bending)
   (b) Maximum voltage sensitivity to changes in resistance (sensitivity to the strain)
2. Label one (or two) strain gauges used, and specify the values of all deployed resistances.
3. Show power supply and DMM connections.

Combined Voltage and Current Dividers

Problem 3.61. For the circuit shown in the figure:

A. Find currents i, i_1, i_2 (show units).
B. Find power P delivered by the voltage source to the circuit.
C. Find voltages V_1, V_2.

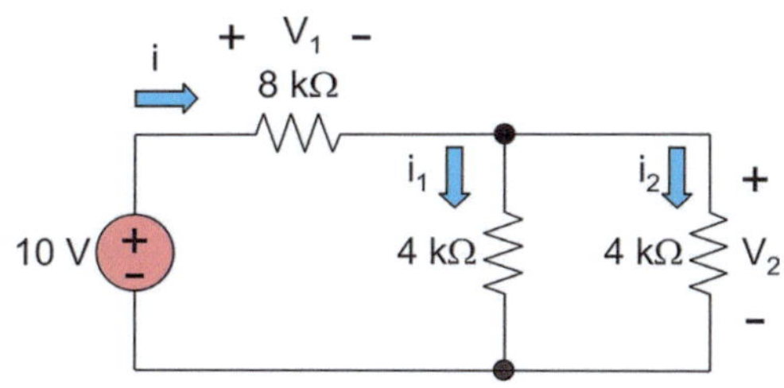

Problem 3.62. Find voltage V across the 20 kΩ resistance and current i for the circuit shown in the figure that follows.

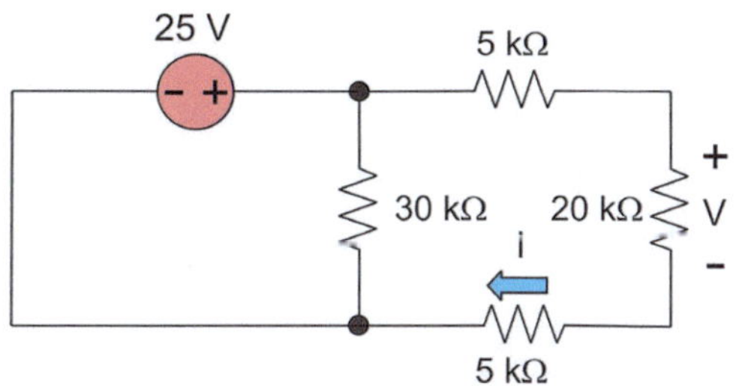

Problem 3.63. For the circuit that follows determine:

A. Current i_2 through the 0.25 kΩ resistance
B. Power P absorbed by the 600 Ω resistance

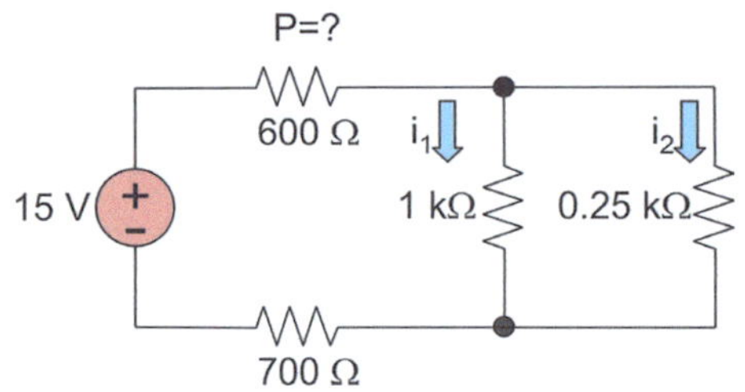

Problem 3.64. For the circuit shown in the figure determine:

A. Current i_2 through the 0.25 kΩ resistance,
B. Power P absorbed by the 0.3 kΩ resistance.

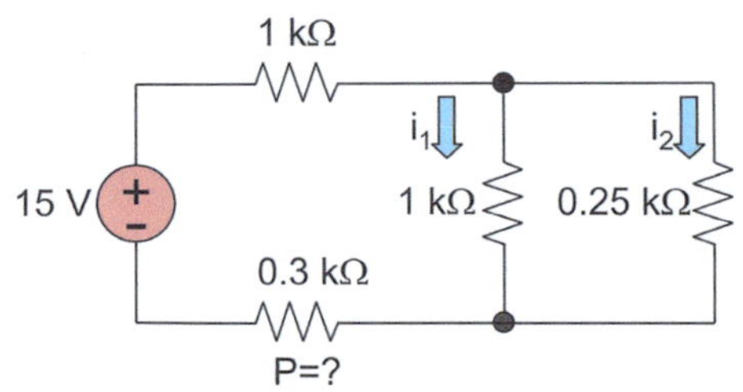

Problem 3.65. Determine current i (show units) in the circuit that follows.

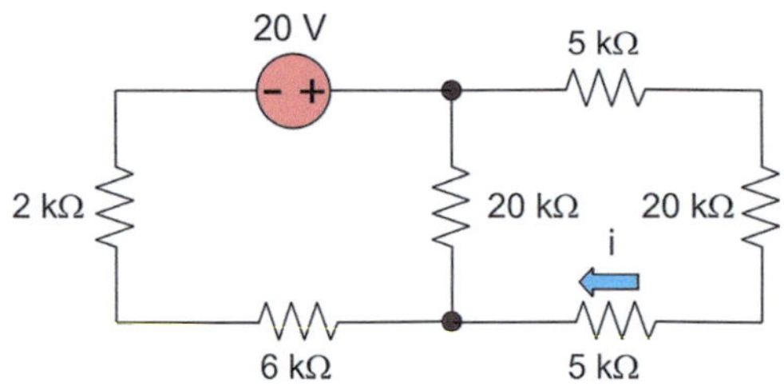

Problem 3.66. Determine current i and voltage υ for the circuit shown in the figure (show units).

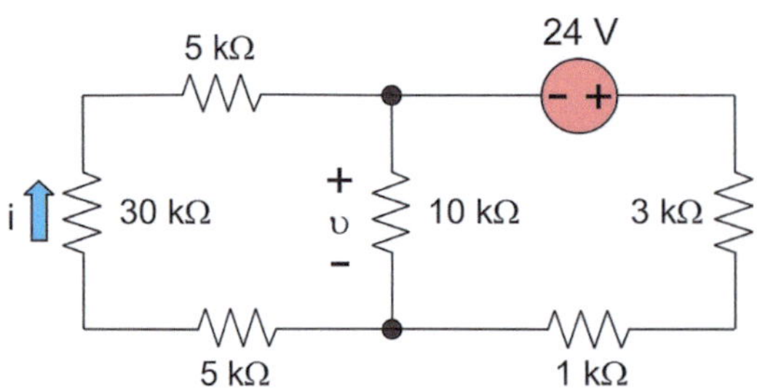

3.3 Superposition Theorem and Its Use

3.3.2 Superposition Theorem or Superposition Principle

Problem 3.67. Solve the circuits shown in the figure that follows (find current I and show units) using superposition theorem.

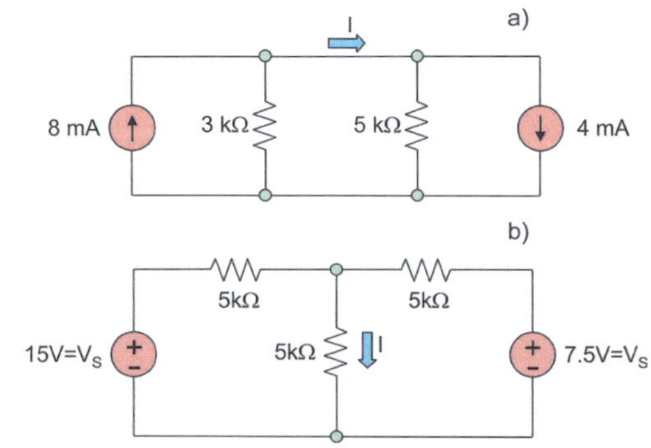

Problem 3.68. Find current I (show units) using the superposition theorem.

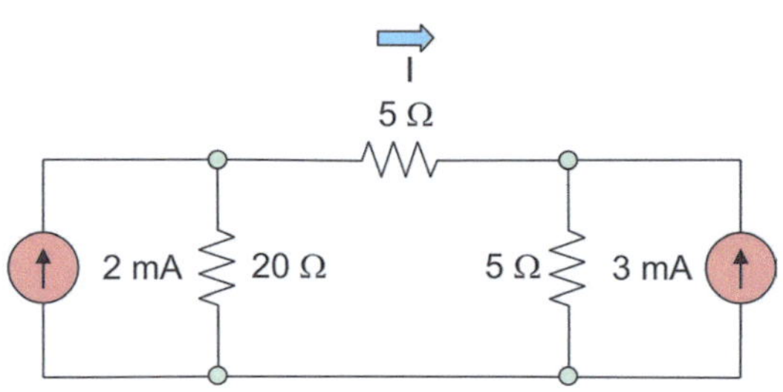

Problem 3.69. Find current I (show units) using the superposition theorem.

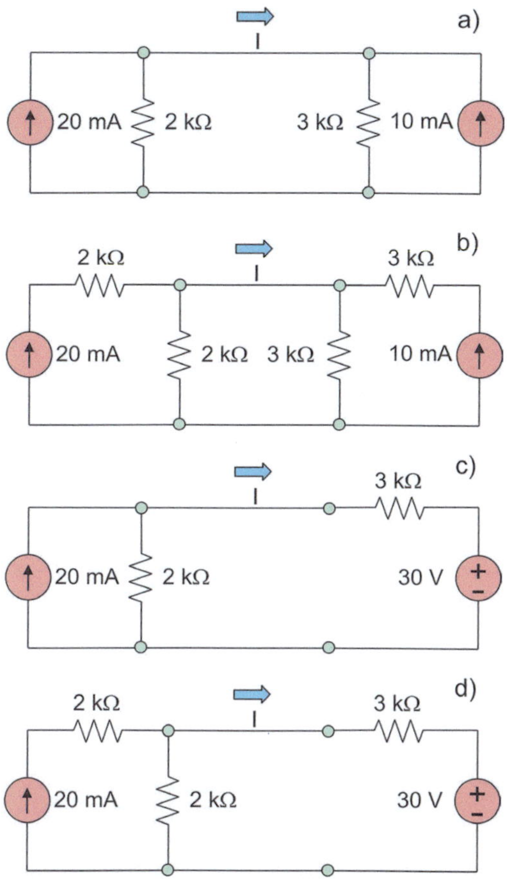

Problem 3.70.

(a) Find current I (make sure to show units) using the superposition theorem.

(b) Find voltage V (make sure to show units) using the superposition theorem.

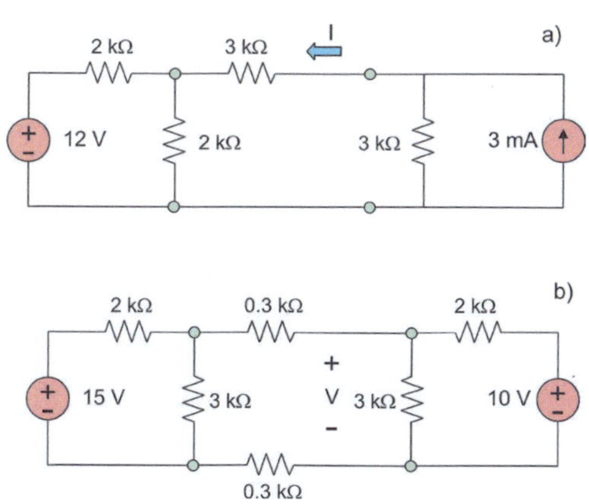

Problem 3.71. For the circuit shown in the figure, determine the absolute voltage (versus chassis ground) and the electric current at the circuit point a.

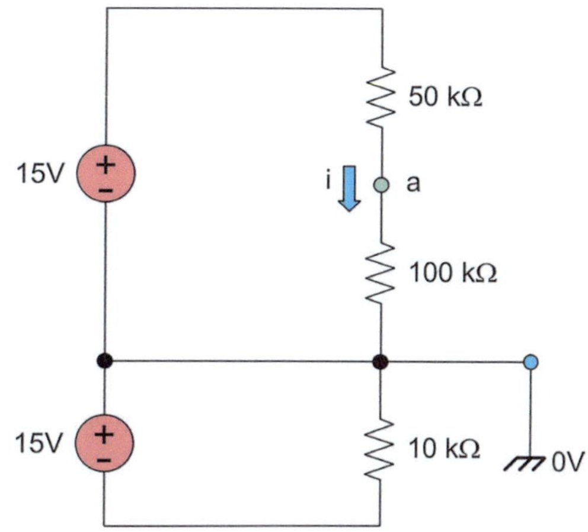

Problem 3.72. Determine the absolute voltage (versus chassis ground) and the current i at the circuit point a.

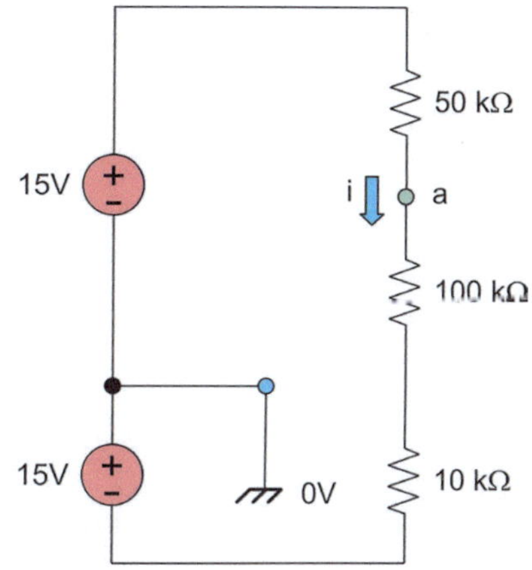

Problem 3.73. For the circuit shown in the figure that follows, find the current across the 10 Ω resistance. Show its direction in the figure.

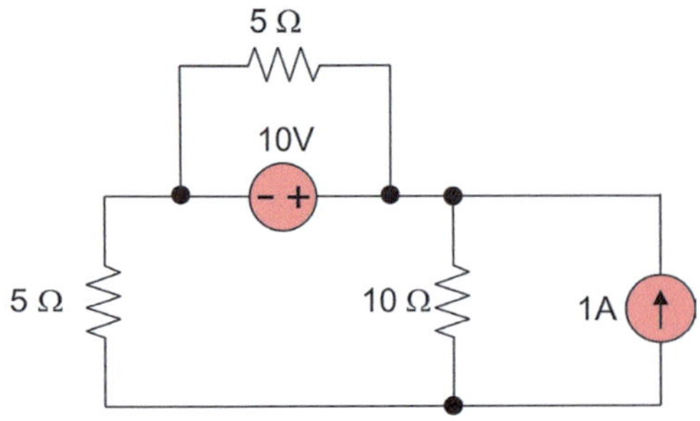

Problem 3.74. Determine the unknown current I in the circuit (solve the problem by superposition).

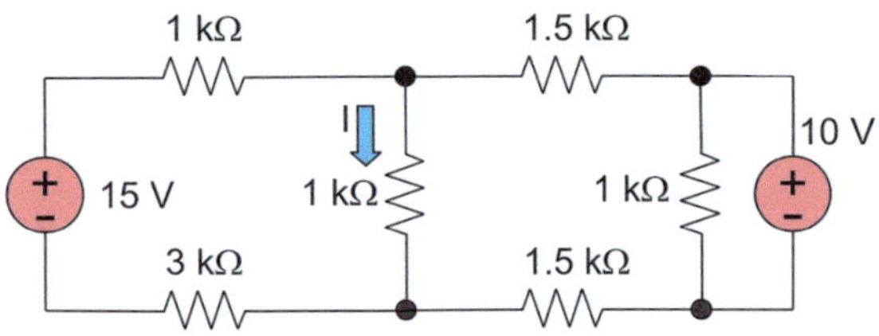

Problem 3.75. Determine the unknown current I in the circuit (solve the problem by superposition).

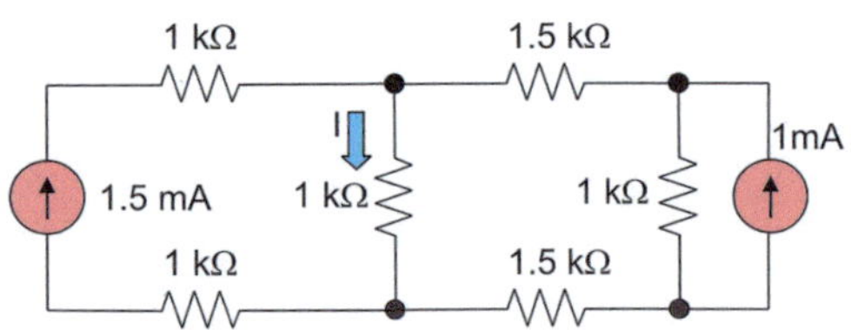

3.3.3 Y (Wye) and Δ (Delta) Networks: Use of Superposition

3.3.4 T and Π Networks: Two-Port Passive Networks

Problem 3.76. Convert the network shown in the figure that follows from Y to Δ:

A. Draw the corresponding Δ network.
B. Label its terminals.
C. Determine and label the corresponding resistance values in the figure.

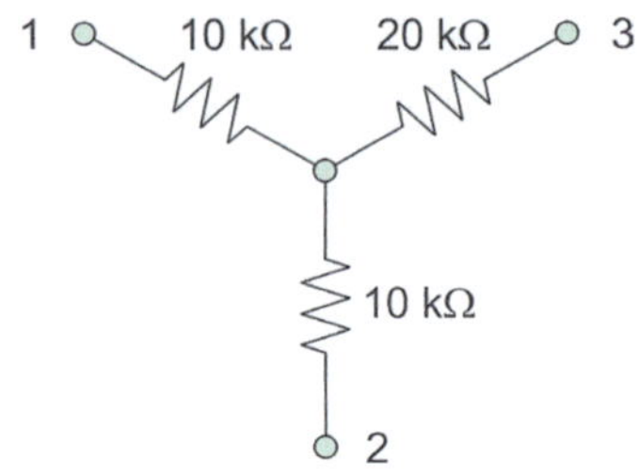

Problem 3.77. Convert the network shown in the figure that follows from Δ to Y:

A. Draw the corresponding Y network.
B. Label its terminals.
C. Determine and label the corresponding resistance values.

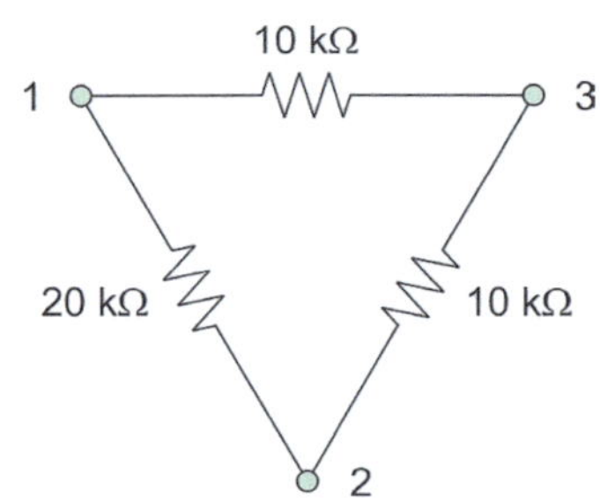

Problem 3.78. Convert the two-port network shown in the figure that follows from T to Π:

1. Draw the corresponding Π network.
2. Label its terminals and ports.
3. Determine and label the corresponding resistance values.

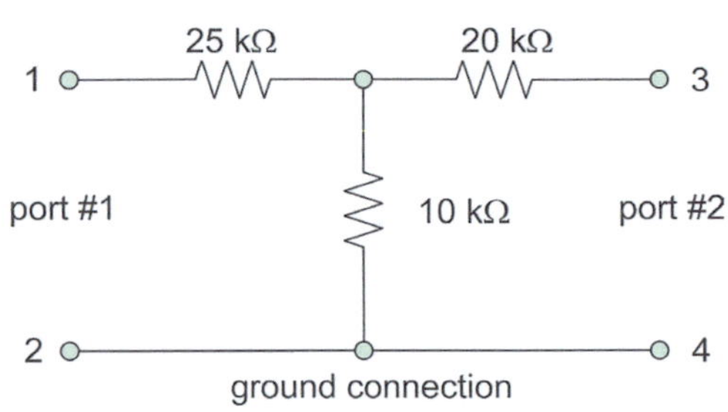

Problem 3.79. Convert the two-port network shown in the figure that follows from Π to T:

A. Draw the corresponding T network.
B. Label its terminals and ports.
C. Determine and label the corresponding resistance values.

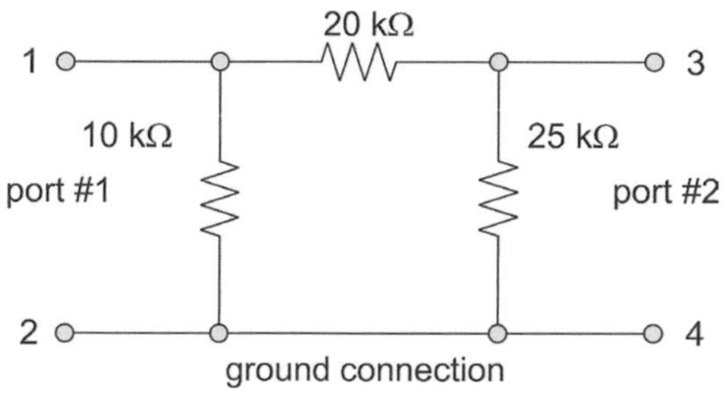

Problem 3.80. For the bridge network, determine the equivalent resistance between terminals *a* and *b*.

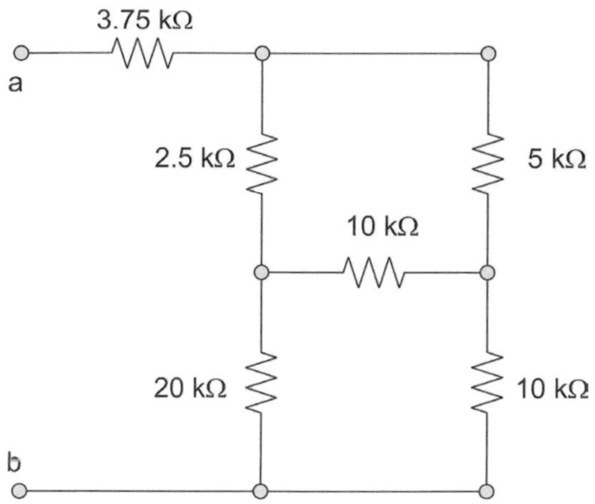

Chapter 4: Circuit Analysis and Power Transfer

Overview

Prerequisites:

- Knowledge of major circuit elements, their v-i characteristics, and Ohm's law (Chap. 2)
- Knowledge of basic networking theorems (Chap. 3)

Objectives of Sect. 4.1:

- Become familiar with the nodal analysis and be able to apply it to solve in arbitrary linear circuits
- Become familiar with the mesh analysis and be able to apply it to solve in arbitrary linear circuits

Objectives of Sect. 4.2:

- Become familiar with the method of short/open circuit
- Establish and prove the source transformation theorem
- Establish and prove Thévenin and Norton's theorems
- Be able to apply Thévenin and Norton's equivalents for circuit solution

Objectives of Sect. 4.3:

- Establish the maximum power theorem and become familiar with the power efficiency concept
- Be able to apply the concepts of Thévenin and Norton's equivalents and maximum power theorem in practice

Objectives of Sect. 4.4:

- Obtain an initial exposure to nonlinear circuit analysis
- Be able to solve in a simple nonlinear circuit

S. N. Makarov et al., *Practical Electrical Engineering*, https://doi.org/10.1007/978-3-319-96692-2_4

Application Examples:
Reading and using data for solar panels
Power radiated by a transmitting antenna
Maximum power extraction from solar panel
Solving the circuit for a generic solar cell

Keywords:
Nodal analysis, Mesh analysis (mesh current analysis), Supernode, Supermesh, Method of short/open circuit (definition of open-circuit network voltage, short-circuit network current), Source transformation theorem, Circuit equivalent (see equivalent circuit), Thévenin's theorem (formulation, proof, special cases), Thévenin equivalent, Norton's theorem, Norton equivalent, R-$2R$ ladder network, negative equivalent (Thévenin) resistance, Maximum power theorem (principle of maximum power transfer), Power efficiency, Analysis of nonlinear circuits, Load line (definition, method of), Iterative method for nonlinear circuits (definition of explicit iterative scheme, implicit iterative scheme), Solar cell (c-Si, open-circuit voltage, short-circuit photocurrent density, fill factor, characteristic equation of), Solar panel (series cell connection, open-circuit voltage, short-circuit photocurrent, fill factor, maximum power load voltage, maximum power load current)

Section 4.1 Nodal/Mesh Analysis

4.1.1 Importance of Circuit Simulators

The series and parallel equivalent circuits along with Y and Δ transformations provide a practical tool for solving simple circuits involving typically only a few elements. However, for more elaborate circuits, circuit simulators such as SPICE (Simulation Program with Integrated Circuit Emphasis) and their various modifications become indispensable tools for the professional engineer. SPICE was developed by the Electronics Research Laboratory at the University of California, Berkeley, and first presented in 1973. These circuit simulators are quite general and allow us to model circuits with passive and active elements including semiconductor devices such as diodes, transistors, and solar cells. Since those elements typically exhibit nonlinear current-voltage behaviors, elaborate solution strategies are needed. The circuit simulators use quite interesting algorithms: they often operate in the time domain, even for DC circuits. For example, a solution for a DC circuit is obtained as the steady-state limit of a transient solution, for voltage and/or current sources turned on at a certain time instance. The key of the time-domain approach is its inherent ability to solve nonlinear problems, with passive and active circuit elements. In this section, we will not discuss the specifics of the numerical circuit simulation. Instead, we will provide the foundation of the *nodal analysis* (or *node analysis*) and the *mesh analysis* (or the *mesh current analysis)*, which are two important features of any professional circuit simulator. The nodal and mesh analyses in their pure form do not involve time-domain methods. They are primarily applicable only to *linear circuits*, also referred to as *linear networks*.

4.1.2 Nodal Analysis for Linear Circuits

The *nodal analysis* is a general method of solving linear networks of arbitrary complexity, which is based on KCL and Ohm's law. Let us consider a circuit shown in Fig. 4.1a, which is a resistive bridge circuit with a bridging resistance. This circuit may be solved using Δ to Y conversion; see, for instance, example 3.16 of Chap. 3. Here, we prefer to use the nodal analysis directly. The nodal analysis operates with the *absolute* values of the *node voltages* in the circuit with respect to *ground reference*. It may be divided into a number of distinct steps:

1. A ground reference needs to be assigned first; this is a node where the voltage is set to 0 V. To this end, we ground the negative terminal of the voltage power supply.
2. Next, we select nontrivial (also called *non-reference*) nodes for which we do not know the voltages. These are nodes 1 and 2 in Fig. 4.1b. The two additional nodes are eliminated from the analysis since the voltages there are already known.

3. We label *absolute* node voltages versus ground reference as V_1, V_2—see Fig. 4.1c.
4. Then currents for every nontrivial node are labeled, assuming that all currents are *outflowing*—see Fig. 4.1c. This condition can be replaced by all inflowing currents.
5. Next, KCL is applied to every nontrivial node. We express the currents as the difference of two absolute voltages: the voltage at the beginning of the current arrow (voltage at the master node) minus the voltage at the end of the current arrow (voltage at any other node) and then divide this difference by the appropriate resistance. Hence, we arrive at a system of linear equations for the nodal voltages. Currents are no longer involved.
6. After the resulting system of linear equations is solved, all circuit parameters can be determined as necessary.

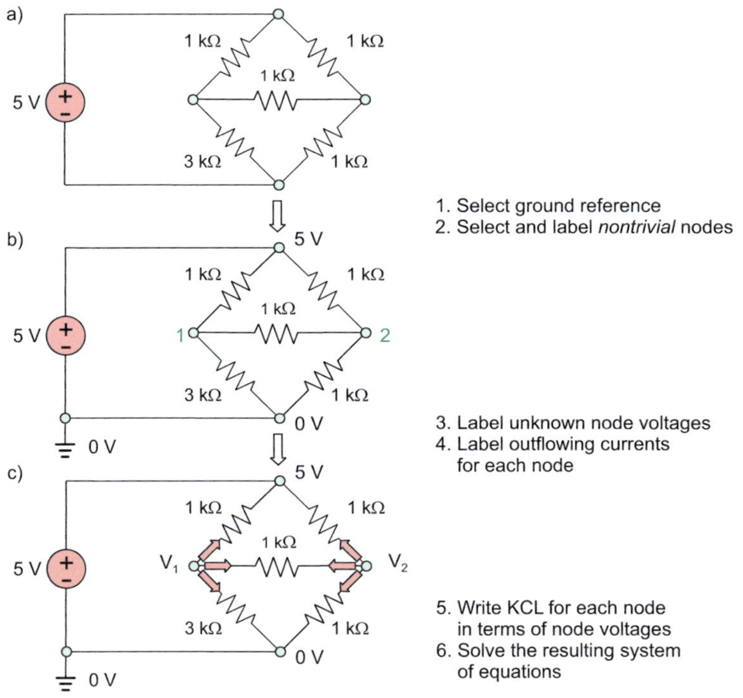

Fig. 4.1. Major steps of the nodal analysis applied to a bridge circuit.

Note that the resulting system of linear (or nodal) equations may have one, two, three, or more equations depending on the complexity of the circuit as well as on your own "smart" choice of the nontrivial nodes. The following two examples will apply the nodal analysis to a circuit with a voltage source.

Example 4.1: Solve the circuit shown in Fig. 4.1a using nodal analysis and find the supply current.

Solution: Steps 1 to 4 are indicated in Fig. 4.1b, c. Applying KCL to node 1 and then to node 2 (the order is not important), one has.

$$\frac{V_1 - 5\ \text{V}}{1\ \text{k}\Omega} + \frac{V_1 - 0\ \text{V}}{3\ \text{k}\Omega} + \frac{V_1 - V_2}{1\ \text{k}\Omega} = 0 \tag{4.1a}$$

$$\frac{V_2 - 5\ \text{V}}{1\ \text{k}\Omega} + \frac{V_2 - 0\ \text{V}}{1\ \text{k}\Omega} + \frac{V_2 - V_1}{1\ \text{k}\Omega} = 0 \tag{4.1b}$$

This is a system of linear equations for two unknown voltages. It simplifies to

$$7/3V_1 - V_2 = 5\ \text{V} \tag{4.2a}$$
$$3V_2 - V_1 = 5\ \text{V} \tag{4.2b}$$

Equations (4.2) are solved, for example, via *Gaussian elimination*. This yields

$$V_1 = 3.33\ \text{V},\quad V_2 = 2.78\ \text{V} \tag{4.3}$$

The circuit current (current of the voltage source) is therefore

$$(5\ \text{V} - V_1)/1\ \text{k}\Omega + (5\ \text{V} - V_2)/1\ \text{k}\Omega = 3.89\ \text{mA}.$$

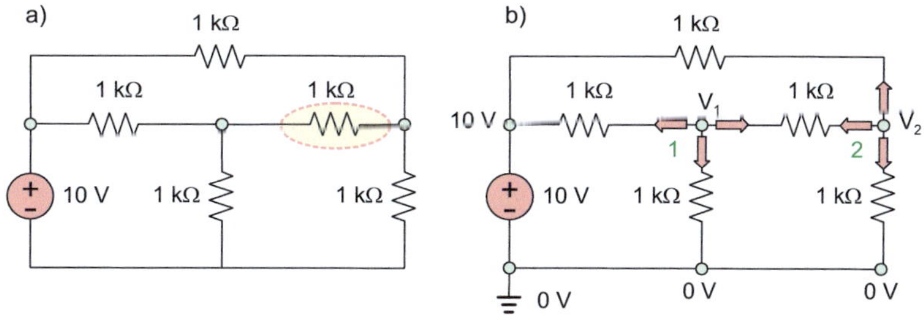

Fig. 4.2. Major steps of the nodal analysis applied to a circuit with a voltage source.

Example 4.2: Solve the circuit shown in Fig. 4.2a employing the nodal analysis.

Solution: Steps 1 to 4 are indicated in Fig. 4.2b. Applying KCL to node 1 and then to node 2, one obtains a system of equations with two unknown voltages:

Example 4.2 (cont.):

$$\frac{V_1 - 10\ \text{V}}{1\ \text{k}\Omega} + \frac{V_1 - 0\ \text{V}}{1\ \text{k}\Omega} + \frac{V_1 - V_2}{1\ \text{k}\Omega} = 0 \quad (4.4a)$$

$$\frac{V_2 - 10\ \text{V}}{1\ \text{k}\Omega} + \frac{V_2 - 0\ \text{V}}{1\ \text{k}\Omega} + \frac{V_2 - V_1}{1\ \text{k}\Omega} = 0 \quad (4.4b)$$

In setting up these equations, it does not matter which sequence of nodes are selected. Simplifying Eqs. (4.4) gives

$$3V_1 - V_2 = 10\ \text{V} \quad (4.5a)$$

$$3V_2 - V_1 = 10\ \text{V} \quad (4.5b)$$

The solution is obtained by symmetry, i.e., $V_1 = V_2 = 5$ V. The circuit current provided by the power supply is 10 mA. All other branch currents can then be found using Ohm's law. An interesting feature of the circuit shown in Fig. 4.2a is that the marked 1 kΩ resistor can be considered as "dead," since there is no current flowing through it (the voltage difference across this resistor is exactly zero). This resistor can be removed from the circuit without affecting the behavior of the circuit in terms of voltages and currents. It might appear at first sight that the circuits shown in Figs. 4.1 and 4.2 have a different network topology. In fact, they do not. To prove this, you should attempt to redraw the circuit in Fig. 4.2a; your redrawn topology should be identical with the circuit in Fig. 4.1a.

Circuits with a Current Source

When a current source is present in a circuit, the solution becomes even simpler: one makes use of the existing current and substitutes its value into the KCL equation written for a particular node. For example, KLC applied to node 1 in Fig. 4.3 includes the outflowing current of −1 mA; the current sign must be taken into account. The same idea is applied to circuits with multiple current power supplies. When only the current sources are present, the ground may be connected to the incoming terminal of a current source.

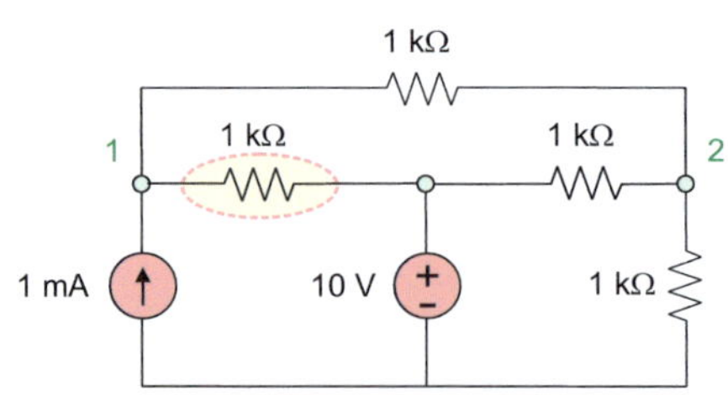

Fig. 4.3. A circuit with a current source solved via nodal analysis.

Exercise 4.1: Determine the voltage across the current source in Fig. 4.3 using the nodal analysis.

Answer: 8.6 V.

4.1.3 Supernode

The nodal analysis requires a "good" eye to see possible simplifications when labeling the nodes. Let us examine a particular case and point out a few useful subtleties.[1] Figure 4.4a depicts a network with two voltage sources. The property of the 5 V source is such that it is not fixed to a particular ground connection—we therefore call it a *floating source*. Setting up the nodal analysis becomes tricky, since we do not know the current through this source. However, a *supernode* may be formed as shown in Fig. 4.4b.

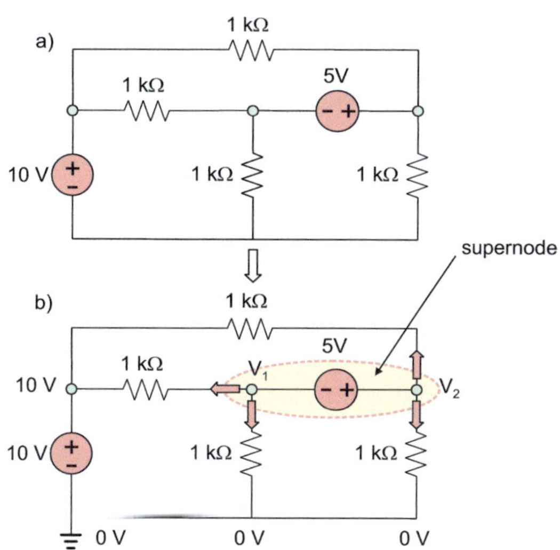

Fig. 4.4. A network with a floating voltage source between nodes 1 and 2.

When KCL is applied to any closed contour around the supernode, the net current must add to zero in such a case. With reference to Fig. 4.4b this yields

$$\frac{V_1 - 10\ \text{V}}{1\ \text{k}\Omega} + \frac{V_1 - 0\ \text{V}}{1\ \text{k}\Omega} + \frac{V_2 - 0\ \text{V}}{1\ \text{k}\Omega} + \frac{V_2 - 10\ \text{V}}{1\ \text{k}\Omega} = 0 \tag{4.6a}$$

[1] Subtleties are often euphemism for "playing" around with the circuit, like redrawing the wire connections and rearranging the circuit elements. This is done to find simpler solution approaches.

What is the second equation for the two unknowns V_1 and V_2? It simply turns out to be the relation between the supernode voltages themselves. Since $V_2 - V_1$ is the voltage of the power source, we obtain

$$V_2 = V_1 + 5\ \mathrm{V} \tag{4.6b}$$

Equations (4.6) can now be solved even without the calculator, eliminating one of the unknowns, which yields

$$V_1 = 2.5\ \mathrm{V}, \quad V_2 = 7.5\ \mathrm{V} \tag{4.7}$$

The circuit is solved. All currents can be found using the nodal voltages and Ohm's law.

Example 4.3: Solve the circuit shown in Fig. 4.4a by using the standard nodal analysis, without the supernode concept.

Solution: We have to specify an unknown current I_x through the 5 V source, which flows, say, from left to right in Fig. 4.4a. This convention results in the following two equations for the two nodes.

$$\frac{V_1 - 10\ \mathrm{V}}{1\ \mathrm{k\Omega}} + \frac{V_1 - 0\ \mathrm{V}}{1\ \mathrm{k\Omega}} + I_x = 0 \tag{4.8a}$$

$$\frac{V_2 - 0\ \mathrm{V}}{1\ \mathrm{k\Omega}} + \frac{V_2 - 10\ \mathrm{V}}{1\ \mathrm{k\Omega}} - I_x = 0 \tag{4.8b}$$

Next, we can add both equations and thereby eliminate I_x. The result is exactly Eq. (4.6a) for the supernode. We must add one more condition to solve this equation. Eq. (4.6b) is the only choice, i.e.,

$$V_2 = V_1 + 5\ \mathrm{V} \tag{4.8c}$$

Interestingly, we arrive at the supernode concept again but in a more complicated way. This is the reason why the supernode approach is such useful tool.

4.1.4 Mesh Analysis for Linear Circuits

The *mesh analysis* (or the *mesh current analysis*) is using loops instead of nodes. Only loops that do not contain any other loops—*the meshes*—are employed. The meshes as elements of the networking topology were defined in Sect. 3.1.1 of Chap. 3. Accordingly, instead of KCL the mesh analysis makes use of KVL. Hence, we need to choose *mesh* currents for every mesh. Figure 4.5 depicts the concept for a circuit with three meshes. Note that this circuit is identical to the circuit from Fig. 4.1. A ground connection does not have to be introduced for the mesh method. Let us denote the mesh current for mesh 1 in Fig. 4.5 by I_1, the mesh current for mesh 2 by I_2, and the mesh current for mesh 3 by I_3.

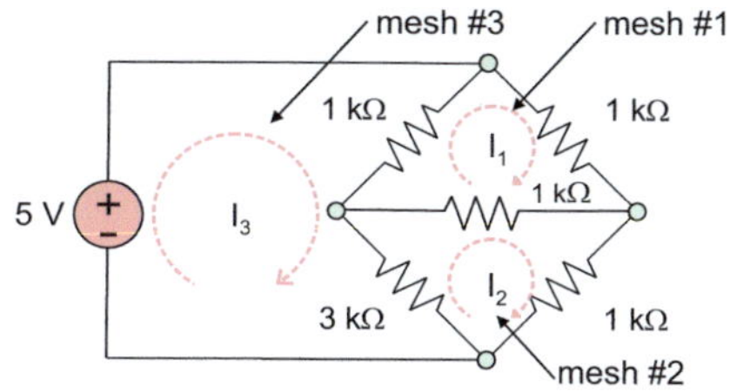

Fig. 4.5. Circuit analysis based on the mesh analysis; circuits in Fig. 4.5 and in Fig. 4.1 coincide.

The KVL equations for the three meshes are based on Ohm's law for the proper reference configuration. We do not need the fourth (large) loop encompassing the entire circuit. For resistors that are shared by two adjacent meshes, we combine either the difference or the sum of the two adjacent mesh currents. The mesh equations become

$$\text{Mesh1}: \quad 1\ \text{k}\Omega \cdot (I_1 - I_3) + 1\ \text{k}\Omega \cdot I_1 + 1\ \text{k}\Omega \cdot (I_1 - I_2) = 0 \tag{4.9a}$$

$$\text{Mesh2}: \quad 3\ \text{k}\Omega \cdot (I_2 - I_3) + 1\ \text{k}\Omega \cdot (I_2 - I_1) + 1\ \text{k}\Omega \cdot I_2 = 0 \tag{4.9b}$$

$$\text{Mesh3}: \quad -5\text{V} + 1\ \text{k}\Omega \cdot (I_3 - I_1) + 3\ \text{k}\Omega \cdot (I_3 - I_2) = 0 \tag{4.9c}$$

We have arrived at a system of *three* equations for the three unknown mesh currents I_1, I_2, and I_3. After division by $1\ \text{k}\Omega$ and combining similar terms it simplifies to

$$\text{Mesh1}: \quad +3I_1 - I_2 - I_3 = 0 \tag{4.10a}$$

$$\text{Mesh2}: \quad -I_1 + 5I_2 - 3I_3 = 0 \tag{4.10b}$$

$$\text{Mesh3}: \quad -I_1 - 3I_2 + 4I_3 = 5\ \text{mA} \tag{4.10c}$$

In contrast, the nodal analysis applied to the same circuit requires only *two* equations for two unknown node voltages, see Example 4.1. The final solution is indeed the same. Thus, the nodal analysis is more beneficial for small networks when a voltage source or sources are present. If, however, instead of a voltage source, a current source were present in Fig. 4.5, the nodal analysis would require three equations. By contrast, the mesh analysis would require only two equations, because I_3 is defined by the current source. Reasoning like this gives us clues which method is most suitable. When mixed power supplies like voltage and current sources are involved, there is usually no real difference between the two methods. The choice often becomes a matter of taste.

Exercise 4.2: Determine the mesh currents for the circuit in Fig. 4.5.

Answer: $I_1 = +2.22$ mA, $I_2 = +2.78$ mA, $I_3 = +3.89$ mA.

4.1.5 Supermesh

Consider the circuit shown in Fig. 4.6. The straightforward mesh analysis should use KVL written for meshes 1 and 2. However, KVL cannot be formulated directly since we do not know the voltage across the current source. A solution is to combine meshes 1 and 2 into a *supermesh* and write KVL around its periphery. The mesh equations become

$$\text{Supermesh}: \quad 1\ \text{k}\Omega \cdot (I_1 - I_3) + 1\ \text{k}\Omega \cdot I_1 + 1\ \text{k}\Omega \cdot I_2 + 3\ \text{k}\Omega \cdot (I_2 - I_3) = 0 \tag{4.11a}$$

$$\text{Mesh3}: \quad -5\ \text{V} + 1\ \text{k}\Omega \cdot (I_3 - I_1) + 3\ \text{k}\Omega \cdot (I_3 - I_2) = 0 \tag{4.11b}$$

$$\text{KCL for the central branch of the source}: \quad I_1 - I_2 = 1\ \text{mA} \tag{4.11c}$$

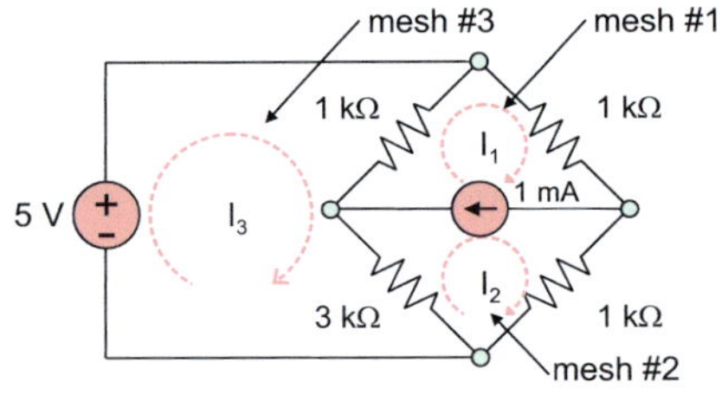

Fig. 4.6. Circuit solved with the supermesh method.

After division by 1 kΩ and combining similar terms, the system of Eq. (4.11) is simplified to

$$\begin{aligned} &+2I_1 + 4I_2 - 4I_3 = 0 \\ &-I_1 - 3I_2 + 4I_3 = 5\ \text{mA} \\ &+I_1 - I_2 = 1\ \text{mA} \end{aligned} \tag{4.12}$$

Exercise 4.3: Determine the mesh currents for the circuit in Fig. 4.6.

Answer: $I_1 = +3$ mA, $I_2 = +2$ mA, $I_3 = +3.5$ mA.

Example 4.4: Outline the solution approach for the circuit shown in Fig. 4.6 using the standard mesh analysis, without the supermesh concept.

Solution: The voltage across the current source is introduced as an extra unknown, V_x. Then, we write *three* KVL equations for *three* meshes in Fig. 4.6, which will contain *four* unknowns: I_1, I_2, I_3, and V_x. An extra equation is needed, which is KCL applied to the central branch: $I_1 - I_2 = 1$ mA. As a result, we need to solve a system of four simultaneous equations. This is considerably more work than in the previous case. This is why the supermesh approach is such a useful tool for the mesh analysis.

Section 4.2 Generator Theorems and Their Use

4.2.1 Equivalence of Active One-Port Networks: Method of Short/Open Circuit

In Chap. 3, we considered *passive linear networks* involving only resistors, which we have transformed into equivalent circuits. *Active linear networks*, which include sources and resistors simultaneously, can be subjected to similar transformations. We know that two electric single-port networks are equivalent when their terminal v-i characteristics are identical. For passive resistive networks studied in Chap. 3, we connected *arbitrary source(s)* across the network terminals and checked the resulting v-i characteristics. For active networks with sources and resistances, we can use the same method. Alternatively, we can connect arbitrary *resistance(s)* across the network terminals and check either the resulting voltage or current. A test resistance to be connected will be denoted here by R. If for two networks the voltages across the resistance R (or currents through it) are identical for *all* values of R, the networks are equivalents.

Method of Short/Open Circuit

However, testing all possible values of resistance R connected to terminals a and b of a network in Fig. 4.7 is not necessary. Note that an active linear network may ultimately have only *two* elements: a source and a resistance. To uniquely determine the *two* elements (their values), only *two* equations are necessary. It is therefore customary to check only *two* (limiting) values of the test resistance:

$$R \to \infty \quad \text{and} \quad R = 0 \tag{4.13}$$

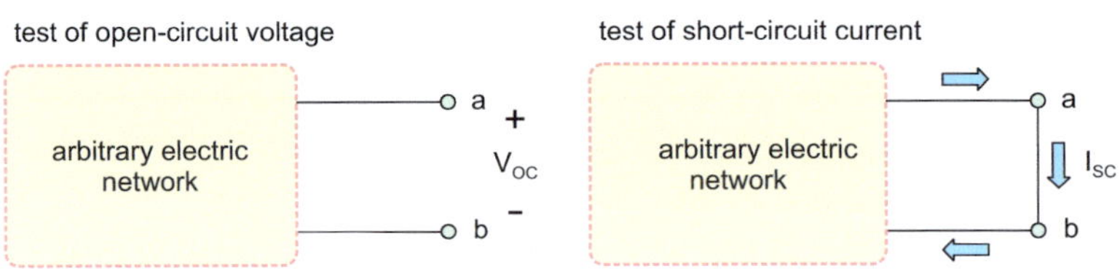

Fig. 4.7. Method of short/open circuit for an active, one-port network.

Conveniently, this corresponds to open- and short-circuit conditions. In the first case, the voltage between terminals *a* and *b* is the *open-circuit network voltage* V_{OC}. In the second case, the current flowing from terminal *a* to terminal *b* is the *short-circuit network current* I_{SC}. The pair V_{OC}, I_{SC} is the key for the *method of short/open circuit*. This method states that two active linear circuits are equivalent when their V_{OC} and I_{SC} coincide. Network equivalency relates not only to the linear active networks with two components, but, as will be shown soon, it is valid for *all* active linear networks.

4.2.2 Application Example: Reading and Using Data for Solar Panels

The method of short/open circuit is also very useful for *active nonlinear networks*, including nonlinear sources. An example is a *solar cell* or a combination thereof, a *solar panel*. Every solar panel has the measured data for V_{OC} and I_{SC} listed on its backside. The short-circuit current is simultaneously the *photocurrent* of the solar cell. Table 4.1 collects this data for common *crystalline silicon* (or c-Si) solar panels. It is organized in such way that V_{OC} is given per one cell in the panel and I_{SC} is given in terms of photocurrent density, J_P, per unit cell area. The cells in the panel are connected *in series*.

Table 4.1. Manufacturers' specified parameters for different c-Si solar panels from five different manufacturers (1–230 W output power range, circa 2012–2015). The cell area is either measured directly or extracted from the datasheet.

Solar panel	Cells, N	V_{OC}/N, V	Cell area A, cm^2	$J_P = I_{SC}/A$ A/cm^2
1 W BSPI-12 power up c-Si panel	36	0.59	2.36	0.030
10 W BSP-1012 power up c-Si panel	36	0.59	~22.0	0.030
65 W BSP-1012 power up c-Si panel	36	0.61	121.7	0.032
230 W sharp ND-U230C1 c-Si panel	60	0.62	241.0	0.034
175 W BP solar SX3175 c-Si panel	72	0.61	156.25	0.033
6 W global solar GSE-6 c-Si panel	44	0.52	16.6	0.027
200 W GE Energy GEPVp-200 c-Si panel	54	0.61	249.3	0.032
Average		**0.593**		**0.0311**

Table 4.1 demonstrates that c-Si solar cells have approximately the same *open-circuit voltage* of 0.6 V per cell. The open-circuit voltage does not depend on the area of the cell. The *short-circuit photocurrent density* is also approximately the same for c-Si solar cells from different manufacturers. On average, it is given by $J_P = 0.03$ A/cm^2. These values corresponds to an incident light intensity of 1000 W/m^2 at $T = 25^\circ$C. The photocurrent density does not depend on the area of the cell. Obviously, the total photocurrent does.

Exercise 4.4: A c-Si solar panel (or *solar module*) has the open-circuit voltage of 23.4 V? How many individual solar cells does it have?

Answer: Approximately 39.

Exercise 4.5: A c-Si solar panel is needed with the open-circuit voltage of 12 V and a short-circuit current of 3 A. Design the panel: give the number of cells to be connected in series and the required unit cell area.

Answer: Twenty cells with the area of 100 cm^2 (10 × 10 cm) each.

4.2.3 Source Transformation Theorem

The most fundamental transformation of active linear networks is the source transformation theorem. The *source transformation theorem* involves the substitution of an independent voltage source V_T in series with resistance R_T for an independent current source I_N with resistance R_N and vice versa, as seen in Fig. 4.8a, b. The meaning of indexes N and T will become apparent soon. The *identical* theorem applies to the dependent sources shown in Fig. 4.8c, d.

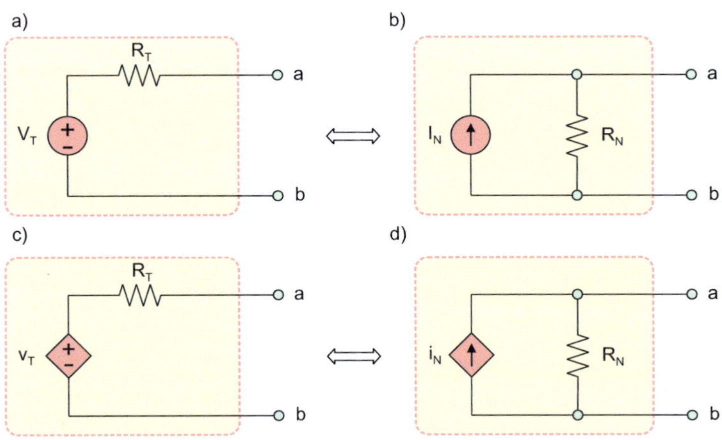

Fig. 4.8. Transformation of dependent and independent sources.

Let us prove this theorem by establishing the circuit equivalence. The pair V_{OC}, I_{SC} is to be found for every network. For the two networks in Fig. 4.8a, b, we have

$$V_{OC} = V_T, \quad I_{SC} = \frac{V_T}{R_T} \tag{4.14a}$$

$$V_{OC} = R_N I_N, \quad I_{SC} = I_N \tag{4.14b}$$

Equations (4.14) have a unique solution in the form of the source transformation theorem

$$R_N = R_T, \quad I_N = \frac{V_T}{R_T} \tag{4.15}$$

If Eqs. (4.15) are satisfied, both networks in Fig. 4.8a, b have equal V_{OC} and I_{SC}. This ensures that their entire *υ-i* characteristics are also the same. To confirm this fact, an arbitrary resistance R could be connected across the port. The resulting voltages is found directly by solving the voltage divider and the current divider circuits, respectively. Both voltages are equal to $V_T R/(R+R_T)$. Thus, the source transformation theorem is proved.

Exercise 4.6: A network has a 10 V voltage source in series with a 20 Ω resistance. It is replaced by a current source I_N in shunt with resistance R_N. Find I_N and R_N.

Answer: $I_N = 0.5$ A, $R_N = 20$ Ω.

Exercise 4.7: A linear active circuit records an open-circuit voltage of 5 V and a short-circuit current of 1 mA. Determine its equivalents in the form of a voltage source in series with a resistance and in the form of a current source in shunt with a resistance.

Answer: $V_T = 5$ V, $R_T = 5$ kΩ and $I_N = 1$ mA, $R_N = 5$ kΩ.

Often, the source transformation theorem allows us to simplify the circuit analysis through simple network manipulations.

Example 4.5: Find current I_1 in the circuit shown in Fig. 4.9a.

Solution: The circuit can be solved using the superposition theorem. Another way is to employ the source transformation theorem. The corresponding steps are outlined in Fig. 4.9b, c. We use the source transformation three times and end up with the parallel combination of two current sources and three resistances. The three resistances in parallel are equivalent to the 0.75 kΩ resistance; the voltage across every element in parallel is then 0.75 kΩ × 2 mA = 1.5 V. Therefore, $I_1 = 0.75$ mA.

Example 4.6: The circuit in Fig. 4.10a includes a current-controlled voltage source with the strength of $4000 i_x$ [V]. Find current i_x by using source transformation.

Solution: The corresponding circuit transformation is shown in Fig. 4.10b. The circuit with the current-controlled current source in Fig. 4.10b is solved using KCL and KVL. KCL applied to the bottom node states that the current of 3 mA+$3i_x$ flows through the rightmost 1 kΩ resistance (directed down). Since, by KVL, the voltages across both resistances must be equal, one has

Example 4.6 (cont.):

$$3\,\text{mA} + 3i_x = i_x \Rightarrow i_x = -1.5\ \text{mA} \tag{4.16}$$

Alternatively, one could convert the independent current source to an independent voltage source. However, this method would hide i_x.

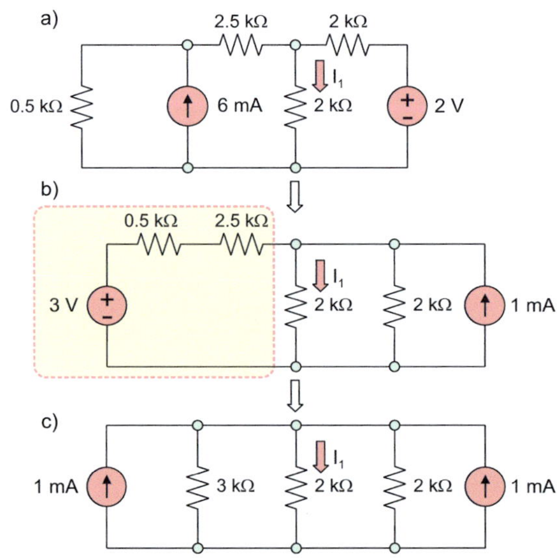

Fig. 4.9. Circuit modifications using the source transformation theorem.

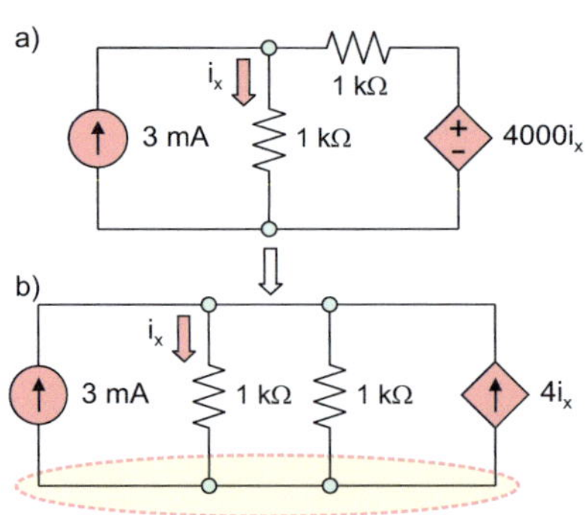

Fig. 4.10. Using source transformation for a circuit with dependent sources.

4.2.4 Thévenin's and Norton's Theorems: Proof Without Dependent Sources

The origin of *Thévenin's theorem* is due to Léon Charles Thévenin, a French engineer (1857–1926). The theorem is illustrated in Fig. 4.11a, b and can be expressed in the following form:

1. Any linear network with independent voltage and current sources, dependent linear sources, and resistances as shown in Fig. 4.11a, can be replaced by a simple equivalent network: a voltage source V_T in series with a resistance R_T.
2. The equivalent network in Fig. 4.11b is called the *Thévenin equivalent.*
3. Voltage V_T is the open-circuit voltage V_{OC} of the original network (with no current flowing into either terminal).
4. When dependent sources are *absent*, Thévenin resistance R_T is the equivalent resistance R_{eq} of the original network with *all* independent sources *turned off* (voltage sources are replaced by short circuits, current sources by open circuits).
5. When *both* dependent and independent sources are present, the independent sources are not turned off. Resistance R_T is given by $R_T = V_{OC}/I_{SC}$, where I_{SC} is the short-circuit current of the original network.
6. When *only* the dependent sources are present, a current (or voltage) source is connected to the network terminals. Thévenin or equivalent resistance is given by $R_T = V/I$ where I is the source current and V is the voltage across the source. Thévenin voltage is strictly speaking not defined in this case.

The *Norton's theorem* is *dual* to the Thévenin's theorem. It was named in honor of Edward L. Norton (1898–1983), an engineer at Bell Labs in New Jersey.[2] Norton's theorem is illustrated in Fig. 4.11c, d. The equivalent circuit (*Norton's equivalent*) is now the current source in shunt with the resistance as shown in Fig. 4.11d. Since the equivalence of both networks in Fig. 4.11b, d has already been established, the Norton's theorem will follow from the Thévenin's theorem and vice versa.

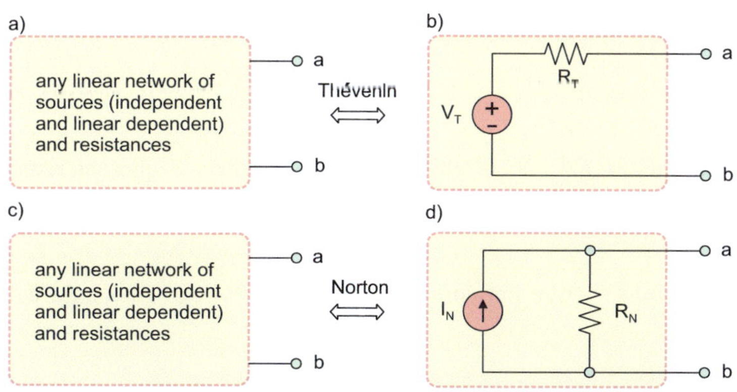

Fig. 4.11. Thévenin's and Norton's theorems: replacing linear active circuits by its Thévenin and Norton equivalents.

[2] The first publication that discusses this equivalent circuit concept is actually due to Hans F. Mayer (1895–1980) who made the discovery in 1926 while a researcher at Siemens Company.

Proof of Thévenin's Theorem for Active Networks Without Dependent Sources

The proof is based on the assumption of circuit linearity. The *υ-i* characteristics of both networks is established using a current source of strength I connected as seen in Fig. 4.12.

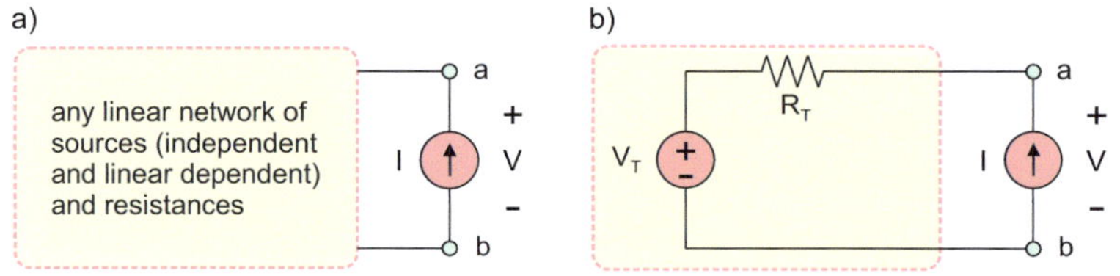

Fig. 4.12. Derivation of Thévenin's theorem by establishing the *υ-i* characteristics.

Since the entire circuit is assumed linear, the *υ-i* characteristic of the current source in Fig. 4.12a must have the form of *linear function*,

$$V = AI + B \tag{4.17}$$

where V is the voltage across the current source. A and B are some "constant" coefficients, which do *not* depend on I, but do depend on the network parameters. Our goal is to find A and B, respectively. First, we check the value $I = 0$ when the external current source is turned off, i.e., replaced by an open circuit. From Eq. (4.17), voltage V equals B. Therefore, it is equals V_{OC} of the original network. Therefore,

$$B = V_{OC} \tag{4.18}$$

Next, let us turn off all internal sources. The network becomes an equivalent resistance R_{eq}. The constant B (its open-circuit voltage) is zero. Equation (4.17) therefore yields $V = AI$ for any value of I. On the other hand, for the current source I connected to the resistance R_{eq}, it must be $V = R_{eq}I$. Comparing the two expressions we obtain,

$$A = R_{eq} \tag{4.19}$$

The simpler network in Fig. 4.12b is also described by the *υ-i* characteristic in the form of Eq. (4.17). In this case, $B = V_{OC} = V_T$, $A = R_{eq} = R_T$. We finally compare the two *υ-i* characteristics,

$$V = R_{eq}I + V_{OC} \text{ for linear active network;} \quad V = R_T I + V_T \text{ for Thé venin equivalent} \tag{4.20}$$

and establish the Thévenin's theorem. A test voltage source could be used in place of the current source in Fig. 4.12, with the same result obtained. The physical background of Thévenin's theorem is thus the fact that the terminal response of any linear network has a linear *υ-i* characteristic—a linear function with only *two* independent coefficients: A and B. A simpler network with exactly *two* independent parameters, V_T and R_T, allows us to model this response.

Equivalence of Arbitrary Linear Networks with Identical V_{OC}, I_{SC}

On the one hand, linear active networks with only two elements (a source and a resistance) are equivalent when their V_{OC}, I_{SC} are identical. On the other hand, any active linear network is equivalent to a linear network with only two elements. Therefore, we conclude that two arbitrary linear networks are equivalent when their V_{OC} and I_{SC} coincide.

4.2.5 Finding Thévenin and Norton Equivalents and Using Them for Circuit Solution

Along with the superposition and source transformation theorems, Thévenin's and Norton's theorems (or their outcome, Thévenin and Norton *equivalent circuits* or simply *equivalents*) are powerful tools to solve more complicated electric circuits. The idea is to replace a larger part of the circuit by its equivalent and, subsequently, find the required voltage/current/power for a remaining small circuit part. However, a bit of practice is required to implement such a solution. We will first establish Thévenin (and implicitly Norton) equivalents for some "large" circuits—one-port networks—shown in Figs. 4.13 and 4.14, and then implement this idea using a few examples.

Example 4.7: Establish Thévenin equivalent circuits in the form of Fig. 4.11b for the one-port networks shown in Fig. 4.13. This first network is a battery bank or a network consisting of series-connected practical voltage sources, with $R_{B1} = R_{B2} = R_{B3} = 1\ \Omega$, $V_{B1} = V_{B2} = V_{B3} = 9\ \text{V}$.

Solution:

Case (a): The Thévenin voltage is the open-circuit voltage between terminals *a* and *b*. Since no current flows through any of the resistors, KVL gives $V_T = V_{ab} = 27\ \text{V}$. The Thévenin resistance is found by turning off the voltage sources, i.e., replacing them by short circuits. Therefore, $R_T = R_{ab} = 3\ \Omega$.

Case (b): The Thévenin voltage is the open-circuit voltage between terminals *a* and *b*. Current cannot flow through open terminals *a* and *b*, but it still can flow within the loop containing the source and the two resistors. Therefore, $V_T = V_{ab} = V_{3\text{k}\Omega} = 9\ \text{V}$. The Thévenin resistance is found by turning off the voltage source (replacing it by a wire). Since the two resistors are combined in parallel, $R_T = R_{ab} = 1.2\ \text{k}\Omega$.

Case (c): Although the source transformation theorem immediately gives us $V_T = 3\ \text{V}$, $R_T = 3\ \text{k}\Omega$, we would like to solve the problem directly. The Thévenin voltage is the open-circuit voltage between terminals *a* and *b*. Current cannot flow through open terminals *a* and *b*, but it still can flow within the loop containing the current source and the resistor. Therefore, $V_T = V_{ab} = V_{3\text{k}\Omega} = 3\ \text{V}$. The Thévenin resistance is found by turning off the current source (replacing it by an open circuit). This gives $R_T = R_{ab} = 3\ \text{k}\Omega$.

Example 4.7 (cont.):

Case (d): The Thévenin voltage is the open-circuit voltage between terminals a and b. Current cannot flow through open terminals a and b and through a 0.2 kΩ resistor, but it still can flow within the loop containing the voltage source and both 2 kΩ and 3 kΩ resistors. Therefore, $V_T = V_{ab} = V_{3k\Omega} = 9$ V. The Thévenin resistance is found by turning off the voltage source (replacing it by a short circuit). This gives us $R_T = R_{ab} = 1.2\ \text{k}\Omega + 0.2\ \text{k}\Omega = 1.4\ \text{k}\Omega$.

Case (e): First, note that the 1 kΩ resistor placed in shunt with the voltage source has no effect on the solution whatsoever; it is either shorted out by a wire (when finding Thévenin resistance) or simply possesses the supply voltage (when finding Thévenin voltage). In practice, such a combination only dissipates power and should be avoided when possible. The rest of the solution follows case (b) with interchanged resistors and yields $V_T = V_{ab} = V_{2k\Omega} = 6$ V, $R_T = R_{ab} = 1.2$ kΩ.

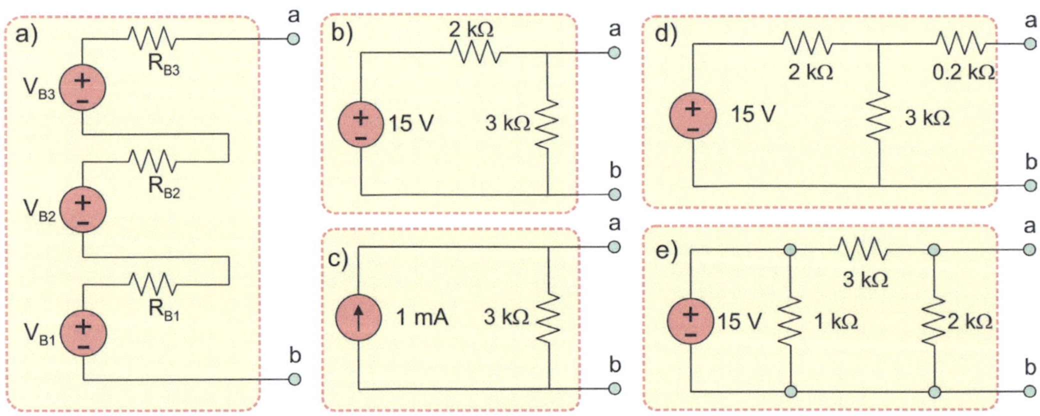

Fig. 4.13. Active linear networks to be converted to their Thévenin equivalents as given in the form of Fig. 4.11b.

Exercise 4.8: Along with the combination of a voltage source in parallel with a resistance as in Fig. 4.13e, give another example of a "meaningless" circuit configuration.

Answer: A current source in series with a resistance.

Example 4.8: A two-terminal network shown in Fig. 4.14a is a two-bit *R*-2*R ladder network* used for digital-to-analog conversion. Express

(i) Thévenin (or equivalent) voltage V_T
(ii) Thévenin (or equivalent) resistance R_T

in terms of (digital) voltages D_0,D_1, and resistance R.

Solution: One way to solve the problem is to find V_T and R_T directly from the circuit in Fig. 4.14a. While the solution for R_T is straightforward, finding V_T requires more work. Yet another method is to apply Thévenin equivalent to the leftmost section of the ladder network first. The result is the circuit shown in Fig. 4.14b. The final Thévenin equivalent has the form $R_T = R,\ \ V_T = \frac{D_1}{2} + \frac{D_0}{4}$.

This method may be applied to ladder networks with multiple sections.

Exercise 4.9: Repeat the previous example for the ladder shown in Fig. 4.14c.

Answer: $R_T = R,\ \ V_T = \frac{D_2}{2} + \frac{D_1}{4} + \frac{D_0}{8}$.

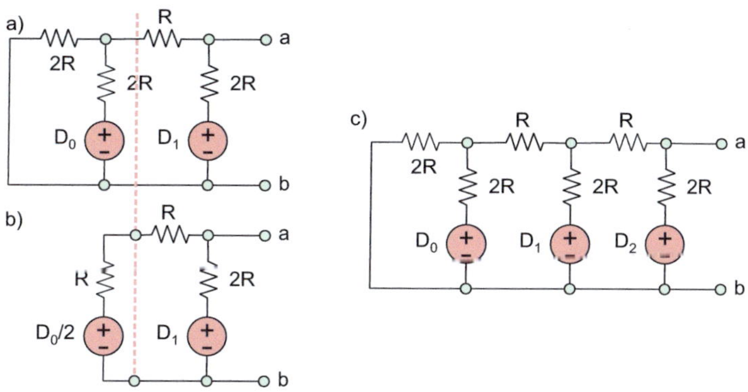

Fig. 4.14. Two-bit and three-bit *R*-2*R* ladder networks.

Using Thévenin and Norton Equivalents for Circuit Solution

In this section, we will use the method of Thévenin equivalent and conduct a circuit solution. Note that, in contrast to a nodal analysis, this is a semi-analytical method which may not always work out for every network. Consider two circuits shown in Fig. 4.15a, b. In either case, the goal is to find the current *I*. We will implement the idea discussed above, i.e., replace a larger part (or parts) of the circuit by its equivalent and then solve for the current in the remaining part.

Example 4.9: For the two circuits shown in Fig. 4.15a, b, find the current *I*.

Solution:

Case (a): One way to solve the problem is to apply the current division principle multiple times. Another method is to replace the circuit to the left of the dashed line (also called "reference plane" in more advanced applications) by its Thévenin equivalent as shown in the figure. The remaining solution becomes trivial and yields a current value of 0.25 mA.

Case (b): One way to solve the problem is to apply the superposition theorem with or without source transformation and then use the current division again. Yet another method is to replace the circuits to the left/right of two dashed lines (two "reference planes") by its Thévenin equivalents as shown in the same figure. The remaining solution is straightforward; it yields the current value of −3.214 mA.

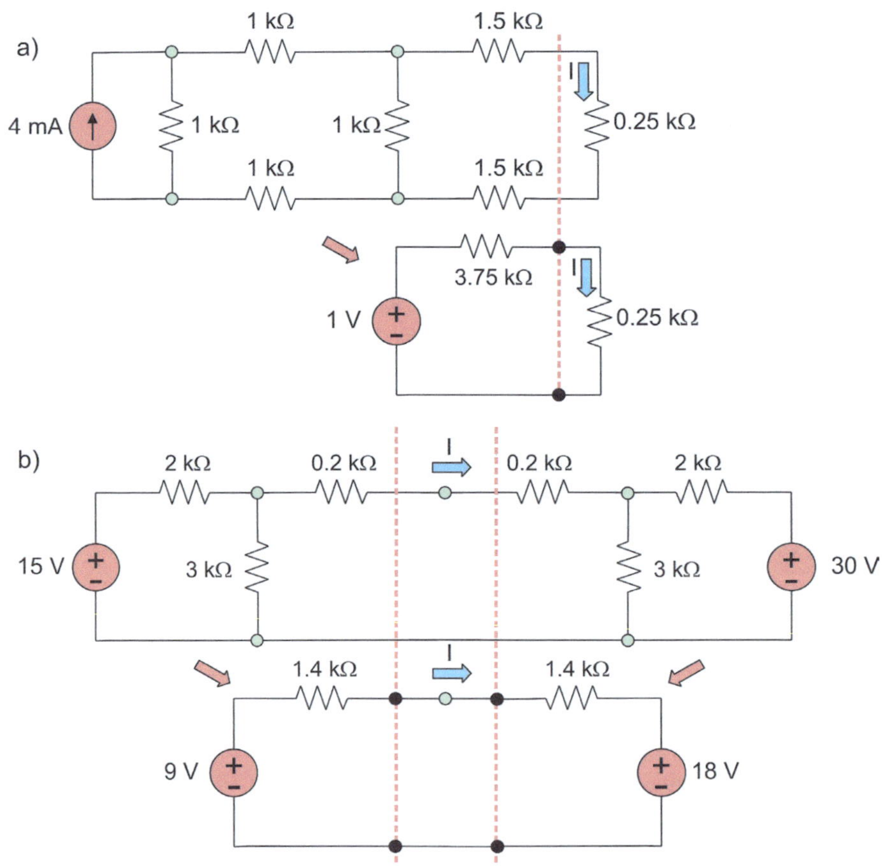

Fig. 4.15. Two circuits solved using the method of Thévenin equivalent.

Thévenin and Norton Equivalents with Dependent Sources

Thévenin and Norton equivalents with dependent sources (amplifiers) are a fascinating subject that can lead to unexpected concepts such as the negative equivalent resistance.

Example 4.10: Find Thévenin and Norton equivalent circuits for the network in Fig. 4.16a.

Solution: The network includes a voltage-controlled voltage source. Therefore, its analysis should be performed in a general form, by finding the pair V_{OC}, I_{SC}. The short-circuit current I_{SC} is found straightforwardly. Since the rightmost resistance is shorted out, $\upsilon_x = 0$, and $I_{SC} = 10$ mA. To find the open-circuit voltage, which is equal to υ_x, we use the source transformation theorem and arrive at the equivalent circuit in Fig. 4.16b. Next, we solve this circuit. Employing KVL, the voltage across the leftmost resistance is equal to $10\ \mathrm{V} - 4\upsilon_x$. By KCL, the currents through both resistances must be the same. Since the resistors are equal, we obtain the equality $10\ \mathrm{V} - 4\upsilon_x = \upsilon_x$ so that $\upsilon_x = 2\ \mathrm{V}$. The open-circuit voltage has the same value. The Thévenin and Norton equivalents are

$$V_T = V_{OC} = 2\ \mathrm{V},\quad R_T = \frac{V_{OC}}{I_{SC}} = 200\ \Omega \tag{4.21a}$$

$$I_N = I_{SC} = 10\ \mathrm{mA},\quad R_N = \frac{V_{OC}}{I_{SC}} = 200\ \Omega \tag{4.21b}$$

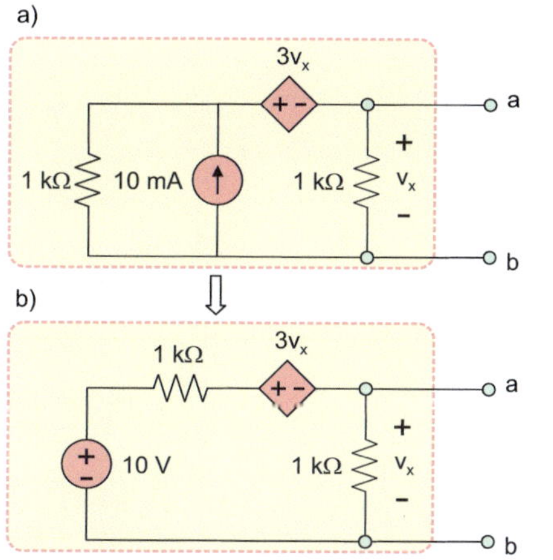

Fig. 4.16. A network with dependent sources to be converted to its equivalent forms.

4.2.6 Application Example: Generating Negative Equivalent Resistance

Consider a circuit shown in Fig. 4.17a. It includes only a dependent source: the voltage-controlled voltage source; it does not have independent sources. Therefore, the circuit analysis has to be done by connecting a current (or voltage) source between terminals *a* and *b* as shown in Fig. 4.16b. The quantity of interest is the Thévenin (or equivalent) resistance. The KVL for the circuit in Fig. 4.17b gives $-Av_x - RI + v_x = 0$. Therefore, by definition,

$$R_T \equiv \frac{v_x}{I} = \frac{R}{1-A} \tag{4.22}$$

As long as the open-circuit voltage gain of the dependent source, A, is greater than one, Eq. (4.22) states the *negative equivalent or Thévenin resistance*. Physically, this means that the Thévenin equivalent circuit is delivering power instead of absorbing it.

Construction and the Use of Negative Equivalent Resistance

A circuit block, which is equivalent to the negative resistance, may be constructed using the operational amplifier studied in the next chapter. This block may be used for different purposes including signal generation. The difference between the negative resistance and the power source is that the negative resistance may supply power of any type (DC, AC, or an arbitrary waveform), i.e., support the *self-oscillating circuits*. Figure 4.17c summarizes Thévenin resistances generated by their basic networks with only dependent sources. The same method of analysis (simultaneous use of KCL and KVL) has been applied to every network. Although all of the networks in principle generate the negative equivalent resistance values, the realization of some particular circuits may be difficult.

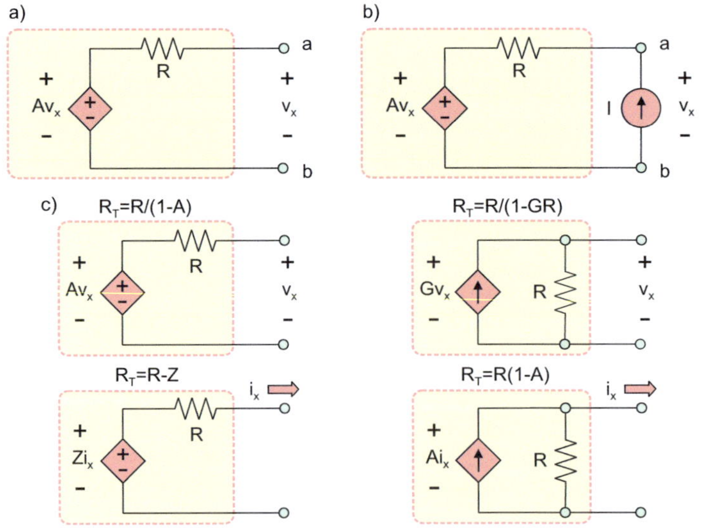

Fig. 4.17. Thévenin equivalent resistances for basic networks with dependent sources.

4.2.7 Short Summary of Circuit Analysis Methods

In summary, a linear circuit can be solved using any of the methods studied in this and in the previous chapters:

- Superposition theorem
- Nodal/mesh analysis
- Source transformation theorem
- Method of Thévenin and/or Norton equivalent circuit(s)

or a combination of those. While the nodal/mesh analysis is always applicable, other methods may turn out to be even more useful since they often provide a physical insight into the circuit behavior.

Exercise 4.10: How could you find the open-circuit voltage V_{ab} in Fig. 4.14a?

Answer:

A. When the superposition theorem is applied, shorting out D_0 gives $V_{ab} = D_1/2$. Shorting out D_1 gives $V_{ab} = D_0/4$.
B. When the nodal analysis is applied, we ground negative terminals of both sources and find the unknown voltage of the upper left node via KCL. Only one equation needs to be solved. This is perhaps the simplest solution method.
C. The source transformation theorem cannot easily be applied.
D. The method of Thévenin equivalent circuits is described in example 4.8.

Exercise 4.11: How could you find current I for the circuit in Fig. 4.15b?

Answer:

A. When the superposition theorem is applied, shorting out the 30 V source gives $I_1 = 3.214$ mA. Shorting out the 15 V source gives $I_2 = -2I_1$(due to symmetry!). The resulting current is $I = I_1 + I_2 = -3.214$ mA.
B. When the nodal analysis is applied, we ground the entire bottom node and combine both 0.2 kΩ resistors in series. There are two nontrivial nodes on top and two nodal equations. After the solution is obtained ($V_1 = 12.86$ V for the left node 1 and $V_2 = 14.14$ V for the right node 2), current I is found as the difference of two nodal voltages divided by 0.4 kΩ.
C. The source transformation theorem followed by superposition could be applied too.
D. The method of Thévenin equivalent circuits is described in example 4.9.

Section 4.3 Power Transfer

4.3.1 Maximum Power Transfer

The *principle of maximum power transfer* from a source to a load will now be quantified. This principle is also known as a *maximum power theorem*. The circuit under study is shown in Fig. 4.18. It involves an arbitrary linear source (a battery, a generator, etc.), which is represented by its Thévenin equivalent, and a load, which is characterized by its equivalent resistance R_L. All other load parameters (dynamic, mechanical, and thermal) are implicitly included into the load's resistance.

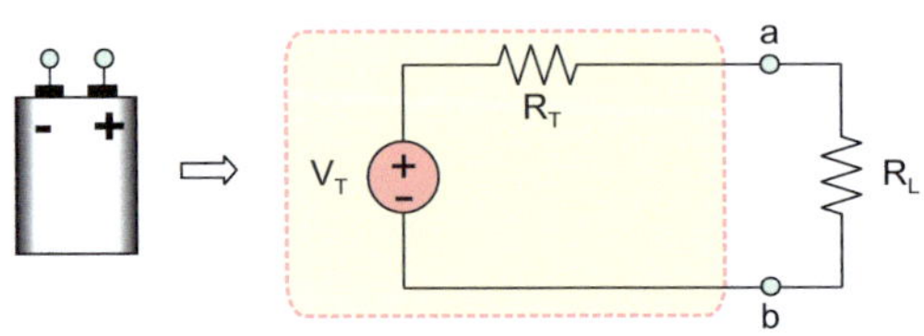

Fig. 4.18. A battery (or another practical voltage source) connected to a load.

The key question you have to ask yourself is this: For a *given* ideal voltage source V_T and a *given* internal resistance R_T, can the electric power delivered to the load be maximized, and at which value of R_L does the maximum occur? The answer is found by solving the circuit in Fig. 4.18. First, the current is determined from the given voltage source V_T and the total resistance using the series equivalent,

$$I = \frac{V_T}{R_T + R_L} \tag{4.23}$$

This allows us to compute the power at the load based on

$$P_L = R_L I^2 = \frac{R_L V_T^2}{(R_T + R_L)^2} \tag{4.24}$$

When V_T and R_T are fixed, the magnitude of the load resistance determines the delivered power P_L. This power tends to zero when either $R_L \to 0$ or $R_L \to \infty$; moreover, it is always positive. Therefore, according to Rolle's theorem of calculus, the power must have a maximum at a certain value of R_L. For example, Fig. 4.19 shows a plot of the load power as a function of R_L for $V_T = 9$ V and $R_T = 5$ Ω.

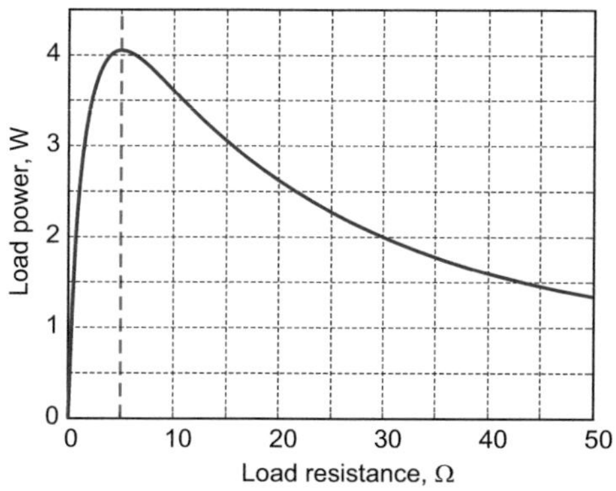

Fig. 4.19. Load power as a function of the load resistance for fixed $V_T = 9$ V, $R_T = 5$ Ω.

We will find the maximum of the load power analytically. We treat P_L in Eq. (4.24) as a function of R_L, i.e., $P_L = P_L(R_L)$. It is known that a function has a maximum when its first derivative is set to zero. Consequently, differentiating P_L with respect to R_L gives

$$\frac{dP_L}{dR_L} = V_T^2 \left[\frac{1}{(R_T + R_L)^2} - 2\frac{R_L}{(R_T + R_L)^3} \right] = V_T^2 \left[\frac{R_T - R_L}{(R_T + R_L)^3} \right] = 0 \tag{4.25}$$

The necessary and sufficient condition for Eq. (4.25) to hold is

$$R_L = R_T \Rightarrow P_L = 0.25 V_T^2 / R_T \tag{4.26}$$

This result is of great practical value despite, or maybe because of, its simplicity. The maximum output power is achieved when the load resistance is equal to the internal resistance of the power source. In other words, the load is *matched* to the source; it is called the *matched load*. In power engineering and in RF and microwave engineering, the problem of load matching is very important. However, it must be clearly stated that *no more* than 50% of the *total* circuit power can be extracted even in the best case. This statement makes sense if we again examine the circuit in Fig. 4.18 and assume two equal resistances. The power is divided equally; half of the total power is spent to heat up the power source. The power maximum in Fig. 4.19 is relatively flat over the domain $R_L > R_T$; however, the power drops sharply when $R_L < R_T$. This last condition should be avoided if at all possible.

Example 4.11: An audio amplifier produces an rms output of 20 V. The amplifier's output resistance is rated at 4 Ω. You are given four 4 Ω speakers. How should you connect the speakers for the maximum acoustic power—in series, parallel, or a single speaker only?

Solution: The rms voltage simply means the equivalent DC voltage that provides the same power to the load as the average power of the primary AC voltage. Hence, the sophisticated AC audio amplifier circuit is essentially replaced by its DC Thévenin equivalent with $V_T = 20$ V and $R_T = 4\ \Omega$. Similarly, the dynamic speakers are replaced by a DC load with $R_L = 16\ \Omega$ if connected in series combination, $R_L = 1\ \Omega$ if connected in parallel, or with $R_L = 4\ \Omega$ if only a single speaker is employed. The output (audio) powers are as follows:

$$P_L = \frac{1 \times 400}{(4+1)^2} = 16\ \text{W} \quad \text{four speakers in parallel} \tag{4.27a}$$

$$P_L = \frac{4 \times 400}{(4+4)^2} = 25\ \text{W} \quad \text{single speaker} \tag{4.27b}$$

$$P_L = \frac{16 \times 400}{(4+16)^2} = 16\ \text{W} \quad \text{four speakers in series} \tag{4.27c}$$

The best (loudest) choice would be surprisingly one single speaker.

4.3.2 Maximum Power Efficiency

A power analysis would be incomplete without discussing the efficiency of the power transfer. Consider an electric boat driven by a marine battery. The optimization of the battery-motor system for maximum power transfer implies that we will move fast, but perhaps not very far. Another optimization is possible for maximum power efficiency. In this case, we could tolerate a smaller speed in order to travel a longer distance. The circuit to be analyzed is again the circuit in Fig. 4.18. The useful power delivered to the load is given by Eq. (4.24). The total power delivered by the source is

$$P = (R_T + R_L)I^2 = \frac{V_T^2}{R_T + R_L} \tag{4.28}$$

The *power efficiency E* is defined as the ratio of the useful power to the total power,

$$E = \frac{P_L}{P} = (R_T + R_L)I^2 = \frac{R_L}{R_T + R_L} \tag{4.29}$$

Thus, the power efficiency is a simple function of the load resistance and the source resistance. It does not depend on the source voltage. The efficiency is zero when the load resistance is zero. It monotonically increases and approaches maximum (the maximum value is unity, which corresponds to an efficiency of 100%) when the load resistance becomes large enough when compared to the source resistance. The *maximum power efficiency* is thus achieved when $R_L >> R_T$. For example, Fig. 4.20 augments the load power graph from Fig. 4.19 with the corresponding

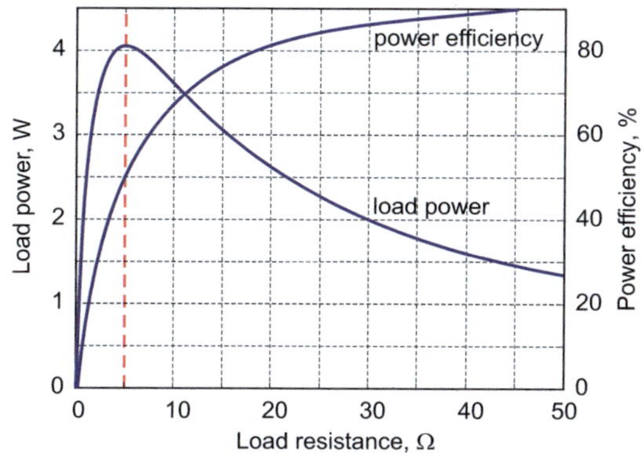

Fig. 4.20. Load power and power efficiency for fixed $V_T = 9V$, $R_T = 5\ \Omega$.

efficiency curve.

Example 4.12: A battery with the stored energy of $E_B = 0.1$ MJ, and $V_T = 12$ V, $R_T = 5\ \Omega$, delivers its entire energy during the time period $0 \le t \le T$ and discharges with a constant output voltage/current. Two loads are used: $R_L = 5\ \Omega$ and $R_L = 50\ \Omega$. Determine discharge time T and total energy delivered to the load in each case.

Solution: The discharge time, $T = E_B/P$, is determined first when the total power P follows based on Eq. (4.28). Assuming constant battery discharge rate, we obtain

$$T \approx 1.9 \text{ hours for the } 5\ \Omega \text{ load}; \quad T \approx 10.6 \text{ hours for the } 50\ \Omega \text{ load} \tag{4.30}$$

The total energy delivered to the load, $E = TP_L$, in each case is given by

$$E_{5\Omega} = 50 \text{ kJ for the } 5\ \Omega \text{ load}; \quad E_{50\Omega} = 91 \text{ kJ for the } 50\ \Omega \text{ load} \tag{4.31}$$

Thus, the total energy extracted from the battery is nearly *twice* as high in the second case. However, it takes about *five* times longer to extract this energy.

4.3.3 Application Example: Power Radiated by a Transmitting Antenna

A transmitting antenna in a radio handset features a monopole antenna. It is connected to a source that has the same basic form as in Fig. 4.18 but with an AC generator instead of the DC source and with an internal (generator) resistance of 50 Ω. The antenna represents a load with a "radiation" resistance of 50 Ω. This resistance describes the power loss in terms of the electromagnetic radiation from the antenna. Thus, the antenna, if properly matched to the power source, will radiate 50% of the total power as electromagnetic waves into space. Now, an inexperienced electrical engineer decides to "modify" the handset by cutting the monopole antenna to only one third of its length. In this case, the antenna's radiation resistance is reduced to one ninth of its original value. How does this affect the radiated signal power? To answer this question, we find the instantaneous load power, which according to Eq. (4.24), is

$$P_L(t) = \frac{R_L V_T^2(t)}{(R_T + R_L)^2} \tag{4.32}$$

The ratio of the power levels for the two antenna configurations does not depend on time and we obtain

$$\frac{P_{L-\text{short}}}{P_{L-\text{original}}} = \frac{50/9}{(50+50/9)^2} / \frac{50}{(50+50)^2} = \frac{0.0018}{0.0050} = 0.36 \tag{4.33}$$

Thus, for the shorter antenna, we will only achieve about 36% of the radiated power compared to the original handset. In practice, this estimate becomes even lower due to the appearance of a very significant antenna capacitance.

4.3.4 Application Example: Maximum Power Extraction from Solar Panel

Every solar panel reports the measured data for V_{OC} and I_{SC} listed on its backside. For linear circuits, $V_T = V_{OC}$, $R_T = V_{OC}/I_{SC}$. If the solar panel were a linear circuit, the maximum extracted power would be exactly equal to $0.25 V_{OC} I_{CS}$ according to Eq. (4.26). Fortunately, this is not the case. The maximum extracted power is significantly greater than this value. However, it is still less than the "best" possible value of $V_{OC} I_{CS}$. To quantify the maximum power output, every solar panel has another set of measured data, V_{MP} and I_{MP}, also listed on its backside. V_{MP} stands for *maximum power load voltage*, and I_{MP} stands for the *maximum power load current*. The maximum extracted power is the product $V_{MP} I_{MP}$, which is always less than $V_{OC} I_{CS}$. The ratio of two powers

$$F = \frac{V_{MP} I_{MP}}{V_{OC} I_{SC}} < 1 \tag{4.34}$$

is known as the *fill factor of the solar panel* (or solar module). We will derive the theoretical value of the fill factor in the next section. Table 4.2 lists some experimental data for crystalline (c-Si) solar panels. The experimental fill factor not only accounts for the nonlinear physics of the cell but it also includes some resistive losses in an individual cell and in the entire solar module. Equation (4.34) approximates the *fill factor of a solar cell* as well.

Exercise 4.12: A REC SCM220 220 Watt c-Si solar panel has the following readings on its back plane: a short circuit current of 8.20 A, an open circuit voltage V_{OC} of 36.0 V, a maximum power voltage of 28.7 V, and a maximum power current of 7.70 A. Estimate the load resistance required for the maximum power transfer to the load.

Answer: 3.72 Ω.

Table 4.2. Manufacturer provided circuit parameters for 12 different c-Si solar panels from 5 different manufacturers (1 W to 230 W output power range, circa 2012–2015).

Solar panel	Cells (series)	V_{MP}/V_{OC}	I_{MP}/I_{SC}	F	\$/Watt (2012)
1 W BSPI-12 power up c-Si panel	36	0.81	0.86	0.70	24.00
10 W BSP-1012 power up c-Si panel	36	0.81	0.88	0.71	8.80
65 W BSP-1012 power up c-Si panel	36	0.80	0.94	0.75	6.35
80 W sharp NE-80EJEA c-Si panel	36	0.80	0.88	0.70	4.29
176 W sharp ND-176U1Y c-Si panel	48	0.80	0.91	0.73	4.68
230 W sharp ND- U230C1 c-Si panel	60	0.82	0.92	0.75	3.51
5 W BP solar SX-305 M c-Si panel	36	0.80	0.90	0.72	15.00
20 W BP solar SX-320 M c-Si panel	36	0.80	0.92	0.74	8.30
175 W BP solar SX3175N c-Si panel	72	0.83	0.92	0.76	4.19
65 W Kyocera KC65T c-Si panel	36	0.80	0.94	0.75	5.22
165 W SolarWorld SW230 c-Si panel	72	0.80	0.90	0.72	4.72
230 W SolarWorld SW230 c-Si panel	60	0.80	0.92	0.74	3.18
Average		**0.806**	**0.908**	**0.73**	

Table 4.2 demonstrates that different c-Si solar cells have approximately the *same* values of V_{MP}/V_{OC}, I_{MP}/I_{SC} and of the fill factor. Using the photocurrent density estimate and the open-circuit cell voltage estimate given at the beginning of this section, we can assume the approximate generic values for c-Si solar *cells* at normal irradiation conditions:

$$V_{MP} = 0.8V_{OC}, \quad I_{MP} = 0.9I_{SC}, \quad F = 0.72, \quad V_{OC} = 0.6\ \text{V(cell)}, \quad J_P = 0.03\ \text{A/cm}^2 \tag{4.35}$$

Unfortunately, these values are not exact; they are meant as a convenient tool for engineering estimates. However, Eq. (4.35) can be used to address an important task: how to identify the proper panel configuration and its approximate size in order to provide sufficient power for a given load.

Example 4.13: A 3 Ω load (for instance, a hot plate in a camp) is rated at 23 V and is to be powered by a solar panel. A c-Si photovoltaic sheet material is your material of choice. Outline parameters of a solar module that is capable of powering the load and estimate the overall module size.

Solution: First, we need to find the required load current. It is given by $I = 23\text{ V}/3\ \Omega = 7.67$ A. Thus, the maximum power parameters of the module must be equal to $V_{MP} = 23$ V and $I_{MP} = 7.67$ A. Next, we find the measurable parameters, V_{OC}, I_{SC}, of the module. According to Eq. (4.35),

$$V_{OC} = \frac{V}{0.8} = 28.75\,\text{V}, \quad I_{SC} = \frac{I}{0.9} = 8.52\,\text{A} \tag{4.36}$$

Finally, we find the number of cells N and the area of an individual cell A, assuming a *series* combination of individual cells:

$$N = \frac{V_{OC}}{0.6\,\text{V}} = \text{round}\,(47.9) = 48, \quad A = \frac{I_{SC}}{0.03\,\text{A}/\text{cm}^2} = 284\,\text{cm}^2 \tag{4.37}$$

The overall module (panel) size for closely spaced cells is then 1.36 m^2.

Example 4.14: Compare the theoretical design from Example 4.13 with a real solar module having nearly the same output power (176 W) and nearly the same maximum power voltage (23 V).

Solution: We choose a Sharp ND-176U1Y, 176 Watt solar panel from Table 4.2 for comparison. Its maximum power voltage is 23.4 V. Table 4.3 lists the parameters of both panels. The designs agree with the number of cells and with the size of the unit cell. The overall panel size for closely spaced cells is also quite similar: 1.36 m^2 versus 1.32 m^2.

Table 4.3. Parameters of a theoretically designed 176 W solar panel versus the corresponding 176 W hardware prototype.

Example 4.3 (theory estimates)	176 Watt Sharp ND-176U1Y panel
$V_{MP} = 23.0$ V, $P_L = 176$ W	$V_{MP} = 23.4$ V, $P_L = 176$ W
# of cells: 48	# of cells: 48
Unit cell area: 284 cm^2	Unit cell area: 275 cm^2

Exercise 4.13: A 9.6 W DC motor in an autonomous robot is rated at 17 V and is to be powered by a solar panel. A c-Si photovoltaic sheet material is your material of choice. Outline parameters of a solar module that is capable of powering the load and estimate the overall module size.

Answer: The module should include 36 cells in series, with the area of $A = 21.0\ \text{cm}^2$ each. The overall module (panel) size for closely spaced cells is then $0.0756\ \text{m}^2$.

Section 4.4 Analysis of Nonlinear Circuits: Load Line Method

Consider a nonlinear passive circuit element which possesses a particular *υ-i* characteristic. It is shown in Fig. 4.21 by a blue rectangle. Examples of such elements were given in Chap. 2. The element's polarity (direction of current inflow for a passive reference configuration) is labeled by a "+" sign. Figure 4.21 presents the four basic nonlinear circuits (networks) encountered in practice: a linear active network given by its Thévenin equivalent and connected to a nonlinear load, a practical nonlinear voltage source connected to a linear load, a linear active network in the form of the Norton equivalent connected to a nonlinear load, and a practical nonlinear current source connected to a linear load. Interchanging the place of the nonlinear element and the resistance, if necessary, and using the source transformation theorem, we can state that all four circuits in Fig. 4.21 are topologically *equivalent*. Therefore, only one of them will be studied, for example, the network shown in Fig. 4.21a. The *υ-i* characteristic of the practical voltage source between terminals *a* and *b* in Fig. 4.21a is given, according to the KVL, by

$$I = \frac{V_T - V}{R_T} \tag{4.38}$$

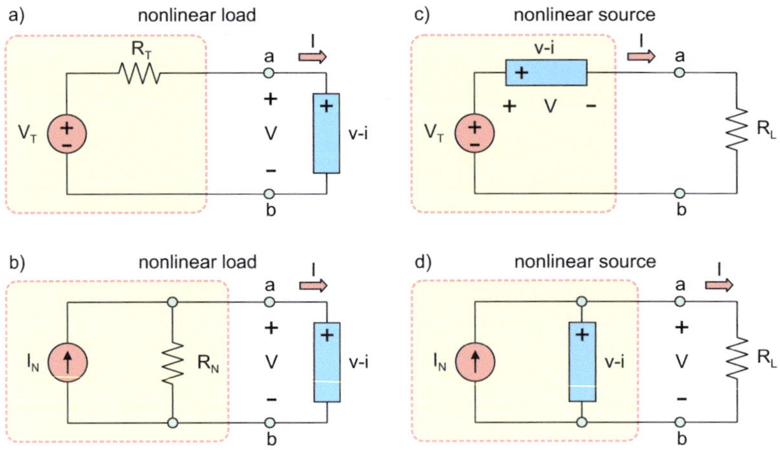

Fig. 4.21. Four basic nonlinear circuits.

This linear function given by Eq. (4.38) is known as the *load line*. It is seen in Fig. 4.22; it intersects the voltage axis at $V = V_T$ and the current axis at $I = V_T/R_T$. The *υ-i* characteristic of a nonlinear element is plotted on the same figure. Both *υ-i* characteristics must give the identical values of voltage and current. Thus, the intersection of the load line with the element's *υ-i* characteristic is the circuit solution. This is the essence of the *load line method*. Though primarily graphical, the load line method provides important insight into the problem under study.

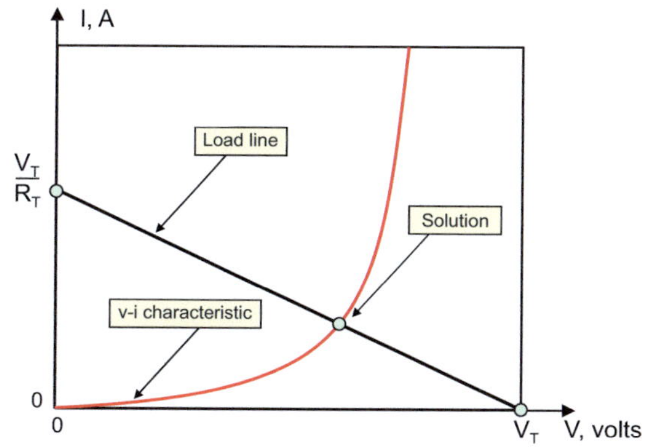

Fig. 4.22. Load line method for a nonlinear circuit.

Example 4.15: The circuit in Fig. 4.21a is characterized by $V_T = 3$ V, $R_T = 1$ kΩ. The *υ-i* characteristic of the nonlinear element (the ideal Shockley diode) is $I = 1 \times 10^{-9}\left[\exp\left(\frac{V}{0.0257\ \mathrm{V}}\right) - 1\right]$ [A]. The goal is to solve the circuit using the load line method.

Solution: Figure 4.23 plots two dependencies: the load line of Eq. (4.38) and *υ-i* characteristic of the ideal diode specified by the present example. Using the visual inspection, the intersection is evaluated as $I \approx 2.6$ mA, $V \approx 0.4$ V. This is the solution for the circuit current and for the load voltage, respectively. The solution accuracy improves when the scale of the plot is adjusted. In particular, we usually do not have to extend the voltage axis all the way from 0 V to the supply voltage; only a small interval may be sufficient.

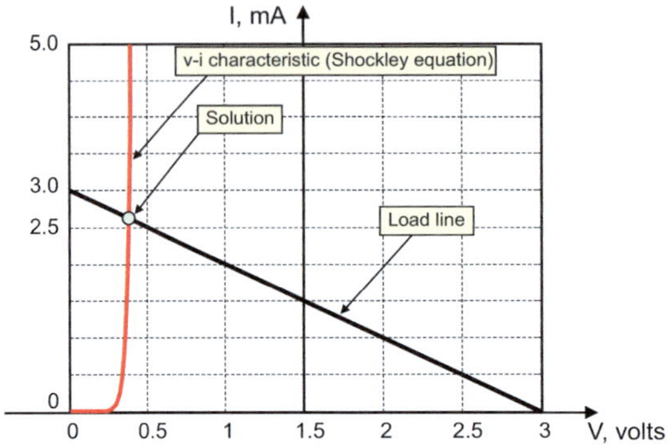

Fig. 4.23. Load line method applied to a nonlinear circuit with an ideal diode.

4.4.1 Iterative Method for Nonlinear Circuits

Assume that the nonlinear circuit element is characterized by an explicit function $I = I(V)$. Then, according to Eq. (4.38), the load line method is equivalent to the graphical solution of the transcendental algebraic equation in the form

$$\frac{V_T - V}{R_T} = I(V) \Rightarrow V = V_T - R_T I(V) \tag{4.39}$$

An alternative is to solve this equation iteratively, starting with an initial guess $V = V^0$. The *iterative method for nonlinear circuits* can be formulated as follows. Two iterative schemes (*explicit* and *implicit*) are formally employed. The first (explicit) scheme follows from the second Eq. (4.39) and the second (implicit) scheme from the first Eq. (4.39),

$$V^{n+1} = V_T - R_T I(V^n), \quad n = 0,\ 1,\ 2,\ \ldots \text{or} \quad V^{n+1} = I^{-1}\left(\frac{V_T - V^n}{R_T}\right), \quad n = 0,\ 1,\ 2,\ \ldots \tag{4.40}$$

where I^{-1} denotes the inverse function of $I(V)$. The first (explicit) scheme is simpler when $I(V)$ is given. However, *only* the second scheme is recommended in practice since the first scheme may not converge for typical nonlinear circuit elements, which model semiconductor devices.

Example 4.16: Solve the previous example using the iterative solution of the transcendental circuit equation.

Solution: We find the inverse *v-i* characteristic of the nonlinear element first. It is good so

Example 4.16 (cont.):

$$V = 0.0257 \times \ln\left[\frac{I}{1 \times 10^{-9}\ \text{A}} + 1\right]\ [\text{V}], \quad I = \frac{3\ \text{V} - V}{1\ \text{k}\Omega} \tag{4.41}$$

The iterative scheme has the form (the second method of Eq. (4.40) is used)

$$V^{n+1} = 0.0257 \times \ln\left[\frac{3\ \text{V} - V^n}{1 \times 10^{-6}\ \text{V}} + 1\right], \quad n = 0,\ 1,\ 2,\ \ldots \tag{4.42}$$

with the initial guess $V^0 = 0$ V. It converges very fast; the corresponding iterations are $V^0 = 0$ V, $V^1 = 0.3833$ V, $V^2 = 0.3798$ V, and $V^3 = 0.3798$ V.
Therefore, only a few iterations are usually sufficient. The final result is $I = 2.62$ mA, $V = 0.380$ V, which improves the solution obtained with the load line method—see the previous example. The initial guess of the iterative solution may vary widely, but it should *not* exceed the source voltage.

Application Example: Solving the Circuit for a Generic Solar Cell

Figure 4.24a shows a simplified physical composition of the solar cell in the form of a pn-junction (a junction of two semiconductor materials), which essentially forms a semiconductor diode.

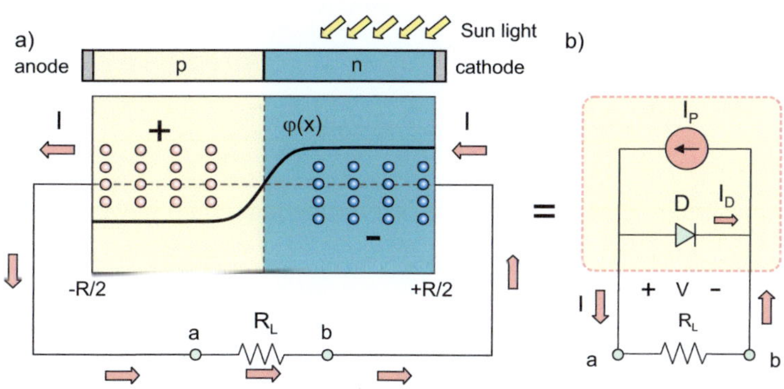

Fig. 4.24. (**a**) Simplified physical composition of the solar cell in the form of a pn-junction—a semiconductor diode and (**b**) simplified (lossless single-diode) equivalent circuit.

Free charge carries generated by sunlight are separated by a built-in voltage or potential $\phi(x)$ within the diode, which is the cause of an equivalent current source: the photocurrent of the solar cell I_P. The photocurrent mostly flows through the load. At the same time, a certain portion of it, I_D, could still flow through the pn-junction diode itself as a *forward diode current*. Therefore, the load current I in Fig. 4.24 is less than the photocurrent. Figure 4.24b shows a simplified equivalent circuit of a solar cell. This circuit coincides with the nonlinear circuit in Fig. 4.21d and can be solved in the same way once the v-i characteristic of the equivalent diode is known. It is often given by

$$I_D = I_S\left[\exp\left(\frac{V}{nV_T}\right) - 1\right], \quad V_T = 0.0257 \text{ V} \tag{4.43}$$

with an *effective* ideality factor n and an *effective* saturation current I_S of the corresponding diode. The *characteristic equation of the cell* is the KCL in Fig. 4.24b,

$$I = I_P - I_S\left[\exp\left(\frac{V}{nV_T}\right) - 1\right] \tag{4.44}$$

Figure 4.25 plots the $I(V)$ dependence of the characteristic Eq. (4.44). The horizontal straight asymptote is the photocurrent I_P or the short-circuit current I_{SC}. The vertical straight asymptote is the open-circuit voltage V_{OC}, $V_{OC} \approx V_T n \ln (I_P/I_S)$ when $I_P/I_S >> 1$. The area of the shaded rectangle is the load power; it is clearly maximized at a certain operating point Q, where $V = V_{MP}$, $I = I_{MP}$.

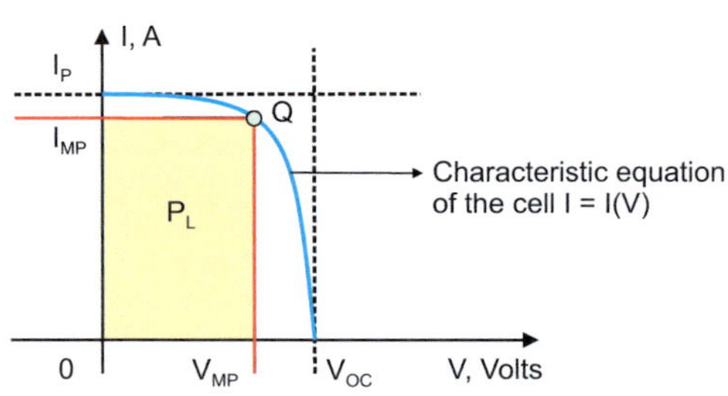

Fig. 4.25. Finding operating point Q of the solar cell for the maximum power transfer.

Finding maximum power parameters is a straightforward but lengthy procedure. The load power is found as a function of V; its derivative must be set equal to zero to maximize the load power. This is essentially the maximum power theorem for nonlinear circuits. The final expressions are

$$V_{MP} = V_{OC} - nV_T \ln\left(1 + \frac{V_{OC}}{nV_T}\right), \quad I_{MP} = I_{SC}\left(1 - \frac{nV_T}{V_{MP}}\right) \approx 90\% \text{of } I_{SC} \tag{4.45}$$

For $n = 1.75$ and $V_{OC} = 0.6$V, $V_{MP} \approx 80$ % of V_{OC}, $I_{MP} \approx 90$ % of I_{SC} which is close to the data in Table 4.2 given that the fill factors for the cell and the panel are approximately the same. Note that the load resistance is found as $R_L = V_{MP}/I_{MP}$.

Summary

Circuit analysis techniques: nodal/mesh analysis		
Nodal analysis	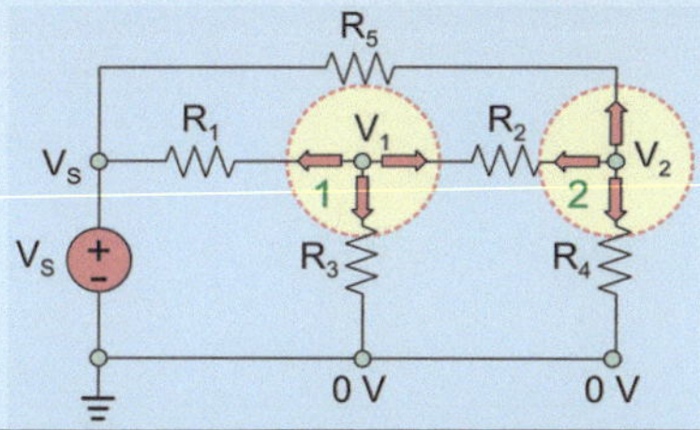	Based on the KCL and Ohm's law: $\frac{V_1-V_S}{R_1}+\frac{V_1-0\ \text{V}}{R_3}+\frac{V_1-V_2}{R_2}=0$ $\frac{V_2-V_S}{R_5}+\frac{V_2-0\ \text{V}}{R_4}+\frac{V_2-V_1}{R_2}=0$
Supernode	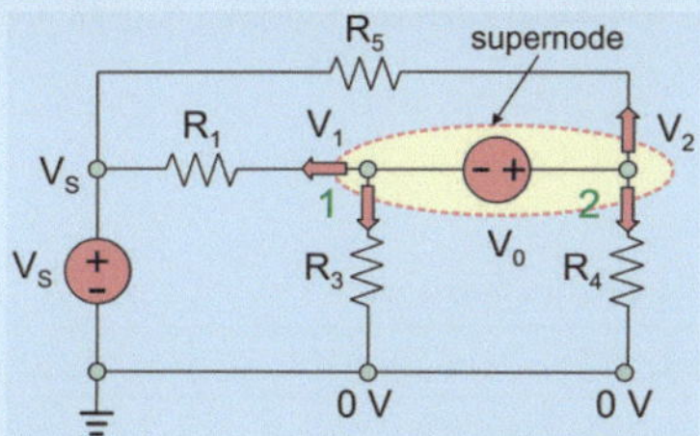	The KCL for the supernode: $\frac{V_1-V_S}{R_1}+\frac{V_1}{R_3}+\frac{V_2}{R_4}+\frac{V_2-V_S}{R_5}=0$ Plus the KVL: $V_2=V_1+V_0$
Mesh analysis	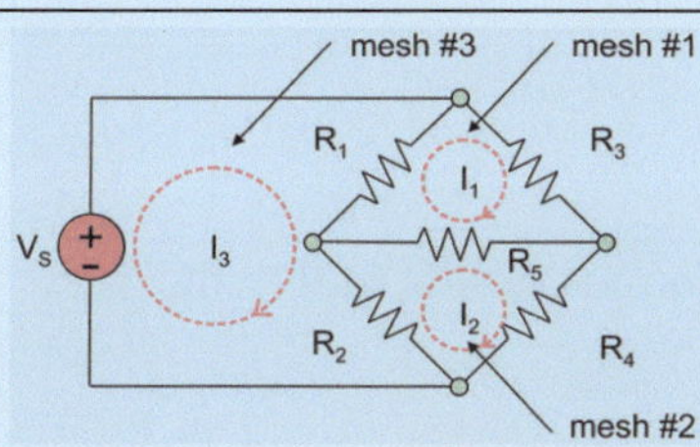	Based on the KVL and Ohm's law: $R_1(I_1-I_3)+R_3I_1+R_5(I_1-I_2)=0$ $R_2(I_2-I_3)+R_5(I_2-I_1)+R_4I_2=0$ $-V_S+R_1(I_3-I_1)+R_2(I_3-I_2)=0$ For meshes 1, 2, and 3.
Supermesh	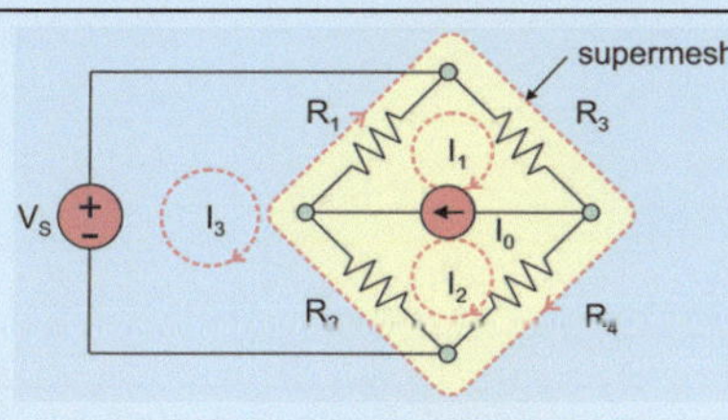	The KVL for the supermesh: $R_1(I_1-I_3)+R_3I_1+R_4I_2+R_2(I_2-I_3)=0$ Plus Eq. for mesh 3 And the KCL: $I_1-I_2=I_S$
Circuit analysis techniques: source transformation theorem		
Source transformation theorem	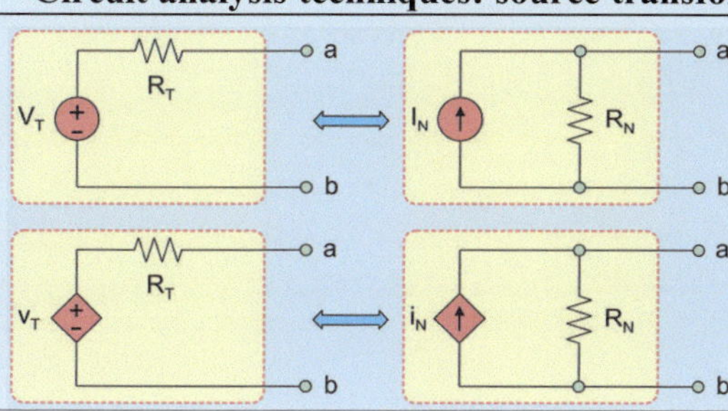	Substitution of voltage source V_T in series with resistance R_T for current source I_N with resistance R_N: $R_N=R_T, I_N=\frac{V_T}{R_T}$ $V_{OC}=V_T, I_{SC}=\frac{V_T}{R_T}$

(continued)

Circuit analysis techniques: Thévenin/Norton theorems/equivalents		
Thévenin and Norton theorems	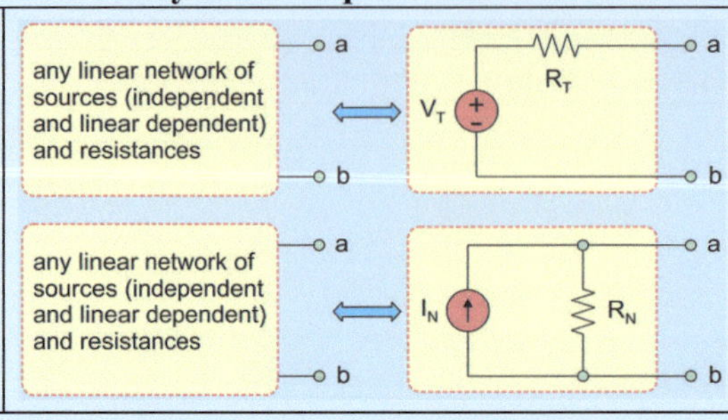	Any linear network with independent sources, dependent linear sources, and resistances can be replaced by a simple equivalent network in the form: i. A voltage source V_T in series with resistance R_T ii. A current source I_N in parallel with resistance R_N
Summary of major circuit analysis methods (linear circuits) - Superposition theorem (pervious chapter) - Nodal/mesh analysis (this chapter) - Source transformation theorem (this chapter) - Thévenin and Norton equivalent circuits (this chapter)		
Linear networks: Measurements/equivalence		
Method of short/open circuit	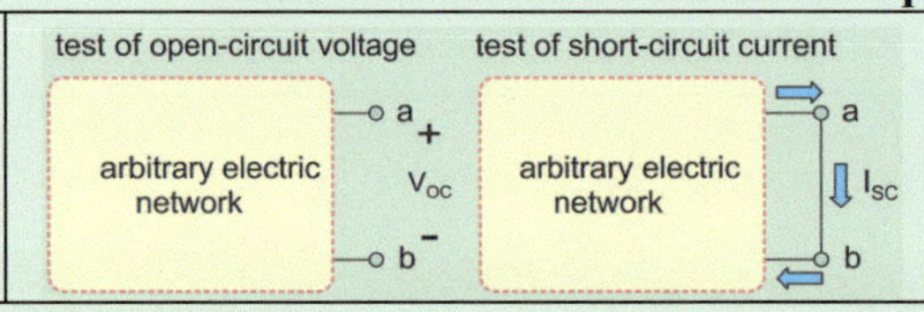	– Two arbitrary linear networks are equivalent when their V_{OC} and I_{SC} coincide – This method is also used for nonlinear circuits
Linear networks: maximum power theorem		
Maximum power theorem (load matching)	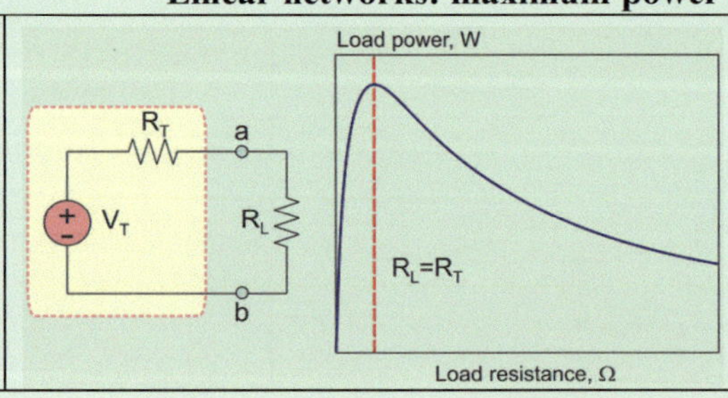	– Power delivered to the load is maximized when $R_L = R_T$ – For high-frequency circuits it also means no "voltage/current wave reflection" from the load
Linear networks: maximum power efficiency		
Power efficiency is maximized when the load resistance is very high (load bridging)	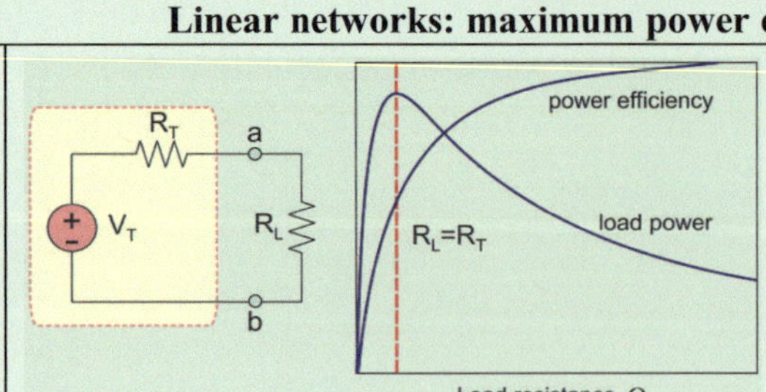	Power transfer efficiency: $E = \frac{P_L}{P} = \frac{R_L}{R_T+R_L}$

(continued)

Linear networks: dependent sources and negative equivalent resistance		
Equivalent resistances of basic linear networks with dependent sources	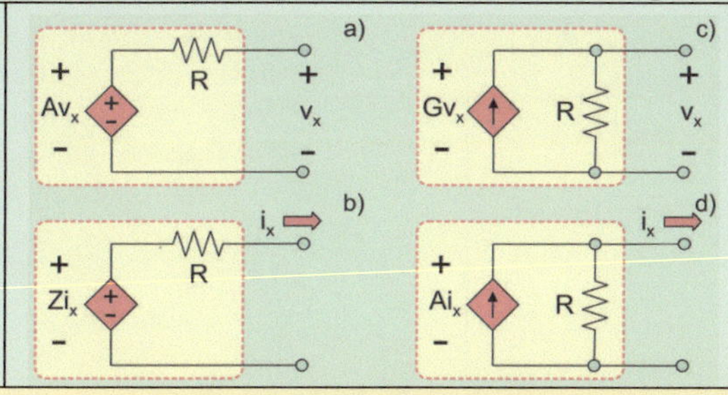	Case (a): $R_T = \frac{R}{1-A}$ Case (b): $R_T = R - Z$ Case (c): $R_T = \frac{R}{1-GR}$ Case (d): $R_T = R(1 - A)$ Equivalent resistance may become negative
Nonlinear networks: four basic topologies		
Basic nonlinear circuits	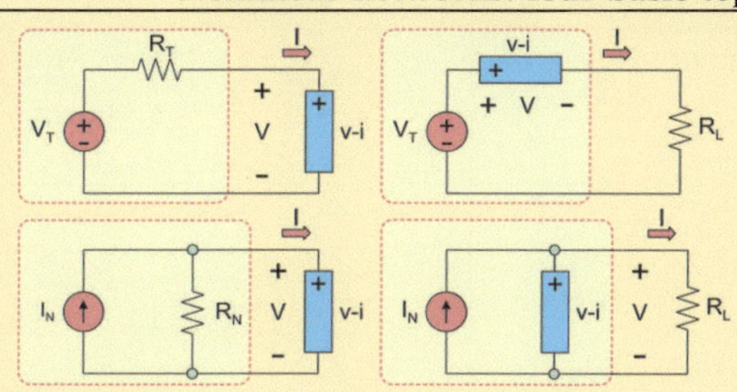	– Linear voltage source connected to a nonlinear load (diode circuit) – Linear current source connected to a nonlinear load – Nonlinear voltage source connected to a linear load; – Nonlinear current source connected to a linear load (photovoltaic circuit)
Nonlinear networks: circuit analysis via load line method		
Load line method	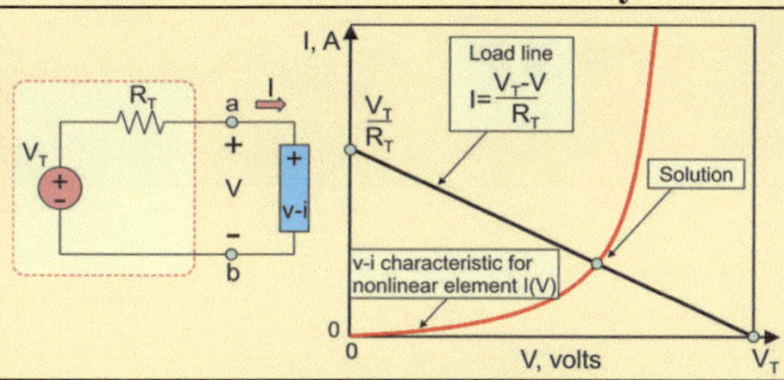	Solution: The load line (v-i characteristic of the linear source, which is $I = \frac{V_T - V}{R_T}$) intersects the v-i characteristic of the nonlinear load, $I(V)$
Nonlinear networks: iterative solution		
Finding the intersection point iteratively	For the circuit in the previous row: $I(V) = \frac{V_T - V}{R_T} \Rightarrow V^{n+1} = I^{-1}\left(\frac{V_T - V^n}{R_T}\right),$ $n = 0, 1, \ldots$	Implicit scheme: Initial guess V^0 may be 0 V
Nonlinear networks: finding resistive load for maximum power extraction		
Finding maximum load power for the equivalent model of a solar cell	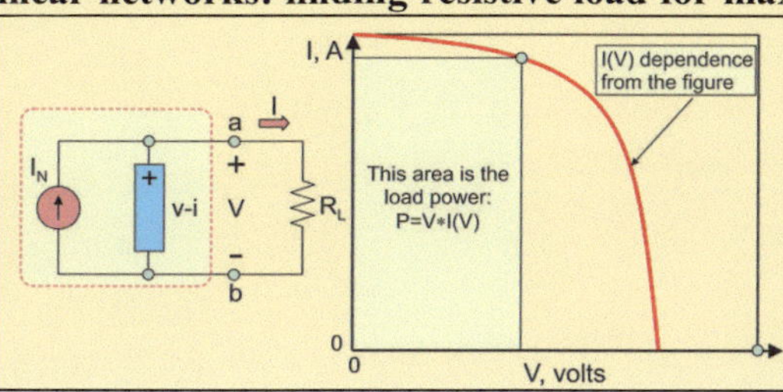	Load power is computed as $P_L = V \times I(V)$. Then, it is maximized, which is equivalent to solving equation: $\frac{dP_L}{dV}(V) = 0$ For unknown voltage V

(continued)

Some useful facts about power extraction from solar cells/modules	
Typical values of open-circuit voltage V_{OC} and photocurrent density J_P of a c-Si cell	Crystalline silicon or c-Si cell: $V_{OC} \approx 0.6$ V; $J_P \approx 0.03$ A/cm^2
– Open-circuit module voltage is N times the cell voltage, $N \cdot V_{OC}$ – Short-circuit module current I_{SC} is the cell short-circuit current $I_{SC} = AJ_P$ where A is cell area	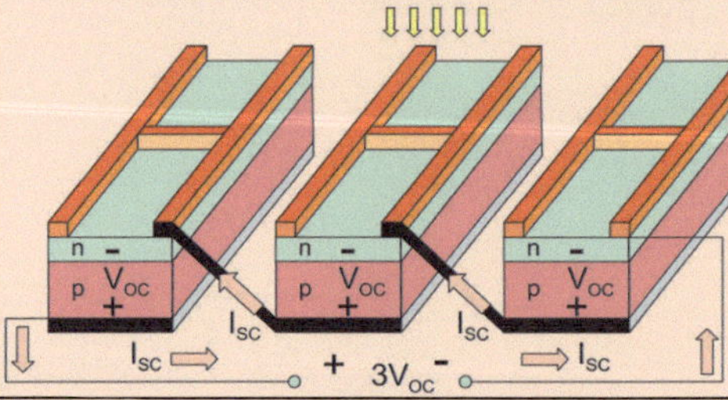
Typical values of maximum power parameters and fill factor for c-Si cells/modules; the load resistance must be $R_L = V_{MP}/I_{MP}$	$F\|_{\text{module}} = \frac{V_{MP}I_{MP}}{V_{OC}I_{SC}} \approx F\|_{\text{cell}}$ (for low-loss modules) $V_{MP} \approx 0.8V_{OC}, \quad I_{MP} \approx 0.9I_{SC}, \quad F \approx 0.72$
Lossless single-diode model of a solar cell: $I_P = AJ_P$—photocurrent (A) A—cell area (cm^2) V_T—thermal volt. (0.0257 V) n—ideality factor ($1 < n < 2$) $I_S \approx I_P \exp\left(-\frac{V_{OC}}{nV_T}\right)$ (A)	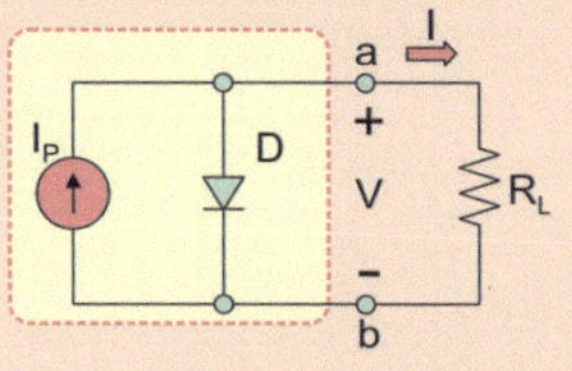 $I = I_P - I_S\left[\exp\left(\frac{V}{nV_T}\right) - 1\right]$
Maximum power analytical solution for lossless single-diode model of a solar cell	$V_{MP} \approx V_{OC} - nV_T \ln\left(1 + \frac{V_{OC}}{nV_T}\right)$, $I_{MP} = I_{SC}\left(1 - \frac{nV_T}{V_{MP}}\right)$

Problems

4.1 Nodal/Mesh Analysis

4.1.2 Nodal Analysis

Problem 4.1. Using the nodal analysis, determine the voltage V_1 in the circuit shown in the following figure.

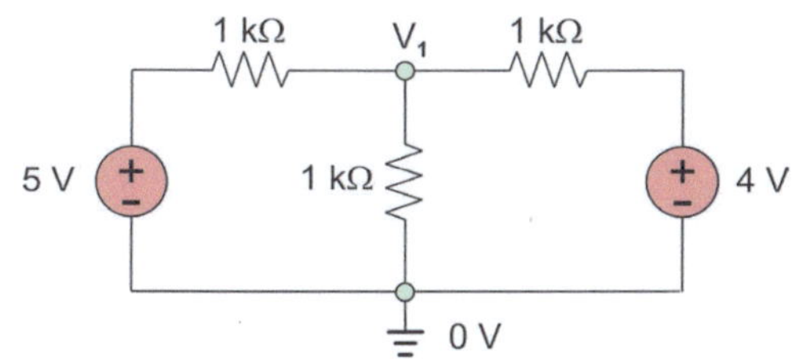

Problem 4.2. Using the nodal analysis, determine the voltage V_1 or current I in the circuits shown in the following figure.

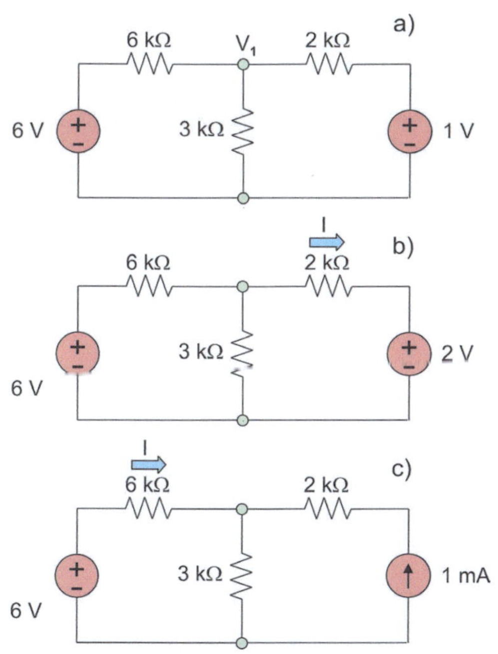

Problem 4.3.

A. Using the nodal analysis, determine the supply current I for the circuit shown in the figure.
B. Show the current directions for every resistance in the circuit.

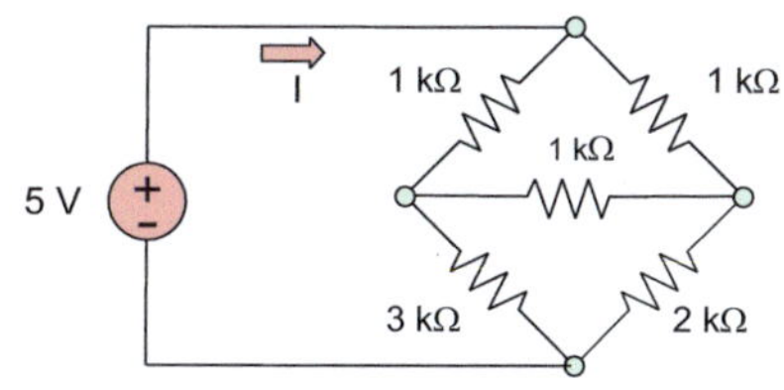

Problem 4.4.

A. Using the nodal analysis, determine the total circuit current I for the circuit shown in the figure below.
B. Show the current directions for every resistance in the circuit.

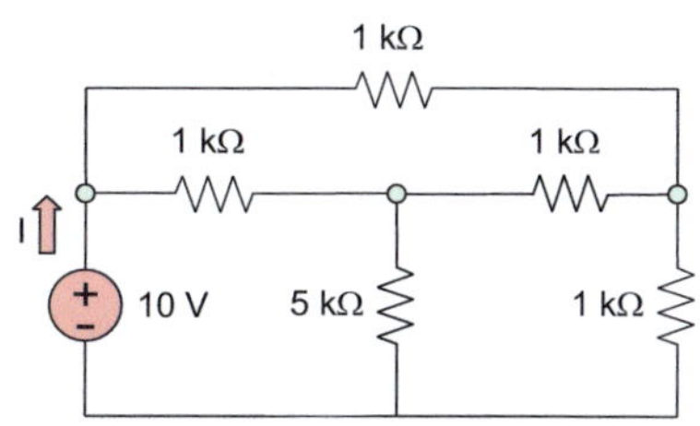

Problem 4.5. Introduce the ground termination, write the nodal equations, and solve for the node voltages for the circuit shown in the figure below. Calculate the current I shown in the figure.

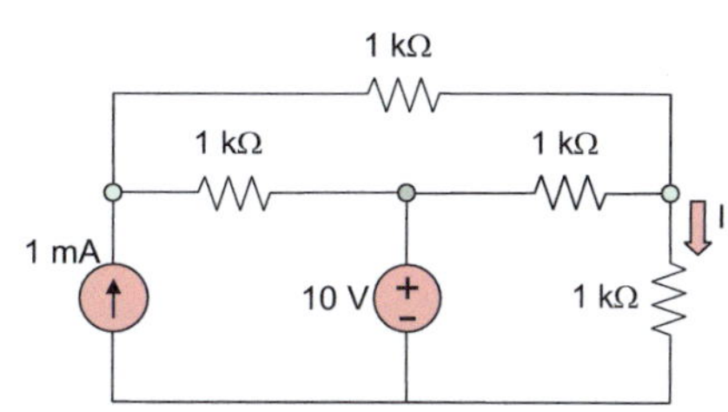

Problem 4.6. Introduce the ground termination, write the nodal equations, and solve for the node voltages for the circuit shown in the figure. Calculate the current through the resistance R_x and show its direction in the figure.

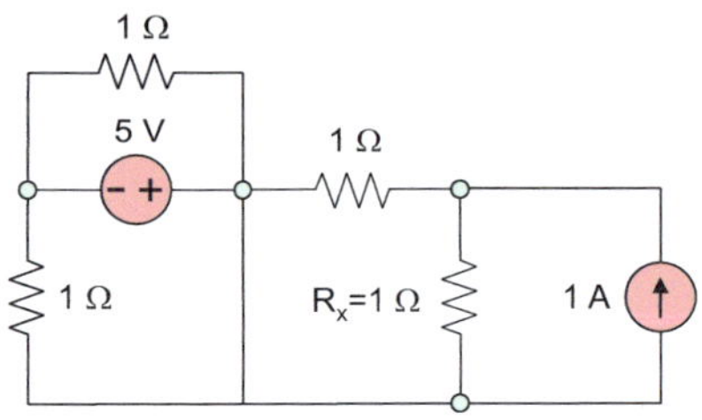

Problem 4.7. For the circuit shown in the figure below

(i) Determine the current i_x through resistance R_x.

(ii) Show its direction on the figure.

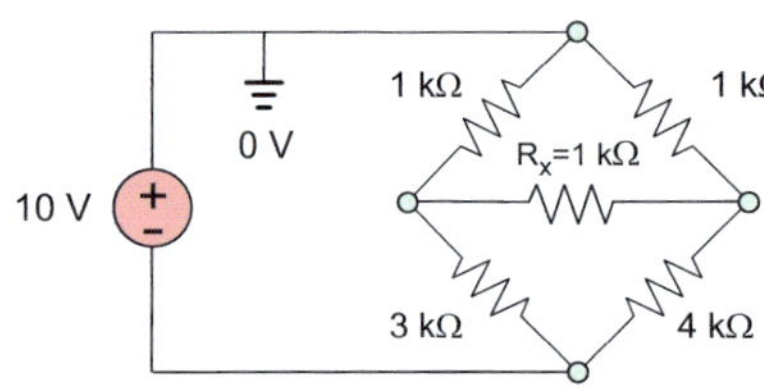

Problem 4.8. For the circuit shown in the figure below

A. Write nodal equations and solve for the node voltages. Then, find the value of i_1.

B. Could this problem be solved in another (simpler) way?

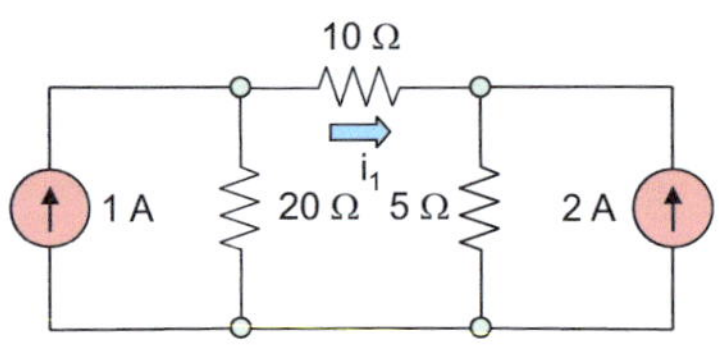

Problem 4.9.

A. Write the nodal equations and solve for the node voltages for the circuit shown in the figure below. Then, find the value of i_1.

B. Use MATLAB or other software of your choice for the solution of the system of linear equations; attach the code to the solution.

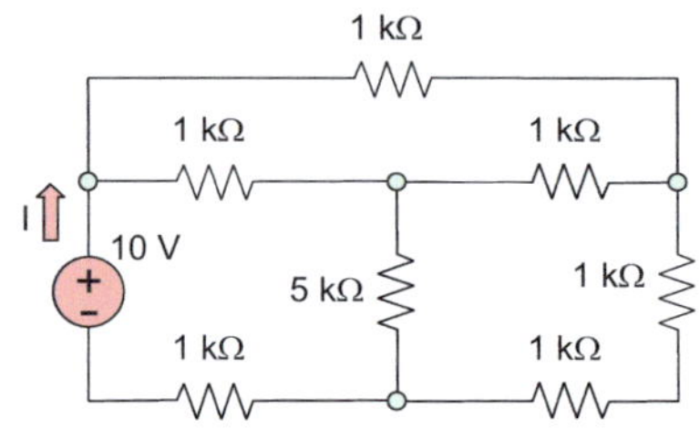

Problem 4.10. The figure below shows the DC equivalent of a residential *three-wire* system, which operates at 240 V rms (do not confuse it with the three-phase system, which carries a higher current). Two 120 V rms power supplies are connected as one dual-polarity power supply, i.e., in series. The 10 Ω load and the 20 Ω are those driven by the two-wire (and one ground) standard wall plug with 120 V rms—the lights, a TV, etc. The 6 Ω load consumes more power, and it is driven with 240 V rms using a separate bigger wall plug (+/– and neutral-not shown)—the stove, washer, dryer, etc. Determine the power delivered to each load. *Hint*: Use a calculator or software of your choice for the solution of the system of linear equations (MATLAB is recommended).

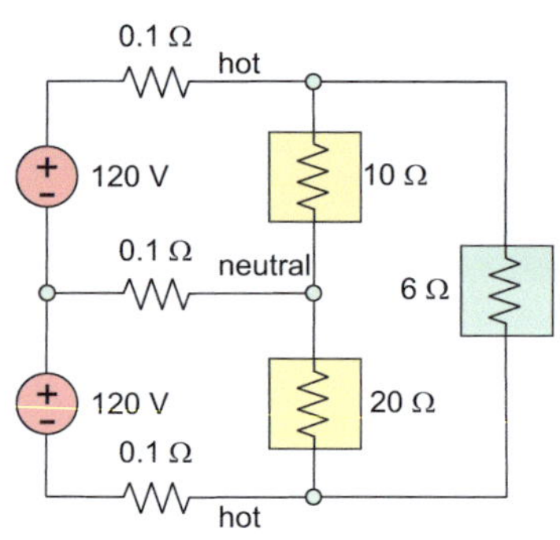

4.1.3 Supernode

Problem 4.11. Introduce the ground terminal, write the nodal equations, and solve for the node voltages for the circuit shown in the figure. Calculate the current I shown in the figure.

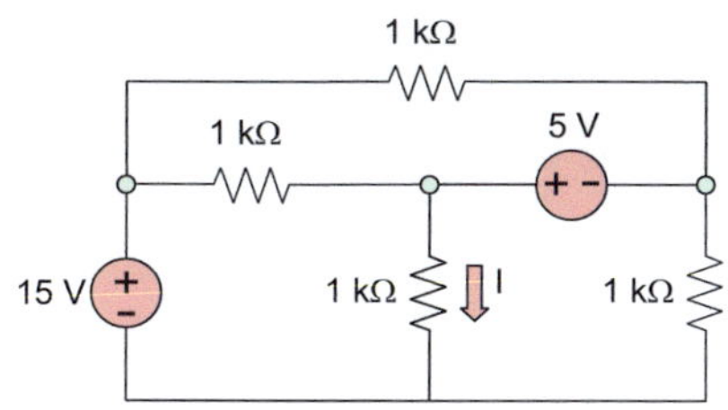

Problem 4.12. Introduce the ground terminal, write the nodal equations, and solve for the node voltages for the circuit shown in the figure. Calculate the current I shown in the figure.

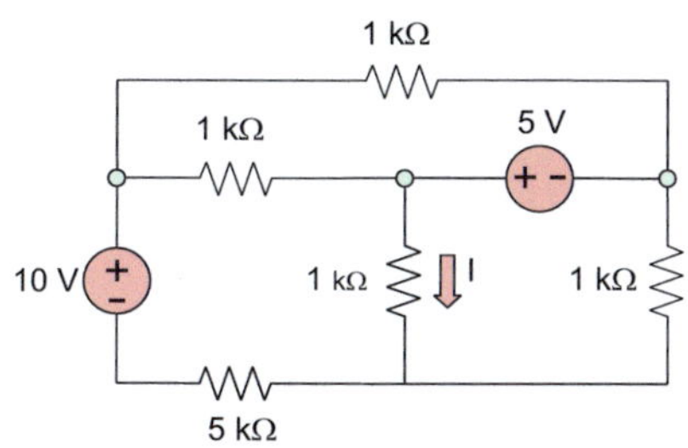

4.1.4 Mesh Analysis

4.1.5 Supermesh

Problem 4.13. For the circuit shown in the figure, determine the current i_1

A. Using the mesh analysis
B. Using the nodal analysis

Which method is simpler?

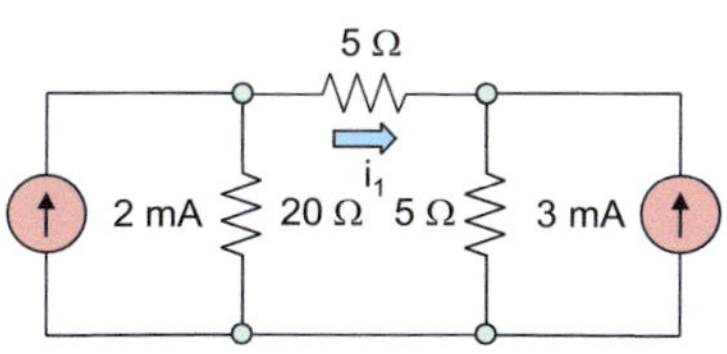

Problem 4.14. For the circuit shown in the figure, determine the voltage V_x

A. Using the mesh analysis
B. Using the nodal analysis

Which method appears to be simpler?

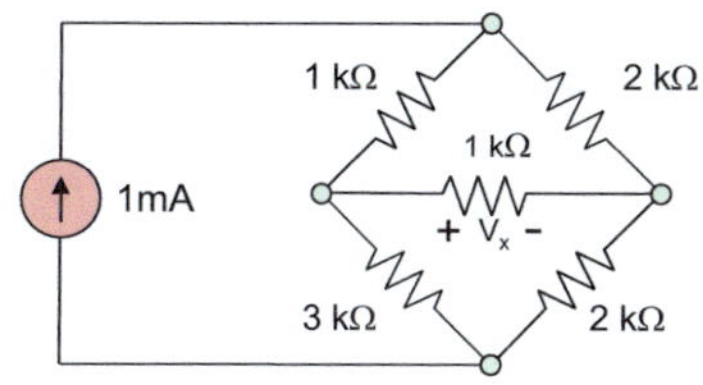

Problem 4.15. For the circuit shown in the figure

A. Determine the circuit current I using either the nodal analysis or the mesh analysis.
B. Explain your choice for the selected method.

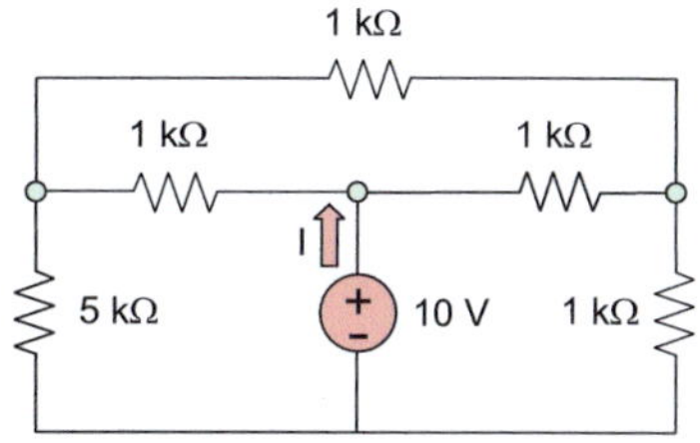

Problem 4.16. For the circuit shown in the figure, determine its equivalent resistance between terminals a and b. *Hint*: connect a power source and use a mesh current analysis or the nodal analysis.

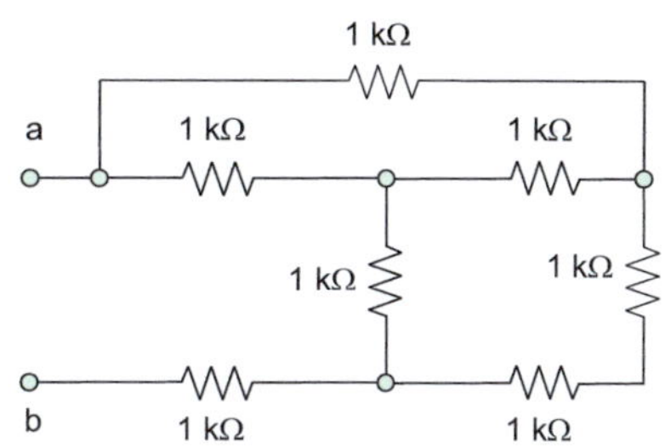

Problem 4.17. For the circuit shown in the figure, determine the current i_1 of the 20 V voltage source.

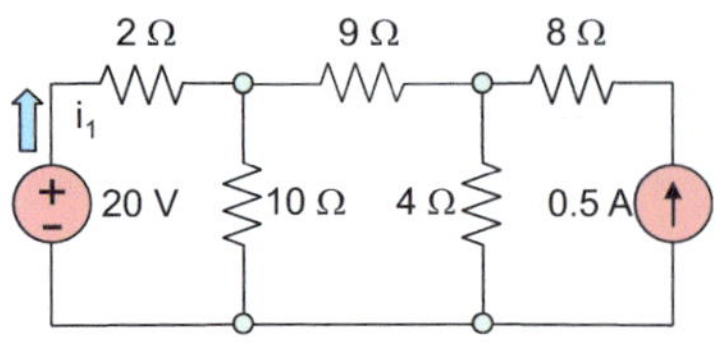

Problem 4.18. Determine the voltage across the current source for the circuit shown in the figure below using the mesh analysis and the supermesh concept.

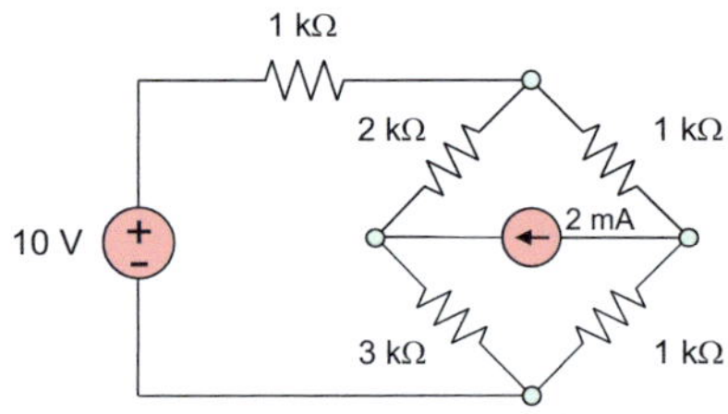

4.2 Generator Theorems

4.2.1 Equivalence of Active One-Port Networks: Method of Short/Open Circuit

Problem 4.19. A linear active network with two unknown circuit elements measures $V_{OC} = 5$ V, $I_{SC} = 10$ mA.

A. Determine the parameters V_T, R_T and I_N, R_N of two equivalent networks shown in the figure below.
B. Could you identify which exactly network is it?

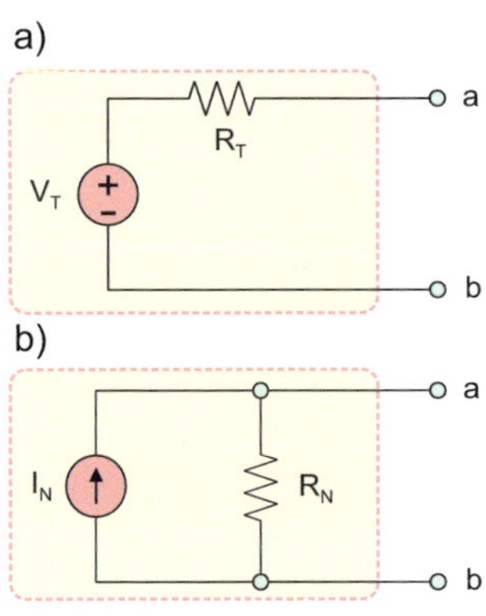

Problem 4.20. Given two networks shown in the figure below:

A. Determine their open-circuit voltage, V_{OC}, and short-circuit current, I_{SC}, for each of them.
B. Are the networks equivalent?

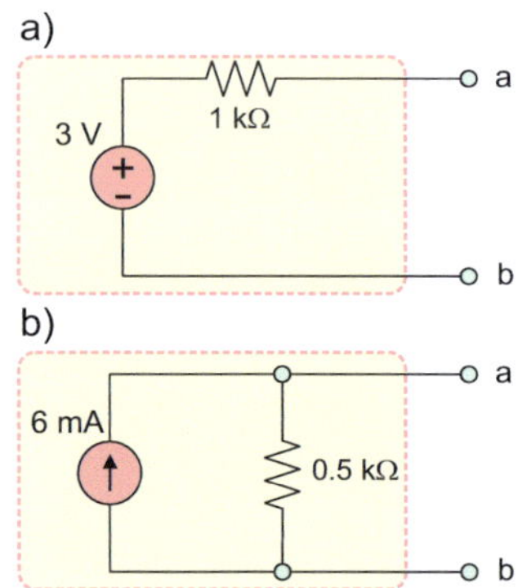

4.2.2 Application Example: Reading and Using Data for Solar Panels

Problem 4.21. The area of a single cell in the 10 W BSP-1012 Power Up c-Si panel is ~22 cm^2. Predict:

A. Open-circuit voltage of the solar cell, V_{OC}
B. Photocurrent density of the cell, J_P
C. Short-circuit current of the cell, I_{SC}

Compare the above value with the value $I_{SC} = 0.66$ A reported by the manufacturer.

Problem 4.22. The area of a single cell in the 175 W BP Solar SX3175 c-Si panel is 156.25 cm^2. Predict:

A. Open-circuit voltage of the solar cell, V_{OC}
B. Photocurrent density of the cell, J_P
C. Short-circuit current of the cell, I_{SC}

Compare the above value with the value $I_{SC} = 5.1\text{A}$ reported by the manufacturer. What value should the photocurrent density have in order to exactly match the short-circuit current reported by the manufacturer?

Problem 4.23. Are the solar cells in the solar module connected in parallel or in series? Why is one particular connection preferred?

Problem 4.24. You are given three c-Si solar cells shown in the figure below, each of area $A/3$. Draw wire connections for a cell bank, which has the performance equivalent to that of a large solar cell with the area A.

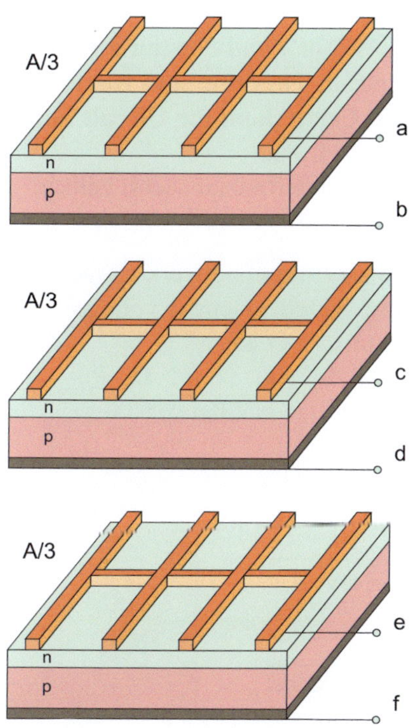

Problem 4.25. Individual solar cells in the figure to the previous problem are to be connected into a standard solar module. Draw the corresponding wire connections.

Problem 4.26. A 10 W BSP1012 PV c-Si module shown in the figure has 36 unit cells connected in series, the short-circuit current of 0.66 A and the open-circuit voltage of 21.3 V.

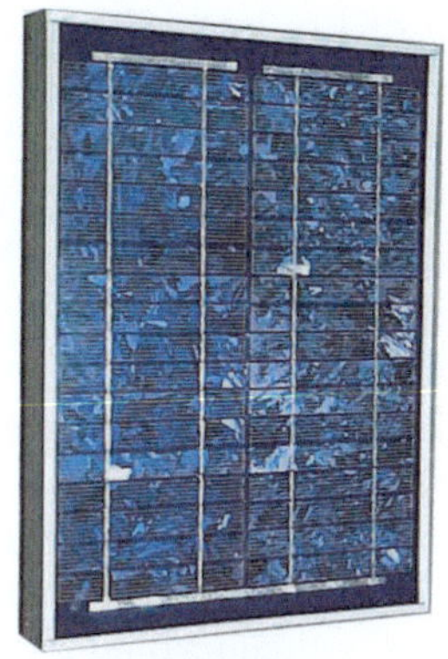

Power Up
10 Watt Module

A. Estimate the area of the single solar cell using the common photocurrent density value for c-Si solar cells.
B. Estimate the open-circuit voltage for the single cell.

Problem 4.27. A 20 W BSP2012 PV c-Si module shown in the figure has 36 unit cells connected in series, the short-circuit current of 1.30 A and the open-circuit voltage of 21.7 V.

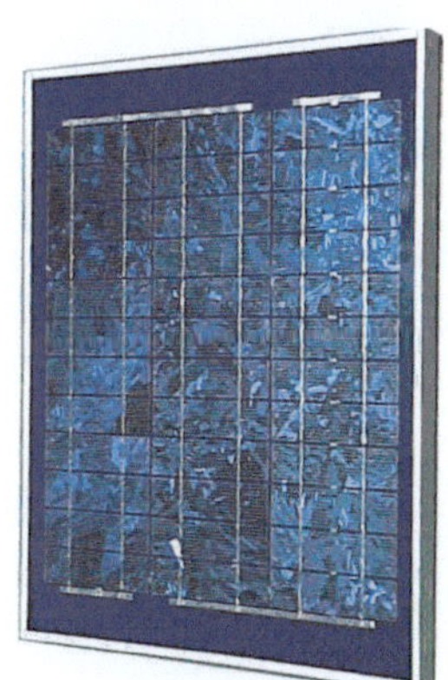

Power Up
20 Watt Module

A. Estimate the area of the single solar cell using the common photocurrent density value for c-Si solar cells.
B. Estimate the open-circuit voltage for the single cell.

Problem 4.28. A c-Si solar module is needed with the open-circuit voltage of 10 V and short-circuit current of 1.0 A. A number of individual solar cells are available, and each has the area of 34 cm^2, open-circuit voltage of 0.5 V, and short-circuit current of 1.0 A. Identify the proper module configuration (number of cells) and estimate module's approximate size.

Problem 4.29. A c-Si solar module is needed with the open-circuit voltage of 12 V and short-circuit current of 3.0 A. A number of individual solar cells are available, and each has the area of 34 cm^2, the open-circuit voltage of 0.5 V, and short-circuit current of 1.0 A. Identify the proper module configuration (number of cells) and estimate module's approximate size.

4.2.3 Source Transformation Theorem

Problem 4.30. Find voltage V in the circuit shown in the figure below using source transformation.

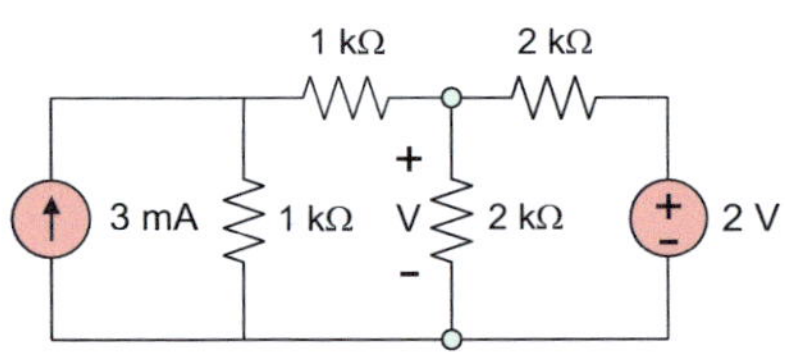

Problem 4.31. The circuit shown in the figure below includes a current-controlled voltage source. Find current i_x using source transformation.

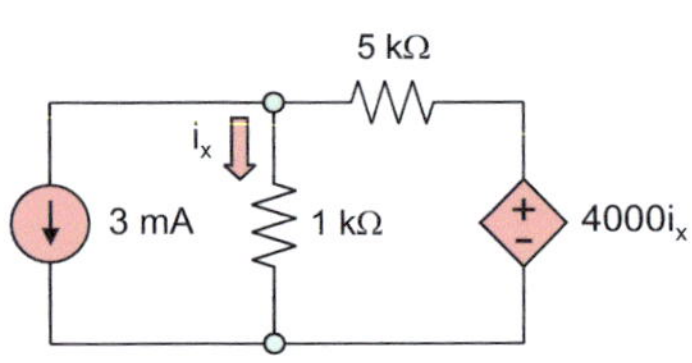

Problem 4.32. The circuit shown in the figure below includes a voltage-controlled voltage source. Find voltage v_x using source transformation.

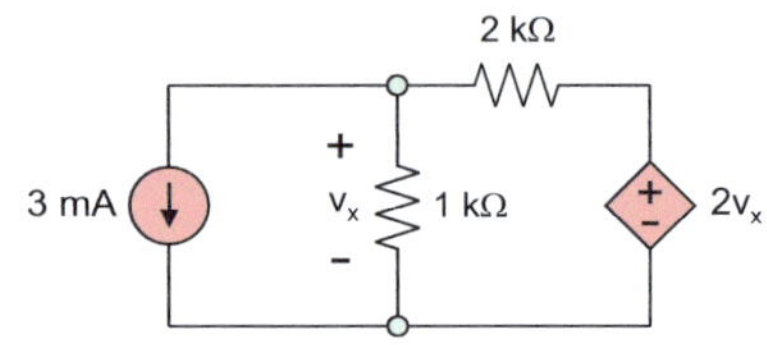

Problem 4.33. Repeat the previous problem for the circuit shown in the figure below.

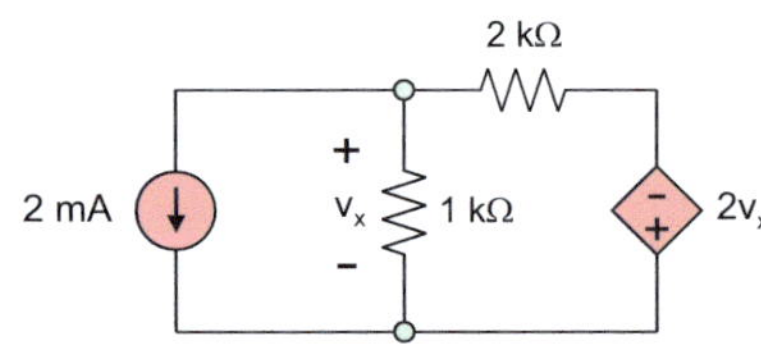

4.2.4 Thévenin's and Norton's Theorems

4.2.5 Finding Thévenin and Norton Equivalents and Using them for Circuit Solution

Problem 4.34. Find

(i) Thévenin (or equivalent) voltage
(ii) Thévenin (or equivalent) resistance

for the two-terminal network shown in the figure below (three practical voltage sources or batteries *in series*) when $R_{B1} = 0.5\ \Omega$, $R_{B2} = 0.5\ \Omega$, $R_{B3} = 0.5\ \Omega$, $V_{B1} = 6$ V, $V_{B2} = 6$ V, $V_{B3} = 6$ V.

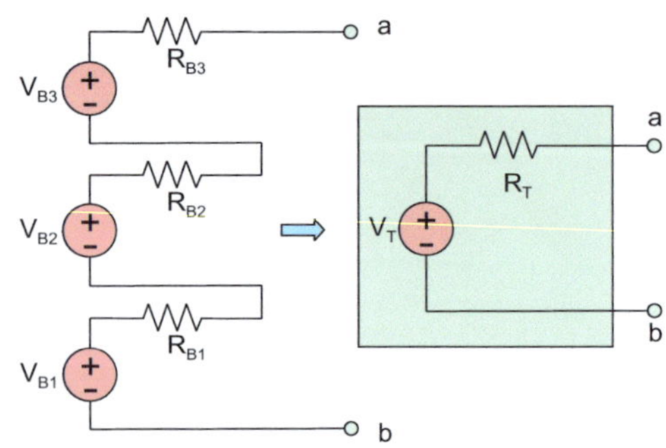

Problem 4.35. Find

(i) Thévenin (or equivalent) voltage

(ii) Thévenin (or equivalent) resistance

for the two-terminal network shown in the figure below (three practical voltage sources or batteries *in parallel*) when
$R_{B1} = 1\ \Omega$, $R_{B2} = 1\ \Omega$, $R_{B3} = 1\ \Omega$,
$V_{B1} = 6$ V, $V_{B2} = 6$ V, $V_{B3} = 6$ V.

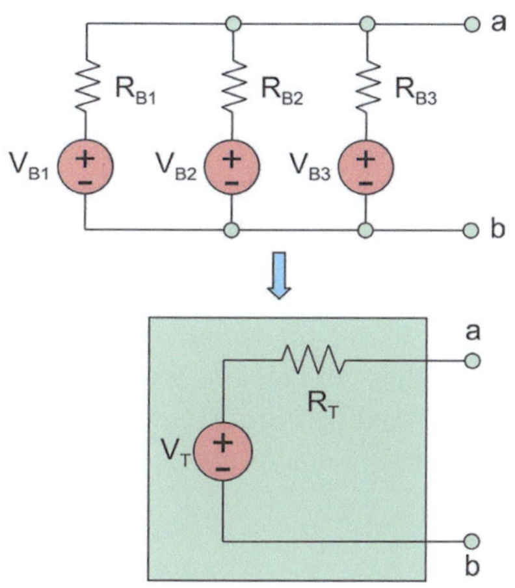

Problem 4.36. Find
(i) Thévenin (or equivalent) voltage
(ii) Thévenin (or equivalent) resistance

for the two-terminal network shown in the figure below. Assume three 9 V sources and resistances of 1 Ω each.

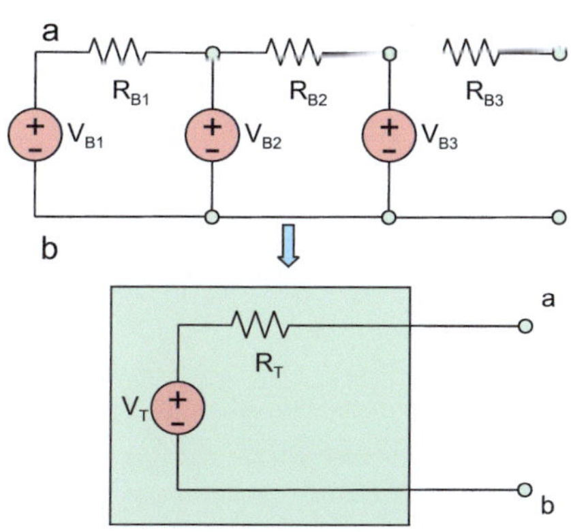

Problem 4.37. For the circuits shown in the figure below, determine their equivalent (Thévenin) circuit by finding V_T and R_T (show units).

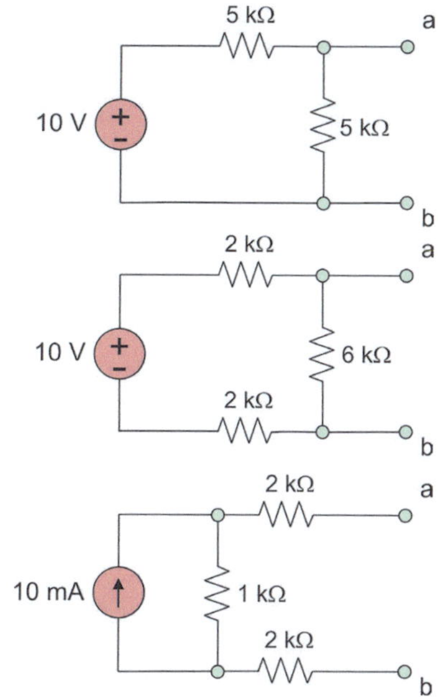

Problem 4.38. For the circuits shown in the figure below, determine their equivalent (Thévenin) circuit by finding V_T and R_T (show units).

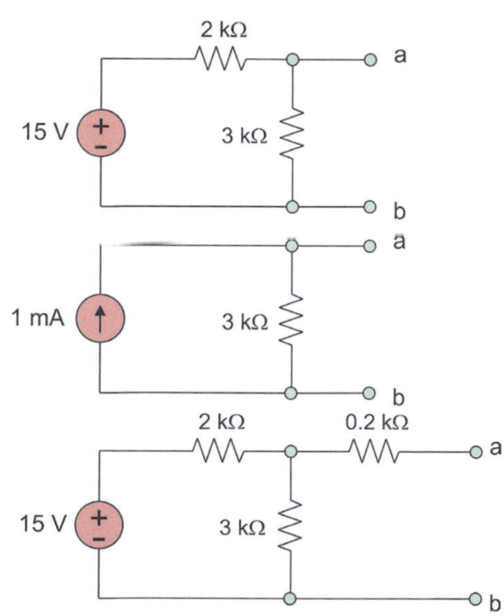

Problem 4.39. Find
(i) Thévenin (or equivalent) voltage

(ii) Thévenin (or equivalent) resistance

for the two-terminal networks shown in the figure below.

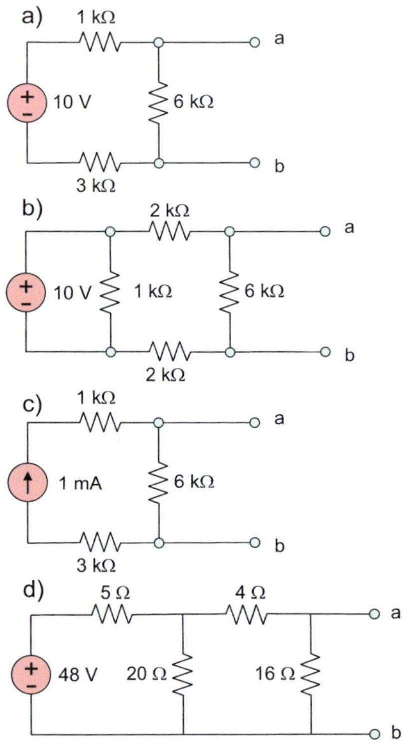

Problem 4.40. Find
(i) Thévenin voltage
(ii) Thévenin resistance

for the two-terminal network shown in the figure below when
$R_1 = R_2 = R_3 = 1\ \text{k}\Omega,\ V_{S1} = V_{S2} = 10\ \text{V}$.

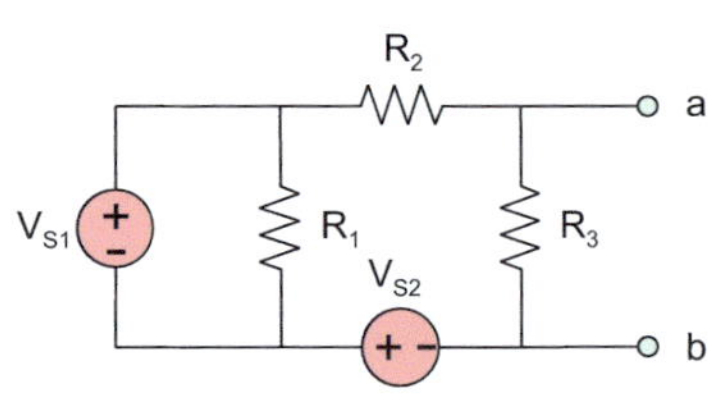

Problem 4.41. A two-terminal network shown in the figure is a starting section of a ladder network used for digital-to-analog conversion. Express
(i) Thévenin (or equivalent) voltage
(ii) Thévenin (or equivalent) resistance

in terms of (digital) voltage $D_0$and resistance R.

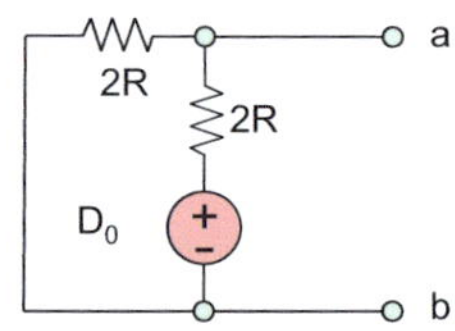

Problem 4.42. A two-terminal circuit network in the figure is a two-bit ladder network used for digital-to-analog conversion. Express parameters of the corresponding Norton equivalent circuit in terms of (digital) voltages D_0,D_1 and resistance R.

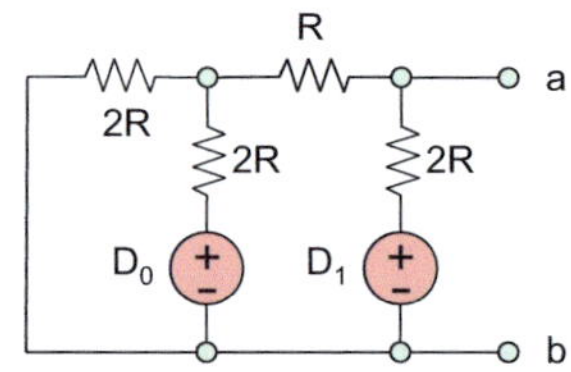

Problem 4.43. A two-terminal network shown in the figure is a four-bit ladder network used for digital-to-analog conversion. Express
(i) Thévenin (or equivalent) voltage
(ii) Thévenin (or equivalent) resistance

in terms of four (digital) voltages D_0,D_1,D_2,D_3 and resistance R.

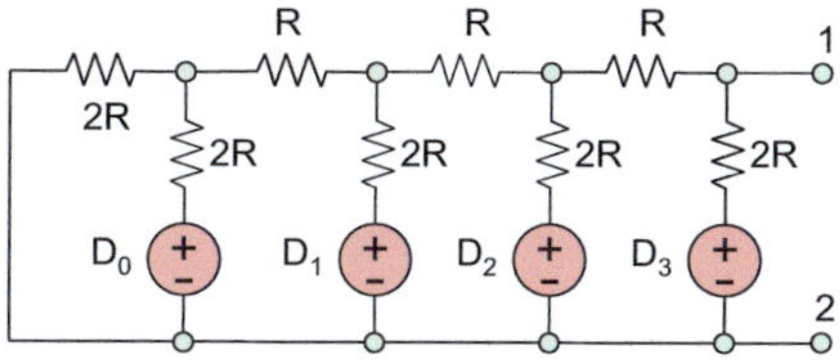

Problem 4.44. Determine the Norton equivalent for the circuit shown in the figure. Express your result in terms of *I* and *R*.

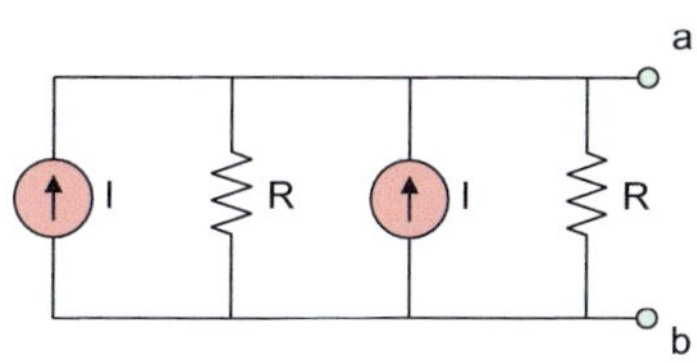

Problem 4.45. Establish Thévenin and Norton equivalent circuits for the network shown in the figure below.

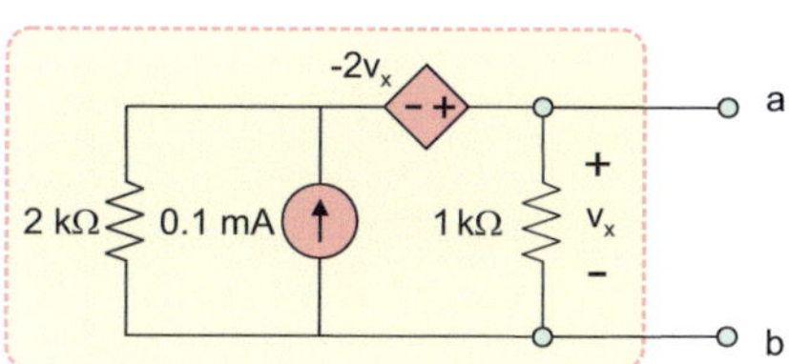

4.2.5 Application Example: Generating Negative Equivalent Resistance

Problem 4.46. Establish the equivalent (Thévenin) resistance for the network shown in the figure below. Carefully examine the sign of the equivalent resistance.

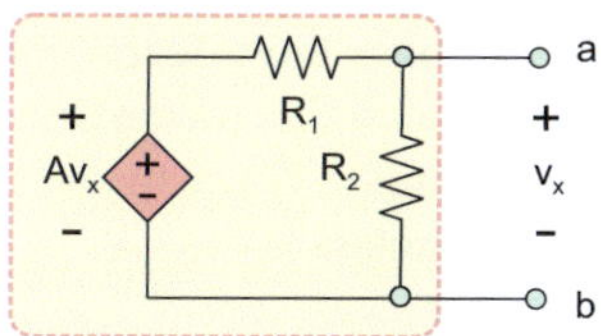

Problem 4.47. Derive the equivalent (Thévenin) resistance for the networks shown in the figure below (confirm Fig. 4.16 of the main text).

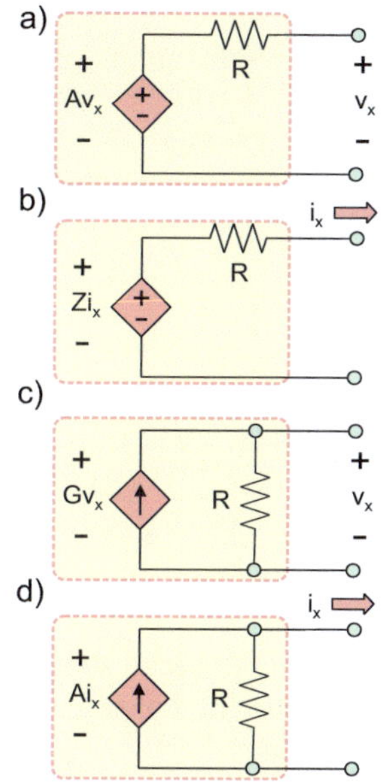

Problem 4.48. Each of three identical batteries is characterized by its Thévenin equivalent circuit with $V_T = 9$ V and $R_T = 1\ \Omega$. The batteries are connected in *series*. The entire battery bank is connected to a 5 Ω load. Find the power delivered to the load.

Problem 4.49. Each of three identical batteries is characterized by its Thévenin equivalent circuit with $V_T = 9$ V and $R_T = 1\ \Omega$. The batteries are connected in *parallel*. The entire battery bank is connected to a 5 Ω load. Find the power delivered to the load.

Problem 4.50. For two circuits shown in the figure below, find the current *I* using the method of Thévenin equivalent.

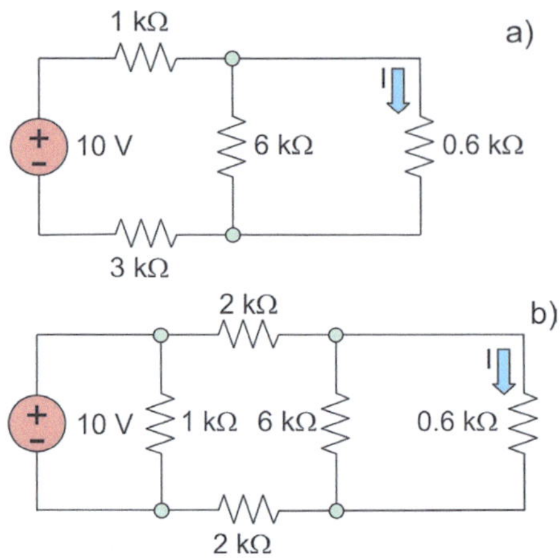

Problem 4.51. For the circuit shown in the figure below, find the current I using the method of Thévenin equivalent.

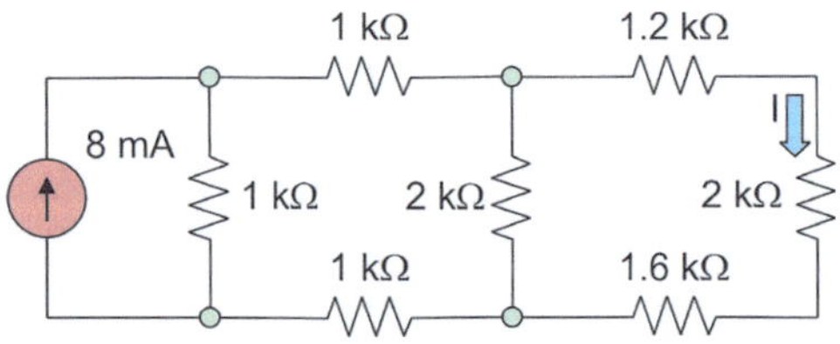

Problem 4.52. For the circuit shown in the figure below, find the current I using the method of Thévenin equivalent.

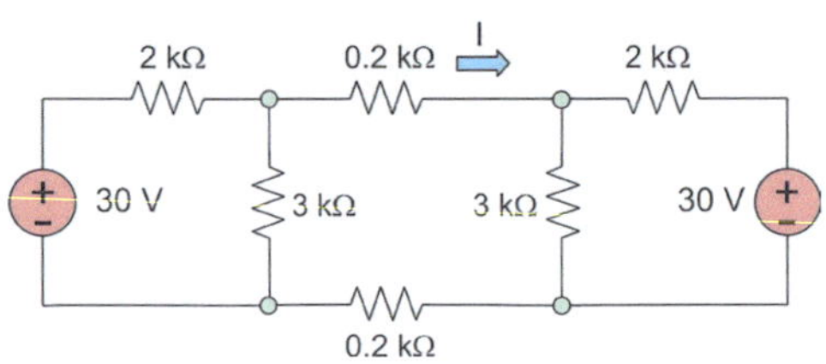

4.2.6 Summary of Circuit Analysis Methods

Problem 4.53. Solve the circuit shown in the figure below and determine the current I (show units). You can use any of the methods studied in class:

- Superposition theorem
- Nodal/mesh analysis
- Source transformation theorem
- Thévenin and Norton equivalents

or their combinations.

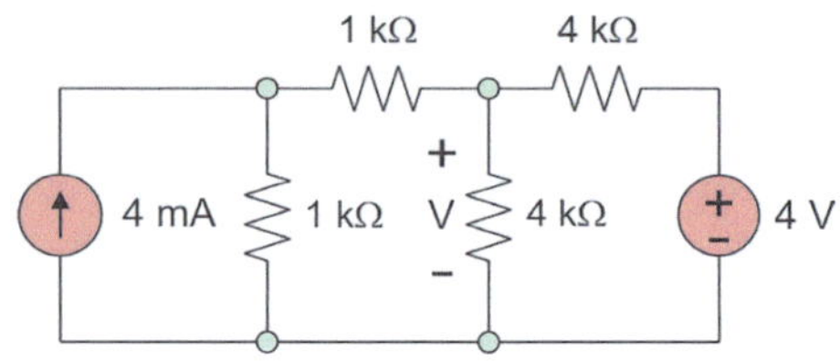

Problem 4.54. Solve the circuits shown in the figure below and determine the current I (show units). You can use any of the methods studied in class:

- Superposition theorem
- Nodal/mesh analysis
- Source transformation theorem
- Thévenin and Norton equivalents

or their combinations.

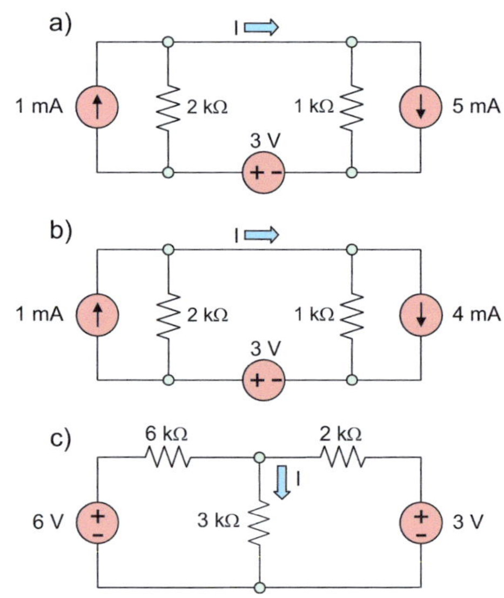

Problem 4.55. Solve the circuits shown in the figure below and determine unknown voltage V or current I (show units). You can use any of the methods studied in class:

- Superposition theorem
- Nodal/mesh analysis
- Source transformation theorem
- Thévenin and Norton equivalents

or their combinations.

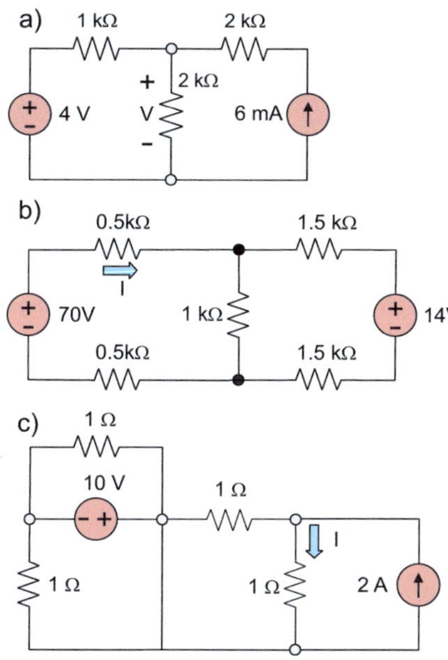

Problem 4.56. Solve the circuit shown in the figure below and determine the current I (show units). You can use any of the methods studied in class:

- Superposition theorem
- Nodal/mesh analysis
- Source transformation theorem
- Thévenin and Norton equivalents

or their combinations.

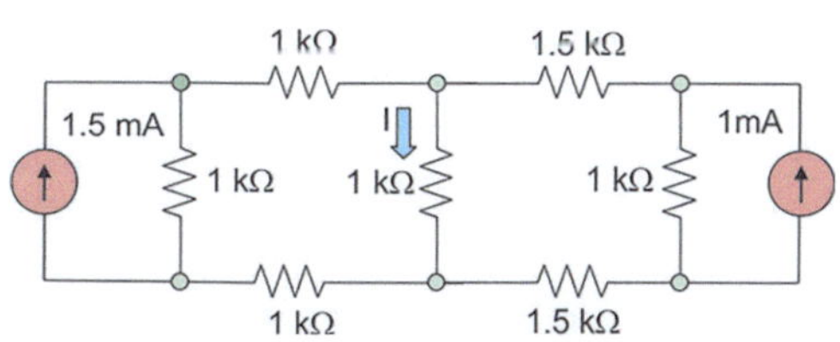

4.3 Power Transfer

4.3.1 Maximum Power Transfer

4.3.2 Maximum Power Efficiency

Problem 4.57. A deep-cycle marine battery is modeled by an ideal voltage source of 24 V in series with a 1 Ω resistance shown in the figure below. The battery is connected to a load, and the load's resistance, R_L, needs to be optimized. Power delivered to the load has a *maximum* at exactly one value of the load resistance. Find that value and prove your answer graphically using software of your choice (MATLAB is recommended).

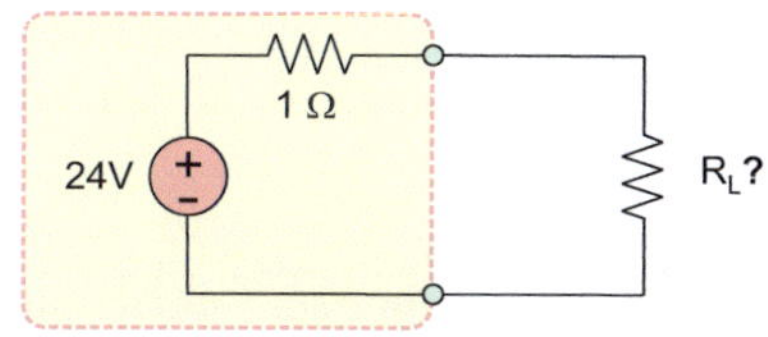

Problem 4.58. A power supply for an electric heater can be modeled by an ideal voltage source of unknown voltage in series with the internal resistance $R_T = 4\ \Omega$.

A. Can you still determine when the power delivered to a load (a heating spiral with resistance R_L) is maximized?

B. Does the answer depend on the source voltage?

Problem 4.59. For the circuit shown in the figure, when is the power delivered to the load maximized?

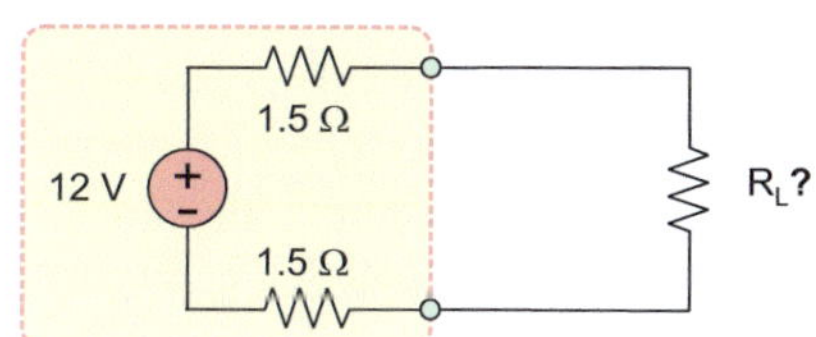

Problem 4.60. A battery can be modeled by an ideal voltage source V_T in series with a resistance R_T.

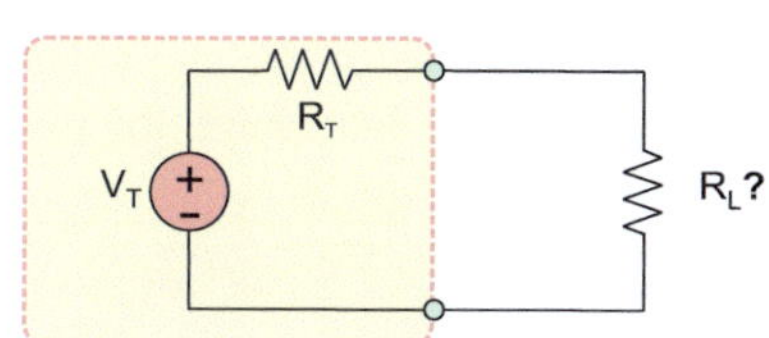

(i) For what value of the load resistance R_L is the power delivered to the load maximized?
(ii) What percentage of the power taken from the voltage source V_T is actually delivered to a load (assuming R_L is chosen to maximize the power delivered)?
(iii) What percentage of the power taken from the voltage source V_T is delivered to a load when $R_L = 0.1R_T$?

Problem 4.61. A certain micro-power photovoltaic device can be modeled under certain conditions as an ideal current power source and a resistance in parallel—see the figure below. At which value of the load resistance, R_L, is the power delivered to the load maximized?

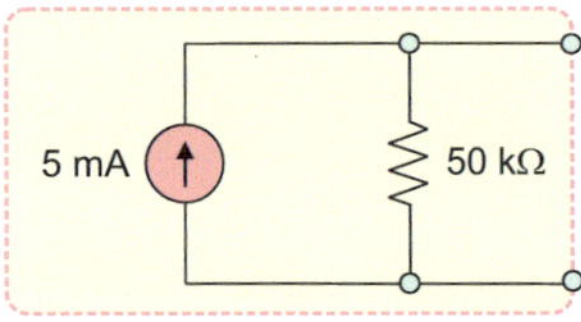

Problem 4.62. A low-cost polycrystalline Power Up BSP1–12 1 W Solar panel lists ratings for the output voltage and current, which give maximum load power: $V_{L\max}$=17.28 V, $I_{L\max}$=0.06A. Based on these cell specifications, which value of the equivalent resistance should the load to be connected to the solar cell have for maximum power output?

Problem 4.63. The solar panel from the previous problem generates a significant voltage of ~13 V in a classroom without direct sun light, but the resulting current is small, ~1 mA.

A. Now, which value of the equivalent resistance should the load have for maximum power output?
B. How is the maximum load power different, compared to the previous problem?

Problem 4.64. The heating element of an electric cooktop has two resistive elements, $R_1 = 50\ \Omega$ and $R_2 = 100\ \Omega$, that can be operated separately, in series, or in parallel from a certain voltage source that has a Thévenin (rms) voltage of 120 V and internal (Thévenin) resistance of 30 Ω. For the highest power output, how should the elements be operated? Select and explain one of the following: 50 Ω only, 100 Ω only, series, and parallel.

Problem 4.65. You are given two speakers (rated at 4 Ω and 16 Ω, respectively) and an audio amplifier with the output resistance(impedance) equal to 8 Ω.

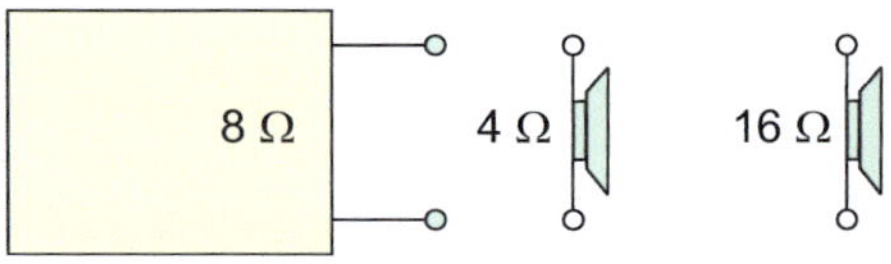

A. Sketch the circuit diagram that gives the maximum acoustic output with the available components. Explain your choice.
B. Sketch the circuit diagram for the maximum power efficiency. Explain your choice.

4.3.4 Application Example: Maximum Power Extraction from Solar Panel

Problem 4.66.

A. Describe in your own words the meaning of the fill factor of a solar cell (and solar module).
B. A 200 W GE Energy GEPVp-200 c-Si panel has the following reading on its back: V_{OC} = 32.9 V, I_{SC} = 8.1 A,

$V_{MP} = 26.3$ V, and $I_{MP} = 7.6$ A. What is the module fill factor? What is approximately the fill factor of the individual cell?

Problem 4.67. Using two Web links,
http://powerupco.com/site/
http://www.affordable-solar.com/,
identify the solar panel that has the greatest fill factor to date.

Problem 4.68. A 10 W BSP1012 PV c-Si module shown in the figure has 36 unit cells connected in series, the short-circuit current of 0.66 A and the open-circuit voltage of 21.3 V. The maximum power parameters are $V_{MP} = 17.3$ V and $I_{MP} = 0.58$ A.

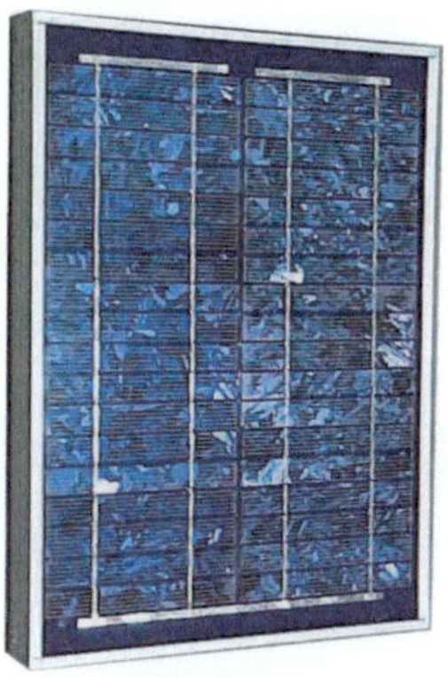

Power Up
10 Watt Module

A. Estimate the area of the single solar cell using the common photocurrent density value for c-Si solar cells (show units).
B. Estimate the open-circuit voltage for the single cell.
C. Estimate the fill factor of the module and of the cell.
D. Estimate the value of the equivalent load resistance R required for the maximum power transfer from the module to the load.

Problem 4.69. A 20 W BSP2012 PV c-Si module shown in the figure has 36 unit cells connected in series: the short-circuit current of 1.30 A, the open-circuit voltage of 21.7 V, the maximum power voltage V_{MP} of 17.3 V, and the maximum power current I_{MP} of 1.20 A. Repeat four tasks of the previous problem.

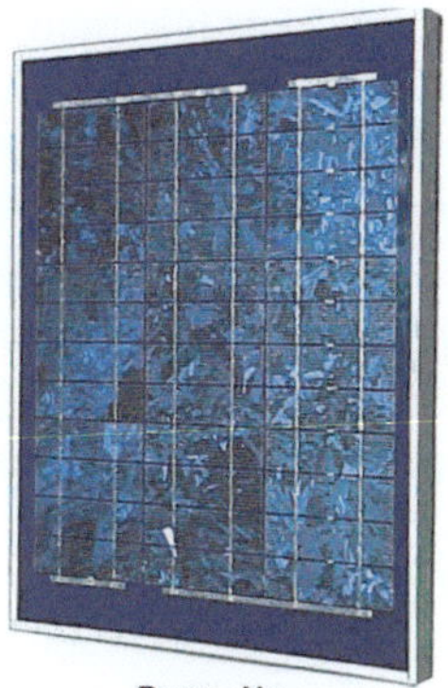

Power Up
20 Watt Module

Problem 4.70. A REC SCM220 220 Watt 20 V c-Si solar panel shown in the figure has the following readings on the back: the short circuit current I_{SC} of ~8.20 A, the open circuit voltage V_{OC} of ~36.0 V, the maximum power voltage V_{MP} of ~28.7 V, and the maximum power current I_{MP} of ~7.70 A. Repeat four tasks of problem 4.57.

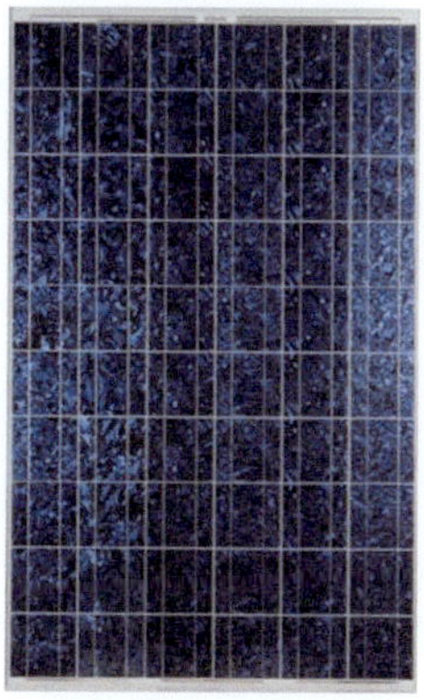

Problem 4.71. A 14.4 W load (a DC motor) rated at 12 V is to be driven by a solar panel. A c-Si photovoltaic sheet material is given, which has the open-circuit voltage of 0.6 V and the photocurrent density of $J_P = 0.03$ A/cm^2.

Outline parameters of a solar module (number of cells, cell area, and overall area) which is capable of driving the motor at the above conditions and estimate the overall panel size.

Problem 4.72. A custom 100 W load (a DC motor) rated at 24 V is to be driven by a solar panel. A c-Si photovoltaic sheet material is given, which has the open-circuit voltage of 0.6 V and the photocurrent density of $J_P = 0.03$ A/cm^2. Outline parameters of a solar module (number of cells, cell area, and overall area) which is capable of driving the motor at the above conditions and estimate the overall panel size.

Problem 4.73. You are given the generic fill factor $F = 0.72$ for the c-Si solar panels, the generic open-circuit voltage $V_{OC} = 0.6$ V of the c-Si cell, and the generic photocurrent density $J_P = 0.03$ W/cm^2.

A. Derive an analytical formula that expresses the total area A_{module} in cm^2 of a solar panel, which is needed to power a load, in terms of the required load power P_L.

B. Test your result by applying it to the previous problem.

Problem 4.74. You are given a low-cost low-power flexible (with the thickness of 0.2 mm) a-Si laminate from PowerFilm, Inc., IA with the following parameters: the fill factor of $F = 0.61$, the single-cell open-circuit voltage, $V_{OC} = 0.82$ V, and the photocurrent density, $J_P = 0.0081$ W/cm^2.

A. Derive an analytical formula that expresses the total module area A_{module} in cm^2, which is needed to power a load, in terms of the required load power P_L.

B. Compare your solution with the solution to the previous problem.

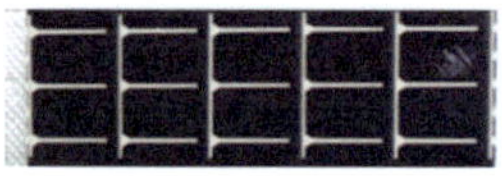

4.4 Analysis of Nonlinear Circuits: Generic Solar Cell

4.4.1 Analysis of Nonlinear Circuits: Load Line Method

4.4.2 Iterative Solution for Nonlinear Circuits

Problem 4.75. A circuit shown in the figure below contains a nonlinear passive element. Using the load line method approximately, determine the voltage across the element and the current through it for the two types of the v-i characteristic, respectively.

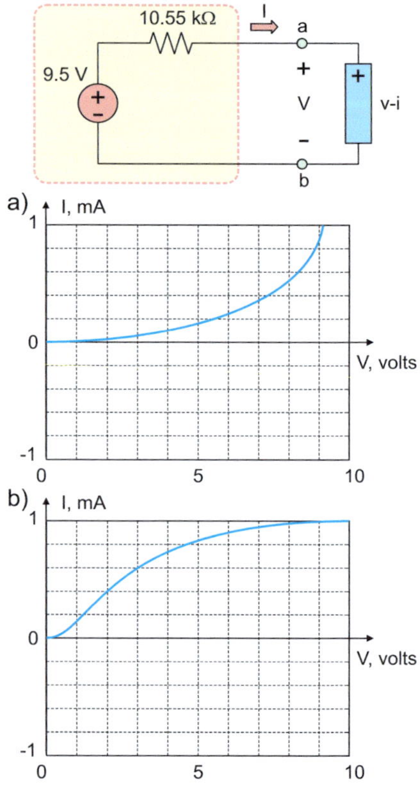

Problem 4.76. A circuit shown in the figure below contains a nonlinear passive element as a part of a current source. Using the load line method approximately, determine the voltage across the element and the current through it for the *υ*-*i* characteristic of the nonlinear element shown in the same figure.

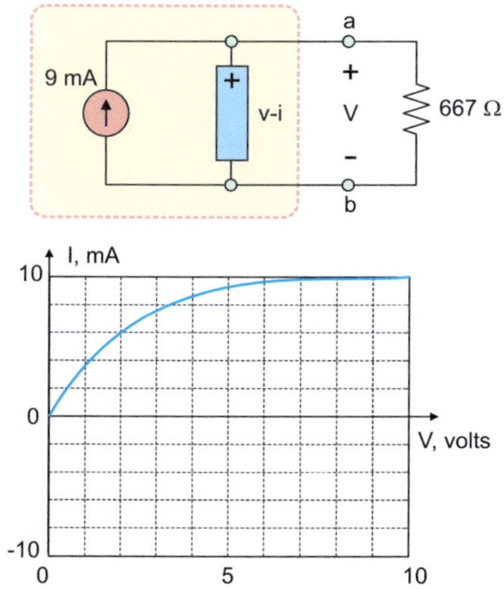

Problem 4.77. Repeat the previous problem for the circuit shown in the figure that follows.

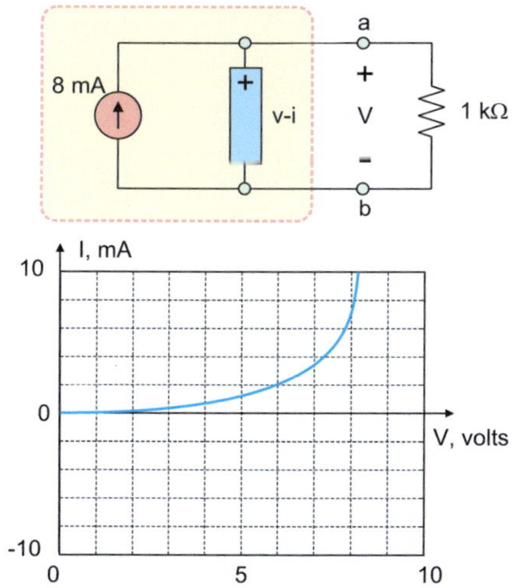

Problem 4.78. A circuit shown in the figure below contains a nonlinear passive element. The *υ*-*i* characteristic of the nonlinear element (the ideal Shockley diode) is

$$I = 5 \times 10^{-10}\left[\exp\left(\frac{V}{0.025\ \mathrm{V}}\right) - 1\right]\ [\mathrm{A}].$$

Using the iterative solution, determine the voltage across the element and the current through it.

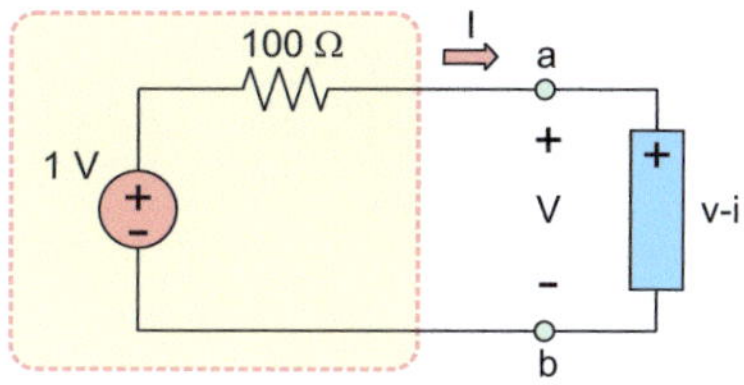

Problem 4.79. Repeat the previous problem for the circuit shown in the figure below. The *υ*-*i* characteristic of the nonlinear element is the same.

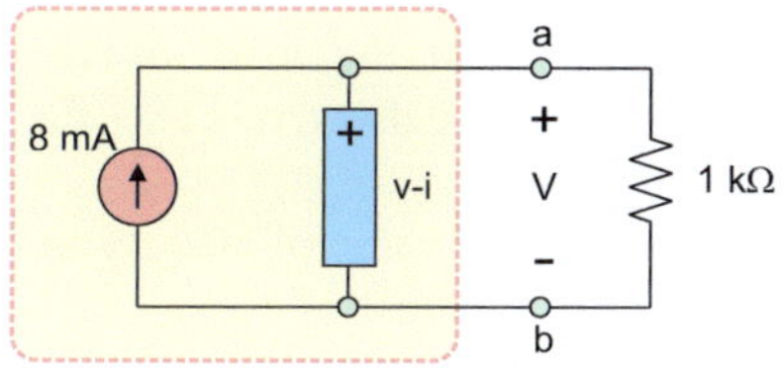

4.4.3 Application Example: Solving the Circuit for a Generic Solar Cell

Problem 4.80. The *I*(*V*)dependence for a resistive load in a circuit is shown in the figure below.

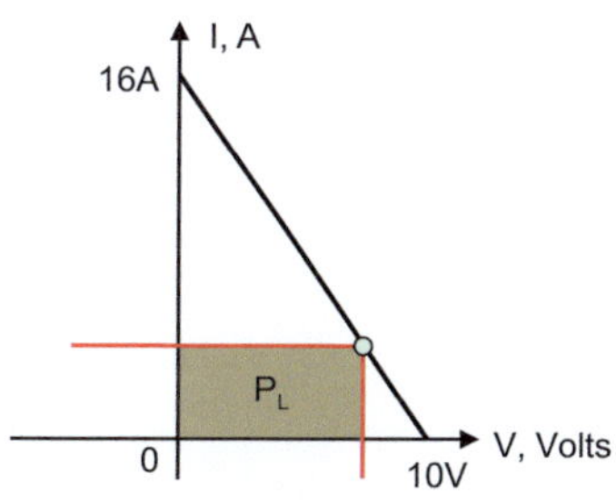

A. At which value of the load voltage is the power delivered to the load maximized?

B. What is the related value of load resistance?

Problem 4.81. A hypothetic thermoelectric engine developed by US Navy has the $I(V)$ dependence shown in the figure below.

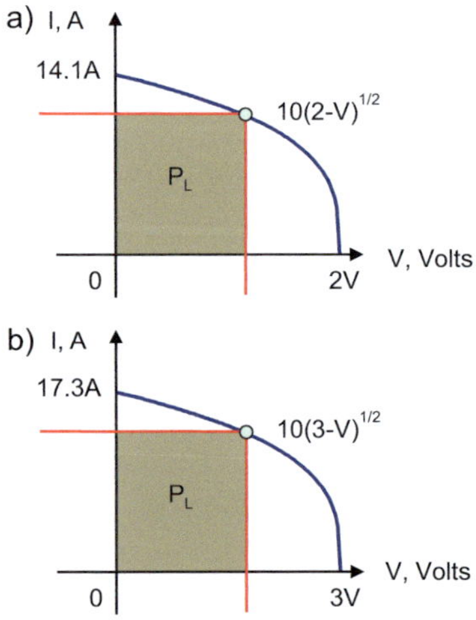

1. At which value of the load voltage is the power, P_L, delivered to the load maximized?
2. What is the related value of load resistance for maximum power transfer?

Problem 4.82. Estimate the values of V_{MP} and I_{MP} versus V_{OC} and I_{SC} for a set of generic c-Si solar cells. Every cell has $V_T = 0.026$V (room temperature of 25° C) and $V_{OC} = 0.6$ V. The ideality factor n in Eqs. (4.43)–(4.45) is allowed to vary over its entire range as shown in the Table below.

n	V_{MP}/V_{OC},%	I_{MP}/I_{OC},%	F
1.00			
1.25			
1.50			
1.75			
2.00			

Chapter 5: Operational Amplifier and Amplifier Models

Overview

Prerequisites:

- Knowledge of major circuit elements (dependent sources) and their v-i characteristics (Chapter 2)
- Knowledge of basic circuit laws (Chapter 3) and Thévenin equivalent (Chapter 4)

Objectives of Section 5.1:

- Learn and apply the model of an operational amplifier including principle of operation, open-circuit gain, power rails, and input and output resistances
- Correlate the physical operational amplifier with the amplifier circuit model
- Establish the ideal-amplifier model
- Learn the first practical amplifier circuit—the comparator

Objectives of Section 5.2:

- Understand and apply the concept of negative feedback to an operational amplifier circuit
- Construct three canonic amplifier circuit configurations with negative feedback: the non-inverting amplifier, the inverting amplifier, and the voltage follower
- Understand the current flow in the amplifier circuit including the power transfer from the power supply to the load

Objectives of Section 5.3:

- Choose the proper resistance values for the feedback loop and learn how to cascade multiple amplifier stages
- Learn about input/output resistances of the amplifier circuit and establish load bridging and load matching conditions important in practice
- Find ways to eliminate the DC imperfections of the amplifier that become very apparent at high amplifier gains
- Use an amplifier IC with a single voltage supply (a battery)

Objectives of Section 5.4:

- Obtain the initial exposure to differential signals and difference amplifiers
- Build an instrumentation amplifier
- Connect an instrumentation amplifier to a resistive sensor

S. N. Makarov et al., *Practical Electrical Engineering*, https://doi.org/10.1007/978-3-319-96692-2_5

Objectives of Section 5.5:

- Learn a general feedback system including closed-loop gain and error signal
- Apply the general feedback theory to voltage amplifier circuits
- Construct current, transresistance, and transconductance amplifiers with the negative feedback

Application Examples:
Operational amplifier comparator
Instrumentation amplifier in laboratory

Keywords:
Operational amplifier: (abbreviation op-amp, integrated circuit, dual in-line package, non-inverting input, inverting input, output terminal, power terminals, offset-null terminals, differential input voltage, open-circuit voltage gain, open-loop voltage gain, open-loop configuration, closed-loop configuration, power rails, voltage transfer characteristic, rail-to-rail, comparator, digital repeater, zero-level detector, circuit model, input resistance, output resistance, ideal amplifier, ideal-amplifier model, marking, summing point, common-mode input signal, differential input signal, summing-point constraints, first summing-point constraint, second summing-point constraint, sourcing current, sinking current, DC imperfections, input offset voltage, input bias current, input offset currents), Negative feedback, Feedback loop, Feedback as a dynamic process, Non-inverting amplifier, Inverting amplifier, Voltage follower (buffer) amplifier, Summing amplifier, Digital-to-analog converter, Binary counter, DC-coupled amplifier, AC-coupled amplifier, Capacitive coupling of an amplifier, Gain tolerance of an amplifier, Circuit model of a voltage amplifier, Input resistance of amplifier circuit, Output resistance of amplifier circuit, Load bridging (impedance bridging), Load matching (impedance matching), Cascading amplifier stages, Virtual-ground (integrated) circuit, Differential voltage of a sensor, Common-mode voltage of a sensor, Differential sensor, Single-ended sensor, Difference amplifier, Differential amplifier circuit gain, Common-mode amplifier circuit gain, Common-mode rejection ratio (CMRR), Unity common-mode gain stage, Instrumentation amplifier, Load cell, Current amplifier using op-amp, Transconductance amplifier using op-amp, Transresistance amplifier using op-amp, Howland current source (Howland current pump)
Linear feedback system: (forward gain? open-loop gain, feedback gain, feedback factor, summing node, difference node, closed-loop gain, error signal)

Section 5.1 Amplifier Operation and Circuit Models

The low-power amplifier *integrated circuit* (IC) is arguably the most widely employed discrete circuit component encountered in common electronic audio, control, and communication systems. Among amplifiers, the differential input, high-gain amplifier called the *operational amplifier* (or simply *op-amp*) has become a popular choice in many circuit applications. At this point, it is impossible for us to understand the internal operation of the amplifier IC without basic knowledge of semiconductor electronics, especially the junction transistor studied in the following courses. Fortunately, the circuit model of an operational amplifier does not require knowledge of the IC fabrication steps, nor does it require an understanding of the internal transistor architecture. Conceptually, operational amplifiers can be introduced early in the book, which enables us to immediately proceed toward our goal of designing and building practical circuits.

5.1.1 Amplifier Operation

Symbol and Terminals

After the amplifier chip is fabricated as an integrated circuit and the bond wires are attached, it is permanently sealed in a plastic package. Often the encasing is done in a *dual in-line* (DIP-N) *package* with N denoting the number of IC pins. Figure 5.1 on the right shows an example of a DIP package. One IC chip may contain several independent individual amplifiers. We start analyzing the amplifier model by first labeling the terminals and introducing the amplifier circuit symbol (a triangle) as shown in Fig. 5.1 on the left. The amplifier is typically powered by a dual-polarity voltage power supply with three terminals: $\pm V_{\mathrm{CC}}$ and common (ground) port of 0 V, see Section 3.2. The index C refers to the collector voltage of the internal transistors.

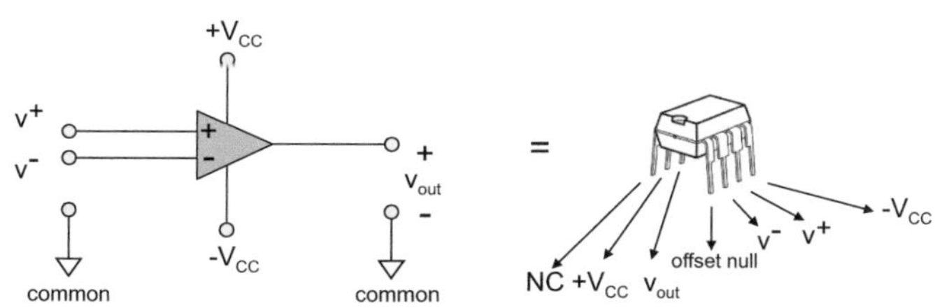

Fig. 5.1. Terminals of the operation amplifier (*left*); they also denote pins of the amplifier IC package (see a common LM 741 chip on the *right*). All voltages are referenced with respect to a common port of the dual-polarity voltage supply.

The amplifier has a total of *five* terminals, notably:

1. A *non-inverting input* with the input voltage υ^+ with respect to common
2. An *inverting input* with the input voltage υ^- with respect to common
3. An *output terminal* with the output voltage υ_{out} with respect to common

4. *Power terminal* $+V_{CC}$ with a positive voltage V_{CC} (+9 V) with respect to common
5. *Power terminal* $-V_{CC}$ with the negative voltage $-V_{CC}$ (−9 V) with respect to common

Each of the five terminals corresponds to a particular metallic pin of the IC package. *All* of the amplifier's terminals are used in an amplifier circuit and *none* of them should be left disconnected. However, the chip itself could have some *not connected* (NC) terminals that maintain symmetry and which are used as heat sinks, see Fig. 5.1 on the right. Also note that a number of amplifier ICs, including the LM74, may have extra terminals or pins, the so-called offset or offset-null terminals. These terminals are used to control the *input offset voltage* (an imperfection) of the amplifier.

Historical: The abbreviation for the operational amplifier is *op-amp*; this abbreviation is not quite official but is used by most practitioners. The term *operational amplifier* first appeared in a 1943 paper by John R. Ragazzini, an American electrical engineer and ECE professor. One of his students introduced the terms *inverting* and *non-inverting inputs*. One of his most notable students was Rudolf Kalman who became famous for the invention of the Kalman filters.

Open-Circuit or Open-Loop Voltage Gain

Once the amplifier chip is properly powered, its operation is quite simple: the output voltage is expressed through the two input voltages in the form

$$v_{\text{out}} = A(v^+ - v^-) \tag{5.1}$$

which is *identical* to the operation of the voltage-controlled voltage source introduced in Section 2.4. Here, $v^+ - v^-$ is the *differential input voltage* to the amplifier. The dimensionless constant A is called the *open-circuit voltage gain* of the amplifier. Quite frequently, the term *open-loop gain* is used and A is replaced by A_{OL}. Equation (5.1), which will be called the *amplifier equation*, is always valid. It does not matter if the amplifier is in the *open-loop configuration*, (i.e., no feedback loop is present) or in a *closed-loop configuration* (a feedback loop is present; see the next section). The amplifier IC is intentionally built in such a way as to provide the highest possible open-circuit gain; it is achieved using transistors connected in series such as the Darlington pair. Typically,

$$A \approx 10^5 - 10^8 \tag{5.2}$$

The exact gain value cannot be controlled precisely due to manufacturing tolerances. The open-circuit gain is often measured in V/mV. For example, the value of 160 V/mV corresponds to the gain value of 160,000. The open-circuit gain is difficult to measure.

Power Rails and Voltage Transfer Characteristic in the Open-Loop Configuration

Two power interconnects of the amplifier are often called *rails*. The term "rail" appears simply because the power interconnections are represented by two long horizontal wires in the circuit diagram connected to $+V_{CC}$ and $-V_{CC}$, respectively, which resemble long metal rails. The positive rail is $+V_{CC}$, and the negative rail is $-V_{CC}$. The power rails are interfaced to a laboratory dual-polarity voltage power supply that also provides a common (ground) port to be used later. The output amplifier voltage can never exceed the positive rail voltage or be less than the negative rail voltage. In other words,

$$-V_{CC} \le v_{out} \le V_{CC} \tag{5.3}$$

Should the output voltage found in Eq. (5.1) exceed V_{CC}, it will be forced to V_{CC}. Likewise, should the output voltage drop to less than $-V_{CC}$, it will be forced to $-V_{CC}$. In view of these physical constraints, Eq. (5.1) may be rewritten in the form

$$\begin{aligned} v_{out} &= A(v^+ - v^-), && |v_{out}| < V_{CC} \\ v_{out} &= +V_{CC}, && A(v^+ - v^-) > +V_{CC} \\ v_{out} &= -V_{CC}, && A(v^+ - v^-) < -V_{CC} \end{aligned} \tag{5.4}$$

Example 5.1: Plot to scale the output voltage of an operational amplifier with an open-circuit gain of $A = 10^5$ when the non-inverting input voltage v^+ changes from -1 mV to $+1$ mV and the inverting input voltage v^- is set to zero. The amplifier is powered by a ± 16-V dual voltage supply. This plot will give us the *voltage transfer characteristic* of the open-loop amplifier.

Solution: Amplifier Eq. (5.4) gives the result shown in Fig. 5.2 by a thick piecewise-linear curve. Due to the extremely high open-loop gain, the amplifier output is almost always *saturated*. This means that, except for a very narrow domain of input voltages on the order of ± 0.2 mV, the output simply follows the power rail voltage, either positive or negative. This is a very remarkable feature of the open-loop amplifier.

Power Rails in Practice

The power rail(s) of the amplifier or the supply voltage is specified in the datasheet. For example, the LM358 amplifier IC operates using a single supply 3 V to 32 V or dual supplies ± 1.5 V to ± 16 V. As we can see from this data, the amplifier does not necessarily operate using a dual voltage supply; a single supply ("single rail") can be used as well. The same amplifier chip (e.g., LM358) can be used either with the single voltage supply or with a dual supply. This question, although less important in theory, is very important in practice. Also note that, in practice, the output never exactly reaches the positive or negative rail voltages; there is always a voltage offset; it can vary from a minimum value of between 0.01 V and 0.05 V for certain special ICs (called the *rail-to-rail amplifiers*) all the way up to 1.8 V for other amplifiers (e.g., LM741).

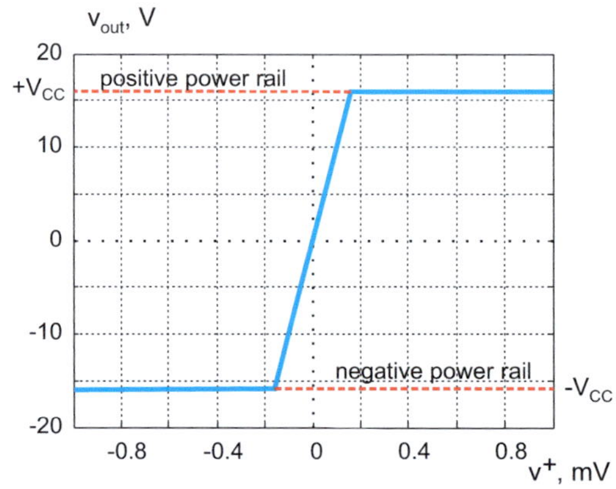

Fig. 5.2. Amplifier output voltage in the open-loop configuration. The open-loop gain is $A_{OL} = 10^5$ and the supply voltage is ± 16 V. Note that the scale for the input voltage is in mV.

5.1.2 Application Example: Operational Amplifier Comparator

A *comparator* is a circuit or a device that compares *two* input voltages and outputs a *digital voltage* (e.g., ± 10 V) as an indication of which input voltage is larger. Due to a very high gain, the operational amplifier in the open-loop configuration shown in Fig. 5.3a may operate as a basic comparator. Figure 5.3b shows one possible application of the comparator: *a digital repeater*. We assume that $v^- = V_{\text{threshold}} = 0$. The *input voltage to the comparator* $v^+(t)$ is a weak noisy digital signal shown in Fig. 5.3b. This signal is compared to a threshold level of zero volts (the threshold voltage in Fig. 5.3).

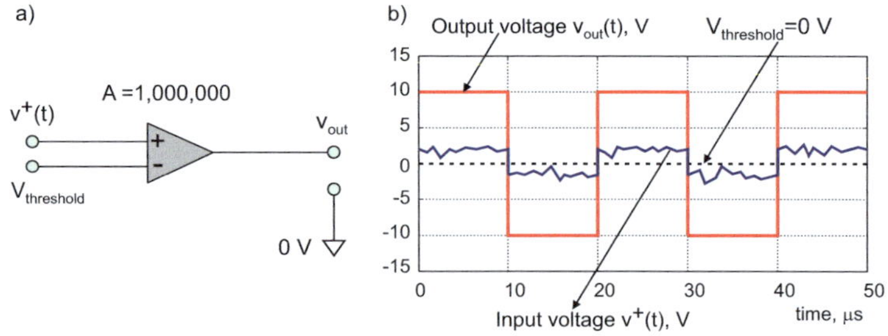

Fig. 5.3. A simple operational amplifier as a voltage comparator.

When the amplifier open-circuit gain tends to infinity (the transfer characteristic in Fig. 5.2 becomes a straight vertical line), Eq. (5.4) applied to the present case is reduced to

$$\begin{aligned} v_{\text{out}} &= +V_{\text{CC}}, & v^+(t) &> 0 \\ v_{\text{out}} &= -V_{\text{CC}}, & v^+(t) &< 0 \end{aligned} \tag{5.5}$$

Figure 5.3b shows the resulting output voltage for $V_{\text{CC}} = 10$ V. The weak input digital signal will thus be amplified and cleaned from noise, which is one major function of a *digital repeater*. In practice, dedicated comparators are used instead of this simple setup, which are much faster and have useful additional features. The comparator amplifier may also be employed for other purposes such as a *zero-level detector*.

Exercise 5.1: In Fig. 5.3, the threshold voltage of the comparator amplifier is changed to +5 V. What will be the output of the comparator circuit?
Answer: -10 V at any time instant.

5.1.3 Amplifier Circuit Model

Circuit Model

An equivalent circuit model of an amplifier is shown in Fig. 5.4. This circuit model is a two-port electric network. It includes three single circuit elements: an ideal voltage-controlled voltage source $A(v^+ - v^-)$ (Section 2.4), *input resistance* R_{in} *of the amplifier*, and *output resistance* R_{out} *of the amplifier*.

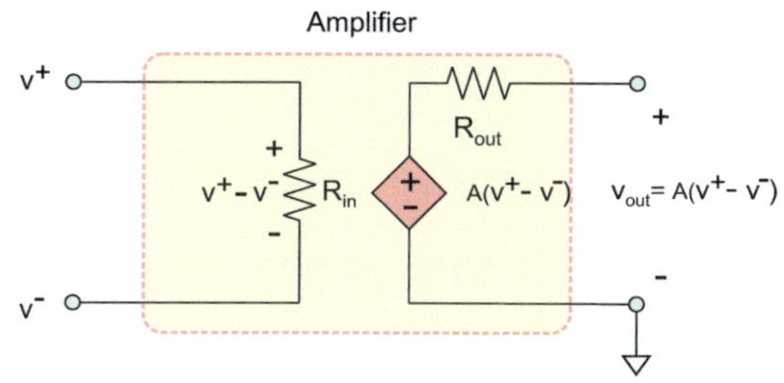

Fig. 5.4. Equivalent circuit model of an amplifier is in the shadow box as a two-port network. No load is connected. The ground of the output terminal is the common port.

Analysis of the Amplifier Circuit Model: Effect of Input/Output Resistances

The input/output resistances of an amplifier in Fig. 5.4 impose rather severe limitations on its desired operation. First, a large but finite input resistance always implies that some *input current*, $i_{\text{in}} = (v^+ - v^-)/R_{\text{in}}$, will flow into the amplifier as long as the input voltage signal is different from zero. Consequently, the amplifier would require not only the input voltage but also a certain amount of input power. As a result, the amplifier may appreciably *load* a sensor connected to its input, i.e., require more power than (a tiny) sensor can actually provide. Second, a finite output resistance limits the *output current* i_{out} to the amplifier; this resistance operates as a current limiting resistor which is studied in Chapter 3. Along with this, it also leads to the fact that the voltage across any load connected to the amplifier's output will *not* be equal to the desired output voltage given by Eq. (5.1), except for an open circuit. These limitations are quantified when we consider a circuit shown in Fig. 5.5. The circuit includes the amplifier model, an arbitrary source represented by its Thévenin equivalent v_S, R_S, and a load represented by its equivalent resistance R_L.

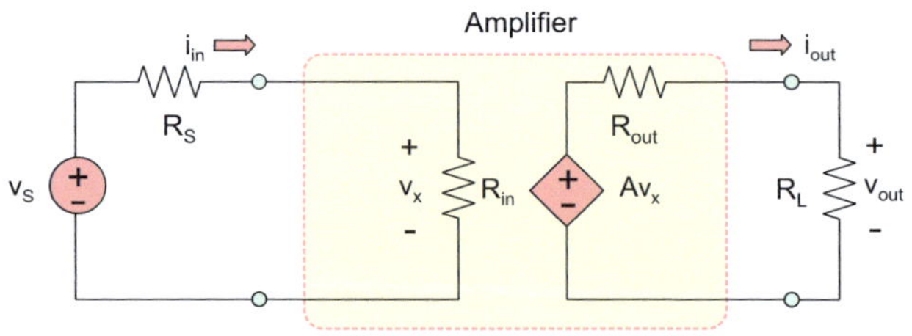

Fig. 5.5. Amplifier circuit model with connected source and load resistances.

Using the voltage division principle twice, the output voltage in Fig. 5.5 is expressed as

$$v_{out} = \underbrace{v_S \times \left(\frac{R_{in}}{R_{in} + R_S}\right)}_{v_X} \times A \times \left(\frac{R_L}{R_L + R_{out}}\right) \tag{5.6}$$

This result is quite different from the ideal behavior of the amplifier described by the perfect amplification of the source signal

$$v_{out} = Av_S \tag{5.7}$$

using the available open-circuit gain A of the amplifier.

Exercise 5.2: For the amplifier circuit in Fig. 5.5 with $A = 1000$, determine the output voltage given that $v_S = 1$ mV, $R_S = 50\ \Omega$, and $R_L = 50\ \Omega$ for two cases:

A. $R_{in} = 1\ \text{M}\Omega$ and $R_{out} = 1\ \Omega$.
B. $R_{in} = 50\ \Omega$ and $R_{out} = 50\ \Omega$.

Answer:
Case A: $v_{out} = 0.98$ V (which is close to the ideal behavior, $v_{out} = 1.00$ V).
Case A: $v_{out} = 0.25$ V (three quarters of the voltage gain are lost).

According to Eq. (5.6), $v_{out} < Av_S$ for any positive finite values of R_{in} and R_{out}. In order to make use of the full available open-circuit gain A of the amplifier, we should:

1. Design R_{in} as large as possible, ideally an open circuit, that is,

$$R_{\text{in}} = \infty \tag{5.8}$$

2. Design R_{out} as small as possible, ideally a short circuit, that is,

$$R_{\text{out}} = 0 \tag{5.9}$$

In this and only this case, the equality $\upsilon_{\text{out}} = A\upsilon_S$ will be satisfied exactly.

5.1.4 Ideal-Amplifier Model and First Summing-Point Constraint

The amplifier IC can then be described with a high degree of accuracy by using the so-called ideal-amplifier model. It is based on the best possible choices for input/output resistances as described by Eqs. (5.8) and (5.9), respectively. It is also based on the assumption that the open-loop gain in Eq. (5.1) is made as high as possible, i.e., equal to infinity. The ideal-amplifier model is an important theoretical and practical tool for the analysis of microelectronic amplifier circuits. This model will be used in the following sections of this chapter and in subsequent chapters. We can summarize the model of an ideal operational amplifier in concise form:

1. No current can flow into the amplifier (into either input terminal).
2. The open-loop gain A is infinitely high.
3. The input resistance R_{in} is infinitely high.
4. The output resistance R_{out} is zero.

Property 1 follows from property 3 and vice versa. One more condition of the ideal-amplifier model could be added, namely, that the power rails $\pm V_{CC}$ are exactly reached when operated in saturation. The ideal-amplifier model does not use the accurate internal amplifier circuit shown in Fig. 5.4 or in Fig. 5.5, respectively. Instead, a simple triangle symbol may be used for the ideal amplifier, which is shown in Figs. 5.1 and 5.3.

Exercise 5.3: Solve the previous exercise for the ideal-amplifier model.

Answer: Case A, B: $\upsilon_{\text{out}} = 1.00$ V.

First Summing-Point Constraint

The *summing point* of an amplifier is the connection of the two inputs to the amplifier. The *common-mode input signal* is the half sum of the two input voltages, $(\upsilon^+ + \upsilon^-)/2$. The *differential input signal* is the input voltage difference, $\upsilon_x = \upsilon^+ - \upsilon^-$. Conditions applied to the amplifier's input are called *summing-point constraints*. The first *summing-point constraint* is applied to the ideal-amplifier model. It states that no current can flow into either of the amplifier terminals as shown in Fig. 5.6.

This is consistent with an infinitely high input resistance. The condition of no input current into the amplifier means that virtually no input power is necessary. For example,

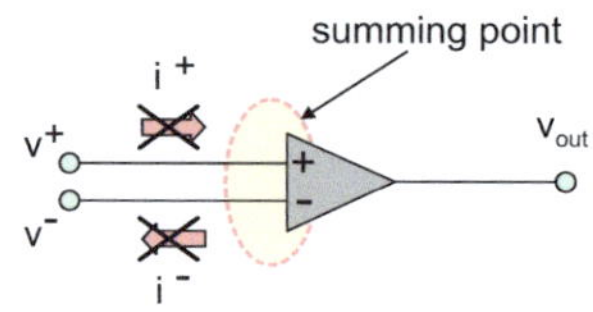

Fig. 5.6. The first summing-point constraint stipulates that no current flows into the ideal amplifier.

a tiny sensor, which does not deliver any appreciable power, could directly be connected to the input. The voltage from the sensor will still be accepted as the input of the amplifier. This condition is a convenient abstraction of the ideal-amplifier model. In reality, a very small input current does exist, typically on the order of nanoamperes (nA) for common amplifier ICs or picoamperes (pA) for ICs with an input JFET stage.

Realistic Values of Input/Output Resistances and Output Current

How far off from reality is the assumption of infinite input resistance? A review of the datasheets reveals that the input resistance of the older amplifier IC (e.g., LM741, LM1458) varies from 0.3 to 6 MΩ. The input resistance of JFET-input stage amplifiers (TL082) is on the order of 1 TΩ (10^{12} Ω). Now, how realistic is the assumption of zero output resistance? Note that if the output resistance were exactly zero, the amplifier would be able to source an infinite current (power) into a low-resistance load. Clearly, we cannot expect a large output power from a physically small amplifier IC. Therefore, we have to introduce a small internal output resistance, which appears to be on the order of 1–100 Ω. The corresponding *output short-circuit current* of the common amplifier ICs (LM741, LM1458, LM358) cannot exceed 40–60 mA; the current into a load is smaller. The output current of faster amplifier ICs (TL082) is even smaller. If the load requires more current than the chip can provide, then the output voltage will notably be clipped.

Section 5.2 Negative Feedback

In most practical circuits, the amplifier IC is not used in open-loop configuration. Engineers have modified the open-loop condition into a negative feedback loop in order to set the gain to a desired value and ensure the amplifier's stability. This section provides you with all the essential knowledge needed to design an amplifier with negative feedback. The mathematical model introduced in this section is based on two conditions imposed at the amplifier's input; we term them the *summing-point constraints*:

a) No electric current flows into or out of the amplifier inputs.
b) The differential voltage at the amplifier's input is zero.

The first summing-point constraint has already been introduced in the previous section. The second summing-point constraint has yet to be derived. We will show that the two summing-point constraints, along with KCL and KVL, will enable us to solve *any* amplifier circuit that involves a negative feedback, no matter what the specific nature of the feedback loop is and regardless of whether it is DC, AC, or a transient circuit.

5.2.1 Idea of the Negative Feedback

The idea of the *negative feedback* goes way back—we may say almost to the Stone Age. Take a wooden rod of 1–2 feet in length. Hold the rod in the vertical position at the tip of your finger. You will probably succeed. Now, close your eyes and try to do the same. You will most likely fail. The reason for the failure is a breakdown of the feedback loop. This loop is created by visual control of the rod's position; you automatically apply a compensating acceleration to the bottom tip of the rod when it begins to fall. Another good example is driving a car and trying to stay in the center of the lane. The negative feedback for electronic amplifiers was first invented and realized by Harold S. Black (1898–1983), a 29-year-old American electrical engineer at Bell Labs. To many electrical engineers, this invention is considered perhaps the most important breakthrough of the twentieth century in the field of electronics because of its wide applicability. We will construct simple amplifier circuits of a given gain, using a resistive feedback loop. Being able to perform this task is already critical from the practical point of view.

5.2.2 Amplifier Feedback Loop: Second Summing-Point Constraint

We construct *the feedback loop*, as shown in Fig. 5.7, by connecting the output to the inverting input terminal. This was exactly the idea of Harold Black. The shadowed box in the feedback loop may represent one or more circuit elements. The feedback loop may be a simple wire, a resistance, a network of circuit elements (resistances, inductances, capacitances), etc. The *negative feedback* simply means that *the output voltage, or rather a portion of it, is returned back to the inverting input.*

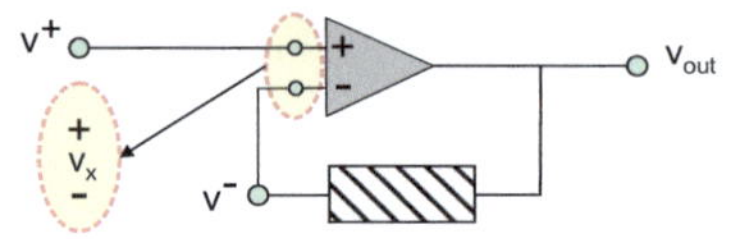

Fig. 5.7. A feedback loop around an amplifier.

Feedback as a Dynamic Process

According to Eq. (5.1) of the previous section, the output voltage is proportional to

$$\upsilon_x = \upsilon^+ - \upsilon^- \tag{5.10}$$

where υ_x is the differential input voltage. Hence, υ_x or υ_x multiplied by a constant is returned to the input during a very short period of time. The feedback effect is inherently a very fast *dynamic process*, which leads to a static solution with quite remarkable properties. In the example that follows, we will attempt to model the effect of the feedback loop using several very simplifying assumptions.

Example 5.2: An amplifier with a feedback loop in Fig. 5.7 has υ^+ fixed at +10 V. υ^- is equal to 0 V at $t = 0$. We shall assume that 50 % of υ^+ is returned back to the input in 1 μs. How does the differential voltage υ_x change with time?

Solution:

1. At $t = 0$, $\upsilon_x = 10\,\text{V} - 0\,\text{V} = 10\,\text{V}$. Next, 50 % of 10 V is returned in 1 μs. The voltage υ^- becomes equal to $0\,\text{V} + 5\,\text{V} = 5\,\text{V}$ after 1 μs.
2. At $t = 1$ μs, $\upsilon_x = 10\,\text{V} - 5\,\text{V} = 5\,\text{V}$. Next, 50 % of 5 V is returned in 1 μs. The voltage υ^- becomes equal to $5\,\text{V} + 2.5\,\text{V} = 7.5\,\text{V}$ after 2 μs.
3. At $t = 2$ μs, $\upsilon_x = 10\,\text{V} - 7.5\text{V} = 2.5\,\text{V}$. Next, 50 % of 2.5 V is returned in 1 μs. The voltage υ^- becomes equal to $7.5\,\text{V} + 1.25\,\text{V} = 8.75\,\text{V}$ after 3 μs.

The process further continues so that voltage υ_x halves every microsecond. The process dynamic is shown in Table 5.1 and visualized in Fig. 5.8.

Table 5.1. Dynamics of the differential input voltage as a function of time for Example 5.2.

Time, μs	υ^+	υ^-	$\upsilon_x = \upsilon^+ - \upsilon^-$
0	10 V	0 V	10 V
1	10 V	5 V	5 V
2	10 V	7.5 V	2.5 V
3	10 V	8.75 V	1.25 V
4	10 V	9.375 V	0.625 V

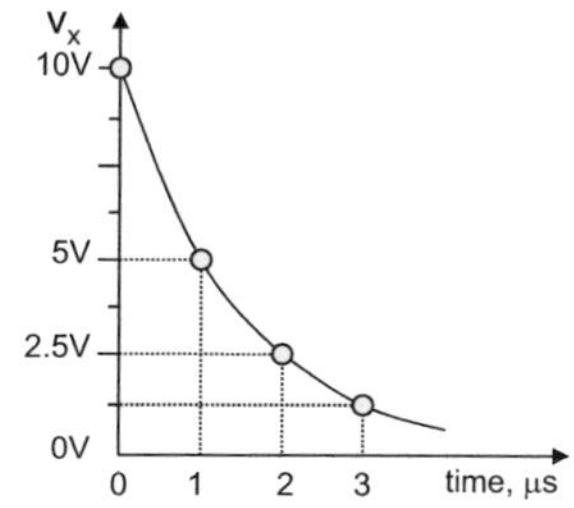

Fig. 5.8. Dynamics of the differential input voltage as a function of time.

Both Table 5.1 and Fig. 5.8 make clear that the differential voltage v_x decays to zero very rapidly, once the feedback loop is introduced. Hence we arrive at the *second summing-point constraint*, which is valid only for the amplifiers with the negative feedback loop: *the differential input voltage to the amplifier is exactly equal to zero.* The second summing-point constraint is a close approximation to reality. Its accuracy depends on the value of the open-loop gain of the amplifier. If the open-loop gain were infinite, the second summing-point constraint would be exact.

5.2.3 Amplifier Circuit Analysis Using Two Summing-Point Constraints

Next, we will solve an amplifier circuit with negative feedback using the two summing-point constraints (SPC): (i) no current into or out of the input amplifier terminals and (ii) the differential input voltage is zero. The method of two summing-point constraints is an accurate solution method for a wide variety of amplifier circuits with the negative feedback. For amplifier circuits with a *single input*, we will denote the *input voltage to the amplifier circuit* by v_{in}. Voltage v_{in} may be equal to v^+ or to v^-, depending on amplifier type to be used.

Non-inverting Amplifier

The first amplifier configuration is the so-called non-inverting amplifier shown in Fig. 5.9. The feedback loop contains one resistance R_2. Another resistance R_1 shunts the inverting input to ground. The input voltage to the amplifier circuit is the voltage v_{in} with respect to ground, or common in this case, which implies the use of the dual-polarity voltage power supply. The output voltage with respect to common is v_{out}. We apply the first summing-point constraint and KCL to the node "*" in Fig. 5.9 and obtain

$$i_1 = i_2 \tag{5.11}$$

Equation (5.11) is further transformed using Ohm's law in the form

$$\frac{v^* - 0}{R_1} = \frac{v_{out} - v^*}{R_2} \tag{5.12}$$

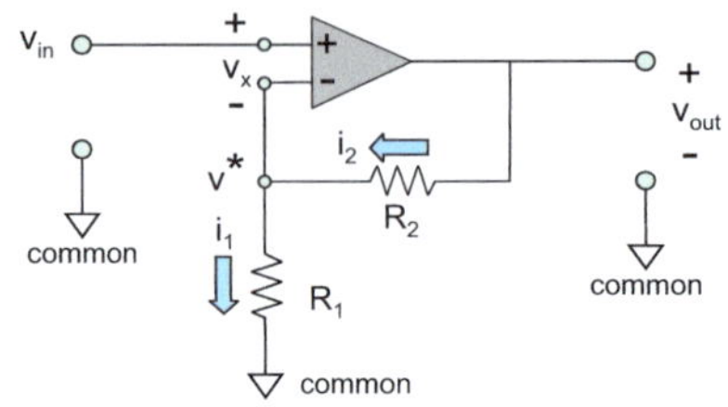

Fig. 5.9. Circuit diagram of the non-inverting amplifier. A dual power supply is not shown.

The second summing-point constraint yields

$$v^* = v_{\text{in}} \tag{5.13}$$

since $v_x = 0$. Equation (5.12) thus reads

$$\frac{v_{\text{in}}}{R_1} = \frac{v_{\text{out}} - v_{\text{in}}}{R_2} \Rightarrow \frac{v_{\text{out}}}{R_2} = \frac{v_{\text{in}}}{R_1} + \frac{v_{\text{in}}}{R_2} \tag{5.14}$$

As a result, we find that the voltage input-to-output relation becomes

$$v_{\text{out}} = \left(1 + \frac{R_2}{R_1}\right) v_{\text{in}} \tag{5.15}$$

The amplifier circuit is solved: we have expressed the output voltage in terms of the input voltage and a resistor ratio. Equation (5.15) is the basic result in amplifier theory. It shows that the feedback loop allows us to precisely control the gain with two arbitrary resistances. One chooses the proper resistance combination to achieve any finite gain between one (setting $R_2 = 0$) and the open-loop (infinite) gain (setting $R_1 = 0$). In the last case, the negative input terminal becomes grounded; the feedback loop is irrelevant and can be replaced by an open circuit so that the amplifier again becomes the *comparator*. The gain expression

$$A_{\text{CL}} = \left(1 + \frac{R_2}{R_1}\right) \geq 1 \tag{5.16}$$

is called the *closed-loop gain* of the amplifier; it clearly relates the output voltage to the input voltage. Equation (5.16) is a dramatic illustration of the negative feedback. We started with an amplifier having a very large yet loosely predictable open-loop gain. Through applying the negative feedback, we arrived at a gain that is much smaller than the open-loop gain; however, it is controllable and stable. Equation (5.16) can be derived more simply using the voltage divider concept. Namely, resistors R_1, R_2 form a voltage divider between 0 V and the output voltage. Hence, the voltage at node (*) may be found. Equating this voltage to the input voltage gives us Eq. (5.16).

Exercise 5.4: Solve the circuit shown in Fig. 5.10, i.e., find the output voltage v_{out} with respect to common.

Answer: $v_{out} = \left(1 + \frac{R_2}{R_1}\right) v_{in} = \left(1 + \frac{1 \times 10^6}{5.1 \times 10^3}\right) \times 1 \text{ mV} = 197 \text{ mV}.$

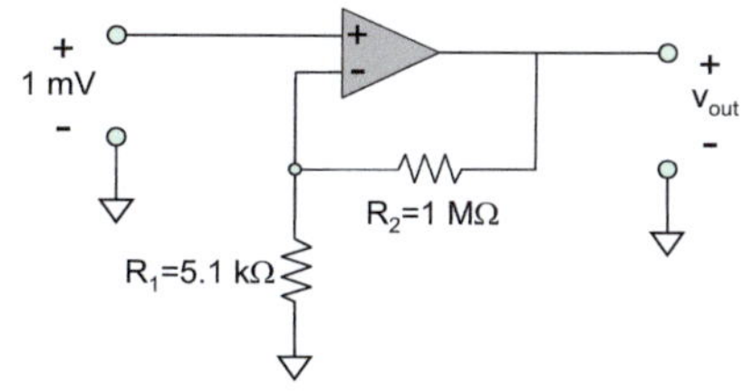

Fig. 5.10. A non-inverting amplifier circuit with an input voltage of 1 mV.

Inverting Amplifier

The next amplifier circuit configuration is the *inverting* amplifier shown in Fig. 5.11. Note that the input terminals are now flipped. The negative feedback loop is still present; it involves resistance R_2. Another resistance, R_1, forms a voltage divider.

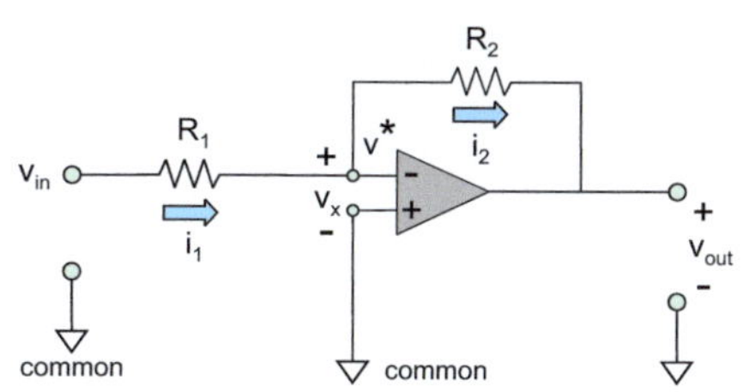

Fig. 5.11. Circuit diagram of the inverting amplifier; a dual power supply is used (not shown).

The input voltage to the amplifier circuit is the voltage v_{in} with respect to ground or common. The output voltage with respect to common is v_{out}. To solve the amplifier circuit, we use the same solution procedure as for the non-inverting amplifier. However, the final result will be quite different. We apply the first summing-point constraint and KCL to the node labeled "*" in Fig. 5.11 and again obtain

$$i_1 = i_2 \tag{5.17}$$

Equation (5.17) is transformed using Ohm's law,

$$\frac{v_{in} - v^*}{R_1} = \frac{v^* - v_{out}}{R_2} \tag{5.18}$$

The second summing-point constraint yields $v^* = 0$ since $v_x = 0$. Equation (5.18) then gives

$$v_{\text{out}} = -\frac{R_2}{R_1} v_{\text{in}} \tag{5.19}$$

The amplifier circuit is solved: we have expressed the output voltage in terms of the input voltage. Equation (5.19) is another key result in amplifier theory. The expression

$$A_{\text{CL}} = -\frac{R_2}{R_1} \tag{5.20}$$

is also called the *closed-loop gain* of the inverting amplifier; the gain again relates the output voltage to the input voltage. It is now negative, which means that the output voltage is inverted. This circumstance is hardly important for the AC signals where the voltage inversion is equivalent to a phase shift of π radians or 180 degrees. The feedback loop of the inverting amplifier also enables us to control the gain of the amplifier with two standard resistors. We can choose the proper resistance combination to achieve any finite gain between zero ($R_2 = 0$) and negative infinity ($R_1 = 0$). In Fig. 5.11 we clearly see how the amplifier gain is controlled by the voltage divider with resistors R_1 and R_2.

Exercise 5.5: Solve the inverting-amplifier circuit shown in Fig. 5.12, i.e., find the output voltage v_{out} with respect to common.

Answer: $v_{\text{out}} = -\frac{R_2}{R_1} v_{\text{in}} = -\frac{1 \times 10^4}{51} \times 1 \text{ mV} = -196 \text{ mV}.$

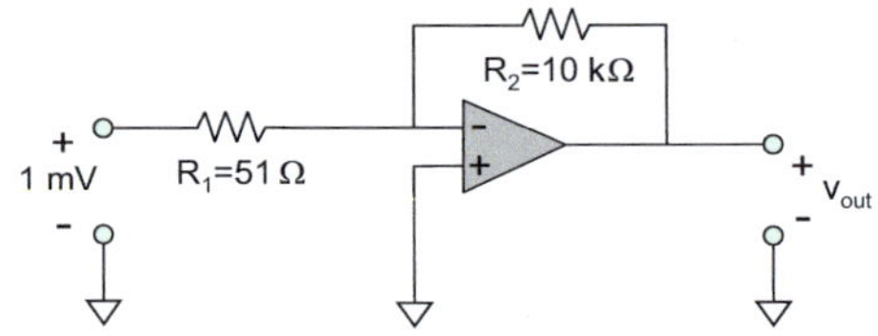

Fig. 5.12. An inverting amplifier circuit with an applied input voltage of 1 mV.

Voltage Follower or Buffer Amplifier

The third important member of the amplifier family is the *voltage follower* or *buffer* amplifier whose circuit is shown in Fig. 5.13. The negative feedback loop is just a wire.

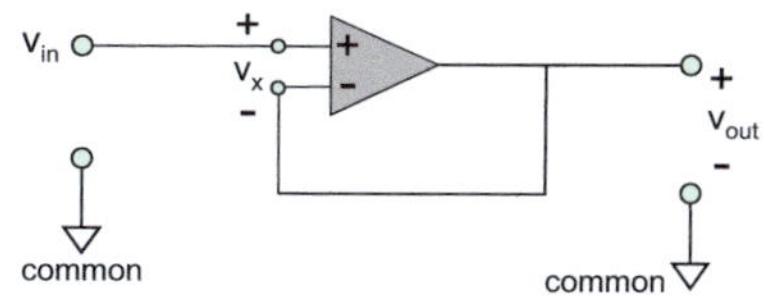

Fig. 5.13. Circuit diagram of the buffer amplifier; a dual power supply is used.

The use of the second summing-point constraint immediately leads to

$$v_{\text{out}} = v_{\text{in}}, \quad A_{\text{CL}} = 1 \tag{5.21}$$

so that the gain of this amplifier type is simply unity. Why do we need a unity-gain amplifier? The reason is that, while the buffer amplifier in Fig. 5.13 passes the voltage without change, it requires virtually no current at the input (virtually no input power) but, at the same time, could source a significant current (on the order of 20–40 mA) at the output, i.e., provide significant output power. In order words, it becomes in a certain sense a *power* amplifier. A simple example would be a capacitive sensor that cannot deliver currents on the order of 10 mA or even smaller currents; otherwise, the corresponding capacitor would immediately discharge. Such a sensor cannot directly be connected to an LED indicator that requires at least 10 mA. However, this sensor may deliver significant voltages, on the order of 1–5 V, which do not need to be amplified. The use of a buffer amplifier can nicely solve this connection problem. The above discussion directly leads us to the concept of *input resistance of the amplifier circuit* studied in the next section.

Exercise 5.6: Solve the voltage-follower circuit shown in Fig. 5.13, i.e., find the output voltage v_{out} with respect to common when the input voltage with respect to common is a) 1 V and b) 10 V. The amplifier is powered by a ±6-V dual supply.

Answer: a) $v_{\text{out}} = 1$ V; b) $v_{\text{out}} = 6$ V.

5.2.4 Mathematics Behind the Second Summing-Point Constraint

The second summing-point constraint might appear to be mysterious, at least at first sight. How does the amplifier accept the input signal if there is no current at the input and the differential input voltage is zero? Can we avoid using the second SPC, and at what cost? We will show that the second SPC is nothing but a handy tool to solve the amplifier circuit with the negative feedback, with a high degree of accuracy. Mathematically, the second SPC gives a leading (and usually very accurate) term of what is known as an *asymptotic expansion* with regard to a small parameter, here the inverse open-loop gain A^{-1}. Let us now ignore the second SPC and derive the gain equation for the buffer amplifier exactly. A similar derivation for the non-inverting amplifier is given as a homework problem. Looking at Fig. 5.13, we conclude that $v^{-} = v_{\text{out}}$ since the negative amplifier terminal is directly connected to the output. According to the amplifier equation, Eq. (5.1) of the previous section, we have

$$\upsilon_{\text{out}} = A(\upsilon^+ - \upsilon^-) = A(\upsilon_{\text{in}} - \upsilon_{\text{out}}) \tag{5.22}$$

Solving Eq. (5.22) for the output voltage yields

$$\upsilon_{\text{out}} = \frac{A}{1+A}\upsilon_{\text{in}} \tag{5.23}$$

Using a Maclaurin series expansion, we obtain with $A >> 1$ the result

$$\frac{A}{1+A} = \frac{1}{1+1/A} \approx 1 - 1/A \approx 1 \tag{5.24}$$

which is consistent with Eq. (5.21) and is very accurate since typically $A > 10^5$. A similar derivation holds for the non-inverting (or the inverting) amplifier configuration. With this in mind, the second SPC is clearly optional. Instead, the amplifier definition Eq. (5.1) may be used, along with the condition of the high open-loop gain. However, it is rather tedious to repeat the asymptotic analysis every time; so we prefer to use the accurate and simple summing-point constraint. The finite value of the open-circuit gain A becomes important for high-speed amplifiers with the feedback loop; see Chapter 10.

5.2.5 Current Flow in the Amplifier Circuit

The current flow in the complete amplifier circuit is illustrated in Fig. 5.14. The output current through the load resistance R_L of the amplifier circuit in Fig. 5.14 is provided by the dual-polarity power supply. In this sense, the amplifier is also a "valve" (similar to its building block, the transistor), which "opens" the power supply in response to the low-power (or virtually no-power) input voltage signal. In Fig. 5.14, you should note that standard resistor values (5 % or 1 % tolerance) may be slightly different from the values used in this figure for convenience. We consider the positive input voltage of 100 mV in Fig. 5.14a first. The non-inverting amplifier has a closed-loop gain A_{CL} equal to 50. The output voltage is thus +5 V, which is the push mode. The load current of 10 mA is found from Ohm's law. The feedback current of 0.1 mA is found using the second SPC and Ohm's law. The power supply current is the sum of both. The feedback current controls the gain of the amplifier, and the load current drives the load. The overall amplifier circuit efficiency (neglecting the loss in the IC itself) depends on the ratio of these two currents. Therefore, we should keep the feedback current small. The power current path is shown in Fig. 5.14a by a thick trace. The current at node A can only enter the upper power supply. Thus, it is the upper power supply being used. The amplifier is operating in the "push" mode, i.e., the *amplifier sources* the current. When the input voltage is negative as in Fig. 5.14b, the lower power supply is delivering power. Now, the *amplifier sinks* the current; it is operating in the "pull" mode.

However, the ratio of the load current and feedback loop current remains the same, at least for the ideal amplifier. Similar results are obtained for the inverting amplifier and the voltage follower, respectively.

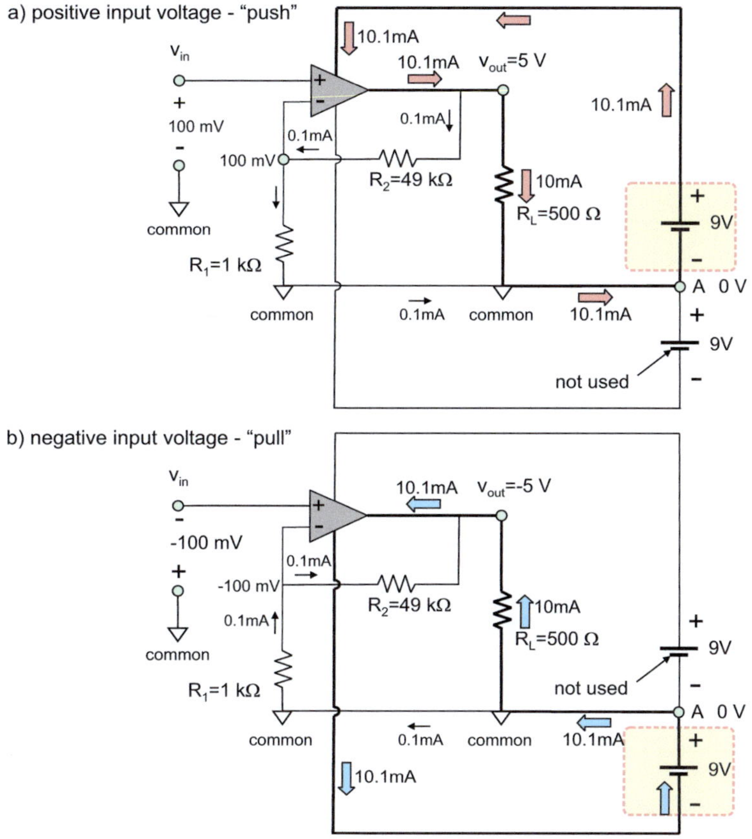

Fig. 5.14. Current flow in the non inverting amplifier circuit operating in the (**a**) push mode and (**b**) pull mode. The path of the (relatively high) load current is marked in bold.

5.2.6 Multiple-Input Amplifier Circuit: Summing Amplifier

Figure 5.15 shows an important amplifier type on the basis of the inverting amplifier—the *summing amplifier*. A summing amplifier performs a simple mathematical operation: it sums several weighted input voltages. The summing amplifier is a prototype of the *binary-weighted-input digital-to-analog converter*. According to KCL and to the first summing-point constraint, one has with reference to Fig. 5.15

$$i_F = i_1 + i_2 + i_3 \tag{5.25}$$

On the other hand, the second summing-point constraint (the differential voltage to the amplifier is zero and the inverting input is the common or virtual ground) yields

$$i_1 = \frac{v_1}{R_1}, \quad i_2 = \frac{v_2}{R_2}, \quad i_3 = \frac{v_3}{R_3} \tag{5.26}$$

in terms of input voltages v_1, v_2, v_3. Therefore, voltage v_{out} in Fig. 5.15 found from Eq. (5.25) is now written in the form

$$i_F = \frac{0 - v_{out}}{R_F} = i_1 + i_2 + i_3 = \left(\frac{v_1}{R_1} + \frac{v_2}{R_2} + \frac{v_3}{R_3}\right) \Rightarrow v_{out} = -\frac{R_F}{R_1}v_1 - \frac{R_F}{R_2}v_2 - \frac{R_F}{R_3}v_3 \tag{5.27}$$

Example 5.3: An input to the amplifier circuit in Fig. 5.15a is a timing sequence shown in Fig. 5.15b. Such a sequence is known as a *binary counter*; it represents all three-bit binary numbers in an ascending order, with the time interval of 1 μs. The amplifier circuit is characterized by $R_F = 2$ kΩ, $R_1 = 40$ kΩ, $R_2 = 20$ kΩ, and $R_3 = 10$ kΩ. Plot the absolute output voltage to scale.

Solution: After plugging in the numbers, Eq. (5.27) is transformed to

$$|v_{out}| = 0.05v_1 + 0.1v_2 + 0.2v_3 \tag{5.28}$$

Figure 5.15c shows the result. This is a staircase approximation of the straight line.

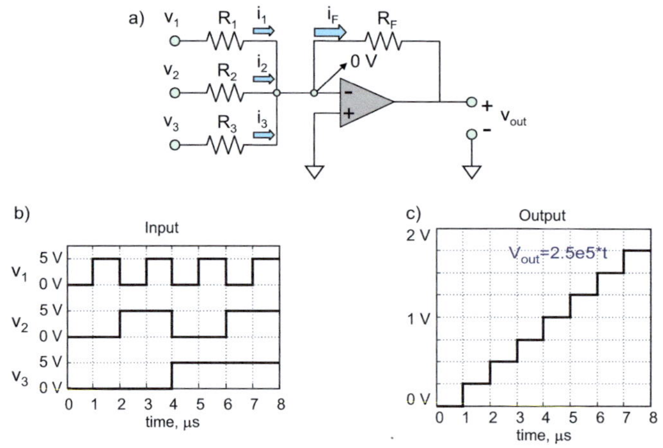

Fig. 5.15. (**a**) Circuit diagram of a summing amplifier. (**b**) and (**c**) Typical input and output voltages.

A large collection of practical amplifier circuits with the negative feedback exists. Some of them are *DC-coupled amplifiers* (considered here), some are intended for amplification of AC voltage signals with zero mean (the so-called *AC-coupled amplifiers*).

Section 5.3 Amplifier Circuit Design

Now that the theory of the negative feedback loop has been established, we can turn our attention to the laboratory. Our hope is to be immediately successful with our designs. However, a number of questions will arise almost instantly. They raise issues such as how to choose the resistor values, how to connect the sensor as part of the input load, and how to use an amplifier chip with a single power supply (a battery).

5.3.1 Choosing Proper Resistance Values

There are several rules regarding how to choose resistances R_1, R_2 controlling the feedback loop in both non-inverting and inverting configurations. They are:

1. Resistances R_1, R_2 cannot be too small. Imagine that in Fig. 5.14 of Section 5.2, the resistor values are changed to $R_1 = 1\ \Omega,\ \ R_2 = 49\ \Omega$. The same non-inverting gain will be achieved and the same output voltage will be obtained. However, the feedback loop current now becomes 100 mA instead of 0.1 mA. The general-purpose op-amp chips are not capable of delivering such large currents. Furthermore, the ohmic losses in the feedback loop become high. Therefore, one should generally use

$$R_1, R_2 \geq 50 - 100\ \Omega \tag{5.29a}$$

2. Resistances R_1, R_2 cannot be too large. Let us assume that resistance R_2 equals 100 MΩ. This means that this physical resistor and the feedback loop represent almost an "open circuit." Unwanted electromagnetic signals may couple into such a circuit through the related electric field difference across its terminals. This effect is known as *capacitive coupling*. Furthermore, the very large resistances increase the parasitic effect of the input offset current. Plus, very large resistances are unstable—their values depend on moisture, temperature, etc. Therefore, one should generally use

$$R_1, R_2 \leq 1\,\mathrm{M}\Omega \tag{5.29b}$$

3. When a precision design is not warranted, inexpensive 5 % tolerance resistors may be used. Otherwise, 1 % or even 0.1 % tolerance resistors are employed. Moreover, in lieu of fixed resistors, we may use one or two potentiometers to make the gain adjustable.
4. The load resistance should be sufficiently large in order not to overdrive the amplifier. A good choice is

$$R_L \geq 100\,\Omega \tag{5.29c}$$

This requires an output current of 20 mA at $v_{out} = 2\,\mathrm{V}$ when R_L is exactly 100 Ω. If $R_L < 100\ \Omega$, the amplifier output voltage may decrease compared to the expected value due to amplifier's inability to source/sink sufficient current.

Example 5.4: A non-inverting amplifier circuit with a gain of 11 is needed in the configuration depicted in Fig. 5.16. Identify one set of proper resistance values.

Solution:
To satisfy Eqs. (5.29), we simply choose the round numbers

$$R_1 = 1\ \mathrm{k}\Omega,\ R_2 = 10\ \mathrm{k}\Omega,\ R_\mathrm{L} = 100\ \Omega \tag{5.30}$$

However, other choices are indeed possible. For example, the set

$$R_1 = 100\ \mathrm{k}\Omega,\ R_2 = 1\ \mathrm{M}\Omega,\ R_\mathrm{L} = 100\ \Omega \tag{5.31}$$

will solve the problem too.

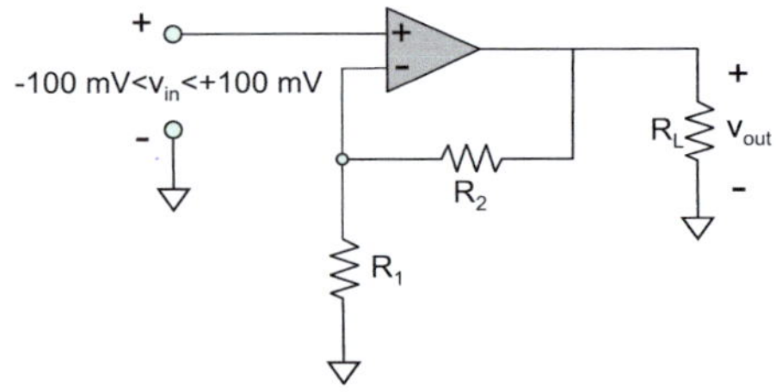

Fig. 5.16. A non-inverting amplifier with unknown resistance values.

Discrete Resistance Values and Potentiometers

To achieve a proper gain, we sometimes have to use "strange" resistor values like 49 kΩ (Fig. 5.14 of Section 5.2). Do such resistors really exist? For 5 % tolerance resistors, they do not. However, for 1 % tolerance resistors, you can find the standard value of 48.7 kΩ, which is close to the above value. When the exact resistor values are not available, a potentiometer can be used as a variable resistor. Furthermore, an externally controlled potentiometer in the feedback loop is also important when a *variable-gain amplifier* is needed, for example, for applications that require *automatic gain control.*

Gain Tolerance

What about the *gain tolerance*? The feedback resistor tolerances indeed determine the gain tolerance. If the resistor tolerance is X, then the gain tolerance is $2X$. This result is valid for both the inverting and the non-inverting amplifier. The corresponding proof uses an asymptotic expansion for the gain about its unperturbed value. We consider the worst-case scenario for the inverting amplifier and obtain

$$A_{CL} = -\frac{R_2(1+X)}{R_1(1-X)} \approx -\frac{R_2}{R_1}(1+X)(1+X) \approx -\frac{R_2}{R_1}(1+2X) \quad \text{for } X << 1 \tag{5.32}$$

The non-inverting gain is treated similarly. For example, if two resistors of an inverting amplifier circuit are $R_1 = 1\ \text{k}\Omega,\ R_2 = 100\ \text{k}\Omega$ and both resistors have 5 % tolerances, then the amplifier gain is equal to -100 with tolerance of 10 %. Similarly the gain of the non-inverting amplifier circuit becomes 101 with tolerance of slightly less than 10 %.

5.3.2 Model of a Whole Voltage Amplifier Circuit

Any *voltage amplifier circuit with the negative feedback* loop may be modeled in a form similar to Fig. 5.5. The corresponding model is shown in Fig. 5.17.

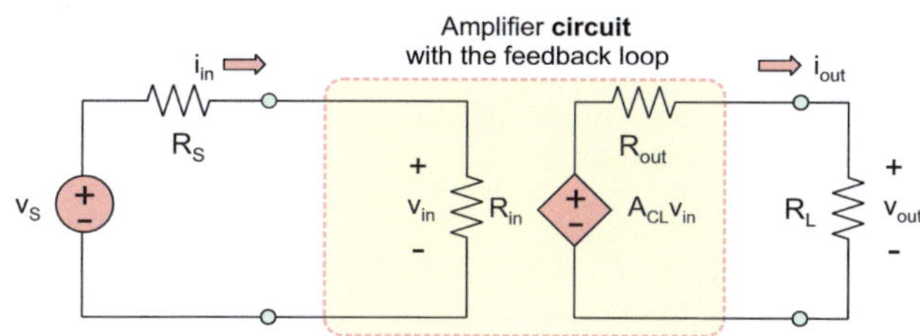

Fig. 5.17. Equivalent model of an amplifier circuit with a negative feedback loop.

First, the open-circuit gain A of the amplifier is replaced by the closed-loop gain $A_{CL} << A$. Second, the differential input voltage v_x is replaced by v_{in}. Third, resistances R_{in} and R_{out} in Fig. 5.17 now become *input and output resistances of the amplifier circuit*, not the amplifier itself. This difference may be quite important in practice.

Input/Output Resistances of Basic Amplifier Circuits Using Ideal-Amplifier Model

The solution simplifies for the ideal-amplifier model; it is shown in Fig. 5.18. We assume ideal operational amplifiers in all three cases. If there were no feedback loop, R_{in} would be exactly equal to the input resistance of the amplifier itself; this is for an ideal operational amplifier $R_{in} = \infty$. When the feedback loop is present, a more general definition should be used, namely,

$$R_{in} \equiv \frac{v_{in}}{i_{in}}, \tag{5.33}$$

where v_{in} is the amplifier's circuit input voltage and i_{in} is now the current into the amplifier's circuit *with the feedback loop*. Figure 5.18 illustrates the corresponding calculation for the three basic amplifier types. For both the non-inverting amplifier and the voltage follower circuits, we have $R_{in} = \infty,\ R_{out} = 0$. However, for the inverting amplifier circuit, $R_{in} = R_1,\ R_{out} = 0$, since an input current can still flow into the feedback loop but not into the amplifier itself. Thus, the inverting amplifier circuit potentially provides greater flexibility in the input resistance simply by varying R_1. To ensure the necessary gain, R_2 has to be chosen accordingly.

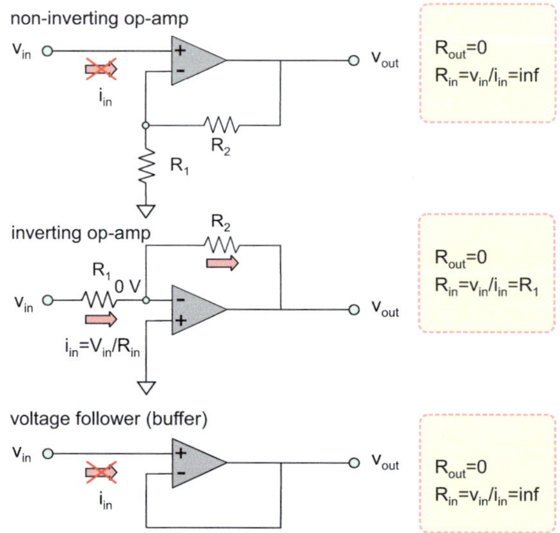

Fig. 5.18. Input and output resistances of amplifier circuits with the ideal operational amplifier.

5.3.3 Voltage Amplifier Versus Matched Amplifier

A sensor (input load to an amplifier circuit) "sees" an amplifier circuit as a simple resistance R_{in} as depicted in Fig. 5.19. The sensor is represented by its Thévenin equivalent circuit.

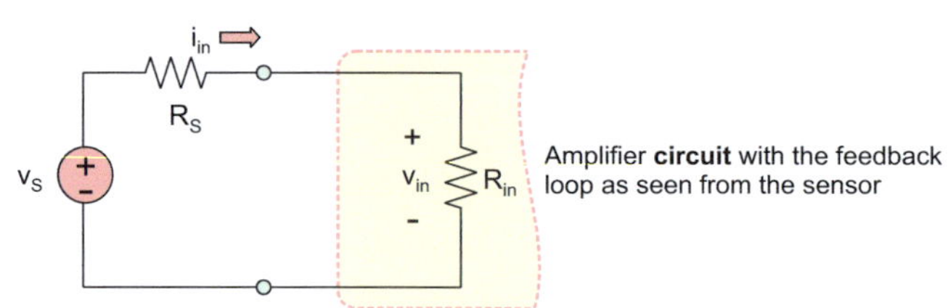

Fig. 5.19. Sensor's equivalent circuit and amplifier's equivalent circuit as seen from the sensor.

Input Load "Bridging"

The general-purpose operational amplifier is a *voltage amplifier* (also called a *signal amplifier*) and not a *power amplifier*. The input voltage matters, not the input power. For the voltage divider circuit in Fig. 5.19, the input voltage v_{in} to the amplifier circuit is *maximized* when R_{in} is maximized. Therefore, R_{in} must satisfy the inequality

$$R_{\mathrm{in}} >> R_{\mathrm{S}} \tag{5.34}$$

An appropriate value would be $R_{\mathrm{in}} = 100R_{\mathrm{S}}$, for example. Equation (5.34) is sometimes called *load bridging* (or *impedance bridging*) *condition*, where the "load" resistance R_{in} seen by the Thévenin source is much larger than the source resistance R_{S}. The load bridging is automatically satisfied for the non-inverting amplifier or for the voltage follower. For the inverting amplifier, one should use large values of R_1, for instance,

$$R_1 = 100R_{\mathrm{S}} \tag{5.35}$$

Bridging connections are used to maximize the voltage transfer from a sensor to an amplifier. Even more importantly, the amplifier does not appreciably load the sensor.

Example 5.5: A sensor is given by its Thévenin equivalent circuit in Fig. 5.19 where the sensor voltage υ_{S} is small. The sensor's equivalent resistance R_{S} is 100 Ω. An inverting amplifier circuit is needed to generate an amplified version of the sensor's voltage. The output voltage should be $\approx -100\upsilon_{\mathrm{S}}$.

Solution: The corresponding circuit is shown in Fig. 5.20. The input voltage to the amplifier circuit is computed by voltage division:

$$\upsilon_{\mathrm{in}} = \frac{R_1}{R_1 + R_{\mathrm{S}}}\upsilon_{\mathrm{S}} \tag{5.36}$$

If $R_{\mathrm{S}} = 100\ \Omega,\ R_1 = 10\ \mathrm{k}\Omega$, then $\upsilon_{\mathrm{in}} \approx \upsilon_{\mathrm{S}}$ and there is almost no loss of voltage signal strength across resistance R_{T}. Therefore, a pair of resistors with $R_1 = 10\ \mathrm{k}\Omega,\ R_2 = 1\ \mathrm{M}\ \Omega$ will solve the problem, with the amplifier voltage gain of -100. If, however, we choose $R_1 = 0.25R_{\mathrm{S}} = 25\,\Omega$, then $\upsilon_{\mathrm{in}} = 0.2\upsilon_{\mathrm{S}}$ and 80 % of available voltage signal strength will be lost! Even if the remaining voltage signal is still appreciable (above the *sensitivity threshold* of the amplifier), the necessary amplifier gain becomes not -100, but -500. An increase in gain leads to an increase in additive voltage noise at the output. Therefore, in the best case, the amplified signal will be a noisier version of the corresponding signal in the previous design.

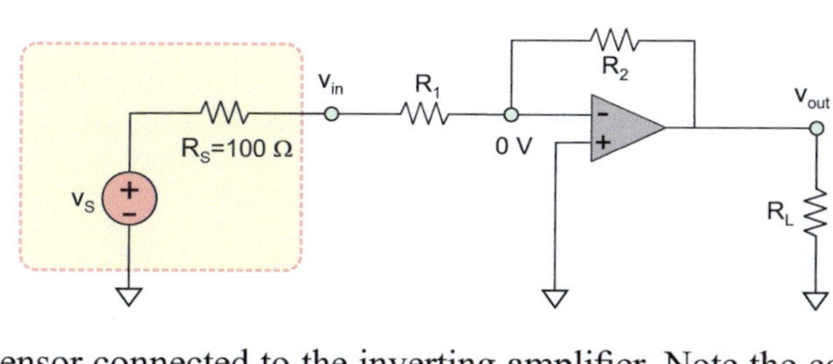

Fig. 5.20. A 100 Ω sensor connected to the inverting amplifier. Note the common connections.

Example 5.6: A sensor is given by its Thévenin equivalent in Fig. 5.19 where the sensor voltage v_S is small. The sensor's equivalent resistance R_S may vary in time but is always less than 100 Ω. An inverting amplifier is needed that generates an amplified version of the sensor's voltage, which is $\approx -100 v_S$.

Solution: With reference to Fig. 5.20, the input voltage to the amplifier is again given by Eq. (5.36). This equation is further transformed to

$$v_{in} = \frac{R_1}{R_1 + R_S(t)} v_S \approx v_S \tag{5.37}$$

if we choose $R_1 = 100\max(R_S(t)) = 10\,\text{k}\Omega$. In other words, not only have we provided amplification but we also *eliminated* the effect of the sensor's resistance variation by proper load bridging. Therefore, a pair of resistors with $R_1 = 10\ \text{k}\Omega$, $R_2 = 1\ \text{M}\Omega$ will solve the problem, with the amplifier output of $\approx -100 v_S$, irrespective of the specific value of R_S.

Input Load Matching

And yet, in many modern applications related to radio-frequency (RF) circuits, the *load matching* (but not the load bridging) may be a critical condition. RF amplifiers are internally designed for matching to a precise 50 Ω load at *both* the input and the output—see Fig. 5.21. The reason is that voltage and current signals in conductors behave like propagating electromagnetic waves at high frequencies. If there is no matching, then multiple wave reflections between the amplifier and its input and/or output loads can occur, resulting in superimposing the previous signal onto the next signal.

Fig. 5.21. A RF amplifier to be matched to a 50 Ω at both the input and output.

Therefore, the amplifier circuit optimized for proper load matching (which also achieves maximum power transfer from the input load to the amplifier circuit, see *the generator theorems*) may still be critical in many high-frequency applications.

Example 5.7: Construct an amplifier circuit matched to an input source with $R_S = 50\ \Omega$. The amplifier's voltage gain is $|A_{CL}| = 20$. The sign of A_{CL} (either positive or negative) is not important since an AC input signal is assumed.

Example 5.7 (cont.):

Solution: Two possible solutions are shown in Fig. 5.22. In the first case, we use the inverting amplifier; in the second case, a smart trick is employed: a non-inverting amplifier with a 50-Ω shunt resistor. You should note that the maximum power transfer will be achieved for the *entire* amplifier circuit, including the shunt resistor.

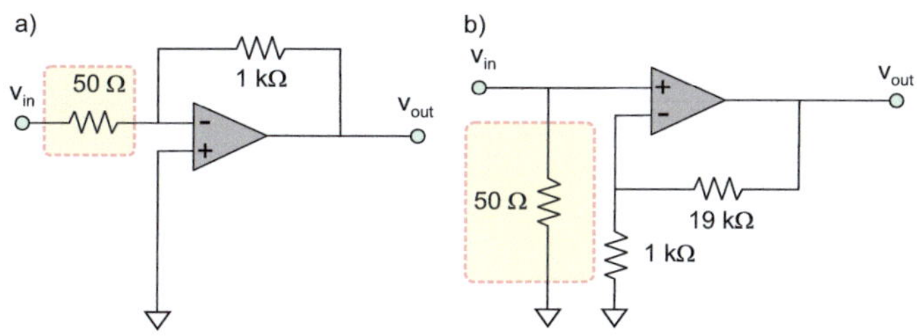

Fig. 5.22. Two possible amplifier configurations matched to a 50-Ω input resistance.

5.3.4 Cascading Amplifier Stages

Assume that we need an inverting amplifier with an overall gain of −1000. The input to the amplifier is a sensor as in Fig. 5.19 with the equivalent resistance given by $R_S = 1\,\text{k}\Omega$. If we require load bridging according to Eq. (5.35), we arrive at

$$R_1 = 100R_S = 100\,\text{k}\Omega \tag{5.38}$$

This yields

$$R_1 = 100\ \text{k}\Omega \Rightarrow R_2 = 100\ \text{M}\Omega \tag{5.39}$$

Such a resistance value is too large to satisfy Eq. (5.29b); it perhaps will not even be included in your laboratory kit (although the ECE shop may still have such resistors). What should we do? The answer lies in *cascading the amplifier stages* as shown in Fig. 5.23. We use the non-inverting amplifier with a gain of 10 as the first stage; this simultaneously provides the load bridging condition. We use the inverting amplifier with a gain of −100 and with the reasonable resistor values as the second stage. The key point of cascading is to realize that the overall gain of the cascade amplifier is given by the *product* (not the sum!) of the individual stage gains, i.e.,

$$A_{CL} = A_{CL1} \times A_{CL2} = 10 \times (-100) = -1000 \tag{5.40}$$

The same result is valid for more than two stages. The proof for two stages is simple:

$$A_{\mathrm{CL}} \equiv \frac{v_{\mathrm{out}}}{v_{\mathrm{in}}} = \frac{A_{\mathrm{CL2}} v^1_{\mathrm{out}}}{v_{\mathrm{in}}} = \frac{A_{\mathrm{CL1}} A_{\mathrm{CL2}} v_{\mathrm{in}}}{v_{\mathrm{in}}} = A_{\mathrm{CL1}} A_{\mathrm{CL2}} \tag{5.41}$$

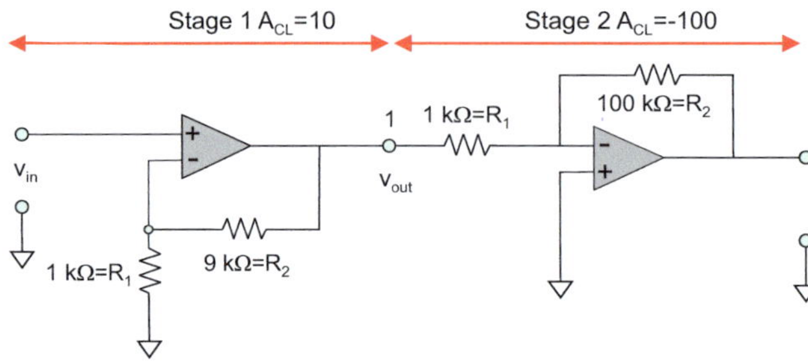

Fig. 5.23. Cascading two amplifier stages into a high-gain circuit. The first stage is a non-inverting amplifier with a gain of 10; the second stage is an inverting amplifier with a gain of −100. The overall gain is therefore −1000.

The cascading of individual amplifier stages is a simple and powerful tool to build high-gain amplifier circuits. Cascading has a number of remarkable features, some of which are studied here:

1. The gains of the individual stages multiply.
2. The gain per stage generally should not exceed 100 (absolute value) in order to avoid instability.
3. The first stage in Fig. 5.23 sees R_1 of the second stage as its output load resistance. Therefore, R_1 should be large enough.
4. The amplifier ICs usually include two (*dual* op-amp) or even four (*quad* op-amp) individual amplifiers in one package. Therefore, they are ideally suited for building multistage amplifiers.
5. The effect of an input offset voltage (an amplifier imperfection studied next) is primarily important for the first stage, but it then loses its significance with every subsequent stage.

Note that cascading is equivalent to a series combination of individual amplifiers. Parallel configurations also exist, particularly in *analog-to-digital converters*.

Example 5.8: The input to the amplifier is a sensor in Fig. 5.19 with an equivalent resistance given by $R_S = 100\ \Omega$ and an equivalent sensor voltage, v_S. An amplifier circuit is needed that generates ~10,000 v_S at its output. The load bridging condition must be satisfied.

Example 5.8 (cont.):

Solution:

A. A single-stage, non-inverting amplifier with $R_1 = 100\ \Omega$, $R_2 = 1\ \text{M}\Omega$ might do the job, including load bridging. However, the gain per stage (10,000) is far too high for stable operation.

B. A series combination of two inverting stages will do a much better job: the first inverting op-amp assures load bridging, and it consists of $R_1 = 10\ \text{k}\Omega$, $R_2 = 1\ \text{M}\Omega$. The second inverting amplifier has exactly the same resistor values: $R_1 = 10\ \text{k}\Omega$, $R_2 = 1\ \text{M}\Omega$. And the overall gain is given by $A_{CL} = (-100) \times (-100) = 10{,}000$.

C. A series combination of two non-inverting stages will perform equally well: we choose $R_1 = 10\ \text{k}\Omega$, $R_2 = 1\ \text{M}\Omega$ for the first stage and $R_1 = 10\ \text{k}\Omega$, $R_2 = 1\ \text{M}\Omega$ for the second stage. The overall gain then yields $A_{CL} = 101 \times 101 = 10{,}201 \approx 10{,}000$.

5.3.5 Amplifier DC Imperfections and Their Cancellation

In general, *DC imperfections of the operational amplifier* can have a severe influence on its performance for high-gain amplifiers. Below, we study two types of imperfections, the *input offset voltage* V_{OS} and the *input (bias and offset) currents*, and provide a simple way of how to cancel the corresponding output offset voltage.

Input Offset Voltage

The input offset voltage results in a nonzero output voltage when the two input terminals of the amplifier are shorted out. It arises due to a small asymmetry in the input differential transistor stage inside the amplifier chip. It is fixed for a certain chip but varies from chip to chip. General-purpose amplifiers have the input offset voltage V_{OS} in the range of $1\ \text{mV} - 6\ \text{mV}$. The offset can be modeled as a small DC source of strength, V_{OS}, in series with one of the input terminals to the amplifier as shown in Fig. 5.24. The input offset voltage produces a similar effect for any amplifier configuration, including the comparator where its effect becomes most dramatic. A large triangle in Fig. 5.24 indicates the actual amplifier chip, whereas the small triangle is the ideal amplifier without DC imperfections. The circuit in Fig. 5.24 is analyzed using the two summing-point constraints. Since the negative feedback is present, v_x must be zero, which yields

$$v_{in} + V_{OS} = v^* = \frac{R_1}{R_1 + R_2} v_{out} \Rightarrow v_{out} = A_{CL}(v_{in} + V_{OS}), \quad A_{CL} = 1 + \frac{R_2}{R_1} \tag{5.42}$$

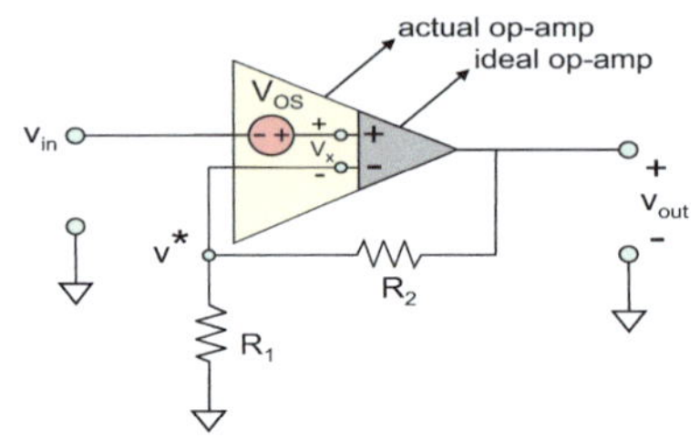

Fig. 5.24. Circuit for the non-inverting amplifier with an input offset voltage.

Exercise 5.7: The amplifier circuit in Fig. 5.24 has the closed-loop gain of $A_{CL} = 100$ and an input offset voltage of $V_{OS} = 5$ mV. What is the general expression for the output voltage?

Answer:

$$v_{out} = 100 v_{in} + 0.5 \text{ V} \tag{5.43}$$

where 0.5 V is the resulting output offset voltage to the amplifier.

Canceling the Output Offset Voltage

In some amplifiers like the LM741, special *offset-null terminals* are available for trimming the output DC voltage to zero. Figure 5.25a shows the concept. The wiper of the potentiometer is to be connected to the negative supply rail. Both inputs to the amplifier should be connected to the common port during the adjusting procedure, which implies that the output voltage is trimmed to zero. If the offset-null terminals are not available, the voltage at the common port, which controls the feedback loop, might be subject to a small offset. Figure 5.25b shows a circuit that can be used to eliminate the DC imperfections for a particular non-inverting amplifier in the laboratory. A compensating voltage offset is introduced by means of an adjustable voltage divider with a potentiometer. To achieve a good degree of accuracy, one should set

$$R_P << R \tag{5.44}$$

Furthermore, R_1 should be considerably larger than R. This condition can be avoided by a further modification of the present voltage divider. The output offset voltage is trimmed to zero when the input to the amplifier in Fig 5.26b is connected to the common port (circuit ground). This ensures that the effect of the input offset voltage will be eliminated entirely, for *any* value of the input voltage. Note that such an adjustment has to be done for every discrete temperature point, since V_{OS} depends on temperature.

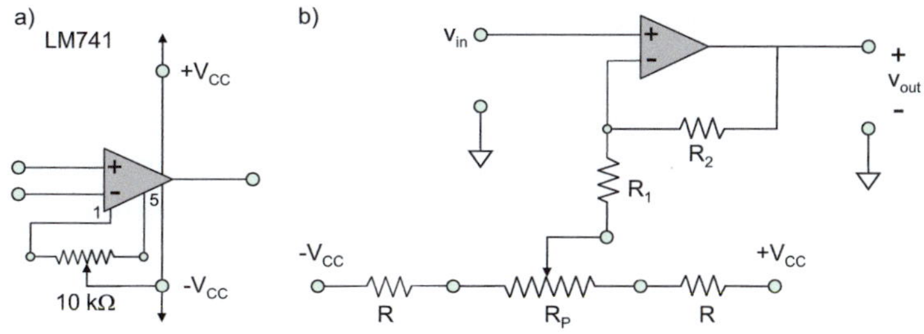

Fig. 5.25. (**a**) Output DC offset voltage for the LM741 is reduced to zero by adjusting the potentiometer placed between its offset-null pins and (**b**) a similar operation performed with the virtual ground of the feedback loop.

Input Bias and Offset Currents

In reality, for the op-amp to operate, there will be very small currents into the input terminals (into transistor bases), typically on the order of 100 nA. When those currents flow through the feedback resistances, they create corresponding voltages which appear as an output offset voltage as well. Let a current of 100 nA flow *into* the negative terminal of the amplifier in Fig. 5.24. If the input to the amplifier is grounded, this current must flow through resistance R_2. Therefore, it will create the extra output DC voltage of

$$v_{\text{out}} = +100 \text{ nA} \times R_2 \tag{5.45}$$

when the input voltage to the amplifier v_{in} is exactly zero. To appreciate its value, we can use a resistance $R_2 = 1\ \text{M}\Omega$ as an example. This yields

$$v_{\text{out}} = 0.1 \text{ V} \tag{5.46}$$

at the output. Fortunately, the currents flowing into the amplifier are nearly the *same* for either terminal. Therefore, their average (the *input bias current*) considerably exceeds their difference (the *input offset current*). There is a way to eliminate the larger effect of the input bias current. It consists of modifying the circuits for the non-inverting and inverting amplifier by adding one extra resistance $R = R_1||R_2$ as shown in Fig. 5.26.

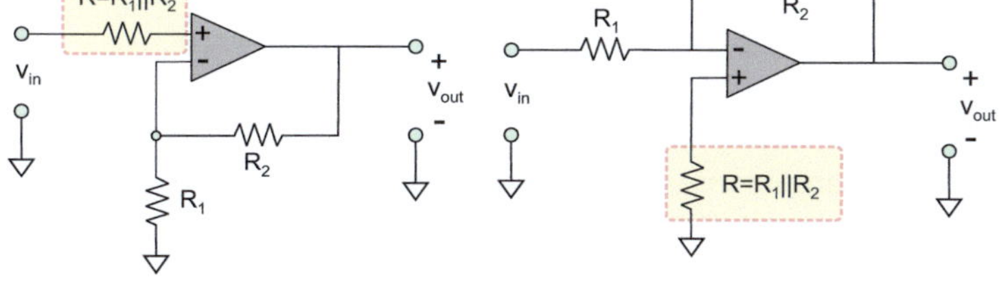

Fig. 5.26. Cancellation of the effect of input bias currents.

From the amplifier's gain point of view, the resistance R has a negligible, if any, effect. The proof of the cancellation effect for the non-inverting amplifier circuit is as follows. When both the input and output to the amplifier in Fig. 5.26a are grounded (connected to the common port), the input current source at the non-inverting input sees resistance R and the input current source at the inverting input sees the parallel combination of R_1, R_2, respectively. Making R equal to $R_1||R_2$ yields an offset differential voltage that is zero.

5.3.6 DC-Coupled Single-Supply Amplifier: Virtual-Ground Circuit

A single voltage supply is a battery. We consider the non-inverting amplifier circuit driven by a 9-V battery. To handle the problem of not being able to generate negative voltages, a *virtual-ground circuit* may be used as shown in Fig. 5.27. We simply divide the voltage of the battery by two, with two large, equal resistances R, and assign this voltage of 4.5 V to the common port. The battery terminal voltages then formally become ±4.5 V versus the common port. The power supply so constructed is unfortunately not exactly the dual-polarity voltage power supply. Namely, both resistances R have to be large to avoid ohmic losses in the virtual-ground circuit. If they are, we cannot source/sink an appreciable output current into the common terminal since these resistances simultaneously operate as current limiters. We can reference the output to ground (which is the negative terminal of the battery), but not to the *virtual ground* of 4.5 V. When referenced to ground, the circuit in Fig. 5.27 has one remarkable property. Namely, if the input vs. ground is the (small) sensor voltage v_{in} plus 4.5 V, then the output vs. ground is the amplified sensor voltage, $A_{CL}v_{in}$, plus the *same* 4.5 V offset. In other words, the offset voltage of the virtual ground is not amplified! This statement is proved in the following example.

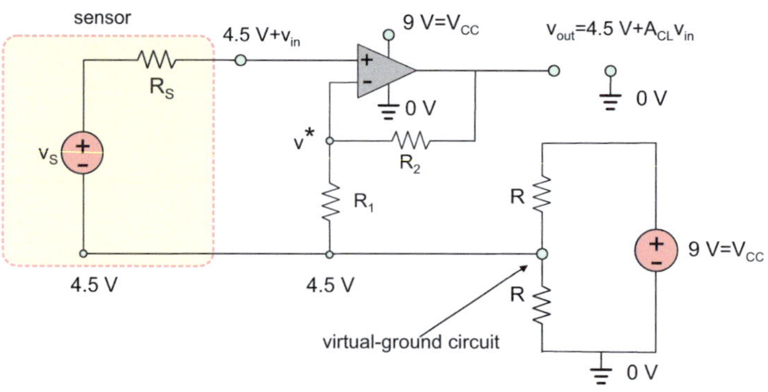

Fig. 5.27. A non-inverting amplifier driven by a single voltage supply, a battery. Absolute voltages versus ground (negative terminal of the battery) are shown.

Example 5.9: Solve the circuit in Fig. 5.27.

Solution: The solution is based on the two summing-point constraints. There is no current into the input terminals so that resistances R_1 and R_2 again form a voltage divider but now between υ_{out} and 4.5 V. By voltage division, the voltage at node (*) becomes

$$\upsilon^* = 4.5\ \text{V} + \frac{R_1}{R_1 + R_2}(\upsilon_{out} - 4.5\ \text{V}) \tag{5.47}$$

At the same time, using the second summing-point constraint for the amplifier with negative feedback, we obtain:

$$\upsilon^* = 4.5\ \text{V} + \upsilon_{in} \tag{5.48}$$

Equating Eqs. (5.47) and (5.48), we arrive at the expected result:

$$\upsilon_{out} = 4.5\ \text{V} + A_{CL}\upsilon_{in}, \quad A_{CL} = 1 + \frac{R_2}{R_1} \tag{5.49}$$

Feedback resistances R_1, R_2 should be much larger than resistance R in Fig. 5.27, in order to assure a flawless circuit operation. Yet another solution is to use (Zener) diodes in the bias circuit. Special *virtual-ground integrated circuits* exist that support single-supply amplifier operation. They generate an output precisely midway between the two supply rails.

Section 5.4 Difference and Instrumentation Amplifiers

5.4.1 Differential Input Signal to an Amplifier

Consider a sensing element that is a variable resistance. The sensor configuration is a Wheatstone bridge (Section 3.2) as seen in Fig. 5.28. Since the amplifier is powered by a dual supply with three terminals, $\pm V_{CC}$, and common (ground) port, the same power supply is connected to the bridge. We will use the positive rail and the ground rail in Fig. 5.28 since the sensor (e.g., the strain gauge) may require lower voltages than the amplifier chip itself.

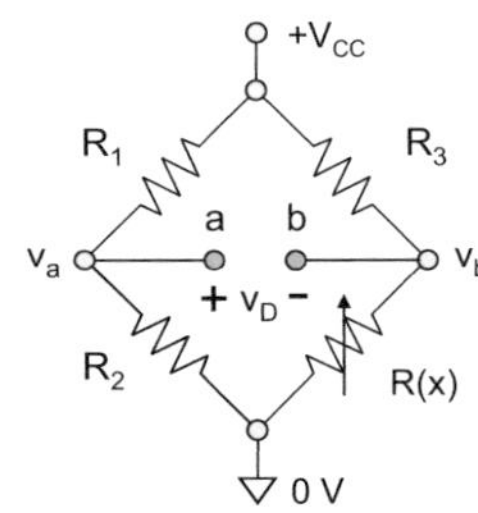

Fig. 5.28. A Wheatstone bridge sensor to be connected to an amplifier circuit.

Both voltages v_a, v_b have to be used when reading sensor information. They may be written in terms of the other two voltages v_D, v_{CM}:

$$v_a = v_{CM} + \frac{1}{2}v_D, \quad v_b = v_{CM} - \frac{1}{2}v_D \tag{5.50a}$$

$$v_D = v_a - v_b, \; v_{CM} = 0.5(v_a + v_b) \tag{5.50b}$$

Here, v_D is the differential component of the combined input signal or the *differential voltage* and v_{CM} is the sum component of the combined input signal or the *common-mode voltage*. Only the differential voltage is usually important for sensor reading; the common-mode voltage does not carry any information. When the bridge is exactly balanced, i.e., when

$$\frac{R_1}{R_2} = \frac{R_3}{R(x)}, \tag{5.51a}$$

the differential voltage is exactly zero:

$$v_D = 0 \tag{5.51b}$$

Still, the common-mode DC voltage given by

$$v_{\mathrm{CM}} = V_{\mathrm{CC}}\left(\frac{R_2}{R_2 + R_1}\right) \tag{5.51c}$$

can have any large positive value. For example, it is $V_{\mathrm{CC}}/2$ when all resistances are equal. Even if the bridge in Fig. 5.28 uses $\pm V_{\mathrm{CC}}$ power rails, it is hardly possible that the common-mode voltage is set to zero since absolutely identical resistors do not exist.

Exercise 5.8: In the Wheatstone bridge in Fig. 5.28, we assume $R_2 = 1.1R_1$, $R(x) = 1.1R_3$, and $V_{\mathrm{CC}} = 6$ V. What are the differential and common-mode voltages?

Answer: $v_{\mathrm{D}} = 0$, $v_{\mathrm{CM}} = 3.14\,\mathrm{V}$.

5.4.2 Difference Amplifier: Differential Gain and Common-Mode Gain

The sensor in Fig. 5.28 is the *differential sensor* with *three* terminals: *a*, *b*, and ground. How do we amplify the differential voltage? Reviewing the inverting and non-inverting amplifier types reveals that they are not appropriate for this purpose: we simply do not have two input terminals to be connected to nodes *a* and *b* in Fig. 5.28. Only *one* input terminal referenced to common (ground) is available. Note that *single-ended sensors* with *two* terminals (plus and ground) indeed exist and may be used. In that case, inverting or non-inverting amplifiers will function well. However, the overall design accuracy may deteriorate compared to the differential design. Therefore, a new amplifier type with two input terminals should be introduced. It is the *difference amplifier* shown in Fig. 5.29.

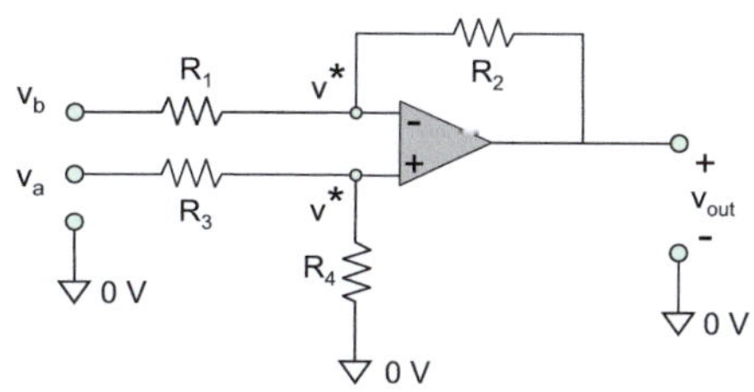

Fig. 5.29. A difference amplifier.

Both inputs to this amplifier are referenced to common (ground) port. First, we see that the difference amplifier is an inverting amplifier with the negative feedback loop. However, the second input signal is now added to its positive terminal through a voltage divider. The analysis of this amplifier type is done using two summing-point constraints. A shortcut is to recognize, with the help of the first SPC, that in Fig. 5.29 we have two voltage dividers: one between resistances R_1, R_2 and another between resistances R_3, R_4. Therefore, for the voltage at node (*), one has, using the first voltage divider,

$$v^* = v_b + \frac{R_1}{R_1 + R_2}(v_{out} - v_b) \tag{5.52a}$$

and, according to the second voltage divider,

$$v^* = v_a + \frac{R_3}{R_3 + R_4}(0\ \text{V} - v_a) \tag{5.52b}$$

Both expressions must be equal to each other due to the second SPC (the differential input voltage in a negative feedback amplifier is zero). Therefore,

$$\begin{aligned} &v_b + \frac{R_1}{R_1 + R_2}(v_{out} - v_b) = v_a - \frac{R_3}{R_3 + R_4}v_a \Rightarrow \\ &\frac{R_1}{R_1 + R_2}v_{out} = \left(\frac{R_1}{R_1 + R_2} - 1\right)v_b - \left(\frac{R_3}{R_3 + R_4} - 1\right)v_a \end{aligned} \tag{5.52c}$$

To create a voltage difference, i.e., $v_a - v_b$, between the input voltages on the right-hand side of Eq. (5.52c), we select

$$\frac{R_2}{R_1} = \frac{R_4}{R_3} \tag{5.52d}$$

as the necessary condition. Then, both factors in parentheses on the right-hand side of Eq. (5.52c) become equal. This yields the basic equation of the difference amplifier,

$$v_{out} = \frac{R_2}{R_1}(v_a - v_b) = \frac{R_2}{R_1}v_D \tag{5.53}$$

Equation (5.53) is a simple, yet highly useful result for amplifier circuit design. Namely, once the amplifier in Fig. 5.29 is connected to the sensor in Fig. 5.28, the differential voltage $v_D = v_a - v_b$ is amplified with the gain of R_2/R_1 (the *differential amplifier circuit gain*). At the same time, the undesired common-mode voltage $v_{CM} = 0.5(v_a + v_b)$ is completely *rejected*, i.e., amplified with a gain of 0, no matter what specific values the input voltages have versus ground. In other words, the *common-mode amplifier circuit gain* is zero. Note that the ratio of two gains (*differential gain* versus *common-mode gain*) is an important characteristic of the difference-amplifier circuit. It is called the *common-mode rejection ratio* (CMRR). In our case, this ratio is clearly infinity. Unfortunately, in reality, this value is finite though quite large. One obvious reason is a possible mismatch in resistance ratios in Eq. (5.52d), which will not allow us to obtain Eq. (5.53) exactly. A certain portion of $v_{CM} = 0.5(v_a + v_b)$ will be present at the output.

Example 5.10: Find the output voltage of the amplifier circuits shown in Fig. 5.30 below. Assume the ideal amplifier and exact resistance values.

Solution: We check Eq. (5.52d) first and conclude that the circuit in Fig. 5.30 is a true difference amplifier: it rejects the common-mode voltage. The differential voltage to the amplifier is -0.01 V. Using Eq. (5.53) gives us an output voltage of $\upsilon_{out} = -0.1$ V. If the resistance ratios were not equal to each other, a common-mode signal would be present at the output. In that case, the complete amplifier equation (5.52c) should to be used.

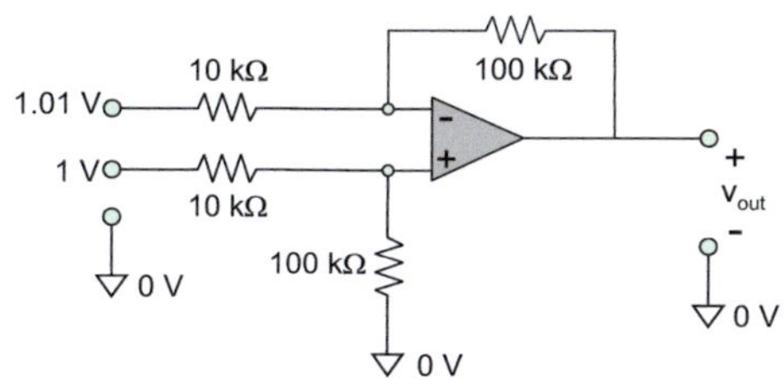

Fig. 5.30. Difference amplifier for Example 5.10.

To minimize the effect of bias currents (Section 5.3), we should choose

$$R_1 = R_3, \quad R_2 = R_4 \tag{5.54}$$

5.4.3 Application Example: Instrumentation Amplifier

Motivation for an Instrumentation Amplifier

Figure 5.31 shows a 700-Ω uniaxial strain gauge. The strain gauge is attached to an aluminum slab. Figure 5.31 also shows a Wheatstone bridge intended for strain measurement with the present device. The dual-polarity supply voltage is ±7.5 V. You may build the Wheatstone bridge according to Fig. 5.31 and connect it to a DMM, with the DMM leads attached to terminals *a* and *b*, respectively. In this configuration, the DMM measures the differential voltage υ_D. Using the potentiometer, you may balance the bridge, i.e., reduce voltage υ_D at no strain to a minimum, which should be within the range $0\ \text{V} \pm 3\ \text{mV}$. The highest DMM resolution should be used. Now, by applying a strong bending force to the slab with two hands, you probably could obtain a maximum voltage change of $\pm 2\ \text{mV}$.

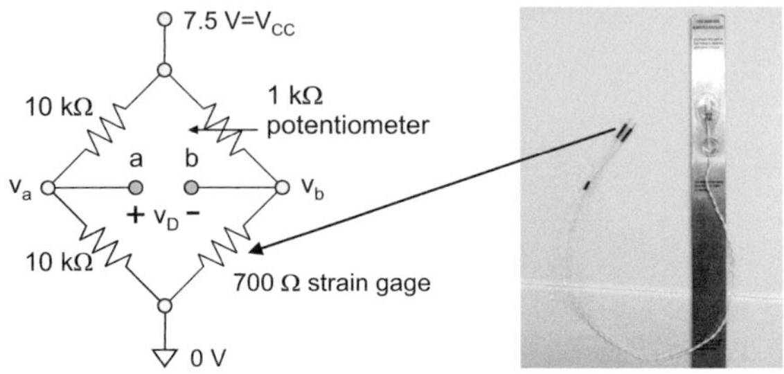

Fig. 5.31. A Wheatstone bridge sensor where the strain gauge forms one of the four resistors.

The positive voltage change corresponds to one bending direction, the negative change, to the opposite direction. Next, you may connect an oscilloscope instead of the DMM and use the DC-coupled settings and the highest voltage resolution of 20 mV per division. You will find that the noisy signal line on the screen hardly changes when you try your best. So are you not strong enough? Well, no. These are the realistic differential voltages for strain gauges which correspond to gauge resistance changes on the order of 1 Ω. Even *smaller* voltage changes are often encountered in practice. Therefore, an accurate amplification of an extremely small differential voltage should be done while rejecting the large common-mode voltage (3.75 V in the present case). This nontrivial task is accomplished by an *instrumentation amplifier.*

How to Build an Instrumentation Amplifier?

The initial guess is probably to use the difference amplifier from Fig. 5.30. For this amplifier, $R_1 = R_3 = 10\ \text{k}\Omega,\ R_2 = R_4 = 100\ \text{k}\Omega$. However, we encounter two problems. The first one is that the amplifier should not perturb the sensor operation. In other words, the amplifier's input resistance must be large compared to any of the resistances in the Wheatstone bridge. The *differential-mode input resistance* to the amplifier in Fig. 5.30 is $R_{\text{in}} = 2R_1 = 2R_3$ (check problem 5.84 at the end of this chapter). Therefore, we may wish to increase $R_1 = R_3$ by a factor of 10, i.e., choose $R_1 = R_3 = 100\ \text{k}\Omega$. The second problem is the amplifier gain. An overall gain of 1000 is required in Eq. (5.53) in order to obtain appreciable output voltages on the order of $\pm$ 2 V. This gain is too high; prohibitively large resistor values $R_2 = R_4 = 100\ \text{M}\Omega$ would be needed in Fig. 5.30.

Concept of an Instrumentation Amplifier

Thus, the solution for at least one problem is clear: we need to add an extra amplifier stage. One way of doing so is shown in Fig. 5.32. Two non-inverting amplifiers are added to both inputs of the difference amplifier. This design achieves two goals simultaneously. First, it isolates the amplifier circuit from the Wheatstone bridge since the non-inverting amplifiers have an infinite input resistance and do not sink any current from the bridge. Second, it adds the extra gain; in other words, it eases the burden on the difference amplifier in the second stage. The difference amplifier becomes mainly responsible for rejecting the common-mode signal and the amplification of the differential signal.

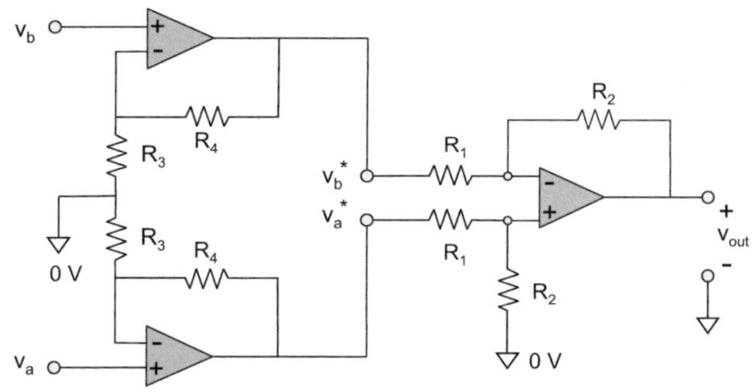

Fig. 5.32. An important step toward the instrumentation amplifier: we add non-inverting amplification stages at the input.

However, another problem arises there. The two non-inverting amplifiers amplify voltages v_a, v_b (close to 3.75 V in the present case). Therefore, at any appreciable gain (say, $A_{CL} = 1 + R_4/R_3 = 10$), they simply saturate and will not function! To avoid this issue, we use a simple yet critical change shown in Fig. 5.33 where we remove the common-port connection from the non-inverting stage. The circuit in Fig. 5.33 behaves completely differently compared to the original circuit in Fig. 5.32. We no longer have the output voltages v_a^*, v_b^* given by

$$v_a{}^* = A_{CL} v_a, \quad v_a{}^* = A_{CL} v_a, \quad A_{CL} = 1 + \frac{R_4}{R_3} \tag{5.55}$$

Instead, those voltages now become

$$v_a{}^* = v_a + \frac{R_4}{2R_3}(v_a - v_b), \quad v_b{}^* = v_b - \frac{R_4}{2R_3}(v_a - v_b) \tag{5.56}$$

The details of the derivation are seen in Fig. 5.33. The currents and voltages labeled in this figure are obtained using two summing-point constraints. The key observation is that the absolute voltages v_a, v_b are no longer amplified but are simply passed through. Only the differential voltage $v_D = v_a - v_b$ is amplified. The circuit in Fig. 5.33 is also a "difference amplifier," and it may be called the *unity common-mode gain stage*. The final step in the construction of the *instrumentation amplifier* is to connect both stages together. Figure 5.34 gives the final circuit that can be employed in conjunction with the Wheatstone bridge for the strain gauge shown in Fig. 5.31. Here, a quad op-amp chip (LM148 series) is used; it has four individual amplifiers inside the chip. For our circuit, we need three of them. The circuit is powered by a ± 7.5 V dual supply. According to Eqs. (5.53) and (5.56), the overall (differential) gain in Fig. 5.34 becomes

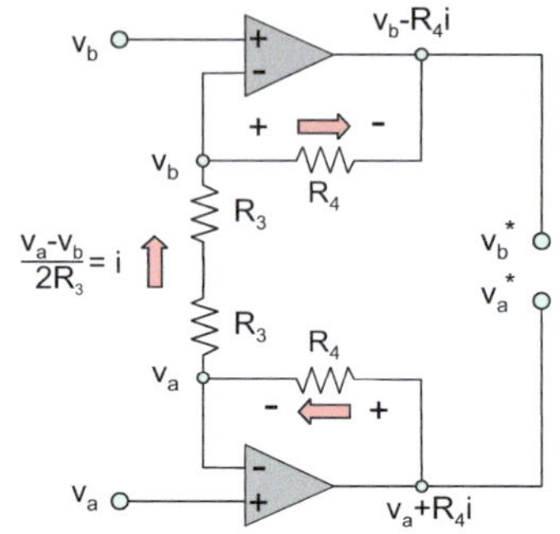

Fig. 5.33. Next step toward the instrumentation amplifier: we convert the non-inverting stage to a unity common-mode gain amplifier.

$$v_{\text{out}} = \frac{R_2}{R_1}\left(1+\frac{R_4}{R_3}\right)(v_{\text{a}} - v_{\text{b}}) = \frac{R_2}{R_1}\left(1+\frac{R_4}{R_3}\right)v_{\text{D}}, \quad A_{\text{CL}} = \frac{R_2}{R_1}\left(1+\frac{R_4}{R_3}\right) \tag{5.57}$$

The overall common-mode gain is exactly zero (ideal resistances). Choosing resistance values from Fig. 5.34 gives us an overall differential gain of 1010. We retain a certain gain (10) of the differential stage in order to have the gain of no more than 100 per stage.

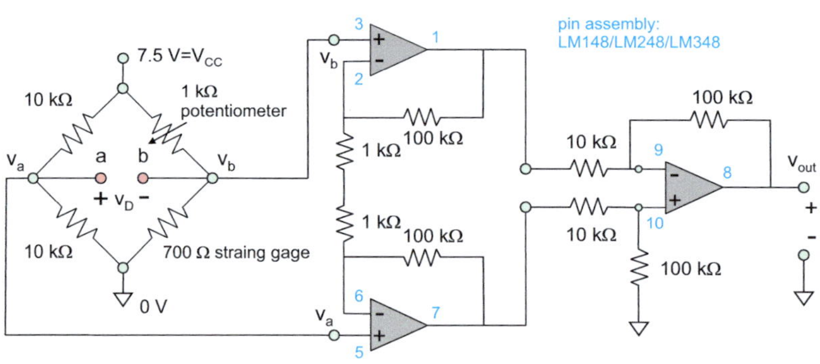

Fig. 5.34. The complete instrumentation amplifier for the strain gauge testing.

5.4.4 Instrumentation Amplifier in Laboratory

The operation of the circuit in Fig. 5.34 is shown in Fig. 5.35. The oscilloscope resolution is 1 V per division. At no applied strain, the oscilloscope connected to the amplifier output (the ground terminal of the oscilloscope is connected to circuit ground) shows a relatively small output voltage signal. When a bending moment is applied, as shown in Fig. 5.35a, the output voltage rises to approximately 2 V. This voltage is sufficient to light up a yellow LED (light-emitting diode) indicator connected between the output port and common port. When the opposite bending moment is applied, as in Fig. 5.35b, the output voltage drops to approximately –2 V. This voltage, taken with a negative sign, is again sufficient to light up a red LED connected from the common port to the output port.

The states in Fig. 5.35 have been achieved by a proper tuning of the potentiometer in the Wheatstone bridge. Thus, we have built a simple, yet useful, uncalibrated, uniaxial, stress-monitoring system. Frequently, the output of an instrumentation amplifier is connected to an analog-to-digital converter (ADC) and then to a computer system.

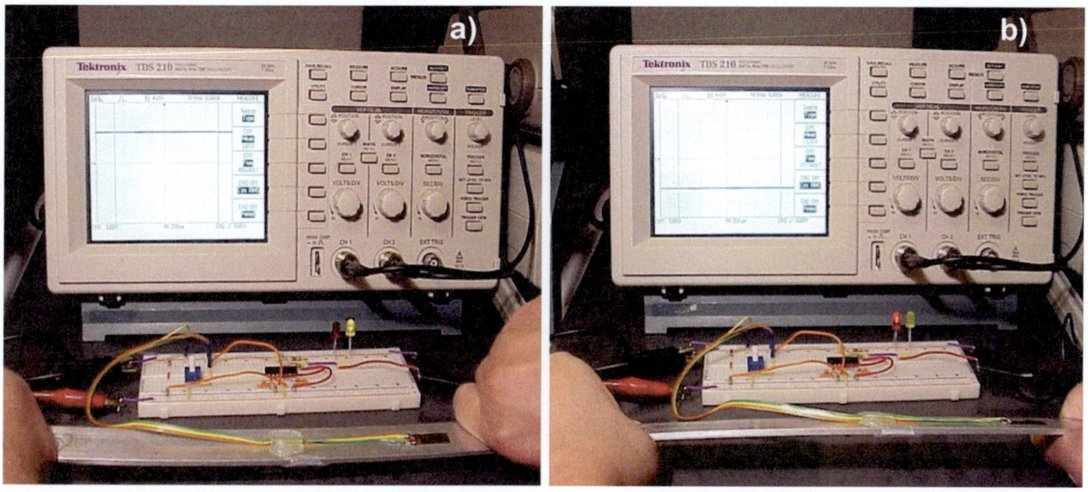

Fig. 5.35. Operation of the instrumentation amplifier with the strain gauge attached to a metal slab. (**a**) A "positive" bending moment is applied and (**b**) a "negative" bending moment is applied. The oscilloscope resolution is 1 V per division in every case.

Load Cell and Other Uses

The circuit in Fig. 5.34 gives us the idea of a commercial *load cell*. Strain gauges are commercially available in prefabricated modules such as load cells that measure force, tension, compression, and torque. All four resistors of the Wheatstone bridge may be strain gauges. Load cells typically use a full-bridge configuration and contain four leads for bridge excitation and measurement. The manufacturers provide calibration and accuracy information. However, the load cells do *not* normally include the instrumentation amplifier itself. Another mechanical engineering example where the differential amplifier is quite useful is a thermocouple. When measuring a thermocouple in a noisy environment, the noise from the environment appears as an offset on both input leads, making it a common-mode voltage signal. Many other examples indeed exist, particularly in biomedical engineering. The instrumentation amplifier is used to amplify an output signal from virtually any analog differential sensor instrument. Also note that instrumentation amplifiers with precision resistors are available as separate integrated circuits. Those ICs have a much better performance than an instrumentation amplifier wired on the protoboard. Other instrumentation amplifier types exist, which are different from the topology of the instrumentation amplifier circuit in Fig. 5.34. In principle, it is possible to design an instrumentation amplifier circuit with only *two* amplifier gain stages. The summary to this chapter provides an example used in practice.

Section 5.5 General Feedback Systems

5.5.1 Signal-Flow Diagram of a Feedback System

Although the negative feedback was first quantified by electrical engineers, it is extensively observed, studied, and employed in many mechanical, biomedical, chemical, and other systems. Figure 5.36 shows a generic structure of a *linear feedback system*.

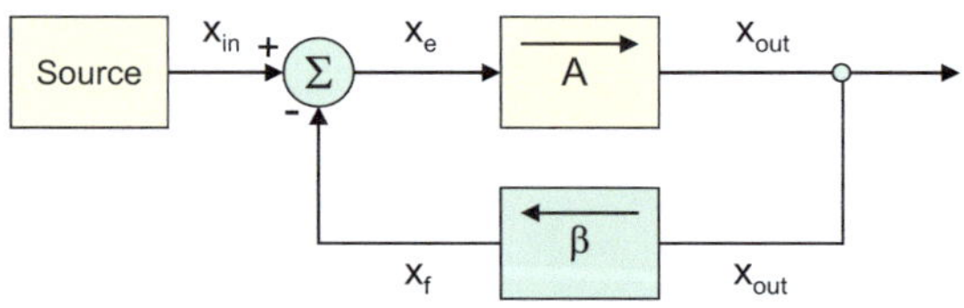

Fig. 5.36. A negative feedback loop for an arbitrary system (signal-flow diagram).

Figure 5.36 is a simplified *signal-flow diagram* or *controls block diagram*. The variable x is a *signal*; it may be voltage, current (electrical engineering) or displacement, velocity (mechanical engineering), etc. These three blocks have special names in control theory:

1. The first block (basic voltage amplifier in terms of ECE) is called the *forward* or *open-loop gain A* (often denoted by A_{OL}). It operates according to the ideal-amplifier rule

$$x_{\mathrm{out}} = Ax_{\mathrm{e}}, \quad A = \mathrm{const} > 0 \tag{5.58}$$

2. The second block is called the *feedback gain*. It operates according to a linear rule

$$x_f = \beta x_{\mathrm{out}}, \quad \beta = \mathrm{const} > 0 \tag{5.59}$$

   where β is called the *feedback factor*.
3. The block where the input signal and feedback signal are compared and subtracted from one another is called the *summing* (or *difference*) *node*.

5.5.2 Closed-Loop Gain and Error Signal

Mathematically, from Eq. (5.59), one has

$$x_{\mathrm{e}} = x_{\mathrm{in}} - x_f = x_{\mathrm{in}} - \beta x_{\mathrm{out}} \tag{5.60}$$

According to Eqs. (5.58) and (5.60), we obtain

$$x_{\mathrm{out}} = Ax_{\mathrm{e}} = A(x_{\mathrm{in}} - \beta x_{\mathrm{out}}) \tag{5.61}$$

Solving for x_{out} gives us the *closed-loop gain* A_{CL} (sometimes denoted by G or A_f)

$$x_{\text{out}} = \frac{A}{1 + A\beta} x_{\text{in}} \Rightarrow A_{\text{CL}} = \frac{A}{1 + A\beta} \tag{5.62}$$

If the *open-loop* gain A is made arbitrarily large (ideally approaching ∞), then the *closed-loop* gain approaches

$$A_{\text{CL}} \approx \frac{1}{\beta} \tag{5.63}$$

The significance of Eq. (5.63) cannot be overstated. It means that as long as the open-loop gain A is large enough, the closed-loop gain A_{CL} will approach the constant value $1/\beta$, which is precisely controlled by an external passive feedback network. In other words, manufacturing uncertainties in A and the potential nonlinear behavior of A are eliminated since A itself is eliminated. The price for this operation is a significant reduction of the overall system gain. Equation (5.63) implies that

$$A\beta >> 1 \Rightarrow A_{CL} = \frac{1}{\beta} << A \tag{5.64}$$

Clearly, the feedback loop reduces the initial gain substantially. And yet, despite this drawback, the closed-loop gain A_{CL} may still be large enough and sufficient for amplification as long as A is made very large. Thus, the feedback loop in Fig. 5.36 is a simple and powerful means to control the operation of an arbitrary high-gain system.

Exercise 5.9: The open-loop gain A in Fig. 5.36 varies between two extreme values of $A = 1000 \pm 200$ (± 20 % gain variation) depending on the system parameters. The forward gain block is used in the closed-loop configuration with the feedback factor β of 0.1. Approximate the two extreme values of the closed-loop gain, A_{CL}.

Answer: $A_{\text{CL}} = A/(1 + A\beta) = 9.90 \pm 0.02$ or $\pm 0.2\%$ closed-loop gain variation.

Error Signal

The second question of interest is finding the *error signal*, x_{e}, in Fig. 5.36, which corresponds to the differential input voltage for an amplifier circuit with the negative feedback. Substitution of Eq. (5.62) into Eq. (5.60) yields

$$x_{\text{e}} = x_{\text{in}} - \beta \frac{A}{1 + A\beta} x_{\text{in}} = \frac{1}{1 + A\beta} x_{\text{in}} \tag{5.65}$$

When the open-loop gain A is large and furthermore $A\beta >> 1$, one obtains

$$x_{\text{e}} \approx 0 \tag{5.66}$$

As applied to the amplifier circuits, Eq. (5.66) is exactly the second summing-point constraint or the condition of the zero differential input amplifier voltage under presence of the negative feedback.

Exercise 5.10: The open-loop gain A in Fig. 5.36 is 10,000. The forward gain block is used in the closed-loop configuration with the feedback factor β of 0.1. Determine the error signal x_e if the input voltage signal is 1 mV.

Answer: $x_e = 0.999\ \mu V \approx 1\ \mu V$.

5.5.3 Application of General Theory to Voltage Amplifiers with Negative Feedback

Two circuits shown in Fig. 5.37 are the buffer amplifier circuit and the non-inverting amplifier circuit, respectively. The goal is to find the closed-loop gain A_{CL} of the amplifier circuit when the given open-circuit voltage gain A is large but finite. In this case, the second summing-point constraint cannot be applied. Therefore, simple Eqs. (5.16) and (5.21) obtained previously need to be modified.

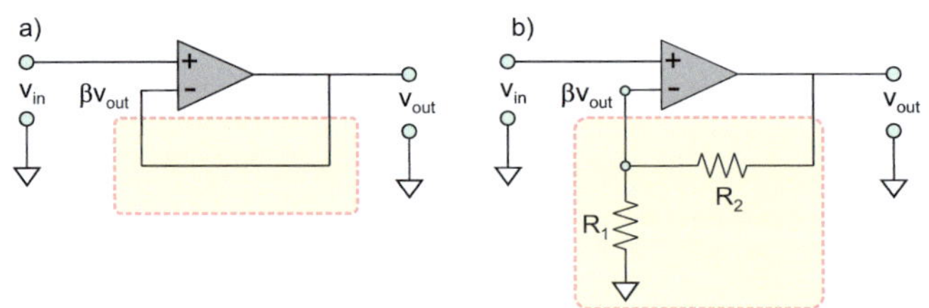

Fig. 5.37. Two amplifier circuits with negative feedback networks indicated by a shaded rectangle.

Both circuits from Fig. 5.37 have the form of the feedback system as in Fig. 5.36. The feedback network is indicated by a shaded rectangle. The signal x is now the voltage. Since A is given, the only problem is to find the feedback factor, β. For the buffer amplifier, the feedback factor is clearly one. For the non-inverting amplifier circuit, the feedback loop is the voltage divider, where β is determined by the resistance ratio. Note that the voltage divider model implies no current into amplifier's input terminals. Specifically, βv_{out} is equal to $R_1/(R_1 + R_2)v_{out}$. Therefore,

$$\beta = 1 \;-\; \text{buffer ampl. circuit;} \quad \beta = \frac{R_1}{R_1 + R_2} \;-\; \text{non-inv. ampl. circuit} \tag{5.67}$$

Substitution into Eq. (5.62) gives us two expressions for the closed-loop gain:

$$A_{CL} = \frac{A}{1 + A} \;-\; \text{buffer ampl. circuit;} \quad A_{CL} = \frac{A}{1 + A\frac{R_1}{R_1 + R_2}} \;-\; \text{non-inv. ampl. circuit} \tag{5.68}$$

The first equation (5.68) coincides with Eq. (5.23) obtained in Section 5.2 using the accurate circuit analysis. So does second equation (5.68) when we repeat the same analysis for the non-inverting amplifier configuration. If $A \to \infty$, then the simple gain expressions—Eqs. (5.16) and (5.21)—derived with the help of the second summing-point constraint are obtained from Eqs. (5.68). The analysis of the inverting amplifier requires more efforts since this amplifier type is not exactly the voltage amplifier but rather a transresistance amplifier considered next.

Exercise 5.11: The open-loop (open-circuit) gain A of a non-inverting amplifier circuit with $R_1 = 1\,\text{k}\Omega, R_2 = 9\,\text{k}\Omega$ is 10,000. Determine the closed-loop gain.

Answer: $A_{CL} = 9.99$, which is by 0.1 % different from $A_{CL} = 1 + \frac{R_2}{R_1} = 10$.

Last but not least, we emphasize another significant advantage of the negative feedback. When the input and output resistances of the amplifier model in Fig. 5.5 have finite values (which occurs in practice), the negative feedback loop effectively *increases* the input resistance and *decreases* the output resistance, i.e., makes the entire amplifier circuit look closer to the ideal-amplifier model.

5.5.4 Voltage, Current, Transresistance, and Transconductance Amplifiers with the Negative Feedback

At the end of this short section, we consider four basic amplifier circuits with negative feedback, which correspond to the four basic dependent sources studied in Chapter 2: *voltage amplifier*, *transconductance amplifier*, *transresistance amplifier*, and *current amplifier*. Figure 5.38 shows the corresponding circuit diagrams. Load resistance R_L is introduced for the transconductance amplifier and the current amplifier, respectively, where the output is the load current. Although every amplifier circuit may be represented in the form similar to the feedback diagram in Fig. 5.36 and analyzed accordingly, only a simplified treatment will be given here. It utilizes the condition $A \to \infty$ and the resulting second summing-point constraint. Using the both summing-point constraints, we may obtain, with reference to Fig. 5.38,

$$v_{out} = \left(1 + \frac{R_2}{R_1}\right) v_{in} \quad \text{voltage amplifier} \tag{5.69a}$$

$$i_{out} = G_F v_{in}, \quad G_F = 1/R_F \quad \text{transconductance amplifier} \tag{5.69b}$$

$$v_{out} = -R_F i_{in} \quad \text{transresistance amplifier} \tag{5.69c}$$

$$i_{out} = \left(1 + \frac{R_2}{R_1}\right) i_{in} \qquad \text{current amplifier} \tag{5.69d}$$

Note that other more elaborate circuits may be considered; some of them are analyzed in the corresponding homework problems.

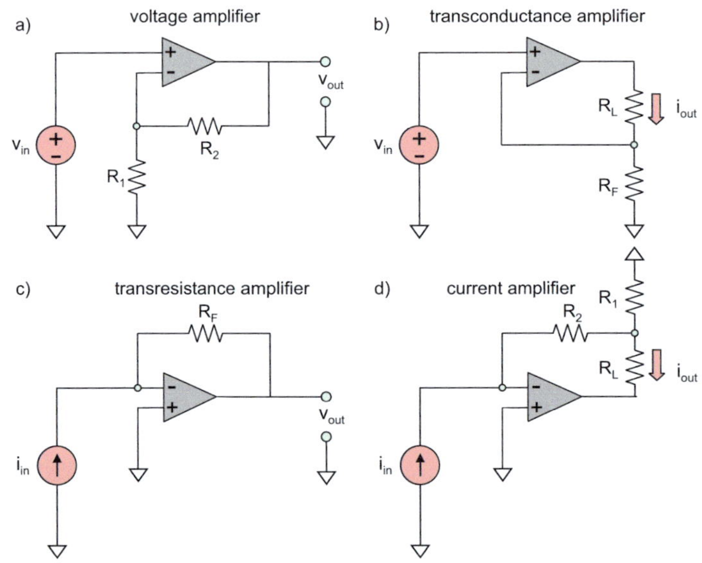

Fig. 5.38. Four basic amplifier types with negative feedback.

Summary

Amplifier circuit	Operation	Formulas
Open-loop operational amplifier (comparator) 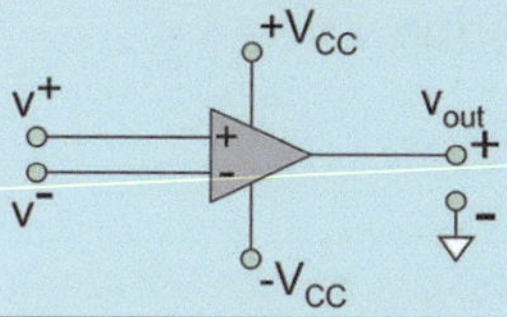	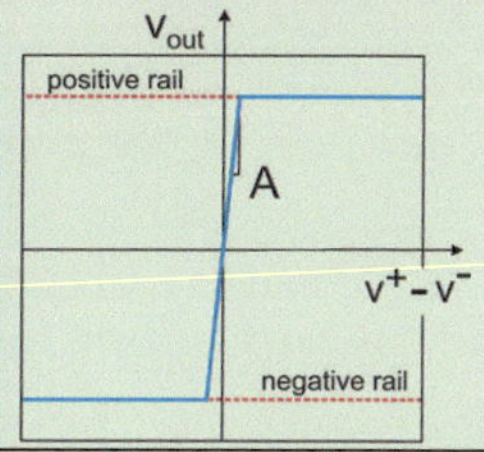	Operation with $\pm V_{CC}$ power rails: $v_{out} = A(v^+ - v^-), \quad \lvert v_{out}\rvert < V_{CC}$ Open-circuit (open-loop) gain A is very high
Amplifier circuit model 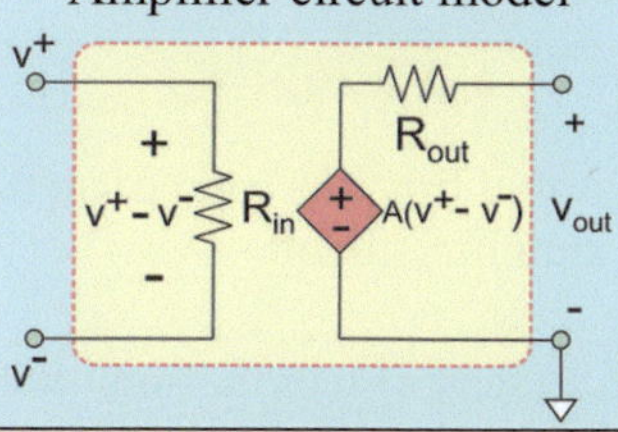	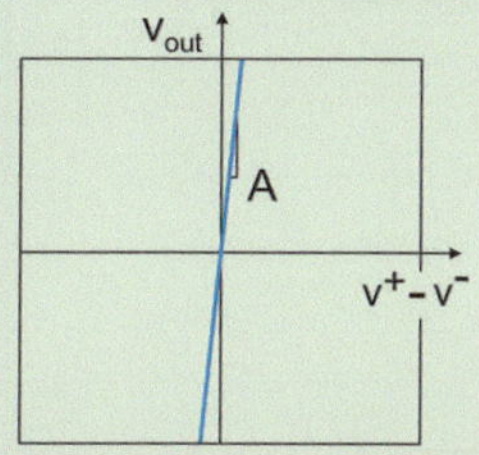	– Valid for *any* voltage amplifier or voltage amplifier circuit, but R_{in}, R_{out}, A are different in every case; – Ideal amplifier model (useful simplification): $R_{in} = \infty$, $R_{out} = 0$, $A = \infty$
Negative feedback for the ideal-amplifier model: differential input voltage is zero (2nd SPC)		
Non-inverting amplifier 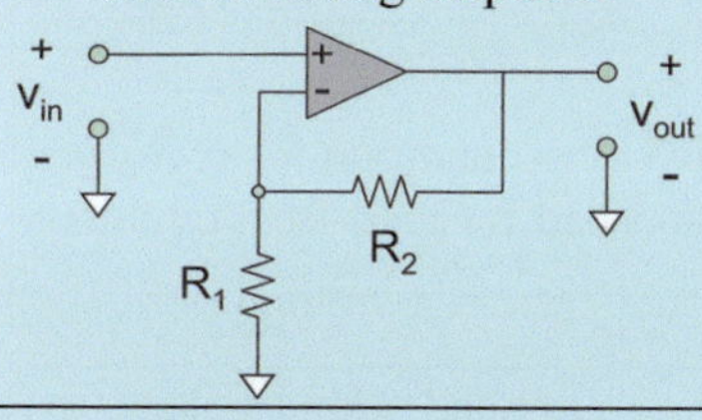	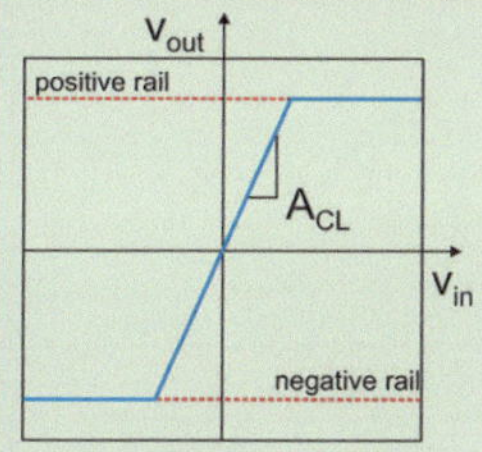	For ideal-amplifier model: $v_{out} = A_{CL} v_{in}, \; A_{CL} = 1 + \frac{R_2}{R_1}$ $R_{in} = \infty$, $R_{out} = 0$ Exact: $A_{CL} = A\left(1 + A\frac{R_1}{R_1 + R_2}\right)^{-1}$
Voltage follower (buffer) amplifier 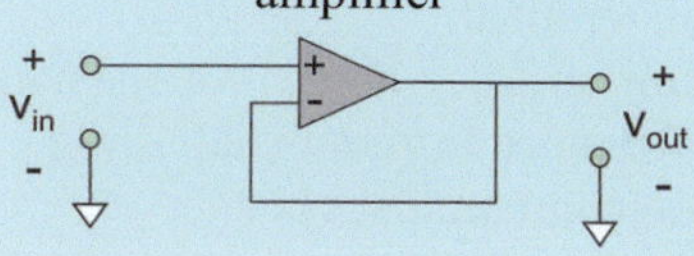	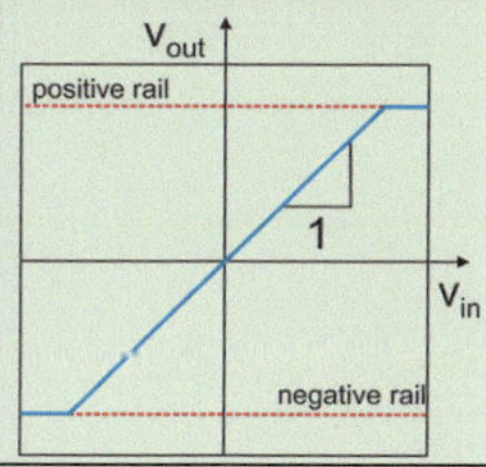	For ideal-amplifier model: $v_{out} = A_{CL} v_{in}, \; A_{CL} = 1$ $R_{in} = \infty$, $R_{out} = 0$ Exact: $A_{CL} = \frac{A}{1 + A}$
Inverting amplifier 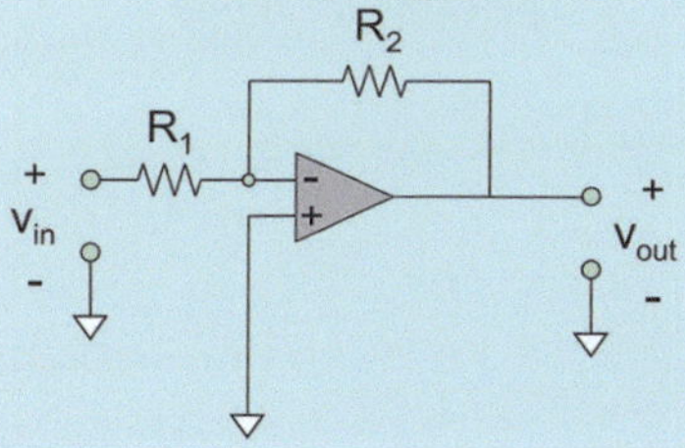	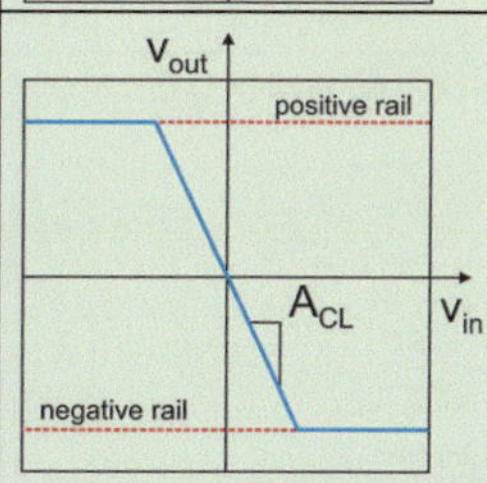	For ideal-amplifier model: $v_{out} = A_{CL} v_{in}, \; A_{CL} = -\frac{R_2}{R_1}$ $R_{in} = R_1$, $R_{out} = 0$ Exact: $A_{CL} = -\frac{R_2}{R_1} A\left(A + 1 + \frac{R_2}{R_1}\right)^{-1}$

(continued)

Cascaded amplifier

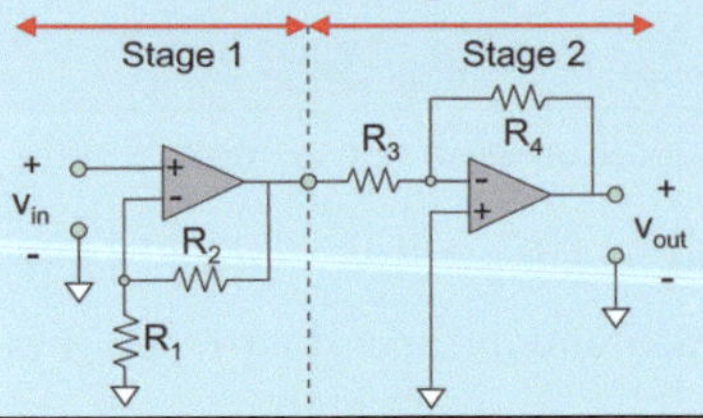

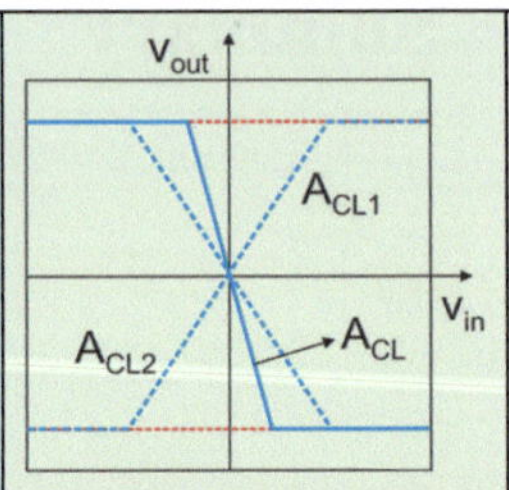

- Gains of individual stages multiply;
- Input resistance of the amplifier circuit is the input resistance of stage 1;
- Individual stage gain should not exceed 100

Input load bridging versus input load matching

Load bridging

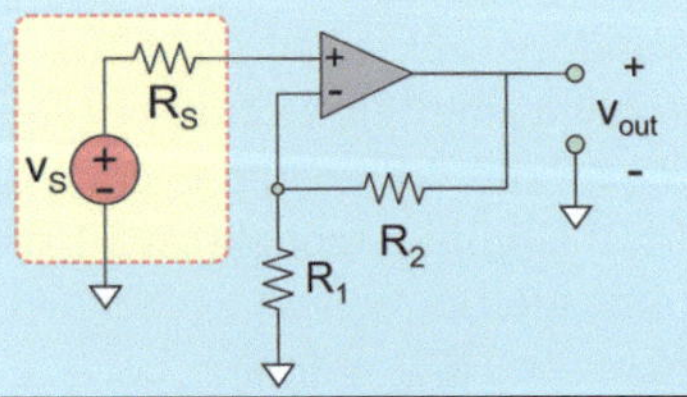

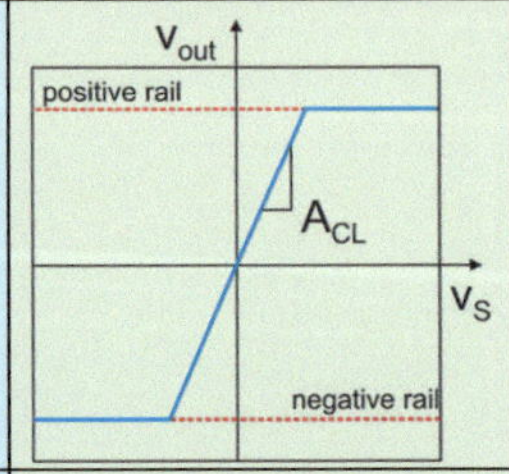

- Source (sensor) sees the amplifier as an open circuit:

$$v_{\text{out}} = v_{\text{S}} \times A_{\text{CL}}$$

- No current from the source can flow into amplifier circuit

Load matching

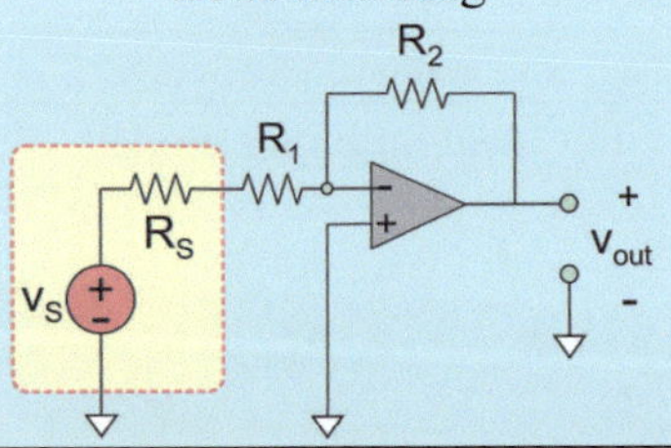

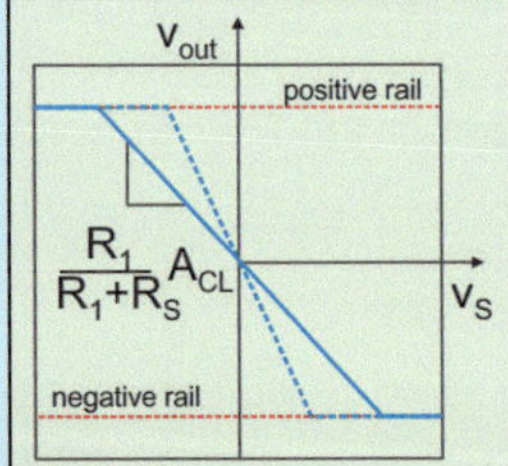

- Source (sensor) sees the amplifier as resistance $R_{\text{in}} = R_1$:

$$v_{\text{out}} = \frac{R_{\text{in}}}{R_{\text{S}} + R_{\text{in}}} v_{\text{S}} \times A_{\text{CL}}$$

- Matching condition $R_{\text{S}} = R_{\text{in}}$ is important for high-freq. circuits

DC imperfections and their cancellation

Non-inverting amplifier

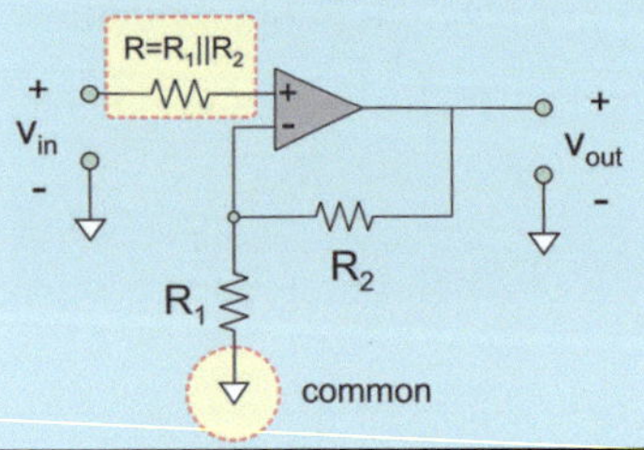

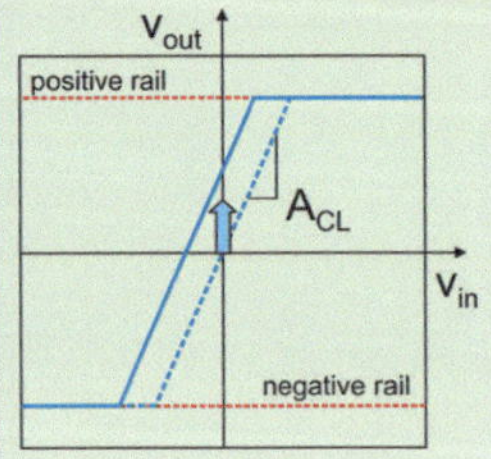

- Short-circuited output voltage may be trimmed to zero using the offset null terminal
- Short-circuited output voltage may be trimmed to zero by adjusting common-terminal voltage;
- Extra resistance R eliminates the effect of the input bias current

Inverting amplifier

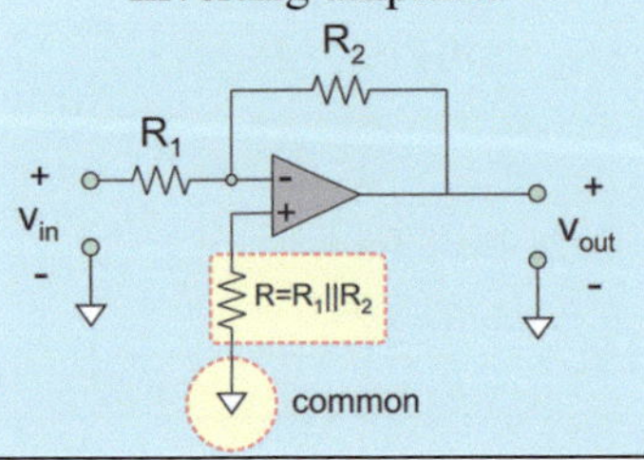

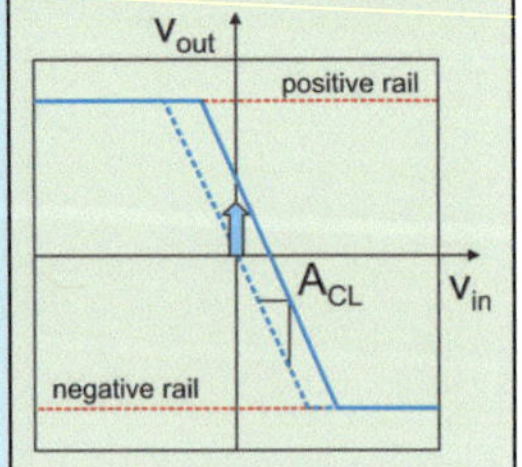

The same as above

(continued)

Multiple-input amplifier circuits		
Summing amplifier 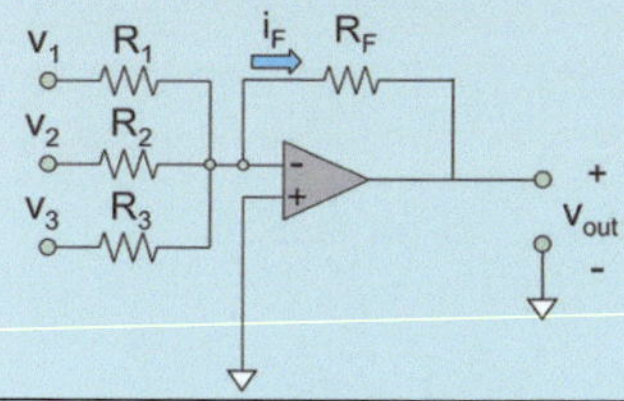	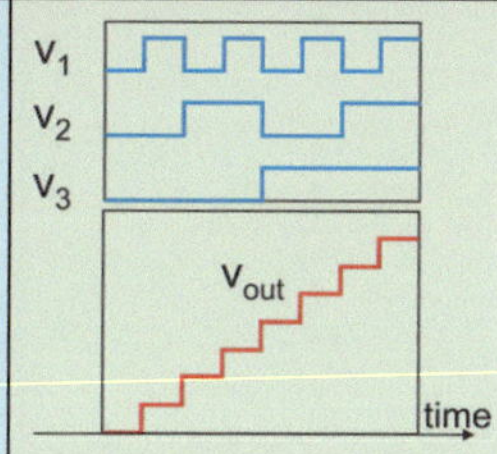	– The summing amplifier sums several weighted input voltages: $v_{\mathrm{out}} = -\frac{R_F}{R_1}v_1 - \frac{R_F}{R_2}v_2 - \frac{R_F}{R_3}v_3$ – Used as a prototype of the digital to analog converter
Difference amplifier 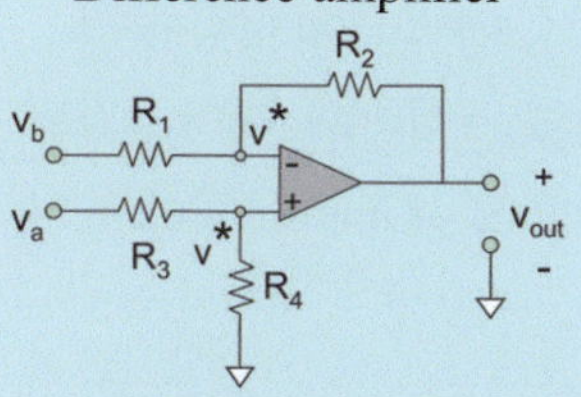	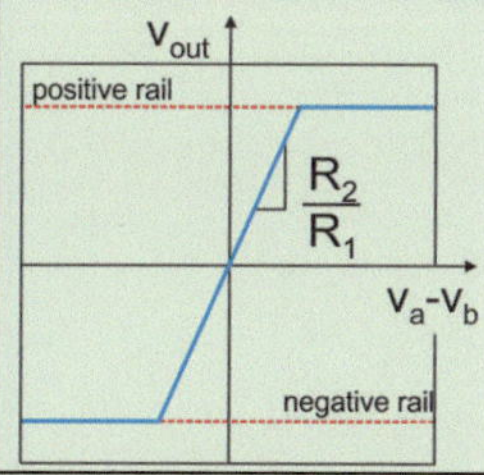	– True difference amplifier: $\frac{R_2}{R_1} = \frac{R_4}{R_3},\ v_{\mathrm{out}} = \frac{R_2}{R_1}(v_a - v_b)$ – Rejects common-mode voltage – For the general difference amplifier circuit see Eq. (5.52c)
Unity common-mode gain stage 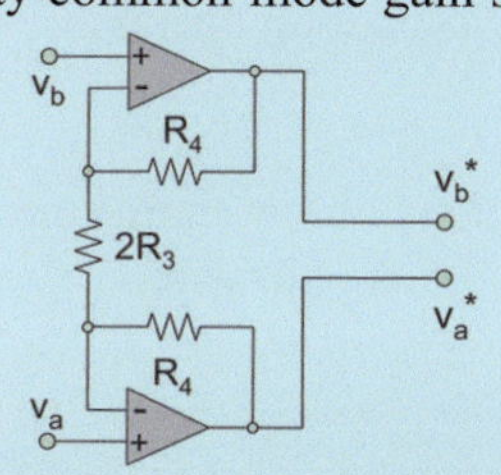	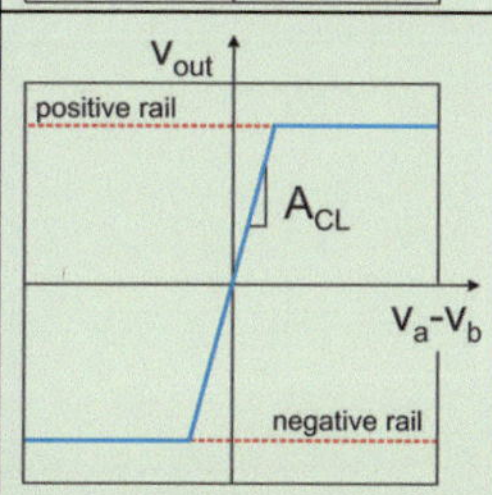	– Differential gain: $v_a{}^* - v_b{}^* = A_D(v_a - v_b)$ $A_D = 1 + \frac{R_4}{R_3}$ – Common-mode gain: $v_a{}^* + v_b{}^* = A_{CM}(v_a + v_b)$ $A_{CM} = 1$
Instrumentation amplifier 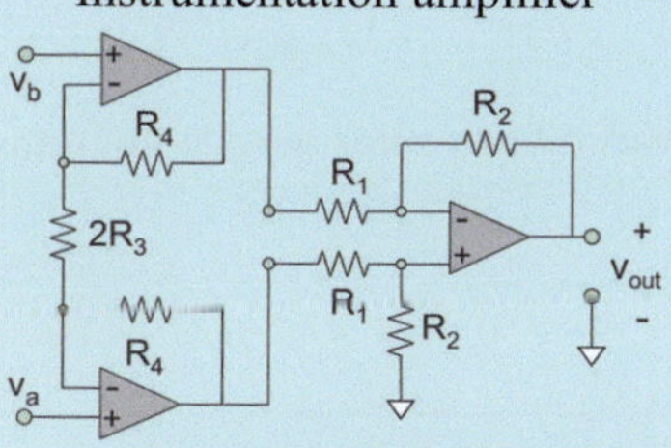	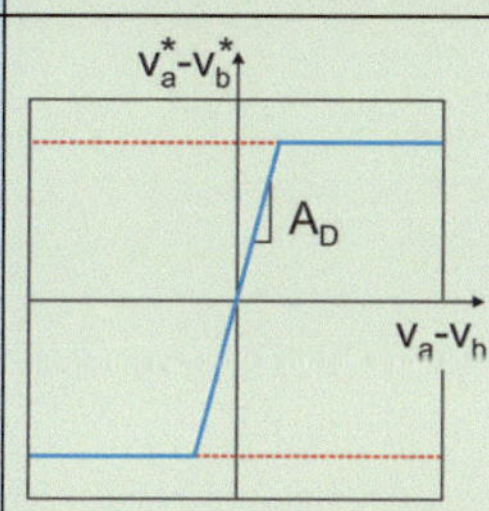	Output voltage: $v_{\mathrm{out}} = \frac{R_2}{R_1}\left(1 + \frac{R_4}{R_3}\right)(v_a - v_b)$ Closed-loop differential gain: $A_{CL} = \frac{R_2}{R_1}\left(1 + \frac{R_4}{R_3}\right)$ – Rejects common-mode voltage
Simpler instrumentation amplifier 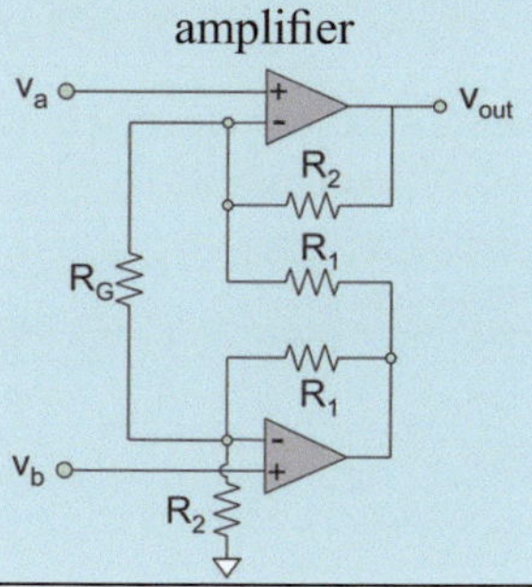	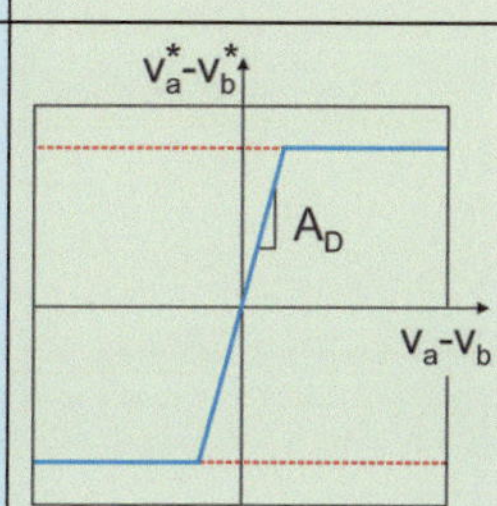	Output voltage: $v_{\mathrm{out}} = \left(2\frac{R_2}{R_G} + \left(1 + \frac{R_2}{R_1}\right)\right)(v_a - v_b)$ – Closed-loop differential gain: $A_{CL} = 2\frac{R_2}{R_G} + \left(1 + \frac{R_2}{R_1}\right)$ – Rejects common-mode voltage

(continued)

General feedback systems and amplifiers with negative feedback		
Signal—flow diagram 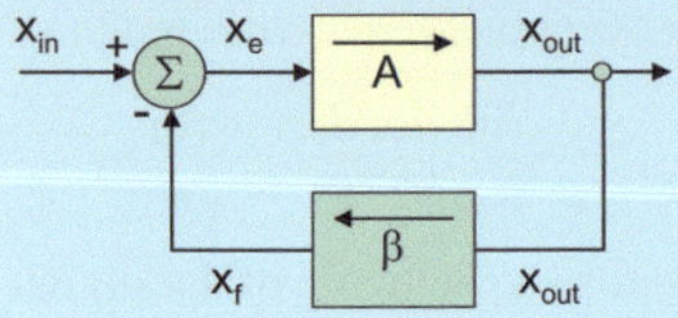	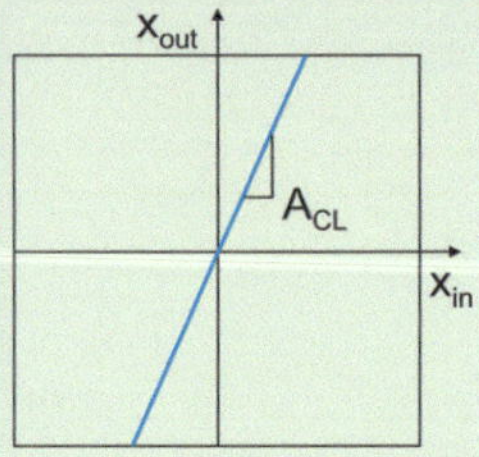	Closed-loop gain: $x_{\text{out}} = A_{\text{CL}} x_{\text{in}}, \ A_{\text{CL}} = \dfrac{A}{1 + A\beta} \approx \dfrac{1}{\beta}$ Error signal: $x_e = \dfrac{1}{1 + A\beta} x_{\text{in}} \approx 0$
Transconductance amplifier 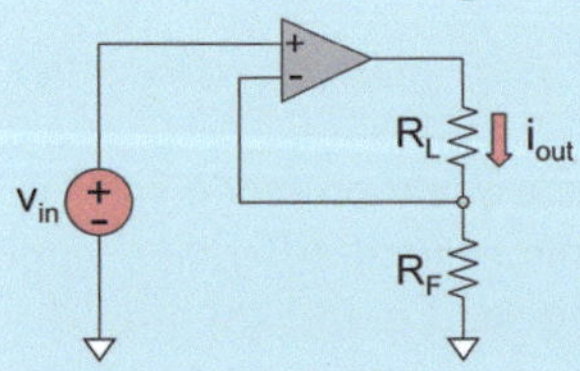	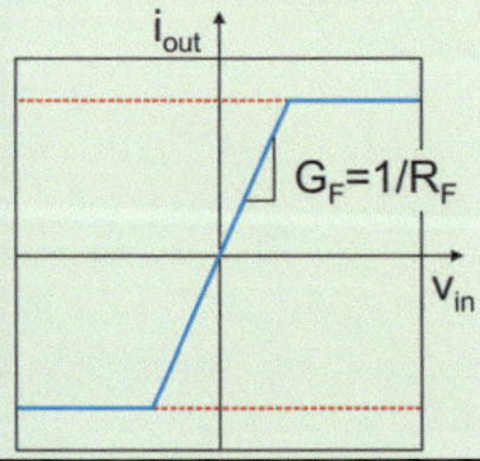	Closed-loop operation: $i_{\text{out}} = G_{\text{F}} v_{\text{in}}, \ G_{\text{F}} = 1/R_{\text{F}}$ Variations of this simple circuit are possible
Transresistance amplifier 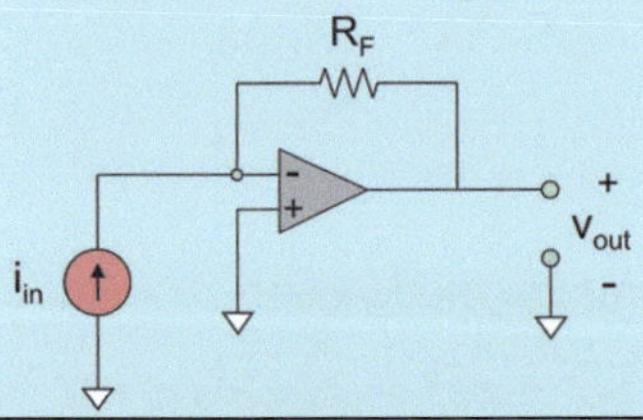	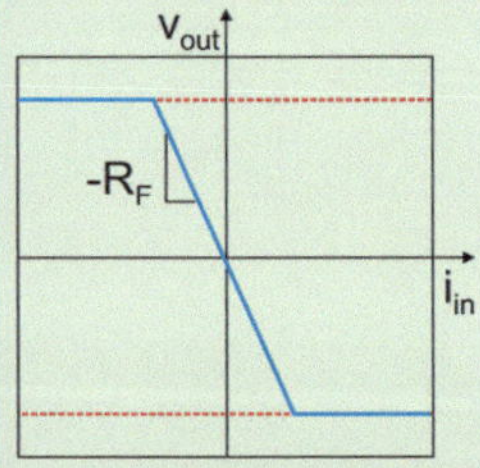	Closed-loop operation: $v_{\text{out}} = -R_{\text{F}} i_{\text{in}}$ Variations of this simple circuit are possible
Current amplifier 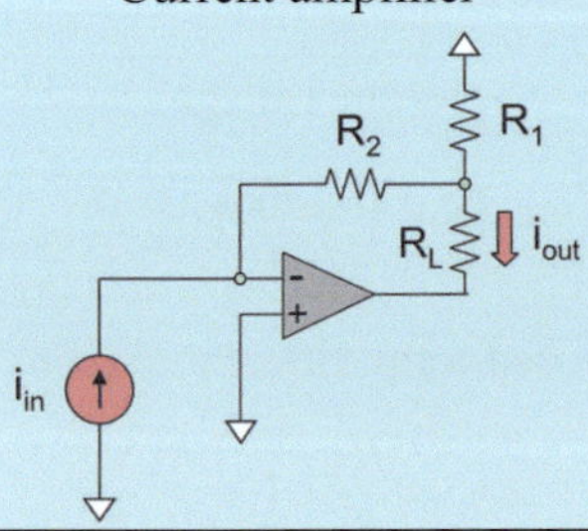	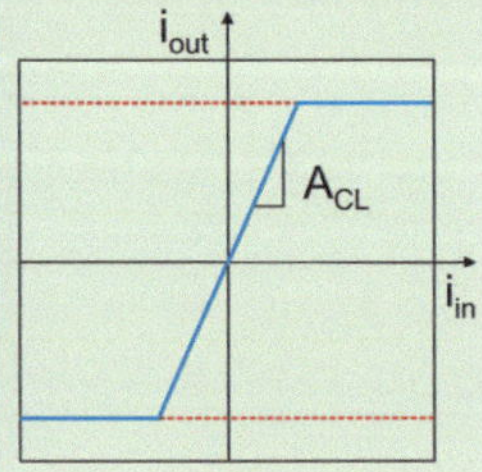	Closed-loop operation: $i_{\text{out}} = \left(1 + \dfrac{R_2}{R_1}\right) i_{\text{in}}$ Variations of this simple circuit are possible
Howland amplifier (current pump) 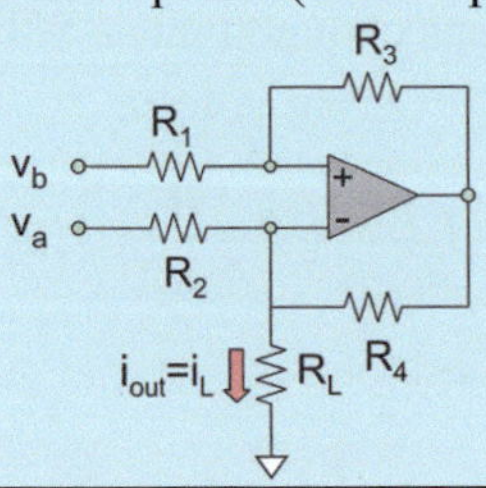	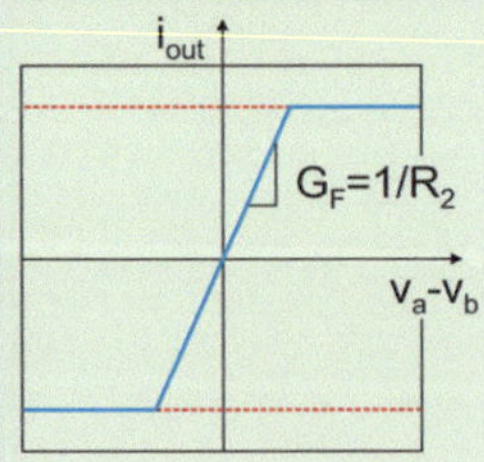	Closed-loop operation: $(R_1/R_3 = R_2/R_4)$ $i_{\text{out}} = G_{\text{F}}(v_a - v_b), \ G_{\text{F}} = 1/R_2$ Variations of this clever circuit are possible

Problems

5.1 Amplifier operation and circuit models

5.1.1 Amplifier Operation

Problem 5.1. An operational amplifier has five terminals.

A. Sketch the amplifier symbol.
B. Name each of the op-amp terminals and describe its function in one sentence per terminal.
C. Can the amplifier IC have more than five terminals? Explain.

Problem 5.2. You may wonder about the meaning of the two letters preceding amplifier marking, e.g., LM741. Each of the semiconductor companies has its own abbreviation, e.g., LM for an amplifier designed and manufactured by the National Semiconductor Corporation (acquired by Texas Instruments in 2011), AD for an amplifier manufactured by Analog Devices, MC for STMicroelectronics, TL for Texas Instruments, etc. The same chip, e.g., LM741, may be manufactured by several semiconductor chip makers. The part number is given by a numerical code that is imprinted on the top of the package. An MC1458 amplifier IC chip is shown in the figure below. This IC is a *dual operational amplifier*. In other words, one such IC package contains two separate operational amplifiers.

N
DIP8
(Plastic Package)

A. Download the amplifier's datasheet from http://www.datasheetcatalog.com
B. Redraw the figure to this problem in your notes and label the pins for the non-inverting input, the inverting input, and the output of the operational amplifier #1.
C. Label pins for $+V_{CC}$ and $-V_{CC}$.

Problem 5.3. What is the minimum number of pins required for:

A. The dual operational amplifier (the corresponding IC package contains two separate operational amplifiers)?
B. The *quad operational amplifier* (the corresponding IC package contains four separate operational amplifiers)?

Problem 5.4. An operational amplifier has an open-circuit gain of $A = 2 \times 10^5$ and is powered by a dual source of ± 10 V. It is operated in the open-circuit configuration. What is the amplifier's open-circuit output voltage v_{out} if

A. $v^+ = 0\text{V},\ v^- = 0\text{V}$
B. $v^+ = +1\text{V},\ v^- = +1\text{V}$
C. $v^+ = +1\text{V},\ v^- = 0\text{V}$
D. $v^+ = 0\text{V},\ v^- = -1\text{V}$
E. $v^+ = +1\text{mV},\ v^- = 0\text{V}$
F. $v^+ = -1\text{mV},\ v^- = 0\text{V}$
G. $v^+ = 10\mu\text{V},\ v^- = 0\text{V}$
H. $v^+ = 0\text{V},\ v^- = 10\mu\text{V}$

Problem 5.5. Based on the solution to Problem 5.4, why do you think the operational amplifier is seldom used in the open-loop configuration, at least in analog electronics?

Problem 5.6. Using the website of the National Semiconductor Corporation, determine the maximum and minimum supply voltages (operating with the dual-polarity power supply) for the following amplifier's ICs:

A. LM358
B. LM1458
C. LM741

Which amplifier IC from the list may be powered by two AAA batteries?

Problem 5.7. Plot to scale the output voltage of the operation amplifier with an open-circuit gain $A = 5 \times 10^4$ when the non-inverting input voltage v^+ changes from -2 mV to +2 mV and the inverting input voltage v^- is equal to -1 mV. The amplifier is powered by a ± 12-V dual voltage supply. Label the axes.

Problem 5.8. Repeat the previous problem for $A = 5 \times 10^5$.

5.1.2 Operational Amplifier Comparator

Problem 5.9. In a circuit shown in the figure below, an operational amplifier is driven by a ±10-V dual power supply (not shown). The open-circuit DC gain of the amplifier is $A = 1{,}000{,}000$. Sketch to scale the output voltage to the amplifier when

a) $V_{\text{threshold}} = -30$ mV
b) $V_{\text{threshold}} = +30$ mV

Assume that the amplifier hits the power rails in saturation.

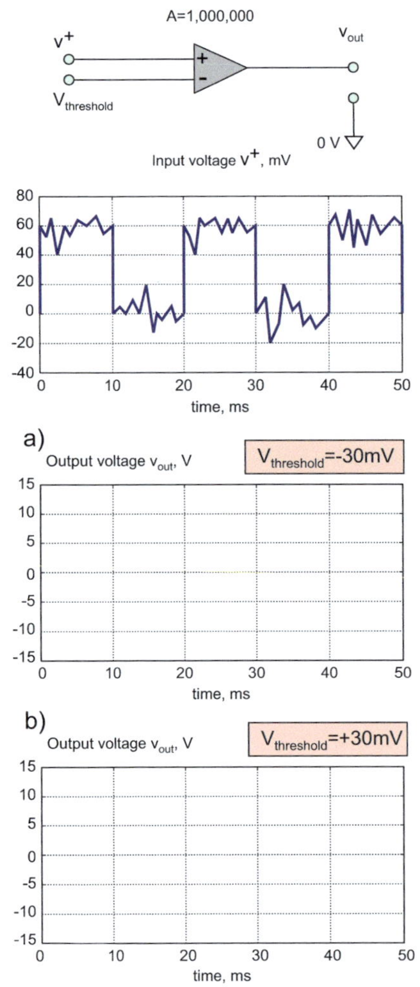

Problem 5.10. Based on the solution to the previous problem, why do you think the operational amplifier in the open-loop configuration may be useful for digital circuits?

Problem 5.11. Solve Problem 5.9 when the input voltage is applied to the inverting input and the threshold voltage is applied to the non-inverting input.

Problem 5.12. The circuit shown in the figure is a *zero-level detector*. An operational amplifier in the open-loop configuration is driven by a ±10-V dual power supply (not shown). The open-circuit amplifier gain is 100,000. Sketch the output voltage to scale. Assume that the amplifier hits the power rails in saturation.

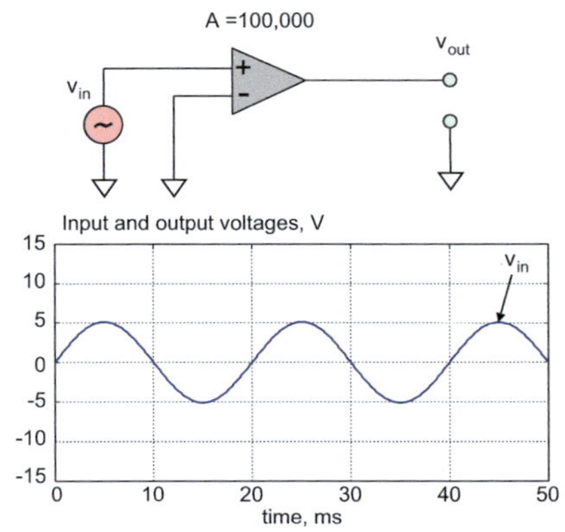

Problem 5.13. In a circuit shown in the figure below, an operational amplifier is driven by a ±15-V dual power supply (not shown). The open-circuit gain of the amplifier is $A = 100{,}000$. Sketch to scale the output voltage to the amplifier when

a) $V_{\text{threshold}} = 0$ mV
b) $V_{\text{threshold}} = +4$ mV

Assume that the amplifier hits the power rails in saturation.

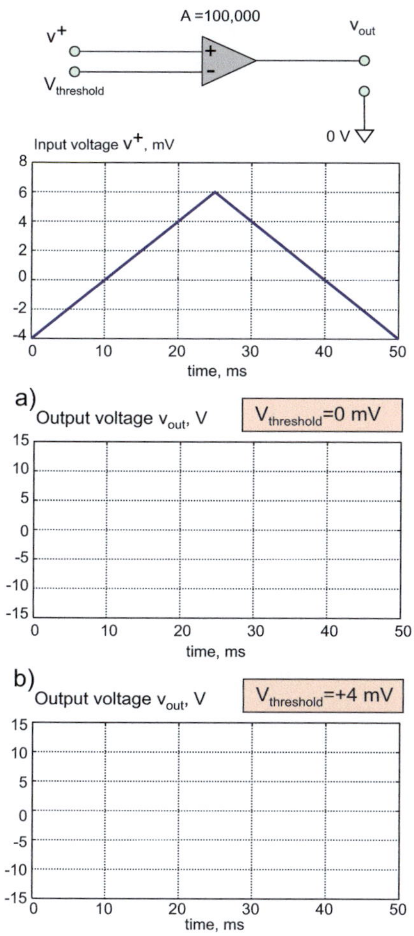

5.1.3 Amplifier Circuit Model
5.1.4 Ideal-Amplifier Model

Problem 5.14

A. Draw the circuit model of an operational amplifier.

B. Describe the meaning of the amplifier as the voltage-controlled voltage source in your own words.

Problem 5.15. For an equivalent amplifier circuit with $A = 1500$ shown in the figure below, determine the output voltage given that $v_S(t) = 1\cos\omega t$ [mV], $R_S = 50\ \Omega$, $R_L = 50\ \Omega$ for three cases:

A. $R_{in} = 100\ k\Omega$ and $R_{out} = 2\ \Omega$.

B. $R_{in} = 50\ \Omega$ and $R_{out} = 25\ \Omega$.

C. $R_{in} = \infty$ and $R_{out} = 0$.

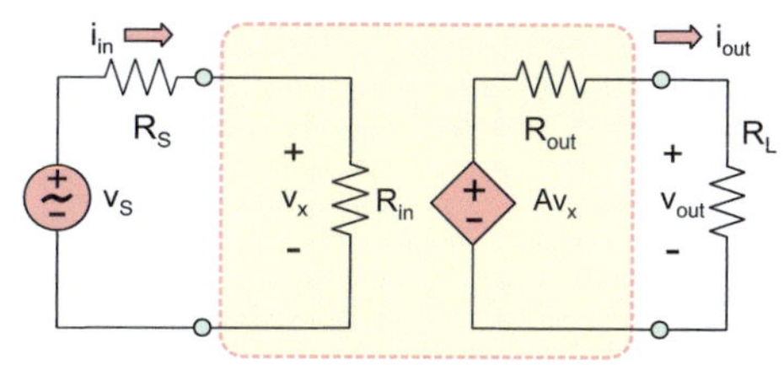

Problem 5.16. Name one reason why we should attempt to:

A. Make the input resistance (impedance) to the amplifier as high as possible.

B. Make the output resistance (impedance) to the amplifier as low as possible.

Problem 5.17

A. List all conditions of the ideal-amplifier model.

B. What is the short-circuit output current of the ideal amplifier?

Problem 5.18

A. What is the first summing-point constraint?

B. What is the equivalent formulation of the first summing-point constraint in terms of the input resistance (impedance) to the amplifier?

Problem 5.19. An amplifier circuit is shown in the figure below. The first summing-point constraint applies. Determine current i_2.

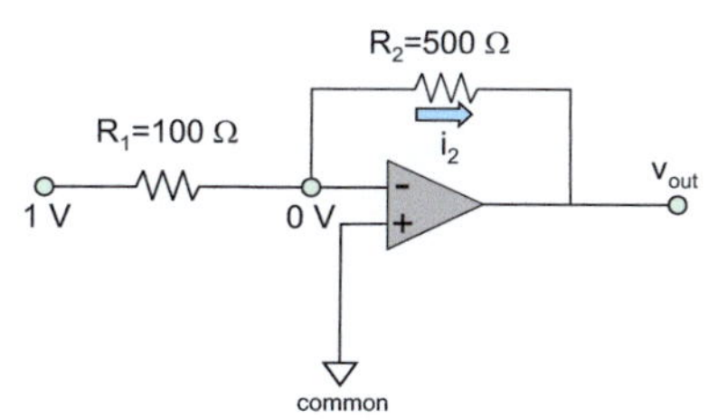

Problem 5.20. An amplifier circuit is shown in the figure below. The first summing-point constraint applies. An ideal operational amplifier has an open-circuit gain of $A = 2 \times 10^5$.

Determine the output voltage, V_{out}. You are not allowed to use any of the materials of the next section!

Hint: Denote the unknown voltage at node * by v^*, express v^* in terms of v_{out}, and then solve for v_{out}.

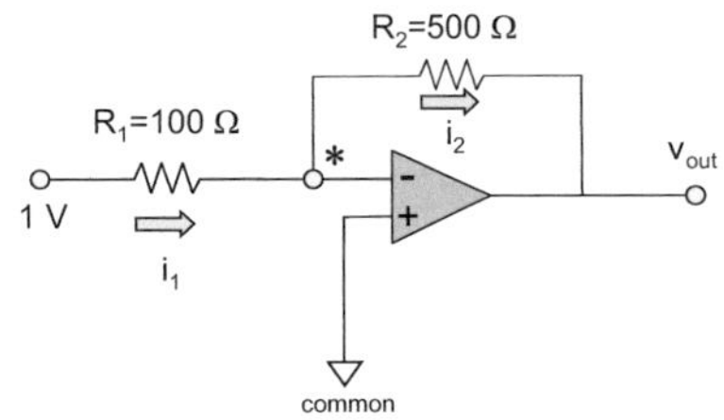

Problem 5.21. An ECE laboratory project uses the LM358 amplifier IC.

A. What semiconductor company has developed this chip?
B. Is the chip from the lab project necessarily manufactured by this company? (See http://www.datasheetcatalog.com/ for manufacturers' datasheets related to this product.)
C. Use the Digi-Key distributor's website and estimate average cost for this amplifier chip (DIP-8 package) in today's market.

Problem 5.22. An ECE laboratory project uses the TL082 amplifier IC.

A. What semiconductor company has developed this chip?
B. Is the chip from the lab project necessarily manufactured by this company? (See http://www.datasheetcatalog.com/ for manufacturers' datasheets related to this product.)
C. Use the Digi-Key distributor's website to estimate the average cost for this amplifier chip (DIP-8 package) in today's market.

5.2 Negative Feedback

5.2.2 Amplifier Feedback Loop. Second Summing-Point Constraint

5.2.3 Amplifier Circuit Analysis Using Two Summing-point Constraints

Problem 5.23

A. Name the two summing-point constraints used to solve an amplifier circuit.
B. Which summing-point constraint remains valid without the negative feedback?

Non-inverting Amplifier

Problem 5.24

A. Draw the circuit diagram of the basic non-inverting amplifier configuration.
B. Accurately derive the expression for the amplifier gain in terms of the resistances, assuming an ideal operational amplifier.

Problem 5.25. Using the two summing-point constraints, solve the ideal-amplifier circuit shown in the figure if the input voltage has the value of 2 mV.

A. Label and determine the currents in the feedback loop.
B. Determine the output voltage of the amplifier versus the common port.

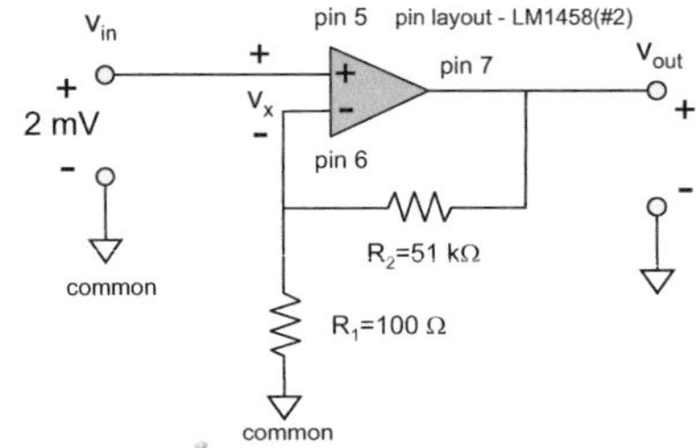

Problem 5.26. Determine the output voltage of the ideal operational amplifier shown in the figure. The amplifier is driven by a ±10-V dual power supply.

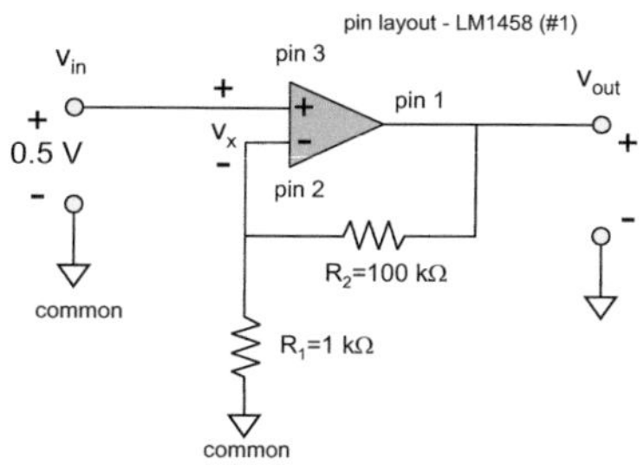

Inverting Amplifier

Problem 5.27

A. Draw the circuit diagram of the basic inverting amplifier configuration.
B. Give the expression for the amplifier gain in terms of the resistances, assuming an ideal operational amplifier.

Problem 5.28. Using the two summing-point constraints, solve the ideal-amplifier circuit shown in the figure that follows if the input voltage is 1 mV.

A. Label and determine the currents in the feedback loop.
B. Determine the output voltage of the amplifier versus the common port.

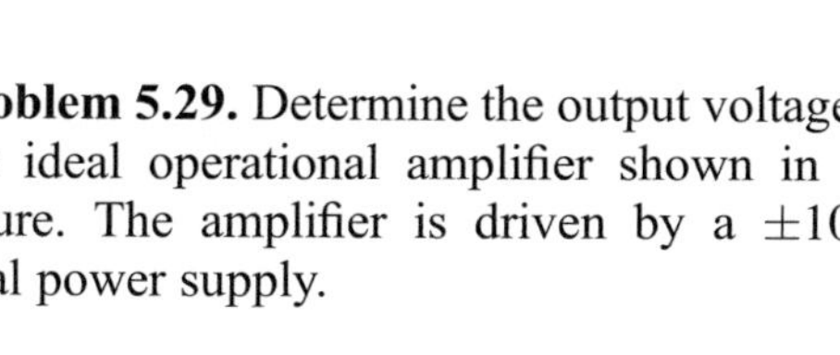

Problem 5.29. Determine the output voltage of the ideal operational amplifier shown in the figure. The amplifier is driven by a ±10-V dual power supply.

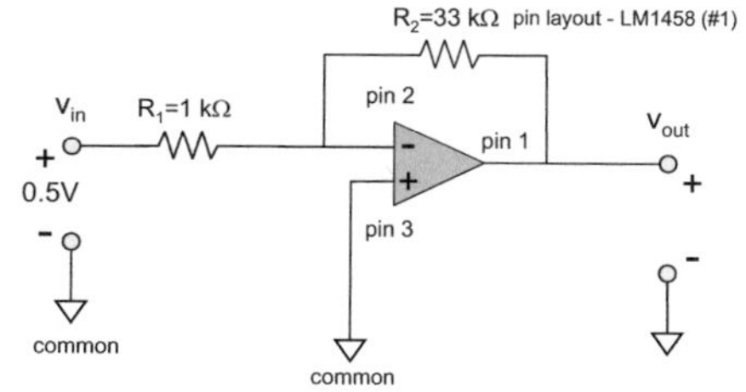

Voltage Follower

Problem 5.30

A. Using only the first summing-point constraint (SPC), solve the circuit shown in the figure, i.e., determine the output voltage of the amplifier versus the common port.
B. What function does this amplifier have? Why is it important?

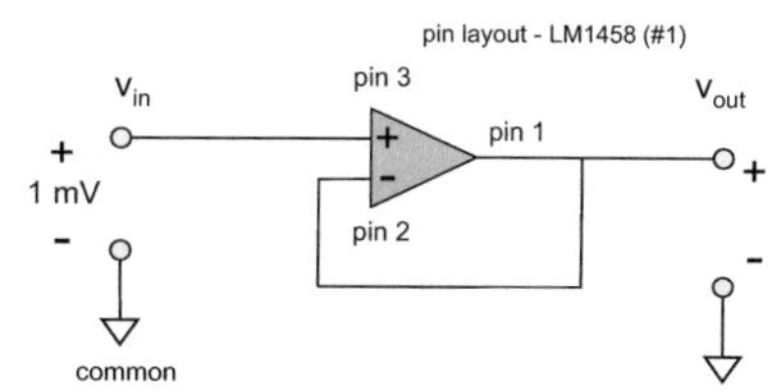

Exercises on the Use of the Negative Feedback

Problem 5.31. (A review problem) For three basic ideal-amplifier circuits:

Inverting amplifier
Non-inverting amplifier
Voltage follower

(each includes negative feedback) present

1. A circuit diagram
2. Expression for the amplifier gain

Problem 5.32. Determine the output voltage of amplifier configurations shown in the figure that follows. The amplifier is powered by a ±9-V dual-polarity voltage power supply. Assume an ideal operational amplifier.

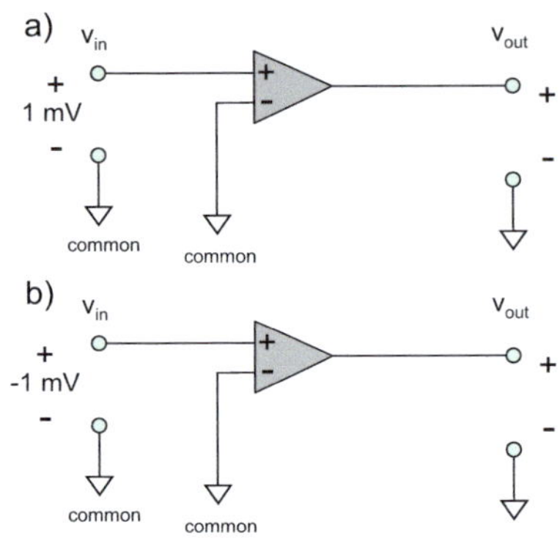

Problem 5.33. Each of the circuits shown in the figures below employs negative feedback. Find the output voltage v_{out} vs. ground (or common). *Hint:* The ground symbol in an amplifier circuit usually has the *same* meaning as the common port.

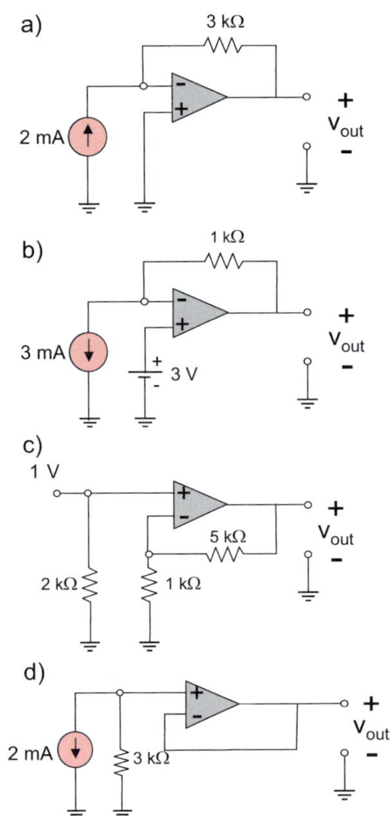

Problem 5.34. Each of the circuits shown in the figures below employs negative feedback. Find the output voltage V_{out} vs. ground (or common). *Hint:* The ground symbol in the amplifier circuit usually has the same meaning as the common port.

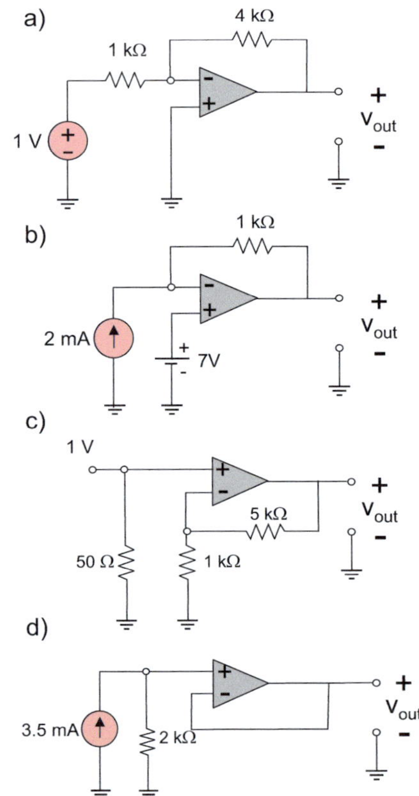

Problem 5.35. Each of the circuits shown in the figures below employs an inverting amplifier.

1. Solve each circuit (find v_{out}) with an input voltage of 1 mV.
2. Based on this solution, find the closed-loop voltage gain A_{CL} of the corresponding amplifier circuit.

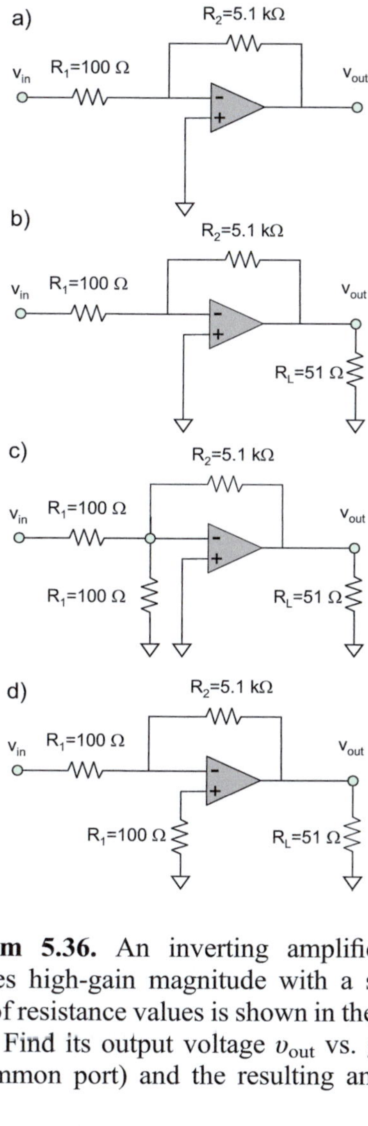

Problem 5.36. An inverting amplifier that achieves high-gain magnitude with a smaller range of resistance values is shown in the figure below. Find its output voltage υ_{out} vs. ground (or common port) and the resulting amplifier gain.

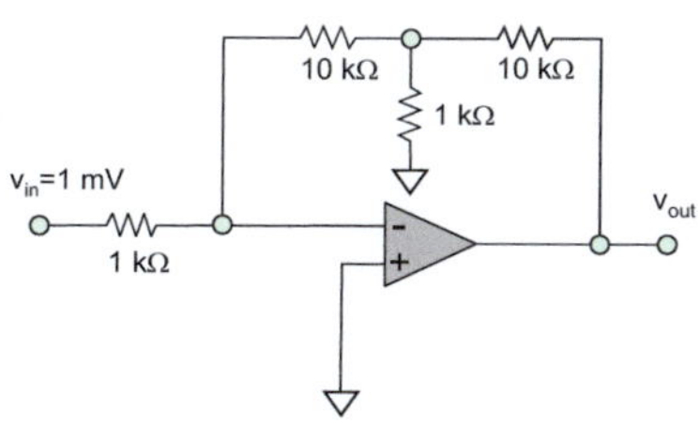

Problem 5.37. The amplifier circuit shown in the figure employs negative feedback.

A. Find the value of the output current i_{out} if the input current is 1 mA, $R_1 = 9\ \text{k}\Omega$, $R_2 = 1\ \text{k}\Omega$.
B. Why do you think this amplifier type is known as the *current amplifier*? To answer this question quantitatively, analytically express the output current i_{out} (current through the load) in terms of the unknown input current i_{out} and two arbitrary resistor values, $R_{1,2}$.

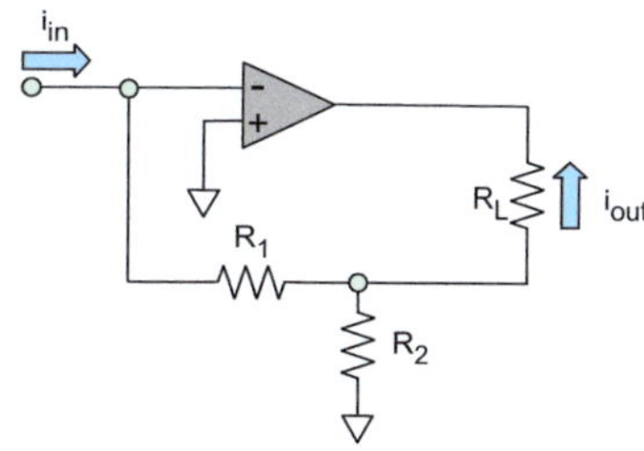

5.2.4 Mathematics Behind the Second Summing-Point Constraint

Problem 5.38

A. Derive an expression for the closed-loop gain of the non-inverting amplifier based only on the definition of the output voltage $\upsilon_{out} = A\left(\upsilon_{in}^{+} - \upsilon_{in}^{-}\right)$, without using the second summing-point constraint.
B. Determine the exact gain value when

$$A = 2 \times 10^5$$

$$R_1 = 1\ \text{k}\Omega, R_2 = 9\ \text{k}\Omega$$

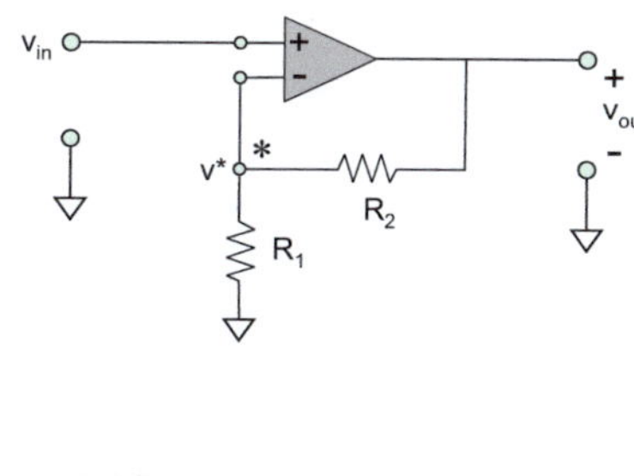

Problem 5.39

A. Derive an expression for the closed-loop gain of the inverting amplifier based only on the definition of the output voltage υ_{out}

$= A\left(v_{\text{in}}^{+} - v_{\text{in}}^{-}\right)$, without using the second summing-point constraint.

B. Determine the exact gain value when

$$A = 2 \times 10^5$$

$$R_1 = 1\ \text{k}\Omega, R_2 = 10\ \text{k}\Omega$$

5.2.5 Current Flow in the Amplifier Circuit

Problem 5.40. The amplifier circuit shown in the figure below is powered by a ±9-V dual-polarity voltage power supply.

A. Redraw the amplifier schematic in your notes.
B. Show the current direction in every wire of the circuit by an arrow and write the corresponding current value close to each arrow.

Hint: Change the polarity of the input voltage and the voltage sign if you have trouble operating with negative values.

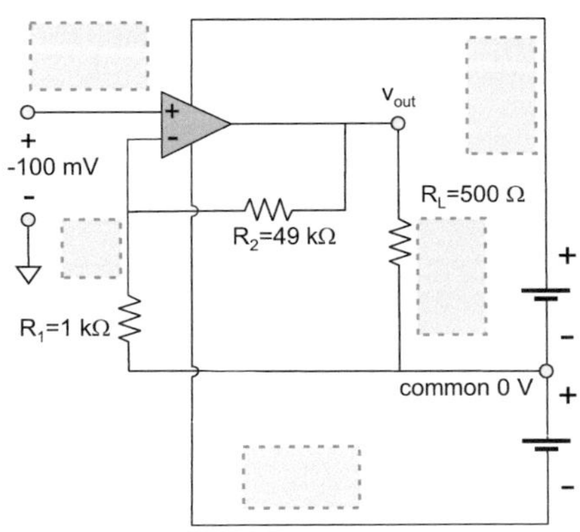

Problem 5.41. Repeat the previous problem for the circuit shown in the figure below.

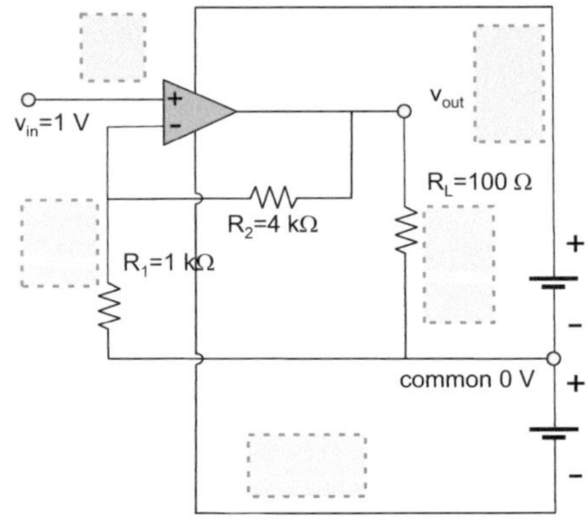

Problem 5.42. The amplifier shown in the figure below is powered by a ±9-V dual-polarity voltage power supply.

A. Redraw the amplifier schematic in your notes.
B. Show the current direction in every wire of the circuit by an arrow and write the corresponding current value close to each arrow.

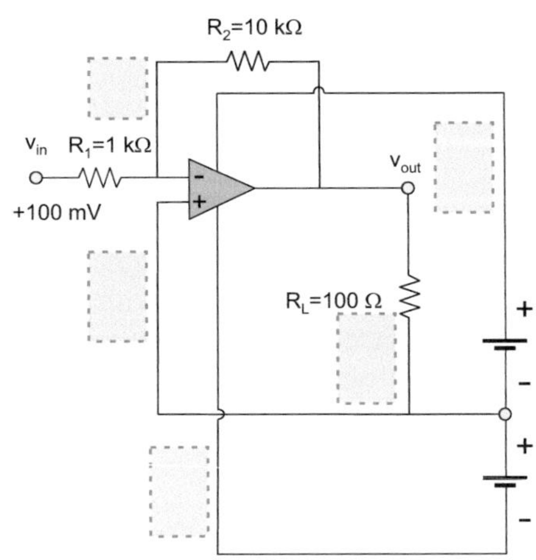

Problem 5.43. Repeat the previous problem when the input voltage to the amplifier is −100 mV.

5.2.6 Multiple-Input Amplifier Circuit: Summing Amplifier

Problem 5.44. By solving the amplifier circuit shown in the figure, fill out the table that follows. Assume that $R_0 = R_F/4$, $R_1 = R_F/2$, and $R_2 = R_F$.

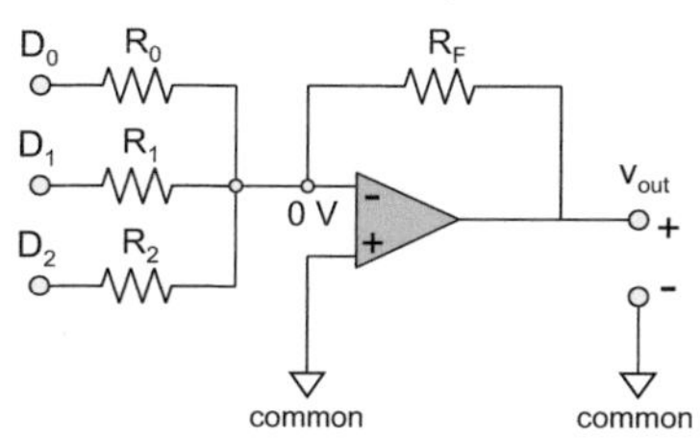

D_2, V	D_1, V	D_0, V	v_{out}, V
0	0	0	
0	0	5	
0	5	0	
0	5	5	
5	0	0	
5	0	5	
5	5	0	
5	5	5	

Problem 5.45. The amplifier circuit shown in the figure below employs negative feedback. This configuration is known as a three-bit *digital-to-analog converter* (DAC) on the base of an R/2R ladder. By solving the amplifier circuit, determine its output voltage in terms of resistances R, R_F, given the input voltages $D_0 = 0$ V, $D_1 = 0$ V, $D_2 = 5$ V.

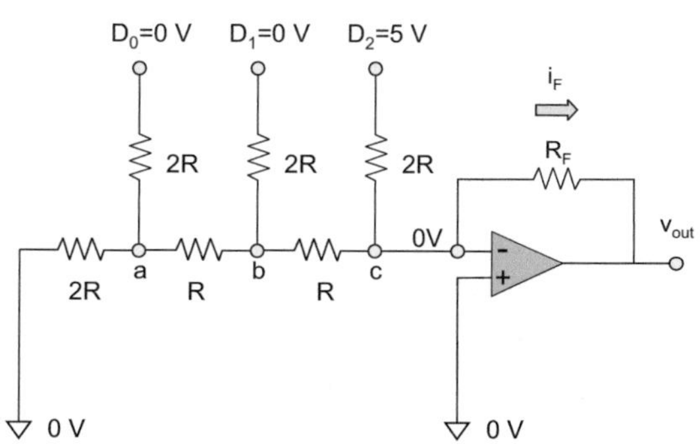

Problem 5.46. By solving the amplifier circuit shown in the following figure, determine its output voltage in terms of resistances R, R_F, given the input voltages $D_0 = 0$ V, $D_1 = 5$ V, $D_2 = 0$ V.

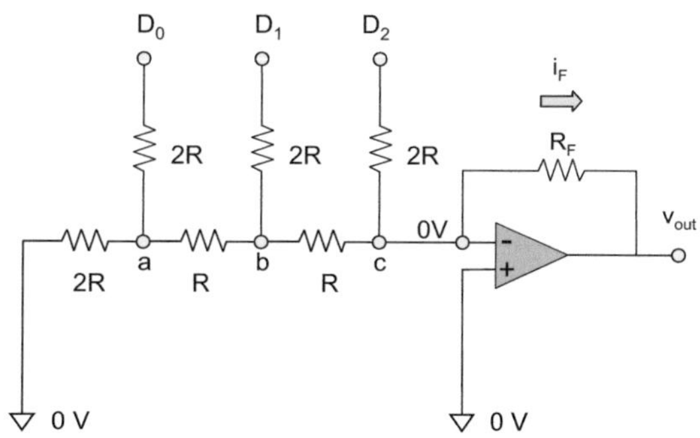

5.3 Amplifier Circuit Design

5.3.1 Choosing Proper Resistance Values

Problem 5.47. State the limitations on the feedback resistances and the output load resistance of an amplifier circuit.

Problem 5.48. The non-inverting amplifier shown in the figure below has been wired in laboratory.

A. Do you have any concerns with regard to this circuit?
B. If you do, draw the corrected circuit diagram.

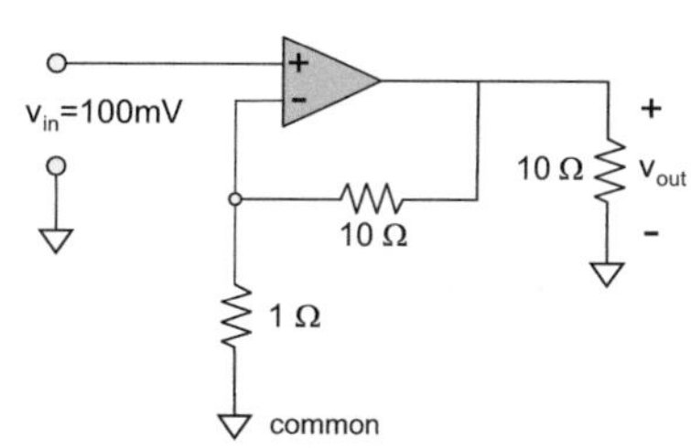

5.3.2 Model of a Whole Amplifier Circuit

5.3.3 Input Load Bridging or Matching

Problem 5.49. For three basic amplifier circuits

Inverting amplifier
Non-inverting amplifier
Voltage follower

(each includes negative feedback), present

1. A circuit diagram
2. An expression for the closed-loop amplifier circuit gain
3. An expression for the input resistance (impedance)
4. An expression for output resistance (impedance)

Problem 5.50

A. Explain in your own words the concept of load bridging (impedance bridging).

B. Which amplifier, the non-inverting or inverting, should be subject to load bridging?

Problem 5.51. An electromechanical sensor is given by its Thévenin equivalent wherein the sensor voltage υ_S is small. The sensor's equivalent resistance R_S may vary in time; but it is always less than 1 kΩ. An inverting amplifier is needed that generates an amplified version of the sensor's voltage. The output voltage should be $\approx -100\upsilon_S$. Draw the corresponding circuit diagram and specify one possible set of resistor values.

Problem 5.52. Construct an amplifier circuit matched to a 100-Ω load. The amplifier's gain is $|A_{CL}| = 100$. The sign of the gain (positive or negative) is not important and the input AC signal.
Hint: Multiple solutions many exist. Present at least two solutions.

Problem 5.53. Find the input resistance (impedance) to the amplifier circuit shown in the figure below.

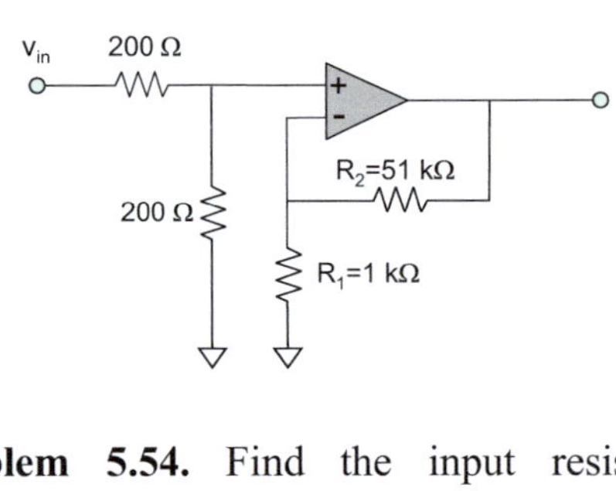

Problem 5.54. Find the input resistance (impedance) to the amplifier circuit shown in the figure below.

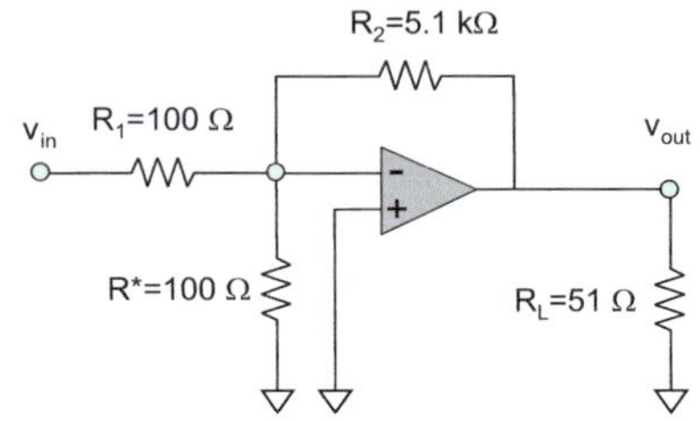

5.3.4 Cascading Amplifier Stages

Problem 5.55. For the amplifier circuit shown in the figure, find the output voltage and the input resistance (show units).

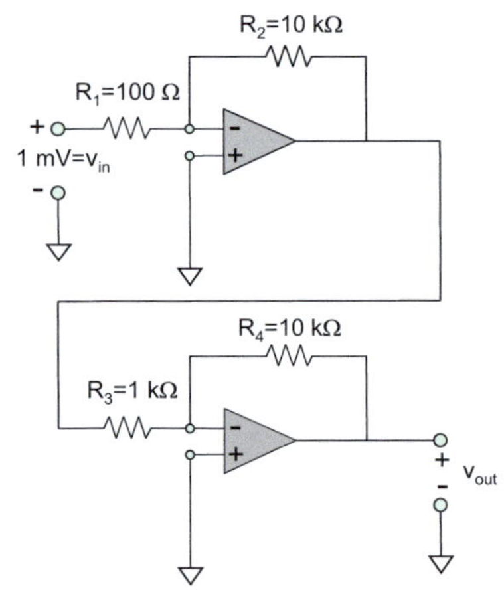

Problem 5.56. A sensor with Thévenin voltage (source voltage) $\upsilon_S = 2.5$ mV shown in the figure is to be connected to an amplifier circuit. An amplified replica of the sensor's voltage, $\upsilon_{out} \approx 1000\,\upsilon_S$, is needed at the circuit's output.

A. Present one possible circuit diagram and specify all necessary resistor values.

B. Present another (distinct) circuit diagram and specify all necessary resistor values.

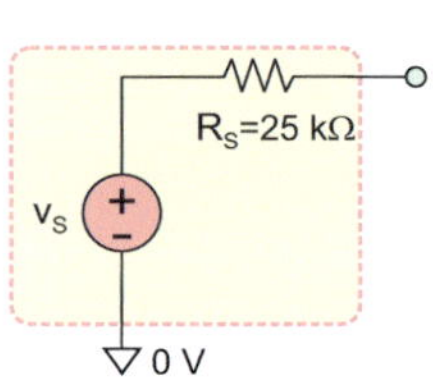

Problem 5.57. An amplifier in the configuration shown in the figure below is connected to a

sensor with Thévenin (source) voltage $\upsilon_S = 25$ mV. An amplified replica of the sensor's voltage, $\upsilon_{out} \approx -100\,\upsilon_S$, is needed at the output.

A. Do you have any concerns with regard to this circuit?
B. If you do, draw an appropriate circuit diagram.

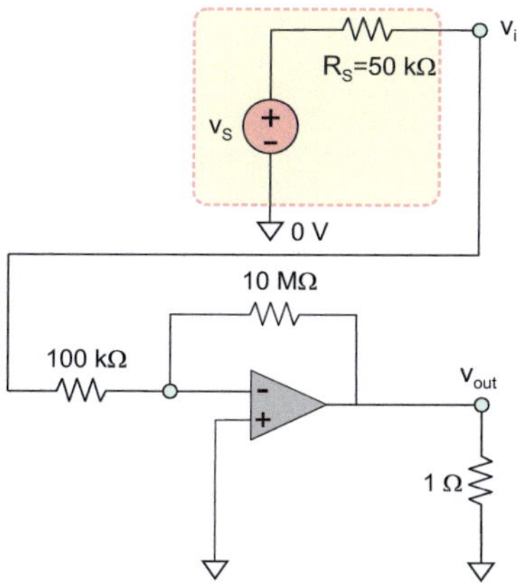

Problem 5.58. An amplifier circuit is needed with the closed-loop gain $A_{CL} = +1000$. The input resistance (impedance) to the circuit should be 5 kΩ. Present two alternative circuit diagrams and specify the necessary resistor values. The first circuit must use inverting amplifiers and the second circuit-non-inverting amplifiers.

Problem 5.59. An amplifier circuit is needed with the closed-loop gain $A_{CL} = +10,000$. The input resistance (impedance) to the circuit should be as high as possible. Present the corresponding circuit diagram and specify the necessary resistor values.

Problem 5.60. An amplifier circuit is needed with a positive gain of 1000 ± 20 %. The input resistance (impedance) to the circuit should be 1 kΩ. Present the circuit diagram and specify the necessary resistor values including tolerance.

Problem 5.61. An amplifier circuit is needed with a positive gain of 5000 ± 5 %. The input resistance (impedance) should be as high as possible. Present one possible circuit diagram and specify the necessary resistor values including tolerance.

5.3.5 Amplifier DC Imperfections and Their Cancellation

Problem 5.62. Determine the output voltage to nonideal operational amplifier circuits (with the nonzero input offset voltage) shown in the figures below.

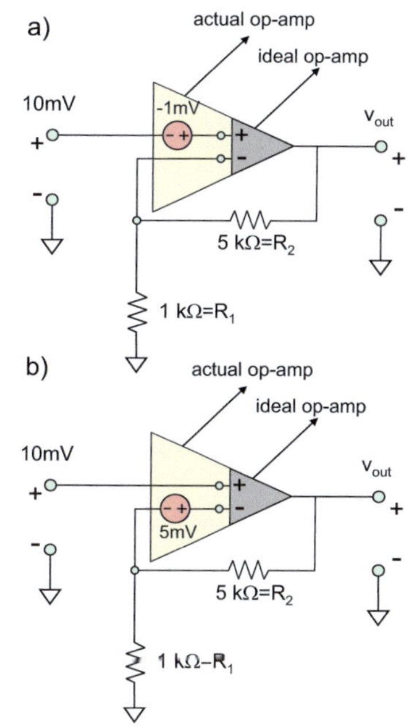

Problem 5.63. Determine the output voltage to nonideal operational amplifier circuits (with nonzero input currents) shown in the figures below. The input terminal is connected directly to the common terminal (grounded). The strength of every bias current source is 100 nA. *Hint:* The upper bias current source does not contribute to the solution.

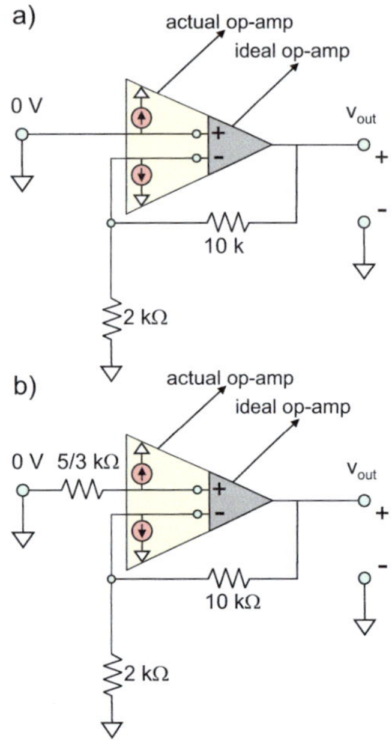

Problem 5.64. For the amplifier circuit shown in the figure below, determine the output voltage. Use $R_1 = R_2 = 1\ \text{k}\Omega$.

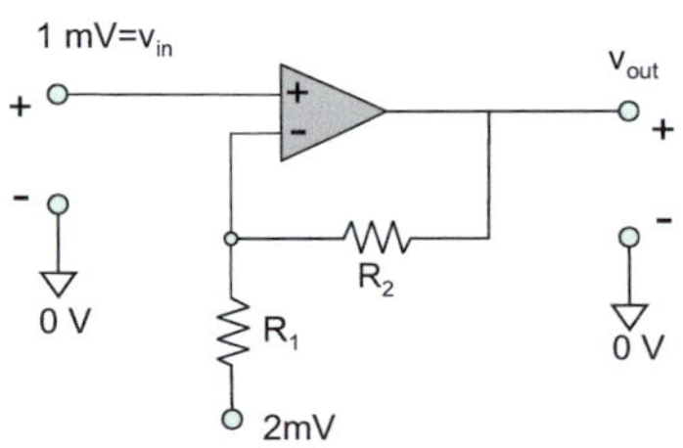

Problem 5.65. In the previous problem, denote the terminal voltage of 2 mV by V_{off}, the input voltage of 1 mV by v_{in}, the output voltage by v_{out}, and the amplifier gain by A_{CL}. Derive an analytical formula that determines V_{off} in terms of V_{in} given that the output voltage v_{out} to the amplifier is exactly zero.

Problem 5.66. For the amplifier circuit shown in the figure below, determine the output voltage. Use $R_1 = 1\ \text{k}\Omega$ and $R_2 = 4\ \text{k}\Omega$.

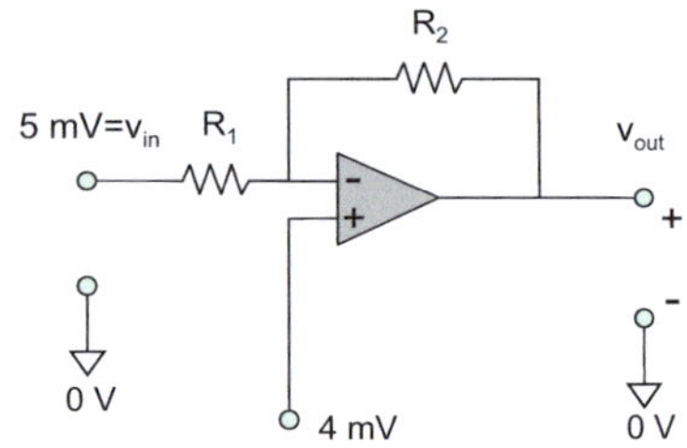

Problem 5.67. In the previous problem, denote the terminal voltage of 4 mV by V_{off}, the input voltage of 5 mV by v_{in}, the output voltage by v_{out}, and the amplifier gain by A_{CL}. Derive an analytical formula that determines V_{off} in terms of v_{in} given that the output voltage to the amplifier is exactly zero.

Problem 5.68. For the circuit shown in the figure below:

A. Determine the output voltage of the nonideal operational amplifier circuit (with the nonzero input offset voltage).
B. Does the amplifier circuit really follow the ideal-amplifier circuit law: $v_{\text{out}} = A_{\text{CL}} v_{\text{in}}$?
C. What happens if the input voltage changes from 10 mV to 20 mV?

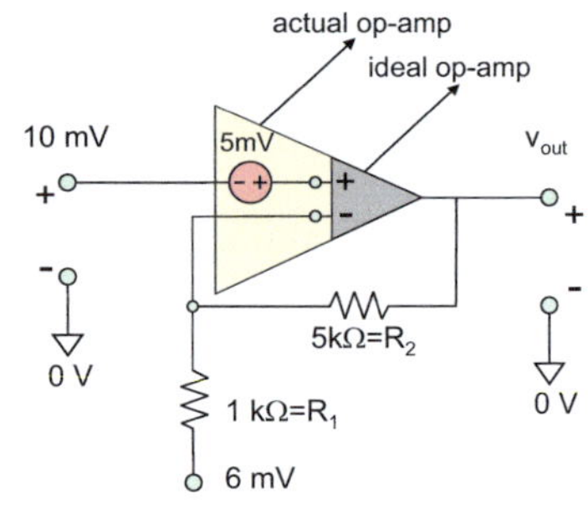

Problem 5.69. Solve the previous problem with the offset voltage in the feedback loop changed from 6 mV to 5 mV.

Problem 5.70. In problem 5.68, denote the terminal voltage of 6 mV by V_{off}, the input voltage of 10 mV by v_{in}, the input offset voltage by V_{OS}, the output voltage by v_{out}, and the amplifier gain by A_{CL}. Derive an analytical formula that determines V_{off} in terms of V_{OS} given that the output voltage v_{out} to the amplifier must exactly follow the ideal-amplifier gain law: $v_{out} = A_{CL} v_{in}$.

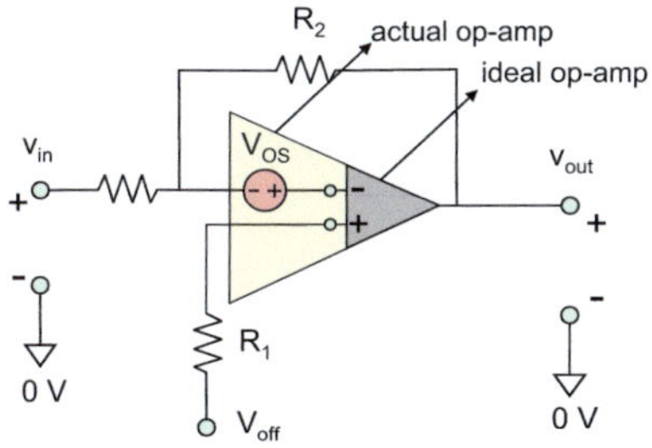

Problem 5.71. For two nonideal operational amplifier circuits (with the nonzero input offset voltage) shown in the figures below, determine the necessary offset voltage, V_{off}, which ensures that the output voltage, v_{out}, to the amplifier exactly follows the ideal-amplifier gain law: $v_{out} = A_{CL} v_{in}$. You need to express this voltage in terms of other circuit parameters that are given in figures a) and b).

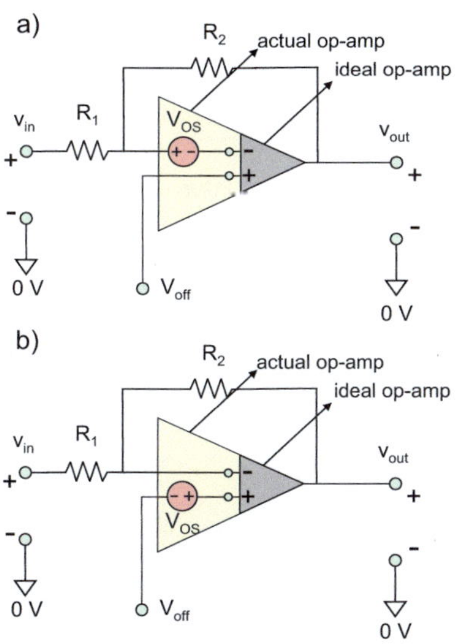

5.3.6 DC-Coupled Single-Supply Amplifier

Problem 5.72. For the single-supply amplifier circuit shown in the figure:

A. Determine the output voltage versus circuit ground (the negative terminal of the voltage power supply).

B. What potential problem do you see with this circuit? How could you fix it?

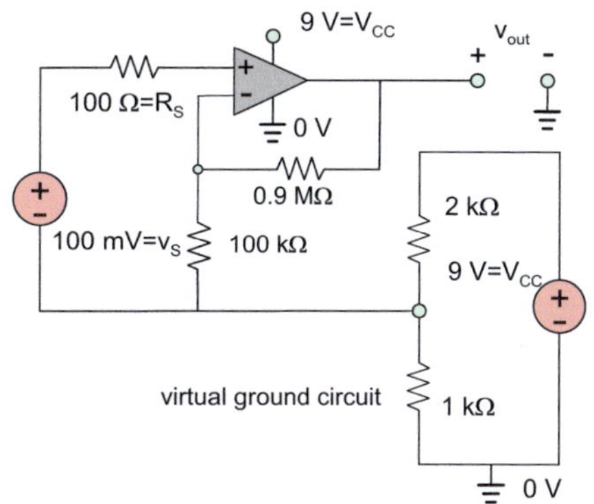

Problem 5.73. For the single-supply amplifier circuit shown in the figure:

A. Determine the output voltage versus circuit ground (the negative terminal of the voltage power supply).

B. Do you see any problem with this circuit?

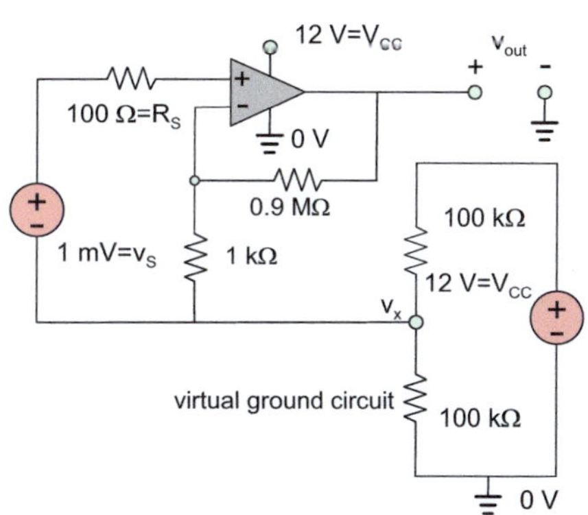

5.4 Difference and Instrumentation Amplifiers

5.4.1 Differential Input Signal to an Amplifier

Problem 5.74. The model of an input signal from a three-terminal sensing device is shown in the figure below. What are the differential and common-mode voltages at terminals a and b?

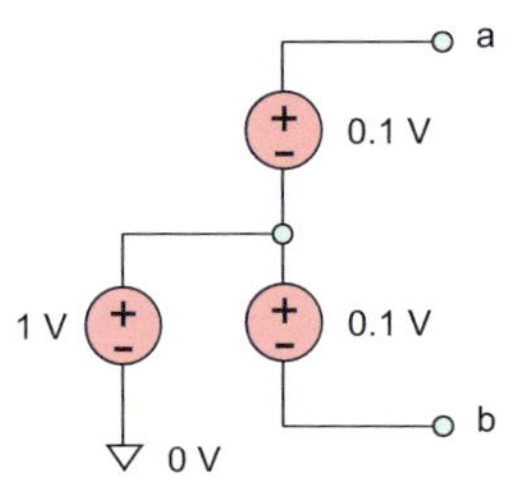

Problem 5.75. The Wheatstone bridge in Fig. 5.27 is connected to $\pm V_{CC}$ rails instead of $+V_{CC}$ and ground. Furthermore, $R_2 = 1.1R_1$, $R(x) = 1.1R_3$, and $V_{CC} = 6\,\text{V}$. What are the differential and common-mode voltages?

Problem 5.76. The Wheatstone bridge in Fig. 5.27 is connected to ground and $-V_{CC}$ rails instead of $+V_{CC}$ and ground. Given that $R_2 = 1.05R_1$, $R(x) = 1.05R_3$, and $V_{CC} = 6$ V, determine the differential and common-mode voltages.

5.4.2 Difference Amplifier

Problem 5.77. Design a difference amplifier with a differential gain of 20. Present the circuit diagram and specify one possible set of resistor values. In the circuit diagram, label the input voltages as v_a, v_b and express the output voltage in terms of v_a, v_b.

Problem 5.78. Repeat the previous problem for a differential gain of 100.

Problem 5.79. Find the output voltage to the difference-amplifier circuits shown in the figures below. Assume the ideal-amplifier model and exact resistance values.

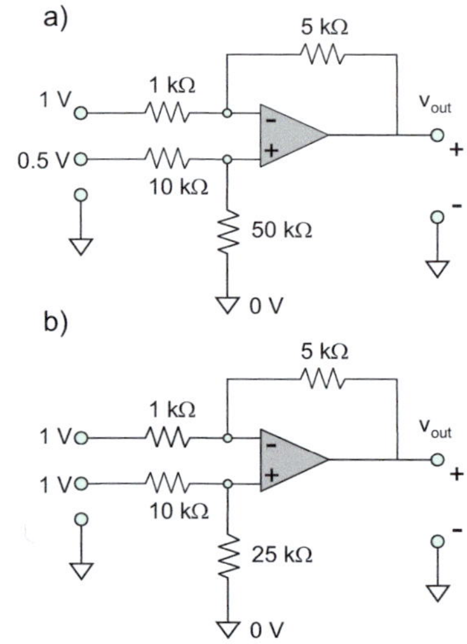

Problem 5.80. Your technician needs to control a process using two sensors with output voltages v_1 and v_2, respectively. The weighted difference in sensor reading, $v = 1v_1 - 0.5v_2$, is critical for the product quality. The technician reads voltage v_1 and then voltage v_2 and then uses a calculator to find v. Help the technician, i.e., sketch for him a difference-amplifier circuit that will directly output v to the DMM. The negative terminal of the DMM is always grounded. Specify one possible set of resistor values.

Problem 5.81. Repeat the previous problem when the weighted difference in sensor reading, $v = 10v_1 - 5v_2$, needs to be processed.

Problem 5.82. For the circuit shown in the figure, find the output voltage if the input voltages are $v_b = 1\,\text{V}$ and $v_a = 1.01\,\text{V}$, respectively. Assume the ideal-amplifier model and exact resistance values.

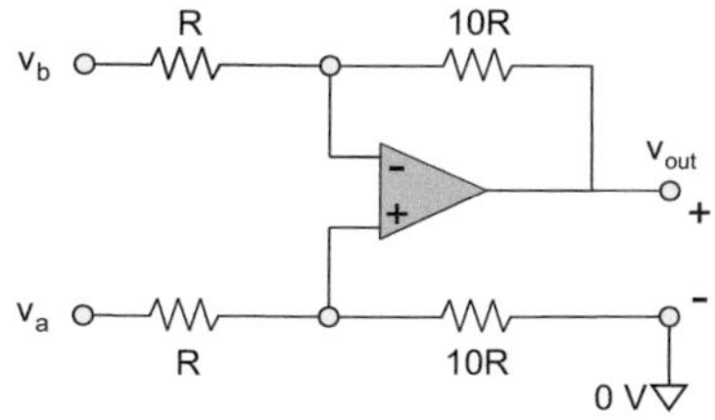

Problem 5.83

A. For the circuits shown in the figures below, find the output voltage if the input voltages are $\upsilon_a = 1\,\text{V}$ and $\upsilon_b = 1\,\text{V}$, respectively Assume the ideal amplifier and exact resistance values.

B. What is the value of the common-mode gain in every case?

a)

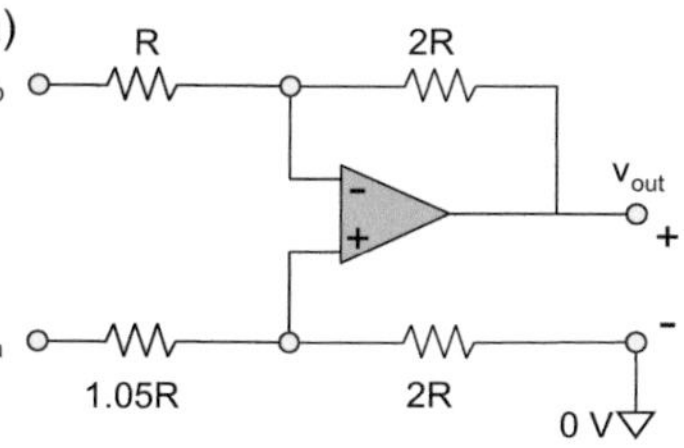

b)

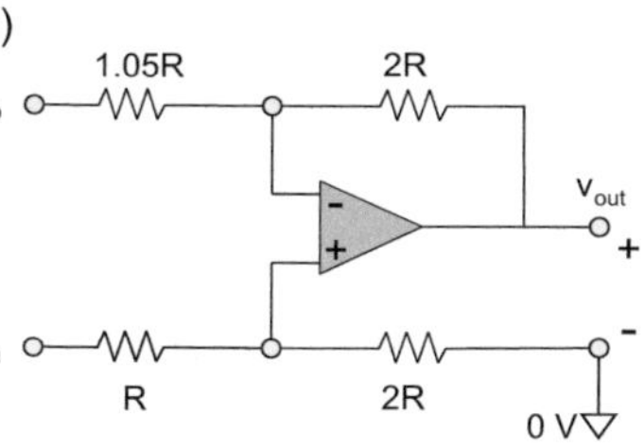

Problem 5.84. For the difference-amplifier circuit shown in the figure below, find the *differential-mode resistance (impedance)* to the amplifier. The differential-mode resistance is defined as the ratio of a voltage of a power supply placed between terminals *a* and *b* to the current that flows through this power supply.

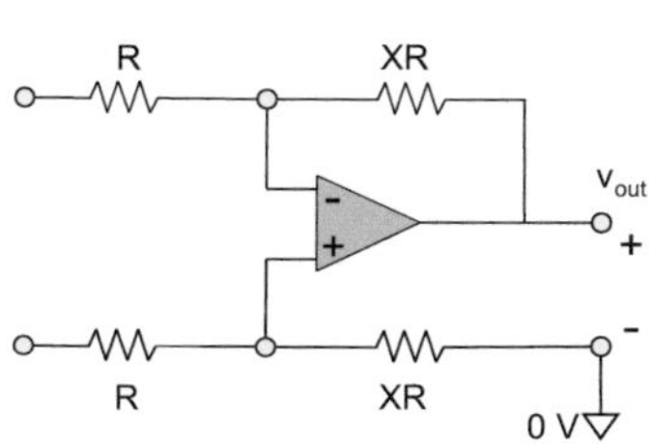

5.4.3 Instrumentation Amplifier

Problem 5.85

A. Why is the original difference amplifier not used as an instrumentation amplifier?

B. Why is the circuit in Fig. 5.31 not used as the instrumentation amplifier?

Problem 5.86

A. Find the differential gain and the common-mode gain for the amplifier circuit shown in Fig. 5.32. The differential output voltage is $\upsilon_a{}^* - \upsilon_b{}^*$, and the common-mode output voltage is $0.5(\upsilon_a{}^* + \upsilon_b{}^*)$.

B. Find the differential gain and the common-mode gain for the amplifier circuit shown in Fig. 5.31.

Problem 5.87. Design an instrumentation amplifier with a differential gain of 210. Present the corresponding circuit diagram and specify one possible set of resistance values. In the circuit diagram, label the input voltages as υ_a, υ_b and express the output voltage in terms of υ_a, υ_b.

Problem 5.88. Design an instrumentation amplifier with a differential gain of 1010. Present the corresponding circuit diagram and specify one possible set of resistor values. In the circuit diagram, label the input voltages as υ_a, υ_b and express the output voltage in terms of υ_a, υ_b.

Problem 5.89. The following voltages are measured: $\upsilon_a = 3.750$ V and $\upsilon_b = 3.748$ V. Find voltages versus circuit ground (common port of the dual supply) for every labeled node in the circuit shown in the figure below. The amplifier circuit is powered by a $\pm 10 - \text{V}$ dual supply. Assume exact resistance values and the ideal-amplifier model.

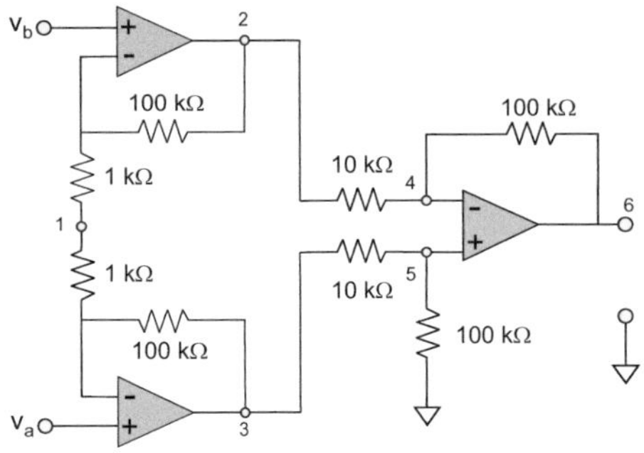

Problem 5.90. Repeat the previous problem when node 1 is grounded.

Problem 5.91. The following voltages are measured: $v_a = 5.000\,\text{V}$ and $v_b = 5.001$ V. Find voltages versus circuit ground (common port of the dual supply) for every labeled node in the circuit shown in the figure below. The amplifier circuit is powered by a $\pm 10 - \text{V}$ dual supply. Assume exact resistance values and the ideal-amplifier model.

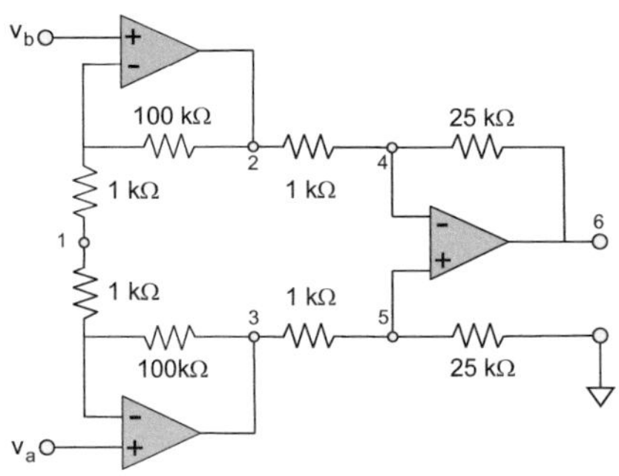

Problem 5.92. Repeat the previous problem when node 1 is grounded.

Problem 5.93. The following voltages are measured: $v_a = 5.000$ V and $v_b = 5.001$ V. Find voltages versus circuit ground (common port of the dual supply) for every labeled node in the circuit shown in the figure below. The amplifier circuit is powered by a $\pm 10 - \text{V}$ dual supply. Assume exact resistance values and the ideal-amplifier model.

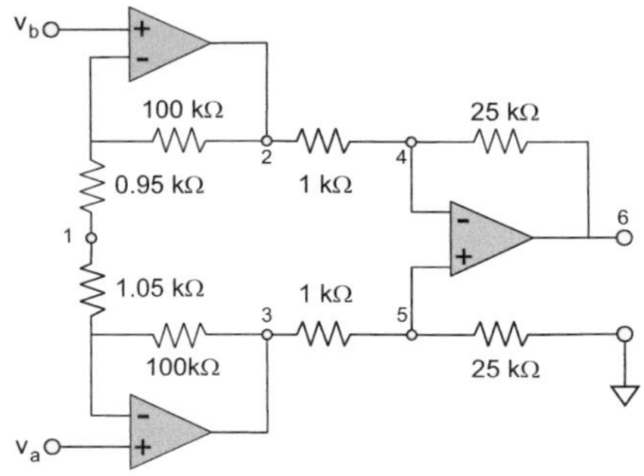

Problem 5.94

A. Find the output voltage for the amplifier circuit shown in the figure below.
B. Denote the input voltage of 0.1 V by v_a, the input voltage of 0.08 V by v_b, the 10-kΩ resistor by R_1, the 40-kΩ resistor by R_2, and the 10-kΩ resistor by R_G. Express the output voltage in the general form, in terms of two input voltages and the resistances.

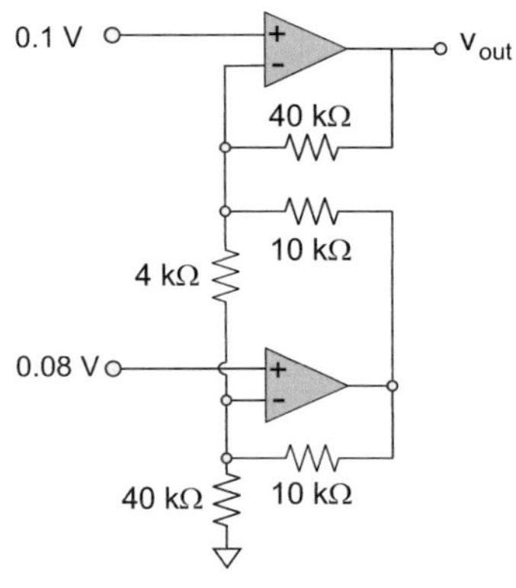

5.5 General Feedback Systems

5.5.1 Signal-flow Diagram of a Feedback System

5.5.2 Closed-Loop Gain and Error Signal

Problem 5.95. The block diagram of Fig. 5.36 is applied to a voltage amplifier.

A. Given the input signal $x_{in} = 10$ mV, the error signal $x_e = 1$ μV, and the output signal $x_{out} = 1$ V, determine the open-loop gain and the feedback factor.
B. Given the ratio of input to error signal $x_{in}/x_e = 100$ and the feedback factor of 0.1, determine the open-loop gain.

Problem 5.96. The open-loop gain A in Fig. 5.36 varies between two extreme values of $A = 10,000 \pm 2,000$ ($\pm 20\,\%$ gain variation) depending on the system parameters. The forward gain block is used in the closed-loop configuration with the feedback factor β of 0.1. Determine the two extreme values of the closed-loop gain, A_{CL}.

Problem 5.97. The open-loop gain A in Fig. 5.36 is 100,000. The forward gain block is used in the closed-loop configuration with the feedback factor β of 1. Determine the error signal, x_e, if the input voltage signal is 1 mV.

Problem 5.98. The closed-loop gain of a non-inverting amplifier circuit with $R_1 = 1$ kΩ, $R_2 = 100$ kΩ is 99. Determine the open-circuit gain A of the amplifier chip.

5.5.3 Application of General Theory to Voltage Amplifiers with Negative Feedback

5.5.4 Voltage, Current, Transresistance, and Transconductance Amplifiers with the Negative Feedback

Problem 5.99. Derive the gain Eqs. (5.69) for the amplifier circuits shown in Fig. 5.38.

Problem 5.100. The circuit shown in the figure that follows is a feedback transconductance amplifier. Express i_{out} in terms of v_{in}.

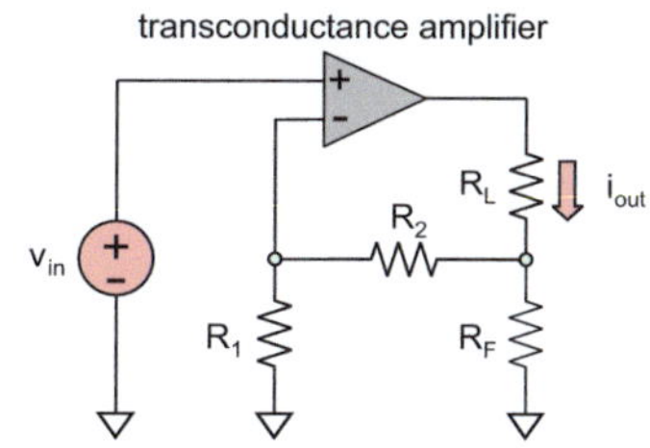

Problem 5.101. The amplifier circuit shown in the figure that follows is the *Howland current source* widely used in biomedical instrumentation; its output is the current through the load resistance.

A. Classify the amplifier circuit in terms of four basic amplifier topologies and mention the most important circuit features.
B. Derive its gain equation $i_{out} = (v_a - v_b)/R_2$ given that $R_1/R_3 = R_2/R_4$.

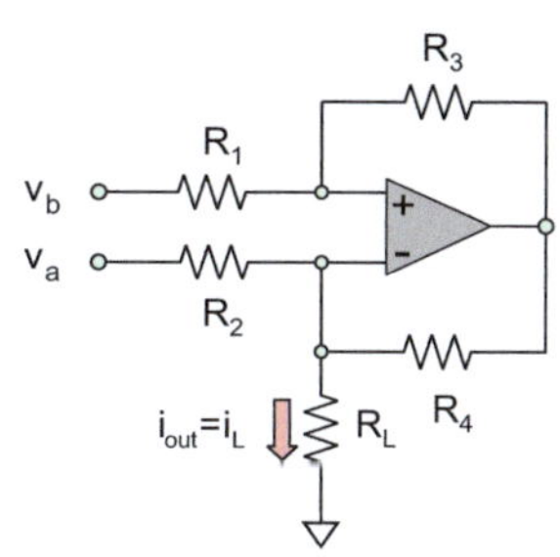

Part II

Transient Circuits

Chapter 6: Dynamic Circuit Elements

Overview

Prerequisites:
- Knowledge of basic circuit theory (Chapters 2 and 3)
- Knowledge of operational amplifiers with negative feedback (Chapter 5)

Objectives of Section 6.1:
- Define types of capacitance encountered in electric circuits
- Define self-inductance and mutual inductance from the first principles
- Define field energy stored in a capacitor/inductor
- Be able to combine capacitances/inductances in series and in parallel
- Understand construction of practical capacitors/inductors
- Understand fringing effect and its use in sensor circuits

Objectives of Section 6.2:
- Derive dynamic equations for capacitance/inductance from the first principles
- Establish how the capacitance may create large transient currents
- Establish how the inductance may create large transient voltages
- Define instantaneous energy and power of dynamic circuit elements
- Establish the behavior of dynamic circuit elements in the DC steady state and at a very high frequency

Objectives of Section 6.3:
- Obtain initial exposure to bypass/blocking capacitor and decoupling inductor
- Obtain initial exposure to amplifier circuits with dynamic circuit elements

Application Examples:
Electrostatic discharge and its effect on integrated circuits
How to design a 1-F capacitor? How to design a 1-mH inductor?
Capacitive touchscreens
Bypassing a DC motor

S. N. Makarov et al., *Practical Electrical Engineering*, https://doi.org/10.1007/978-3-319-96692-2_6

Keywords:
Capacitance, Capacitance of two conductors, Self-capacitance, Capacitance to ground, Capacitance of two equal conductors separated by large distances, Energy stored in a capacitance, Electrostatic discharge (ESD), ESD effect on integrated circuits, Device under test (DUT), Parallel-plate capacitor (base formulas, fringing effect, fringing fields), Capacitor (absolute dielectric permittivity, relative dielectric permittivity, dielectric strength, normalized breakdown voltage, electrolytic, tantalum, ceramic, marking, set of base values), Capacitive touch screens (self-capacitance method, mutual-capacitance method), Magnetic flux density, Magnetic field, Absolute magnetic permeability, Relative magnetic permeability, Magnetic induction, Magnetic flux, Self-inductance, Inductance, Mutual inductance, Energy stored in an inductance, Solenoid (air core, toroidal magnetic core, straight magnetic core, short, fringing fields), Inductor (marking, set of base values, also see *solenoid*), Dynamic equation for capacitance (definition, derivation, fluid mechanics analogy), Capacitance (instantaneous energy, instantaneous power, behavior in the DC steady state, behavior at very high frequencies), Dynamic equation for inductance (definition, derivation, fluid mechanics analogy), Inductance (instantaneous energy, instantaneous power, behavior in the DC steady state, behavior at very high frequencies), Bypass capacitor, Decoupling capacitor, Shunt capacitor, Snubber RC circuit, Decoupling inductor, Inductor choke, Transient circuit, Amplifier circuits with dynamic circuit elements, Active filters, Miller integrator (circuit, DC gain, compensation, time constant), Analog pulse counter, Analog computer, Differentiator amplifier (circuit, gain at very high frequencies), Active differentiator

Section 6.1 Static Capacitance and Inductance

6.1.1 Capacitance, Self-Capacitance, and Capacitance to Ground

Capacitance reflects the ability of arbitrary conductors to store electric charge and, simultaneously, the store energy of the electric field in the surrounding space. When no dielectric is present, capacitance is determined *entirely* by the geometry of conductors. When a dielectric material is present, its permittivity becomes important. Capacitance definitions will be given with reference to Fig. 6.1.

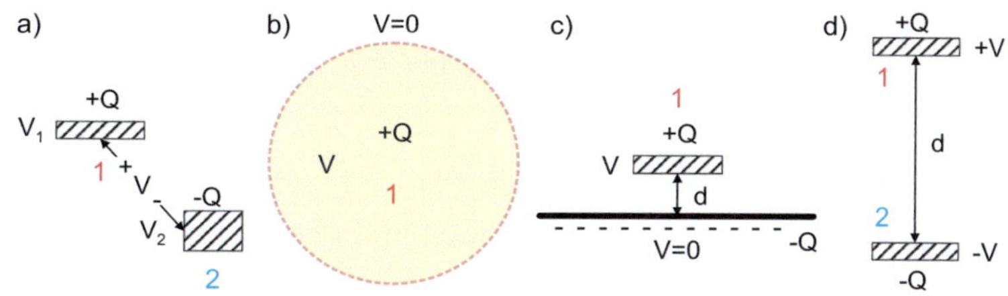

Fig. 6.1. Conductor geometry for capacitance definitions.

1. *Capacitance, C, of Two Conductors*. Two arbitrary insulated conductors near together in Fig. 6.1a constitute a simple *capacitor*. Its capacitance, *C*, is found with the help of electrostatic theory. Further, it is used in various dynamic models. Capacitance *C* of two insulated conductors 1 and 2 is defined by the ratio

$$C \equiv \frac{Q}{V} > 0 \qquad (6.1a)$$

where $Q > 0$ is the (absolute) net charge of *either* conductor given that the net charge of the system with both conductors is *zero* and *V* is the potential difference or voltage *between* two conductors 1 and 2, i.e., $V = V_1 - V_2$. This ratio does not depend on *V*; it is always taken so as to make the capacitance *positive*.

2. *Self-Capacitance, C_{self}, of a Conductor*. When an electric charge *Q* is added to a single isolated conductor in Fig. 6.1b, its surface will possess a certain absolute voltage *V* versus 0 V at infinity. The ratio

$$C_{\text{self}} \equiv \frac{Q}{V} > 0 \qquad (6.1b)$$

is the *self-capacitance* of the conductor. The self-capacitance is the capacitance when the second conductor is a hollow conducting sphere of infinite radius subject to 0 V.

3. *Capacitance to Ground, C, of a Conductor*. For conductor 1 in Fig. 6.1c with charge $+Q$, its capacitance to ground is the capacitance when the second conductor is an infinite conducting ground plane in Fig. 6.1c subject to 0 V (and charged to $-Q$).

Capacitance to ground, C, is always *greater* than the self-capacitance, C_{self}; their ratio becomes quite large when the separation distance d from the plane is small. On the other hand,

$$C \to C_{self} \text{ when } d \to \infty \text{ in Fig. 6.1c.} \tag{6.1c}$$

4. *Capacitance, C, of Two Equal Conductors Separated by Large Distances*. For conductors 1 and 2 in Fig. 6.1d, the capacitance approaches

$$C \to \frac{1}{2}C_{self} \text{ when } d \to \infty \text{ in Fig. 6.1d.} \tag{6.1d}$$

The separation distance d must be large compared to the conductor's size. Equation (6.1d) will be proved shortly.

The capacitance is recorded in units of farads or F. This unit is named in honor of Michael Faraday (1791–1867), a British physicist and chemist, who was known to many as "the best experimentalist in the history of science." Typical capacitance values in electronics are pF (picofarad or simply *paf*), nF (nanofarad), and μF (microfarad). In power electronic circuits, larger capacitances might be used. The capacitance unit is linked to other MKS units as follows:

$$1\ \mathrm{F} = 1\frac{\mathrm{A \cdot s}}{\mathrm{V}} = 1\frac{\mathrm{J}}{\mathrm{V}^2} = 1\frac{\mathrm{C}}{\mathrm{V}} \tag{6.2}$$

where C is the unit of coulomb. The total *electric field energy* stored between two conductors and in the surrounding space is given by

$$E = \frac{1}{2}CV^2 \tag{6.3}$$

This result can be derived from the definition of the electric potential (voltage). The energy is equal to work, which is necessary to put all charges of the capacitor in place. Equation (6.3) is valid for any configuration shown in Fig. 6.1.

Example 6.1: Prove Eq. (6.1d).

Solution: For a conductor in Fig. 6.1b, the stored electric energy is given by

$$E_{self} = \frac{1}{2}C_{self}V^2 \tag{6.4a}$$

For two conductors in Fig. 6.1d separated by a very large distance d, the stored electric energy is approximately given by

Example 6.1 (cont.):

$$E = 2E_{\text{self}} = \frac{1}{2}C(2V)^2 \tag{6.4b}$$

Comparing Eqs. (6.4a) and (6.4b) we obtain the necessary result.

Exercise 6.1: A metal circle or radius $r = 0.1\,\text{m}$ has the self-capacitance $C_{\text{self}} = 8\varepsilon_0 r$ where $\varepsilon_0 = 8.85419 \times 10^{-12}\ \text{F/m}$ is the permittivity of vacuum. Estimate capacitance of a capacitor formed by two coaxial circles separated by 1 m.

Answer: ~3.54 pF from Eq. (6.1d). A precise numerical solution predicts 3.77 pF.

Exercise 6.2: How large is the stored energy in a 100-μF laboratory capacitor at 10 V?

Answer: 0.005 J or, which is the same on the power basis, 5 mW of power delivered during one second. However, this power will not be delivered uniformly.

Exercise 6.3: How large is the stored energy in a 20-F *ultracapacitor* charged to 25 V?

Answer: 6250 J or 6.25 kJ. This is certainly a significant value. At the same time, the discharge rate (available current or power) is much less in this case than the current or power delivered by laboratory electrolytic capacitors.

6.1.2 Application Example: ESD

Self-capacitance results may be applied for the prediction of ESD (*electrostatic discharge*) effects on *integrated circuits* (ICs). One of the most common causes of electrostatic damage is the direct transfer of electrostatic charge through a significant series resistor from the human body or from a charged material to the electrostatic discharge-sensitive (ESDS) device. The concept is shown in Fig. 6.2a. A metal ground plane and the highly conducting human body naturally form a capacitor. The *body capacitance C* in Fig. 6.2b is defined as the capacitance between the body, assumed to be a conductor, and the large (ideally infinite) ground plane. Its value depends significantly on the posture of the body with respect to the ground surface. The typical separation distance is 2 cm. It may be shown that, at such distances, $C \approx 2C_{\text{self}}$. Therefore, instead of calculating C directly, we can find the self-capacitance of the human body, C_{self}, and then multiply it by 2.

Typical self-capacitances of a 177-cm-tall male student are indicated in Fig. 6.3. These values were obtained by the numerical method described in Chapter 1.

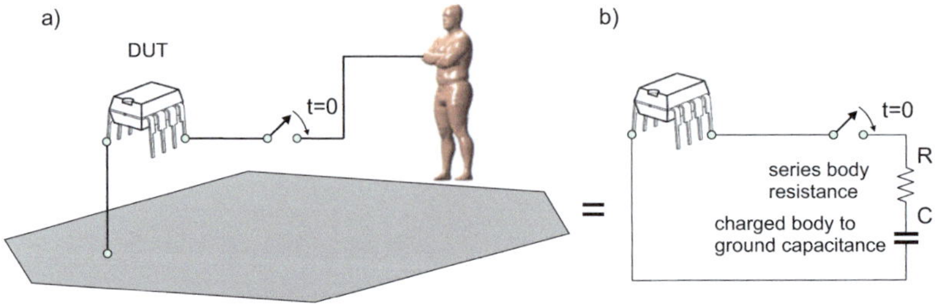

Fig 6.2. Equivalent circuit for understanding ESD and its effect on a *device under test* (DUT).

The simplifying assumption $C \approx 2C_{\text{self}}$ and Fig. 6.3 predict body capacitances in the range 86–95 pF. These values are in a good agreement with the generally accepted *human body model* (HBM), which, with reference to Fig. 6.2b, uses

$$R = 1.5\ \text{k}\Omega,\quad C = 100\ \text{pF} \tag{6.5}$$

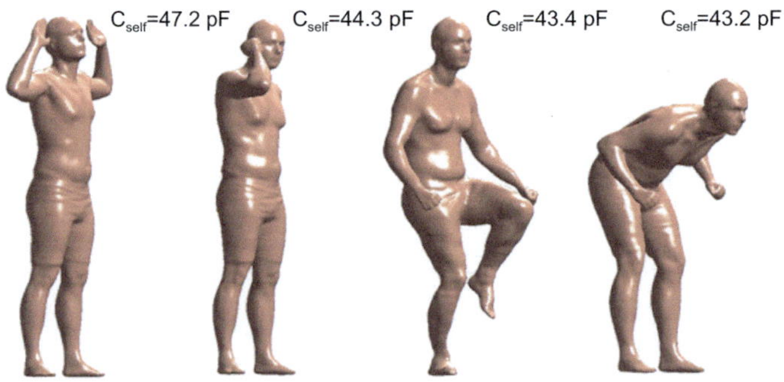

Fig 6.3. Typical self-capacitance values for a 177-cm-tall male person. Note how the self-capacitance changes when the body poses change.

6.1.3 Parallel-Plate Capacitor

Consider a parallel-plate capacitor shown in Fig. 6.4a, b. Both infinitely thin conducting square plates with the side a and area $A = a^2$ are separated by distance d. The upper plate has a total charge $+Q$; the lower plate has the opposite charge $-Q$; the net charge of the capacitor is zero. Feeding conductors are implied to be disconnected; they are excluded from consideration. Assuming that the entire electric field is concentrated *only* within the capacitor and that it is uniform in space (equal to V/d), the approximate capacitance is established as

$$C = \frac{\varepsilon_0 A}{d} \tag{6.6}$$

where ε_0 is the *dielectric permittivity of vacuum* if the capacitor is situated in vacuum. For Eq. (6.6) to hold, the plates do not have to be square. If the capacitor does not have a high-ε dielectric inside, Eq. (6.6) is a good approximation only if d is *very small* compared to the dimensions of the plates. Otherwise, the *fringing effect* must be taken into account. The fringing effect is illustrated in Fig. 6.4a, b. Fringing means that the electric field extends outside the physical capacitor. The electric field outside the capacitor possesses certain extra energy. Therefore, according to Eq. (6.3) where voltage V is fixed, the capacitance must *increase* compared to the non-fringing case. Figure 6.4c given below illustrates numerically found capacitance values C_{exact} for the parallel-plate capacitor with fringing. These values have been accurately computed using a rigorous numerical adaptive procedure. Figure 6.4c predicts a nearly linear increase of the ratio C_{exact}/C as a function of the separation distance. Therefore, the wrong result, $C \to 0$ when $d \to \infty$, which is predicted by Eq. (6.6), is corrected. Instead, one will have $C \to 0.5\, C_{\text{self}}$ when $d \to \infty$.

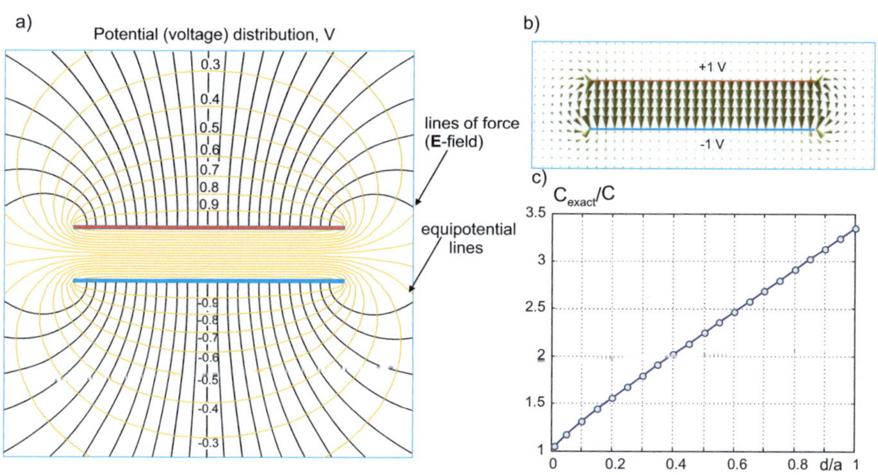

Fig. 6.4. (**a**) Equipotential lines and lines of force for a capacitor with $d/a = 0.2$ in the central cross-sectional plane (the plates are at ± 1 V). (**b**) Fringing electric field for the same capacitor observed in the central cross-sectional plane (the plates are at ± 1 V). (**c**) Ratio of the accurate capacitance values (found numerically) to the values predicted by Eq. (6.6).

The fringing field of capacitors is utilized in capacitive *touch screens*. In this case, the significant fringing field is a desired effect. Therefore, configurations other than the parallel-plate capacitor are used. These configurations will be studied later in this section.

When a dielectric material of relative permittivity $\varepsilon_r > 1$ is inserted between the capacitor plates, a substitution $\varepsilon_0 \to \varepsilon_r \varepsilon_0$ has to be made in Eq. (6.6). The fringing effect is less apparent for higher values of ε_r.

Exercise 6.4: Estimate the static capacitance of a parallel-plate capacitor with $a = 1\ \text{cm}^2$ and $d = 1.57$ mm using the basic formula. The substrate material is Rogers 4003 laminate with $\varepsilon_r = 3.55$.

Answer: 2.00 pF.

6.1.4 Circuit Symbol: Capacitances in Parallel and in Series

Figure 6.5 shows the capacitances in parallel and in series, along with the capacitance circuit symbol. This symbol is reserved for the capacitance as a *circuit element*. Such an element is an *ideal* capacitor excluding manufacturing imperfections (parasitic resistance and inductance). In the following text, we will frequently employ both words—capacitance and capacitor—to denote the same ideal circuit element. The parallel and series connections of capacitances are *opposite* when compared to the resistances. To establish this fact we consider two combinations in Fig. 6.5.

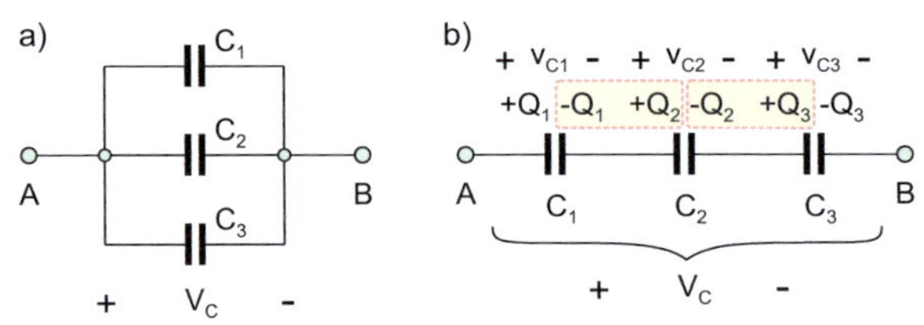

Fig. 6.5. Capacitances in parallel and in series: (**a**) capacitances in parallel are added; they behave similarly to resistances in series and (**b**) capacitances in series are combined in the same way as resistances in parallel.

For the parallel configuration in Fig. 6.5a, the same voltage V_C is applied to every capacitance. One has for the charges on the capacitor plates,

$$
\begin{aligned}
Q_1 = C_1 V_\text{C},\ Q_2 = C_2 V_\text{C},\ Q_3 = C_3 V_\text{C} &\Rightarrow Q_1 + Q_2 + Q_3 = Q_\text{total} \\
&= (C_1 + C_2 + C_3)\ V_\text{C}
\end{aligned}
\tag{6.7}
$$

Thus, the capacitances connected in parallel behave like a single capacitance C_{eq},

$$C_{eq} = C_1 + C_2 + C_3 \tag{6.8}$$

Equation (6.8) also makes intuitive sense if we take a closer look at Fig. 6.5a where the three individual capacitors visually form a bigger capacitor comprised of larger plate areas. This clearly increases the capacitance accordingly. In Fig. 6.5b, however, the situation is different. The thicknesses of each capacitor add together, which decreases the overall capacitance since thickness varies inversely with capacitance. Assume that every capacitor was initially uncharged and apply voltage V_C between terminals A and B. Since each pair of inner conductors in Fig. 6.5b has remained insulated, $Q_1 = Q_2 = Q_3 = Q$. Next, by KVL,

$$V_C = V_1 + V_2 + V_3 = \frac{Q_1}{C_1} + \frac{Q_2}{C_2} + \frac{Q_3}{C_3} = \left(\frac{1}{C_1} + \frac{1}{C_2} + \frac{1}{C_3}\right)Q = \frac{Q}{C_{eq}} \tag{6.9}$$

and for the series combination of the capacitances, one has

$$\frac{1}{C_{eq}} = \frac{1}{C_1} + \frac{1}{C_2} + \frac{1}{C_3} \tag{6.10}$$

Exercise 6.5: Find the equivalent capacitance of the circuit shown in Fig. 6.6.

Answer: $C_{eq} = 44\ \mu F$

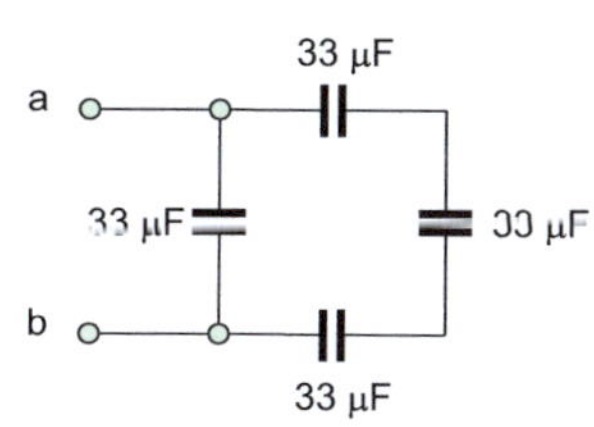

Fig. 6.6. A capacitive network which includes series and parallel combinations of capacitances.

6.1.5 Application Example: How to Design a 1-μF Capacitor?

Let us to design a capacitor. Our modest goal is a 1-μF capacitor. We consider two aluminum plates separated by 1 mm. Equation (6.6) then allows us to predict the required plate area:

$$A = \frac{dC}{\varepsilon_0} = \frac{10^{-3} \times 10^{-6}}{8.854 \times 10^{-12}} = 113\ \text{m}^2! \tag{6.11a}$$

Well, such a capacitor will certainly occupy a significant fraction of a lecture hall and is hardly practical. How, then, do manufactures design a capacitor of 1 μF? The first step is to use a dielectric material sandwiched within the capacitor. A dielectric medium increases charges stored on the two metal capacitor plates depending on $\varepsilon_r \geq 1$, the relative dielectric permittivity of the dielectric medium. Table 6.1 gives us a list of permittivities for a number of dielectric materials. For each material, a *dielectric strength*, or *normalized breakdown voltage*, is also given. This is actually the maximum electric field (notice the unit of V/m) that the capacitor can handle. It is for this reason that capacitors carry a voltage rating that you should not exceed in your circuit. From a practical point of view, the higher the capacitance, the lower the voltage rating. The well-known dilemma with the capacitor is that a decrease in the separation distance increases the capacitance and the stored energy. However, as already mentioned, it simultaneously decreases the maximum applied voltage due to the *dielectric breakdown effect*. For our capacitor, we will again use the mica dielectric material listed in Table 6.1. Equation (6.6) now transforms to

$$A = \frac{dC}{\varepsilon_0 \varepsilon_r} = \frac{10^{-3} \times 10^{-6}}{8.854 \times 10^{-12} \times 7} = 16\ \text{m}^2 \tag{6.11b}$$

Table 6.1. Relative dielectric permittivity and dielectric strength of some common materials.

Material	Relative permittivity	Dielectric strength in V/m
Air	1.0	0.4–3.0 × 10^6
Aluminum oxide	8.5	Up to 1000 × 10^6
Fused silica (glass)	3.8	470–670 × 10^6 (or lower)
Gallium arsenide (GaAs)	13	
Germanium (Ge) crystal	16	~10 × 10^6
Mica	7.0	Up to 400 × 10^6
Nylon	3.8	~20 × 10^6
Plexiglas	3.4	~30 × 10^6
Polyester	3.4	
Quartz	4.3	8 × 10^6 (fused quartz)
Rutile (titanium dioxide)	100–200	10–25 × 10^6
Silicon (Si) crystal	12	~30 × 10^6
Styrofoam	1.03–1.05	
Teflon	2.2	87–173 × 10^6
Water (distilled, deionized)	~80	65–70 × 10^6

Even though the result looks a bit better, it is still far from practical. However, what if we try to make the dielectric layer very thin? An oxide is a dielectric, so could we just oxidize one top aluminum plate with a very thin (i.e., $d = 10\,\mu\text{m}$) oxide layer and press-fit it to the second plate? The result becomes (the relative dielectric constant of 8.5 is now that for aluminum oxide from Table 6.1)

$$A = \frac{hC}{\varepsilon_0 \varepsilon_r} = \frac{10^{-5} \times 10^{-6}}{8.854 \times 10^{-12} \times 8.5} = 13 \text{ cm}^2 \quad (6.11c)$$

Electrolytic Capacitors

Once such a thin film is rolled into a cylinder, it will clearly become a compact design, similar in size to a 1-μF electrolytic capacitor routinely used in the laboratory. Unfortunately, one problem still remains: the permanent oxide layer is fragile and rough in shape. A better idea is to chemically grow such a layer using a so-called anodization process. This process occurs when the aluminum foil is in contact with an electrolyte as a second conductor and an appropriate *voltage* is applied between them. This is the smart idea behind an electrolytic capacitor. And this is also the reason why an electrolytic capacitor is *polarized*. The term *electrolytic capacitor* is applied to any capacitor in which the dielectric material is formed by an electrolytic method; the capacitor itself does not necessarily contain an electrolyte. Along with aluminum capacitors, *tantalum capacitors* (both *wet* and *dry*) are also electrolytic capacitors.

Ceramic Capacitors

A competitor to the electrolytic capacitor is a *non-polarized ceramic* capacitor. Ceramic capacitors consist of a sandwich of conductor sheets alternated with ceramic material. In these capacitors the dielectric material is a ceramic agglomerate whose relative static dielectric permittivity, ε_r, can be changed over a very wide range from 10 to 10,000 by dedicated compositions. The ceramic capacitors with lower ε_r values have a stable capacitance and very low losses, so they are preferred in high-precision circuits and in high-frequency and RF electronic circuits. Typically, these "fast" ceramic capacitors have very small capacitances, on the order of pF and nF, and they can hold a high voltage. At the same time, the "slow" ceramic capacitors may have values as high as 1 μF. Therefore, the task of the above example can be solved with the ceramic capacitor as well.

Capacitor Marking

Figure 6.7 shows two examples of ceramic capacitors, with 100-pF and 1.0-μF capacitance from two different companies. To read the capacitance in the figure, we use the following rule: $101 = 10 \times 10^1$ pF $= 100$ pF, and $105 = 10 \times 10^5$pF $= 1$ μF. Indeed, $473 = 47 \times 10^3$pF $= 47$ nF, and so forth. The tolerance letters may be present: $F = 1$ %, $G = 2$ %, $J = 5$%, $K = 10$ %, and $M = 20$ %. Also, the voltage rating should be given.

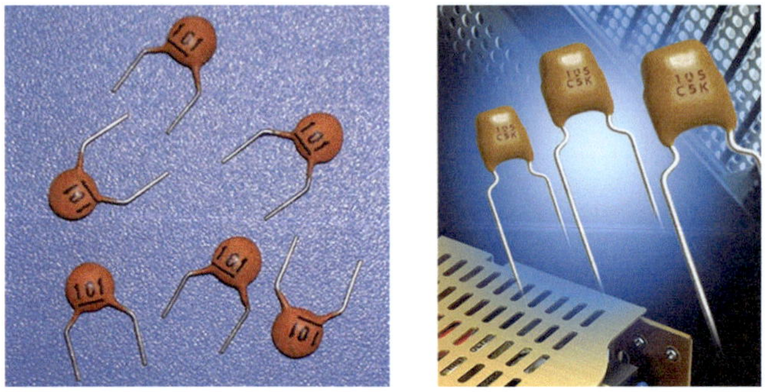

Fig. 6.7. *Left*, ceramic capacitors of 100 pF. *Right*, radial leaded ceramic capacitors of 1.0 μF.

A standardized *set of capacitance base values* is defined in the industry. The capacitance of any (electrolytic or not) capacitor can then be derived by multiplying one of the base numbers **1.0**, **1.5**, **2.2**, **3.3**, **4.7**, or **6.8** by powers of ten. Therefore, it is common to find capacitors with capacitances of 10, 15, 22, 33, 47, 68, 100, 220 μF, and so on. Using this method, values ranging from 0.01 to 4700 μF are customary in most applications. The value of the capacitance and the allowed maximum voltage are prominently written on the case of the electrolytic capacitor so reading those does not constitute any difficulties.

6.1.6 Application Example: Capacitive Touchscreens

Capacitive touchscreens use the fringing field of a capacitor studied previously. Many small capacitors with a significant fringing field are involved. If a conducting finger (an extra conductor) is placed in the fringing field, the corresponding capacitance changes. There are two possible solutions called the *self-capacitance method* and the *mutual-capacitance method*, respectively. The difference is in the measurement nodes for the capacitance. In the first case, the capacitance is measured between the touch pad electrode and a ground. In the second case, the capacitance is measured between two pad electrodes, neither of which is grounded. Both methods may be combined.

Self-Capacitance Method

Consider a human finger in the proximity of a touchscreen as shown in Fig. 6.8a. The touchscreen itself may be a lattice of circular touch pads surrounded by a ground plane and separated from it by an air-gap ring—see Fig. 6.8b. When the finger is not present, each pad has capacitance C_P to ground, which is called a *parasitic capacitance*. When the (grounded) finger appears in the vicinity of the touchpad, there appears another capacitance, C_F, which is called the *finger capacitance*. Figure 6.8a indicates that both capacitances are in parallel so that the resulting ground capacitance *increases* as

$$C_P \to C_P + C_F > C_P \tag{6.12}$$

This change in capacitance is recorded. Physically, the presence of the finger (or hand) increases the size of the ground conductor and thus increases the resulting capacitance. Typical values of C_P are on the order of 100 pF; C_F is on the order of 1–0.1 pF. Now assume that the desired resolution along one dimension of the screen is N. Then, N^2 individual touch pads are needed including the corresponding sensing circuitry. This may be a significant disadvantage of the self-capacitance method.

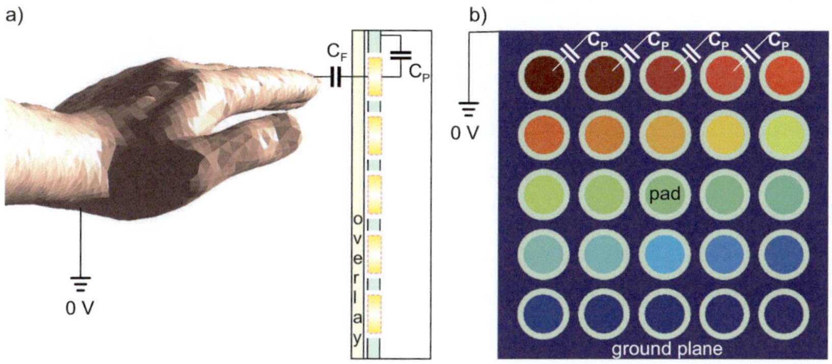

Fig. 6.8. Self-capacitance method for a capacitive touchscreen. The touchscreen is enlarged.

Mutual-Capacitance Method

The electrodes are typically interleaving rows and columns of interconnected square patches, which are shown in Fig. 6.9. Neither of them is connected to circuit ground (the third conductor) or to each other. When a finger touches the panel, the mutual capacitance C_M between the row and column, which mostly concentrates at the intersection, *decreases*, in contrast to the previous case. This change in capacitance is recorded. Assume again that the desired resolution along one dimension of the screen is N. Then, only $2N$ individual touch pads (electrodes) are needed including the corresponding sensing circuitry. This is a significant advantage of the mutual-capacitance method.

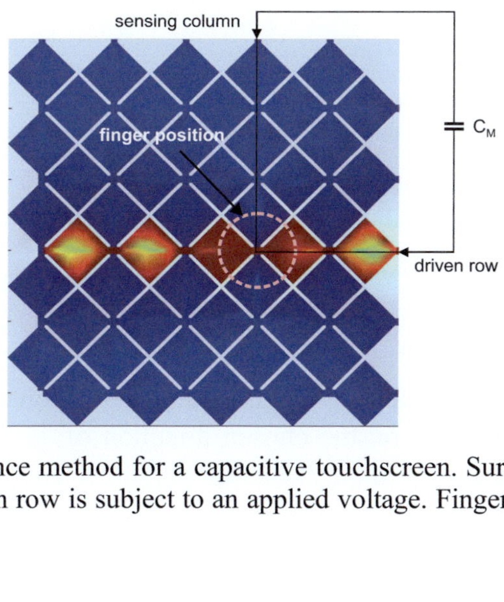

Fig. 6.9. Mutual-capacitance method for a capacitive touchscreen. Surface charge distribution is illustrated when the driven row is subject to an applied voltage. Finger projection is a circle.

6.1.7 Self-Inductance (Inductance) and Mutual Inductance

As long as the physical capacitor stores energy of the electric field, a physical *inductor* stores energy of the magnetic field. The inductor stores magnetic-field energy only when an electric current I flows through it. This is in contrast to a capacitor, which, once charged, stores the electric field energy in vacuum indefinitely, even when disconnected from the charging circuit. To use the inductor as an energy-storage element, one therefore needs to maintain a current in the circuit. The primary quantity is the *magnetic flux density*, $\vec{B}$, which is measured in *webers per* m^2, or *tesla* ($1\ \text{T} = 1\ \text{Wb/m}^2 = 1\ \text{V}\cdot\text{s/m}^2$). The magnetic flux density is related to the *magnetic field*, $\vec{H}$, by $\vec{B} = \mu\vec{H}$ where μ is *magnetic permeability*. In older power electronics texts, $\vec{B}$, may be called *magnetic induction*. In Fig. 6.10, the magnetic flux density $\vec{B}$ is created by circuit #1 (a closed loop of current I). Instead of the vector field $\vec{B}$, it is convenient to use a simpler scalar quantity known as *magnetic flux* or simply *flux*, Φ. For a constant $\vec{B}$, which is strictly perpendicular to the plane of circuit #1 with area A, the magnetic flux would be equal to

$$\Phi = AB \tag{6.13}$$

where B is the magnitude (length) of vector $\vec{B}$. The flux is measured in *webers* or in $\text{V}\cdot\text{s}$ ($1\ \text{Wb} = 1\ \text{V}\cdot\text{s}$).

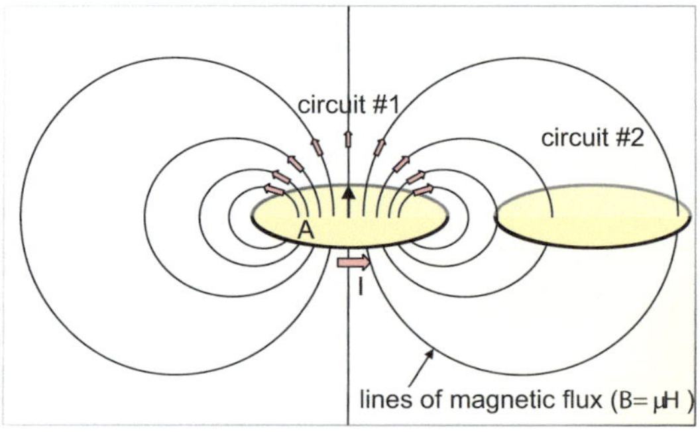

Fig. 6.10. Magnetic flux density generated by circuit #1.

Generally, Eq. (6.13) is only approximately valid for circuit #1 in Fig. 6.10. The exact flux is given by a surface integral over the area of the circuit,

$$\Phi = \iint_A \vec{B}\cdot\vec{n}da \tag{6.14}$$

where $\vec{n}$ is a unit normal to the surface A. Flux is an algebraic quantity and hence could be positive or negative depending on the chosen direction of $\vec{n}$. We assume that *positive current in the circuit produces positive flux*—see Fig. 6.10. The *self-inductance* of circuit

#1 in Fig. 6.10 is its *inductance*; both terms have the same meaning. The inductance, L, of circuit #1 is given by

$$L \equiv \frac{\Phi}{I} > 0 \tag{6.15}$$

Thus, the inductance is the magnetic flux through circuit #1 produced by a unit current in the same circuit. The *mutual inductance*, M, between circuits #2 and #1 in Fig. 6.10 is the magnetic flux, Φ', through circuit #2 produced by unit current in circuit #1, i.e.,

$$M \equiv \frac{\Phi'}{I} \tag{6.16}$$

Both L and M have the units of henry, or H. This unit is named in honor of Joseph Henry (1797–1878), an American scientist. Typical inductance values in electronics are nH (nanohenries) and μH (microhenries). In power electronics, larger inductances may be used. Henry is converted to V, A, and energy, J, as follows:

$$1\ \mathrm{H} = 1\frac{\mathrm{V}\cdot\mathrm{s}}{\mathrm{A}} = 1\frac{\mathrm{J}}{\mathrm{A}^2} \tag{6.17}$$

One may observe a close similarity between Eqs. (6.17) and (6.2). Both equations become identical if we interchange V and A. Equation (6.17) also has a number of simple and important implications related to energy and power. Total *magnetic-field energy* stored in space surrounding circuit #1 in Fig. 6.10 is given by

$$E = \frac{1}{2}LI^2 \tag{6.18}$$

Equation (6.18) may be considered as another definition of self-inductance (or inductance). As such, it is frequently used in practice.

Exercise 6.6: A flux linking the circuit is 0.1 Wb. Find the circuit's inductance and magnetic-field energy stored if the circuit current is 1 A.

Answer: $L = 100$ mH, $E = 0.05$ J.

6.1.8 Inductance of a Solenoid With and Without Magnetic Core

Consider a *solenoid* (a long helical coil of length l) with applied current I shown in Fig. 6.11. The case of air-filled coil in Fig. 6.11a is studied first. Magnetic flux density $\vec{B}$ within the solenoid is nearly uniform and is directed along its axis. Therefore, the flux through one turn of the coil (one loop) is given by Eq. (6.13). It is equal to AB, where A is the loop area. The net flux Φ through the entire solenoid is AB times the number of turns comprising the coil, N. The inductance is therefore obtained from Eq. (6.15) as

$$L = N\frac{AB}{I} \tag{6.19}$$

The magnetic flux, B, within the solenoid is found in physics courses:

$$B = \frac{\mu_0 N I}{l} \tag{6.20}$$

where the *natural constant* $\mu_0 = 4\pi \times 10^{-7}\ \mathrm{H/m}$ is the *magnetic permeability of vacuum* (or air, which is very close to a vacuum with regard to magnetic properties). Substitution of Eq. (6.20) into Eq. (6.19) yields a simple equation for the inductance

$$L = \frac{\mu_0 A N^2}{l} \quad [\mathrm{H}] \tag{6.21}$$

Thus, strong inductances can be created by a large number of turns (a quadratic dependence), a large coil cross section, and a smaller coil length. Equation (6.21) also holds for various bent solenoids (such as toroidal coils). Equation (6.21) makes clear that the inductance, like capacitance and resistance, is independent of externally applied circuit conditions.

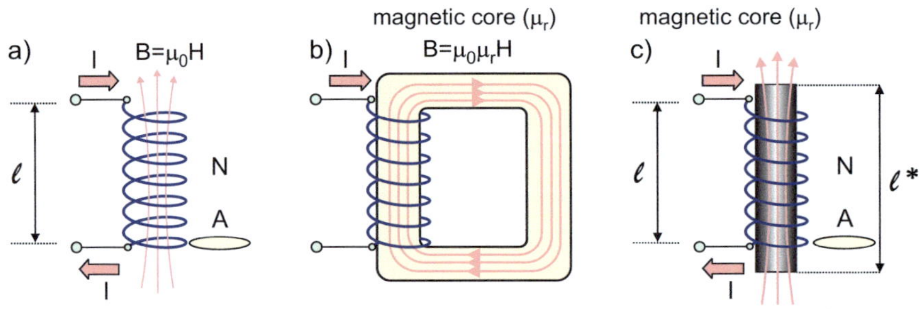

Fig. 6.11. Three types of a solenoid.

The above derivation is only valid for a solenoid that is long compared to its diameter. When this is not the case, a modification to Eq. (6.22) is made, namely,

$$L = \frac{\mu_0 A N^2}{l}\left(1 - \frac{8w}{3\pi} + \frac{w^2}{2} - \frac{w^4}{4}\right), \quad w = \frac{r}{l} < 1 \tag{6.22}$$

where r is the radius of the coil. Other handy formulas for *short solenoids* exist. Equation (6.22) describes the *fringing effect* in practical inductors, which may also be used for sensor purposes, similar to the fringing fields for the capacitor. The inductance of the solenoid increases when it has a core with a magnetic material within the coil, as shown in Fig. 6.11b, c. This material is called a magnetic *core* and inserting it into the coil increases the magnetic-field energy stored in the inductor. When the magnetic core is closed, i.e., has the form of a toroid as in Fig. 6.11b, Eq. (6.21) is transformed to

$$L = \frac{\mu_0 \mu_r A N^2}{l} \tag{6.23}$$

where $\mu_r \geq 1$ is known as the *relative magnetic permeability* of a magnetic material. Emphasize that Eq. (6.23) is not valid for the straight cylindrical core in Fig. 6.11c. The calculation of inductance for the straight core becomes a nontrivial theoretical exercise. We present here a useful theoretical result, which is only valid for a high-permeability magnetic core, with approximately $\mu_r \geq 100$. The resulting inductance has the form

$$L \approx \frac{0.5\pi\mu_0 l^* N^2}{\ln\left[\frac{l^*}{r} - 1\right]}\left(1 - \frac{l}{2l^*}\right) \quad [\mathrm{H}] \tag{6.24}$$

where l, l^* are indicated in Fig. 6.11c and r is the radius of the coil (core). The resulting inductance does not explicitly depend on the specific value of μ_r as long as this value is sufficiently large. Equation (6.24) holds only for the situations where the core length-to-diameter ratio is considerably *smaller* than the relative magnetic permeability, μ_r. It predicts inductances that are much lower than those found by using Eq. (6.23).

Exercise 6.7: A solenoid coil in Fig. 6.11a, b, and c has $r = 0.45\,\mathrm{cm}$, $N = 110$, $l = 2.15$ cm. Determine the coil inductance in all three cases. In cases b and c, respectively, $\mu_r = 100$. Furthermore, $l^* = 8.90$ cm in Fig. 6.11c.

Answer: $L = 45$ μH (air-core coil); $L = 4500$ μH (toroidal coil); $L = 640$ μH (straight-magnetic-core coil).

6.1.9 Circuit Symbol: Inductances in Series and in Parallel

Figure 6.12 shows the inductances in series and in parallel, along with inductance circuit symbol. This symbol is reserved for the inductance as a *circuit element*. Such an element is an *ideal* inductor excluding manufacturing imperfections (parasitic resistance and capacitance). In the following text, we will frequently employ both words—inductance and inductor—to denote the same ideal circuit element. The series and parallel connections of inductances are *identical* when compared to the connections of resistances. To establish this fact we consider two combinations shown in Fig. 6.12.

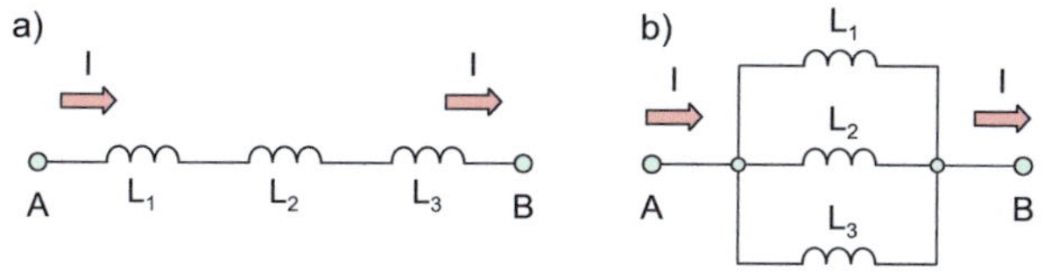

Fig. 6.12. Inductances in series and in parallel.

For the series configuration in Fig. 6.12a, the same current I is applied to every inductance. Given that the equivalent inductance is also subject to current I and must possess the same magnetic-field energy defined by Eq. (6.18), one has

$$E = \frac{1}{2}L_{eq}I^2 = \frac{1}{2}L_1I^2 + \frac{1}{2}L_2I^2 + \frac{1}{2}L_3I^2 \Rightarrow L_{eq} = L_1 + L_2 + L_3 \tag{6.25}$$

The parallel configuration in Fig. 6.12b may be analyzed given the condition of equal magnetic flux through each inductance. Since this condition is related to Faraday's law of induction, we postpone the corresponding discussion until the next section.

6.1.10 Application Example: How to Design a 1-mH Inductor?

Again, after this theoretical excursion, let us design a practical inductor. Our goal is to construct a 1-mH inductance. According to Eq. (6.21) the required number of turns is

$$N = \sqrt{\frac{Ll}{\mu_0 A}} \tag{6.26}$$

We will select a coil length of $l = 5$ cm and an (average) coil radius of $r = 1$ cm. Equation (6.26) then yields

$$N = \sqrt{\frac{10^{-3} \times 5 \times 10^{-2}}{4\pi^2 \times 10^{-7} \times 10^{-4}}} = 356 \tag{6.27}$$

Such a coil can be wound on a former in the laboratory with a sufficiently thin wire, say AWG 28. A different approach to reducing the number of turns while maintaining, or even increasing, the inductance is to use a magnetic material within the coil, as shown in Fig. 6.11. Table 6.2 lists the magnetic permeability for a number of magnetic materials. The simplest magnetic core is an iron core. However, it is lossy since an alternating magnetic field creates so-called eddy currents in the conducting core, which are dissipated into heat. One solution to this problem is to use thin insulated sheets of iron, or *laminations*. Various *ferrites* (oxides of iron, or other metals) are an alternative to iron, which are ceramics and known as good electric insulators. Other types of losses may occur there, explanations of which go beyond the scope of our text. Once a soft ferrite with $\mu_r \sim 100$ is used in the design of the 1-mH inductor, the number of turns necessary to achieve the same inductance decreases by $\sqrt{\mu_r} = 10$. In our example, it becomes equal to 36 turns instead of over 300 turns. However, the magnetic core *cannot* be a short rod like that shown in Fig. 6.11c; it must form a closed loop shown in Fig. 6.11b.

Table 6.2. Relative permeability of some common materials.

Material	Relative static permeability
Air	1.0
Magnetic iron	200
Iron powder	2–75
Nickel	100
Permalloy (78.5 % nickel + 21.5 % iron)	8000
Soft ferrites with low losses at frequencies up to 100 MHz	20–800
Hard ferrites with low losses up to 1 MHz	1000–15,000

Inductor Marking

Leaded inductors have color codes, similar to resistors. A standardized *set of inductance base values* is defined in the industry. The inductance of any inductor can then be derived by multiplying one of the base numbers **1.0**, **1.1**, **1.3**, **1.5**, **1.9**, **2.2**, **2.7**, **3.3**, **3.9**, **4.7**, **5.8**, **6.8**, or **8.2** by powers of ten. Therefore, it is common to find inductances with values of 1.0, 2.7, 6.8 μH, and so on. Using this method, values ranging from 0.01 to 100 μH are customary in most applications.

Section 6.2 Dynamic Behavior of Capacitance and Inductance

6.2.1 Set of Passive Linear Circuit Elements

The three elements, resistance, capacitance, and inductance, constitute the fundamental set of passive circuit elements for any linear electric circuit. This is very similar to a mechanical system consisting of dashpot, spring, and mass, which form the basic set of any linear kinematic system. Having discussed the underlying DC concepts of capacitance and inductance, we now turn our attention to their dynamics. A simple example of dynamic behavior is given by a vacuum cleaner. If one manually unplugs the working vacuum cleaner from the wall outlet (please avoid doing so), a profound spark may appear. On the other hand, turning off the vacuum cleaner normally produces no spark. The reason for the spark is that breaking the current through a dynamic circuit element—an inductance—which models the coil of the motor, creates very large transient voltages. One reason for studying transients is the wish to avoid such sparks and to properly design the electric switch. The use of two dynamic circuit elements—the capacitance and inductance—is enormous, especially in power systems. Every electric motor is basically an inductance; most power motors need a power correction circuit that in turn requires a shunt capacitance. Some motors need starting capacitors or surge capacitors for large motors. On the other hand, the capacitance of logic gates is responsible for the so-called propagation delay. This delay determines a very important measure of the performance of a digital system, such as a computer, which is the maximum speed of operation. Thus, the capacitances and inductances are just everywhere, like mass and spring systems present everywhere in mechanical engineering. However, they are becoming most apparent when we consider a transient behavior, an alternating current, or high-frequency digital and communication circuits.

6.2.2 Dynamic Behavior of Capacitance

Both capacitance and inductance are *passive* circuit elements, which means that, like resistance, they do not deliver a net power increase to the circuit. Indeed, after charging, the capacitor is able to power a circuit, usually for a short period of time. At the end of the discharge cycle, it needs to be recharged. Therefore, we use the passive reference configuration for the capacitance in Fig. 6.13a. Lowercase letters denote time-varying voltage and current. The dynamic behavior of the capacitance is described by the well-known voltage-to-current relation (dynamic equation), which plays the role of "Ohm's law" for the capacitance

$$i_C = C\frac{dv_C}{dt} \tag{6.28}$$

Equation (6.28) follows from the capacitance definition, $q_C = Cv_C$, given in the previous section. It is obtained after differentiation and using the equality

$$\frac{dq_C}{dt} = i_C \tag{6.29}$$

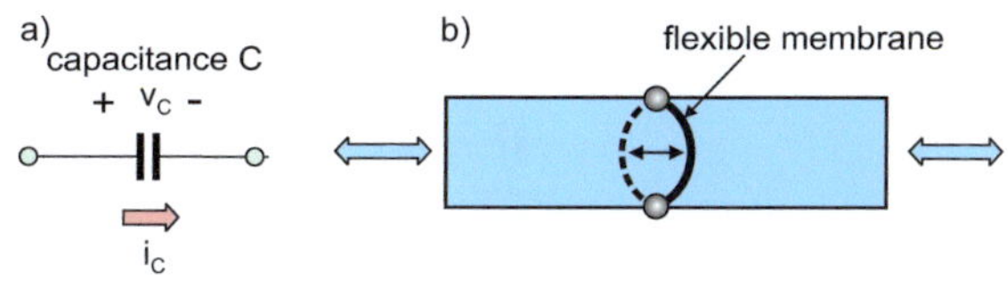

Fig. 6.13. Passive reference configuration for capacitance and its fluid mechanics analogy.

A fluid mechanics analogy of the dynamic capacitance effect corresponds to alternating fluid motion in Fig. 6.13b. A capacitance is represented by a *flexible membrane*. The capacitance value, C, corresponds to the *inverse stiffness*, $1/k$, of the membrane. When $k \to \infty$ and $C \to 0$ (a rigid membrane), the capacitance value tends to zero. The membrane becomes a solid wall, which blocks the alternating flow entirely. In another limiting case ($k \to 0$ or $C \to \infty$), the membrane moves with fluid and has no effect on the fluid flow. Intermediate cases correspond to a partial blocking. The electric current in Eq. (6.28) is not the DC conduction current, but the *displacement* current which was first introduced by Sir James Clerk Maxwell to complete Maxwell's equations. This current can flow through empty (or free) space between two capacitor plates. It is not supported by directional motion of free charges, in contrast to the DC conduction current. The capacitor itself was invented yet in 1745 by Ewald Georg von Kleist (1700–1748), German lawyer and physicist, and by Dutch scientist Pieter Van Musschenbroek (1692–1761). Some also believe the Biblical Ark of the Covenant was protected by a first capacitor—the Leyden jar—capable of producing thousands of volts of static electricity.

Example 6.2: The voltage across a 100-μF capacitor is shown in Fig. 6.14 that follows by a solid curve. At $t = 0$, the voltage is zero. Sketch the current through the capacitor to scale versus time.

Solution: We use Eq. (6.28) to find the current. In Fig. 6.14, $v_C(t) = 10^5(t - 10^{-6})$ V when t changes from 1 to 2 μs, and $v_C(t) = 0$ at $t < 1$ μs. At $t > 2$ μs, the voltage is 100 mV. Therefore, the current is found in the form: $i_C(t) = 10^{-4} \times 10^5 = 10$ A when t changes from 1 to 2 μs and $i_C(\mathrm{t}) = 0$ otherwise. The result is shown in Fig. 6.14 by a dashed curve. We observe a strong current spike when the voltage across the capacitor changes rapidly, and we observe no current flow when the voltage across the capacitor remains constant.

Note that the relatively small voltage on the order of 100 mV in Fig. 6.14 leads to a very large current spike of 10 A(!) through the capacitance. The key point here is that the current increase is due to the *rapid* change in voltage. Such a change can be created when the capacitor discharges through a small resistance. This is the reason why capacitors are routinely employed to deliver large currents, or high power levels, for a very short period of time. The high currents are common in motor starting circuits, in electronic flashes, in solenoids, and in various electromagnetic propulsion systems.

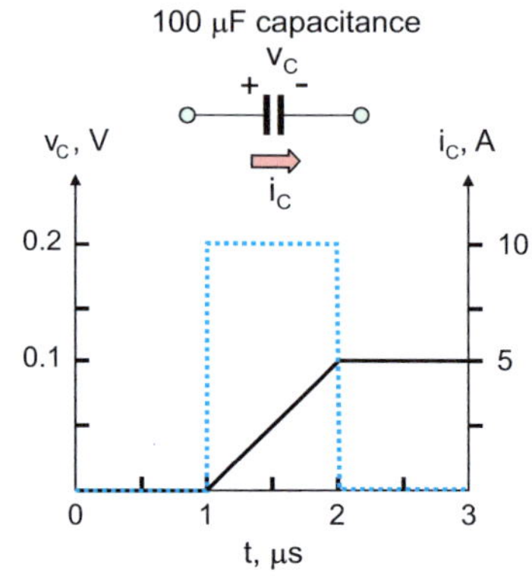

Fig. 6.14. Applied voltage across the capacitor (*solid curve*) and resulting current (*dashed curve*).

The capacitor is charged with an electric current. The voltage across the capacitor, from Eqs. (6.28) and (6.29), is given by

$$v_C = \frac{1}{C}\int_0^t i_C(t')dt' = \frac{q_C(t)}{C} \tag{6.30}$$

Equation (6.30) tells us that the capacitor voltage is equal to zero at the initial time, i.e., at $t = 0$. Once the current $i_C(t)$ is known, the voltage at any point in time is obtained by carrying out the integration in Eq. (6.30). At any time instant, the voltage is equal to the instantaneous stored charge $q_C(t)$ divided by capacitance. The current in Eq. (6.30) is either predefined or found from circuit considerations. The example that follows illustrates voltage calculations.

Example 6.3: A 1-μF capacitor is charged with an electric current, $i_C(t) = 2 \cdot t$ [mA]. The capacitance voltage is equal to zero at the initial time instance $t = 0$. When will the capacitor be charged to 10 V?

Solution: The integration yields

$$v_C = \frac{1}{C}\int_0^t i_C(t')dt' = \frac{0.001}{C}t^2 = 10 \Rightarrow t = 0.1\ \mathrm{s} \tag{6.31}$$

6.2.3 Dynamic Behavior of Inductance

We use the passive reference configuration for the inductance in Fig. 6.15a. Lowercase letters denote time-varying voltage and current. The dynamic behavior of the inductance is described by the well-known voltage-to-current relation (*dynamic equation*), which plays the role of "Ohm's law" for the inductance:

$$v_L = L\frac{di_L}{dt} \tag{6.32}$$

Equation (6.32) follows from the inductance definition, $\Phi = Li_L$, given in the previous section. We obtain Eq. (6.32) after differentiation and using Faraday's law of induction for the time derivative of the magnetic flux (the *plus* sign is used in Faraday's law to be consistent with the passive reference configuration):

$$\frac{d\Phi(t)}{dt} = v_L \tag{6.33}$$

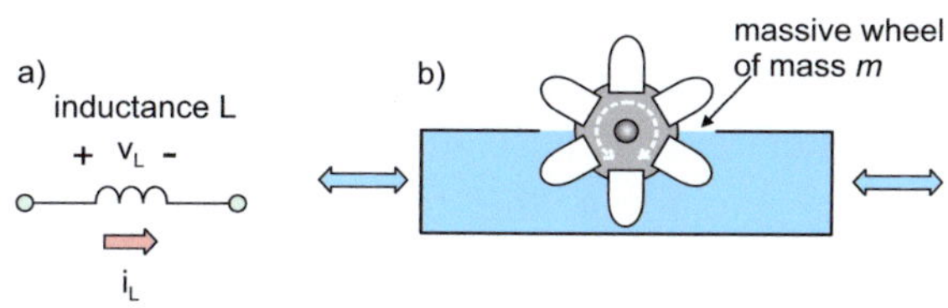

Fig. 6.15. Passive reference configuration for inductance and its fluid mechanics analogy.

A fluid mechanics analogy of the dynamic inductance effect is given here in terms of alternating current, which corresponds to alternating fluid motion in Fig. 6.15b. A massive wheel with rotational inertia in Fig. 6.15b represents inductance. The inductance value, L, corresponds to the *mechanical mass* m of the wheel. When $m \to \infty$ or $L \to \infty$, the wheel does not responds to fluid oscillations and blocks the alternating fluid flow entirely. In the opposite case ($m \to 0$ or $L \to 0$), the wheel has no effect on the fluid flow. Intermediate cases correspond to a partial blocking.

Example 6.4: The current through a 2-mH inductor is shown in Fig. 6.16 by a solid curve. At $t = 0$, the current is zero. Sketch the voltage across the inductance to scale versus time.

Solution: We use Eq. (6.32) to find the voltage across the inductance. In Fig. 6.16, $i_L(t) = 10^3(t - 10^{-6})$ A when t changes from 1 to 2 μs and $i_L(t) = 0$ at $t < 1$ μs. At $t > 2$ μs, the current is 1 mA. Therefore, the voltage is found in the form: $v_L(t) = 2 \times 10^{-3} \times 10^3 = 2$ V when t changes from 1 to 2 μs and $v_L(t) = 0$ otherwise. The result is shown in Fig. 6.16 by a dashed curve.

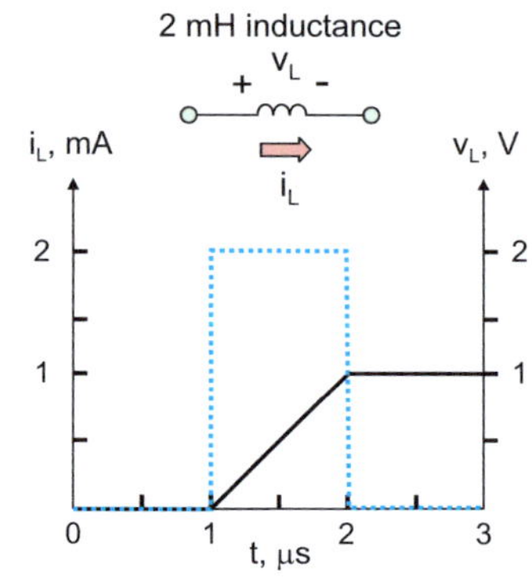

Fig. 6.16. Impressed current through an inductance (*solid curve*) and resulting voltage across the inductance (*dashed curve*).

Note that relatively small current, on the order of 1 mA, leads to a large voltage spike of 2 V across the inductance in Fig. 6.16. The key point here is again the *rapid* change in the current. If the current in the present example were on the order of 1 A, a voltage spike of 2000 V would be observed. This is the reason why inductors are routinely used to boost the voltage to a higher level. These high voltages are common in electric and electronic ignition systems including the most common car ignition plug. From Eqs. (6.32) and (6.33) the current through the inductance is given by

$$i_{\mathrm{L}} = \frac{1}{L}\int_0^t v_L(t')dt' = \frac{\Phi(t)}{L} \tag{6.34}$$

Equation (6.34) implies that the current is equal to zero at the initial time, i.e., at $t = 0$. Once the voltage is known as a function of time, the current through the inductance at any time moment is obtained by the calculation of the integral in Eq. (6.34). At any time instant, the current is equal to the instantaneous magnetic flux $\Phi(t)$ divided by inductance.

Example 6.5: A 1-mH inductor is subject to applied voltage, $v_L(t) = 2 \cdot t$ [mV]. The inductance current is equal to zero at the initial time instance $t = 0$. When will the magnetic-field energy stored in the inductance reach 1 J?

Solution: The integration in Eq. (6.34) and using Eq. (6.18) for the energy stored in the inductor yield

$$i_{\mathrm{L}} = \frac{1}{L}\int_0^t v_L(t')dt' = \frac{0.001}{L}t^2 = t^2 \text{ [A]} \Rightarrow \frac{1}{2}L(t^2)^2 = 1 \text{ J} \Rightarrow t = 6.7 \text{ s} \tag{6.35}$$

So, if the capacitance is associated with a spring, the inductance is associated with the mass, and the resistance is associated with a dash pot (damping element), then the entire electric circuit containing dynamic elements is nothing else but a mechanical system. Is this correct? Clearly the same analysis methods are applicable to both systems, electrical and mechanical! The model of an entire building in terms of lumped mechanical elements is in theory the same as the model of a complicated electric circuit. Both models can be analyzed by using the theory of linear systems, and both models follow the same control theory. A more difficult issue is related to nonlinear circuit elements.

6.2.4 Instantaneous Energy and Power of Dynamic Circuit Elements

An elegant derivation of the energy stored in a capacitance can also be obtained by integrating the power delivered (or taken) by the capacitance. The instantaneous electric power $p_C(t)$ can be written in the form

$$p_C(t) = v_C i_C = v_C C \frac{dv_C}{dt} = \frac{1}{2} C \frac{dv_C^2}{dt} \tag{6.36}$$

The stored energy is then the time integration of the power, i.e.,

$$E_C(t) = \int_0^t p_C(t')dt' = \int_0^t \frac{1}{2} C \frac{dv_C^2}{dt'} dt' = \frac{1}{2} C \left[v_C^2(t) - v_C^2(0) \right] \tag{6.37}$$

where the lower limit, $v_C(0)$, is the initial state of the capacitance. Suppose that $v_C(0) = 0$, i.e., the capacitance is initially uncharged and has zero stored energy. Then,

$$E_C(t) = \frac{1}{2} C v_C^2(t) \tag{6.38}$$

Equation (6.38) is the formal proof of the corresponding static result, Eq. (6.3), postulated in the previous section. We can derive the energy stored in the inductance using the same method—by integrating the power. The instantaneous power supplied to or obtained from the inductance has the form

$$p_L(t) = v_L i_L = i_L L \frac{di_L}{dt} = \frac{1}{2} L \frac{di_L^2}{dt} \tag{6.39}$$

The energy stored in the inductance is the integral of Eq. (6.39), i.e.,

$$E_L(t) = \int_0^t p_L(t')dt' = \int_0^t \frac{1}{2} L \frac{di_L^2}{dt'} dt' = \frac{1}{2} L \left[i_L^2(t) - i_L^2(0) \right] \tag{6.40}$$

Suppose that $i_L(0) = 0$, i.e., the inductance is initially "uncharged" or has no stored energy. Then,

$$E_L(t) = \frac{1}{2} L i_L^2(t) \tag{6.41}$$

Equation (6.41) is the formal proof of the corresponding static result, Eq. (6.18), postulated in the previous section. The series and parallel combinations of inductances and capacitances may also be analyzed using the dynamic element equations; the laws obtained in the previous section will be confirmed.

Exercise 6.8: Determine instantaneous power supplied to the capacitance in Fig. 6.14 at A. $t = 2\ \mu s$ and B. $t = 1\ \mu s$.

Answer: 1 W and 0 W, respectively.

Exercise 6.9: Repeat the previous exercise for the inductance shown in Fig. 6.16.

Answer: 0.002 W and 0 W, respectively.

6.2.5 DC Steady State

According to Eq. (6.28) when voltage across the capacitance does not change with time, the capacitance becomes an *open circuit* (no current) under DC steady-state condition, i.e.,

$$\frac{dv_C}{dt} = 0 \Rightarrow i_C = C\frac{dv_C}{dt} = 0 \tag{6.42}$$

This is to be expected since a DC current cannot flow through empty space between two capacitor plates. Similarly, according to Eq. (6.32) the inductance becomes a *short circuit* for the DC steady state when current across the inductance does not change in time, i.e.,

$$\frac{di_L}{dt} = 0 \Rightarrow v_L = L\frac{di_L}{dt} = 0 \tag{6.43}$$

In other words, there is no voltage drop across a (long) bent piece of wire, which is the inductor, for DC currents. Equations (6.42) and (6.43) allow us to establish the behavior of *any transient electric circuit* in the long run, after the circuit behavior has been stabilized. The *transient circuit* is a circuit with dynamic elements and a switch. Figure 6.17 shows one such circuit that consists of a number of dynamic (and static) elements and a switch. The switch connects the voltage source to the rest of the circuit as the switch closes at $t \to 0$.

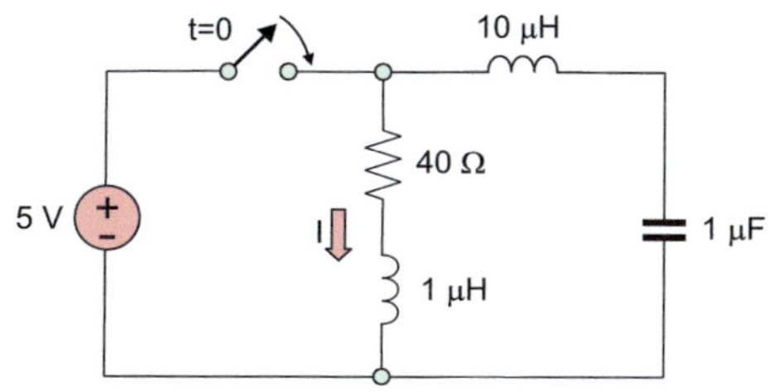

Fig. 6.17. A circuit is used to study the DC steady state. The switch closes at $t = 0$. The behavior of the circuit at $t \to \infty$ is sought after the circuit has stabilized.

Example 6.6: Find current I in Fig. 6.17 at $t \to \infty$, i.e., under DC steady state.

Solution: Immediately after the switch in Fig. 6.17 closes, the voltages and currents in the circuit may be subject to a complicated response. In particular the voltage across certain dynamic elements may be higher than the voltage of the power supply of 5 V. However, in the long run as $t \to \infty$, the circuit behavior stabilizes and we reach the DC steady state. The capacitance in Fig. 6.17 becomes an open circuit and may be ignored. Both inductances can be replaced by a wire (short circuit). The resulting DC circuit is shown in Fig. 6.18. Thus, we obtain

$$I = \frac{5 \ \text{V}}{40 \ \Omega} = 125 \ \text{mA} \tag{6.44}$$

Other more complicated circuits can be analyzed in exactly the same way.

Fig. 6.18. A DC equivalent of the circuit in Fig. 6.17 under steady-state conditions.

6.2.6 Behavior at Very High Frequencies

At very high frequencies, the behavior of the two dynamic circuit elements is exactly the *opposite*: the capacitance becomes a *short* circuit, whereas the inductance becomes an *open* circuit. To establish this fact, we can either use the fluid mechanics analogies or the dynamic equations themselves. For example, the inertia of a massive wheel (inductance) will prevent any very fast movements of it so that the oscillating fluid flow will be entirely blocked when the oscillation frequency tends to infinity. On the other hand, the forces on a flexible membrane of zero mass will be so high for a rapidly oscillating fluid flow that its finite stiffness no longer matters. The membrane will simply be moving along with the fluid, which means the full transmission through its counterpart—the capacitance.

The behavior of dynamic elements at very high frequencies is exactly as important as the behavior at DC; it will be studied quantitatively in Chapter 9.

Example 6.7: Illustrate how is the capacitance becoming a short circuit at very high frequencies using the capacitor's dynamic equation as a starting point.

Solution: Assume that there is a periodic current with the amplitude of 1 A, $i_C(t) = 1\ \mathrm{A} \cos \omega t$ through a 10-μF capacitance. The resulting capacitor voltage is given by Eq. (6.30), $v_C(t) = 1/(\omega C) \sin \omega t$. When $\omega = 10^8$ rad/s, the capacitor voltage has the amplitude of 1 mV. This small voltage approximately corresponds to a short circuit. When ω increases, the voltage amplitude is reduced even further.

Section 6.3 Application Circuits Highlighting Dynamic Behavior

6.3.1 Bypass Capacitor

Let us consider the circuit shown in Fig. 6.19a. It includes a voltage source represented by its Thévenin equivalent and a load represented by its equivalent resistance R_L. The source generates a voltage in the form of a (large) DC component V_S and superimposed (small) AC signal $v_S(t)$. This setup could model a nonideal DC voltage power supply, which does not create the exactly DC voltage. In fact, a weak AC component may be present. This AC component (also called the *noise component*) has a frequency of either $60 \times n$ Hz, where n is an integer (USA, Canada, parts of South America, Saudi Arabia, etc.), or $50 \times n$ Hz (the rest of the world) and appears due to a not quite perfect rectification of the primary AC power. As an aside, switching power supplies create noise spikes at much higher frequencies. The weak AC component may lead to circuit oscillations, especially when dealing with high-gain amplifiers. It therefore should be removed from the load, or "filtered out" as engineers often say. The idea is to use a capacitor C in *parallel* with the (imperfect) power supply and in parallel with the load, the so-called *bypass capacitor*. This capacitor ideally becomes a short circuit for the high-frequency noise component of the source and shorts it out (or bypasses). The corresponding circuit diagram is shown in Fig. 6.19b. An electrolytic capacitor is typically used as the bypass capacitor.

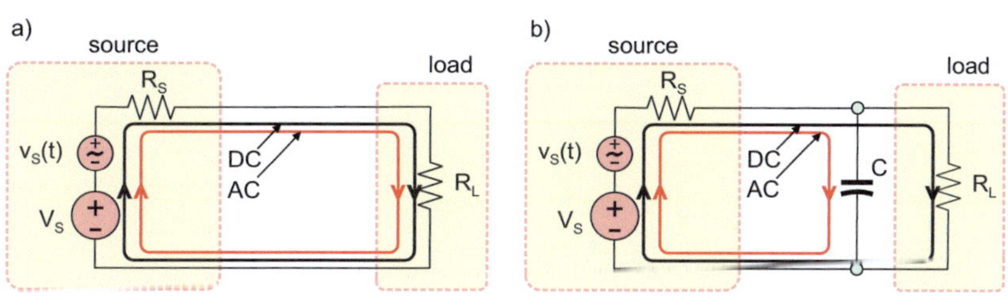

Fig. 6.19. Model of a voltage source connected to a load with a bypass capacitor.

In many cases, the undesired noise source in a circuit is not a low-frequency noise source of a nonideal power supply, but rather a high-frequency noise generator. Examples include high-speed DC motors, analog-to-digital converters, and other digital circuits. Radio-frequency (RF) high-speed amplifiers are also very sensitive to RF noise that is created by connectors and wires which can act like antennas. By placing a bypass capacitor as closely as possible to the power supply pins of every chip, such RF noise sources may be eliminated. Bypass capacitors are so prevalent that they are encountered in virtually *every* working piece of electronic equipment.

Example 6.8: Explain the operation of bypassing a DC motor.

Solution: A DC motor with a *bypass capacitor* is also described by the model shown in Fig. 6.19b. In this case, the source voltage $V_S + v_S(t)$ becomes the induced electromotive force (*emf*), *E*, of the motor, The induced emf is still a DC voltage but with quite a significant high-frequency noise component created by the spinning rotor comprised of a finite number of individual switched coils. The source resistance becomes the armature and brush resistance R_M, that is, $R_S = R_M$. The load resistance R_L may, for example, be the oscilloscope resistance. We consider a small DC fan motor directly connected to a 5-V power supply shown in Fig. 6.20. The motor creates a substantial high-frequency noise seen on the oscilloscope in Fig. 6.20, left (with 100 mV per division resolution). The oscilloscope measures the voltage across the motor, which is the 5-V DC component plus the noise component. Once a 1000-μF capacitor is connected in parallel with the motor (one may call it a *shunt capacitor*), the resulting voltage becomes a highly stable 5-V DC (see Fig. 6.20, right).

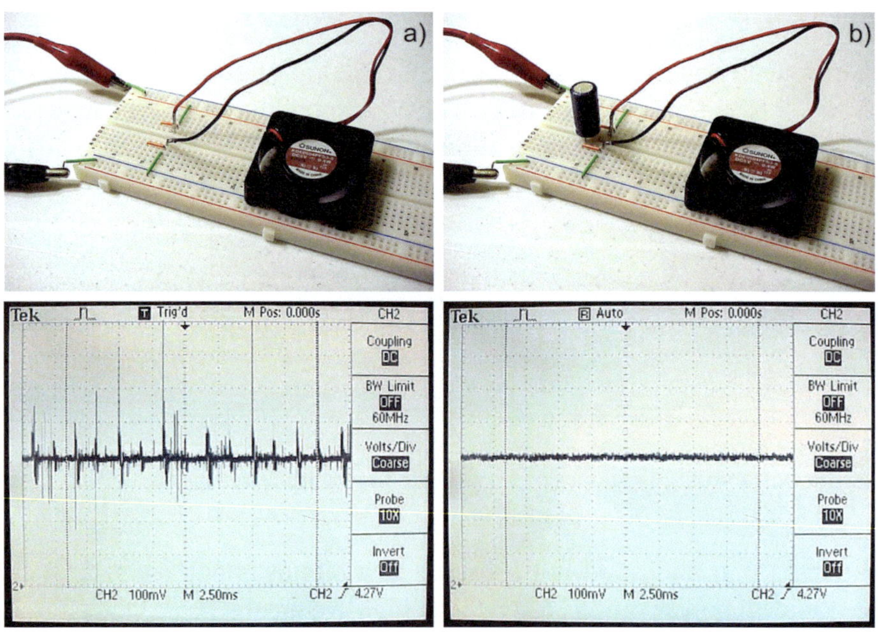

Fig. 6.20. Effect of bypass capacitor on the high-frequency noise created by a DC motor.

The bypass capacitor in Fig. 6.20 may be considered as a part of the *snubber RC circuit*, which includes a capacitance and a small series resistance. The snubber circuits are used to suppress high-voltage spikes in inductive switching systems like electric motors.

6.3.2 Blocking Capacitor

Quite often, an opposite scenario is desired—we want to block a DC component at the load—see Fig. 6.21. An example is an audio amplifier (the source) connected to a speaker (the load). The audio amplifier may generate an unwanted DC component, which may overheat the speaker coil made of a very thin wire. The idea is to use a capacitor C in *series* with the (imperfect) amplifier and in series with the load. This *blocking* or *decoupling capacitor* will block the DC current at the load as shown in Fig. 6.21b.

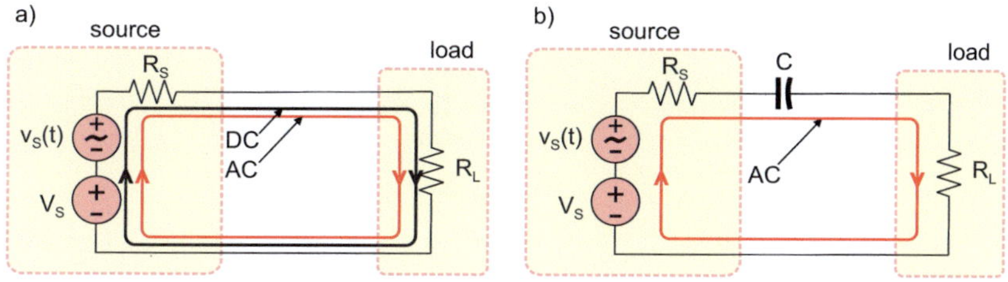

Fig. 6.21. Model of a voltage source connected to a load with a blocking capacitor.

6.3.3 Decoupling Inductor

A *decoupling inductor* is the complement of the bypass capacitor. Consider the circuit shown in Fig. 6.22a. It includes a source represented by its Norton equivalent $I_S + i_S(t)$, R_S and a load with an equivalent resistance R_L. The source generates an electric current in the form of a DC component and a superimposed AC signal. Assume that we would like to have only the direct current at the load. The idea is to use an inductor L *in series* with the load, the so-called decoupling inductor as shown in Fig. 6.22b. At a sufficiently high frequency, this inductor will block the AC component at the load.

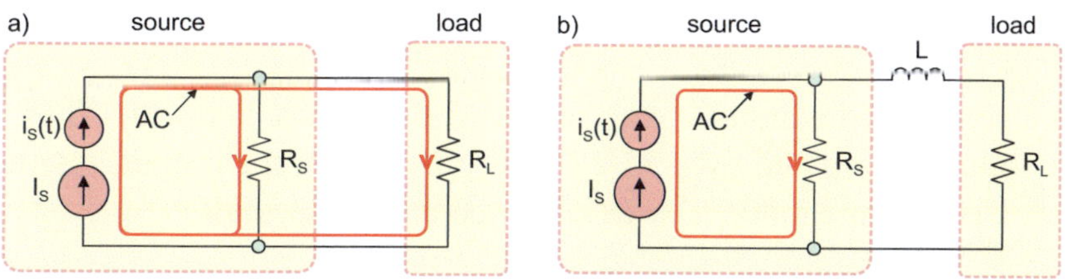

Fig. 6.22. Model of a current source connected to a load with a decoupling inductor.

A common application of the decoupling inductor is the so-called radio-frequency (RF) *inductor choke*. Here, we'd rather intend to redirect the alternating current. The inductor choke prevents the very weak alternating current received by an antenna to be lost in the low-resistance DC power supply, which powers an amplifier. Instead, it forces the current to flow directly into the input port of the amplifier. In order to model the choke effect in Fig. 6.22, we should in fact interchange the role of two resistances: we consider the load resistance as the DC supply resistance and the source resistance as the desired input resistance of the amplifier.

6.3.4 Amplifier Circuits With Dynamic Elements: Miller Integrator

Amplifier circuits with the *dynamic elements* in the *negative feedback loop* can serve different purposes. In particular, they operate as *active filters*. Here, we will introduce the operation concept and present simple examples. The *Miller integrator circuit* is an inverting-amplifier circuit considered in Chapter 5, but with the feedback resistance R_2 replaced by a capacitance C—see Fig. 6.23. Given a time-varying input voltage signal, the capacitor will conduct a current. Therefore, the negative feedback is still present, even though we now have a capacitance instead of the resistance in the feedback loop.

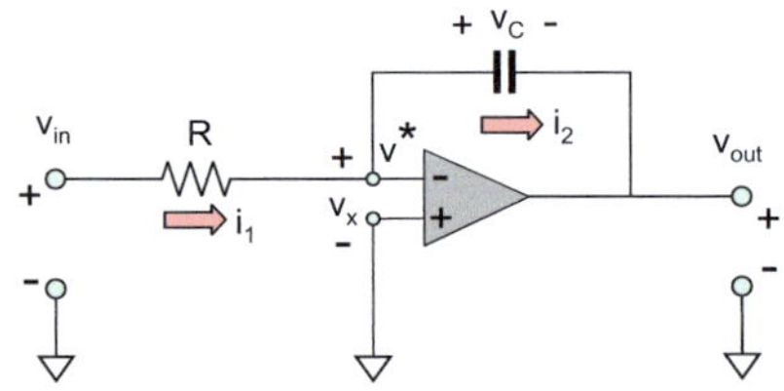

Fig. 6.23. Miller integrator circuit.

The circuit analysis uses two summing-point constraints: no current into the amplifier and zero differential input voltage. Therefore, the node voltage v^* in Fig. 6.23 is also zero. The currents $i_1 = i_R$ and $i_2 = i_C$ in Fig. 6.23 are equal to each other. This yields

$$\overbrace{C\frac{dv_C}{dt}}^{i_C} = C\frac{d(v^* - v_{out})}{dt} = -C\frac{dv_{out}}{dt} = \overbrace{\frac{v_R}{R}}^{i_R} = \frac{v_{in} - v^*}{R} = \frac{v_{in}}{R} \Rightarrow$$

$$-C\frac{dv_{out}}{dt} = \frac{v_{in}}{R} \Rightarrow v_{out} = -\frac{1}{RC}\int_0^t v_{in}(t')dt' - V_C \tag{6.45}$$

where V_C is a constant (the initial voltage across the capacitor at $t = 0$). Thus, an integral of the input voltage (weighted by $-1/(RC)$) is provided at the output. Interestingly, the *time constant* τ of the integrator, $\tau = RC$, has the unit of seconds.

Example 6.9: The analog *pulse counter* is an integrator circuit shown in Fig. 6.23 that counts monopolar voltage pulses simply by integrating the input voltage as time progresses. Assume that the input to the amplifier is the voltage shown in Fig. 6.24, where every rectangular voltage pulse of 10-ms duration corresponds to a car passing through a gate. Given that $R = 10\ \text{k}\Omega$, $C = 0.1\ \mu\text{F}$, and that the initial value of the output voltage is reset to zero, how many cars should pass the gate in order to reach the output voltage threshold of -6 V?

Solution: The time constant τ of the integrator, $\tau = RC$, is equal to 1 ms.

Example 6.9 (cont.):

According to Eq. (6.45),

$$v_{out} = -1000 \ \mathrm{s}^{-1} \sum_{n=1}^{N} (0.06 \ \mathrm{V} \times 0.01 \ \mathrm{s}) = -N \times 0.6 \ \mathrm{V} \tag{6.46}$$

where N is the number of pulses (cars). Equating the above expression to -6 V gives $N = 10$. The time interval between passing cars is not important.

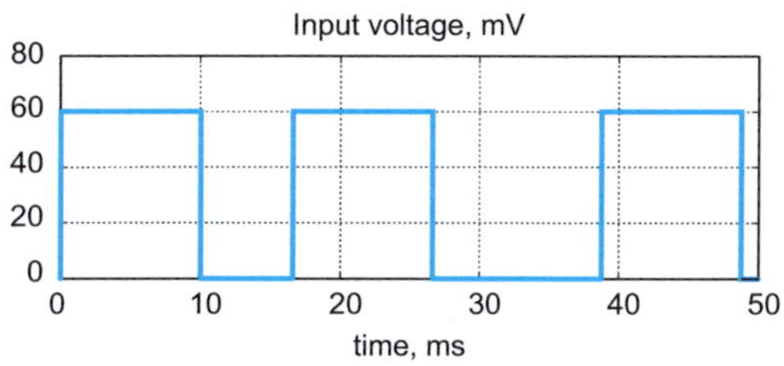

Fig. 6.24. Input signal to the amplifier in Example 6.9.

Along with Example 6.9, other applications of the Miller integrator include various wave-shaping circuits.

6.3.5 Compensated Miller Integrator

The circuit in Fig. 6.23 will not function in the laboratory, when a realistic amplifier chip is used that is different from the ideal-amplifier model. The reason is that the capacitance is equivalent to an open circuit at DC. Therefore, the feedback loop is simply missing in the Miller integrator at DC, and the entire amplifier circuit becomes a comparator with a very high open-loop gain. A small random *input offset voltage*, V_{OS}, which is present for

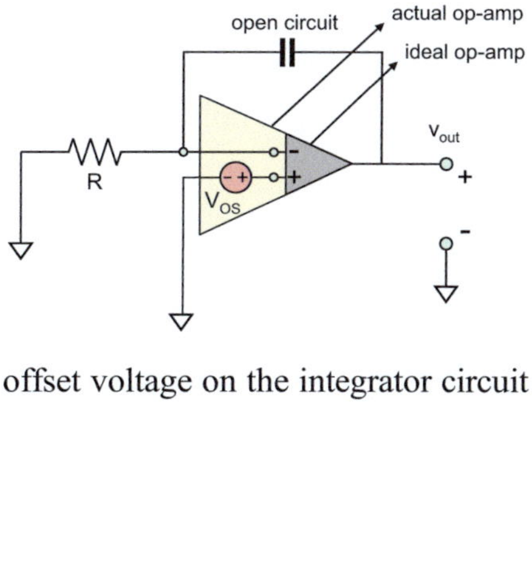

Fig. 6.25. Effect of the input offset voltage on the integrator circuit at DC.

any realistic amplifier IC, will saturate the amplifier toward one of the power rails (depending on the sign of V_{OS}) even if its input is at zero volts (grounded)—see Fig. 6.25.

Note that the input offset voltage source, V_{OS}, may be added to either amplifier terminal. Hence, the voltage across the capacitance will approach the rail voltage and the capacitance itself will become permanently charged. The *dynamics* of this process can be analyzed explicitly, starting with some initial voltage value, say $\upsilon_C = 0\,\text{V}$. In this case, we are allowed to use the negative feedback. For example, given that $R = 10\,\text{k}\Omega$, $C = 0.1\,\mu\text{F}$, $V_{OS} = 5\,\text{mV}$, it takes exactly 1 s to reach the output voltage of 5 V! A similar effect is created by input bias currents to the amplifier. To overcome this issue, a large resistance, R_F, is introduced in parallel with C in order to maintain the negative feedback at DC and discharge the capacitance as needed—see Fig. 6.26. If, for example, $R_F = 10\,\text{M}\Omega$, then the capacitance will discharge over time on the order of $R_F C = 1$ s. This estimate is comparable with the estimate for the charging time. As a result, a balance will be established that results in a certain nonzero υ_{out} with the output of the amplifier grounded. A further quantitative discussion may be carried out.

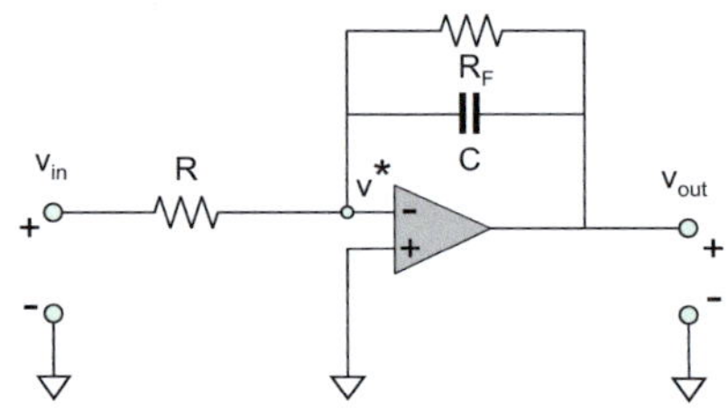

Fig. 6.26. Miller integrator improved with a large resistance, R_F, in the feedback loop.

6.3.6 Differentiator and Other Circuits

When the resistance and the capacitance in Fig. 6.23 are interchanged, a *differentiator amplifier circuit* (or *active differentiator*) is obtained; its output signal is a derivative of the input signal. The corresponding solution is studied in one of the homework problems. The differentiators are rarely used in practice since they attempt to amplify any input noise (they become "noise magnifiers"). The reason for this is an *infinitely high gain* of the amplifier circuit at *high* frequencies, when the capacitance becomes a short circuit. A small resistance added *in series* with the capacitance reduces this effect and assures the finite gain similar to the standard inverting amplifier.

Exercise 6.10: Draw an integrator circuit with an inductance instead of the capacitance.

Answer: The circuit in Fig. 6.23 but with the resistance replaced by an inductance and with the capacitance replaced by the resistance.

Inductances might be used instead of capacitances; the amplifier circuits so constructed will be either a differentiator or an integrator. We again pass the corresponding analysis to the homework exercises. However, the physical inductors tend to have a significant series resistance and are more bulky. Last but not least, we may ask ourselves a question: as long as the amplifier circuits can perform multiplication, addition (or subtraction), and integration (or differentiation), can we now build an *analog computer*, which operates with analog voltages and replaces its digital counterpart at least for simple computational tasks? The answer is yes, we can. In fact, this was done a long time ago, in the mid-1960s.

Summary

<table>
<tr><th colspan="3">Static capacitance and inductance</th></tr>
<tr><th>Property</th><th>Capacitance
</th><th>Inductance
</th></tr>
<tr><td>Definition</td><td>$C = \frac{Q}{V} > 0$
Q—charge of either conductor;
V—volt. between two conductors
Units: $\mathrm{F} = \mathrm{C}/\mathrm{V}$</td><td>$L = \frac{\Phi}{I} > 0$
Φ—magnetic flux through the circuit; I—circuit current
Units: $\mathrm{H} = (\mathrm{V} \cdot \mathrm{s})/\mathrm{A}$</td></tr>
<tr><td>Physical meaning</td><td>Charge on either conductor produced by 1 V voltage difference between the two conductors</td><td>Magnetic flux through the circuit produced by 1 A of current in the same circuit</td></tr>
<tr><td>Stored energy, J (static or dyn.)</td><td>$E = \frac{1}{2} CV^2$ or $E(t) = \frac{1}{2} Cv(t)^2$</td><td>$E = \frac{1}{2} LI^2$ or $E(t) = \frac{1}{2} Li(t)^2$</td></tr>
<tr><td>Series/parallel combinations</td><td>$\frac{1}{C_{\mathrm{eq}}} = \frac{1}{C_1} + \frac{1}{C_2} + \frac{1}{C_3}$ in series
$C_{\mathrm{eq}} = C_1 + C_2 + C_3$ in parallel</td><td>$L_{\mathrm{eq}} = L_1 + L_2 + L_3$ in series
$\frac{1}{L_{\mathrm{eq}}} = \frac{1}{L_1} + \frac{1}{L_2} + \frac{1}{L_3}$ in parallel</td></tr>
<tr><td>Basic models (no fringing fields)</td><td>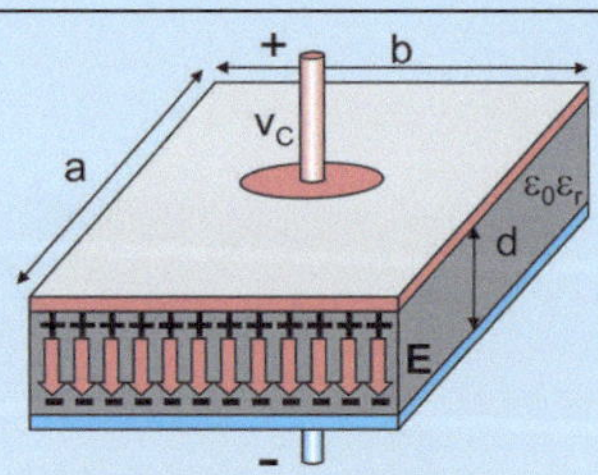

$C = \frac{\varepsilon_0 A}{d}$ or $C = \frac{\varepsilon_r \varepsilon_0 A}{d}$ (diel. material inside)
$A = ab, \quad \varepsilon_0 = 8.854187 \times 10^{-12}\ \mathrm{F/m}$</td><td>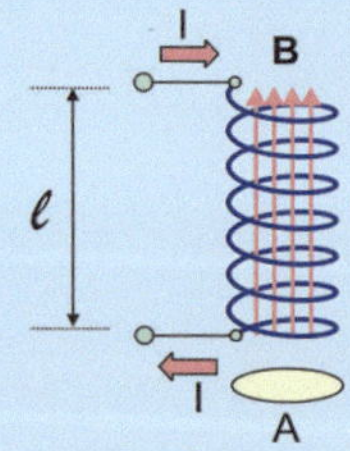

$L = \frac{\mu_0 A N^2}{l}$ or $L = \frac{\mu_0 \mu_r A N^2}{l}$
(closed magnetic core)
$\mu_0 = 4\pi \times 10^{-7}\ \mathrm{H/m}$</td></tr>
<tr><th colspan="3">Models with fringing</th></tr>
<tr><td>Capacitance of a parallel-plate square capacitor (Table 6.1)</td><td colspan="2"><table>
<tr><td>d/a</td><td>0.01</td><td>0.1</td><td>0.2</td><td>0.3</td><td>0.4</td><td>0.5</td></tr>
<tr><td>C_{exact}/C</td><td>1.043</td><td>1.307</td><td>1.554</td><td>1.787</td><td>2.013</td><td>2.237</td></tr>
<tr><td>d/a</td><td>0.6</td><td>0.7</td><td>0.8</td><td>0.9</td><td>1.0</td><td></td></tr>
<tr><td>C_{exact}/C</td><td>2.459</td><td>2.680</td><td>2.901</td><td>3.122</td><td>3.343</td><td></td></tr>
</table></td></tr>
<tr><td>Inductance of a finite-radius solenoid</td><td colspan="2">$L_{\mathrm{exact}} = L\left(1 - \frac{8w}{3\pi} + \frac{w^2}{2} - \frac{w^4}{4}\right), \quad w = \frac{r}{l} < 1$; r is the radius of the coil</td></tr>
<tr><td>Inductance of a solenoid with a finite core</td><td colspan="2">$L \approx \frac{0.5\pi\mu_0 l^* N^2}{\ln\left[\frac{l^*}{r} - 1\right]}\left(1 - \frac{l}{2l^*}\right)$, $\mu_r \geq 100$, see Fig. 6.12c</td></tr>
</table>

(continued)

Dynamic behavior		
Property	**Capacitance**	**Inductance**
Dynamic model-Passive ref. conf.	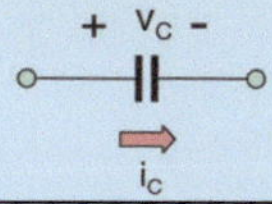	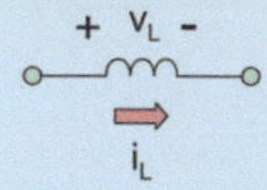
Dynamic model-υ–i characteristic	$i_C = C\frac{d\upsilon_C}{dt}$	$\upsilon_L = L\frac{di_L}{dt}$
Dynamic model-charge and flux	$\frac{dq_C}{dt} = i_C$	$\frac{d\Phi(t)}{dt} = \upsilon_L$ (passive ref. conf.)
Behavior at DC	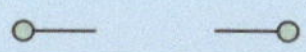	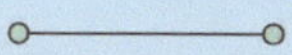
Behavior at very high frequencies		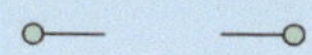
Bypassing/Decoupling		
Bypass capacitor	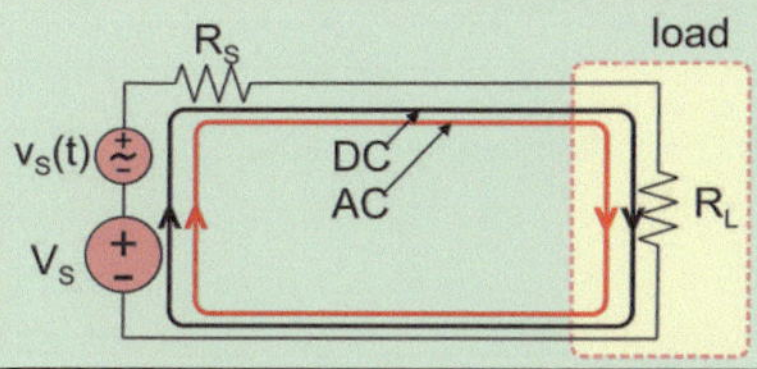	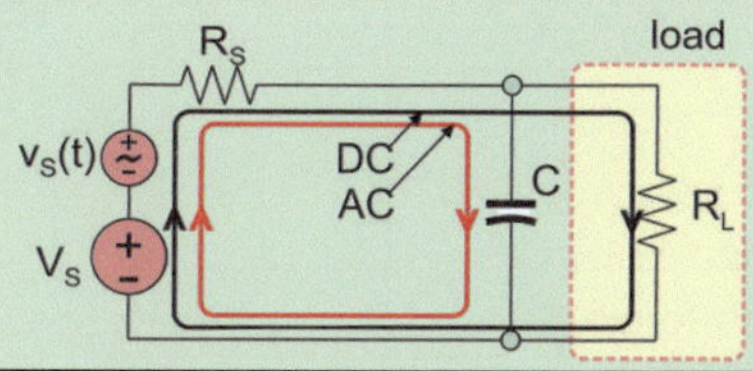
Blocking capacitor	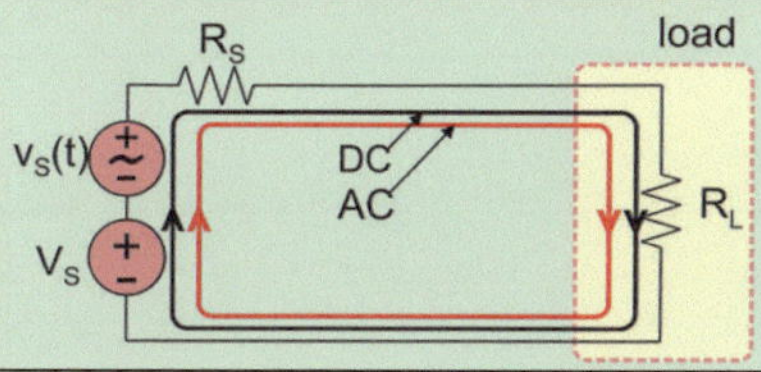	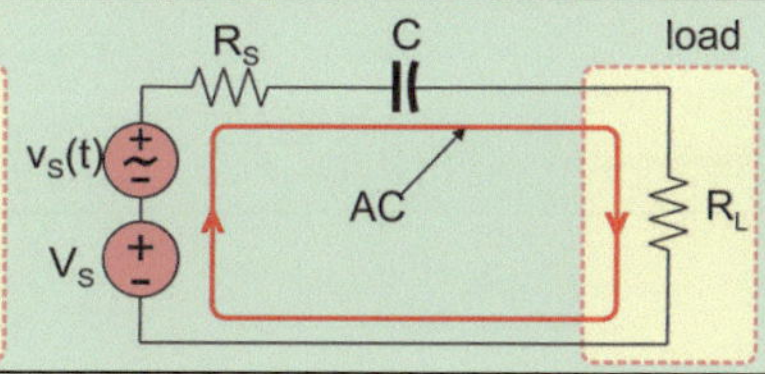
Decoupling inductor	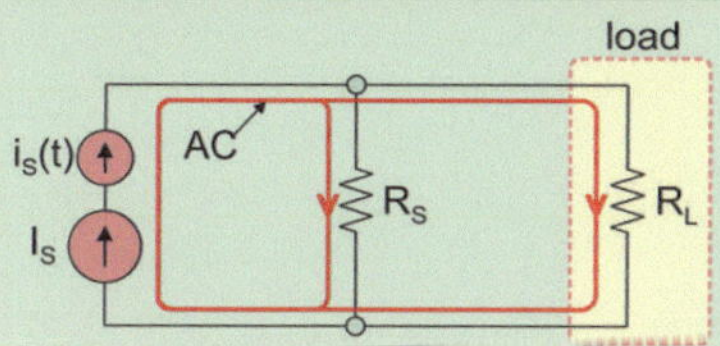	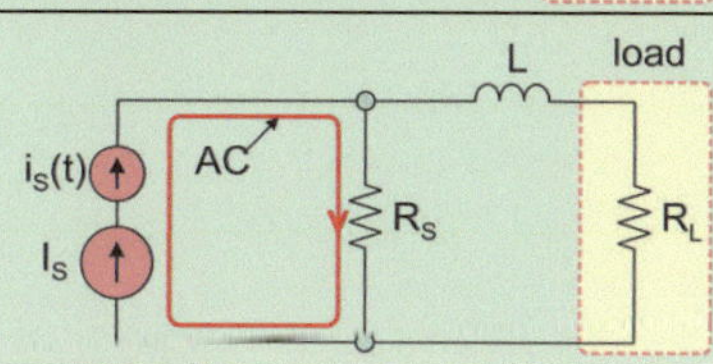
Amplifier circuits with capacitor/inductor		
Property	**Capacitance**	**Inductance**
Miller Integrator (open-loop amplifier at DC)	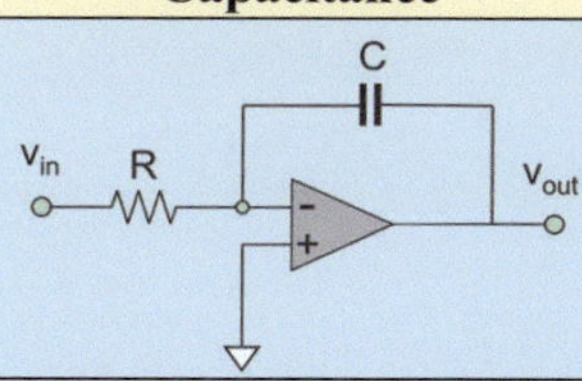	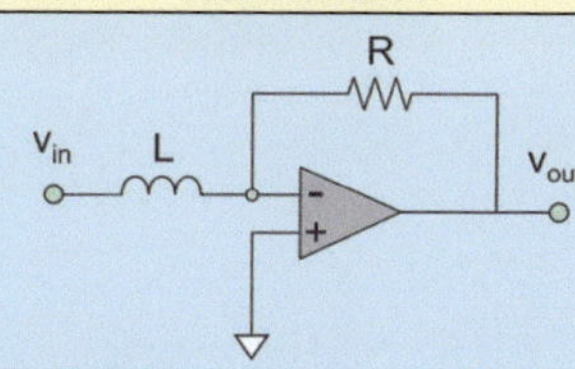
Time-domain operation	$\frac{d\upsilon_{out}}{dt} = -\frac{\upsilon_{in}}{\tau},\ \tau = RC$	$\frac{d\upsilon_{out}}{dt} = -\frac{\upsilon_{in}}{\tau},\ \tau = \frac{L}{R}$

(continued)

Compensated integrator with a finite gain at DC	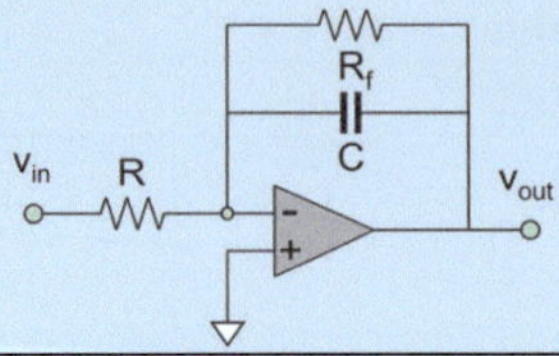	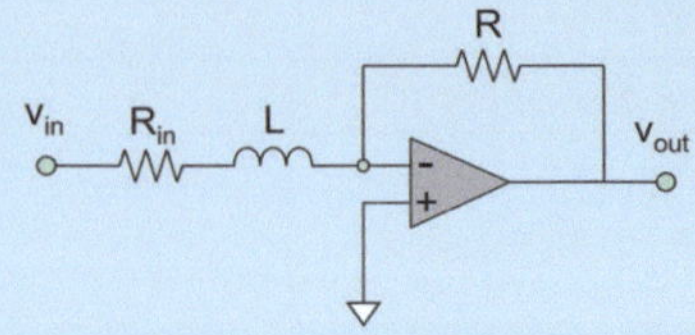
Time-domain operation	$\frac{dv_{\text{out}}}{dt}+\frac{v_{\text{out}}}{R_f C}=-\frac{v_{\text{in}}}{\tau},\quad \tau=RC$	$\frac{dv_{\text{out}}}{dt}+\frac{R_{\text{in}}}{L}v_{\text{out}}=-\frac{v_{\text{in}}}{\tau},\quad \tau=\frac{L}{R}$
Differentiator circuit (infinite gain at very high frequencies)	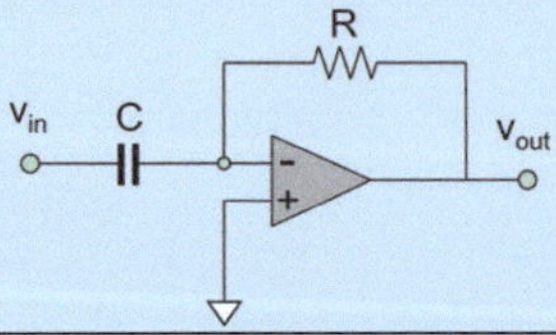	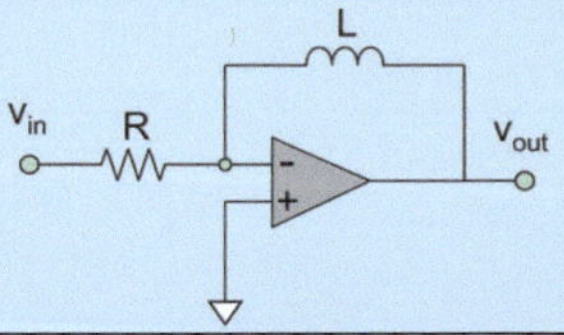
Time-domain operation	$v_{\text{out}}=-\tau\frac{dv_{\text{in}}}{dt},\quad \tau=RC$	$v_{\text{out}}=-\tau\frac{dv_{\text{in}}}{dt},\quad \tau=\frac{L}{R}$
Differentiator with a finite gain at very high frequencies	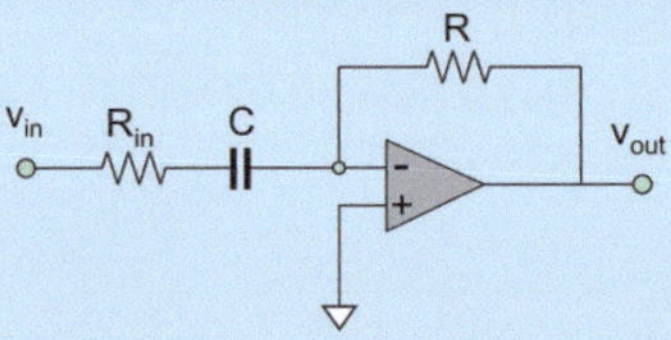	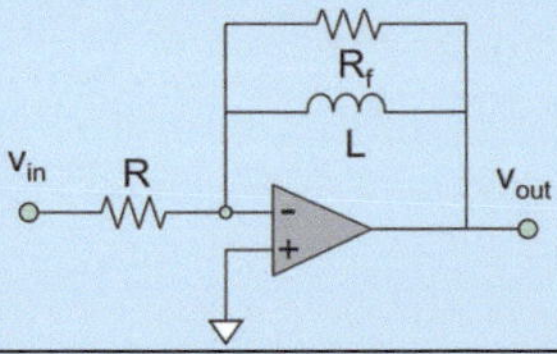
Time-domain operation	$R_{\text{in}}C\frac{dv_{\text{out}}}{dt}+v_{\text{out}}=-\tau\frac{dv_{\text{in}}}{dt},\tau=RC$	$\frac{L}{R_f}\frac{dv_{\text{out}}}{dt}+v_{\text{out}}=-\tau\frac{dv_{\text{in}}}{dt},\tau=L/R$

Problems

6.1 Static Capacitance and Inductance

6.1.1 Capacitance, Self-capacitance, and Capacitance to Ground

6.1.2 Application Example: ESD

Problem 6.1.

A. Describe in your own words the physical meaning of capacitance.
B. Suggest a way to memorize the expression for the capacitance of two conductors.
C. What is approximately the self-capacitance of a human body?
D. What is approximately the capacitance of a human body (to ground)?
E. How does the human-body self-capacitance change in embryo pose (yoga)?

Problem 6.2. A metal square plate with the side of 10 mm has the self-capacitance $C_{\mathrm{self}} = 0.41$ pF.

A. Estimate the capacitance of a capacitor formed by two such parallel plates separated by 30 mm. Compare this value to the exact result of 0.23 pF.
B. Estimate the capacitance of the plate to ground when the separation distance is 30 mm. Compare this value to the exact result of 0.43 pF.

Problem 6.3. Draw the basic electric-circuit model of a human body and specify the generic element values.

6.1.3 Parallel-Plate Capacitor

6.1.4 Capacitances in Parallel and in Series

Problem 6.4. For the parallel-plate capacitor schematically shown in the figure,

$$a = 10 \text{ cm}, \ b = 20 \text{ cm},$$
$$h = 1 \text{ mm}, \ \varepsilon_r = 12$$

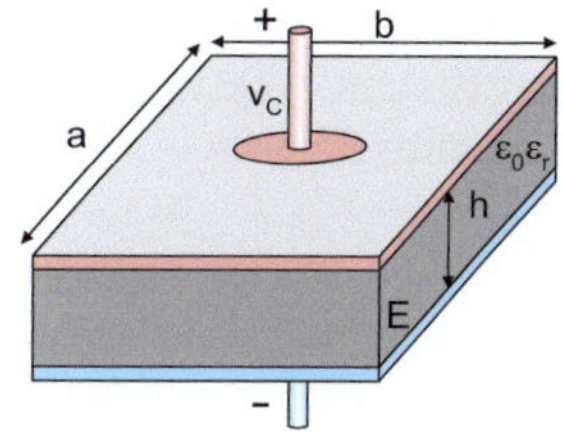

A. Determine capacitance of the capacitor, C.
B. Determine the electric field strength, E (in V/m), within the capacitor volume and total charge, Q, on either capacitor plate if the applied voltage is 25 V.
C. Determine the electric field energy stored in the capacitor if the applied voltage is 25 V.

Problem 6.5. Solve the previous problem when the separation distance, h, between the plates is reduced to 100 μm.

Problem 6.6. For the enclosed-cylinder capacitor shown in the figure, $a = 10\,\text{cm}$, $b = 9.99$ cm, $H = 5\,\text{cm}$, $\varepsilon_r = 16$, the electrodes are the inner and outer cylinder surfaces, respectively.

A. Determine the capacitance of the capacitor, C.
B. Determine the electric field strength, E (in V/m), within the capacitor volume and total charge, Q, on either capacitor plate if the applied voltage is 50 V.
C. Determine the electric field energy stored in the capacitor if the applied voltage is 25 V.

Hint: The capacitance per unit area of the device is that of the parallel-plate capacitor.

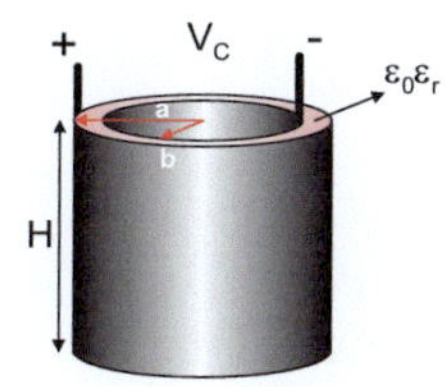

Problem 6.7. Solve the previous problem when the separation distance between the two electrodes is reduced to 25 μm.

Problem 6.8. A cross section of the rolled capacitor is approximated by a rational spiral shown in the figure. Here, $a = 0.25$ cm, $\varepsilon_r = 10$, the separation distance between the two conductors is 20 μm, and the height of the entire roll is 1.5 cm.

A. Determine capacitance of the capacitor, C.
B. Determine the electric field strength, E (in V/m), within the capacitor and total charge, Q, on either capacitor plate if the applied voltage is 12 V.
C. Determine the electric field energy stored in the capacitor if the applied voltage is 12 V.

Hint: The capacitance per unit area of the device is that of the parallel-plate capacitor.

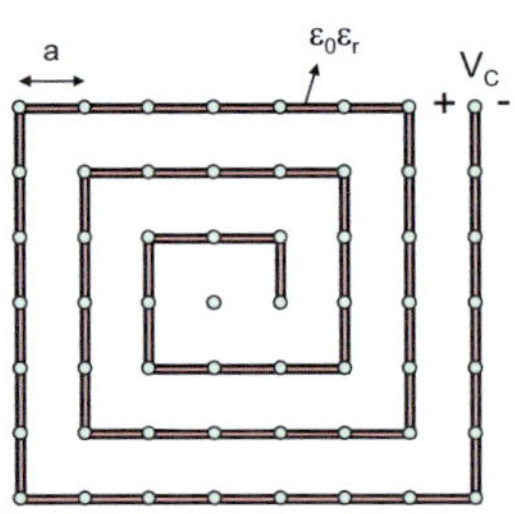

Problem 6.9. Determine the capacitance of the three leaded capacitors shown in the figure (from left to right).

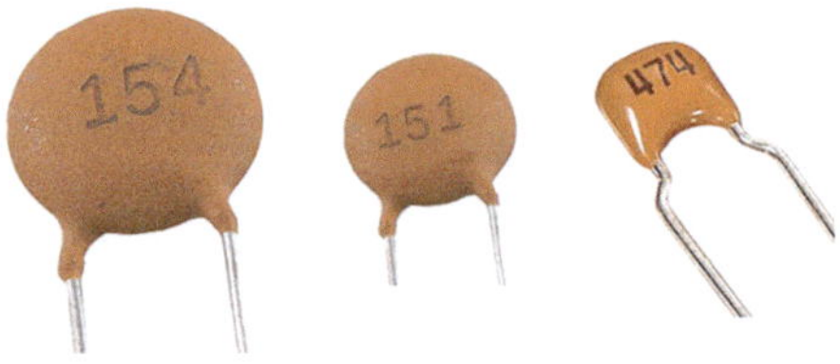

Problem 6.10. Find the equivalent capacitance for the circuit shown in the following figure.

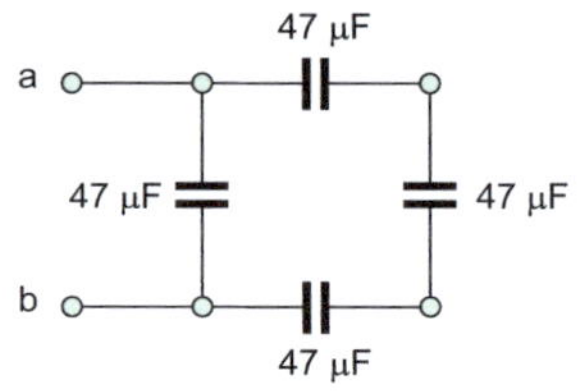

Problem 6.11. Find the equivalent capacitance for each circuit shown in the figure below.

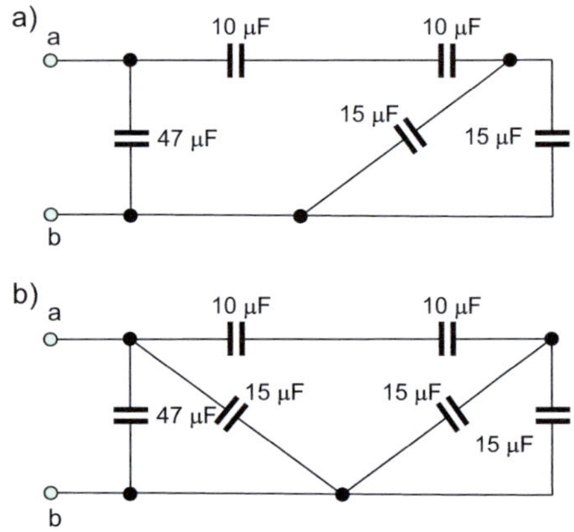

Problem 6.12. Find the equivalent capacitance for each circuit shown in the following figure.

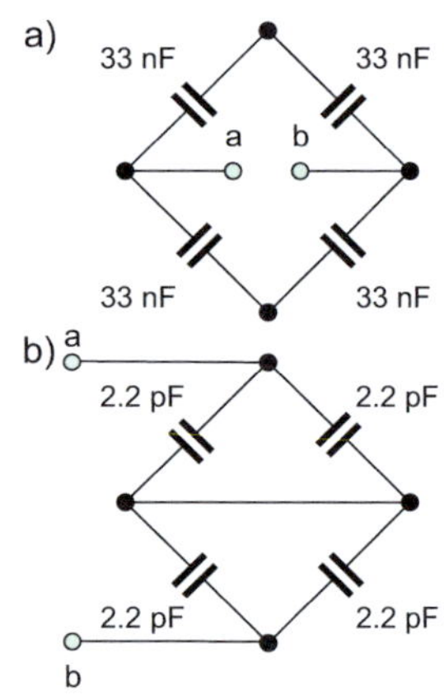

Problem 6.13. Two 1-μF capacitances have an initial voltage of 10 V and 0 V, respectively (before the switch is closed), as shown in the figure. Find the total electric energy stored in the system before the switch is closed. Find the

voltage across each capacitance and the total stored energy after the switch is closed. What could have happened to the missing energy?

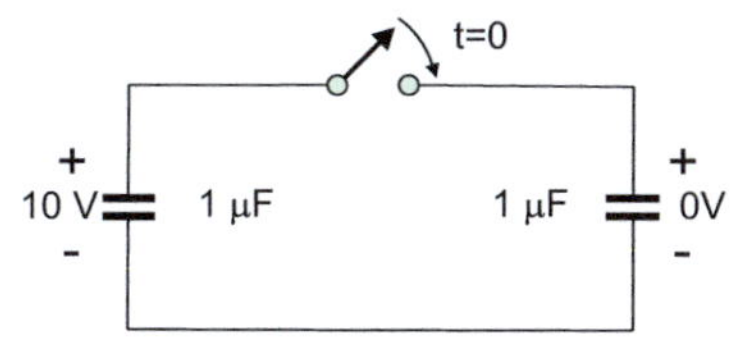

6.1.7 Self-inductance (Inductance) and Mutual Inductance

6.1.8 Inductance of a Solenoid With and Without Magnetic Core

6.1.9 Circuit Symbol. Inductances in Series and in Parallel

Problem 6.14.

A. Describe in your own words the physical meaning of inductance.

B. Do you think a straight wire has a certain inductance per unit length? You might want to ask the TA and/or browse the Web and present the corresponding expression (if any).

Problem 6.15. Three air-core inductors of the same cross section are shown in the following figure. The inductor length is proportional to the number of turns. Find the ratios of inductances: $L_2/L_1, L_3/L_2, L_3/L_1$.

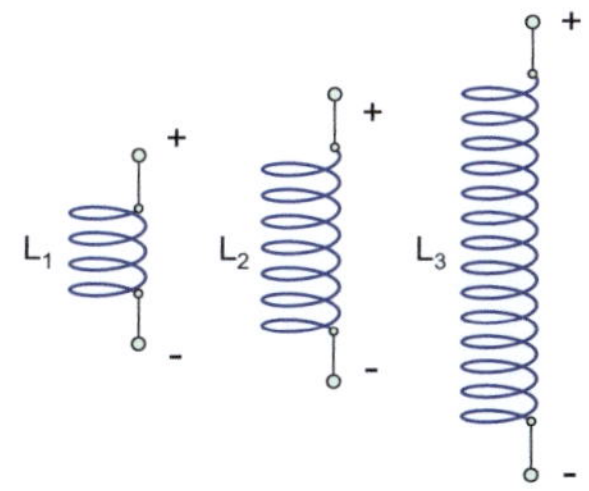

Problem 6.16. The solenoid shown in the figure has a diameter $d = 1$ cm and a length $l = 10$ cm.

A. Find the solenoid's inductance, L, using the common assumption $d/l << 1$.

B. Determine the magnetic field energy stored within the inductance and in the surrounding space if the applied current, I_L, is 0.5 A.

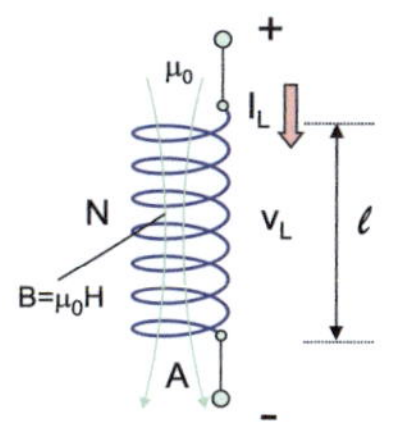

Problem 6.17.

A. Obtain more accurate answers to the previous problem using the precise expression for the inductance of a solenoid, which is Eq. (6.22).

B. Estimate the relative error of the common expression for solenoid's inductance, Eq. (6.21).

Problem 6.18.

A. Design a 100-μH air-core inductor (determine the necessary number of turns) with a radius of 1 cm and a length of 10 cm.

B. Determine the magnetic field energy stored in the inductor if the applied current, I_L, is 1.0 A.

Problem 6.19.

A. Design a 0.5-mH air-core inductor (determine the necessary number of turns) having a radius of 0.5 cm and a length of 10 cm.

B. Determine the magnetic field energy stored in the inductor if the applied current, I_L, is 1.0 A.

Problem 6.20. The solenoid shown in the figure has diameter, $d = 1$ cm, length, $l = 10$ cm, and a closed magnetic core of the same circular cross section with $\mu_r = 1000$.

A. Find the solenoid's inductance, L, using the common assumption $d/l << 1$.

B. Determine the magnetic field energy stored in the inductor if the applied current, I_L, is 0.5 A.

C. Compare two of your answers to the answers in Problem 6.16.

D. Determine voltage across the inductor assuming the ideal (zero-resistance) wire.

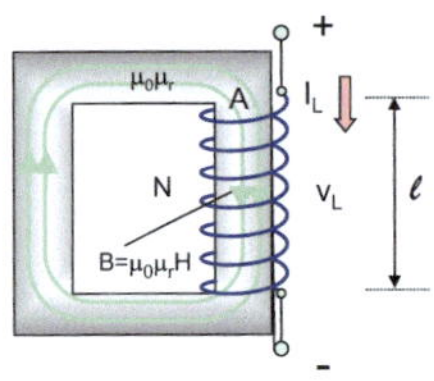

Problem 6.21.

A. Obtain a more accurate solution to the previous problem using the precise expression for the inductance of a solenoid, Eq. (6.22). To do so, propose your own modification of this expression to include the effect of the magnetic core.

B. Estimate the relative error of the simplified expression for the inductance, Eq. (6.21).

Problem 6.22. Find the equivalent inductance for the circuit shown in the following figure.

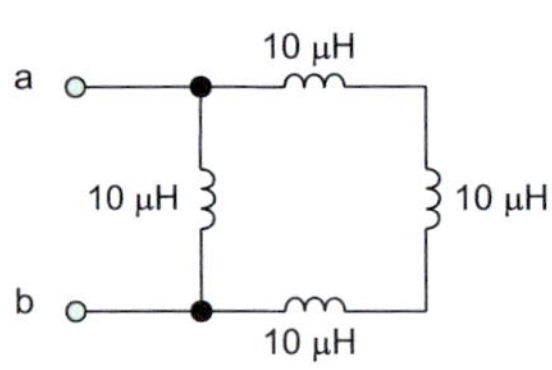

Problem 6.23. Find the equivalent inductance for circuits shown in the figure that follows.

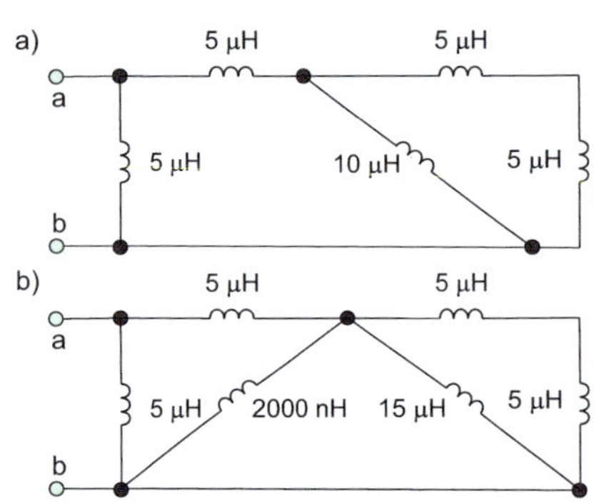

Problem 6.24. Find the equivalent inductance for circuits shown in the figure.

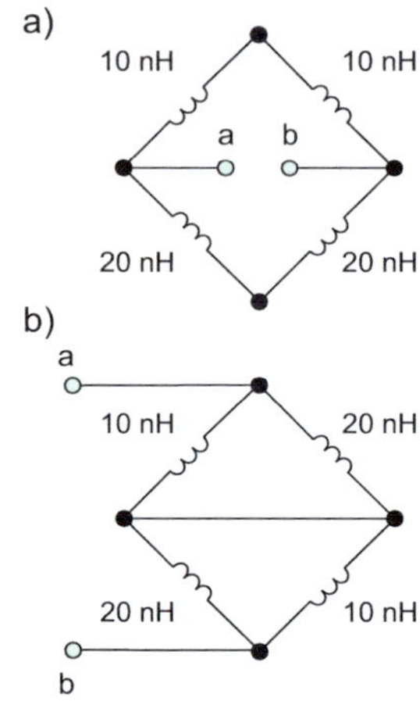

Problem 6.25. For the inductor shown in Fig. 6.11, $l^* = 2l = 10$ cm and $r = 0.75$ cm. How do the inductances with the magnetic core (with $\mu_r \geq 100$) and without the core compare to each other?

Problem 6.26. Repeat the previous problem with $l^* = 15$ cm, but still $2l = 10$cm. The other parameters are the same.

6.2 Dynamic Behavior of Capacitance and Inductance

6.2.2 Dynamic Behavior of Capacitance

6.2.3 Dynamic Behavior of Inductance

6.2.4 Instantaneous Energy and Power of Dynamic Circuit Elements

Problem 6.27.

A. Sketch the circuit symbol for the capacitance and the inductance; label the corresponding voltages and currents.

B. Write two dynamic equations for the capacitance and the inductance which relate the voltages and currents.

C. Express the units for the inductance and capacitance through volts and amperes.

D. Which capacitance and inductance values are typical in electronic circuits?

Problem 6.28.

A. Sketch the hydrodynamic analogies for the capacitance and the inductance.
B. Which mechanical quantities are associated with the capacitance and inductance?

Problem 6.29. The voltage across a 10-μF capacitance is shown in the following figure. At $t = 0$, the voltage is zero. Sketch the current through the capacitance to scale versus time.

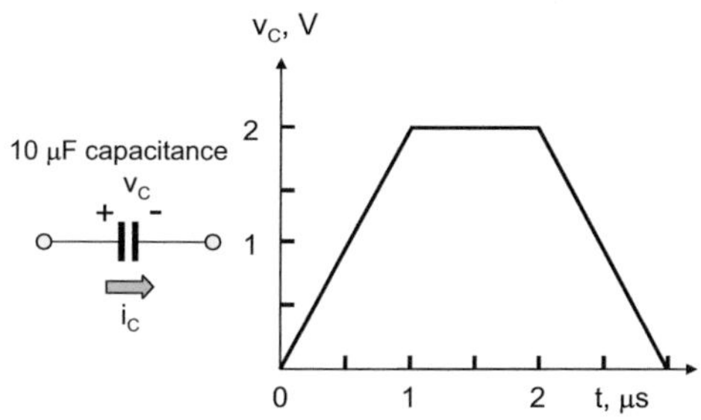

Problem 6.30. Repeat the previous problem for the voltage shown in the following figure.

A. How is the solution different from the previous problem?
B. For creating large currents, should the voltage across the capacitance change slowly or quickly?

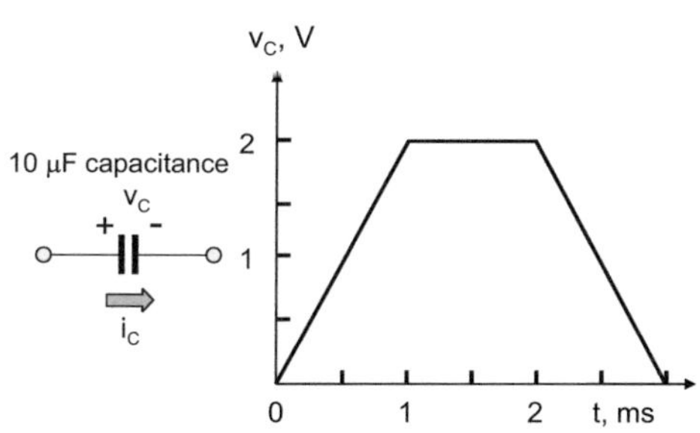

Problem 6.31. A 10-μF capacitance is charged by the current $i_C = 1\,\text{mA} \cdot \cos^2(1000t)$. At $t = 0$, the capacitance voltage is zero.

A. Using software of your choice (MATLAB is recommended), sketch the capacitance voltage to scale versus time over the interval from 0 to 0.05sec.
B. How much time is approximately necessary to charge the capacitance to 1.5 V?
C. Solve the same problem analytically.

Problem 6.32. Repeat the previous problem with the capacitance changed to 1 μF. Does the 10-V charging time increase or decrease?

Problem 6.33. The current through a 5-mH inductance is shown in the following figure. At $t = 0$, the current is zero. Sketch the voltage across the inductance to scale versus time.

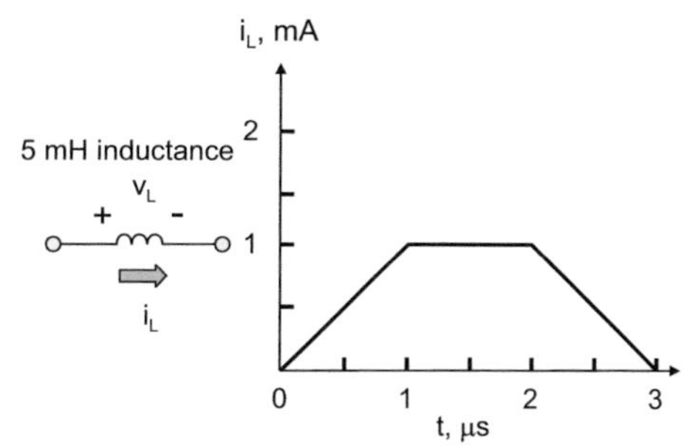

Problem 6.34. Repeat the previous problem for the current shown in the figure.

A. How is the solution different from the previous problem?
B. For creating large voltages, should the current through the inductance change slowly or quickly?

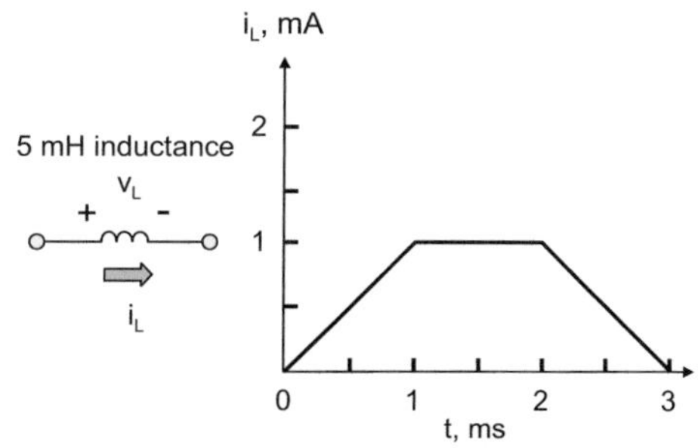

Problem 6.35. The voltage across a 3-mH inductance is given by $v_L(t) = 10\,\text{mV} \cdot \cos^2(1000t)$. The current through the

inductance is equal to zero at the initial time $t = 0$. Using software of your choice (MATLAB is recommended), plot the current through the inductance for $t \leq 50\,\text{ms}$.

6.2.5 DC Steady State

Problem 6.36.

A. Draw the equivalent circuit for the inductance shown in the figure at DC.

B. Repeat for the capacitance.

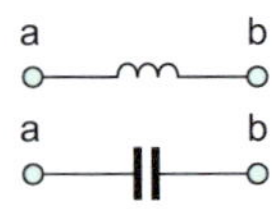

Problem 6.37. Find current I in the circuit shown in the figure that follows at $t \to \infty$, in the DC steady state. The switch closes at $t = 0$.

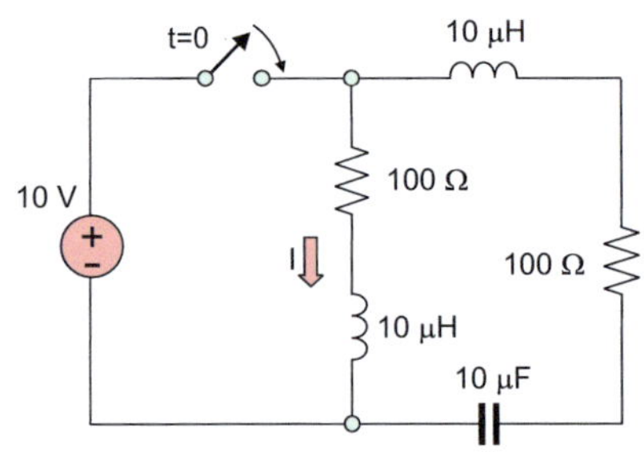

Problem 6.38. Find current I in the circuit shown in the figure at $t \to \infty$, in the DC steady state. The switch closes at $t = 0$.

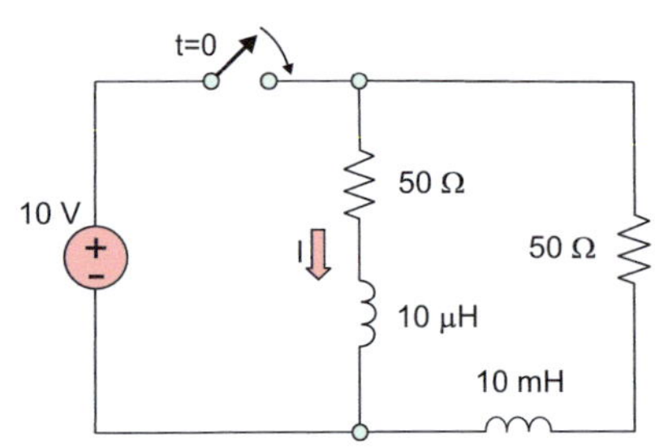

Problem 6.39. Find current I in the circuit shown in the following figure at $t \to \infty$, in the DC steady state. The switch closes at $t = 0$.

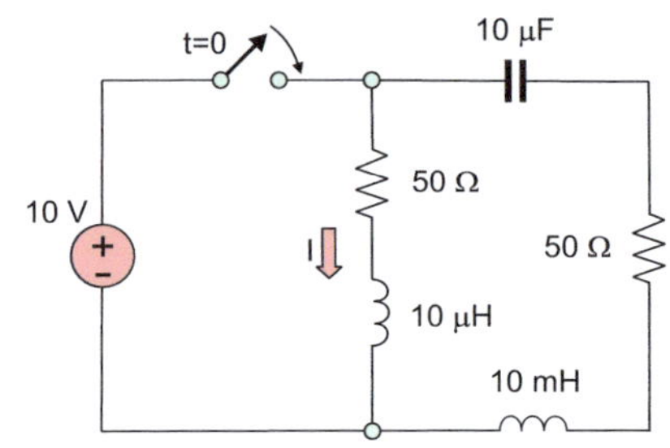

Problem 6.40.

A. Find current I in the circuit shown in the figure at $t \to \infty$, in the DC steady state. The switch closes at $t = 0$.

B. Find voltage across the capacitance in the DC steady state and label its polarity in the figure.

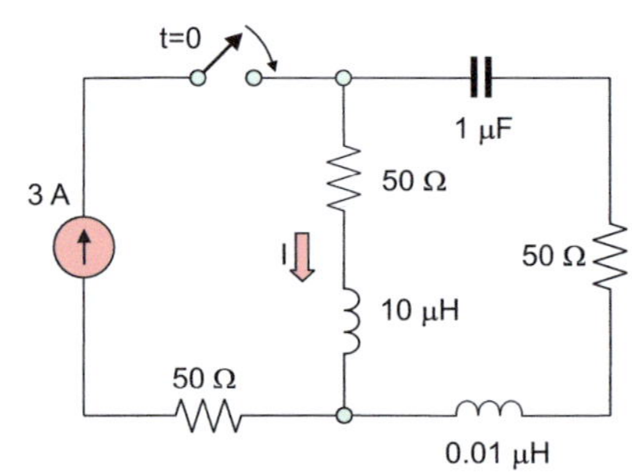

Problem 6.41. Find the voltage across the capacitance in the DC steady state and label its polarity in the figure at $t \to \infty$, in the DC steady state. The switch opens at $t = 0$.

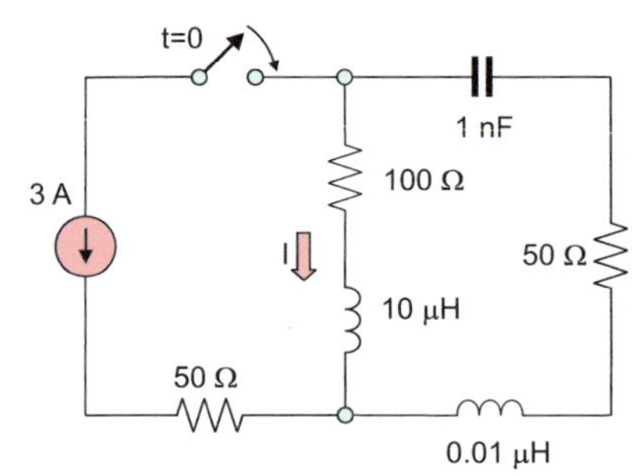

Problem 6.42. Determine the equivalent resistance between terminals a and b for the circuit shown in the following figure in the DC steady state.

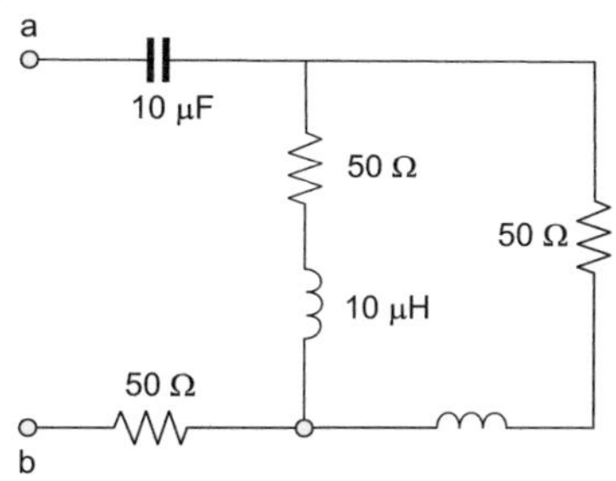

Problem 6.43. Determine the equivalent resistance between terminals a and b for the circuit shown in the following figure in the DC steady state.

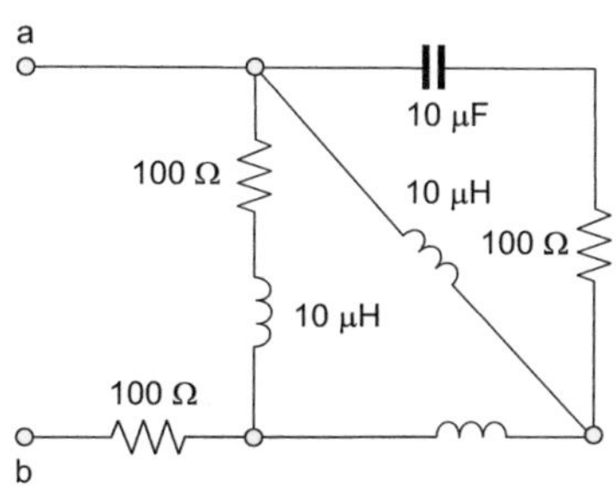

6.3 Application Circuits Highlighting Dynamic Behavior

6.3.1 Bypass Capacitor

6.3.2 Blocking Capacitor

6.3.3 Decoupling Inductor

Problem 6.44. Describe the purpose of a

A. Bypass capacitor
B. Blocking capacitor
C. Decoupling inductor

in your own words. Specify the placement of each component: in series or in parallel with the source.

6.3.4 Amplifier Circuits with Dynamic Elements: Miller Integrator

6.3.5 Compensated Miller Integrator

6.3.6 Differentiator and Other Circuits

Problem 6.45. The input voltage to the Miller integrator circuit with the ideal amplifier shown in the figure is a series of rectangular voltage pulses. Each is 50 mV tall and 8 ms wide. Given that the initial value of the output voltage is zero, how many voltage pulses are necessary to reach the negative output voltage threshold of −8 V?

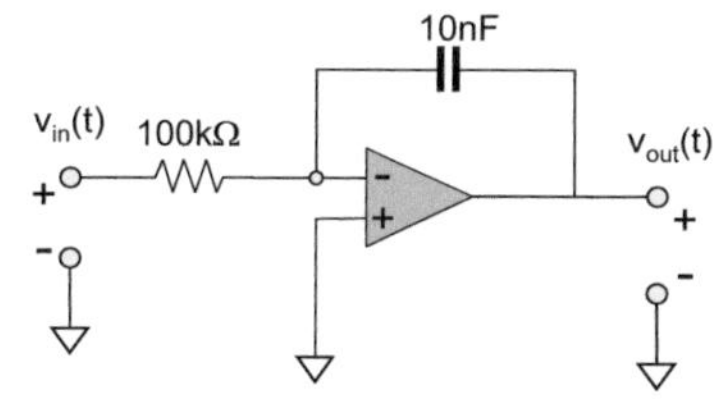

Problem 6.46. The input voltage to the Miller integrator circuit with the ideal amplifier shown in the figure is a series of rectangular voltage pulses. Each is 50 mV tall and 16 ms wide. Given that the initial value of the output voltage is zero, how many voltage pulses are necessary to reach the negative output voltage threshold of −9.2 V?

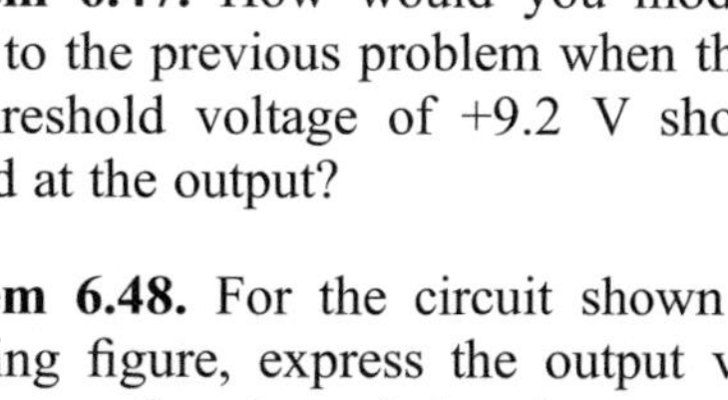

Problem 6.47. How would you modify the circuit to the previous problem when the positive threshold voltage of +9.2 V should be reached at the output?

Problem 6.48. For the circuit shown in the following figure, express the output voltage, $v_{\text{out}}(t)$, as a function of time in terms of the input voltage, $v_{\text{in}}(t)$, and circuit parameters R, C. Assume ideal amplifiers.

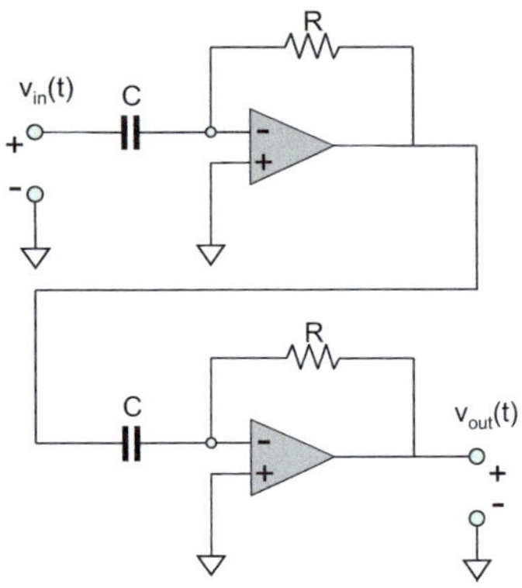

Problem 6.49. For the circuit shown in the following figure, express the output voltage, $v_{out}(t)$, as a function of time in terms of the input voltage, $v_{in}(t)$, and circuit parameters R, C. Assume the ideal amplifier.

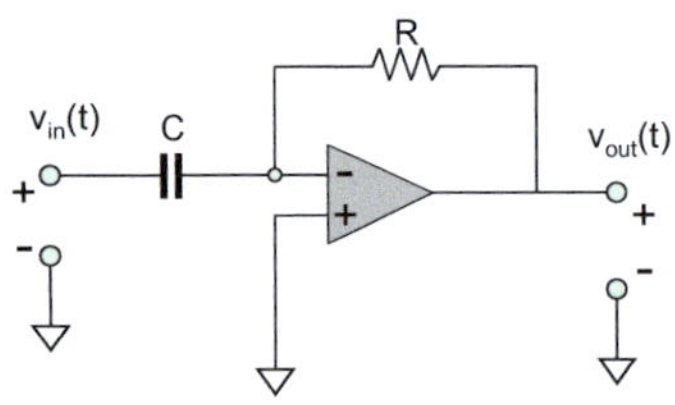

Problem 6.50. For the circuit shown in the following figure, express the output voltage, $v_{out}(t)$, as a function of time in terms of the input voltage, $v_{in}(t)$, and circuit parameters R, L. Assume ideal amplifier.

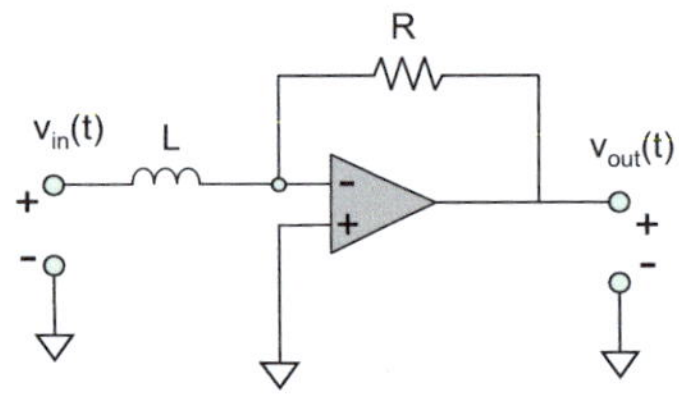

Problem 6.51. For the circuit shown in the following figure, express the output voltage, $v_{out}(t)$, as a function of time in terms of the input voltage, $v_{in}(t)$, and circuit parameters R, L. Assume ideal amplifier.

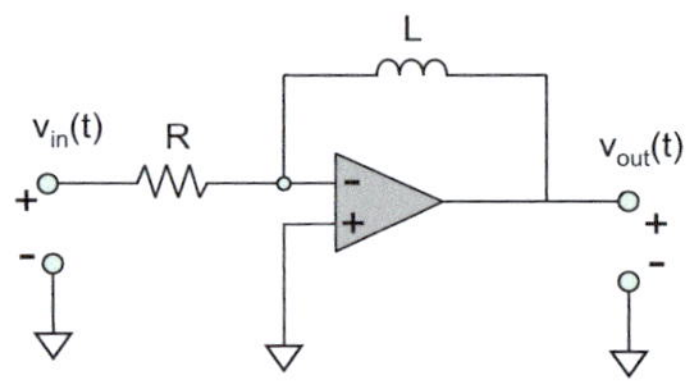

Problem 6.52. Explain why is the Miller integrator typically used with a shunt resistance, R_F.

Problem 6.53. For the circuit shown in the figure, assume that the voltage across the capacitance at $t = 0$ is zero. Also assume that the negative feedback is present. Derive the dynamic expression for $v_C(t)$ at any positive time instance.

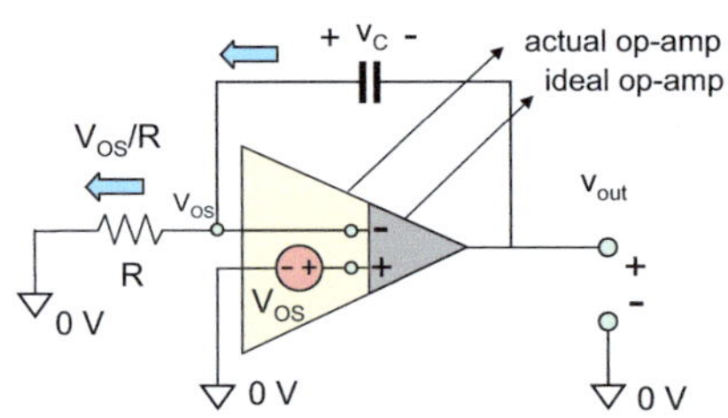

Problem 6.54. Establish the gain of amplifier circuits shown in the figure that follows.

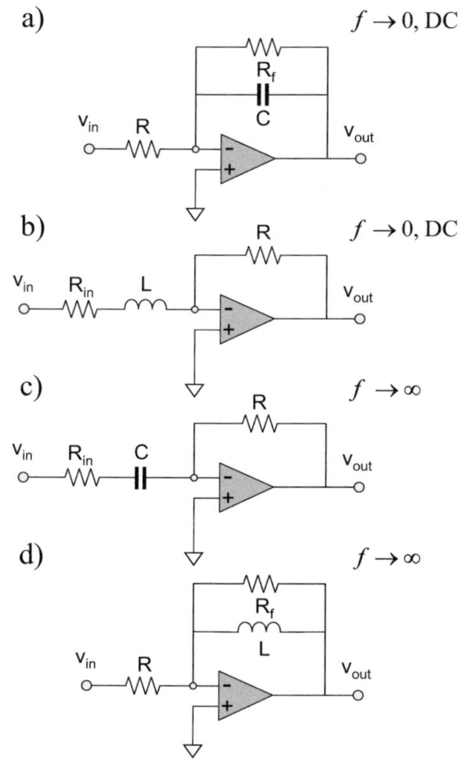

Problem 6.55. For two circuits shown in the figure that follows, obtain an analytical expression for the output voltage as a function of time and circuit parameters when the input voltage has the form $v_{\text{in}}(t) = 1\exp(-\alpha t)$ [mV].

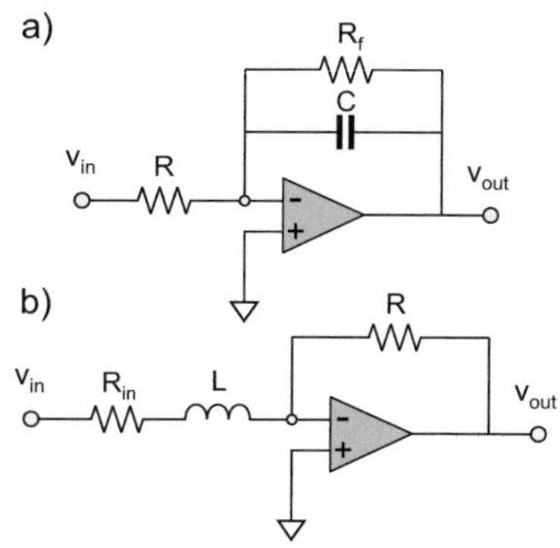

Chapter 7: Transient Circuit Fundamentals

Overview

Prerequisites:

- Knowledge of first-order ordinary differential equations (calculus)
- Knowledge of Thévenin/Norton equivalent circuits (Chapter 4)
- Knowledge of constitutive relations for dynamic circuit elements (Chapter 6)
- Knowledge of basic amplifier theory (Chapter 5)

Objectives of Section 7.1:

- Demonstrate the universal character of the KVL/KCL as applied to any electric circuit including transient circuits
- Establish the general character of the time constant $\tau = RC$ for RC circuits
- Establish the continuity of the capacitor voltage and its role in circuit ODEs
- Solve *any* first-order transient RC circuit configuration and understand the practical meaning of the RC circuit using different application examples

Objectives of Section 7.2:

- Demonstrate the universal character of the KVL/KCL as applicable to any electric circuit including transient circuits
- Establish the general character of the time constant $\tau = L/R$ for RL circuits
- Establish the continuity of the inductor current and its role in circuit ODEs
- Solve *any* first-order transient RL circuit configuration and understand the practical meaning of the RL circuit using an application example

Objectives of Section 7.3:

- Obtain initial exposure to a bistable amplifier circuit with positive feedback
- Understand the principle of operation of a relaxation oscillator—RC timer—on the base of the bistable amplifier circuit
- Establish oscillation frequency and voltage amplitudes from the relaxation oscillator; demonstrate the corresponding laboratory setup
- Briefly discuss the 555 timer IC

S. N. Makarov et al., *Practical Electrical Engineering*, https://doi.org/10.1007/978-3-319-96692-2_7

Objectives of Section 7.4:
- Define the single-time-constant (STC) transient circuit
- Be able to classify any transient circuit with dynamic elements of the same type as either a single-time-constant circuit or a more complicated circuit
- Solve an example of a non-STC circuit
- Convert an arbitrary transient circuit with one capacitance or one inductance to the basic RC/RL first-order circuit
- Solve a first-order transient circuit with a harmonic forcing function

Objectives of Section 7.5:
- Understand topology and classification for the second-order transient circuits
- Convert a transient circuit with a series/parallel LC block to the standard second-order RLC series/parallel transient circuits
- Introduce two major RLC circuit parameters: damping coefficient and undamped resonant frequency
- Introduce the step response of a second-order transient circuit as a solution with a DC source and a switch. Understand the general value of the step response
- Properly select the independent function (capacitor voltage or inductor current) for the standard form of the step response with zero initial conditions

Objectives of Section 7.6:
- Use the method of characteristic equation for second-order transient circuits
- Understand the meaning of overdamped, critically damped, and underdamped circuits
- Use the value of damping ratio ς to distinguish between three different cases of circuit behavior
- Obtain the complete analytical solution for the step response of the RLC circuit
- Apply this solution for modeling a nonideal (realistic) digital waveform

Application Examples:
- Electromagnetic railgun
- Electromagnetic material processing
- Digital memory cell
- Laboratory ignition system
- RC timer or clock circuit in laboratory
- Transient circuit with a bypass capacitor
- Modeling and origin of the nonideal digital waveform

Keywords:
Transient RC circuit, Transient RL circuit, Energy-release RC/RL circuit, Energy-accumulating RC/RL circuit, Time constant of RC circuit, Time constant of RL circuit, Relaxation time, Voltage continuity across the capacitor, Fluid mechanics analogy of transient RC circuit, Lorentz force, Self-induced Lorentz force, Railgun, Electromagnetic material processing, Electromagnetic forming, Current continuity through the inductor, Fluid mechanics analogy of transient RL circuit, Forced response, Electronic ignition system, Piezoelectric effect, Clock frequency, Clock signal, Positive feedback, Linear oscillators, Switching oscillators, Switching RC oscillator, Astable multivibrator, Relaxation oscillator, Bistable amplifier circuit (operation, threshold voltage, mechanical analogy, triggering, trigger signal), Digital memory element, Inverting Schmitt trigger, Non-inverting Schmitt trigger, 555 timer IC, Single-time-constant circuits (definition, classification of, examples of, with general sources), STC circuits, Non-STC circuits (definition, examples of), Series RLC circuit (generic representation, qualitative description, mechanical analogy, step response, duality), Parallel RLC circuit (generic representation, qualitative description, mechanical analogy, step response, duality), Second-order ODE (homogeneous, nonhomogeneous, initial conditions, in terms of current, in terms of voltage, forcing function, general solution, forced response, particular solution, complementary solution natural response, step response), Damping coefficient, Neper, Time constant of the decay envelope, Undamped resonant frequency, Step response, Impulse response, Damping ratio, Natural frequency, Overdamped circuit, Critically damped circuit, Underdamped circuit, Overshoot, Undershoot, Rise time, Fall time, Ringing, Nonideal digital waveform

Section 7.1 RC Circuits

The first-order RC circuits explored in this section involve the process of discharging or charging a capacitor. This is a time-dependent, or transient, circuit behavior, and to understand it, we are required to solve dynamic circuit equations. Mathematically, this implies the solution of first-order ordinary differential equations (ODEs) with time as one independent variable. Fortunately, KVL and KCL remain valid for any static or dynamic circuit. These laws can be employed to derive the circuit equations. After that, it is either solved analytically for simple circuits or numerically for realistic RC circuits.

7.1.1 Energy-Release Capacitor Circuit

The circuit in Fig. 7.1 depicts a capacitor, *C,* that has been charged to a certain voltage

$$V_0 = v_C(t \leq 0) \tag{7.1}$$

prior to use. Through a switch, the capacitor is connected to a load, represented by a resistor $R = 10\ \Omega$. The switch shown in Fig. 7.1 may be a transistor switch. We assume that the switch closes and thereby connects the load to the capacitor at $t = 0$. Our goal is to find all circuit parameters, plus the power delivered to the load as functions of time.

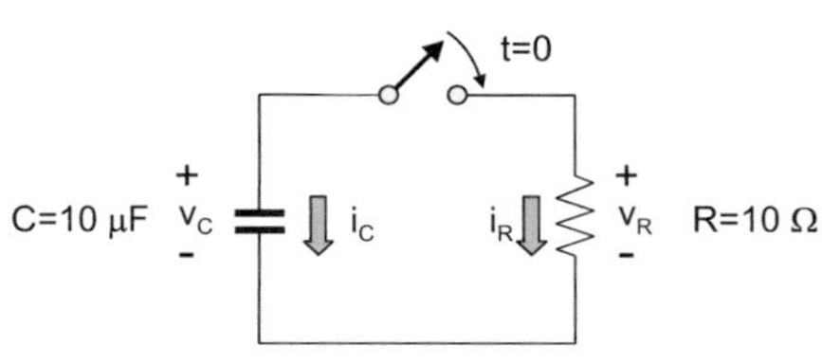

Fig. 7.1. Discharging a capacitor through a load resistor.

The solution to this dynamic circuit is based on applying KVL and KCL, which are valid for all electric circuits. Using KCL gives the result:

$$i_C = -i_R \tag{7.2}$$

at *any* instance of time, *t*. Since both circuit elements in Fig. 7.1 are passive, we can apply the constitutive relations between currents and voltages without changing the sign:

$$\overbrace{C\frac{dv_C}{dt}}^{i_C} = -\overbrace{\frac{v_R}{R}}^{i_R} = -\frac{v_C}{R} \tag{7.3}$$

This is true because KVL states for any *positive* time, $t > 0$,

$$v_{\mathrm{R}} = v_{\mathrm{C}} \tag{7.4}$$

Equation (7.3) therefore yields

$$C\frac{dv_{\mathrm{C}}}{dt} + \frac{v_{\mathrm{C}}}{R} = 0 \Rightarrow \frac{dv_{\mathrm{C}}}{dt} + \frac{v_{\mathrm{C}}}{\tau} = 0, \quad \tau = RC \tag{7.5}$$

This is the famous first-order *transient* circuit equation. Here, τ carries units of seconds since R is recorded in Ω and C is given in $\mathrm{F} = \mathrm{A} \times \mathrm{s/V}$ and is called the *time constant* or the *relaxation constant* of the circuit. It is the *only* constant that is present in the first-order differential equation. The solution of an ODE of this type has the generic form

$$v_{\mathrm{C}}(t) = K\exp\left(-\frac{t}{\tau}\right) \tag{7.6}$$

This fact is proven by direct substitution. The constant K is determined from the initial condition, Eq. (7.1), which yields

$$K = V_0 \tag{7.7}$$

Thus, the circuit voltages have the same form

$$v_{\mathrm{C}}(t) = v_{\mathrm{R}}(t) = V_0\exp\left(-\frac{t}{\tau}\right); \quad t \geq 0 \tag{7.8a}$$

for nonnegative values of t. However, although the capacitor voltage is equal to V_0 at $t < 0$, the resistor voltage is exactly zero at $t < 0$, since the switch was open. The current through the load resistor is

$$i_{\mathrm{R}}(t) = \frac{v_{\mathrm{R}}(t)}{R} = \frac{V_0}{R}\exp\left(-\frac{t}{\tau}\right); \quad t \geq 0 \tag{7.8b}$$

and is zero for negative t. We recall that the capacitor current is the negative of the load current. The instantaneous power delivered to the load resistance is expressed in the form

$$p_{\mathrm{R}}(t) = v_{\mathrm{R}}(t)i_{\mathrm{R}}(t) = \frac{V_0{}^2}{R}\exp\left(-2\frac{t}{\tau}\right); \quad t \geq 0 \tag{7.8c}$$

Equations (7.8a–c) provide the complete solution of the circuit shown in Fig. 7.1. What is the most remarkable and perhaps most important property of the solution? The answer to this question is linked to the amount of power that can be discharged in a finite amount of time. Let us examine Eq. (7.8c) more closely. When the load resistance, R, becomes small, the delivered power can reach an arbitrarily high value at small positive t. Expressed in another way, when discharged through a small resistance, the (ideal) capacitor delivers an extremely high power pulse during a short period of time!

This conclusion is not affected by the specific capacitance value; the capacitance value only affects the discharge duration. In reality, however, an infinitely small resistance cannot be achieved. How can we use the ability of the charged capacitor to create a large current and, consequently, supply a large power for a short period of time? There are a number of well-known applications such as an electronic photoflash or drivers for the light-emitting diodes (LEDs) or even *electromagnetic material processing*.

7.1.2 Time Constant of the RC Circuit and Its Meaning

To appreciate the value of the time constant as a fundamental property of an *RC* circuit, we consider two examples with explicit component values. Our objective is to find the dynamic voltage and current responses of the circuit as the capacitor discharges. You should note that the time constant τ determines the duration at which the capacitor voltage will have dropped to $1/e$ or 0.368 (36.8 %) of the initial voltage V_0. This number arises from the fact that, at the time instance $t = \tau$, we obtain from Eq. (7.8a)

$$v_C(t) = v_R(t) = V_0 \exp\left(-\frac{\tau}{\tau}\right) = V_0 e^{-1} = V_0/e = 0.368\ V_0 \tag{7.9a}$$

It is sometimes useful to study the dynamic response at $t = 2\tau$, in which case we obtain

$$v_C(t) = v_R(t) = V_0 e^{-2} = 0.135\ V_0 \tag{7.9b}$$

or 13.5 % of its original value, V_0. At 3 τ, we already see the voltage drop less than 5 %.

Example 7.1: In Fig. 7.1, a 10-μF capacitor discharges into a 10-Ω load. The capacitor is initially charged to $V_0 = 10$ V. Plot the capacitor voltage v_C, load current i_R, load voltage v_R, and load power p_R, over the interval from −0.2 ms to 0.5 ms.

Solution: First, we determine the time constant τ. According to Eq. (7.5),

$$\tau = RC = 10^{-5}\text{F} \times 10\Omega = 10^{-4}\text{s} = 0.1\ \text{ms} \tag{7.9c}$$

The solution then relies on Eqs. (7.8a) through (7.8c) based on $V_0 = 10$ V. Figure 7.2 shows the behavior of voltage, load current, and load power. The vertical line is the time constant τ. This constant determines how fast the capacitor discharges. At $t = \tau$, the voltage is equal to $1/e$ or 0.368 of the initial capacitor voltage, V_0. Note that a rather low capacitance value of 10 μF is used. We can purchase a 10-μF electrolytic capacitor of 5 mm diameter and 12 mm height and rated at 25 V or 50 V. As seen in Fig. 7.2c, an appreciable load power of 10 W (!) can be created. Unfortunately, it is created for only a very short period of time, on the order of τ.

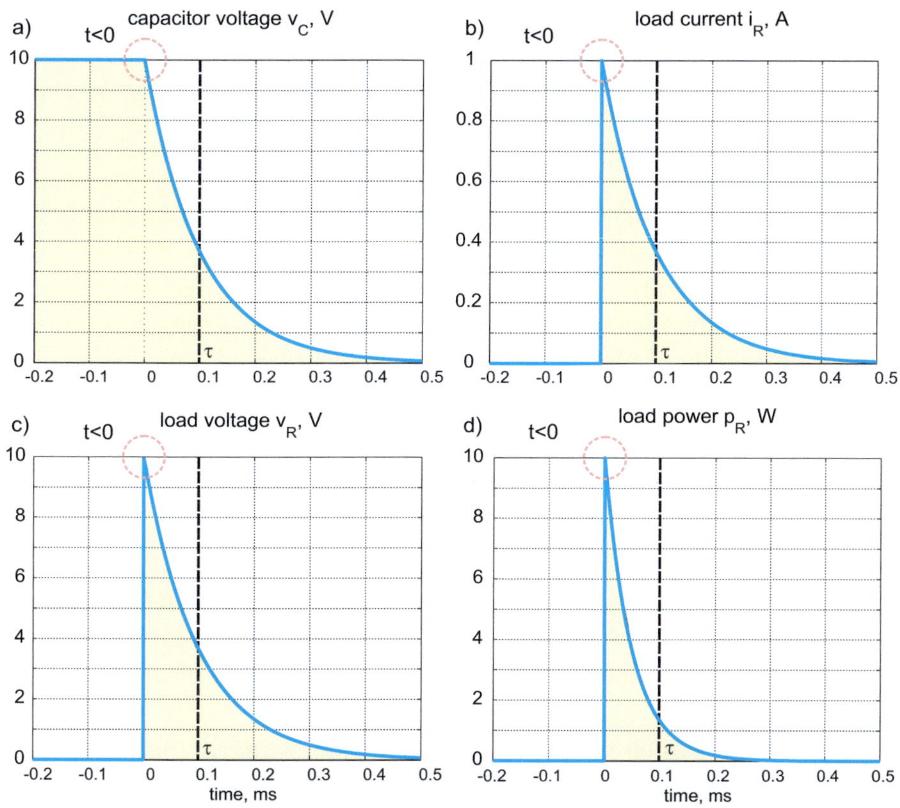

Fig. 7.2. (**a**) Capacitor voltage, (**b**) load current (**c**) load voltage, and (**d**) load power for a 10-μF capacitor discharging into a 10-Ω load resistor.

Exercise 7.1: The capacitor in Fig. 7.1 is initially charged to $V_0 = 20$ V. Determine the capacitor voltage and the instantaneous power delivered to the load resistance at (i) $t = 50$ μs and (ii) $t = 1$ ms.

Answer: (i) 12.13 V and 14.72 W; (ii) 0.91 mV and 0.08 μW.

7.1.3 Continuity of the Capacitor Voltage

Energy Consideration

The voltage across the discharging capacitor remains a *continuous* function of time over the breakpoint $t = 0$. On the other hand, *all* other quantities in Fig. 7.2 such as the circuit current, the load voltage, and the load power are subject to a sudden jump when the switch closes. Why is that so? The electric field energy stored in the capacitor is given by

$$E_C = \frac{1}{2} C v_C^2(t) \tag{7.10a}$$

at any time instant. Any energy cannot be released instantaneously. For instance, a vehicle with mass m and speed υ possesses the kinetic energy

$$E_{\mathrm{T}} = \frac{1}{2} m\upsilon^2 \tag{7.10b}$$

and cannot be stopped instantaneously. The kinetic energy must be a continuous function of time, as does the vehicle speed. Similarly, the capacitor energy must be a continuous function of time and so does the capacitor voltage. Such an effect might be called the "capacitor inertia" in reference to mechanical inertia. Thus, the capacitor voltage $\upsilon_{\mathrm{C}}(t)$ is the *only* variable which is always a continuous function of time in an RC circuit. Therefore, it must be used as an *independent* function in the ODEs for RC circuits. Using any other function (circuit current or resistor voltage) is prohibited since we cannot specify the initial conditions for a noncontinuous function.

Fluid Mechanics Analogy

The continuity of the capacitor voltage may be illustrated by a fluid-flow analogy of the discharging capacitor shown in Fig. 7.3. The voltage corresponds to the fluid level in the water-filled tank, which gradually decreases, but cannot jump instantaneously. On the other hand, the fluid acquires a certain velocity (the equivalence to electric current) immediately after the switch in Fig. 7.3 opens. Interestingly, the value of the load resistance in Fig. 7.1 is the reciprocal of the cross section of the pipe in Fig. 7.3. The smaller the cross section of the pipe (i.e., the greater the resistance), the slower the observed fluid flow from the tank (i.e., the smaller load current). At the same time, the leakage time (or the discharge time) increases accordingly. In Fig. 7.3, we actually need to *open* the mechanical valve, whereas in Fig. 7.1 we *close* the electric switch. There is no real contradiction though since both operations really enable the flow of a substance: either the flow of electric current in Fig. 7.1 or the water flow in Fig. 7.3.

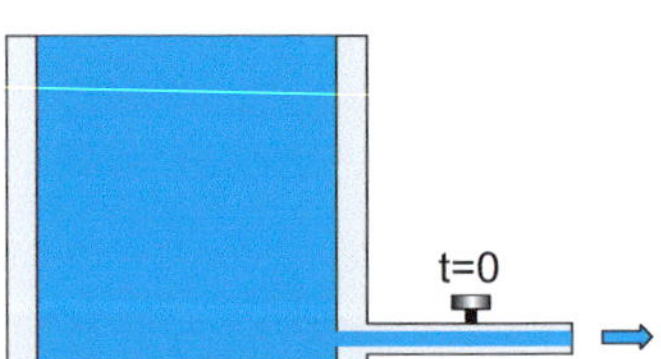

Fig. 7.3. A fluid-flow analogy for the circuit shown in Fig. 7.1.

7.1.4 Application Example: Electromagnetic Railgun

Figure 7.4 shows a generic structure of an electromagnetic accelerator, sometimes called an electromagnetic *railgun*. Apart from potential high-power applications, this setup helps us to visualize the operation of linear motors and generators. The discharging capacitor is connected via a resistor to the rails, as shown on the left in Fig. 7.4. Resistor R models ohmic losses in the metal rails and a (typically small) series resistance of the capacitor. The capacitor current flows through two *metal* rails and through a sliding or rolling *metal* rod to be accelerated. Also shown in Fig. 7.4 are two permanent magnets responsible for creating a magnetic flux emanating from the north pole (N) and terminating at the south pole (S). When this perpendicular magnetic flux density B, measured in tesla (T), is applied between the rails, the *Lorentz force* will act on the moving object of length $l = |\vec{l}|$ and accelerate this object in the direction of the rails. This force is given by

$$\vec{F} = i_C\left(\vec{l} \times \vec{B}\right) \quad [\mathrm{N}] \tag{7.11}$$

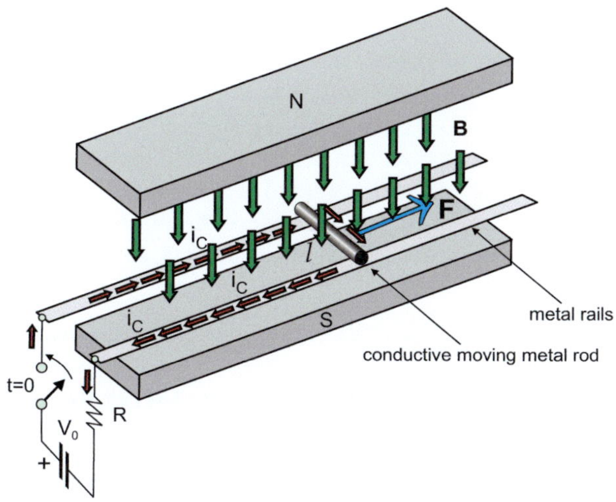

Fig. 7.4. An electromagnetic accelerator based on the Lorentz force effect in the magnetic field created by two permanent magnets and the capacitor discharge current.

The Lorentz force was named after Hendrik Antoon Lorentz (1853–1928), a Dutch physicist and Nobel Prize laureate. Only 24 years of age, Lorentz was appointed to the newly established chair in theoretical physics at the University of Leiden, the oldest university in the Netherlands founded by William, Prince of Orange. Lorentz made significant contributions to field theory ranging from hydrodynamics to general relativity. The Lorentz force is the driving force of electric motors/generators you are using. The cross product in Eq. (7.11) is consistent with the right-hand rule: the current direction of the moving object represents the fingers of your right hand, and they are turned into the magnetic flux direction so that the thumb points in the direction of the Lorentz force. Another way is to picture a screw whose body points along the force and which is turned in the plane spanned by $\vec{l}$ and $\vec{B}$ such that $\vec{l}$ is rotated into $\vec{B}$. From a circuit point of view, the construction in Fig. 7.4 could be replaced by an *RC* circuit, with the capacitor connected to the rails through a switch. Resistance *R* in Fig. 7.4 is formed by the rail resistance combined with the object's resistance and with the contact resistance between the object and the rails. The force is directly proportional to the discharge current i_C.

Example 7.2: The capacitor in Fig. 7.4 is initially charged to a voltage $V_0 = 100\ \text{V}$ and has a capacitance of 1000 μF. The total system resistance *R* is 1 Ω. For the above example, what force in N is to be expected and for how long? The field is 0.1 T; the rod is 6 cm long.

Solution: According to Eq. (7.8b), the maximum current value, which occurs when the switch has just closed, is $V_0/R = 100\ \text{A}$. Using Eq. (7.11), we calculate the initial force value, which is 0.6 N.

As time progresses, the current and the force both quickly decrease. Over a time duration of $\tau = RC = 1\,\text{ms}$, both of these values decrease to 36.8 % of their initial values. For simplicity, we assume that an average force acts over the time duration τ. Its value is estimated as approximately 60 % of the initial force value. We then obtain an average of 0.36 N over the time interval $\tau = 1$ ms, which is a rather modest result. Realistic capacitors used for electromagnetic (EM) acceleration are the so-called pulsed capacitors. They have a high charge voltage of $V_0 \geq 10{,}000\ \text{V}$ and capacitances on the order of 100 μF. Therefore, a high-voltage power supply is needed. A number of capacitors are put in parallel to increase the overall capacitance. Large currents, on the order of 10,000 A, into the 1-Ω load may then produce much higher force values.

7.1.5 Application Example: Electromagnetic Material Processing

Electromagnetic Forming

The moving object in Fig. 7.4 may be implemented in various forms. For example, it could be replaced by a liquid metal such as molten aluminum. In principle, an electromagnetic "die casting" machine could be constructed that creates a high-speed liquid metal jet. The key point is the small mass of the object in order to enable a fast acceleration. *Electromagnetic forming* is used to accelerate solid metal sheets at velocities up to a few hundred meters per second, which are 100–1000 times greater than the deformation rates of conventional forming such as sheet metal stamping. The noncontact electromagnetic forming of metals is a process that has been applied since the 1960s but has not seen extensive use. Its common application is to expand, or compress, axisymmetric metal parts as shown in Fig. 7.5a. It has been commercially applied for the joining and assembly of concentric parts and compression crimp seals. Figure 7.5b shows a more recent experiment at the Ohio State University and made with aluminum car door panels.

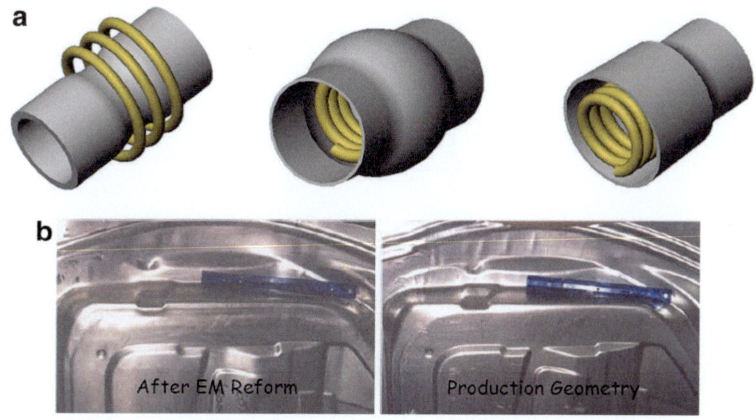

Fig. 7.5. (**a**) Electromagnetic forming of metal joints. The current in the windings generates a magnetic flux that induces eddy currents in the metal. Their product is the Lorentz force. (**b**) Electromagnetically reformed door panel compared with the production geometry.

Self-Induced Lorentz Force

Electromagnetic forming processes shown in Fig. 7.5 do not use permanent magnets. Instead, the so-called *self-induced* Lorentz force is employed. The idea is to generate the magnetic flux B with the same current i_C. Figure 7.5a shows the related concept used in noncontact electromagnetic forming of metal joints. The high discharge current, i_C, creates a strong time-varying magnetic field, both inside and outside of the coils in Fig. 7.5a. In turn, the time-varying magnetic field induces so-called eddy currents in the metal sample. The product of these eddy currents and the magnetic field gives rise to a Lorentz force according to Eq. (7.11). This force is strong enough to deform the joints.

7.1.6 Application Example: Digital Memory Cell

This completely different example investigates a digital circuit that stores binary information. Figure 7.6 shows a schematic of a *dynamic random-access memory* (DRAM) memory cell. The cell stores its bit of information as charge deposited on the cell capacitor C. When the cell is storing a logic 1, the capacitor is charged to a positive voltage V_0; when a logic zero is stored, the capacitor is discharged to a zero voltage. Because of leakage effects, there is always a nonzero resistance R to ground (not shown in the figure). Thus, the cell circuit becomes that of Fig. 7.1. The capacitor will discharge and must be refreshed periodically. During refresh, the capacitor voltage is restored to V_0 if necessary. The refresh operation is in fact performed every 5–10 ms!

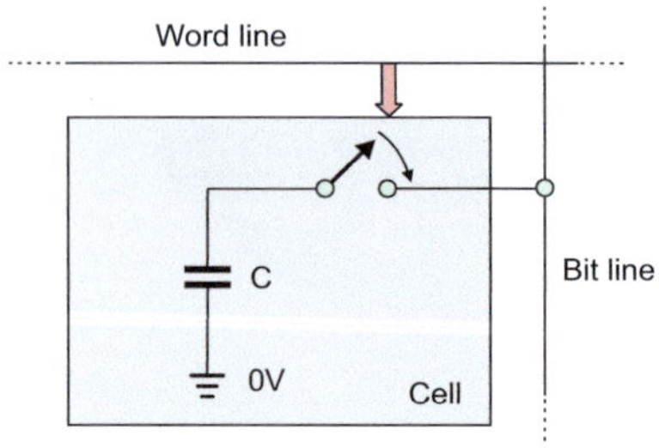

Fig. 7.6. Dynamic RAM memory cell. The bit line carries either logic 1 or 0 information.

7.1.7 Energy-Accumulating Capacitor Circuit

Charging is the inverse process of discharging a capacitor and it involves a power supply. The corresponding circuit is shown in Fig. 7.7. The switch closes at $t = 0$. The resistor R can be either the Thévenin resistance of the practical voltage source or the series parasitic resistance of the capacitor itself or even a combination of both.

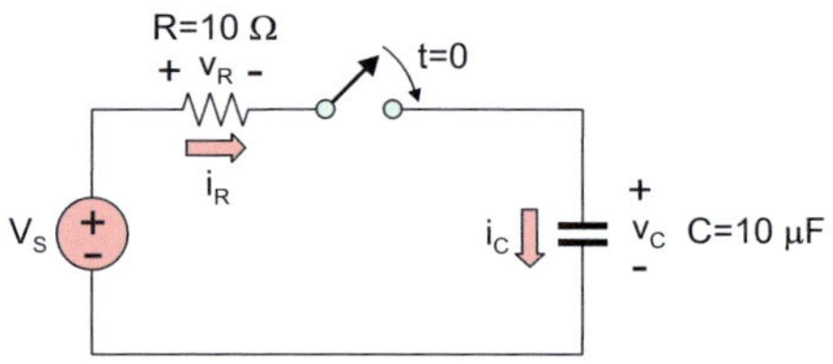

Fig. 7.7. Charging a capacitor with a DC voltage source as an example of another RC circuit.

To support this conclusion, we note that the positions of the switch and the resistor in Fig. 7.7 may be interchanged without affecting the circuit analysis. Similarly, two resistors may be placed on both sides of the switch; the circuit solution will display their series equivalent. The solution to the circuit is once again based on KVL and KCL. By KCL,

$$i_C = i_R \tag{7.12}$$

at *any* time instance t. Since both R and C in Fig. 7.7 are passive elements, we can apply the constitutive relations between currents and voltages without changing the sign:

$$\overbrace{C\frac{dv_C}{dt}}^{i_C} = \overbrace{\frac{v_R}{R}}^{i_R} = \frac{V_S - v_C}{R} \tag{7.13}$$

since KVL states

$$v_R = V_S - v_C \tag{7.14}$$

at any *positive* time instance t. Therefore, Eq. (7.13) yields

$$C\frac{dv_C}{dt}+\frac{v_C}{R}=\frac{V_S}{R}\Rightarrow\frac{dv_C}{dt}+\frac{v_C}{\tau}=\frac{V_S}{\tau},\quad \tau=RC \tag{7.15}$$

Equation (7.15) has now an excitation term (the power supply voltage) on its right-hand side. Consequently, the solution to this equation is called a *forced response*. Equation (7.15) is known as an *inhomogeneous* first-order differential equation. This is in contrast to the *homogeneous* differential equation (7.5). Nonetheless, Eq. (7.15) still remains a first-order transient equation with the same time constant τ. The solution to any first-order ordinary differential equation of that type has the generic form

$$v_C(t)=K_1\exp\left(-\frac{t}{\tau}\right)+K_2 \tag{7.16a}$$

This fact can be checked by direct substitution. The two terms containing the exponential factor will cancel out after differentiation. The remaining terms in Eq. (7.15) yield

$$\frac{K_2}{\tau}=\frac{V_S}{\tau}\Rightarrow K_2=V_S \tag{7.16b}$$

The constant parameter K_1 can be determined from the initial condition, $v_C(t=0)=0$. Since $\exp(0)=1$ in Eq. (7.16a), we conclude

$$K_1+K_2=0\Rightarrow K_1=-V_S \tag{7.16c}$$

Thus, the circuit voltages have the form

$$v_C(t)=V_S\left[1-\exp\left(-\frac{t}{\tau}\right)\right],\quad v_R(t)=V_S\exp\left(-\frac{t}{\tau}\right) \tag{7.16d}$$

The resistor voltage has the same form as the resistor voltage in Eq. (7.8a) for the discharging capacitor. However, the capacitor voltage has not. The capacitor current is

$$i_C(t)=C\frac{dv_C(t)}{dt}=\frac{V_S}{R}\exp\left(-\frac{t}{\tau}\right) \tag{7.16e}$$

It is equivalent to Eq. (7.8b), the discharge current for the RC circuit. What is the most remarkable property of the solution given? According to Eq. (7.16d), we always need a certain amount of time to charge the capacitor. It is clear that this time will be on the order of the time constant τ. Moreover, from a formal point of view, the capacitor voltage will never exactly reach the source voltage (the exact equality only occurs at $t\to\infty$), see Fig. 7.8.

Example 7.3: A 10-μF capacitor in Fig. 7.7 is charged by a 10-V voltage source. The switch closes at $t = 0$, and the system resistance is 10 Ω. Plot capacitor voltage v_C and capacitor current i_C to scale over the time interval from −0.2 ms to 0.5 ms.

Solution: First, we find the time constant τ. According to Eq. (7.15) or (7.5), $\tau = RC = 10^{-5}\text{F} \times 10\ \Omega = 0.1\,\text{ms}$. The solutions for this example are given by Eqs. (7.16d) and (7.16e) with $V_S = 10\,\text{V}$. Solutions for the capacitor voltage and capacitor current are plotted in Fig. 7.8. The vertical line denotes the time constant τ so that you can see how fast the capacitor charges. At one time constant, i.e., $\tau = 0.1$ ms, the capacitor is charged to $(1 - 1/e)V_S$ or to 63.2 % of the source voltage V_S.

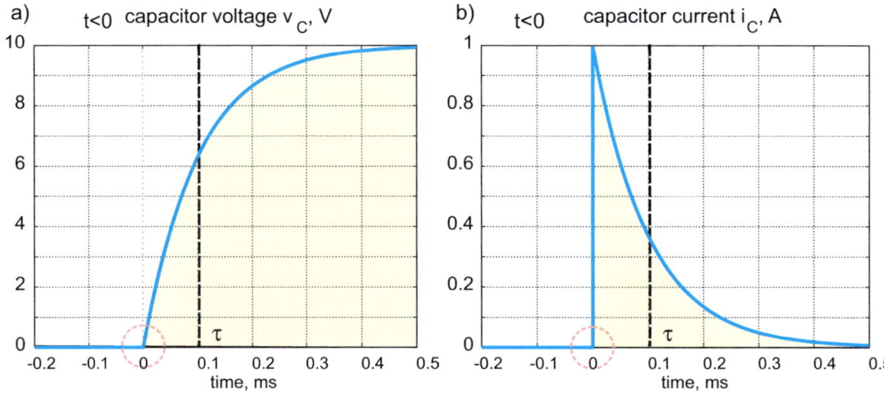

Fig. 7.8. Voltage/current plots for charging a 10-μF capacitor in series with a 10-Ω resistor. We again observe continuity of the capacitor voltage (the "capacitor inertia").

Exercise 7.2: The source voltage in Fig. 7.7 is 20 V. Determine the capacitor voltage and the circuit current at (i) $t = 5$ μs and (ii) $t = 1$ ms.

Answer: (i) 0.98 V and 1.90 A; (ii) 19.999 V and 91 μA.

Section 7.2 RL Circuits

The first-order *RL* circuits studied in this section lay the foundation of understanding the behavior of any transient circuit containing inductances. The transient circuit response is a dynamic process; we are once again required to solve dynamic circuit equations that can be formulated as ordinary differential equations. You can again rely on KVL and KCL since they apply to any static (DC) or dynamic (transient and AC) circuit. Along with the inductor's voltage/current relation, they are used to derive the circuit ODE.

7.2.1 Energy-Release Inductor Circuit

The inductor stores the magnetic-field energy created by an electric current. Thus, in order to be "charged," the inductor must carry some current. A natural choice is therefore a circuit with the *current source* shown in Fig. 7.9a. This is in contrast to the charged capacitor, which does not need a voltage supply to stay charged. If the switch in Fig. 7.9a is open ($t < 0$), the entire current I_S flows through the inductor. The inductor is thus "charged." When the switch closes at $t = 0$, the current source still generates the same current I_S at its terminals. However, the supply is now shorted out, i.e., no current flows into the circuit. In other words, the current supply is effectively disconnected so that the RL circuit becomes a stand-alone circuit in Fig. 7.9b, with the initial current I_S still flowing in the inductor. As time progresses, the inductor releases its energy to the load.

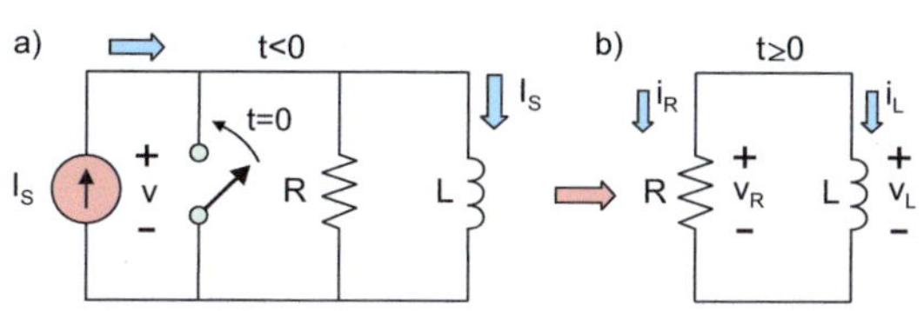

Fig. 7.9. The concept of "discharging" the previously charged inductor through a load resistor *R*.

The solution of the dynamic circuit in Fig. 7.9b is again based on KVL and KCL. With the voltage polarities shown in Fig. 7.9b, the use of KVL yields

$$v_L = v_R \tag{7.17}$$

at *any* time instance, *t*. Since both circuit elements in Fig. 7.9b are passive, we can directly apply the constitutive relations between voltages and currents without changing the sign:

$$\overbrace{L\frac{di_L}{dt}}^{v_L} = \overbrace{Ri_R}^{v_R} = -Ri_L \tag{7.18}$$

According to KCL, $i_R = -i_L$, at any *positive* time instance *t*, and Eq. (7.18) yields

$$L\frac{di_L}{dt} + Ri_L = 0 \Rightarrow \frac{di_L}{dt} + \frac{i_L}{\tau} = 0, \quad \tau = L/R \tag{7.19}$$

Here again we encounter a first-order transient equation. The constant τ carries the unit of seconds (since *R* is in ohms and *L* is in henrys) and is known as the *time constant*, or *relaxation constant,* of the *RL* circuit. It is the only constant that is present in the first-order transient differential equation. We can observe a remarkable similarity between the transient *RL* circuit and the RC circuit of discharging a capacitor. The mathematics is the same, but the capacitor voltage is replaced by the inductor current, and the value of the time constant changes from *RC* to *L/R*. The initial condition $i_L(t=0) = I_S$ includes the past inductor current instead of the past capacitor voltage. The solution of Eq. (7.19) is

$$i_L(t) = K\exp\left(-\frac{t}{\tau}\right), \quad t \geq 0 \tag{7.20}$$

The validity of Eq. (7.20) is seen by direct substitution. The constant *K* is determined from the initial condition. Setting $t = 0$ yields $K = I_S$. Both currents in Fig. 7.9b are

$$i_L(t) = -i_R(t) = I_S\exp\left(-\frac{t}{\tau}\right) \quad t \geq 0 \tag{7.21a}$$

At $t < 0$, the inductor current maintains its value I_S but the resistor current is zero, as shown in Fig. 7.10c. The resistor (or load) voltage is given by

$$v_R(t) = Ri_R(t) = -RI_S\exp\left(-\frac{t}{\tau}\right) \tag{7.21b}$$

At $t < 0$, the load voltage is zero. The instantaneous power delivered to the load is

$$p_R(t) = v_R(t)i_R(t) = RI_S^2\exp\left(-2\frac{t}{\tau}\right) \tag{7.21c}$$

Equations (7.21a)–(7.21c) provide the complete solutions for the circuit depicted in Fig. 7.9.

Example 7.4: A 1-mH inductor in Fig. 7.9 is connected to a 1-kΩ load. The supply current (disconnected at $t = 0$) is 1 A. Plot inductor current i_L, inductor (or load) voltage v_R, load current, i_R, and load power, p_R, over the interval from -2τ to 5τ.

Example 7.4 (cont.):

Solution: First, we find the time constant τ. According to Eq. (7.19),

$$\tau = L/R = 10^{-3}\mathrm{H}/1000\ \Omega = 10^{-6}\mathrm{s} = 1\ \mu\mathrm{s} \tag{7.22}$$

which is a rather small value. The solution for this example is given by Eqs. (7.21a) through (7.21c), with $I_S = 1$ A, and is shown in Fig. 7.10. The vertical line in all plots is the time constant τ. One can see that the time constant determines how quickly the inductor current and the load voltage decrease. At 1τ, the load voltage is equal to $1/e$ or 0.368 (36.8 %) of the initial voltage RI_S.

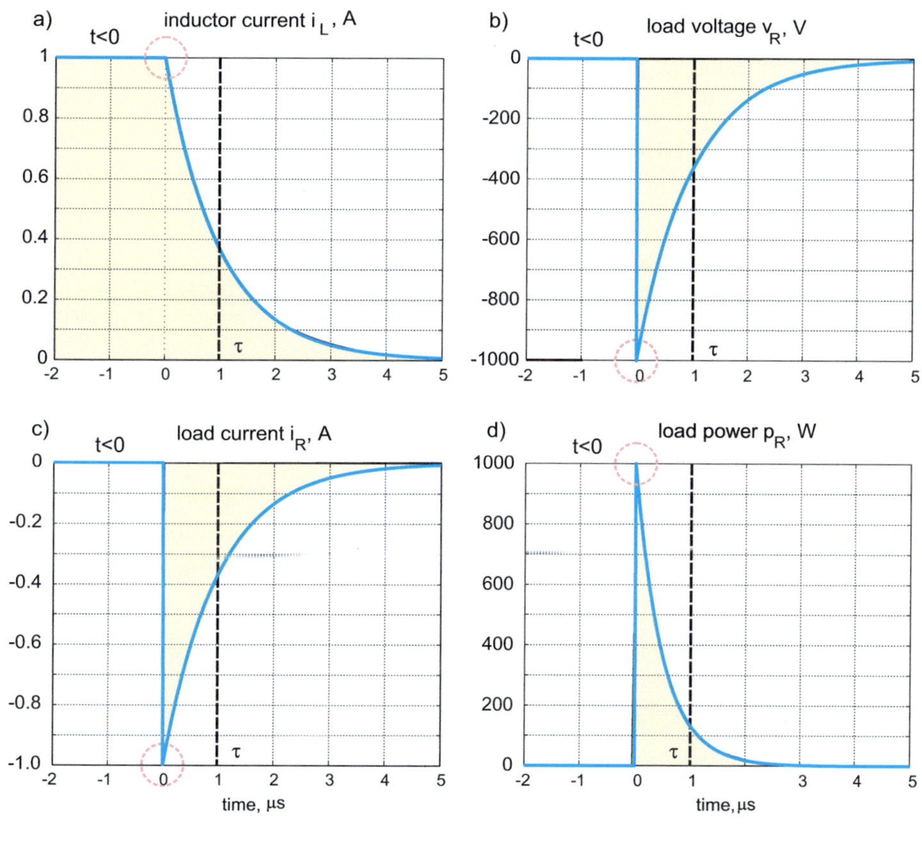

Fig. 7.10. (**a**) Inductor current, (**b**) load voltage, (**c**) load current, and (**d**) load power for a 1-mH inductor connected to a 1000-Ω load resistor.

Exercise 7.3: The supply current in Fig. 7.9 is 0.5 A. Given $L = 1$ mH and $R = 1$ kΩ, determine the load voltage and the instantaneous power delivered to the load resistance at (i) $t = 0.2$ μs and (ii) $t = 10$ μs.

Answer: (i) –409.4 V and 167.6 W; (ii) –23 mV and 0.52 μW.

What is the most remarkable property of the solution? According to Fig. 7.10, the high voltage spike across the inductor is created in an RL switching circuit when the load resistance, R, is large. For example, an air gap has a very high resistance. When used as a load, it may possess a very high voltage drop of several kV and more. This is the idea behind any medium-to-high-power *electronic ignition* system, including the 12-V-powered car ignition system, a missile ignition system, etc. Such a circuit must include at least three basic elements: (a) a voltage or current power supply, (b) a switch, and (c) an inductor (coil). The switch can be a transistor switch controlled by a sensor.

7.2.2 Continuity of the Inductor Current

The current through the inductor remains a *continuous* function of time over the breakpoint $t = 0$. However, *all* other quantities in Fig. 7.10, the load voltage, the load current, and the load power are subject to a sudden jump when the switch closes. The reason for such a continuity is the *finite* magnetic-field energy stored in the inductor:

$$E_{\mathrm{L}} = \frac{1}{2} L i_{\mathrm{L}}^2(t), \tag{7.23}$$

at any time instant. This energy cannot be released instantaneously. Such an effect might be called "inductor inertia" in reference to mechanical inertia of a vehicle with mass m and speed v and kinetic energy $E_{\mathrm{T}} = 0.5\, mv^2$, which cannot be stopped instantaneously. The kinetic energy is a continuous function of time, as is the vehicle speed. Similarly, the inductor energy is a continuous function of time, as is the inductor current. The inductor current is the only variable which is always a continuous function. Therefore, it must be used as an *independent* function in the ODEs for RL circuits. Using any other function is prohibited since we cannot state the initial conditions for a noncontinuous function.

Fluid Mechanics Analogy

The continuity of the inductor current may be illustrated by a fluid-flow analogy of the energy-releasing inductor circuit shown in Fig. 7.11. The electric current corresponds to the velocity of the fluid. The inductance is a massive wheel of mass m. When subject to a DC current (constant water flow), it acquires a certain angular velocity. This is the case of Fig. 7.9a at $t < 0$. When the water pump is suddenly turned off, the wheel inertia will still support the same water flow, at least at the initial time moment. After that, the wheel slowly decelerates. This is the case of Fig. 7.9b.

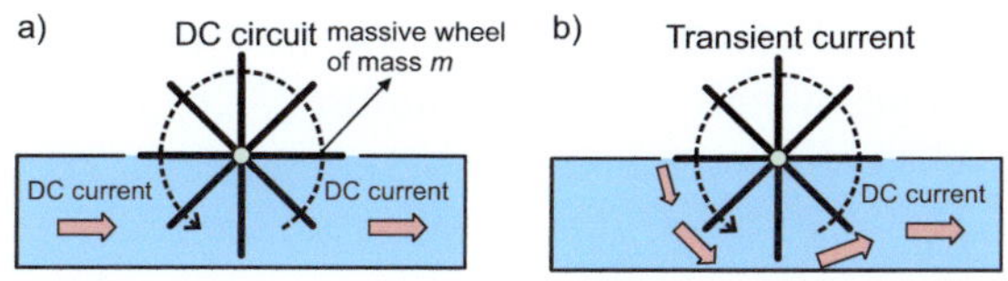

Fig. 7.11. A fluid-flow analogy for the two circuits shown in Fig. 7.9.

7.2.3 Energy-Accumulating Inductor Circuit

The circuit behavior is now exactly the opposite of the circuit shown in Fig. 7.9. For instance, when the switch in Fig. 7.12 is closed, the current supply is shorted out. No current flows into the circuit. However, when the switch opens, the supply current I_S starts to flow into the circuit, and as time approaches infinity, the entire supply current I_S flows through the inductor. Thus, the inductor becomes "charged."

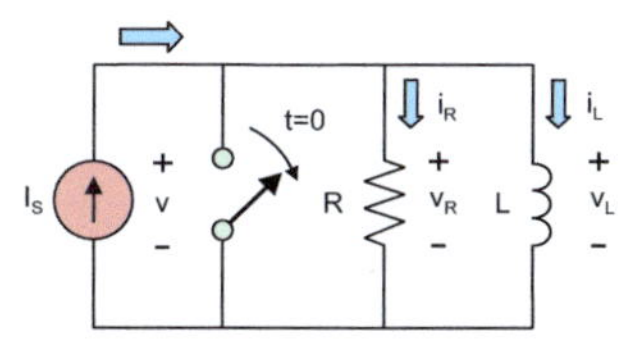

Fig. 7.12. The concept of "charging" an inductor using a current source.

The solution of the dynamic circuit in Fig. 7.12 is based on KVL and KCL. With the voltage polarities in Fig. 7.12, the use of KVL yields

$$v_L = v_R \tag{7.24}$$

at *any* time instance, *t*. Since both circuit elements in Fig. 7.12 are passive, we can apply the constitutive relations between voltages and currents without changing the sign:

$$\overbrace{L\frac{di_L}{dt}}^{n_L} = \overbrace{Ri_R}^{v_R} = RI_S - Ri_L \tag{7.25}$$

because, according to KCL, $i_R = I_S - i_L$, at any *positive* time instance, *t*. Equation (7.25) yields

$$\frac{di_L}{dt} + \frac{i_L}{\tau} = \frac{I_S}{\tau}, \quad \tau = L/R \tag{7.26}$$

which is the inhomogeneous first-order transient equation with the forcing function (right-hand side) equal to I_S/τ. Here, τ is the generic time constant of the *RL* circuit. Once again, there is a close similarity between the present *RL* circuit and the series *RC* circuit for charging the capacitor. The mathematics is the same, but the capacitor voltage is replaced by an inductor current and the voltage supply is replaced by the current supply.

The initial condition for Eq. (7.26) now includes the past inductor current of 0 A instead of the past capacitor voltage, which is 0 V. Equation (7.26) has the solution

$$i_L(t) = K_1 \exp\left(-\frac{t}{\tau}\right) + K_2, \quad K_2 = I_S \tag{7.27}$$

This fact is seen by direct substitution. The constant K_1 is found from the initial condition of zero inductor current at $t = 0$, which yields $K_1 = -K_2$. Therefore,

$$i_L(t) = I_S\left[1 - \exp\left(-\frac{t}{\tau}\right)\right], \quad i_R(t) = I_S \exp\left(-\frac{t}{\tau}\right), \quad t \geq 0 \tag{7.28a}$$

Both currents are zero at $t < 0$. However, the inductor current is continuous over the breakpoint while the resistor current is not. The inductor/resistor voltages are given by

$$v_L(t) = L\frac{di_L(t)}{dt} = RI_S \exp\left(-\frac{t}{\tau}\right), \quad v_R(t) = v_L(t) \quad t \geq 0 \tag{7.28b}$$

Both voltages are zero at $t < 0$. This completes our circuit analysis.

Example 7.5: A 1-mH inductor in Fig. 7.12 is connected to a 1-A current power supply. The resistor value is $R = 1\ \text{k}\Omega$. Plot the inductor current, i_L, and the inductor voltage, v_L, to scale versus time over the interval from -2τ to 5τ.

Solution: First, we find the time constant τ. According to Eq. (7.26), we get $\tau = L/R = 10^{-3}\text{H}/1\,\text{k}\Omega = 10^{-6}\text{s} = 1\,\mu\text{s}$. The solution to the example is given by Eqs. (7.28a, b) with $I_S = 1$ A; see Fig. 7.13. The vertical line in both plots is the time constant τ. One can see that this time constant determines how fast the circuit stabilizes. At 1τ, the inductor current reaches $(1 - 1/e)I_S$, i.e., 63.2 % of the expected DC value. Note that Fig. 7.13 of this section and Fig. 7.8 of the previous section are identical to within interchanging voltage and current terms!

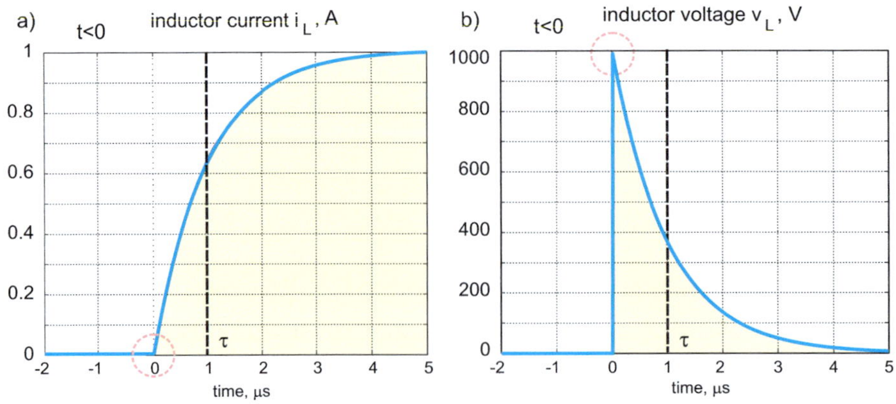

Fig. 7.13. Voltage/current plots for "charging" the 1-mH inductor in parallel with the 1-kΩ resistor.

Exercise 7.4: The supply current in Fig. 7.12 is 0.5 A. Given $L = 1$ mH and $R = 1$ kΩ, determine the inductor voltage and the inductor current at (i) $t = 0.2$ μs and (ii) $t = 10$ μs.

Answer: (i) 409.4 V and 91 mA; (ii) 23 mV and 0.49999 A.

The final question to ask is what is the most remarkable property of the solutions given by Eqs. (7.28a, b)? According to Eq. (7.28a), we always need a certain period of time to create a given current through the inductor. The elapsed time will be on the order of the time constant τ. Moreover, the inductor current will never exactly reach the supply current (the exact equality only occurs as $t \to \infty$). In practice, this effect is masked by noise and by other factors. Interestingly, the resistor carries most of the circuit current when the solution changes rapidly, i.e., close to the initial time $t = 0$. At the same time, when the circuit stabilizes, i.e., when t becomes large compared to τ, the influence of the inductor dominates. This observation leads us to the concept of *impedance* (the "resistance" of dynamic circuit elements) that is considered next. The impedance is similar to a resistance (and has the same unit), but it depends on how fast circuit current and voltage change. When the changes are very fast, the inductor exhibits a much greater "resistance" than the resistor; it becomes virtually an open circuit with no current flow.

7.2.4 Energy-Release RL Circuit with the Voltage Supply

The combination of the current supply I_S and resistor R in Figs. 7.9 and 7.12 is in fact the Norton equivalent circuit of any network of power supplies and resistors. The RL circuit may be modeled a Thévenin equivalent too. The concept is shown in Fig. 7.14.

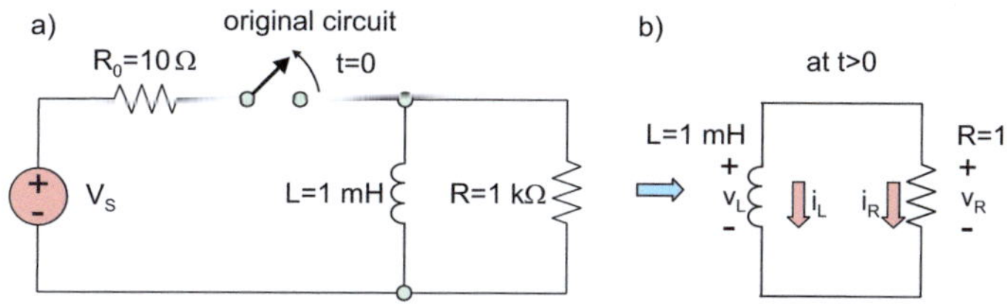

Fig. 7.14. "Discharging" an inductor with a voltage supply, as an example of an RL circuit.

Prior to opening the switch in Fig. 7.14a, the inductor current is found to be V_S/R_0. When the switch opens, the supply is disconnected from the RL circuit; see Fig. 7.14b. As time progresses, the inductor releases its energy into the load. The circuit in Fig. 7.14b is identical to the circuit in Fig. 7.9b. Therefore, all prior results related to the energy-release RL circuit will remain valid if we replace the initial inductor current I_S by V_S/R_0. According to Eq. (7.21b), the load voltage is given by

$$\upsilon_R(t) = Ri_R(t) = -\frac{R}{R_0}V_S\exp\left(-\frac{t}{\tau}\right) \tag{7.29}$$

If the ratio R/R_0 is large, the initial voltage spike of the inductor is large too. The magnitude of the initial voltage spike for the circuit in Fig. 7.14 is a hundred times the supply voltage V_S! Can we model an *electronic ignition system* in the laboratory? Yes, according to Fig. 7.14 this can be accomplished relatively easily. The key, however, is the construction of a fast switch. A proper choice may be a power transistor switch.

7.2.5 Application Example: Laboratory Ignition Circuit

A circuit rated at 6 V for safety purposes is shown in Fig. 7.15. The laboratory DC voltage source usually delivers up to 3 A of current, if not current limited. The electric step-up transformer with two coils is a 6-V car ignition coil. Instead of a simple coil, a transformer is used to further boost the inductor voltage spike of the *RL* circuit. The small resistance R_0 is the transistor/wire resistance. The very large resistance R is the resistance of the spark plug in series with the large resistance of the spark plug cable R_{CABLE}, which is a carbon core wire. This carbon core wire is used to prevent higher EM radiation and its possible influence (we would hear it as noise) on the car audio receiving equipment. Figure 7.16 shows the operating circuit. The sparking frequency ranges from 2 Hz to 100 Hz.

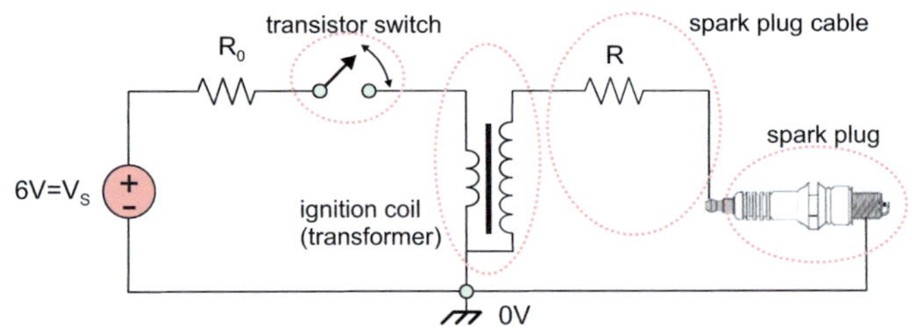

Fig. 7.15. Modeling the ignition system as in a laboratory. The spark plug voltage is about 3–10 kV.

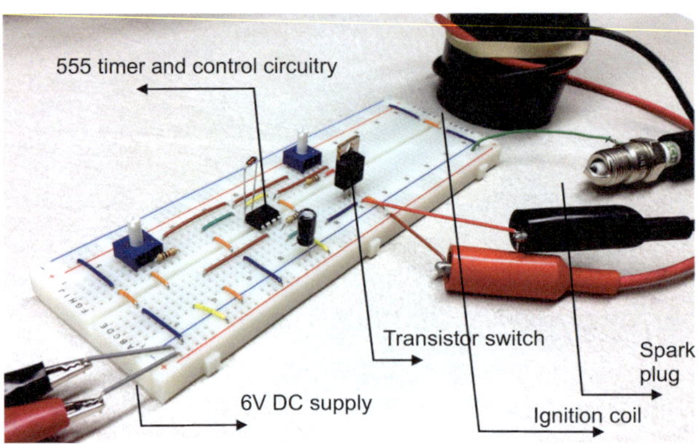

Fig. 7.16. A constructed ignition system.

Spark Gap Radio

The spark gap in the circuit in Fig. 7.15 and in Fig. 7.16 is a powerful and broadband source of electromagnetic radiation. One favorite story in science fiction novels is that, after crashing his or her spaceship on an inhabited planet, the commander can quickly construct a spark generator out of remaining parts of the ship and send an SOS signal out to space. An AM radio can "listen" to the circuit shown in Fig. 7.16 over the entire AM band from 540 to 1610 kHz used in the USA. Even better results are observed for the long-wave AM band from 153 to 279 kHz used in Europe, Africa, and parts of Asia.

Exercise 7.5: The supply voltage in Fig. 7.14 is 10 V. Determine the inductor voltage and the inductor current at (i) $t = 0.2$ μs and (ii) $t = 10$ μs.

Answer: (i) –818.7 V and 0.82 A; (ii) –45 mV and 45 μA.

Section 7.3 Switching RC Oscillator

The time constant τ of an RC circuit provides a natural time scale. It is widely employed for timing purposes. Some of you may have already used microcontroller starter kits. The microcontroller manual discussing the various settings will likely feature a topic entitled "RC oscillator." Here you will discover that the microcontroller *clock frequency* can be controlled by an external resistor R and capacitor C. How is this possible? We have just seen that the RC circuit discharges the capacitor through the resistor, but how can it be used to create a periodic *clock signal* at a given frequency? The present section aims to augment an RC circuit with an amplifier circuit and establish a *clock circuit.*

7.3.1 About Electronic Oscillators

An *electronic oscillator* is a circuit that has an output, but no input in the common sense. It generates a certain periodic waveform at the output. The period, amplitude, and shape of this waveform are determined by the circuit topology. The "heart" of any oscillator circuit is an amplifier block with some sort of a *positive feedback*. The positive feedback is the opposite of the negative feedback. A part of the output amplifier's voltage is fed back into the input with the sign plus. All oscillator circuits may be divided into *linear oscillators*, which create sinusoidal waveforms, and *switching oscillators*, which create square and other periodic nonharmonic waveforms. The subject of this section is a switching oscillator circuit, which is called an *astable multivibrator* or a *relaxation oscillator*. This circuit uses the comparator amplifier but with a positive feedback loop and a transient RC block. It is perhaps the simplest and yet efficient oscillator circuit.

7.3.2 Bistable Amplifier Circuit with the Positive Feedback

Saturation Mechanism

Consider the circuit shown in Fig. 7.17. At first sight, it is similar to the inverting amplifier configuration. However, the amplifier polarity is interchanged, which means that the feedback is now positive. The circuit has no input: both potential inputs are grounded. Since there is no current into the amplifier itself (the first summing-point constraint still applies), two resistors of the feedback loop form a voltage divider between the output voltage v_{out} and ground. Therefore, the voltage at node (+) becomes

$$v^{+} = \frac{R_1}{R_1 + R_2} v_{out} \tag{7.30}$$

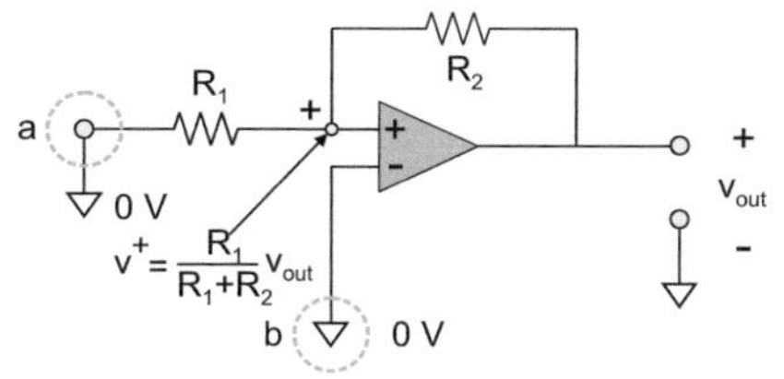

Fig. 7.17. A bistable amplifier circuit.

To analyze the circuit in Fig. 7.17, we again consider the feedback as a fast dynamic process with a very short delay in the feedback loop. We assume that $R_1 = R_2$ for simplicity. It means that 50 % of v_{out} is returned back to the non-inverting input in, say, 1 µs. The open-loop amplifier's gain will be $A = 10^6$; the amplifier hits the power rails $v_{out} = \pm V_{CC}$ in saturation. The initial value of v^+ will be 0 V, and the initial value of v_{out} will be 1 µV (at the noise level). Table 7.1 shows the dynamics of the feedback process where the amplifier operates as $v_{out} = Av^+$, but it takes 1 µs to return 50 % of v_{out}. It follows from Table 7.1 that the amplifier will be very quickly saturated; its output will be the positive rail voltage $v_{out} = +V_{CC}$. All other positive initial values of v_{out} will lead to the same result. Simultaneously, all negative initial values of v_{out} will lead to the saturation at the negative rail $v_{out} = -V_{CC}$.

Table 7.1. Dynamics of the output voltage for the bistable amplifier circuit.

Time, µs	v^+	v_{out}
0	0 V	10^{-6} V
1	0.5×10^{-6} V	0.5 V
2	0.25 V	$+V_{CC}$

Two Stable States

The key point is that once the saturation state

$$v_{out} = +V_{CC},\ v^+ = +\frac{R_1}{R_1 + R_2} V_{CC} \tag{7.31}$$

has been reached, the amplifier circuit will exist in this state *indefinitely*, despite all the subsequent electric noise. To prove this fact, we may introduce a small perturbation in v_{out} and/or in v^+; the circuit will quickly return to the solution given by Eq. (7.31). Quite similarly, once the opposite saturation state

$$v_{out} = -V_{CC}\ \ v^+ = -\frac{R_1}{R_1 + R_2} V_{CC} \tag{7.32}$$

has been reached, the amplifier circuit will exist in this state *indefinitely*. Thus, the positive feedback always forces the comparator to operate in saturation, i.e., in either of the two *stable* states, $v_{\text{out}} = \pm V_{\text{CC}}$, where $\pm V_{\text{CC}}$ is the supply voltage of the amplifier. This result is valid for any pair of the resistances R_1, R_2. The resistance pair specifies values of v^+ in Eqs. (7.31) and (7.32), respectively, known as *threshold voltages*. Figure 7.18 shows the corresponding *mechanical analogy* of the bistable amplifier circuit. Note that a grounded comparator amplifier *without* the positive feedback loop would also be always saturated due to inherent electric noise. However, there are *no* stable states whatsoever; the switching between the rails is random; it is controlled by random noise.

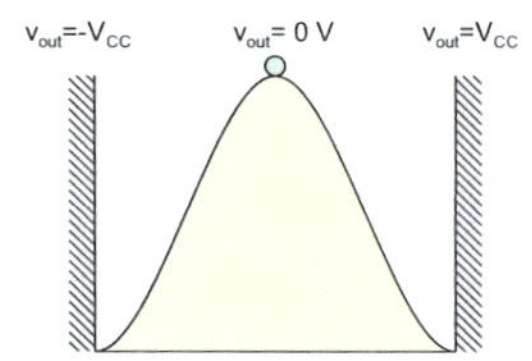

Fig. 7.18. Mechanical analogy of two stable states for the bistable amplifier circuit.

7.3.3 Triggering

The amplifier circuit in Fig. 7.17 can be in either of two stable states defined by the initial conditions. As such, it is useless as long as we do not have a mean to change the state. A *trigger signal* (an input voltage signal) may be applied to switch between the states. After introducing an *external* trigger signal in the form of short pulses, the bistable amplifier circuit becomes a basic *digital memory element* capable of saving and retrieving one bit of data. When the input voltage signal is applied to node (a) in Fig. 7.17 instead of grounding it, the corresponding circuit becomes the *non-inverting Schmitt trigger*. When the input voltage is applied to node (b) in Fig. 7.17 instead of grounding it, the corresponding circuit becomes the *inverting Schmitt trigger*. The Schmitt triggers are used as zero-level detectors in analog electronics and for many other purposes. When triggered, the bistable amplifier circuit operates as a *comparator*.

Exercise 7.6: The bistable amplifier circuit with $R_1 = R_2$ in Fig. 7.17 exists in the positive stable state with $v_{\text{out}} = +V_{\text{CC}}$. A trigger signal is applied to node (b) in Fig. 7.17. Determine output voltage when the applied trigger signal is (i) $-V_{CC}$, (ii) $+0.4V_{\text{CC}}$, and (iii) $+0.6V_{\text{CC}}$, where $\pm V_{\text{CC}}$ is the supply voltage of the amplifier.

Answer: (i) $+V_{\text{CC}}$; (ii) $+V_{\text{CC}}$; (iii) $-V_{\text{CC}}$.

7.3.4 Switching RC Oscillator

The idea of the *switching RC oscillator* (or the *relaxation oscillator*) is not to use an external trigger, but rather to derive the trigger signal from an RC circuit connected to the output of the amplifier itself. The circuit diagram of the relaxation oscillator is shown in Fig. 7.19. The RC circuit with the source voltage v_{out} forms a negative feedback loop. The capacitor may either charge or discharge depending on the source voltage.

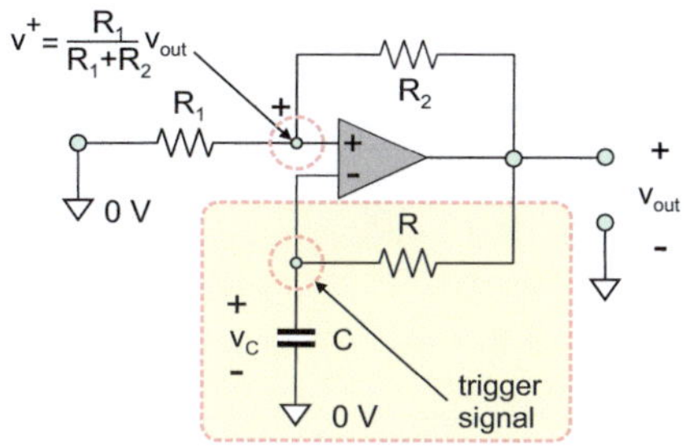

Fig. 7.19. Bistable amplifier circuit with an RC circuit in the negative feedback loop.

To analyze the amplifier circuit in Fig 7.19, we assume an infinitely high open-loop DC gain A and use basic amplifier equations (5.4) with $v^- = v_C$. This yields

$$v_{out} = +V_{CC} \quad \text{if} \, \frac{R_1}{R_1 + R_2} v_{out} > v_C \tag{7.33a}$$

$$v_{out} = -V_{CC} \quad \text{if} \, \frac{R_1}{R_1 + R_2} v_{out} < v_C \tag{7.33b}$$

Thus, the circuit in Fig. 7.19 becomes equivalent to a simple RC circuit given in Fig. 7.20 where the dependent (or rather switching) voltage source is defined by Eqs. (7.33a, b). The corresponding transient analysis is performed starting with some initial conditions.

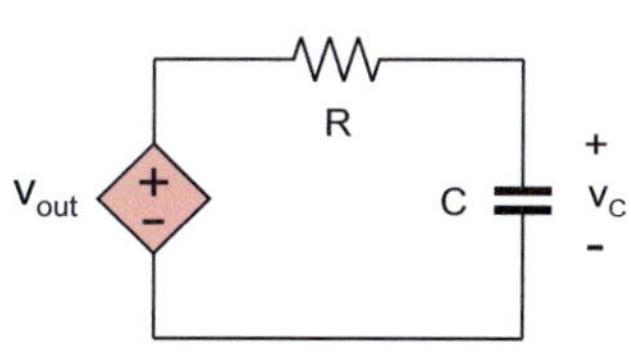

Fig. 7.20. Equivalent circuit for finding oscillation behavior. The dependent voltage source is controlled by the capacitor voltage.

Example 7.6: Given initial conditions $v_{out} = 10$ V and $v_C = 0$ V at $t = 0$, solve the circuit in Fig. 7.19 and Fig. 7.20. Assume that $V_{CC} = 10$ V and $R_1 = R_2$.

Solution: Step 1. The source voltage in Fig. 7.20 is $v_{out} = 10$ V and the capacitor voltage is zero, i.e., $v_C = 0$ V at the initial time moment $t = 0$.

Step 2. The capacitor starts to charge. When its voltage reaches $v_C = 5$ V, the differential voltage at the amplifier's input becomes negative so that Eq. (7.33a) is no longer valid. The source voltage in Fig. 7.20 switches to
$v_{out} = -10$ V according to Eq. (7.33b).

Step 3. The capacitor starts to discharge. When it reaches $v_C = -5$ V, the differential voltage at the amplifier's input becomes positive, and the output voltage therefore switches back to $v_{out} = 10$ V according to Eq.(7.33a).
After Step 3, the circuit returns to Step 1, and the process continues periodically. This results in the output voltage shown in Fig. 7.21.

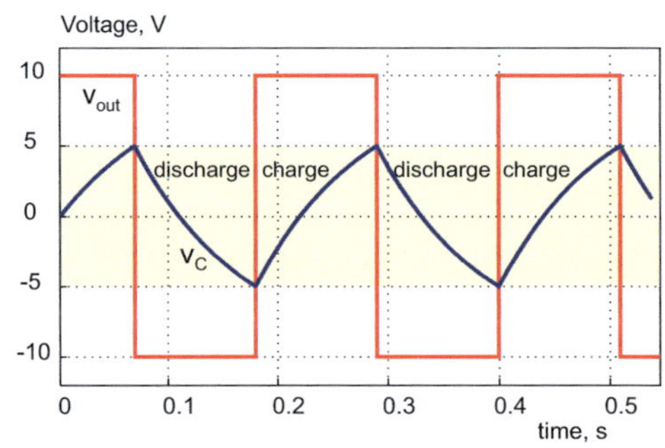

Fig. 7.21. Relaxation oscillator operation for the idealized amplifier model (amplifier hits the rails in saturation); the amplifier supply voltage is 10 V.

Exercise 7.7: The relaxation oscillator circuit in Fig. 7.19 uses $R_1 = 1$ kΩ and $R_2 = 4$ kΩ. The supply voltage of the amplifier is ± 15 V. Determine the amplitude (peak value) of the oscillating capacitor voltage and the oscillating output voltage.

Answer: 3 V and 15 V, respectively.

7.3.5 Oscillation Frequency

Consider the positive half cycle in Fig. 7.21. The solution for the RC circuit in Fig. 7.20 is given by Eqs. (7.16a, b) with $V_S = V_{CC}$. The constant K_1 is found from the initial condition of $v_C(t = 0) = -\beta V_{CC}$ where $\beta = R_1/(R_1 + R_2)$ is the amount of the positive feedback. We assume that the initial time instance has been switched to the start of the half cycle. Therefore, during the entire positive half cycle

$$\upsilon_C(t) = V_{CC} - (1+\beta)V_{CC}\exp\left(-\frac{t}{\tau}\right), \quad \tau = RC \tag{7.34}$$

At the end of the positive half cycle, the capacitor voltage becomes βV_{CC}. This allows us to find the half cycle duration $T/2$ and the oscillation period T. Solving Eq. (7.34) with $\upsilon_C = \beta V_{CC}$ and $t = T/2$, one has (f is the oscillation frequency in hertz)

$$T = 2\tau \ln\frac{1+\beta}{1-\beta}, \quad f = \frac{1}{T} = \frac{1}{2\tau}\left(\ln\frac{1+\beta}{1-\beta}\right)^{-1} \tag{7.35}$$

Exercise 7.8: For the relaxation oscillator with $R_1 = R_2$, express the oscillation frequency in terms of its time constant τ.

Answer: $f = \frac{0.455}{\tau}$ [Hz].

7.3.6 Circuit Implementation: 555 Timer

Figure 7.22 shows the relaxation oscillator circuit implemented in a laboratory and its output voltages when the supply voltages are ±5 V. Since the realistic amplifier might not reach the supply rails, the peak-to-peak (Pk-Pk) value of the output voltage might be less than 10 V. The output current limitations of the amplifier IC may severely affect circuit performance. Also, the oscillation frequency only *approximately* follows Eq. (7.35).

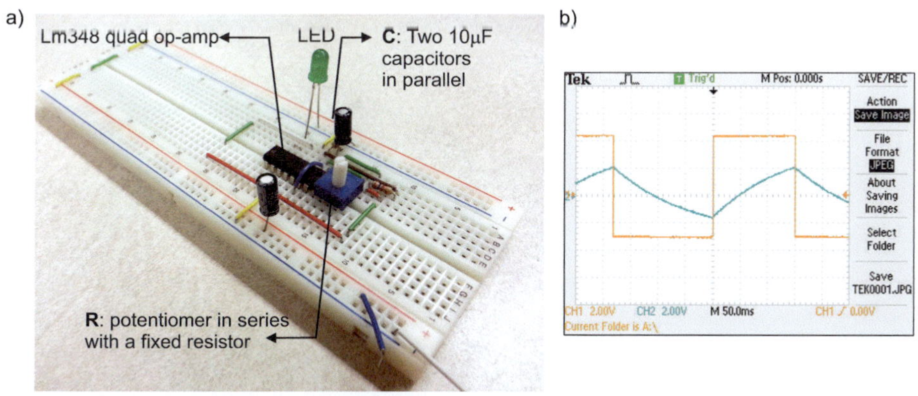

Fig. 7.22. (a) Timer circuit operation in laboratory. The square waveform is the output voltage $\upsilon_{out}(t)$. The curved waveform is the capacitor voltage $\upsilon_C(t)$. A variable resistance R makes it possible to visually control the oscillation frequency using an LED connected to the output through a *buffer amplifier* . The oscillation frequency changes from 0.5 to about 100 Hz.

555 Timer Integrated Circuit

Rectangular pulse forms at lower frequencies are routinely created by the well-known *555 timer IC* (integrated circuit). The 555 timer operates conceptually similarly to the relaxation oscillator circuit described above; it is more versatile though. The 555 timer creates a waveform of relatively sharp and clean rectangular voltage pulses whose frequency is controlled by an external capacitor and resistor. The *duty cycle* (ratio of the positive phase duration to the signal period) can also be controlled. The 555 timer is perhaps one of the most popular integrated circuits ever built.

Section 7.4 Single-Time-Constant (STC) Transient Circuits

7.4.1 Circuits with Resistances and Capacitances

Consider a transient circuit with an arbitrary number of capacitances and resistances. The circuit has an independent voltage source (or sources) and a switch. Instead of the voltage source, some capacitors may be charged prior to closing the switch. A *single-time-constant transient circuit* (*STC circuit*) is that which solution has the form

$$v(t) = K_1 \exp\left(-\frac{t}{\tau}\right) + K_2 \tag{7.36}$$

for *any* branch voltage in the circuit. In other words, only *one* exponential function is involved, similar to the basic RC circuits studied previously. Here, τ is the only *time constant* of the circuit. The STC transient circuits are frequently encountered in practice, in particular, in the study of transistor amplifiers. The STC transient circuits include:

1. Transient circuits with only one capacitance C. According to Thévenin's theorem, the network of resistances and source(s) seen by the capacitor is reduced to its Thévenin equivalent. As a result, we obtain the circuit shown in Fig. 7.7. Its time constant is given by

   $$\tau = R_T C \tag{7.37}$$

   where R_T is Thévenin resistance – the equivalent resistance of the network with the independent voltage source(s) shorted out.
2. Transient circuits with only one resistance R. Thévenin's theorem may be applied again, this time to the network of capacitances and source(s) seen by the resistance. As a result, we again obtain the circuit from Fig. 7.7. Its time constant is given by

   $$\tau = R C_T \tag{7.38}$$

   where C_T is the equivalent capacitance of the network with the independent voltage source(s) shorted out.
3. Transient circuits with an arbitrary number of capacitances and resistances given that the solutions for different capacitor voltages obtained by the simultaneous use of KVL and KCL are all *linear functions* of each other. Consider a circuit with multiple capacitances. Assume that N is the final number of capacitances after all possible series/parallel combinations. For the STC condition to hold, there should be $N-1$ independent closed loops that include *only* capacitances (and possibly independent voltage source(s)) but do not include resistances. This useful result has been confirmed by the authors based on an extensive circuit analysis.

As an example, we consider here simple transient circuits shown in Fig. 7.23.

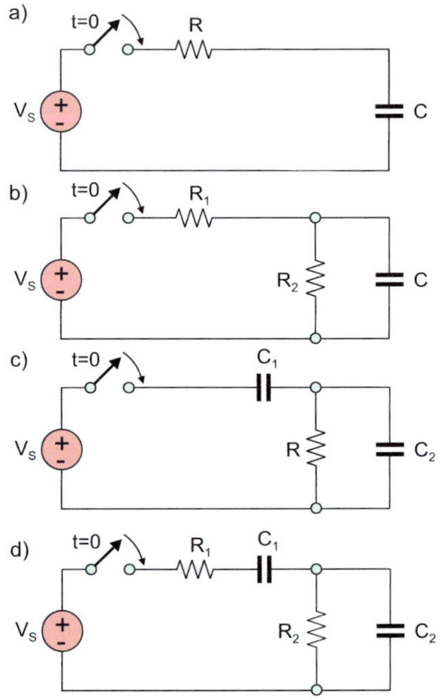

Fig. 7.23. Transient circuits with multiple resistances and capacitances.

Exercise 7.9: For the circuits in Fig. 7.23, establish the number of independent closed loops that include only capacitance(s) and independent voltage source but do not include resistances.

Answer: (a) Zero; (b) zero; (c) one; (d) zero.

The first case in Fig. 7.23 is the transient circuit of Fig. 7.7. The positions of the switch and the resistor may be interchanged without affecting the circuit solution, which has the form of Eq. (7.36)—see also Eqs. (7.16). The second case is again the STC circuit; the solution for the capacitor voltage (and any other voltage in the circuit) is given by Eq. (7.36) with $\tau = \left(R_1 \| R_2\right)C$. The constants K_1 and K_2 in this equation will be different for voltages across different circuit elements. The third case is also the STC circuit; the solution for the capacitor voltage (and any other voltage in the circuit) is still given by Eq. (7.36) with $\tau = R(C_1 + C_2)$. This case requires extra care since two capacitances and the source form a closed loop. Therefore, according to KVL, capacitor voltages cannot be both equal to zero at the initial time moment (and at any other time moment). Finally, consider the last case in Fig. 7.23. For this circuit with two capacitances and two resistances, there is no closed loop that includes only the capacitances but does not include resistances. Therefore, this circuit is *not* a STC circuit. This result will be confirmed shortly.

Exercise 7.10: Given initially uncharged capacitor(s), write solutions for the capacitor voltage for the circuits shown in Fig. 7.23a–c. For the circuit in Fig. 7.23c, assume that that C_1 was initially uncharged but C_2 was initially charged to V_S and connected to the rest of the circuit at $t = 0$ via a second switch.

Answer:

$$v_C(t) = V_S\left[1 - \exp\left(-\frac{t}{\tau}\right)\right], \quad \tau = RC \quad \text{in Fig. 7.23a} \tag{7.39}$$

$$v_C(t) = V_S\frac{R_2}{R_1 + R_2}\left[1 - \exp\left(-\frac{t}{\tau}\right)\right], \quad \tau = (R_1||R_2)C \quad \text{in Fig. 7.23b} \tag{7.40}$$

$$\begin{aligned} &v_{C1}(t) = V_S\left[1 - \exp\left(-\frac{t}{\tau}\right)\right], \quad v_{C2}(t) = V_S\exp\left(-\frac{t}{\tau}\right), \\ &\tau = R(C_1 + C_2) \quad \text{in Fig. 7.23c} \end{aligned} \tag{7.41}$$

7.4.2 Circuits with Resistances and Inductances

The forthcoming analysis is quite similar to the analysis performed previously for the capacitances and resistances. The *single-time-constant transient circuit* (*STC circuit*) with resistances and inductances is that which solution has the form

$$i(t) = K_1\exp\left(-\frac{t}{\tau}\right) + K_2 \tag{7.42}$$

for *any* branch current in the circuit. The list of the corresponding STC circuits also becomes identical to the previous case with a few modifications:

1. Capacitances are replaced by inductances.
2. Independent current source(s) will be used. They may be converted to voltage sources according to the source transformation theorem.
3. In condition #3, instead of loops, we use circuit *node*s. This condition is now formulated as follows. Consider a circuit with multiple inductances. Assume that N is the final number of inductances after all possible series/parallel combinations. For the STC condition to hold, there should be $N - 1$ independent (single) nodes, every branch of which is either an inductance or an independent current source.

As an example, we consider here simple transient circuits shown in Fig. 7.24.

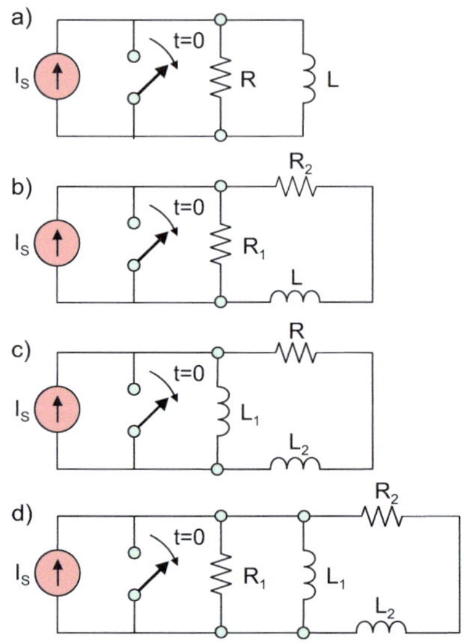

Fig. 7.24. Transient circuits with multiple resistances and inductances.

Exercise 7.11: For the circuits in Fig. 7.24, establish the number of independent nodes, every branch of which is either an inductance or an independent current source.

Answer: (a) Zero; (b) zero; (c) one; (d) zero.

The first case in Fig. 7.24 is the transient circuit of Fig. 7.12. The corresponding solution for the inductor current is given by Eq. (7.42); see also Eqs. (7.28). The second case is again the STC circuit; the solution for the inductor current (and any other current in the circuit) is given by Eq. (7.42) with $\tau = L/(R_1 + R_2)$. The constants K_1 and K_2 in this equation will be different for voltages across different circuit elements. The third case is also the STC circuit; the solution for the inductor current (and any other current in the circuit) is still given by Eq. (7.42) with $\tau = (L_1 + L_2)/R$. This case requires extra care since two inductances and the current source are three branches of the same node. Therefore, according to KCL, inductance currents cannot be both equal to zero at the initial time moment (and at any other time moment). We may assume, for example, that inductance L_2 carried initial current I_S and employs a second switch. Finally, consider the last case in Fig. 7.24. For this circuit with two inductances and two resistances, there is no nontrivial *single* node that includes only the inductances but does not include resistances. Therefore, this circuit is not the STC circuit.

Exercise 7.12: Given zero inductor current at the initial time moment, write solutions for the inductor current for the circuits shown in Fig. 7.24a–c. For the circuit in Fig. 7.24c, additionally assume that that inductor L_1 was carrying zero initial current but inductor L_2 was initially carrying current I_S and connected to the rest of the circuit at $t = 0$ via a second switch.

Answer:

$$i_L(t) = I_S\left[1 - \exp\left(-\frac{t}{\tau}\right)\right], \quad \tau = L/R \quad \text{in Fig. 7.24a} \tag{7.43}$$

$$i_L(t) = I_S\frac{R_1}{R_1 + R_2}\left[1 - \exp\left(-\frac{t}{\tau}\right)\right], \quad \tau = L/(R_1 + R_2) \text{ in Fig. 7.24b} \tag{7.44}$$

$$i_{L1}(t) = I_S\left[1 - \exp\left(-\frac{t}{\tau}\right)\right], \quad i_{L2}(t) = I_S\exp\left(-\frac{t}{\tau}\right), \quad \tau = (L_1 + L_2)/R \quad \text{in Fig. 7.24c} \tag{7.45}$$

7.4.3 Example of a Non-STC Transient Circuit

Figure 7.25a shows the last circuit from Fig. 7.23 to be analyzed here in detail. With reference to Fig. 7.25a, KCL and KVL give

$$\begin{aligned} i = i_1 + i_2 \Rightarrow C_1\frac{dv_1}{dt} = C_2\frac{dv_2}{dt} + \frac{v_2}{R_2}; \\ -V_S + v_1 + v_2 = 0 \Rightarrow R_1C_1\frac{dv_1}{dt} = V_S - v_1 - v_2 \end{aligned} \tag{7.46}$$

Expressing either v_1 in terms of v_2 or vice versa, we obtain from Eq. (7.46) a *second-order ODE* for either of the capacitor voltages. For simplicity, we will assume that $C_1 = C_2 = C$, $R_1 = R_2 = R$, and $\tau_0 = RC$. Then, the corresponding ODE for v_1 has the form

$$\frac{d^2v_1}{dt^2} + \frac{3}{\tau_0}\frac{dv_1}{dt} + \frac{v_1}{\tau_0^2} = \frac{V_S}{\tau_0^2} \tag{7.47}$$

The analysis of the second-order ODEs like Eq. (7.47) is thoroughly explained in the last section of this chapter. Here, we present its succinct version suitable for our immediate purposes. The solution of the *homogeneous second-order ODE* is sought in the form $\exp(-\alpha t/\tau_0)$ with α being a dimensionless constant. Substitution of this expression into the homogeneous ODE gives a quadratic equation for α, $\alpha^2 + 3\alpha + 1 = 0$, with the two positive roots $\alpha_1 = 2.62$ and $\alpha_2 = 0.38$. Therefore, the solution for Eq. (7.47) should have the form

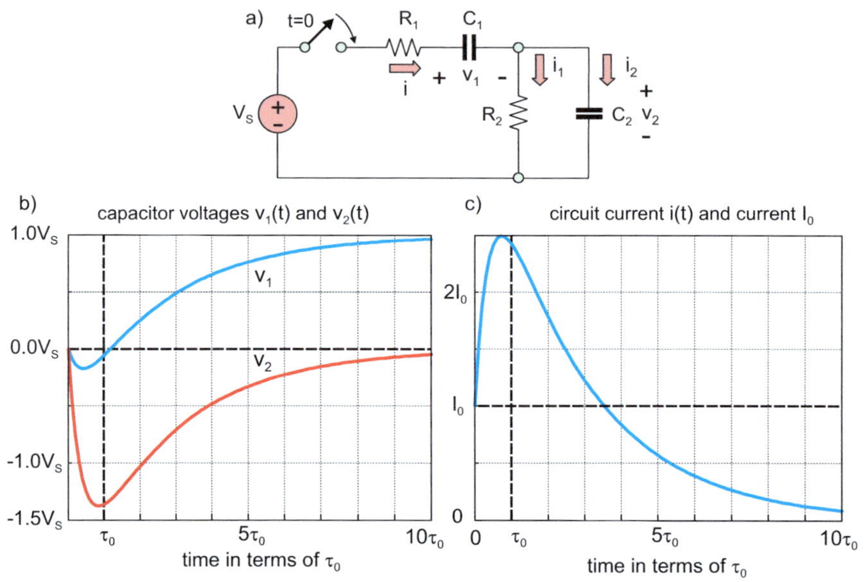

Fig. 7.25. A non-STC circuit with two resistances and two capacitances and its solution behavior.

$$v_1(t) = V_S + K_1\exp\left(-\frac{\alpha_1 t}{\tau_0}\right) + K_2\exp\left(-\frac{\alpha_2 t}{\tau_0}\right) \tag{7.48}$$

where K_1, K_2 are two constants determined by the initial conditions The solution for voltage $v_2(t)$ is found using the second of equations (7.46):

$$v_2(t) = (1+\alpha_1)K_1\exp\left(-\frac{\alpha_1 t}{\tau_0}\right) + (1+\alpha_2)K_2\exp\left(-\frac{\alpha_2 t}{\tau_0}\right) \tag{7.49}$$

The initial conditions imply that both capacitors are uncharged prior to closing the switch. This gives $K_1 = (1+\alpha_2)/(\alpha_1 - \alpha_2)V_S$ and $K_2 = (1+\alpha_1)/(\alpha_2 - \alpha_1)V_S$. With this in mind, the solution is complete. Figure 7.25b, c shows the behavior of the two capacitor voltages. A truly remarkable point is that instantaneous voltage across the second capacitor in Fig. 7.25b *exceeds* the (absolute) source voltage. Moreover, the instantaneous circuit current in Fig. 7.25c *exceeds* the initial circuit current $I_0 = V_S/R_1$ by 2.5 times. Those distinct features are observed for other *second-order* transient circuits studied further.

7.4.4 Example of an STC Transient Circuit

In Fig. 7.26, we present an example of a rather complicated circuit, which still follows the STC circuit model. The proof is based on the observation that node (*) in Fig. 7.26 connects three branches: two of which are exactly the inductances and the remaining one is the current source.

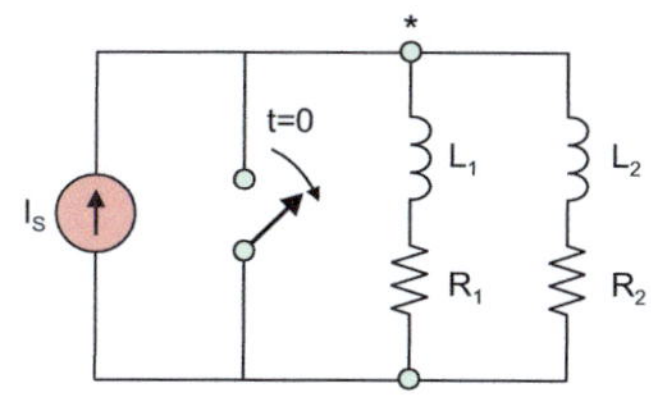

Fig. 7.26. An STC circuit with two inductances and two resistances.

7.4.5 Method of Thévenin Equivalent and Application Example: Circuit with a Bypass Capacitor

Let us consider a transient circuit with a bypass capacitor shown in Fig. 7.27a. It includes a voltage source represented by its Thévenin equivalent $\upsilon_S(t)$, R_S and a load represented by its equivalent resistance R_L. The source is turned on at $t = 0$ and it generates a voltage in the form of a (large) DC component V_S and a superimposed (small) AC signal:

$$\upsilon_S(t) = V_S + V_m \cos \omega t, \quad \omega = 2\pi f \tag{7.50}$$

where f is frequency in Hz. Generally, $V_S >> V_m$.

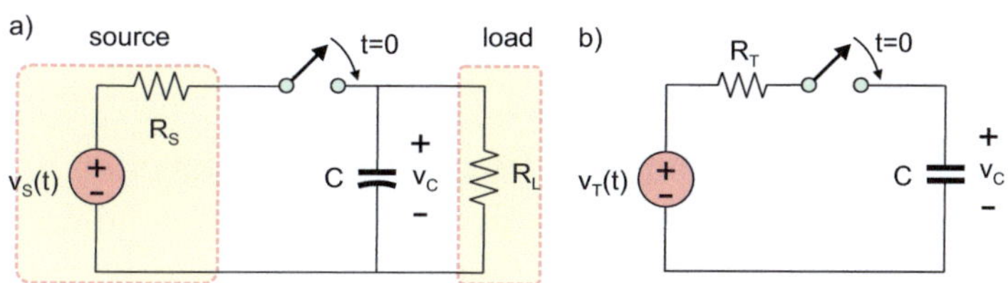

Fig. 7.27. Model of a voltage source connected to a load with a bypass capacitor.

We first interchange positions of capacitance C and resistance R_L in Fig. 7.27. With the switch closed, the voltage source $\upsilon_S(t)$ with two resistances R_S, R_L is then converted to its Thévenin equivalent circuit with

$$\upsilon_T(t) = \frac{R_L}{R_S + R_L}\upsilon_S(t), \quad R_T = R_S \| R_L \tag{7.51}$$

The resulting circuit is the simple RC circuit shown in Fig. 7.27b. The method of a Thévenin (or Norton) equivalent circuit is common for all transient circuits with one capacitor or inductor as explained in the previous section. According to this method, a circuit with *one* capacitance or *one* inductance is *always* converted to the basic RC or RL transient circuit. Therefore, it is always *a first-order* transient circuit which is described by a *first-order* ODE. Once the solution for the capacitor voltage $\upsilon_C(t)$ in Fig. 7.27a is found, the voltage across the load resistor is then simply expressed as $\upsilon_L(t) = \upsilon_C(t)$.

Forced Response and Natural Response

Using KVL, KCL, and the capacitor equation, we obtain the first-order ODE for the circuit in Fig. 7.27b in the form

$$\frac{dv_C(t)}{dt} + \frac{v_C(t)}{\tau} = f(t), \quad f(t) = \frac{1}{\tau}\frac{R_L}{R_S + R_L}(V_S + V_m \cos\omega t), \quad \tau = R_T C \tag{7.52}$$

at any *positive* time instance, $t > 0$. The initial condition is $v_C(t = 0) = 0$. The time-dependent source term $f(t)$ is called a *forcing function*, whether or not the switch is present. The action of the switch may always be included into $f(t)$ by setting $f(t) = 0$ at $t < 0$. A *general solution* of the *homogeneous* Eq. (7.52), let us call it $x_c(t)$, is known as the *complementary solution* (or *natural response*). The natural response is given by $x_c(t) = K\exp(-t/\tau)$ where K is an arbitrary constant. The natural response carries information *about the circuit*. A *particular solution* of the *inhomogeneous* Eq. (7.52), let us call it $x_p(t)$, is known as the *forced response*. The forced response carries information *about the excitation* $f(t)$. It does not need to contain an arbitrary integration constant. By linearity, the complete solution for Eq. (7.53) is the *sum* of both responses, i.e.,

$$v_C(t) = x_p(t) + x_c(t) \tag{7.53}$$

We already know the natural response. The forced response is sought in the form

$$x_p(t) = a\cos\omega t + b\sin\omega t + c \tag{7.54}$$

where constants a, b, c are to be uniquely determined. If the function $f(t)$ is sinusoidal, a combination of sine and cosine would suffice under certain conditions for *any* linear differential equation, not necessarily of the first order. The particular solution in the form of Eq. (7.54) is even useable for second-order circuits considered next.

Example 7.7: Find the forced response for Eq. (7.52).

Solution: Equation (7.54) is substituted into Eq. (7.52); then all terms are pulled to the left-hand side, and all terms with $\cos\omega t$ and $\cos\omega t$ are combined. We obtain:

$$\cos\omega t\left[\frac{a}{\tau} + b\omega - \frac{V_m}{\tau}\frac{R_L}{R_S + R_L}\right] + \sin\omega t\left[\frac{b}{\tau} - a\omega\right] + \left[\frac{c}{\tau} - \frac{V_S}{\tau}\frac{R_L}{R_S + R_L}\right] = 0 \tag{7.55}$$

In order to satisfy Eq. (7.55), we require all three expressions in the square brackets to be zero. This operation yields a system of two equations for a and b, while the constant c is found directly. Working out the algebra produces the results

$$a = \frac{R_L V_m}{R_S + R_L}\frac{1}{1 + (\omega\tau)^2}, \quad b = \frac{R_L V_m}{R_S + R_L}\frac{\omega\tau}{1 + (\omega\tau)^2}, \quad c = \frac{R_L V_S}{R_S + R_L} \tag{7.56}$$

Example 7.8: Obtain the complete solution for Eq. (7.52) and plot it to scale.

Solution: The final solution is based on Eqs. (7.54)–(7.56). The initial condition $v_C(t=0) = 0$ is satisfied if $K = -a - c$. The required load voltage has the form

$$v_L(t) = v_C(t) = \frac{R_L V_S}{R_S + R_L}(1 - \exp(-t/\tau)) + \frac{R_L V_m}{R_S + R_L}\frac{1}{1 + (\omega\tau)^2}[\cos\omega t + \omega\tau\sin\omega t - \exp(-t/\tau)] \quad (7.57)$$

This solution is compared with the load voltage without bypass capacitor:

$$v_L(t) = \frac{R_L V_S}{R_S + R_L} + \frac{R_L V_m}{R_S + R_L}\cos\omega t \quad (7.58)$$

Given that $\omega\tau$ and t/τ are both large, the dominant AC term in Eq. (7.57) is the sine function. Comparing Eqs. (7.57) and (7.58) with each other, we can therefore state that the bypass capacitor reduces the amplitude of the unwanted AC component at the load by a factor $1/(\omega\tau)$ while keeping the DC component unchanged! Figure 7.28 plots the load voltage with and without the bypass capacitor. The circuit parameters are $R_S = 5\ \Omega$, $R_L = 1\ \text{k}\Omega$, $C = 1000\ \mu\text{F}$. The source parameters are $V_S = 10$ V, $V_m = 1$ V, $f = 500$ Hz. One hidden yet critical solution parameter is the source resistance R_S. When this parameter is very small, the bypass capacitor has little if any effect on the solution.

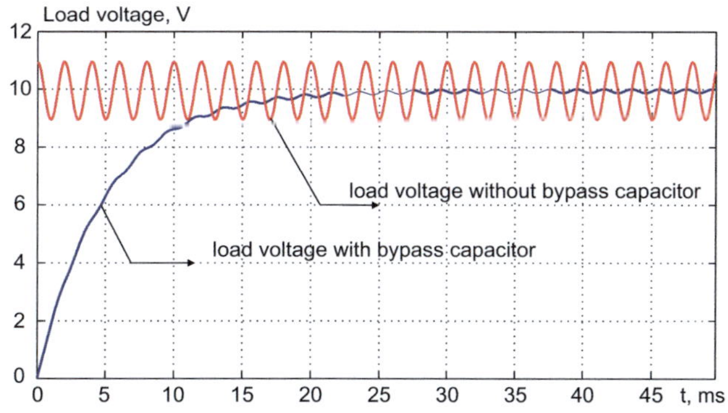

Fig. 7.28. Load voltage with and without the bypass capacitor predicted by Eqs. (7.57) and (7.58).

Section 7.5 Description of the Second-Order Transient Circuits

7.5.1 Types of Second-Order Transient Circuits

A *first-order transient circuit* is described by a *first-order* differential equation. All single-time-constant (STC) circuits considered thus far are the first-order transient circuits. A *second-order transient circuit* is described by a *second-order* differential equation. Figure 7.29 outlines two major types of the second-order order transient circuits.

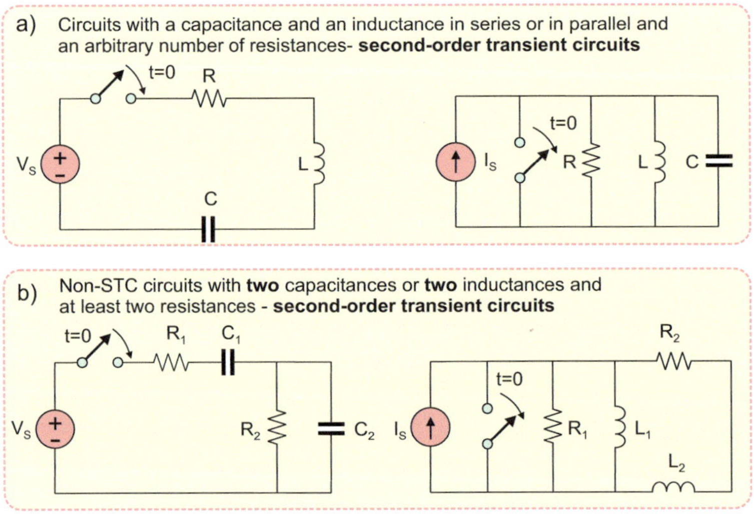

Fig. 7.29. Classification of second-order transient circuits.

The first type involves *two* nonidentical dynamic elements (capacitance and inductance) assembled as a series or parallel two-terminal *LC network*—see Fig. 7.29a. This figure shows two such connections. The number of resistances may be arbitrary. As long as the LC network sees only the combination of source(s) and resistances, the equivalent circuit (obtained with the help of Thévenin or Norton equivalents) will have the form shown in Fig. 7.29a. The second type includes non-STC circuits with two capacitances or two inductances—see Fig. 7.29b. In this section, we will study the second-order circuits on the base of the LC networks.

7.5.2 Series-Connected Second-Order RLC Circuit

Generic Representation of a Series RLC Circuit and Qualitative Operation

Consider a series LC network shown in the shaded box of Fig. 7.30. Using Thévenin's theorem, any network of resistances/independent DC sources connected to this block may be represented as the series combination of the voltage source and the resistance R. This is why we arrive at the very generic series RLC circuit shown in Fig. 7.30. The switch implies that the ideal voltage source is to be connected to the circuit at time $t = 0$. Prior to $t = 0$, this source has no effect, it is disconnected. Qualitatively, just after closing the switch the entire voltage drop will be acquired by the inductance. Then, the capacitor voltage will increase. Finally, the capacitor voltage will assume the supply voltage V_S, and no current will flow in the circuit under DC steady-state condition.

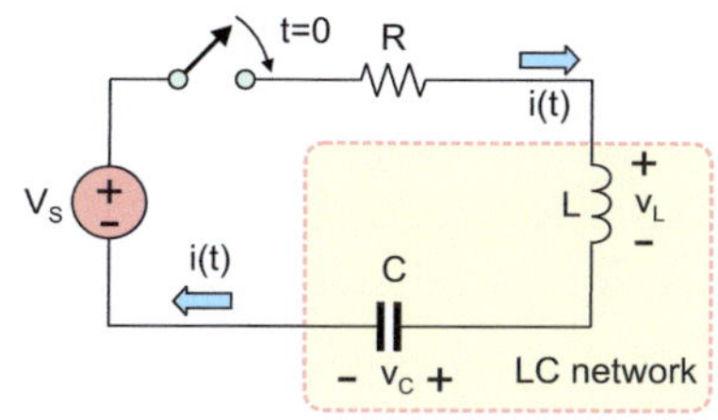

Fig. 7.30. Series RLC second-order transient circuit; the switch closes at $t = 0$.

Mechanical Analogy

Figure 7.31 shows an intuitive analogy between a mechanical mass-spring-damping system and an electric (or electronic) RLC transient circuit depicted in Fig. 7.30.

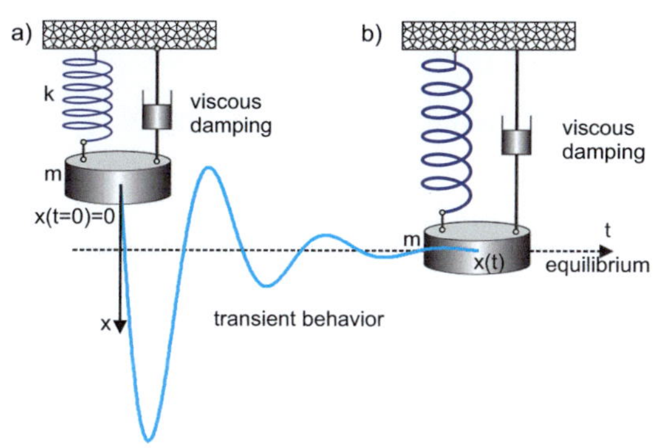

Fig. 7.31. Mechanical counterpart of an RLC circuit; at $t = 0$ the gravity force is applied.

The inductance L corresponds to the mechanical mass m. The capacitance C is the inverse stiffness, $1/k$, of the spring. The resistance R corresponds to viscous damping. At $t = 0$, the gravity force is applied to mass m in Fig. 7.31 initially located at $x = 0$. As a result, the mass reaches a new equilibrium position $x(t \to \infty)$ with or without intermediate oscillations depending on damping. The following correspondences may be established between mechanical and electrical quantities ($q(t)$ is the capacitor charge):

$$i(t) \quad \to \quad dx(t)/dt, \quad q(t) \quad \to \quad x(t), \quad mg \quad \to \quad V_S \tag{7.59}$$

Solution in Terms of Circuit Current

A solution to the dynamic circuit in Fig. 7.30 is based on KVL and KCL. KCL prescribes the same current $i(t)$ through every circuit element at *any* time instant t. KVL states

$$v_L + v_R + v_C = V_S \tag{7.60}$$

at any *positive* time $t > 0$. Indeed, v_L, v_R, v_C are all functions of time. Using constitutive relations for the inductance, resistance, and capacitance (integral form), we obtain

$$L\frac{di(t)}{dt} + Ri(t) + \frac{1}{C}\int_0^t i(t')dt' = V_S \tag{7.61}$$

where we assumed that the initial capacitor voltage is zero. Differentiation over time yields the expected *homogeneous* second-order ODE:

$$L\frac{d^2i(t)}{dt^2} + R\frac{di(t)}{dt} + \frac{1}{C}i(t) = 0 \tag{7.62}$$

which, after division by L, can be cast in the *standard form*:

$$\frac{d^2i(t)}{dt^2} + 2\alpha\frac{di(t)}{dt} + \omega_0^2 i(t) = 0 \tag{7.63}$$

The two constants present in this equation are given by

$$\alpha = \frac{R}{2L}, \quad \omega_0 = \frac{1}{\sqrt{LC}} \tag{7.64}$$

Both constants in Eq. (7.64) have a general mathematical meaning that should be remembered. The first constant α with the units of neper/sec is the *damping coefficient*. It generally characterizes how fast oscillations in Fig. 7.31 decay and reach a steady state. The neper (Np) is a *dimensionless* unit named after John Napier (1550–1617), a Scottish mathematician. The constant α is similar to the inverse time constant $1/\tau$ for the first-order transient circuits. The second constant ω_0 with the units of rad/s is the *undamped resonant frequency* of the RLC circuit. This constant characterizes frequency of oscillations—see again the mechanical analogy in Fig. 7.31. The meaning of the undamped resonant frequency remains the *same* for any LC circuit block, either in transient analysis or in the AC circuit analysis, either for series or parallel configurations.

Exercise 7.13: When is $i(t) = K\cos\omega_0 t$ a solution to Eq. (7.63)?

Answer: At $R = 0$.

7.5.3 Initial Conditions in Terms of Circuit Current and Capacitor Voltage

In contrast to the first-order transient circuits, there are two equally possible choices of the independent function for the RLC circuit in Fig. 7.30:

- Circuit (or inductor) current $i(t)$
- Capacitor voltage, $v_C(t)$

The circuit current remains continuous over time (inductor "inertia") and so does the capacitor voltage (capacitor "inertia"). We have chosen the circuit current and obtained the second-order ODE Eq. (7.63). You may wonder if this is really the best choice. The answer is nontrivial and is hidden in the initial conditions. Any first-order ODE needs *one initial condition*. Any second-order ODE needs *two initial conditions*. Let us establish these initial conditions for the circuit current first. Following the current continuity through the inductor, the circuit current must be zero at $t = 0$, that is,

$$i(t = 0) = 0 \tag{7.65a}$$

Hence, the first initial condition is established. The second one is that of the initial capacitor voltage equal to zero. According to Eq. (7.60) and Eq. (7.65a),

$$v_L(t = 0) + \underbrace{v_R(t = 0)}_{Ri(t=0)=0} + \underbrace{v_C(t = 0)}_{0} = V_S \Rightarrow L\frac{di}{dt}(t = 0) = V_S \Rightarrow \frac{di}{dt}(t = 0) = \frac{V_S}{L} \tag{7.65b}$$

Thus, the initial circuit current is zero, whereas its first derivative is *not*. This is a drawback of the electric current formulation given by Eq. (7.63) for the series RLC circuit. On the other hand, the capacitor voltage and its derivative (which is proportional to the capacitor/inductor/circuit current) are both zero at $t = 0$, which leads to a simpler "universal" homogeneous formulation of the initial conditions, i.e.,

$$v_C(t = 0) = 0, \quad \frac{dv_C(t = 0)}{dt} = 0 \tag{7.66}$$

At the same time, the second-order ODE for the capacitor voltage becomes *inhomogeneous*, i.e., at $t \geq 0$

$$\frac{d^2 v_C(t)}{dt^2} + 2\alpha\frac{dv_C(t)}{dt} + \omega_0^2 v_C(t) = \omega_0^2 V_S \tag{7.67}$$

It is worth noting that this equation has exactly the same form as Eq. (7.63), but with the nonzero right-hand side. The derivation of Eq. (7.67) is similar to the derivation of Eq. (7.63); it is suggested as a homework problem.

7.5.4 Step Response and Choice of the Independent Function

The selection between current and voltage formulations reflects our desire to convert the circuit differential equation into a standard form, which will allow us to use powerful tools of signals and systems theory. What should be preferred: the homogeneous second-order ODE Eq. (7.63) for the circuit current augmented with inhomogeneous initial conditions Eqs. (7.65a, b), or the inhomogeneous second-order ODE Eq. (7.67) for the capacitor voltage augmented with the homogeneous initial conditions Eq. (7.66)? The answer is as follows. If the initial conditions are all zero, the only remaining excitation is *the forcing function*: the right-hand side of Eq. (7.67). We consider a unit *step function*, $u(t)$, defined by (see Fig. 7.32a)

$$u(t) = \begin{cases} 0 & t < 0 \\ 1 & t \geq 0 \end{cases} \tag{7.68}$$

Equation (7.67) may be conveniently written in the form

$$\frac{d^2 v_C(t)}{dt^2} + 2\alpha \frac{dv_C(t)}{dt} + \omega_0^2 v_C(t) = Au(t) \tag{7.69}$$

at *any* time instant where $A = \omega_0^2 V_S$. The forcing function is thus the product of the constant A and the unit step function $u(t)$, as seen in Fig. 7.32b. The solution to Eq. (7.67) or Eq. (7.69), after division by A, is the *normalized step response* of a second-order system. We call it the response to a unit step voltage excitation. It is generally accepted in signals and systems theory that for the unit step response the initial conditions should be homogeneous or *zero*. Therefore, Eq. (7.67) or Eq. (7.69) with zero initial conditions Eq. (7.66) is preferred when dealing with future applications of the unit step response.

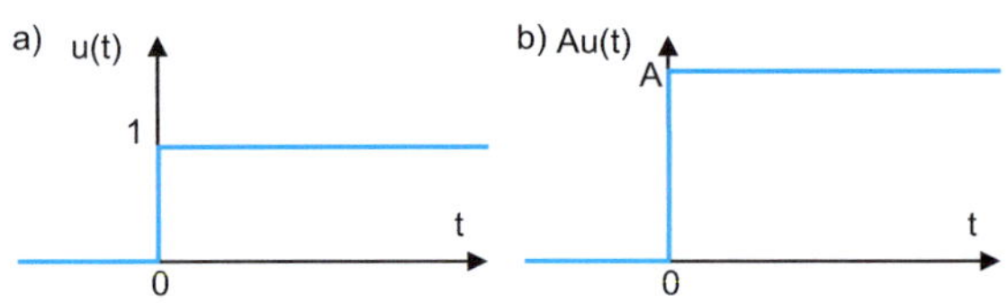

Fig. 7.32. (**a**) Unit step function $u(t)$ and (**b**) the scaled right-hand side of Eq. (7.69).

The step response is the "business card" of the circuit, which actually contains the complete information about its behavior. If the circuit in Fig. 7.30 with an arbitrary time-varying voltage source $V_S \to v_S(t)$ is considered, this source may be represented as a number of "steps" in time. Hence, the complete solution may be constructed as a sum (or integral) of the elementary unit step response solutions, properly scaled and shifted in time. As an example, we consider a voltage source $v_S(t)$ in Fig. 7.30, which generates a pulse (think of one bit) with the duration T and a 5-V peak value, as depicted in Fig. 7.33.

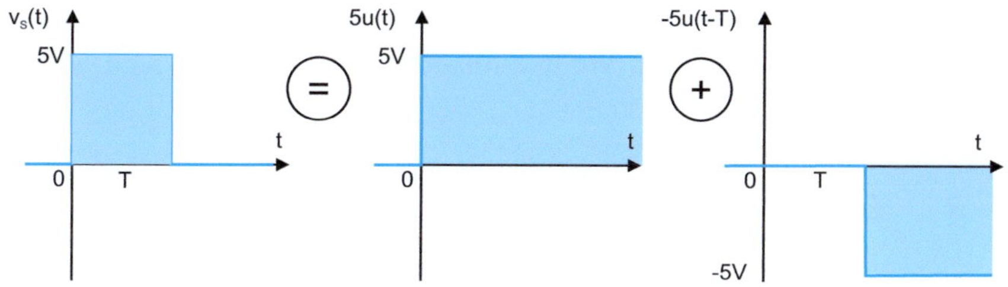

Fig. 7.33. A one-bit excitation voltage constructed as the sum of two step responses.

The switch in Fig. 7.30 is replaced by a short circuit. According to Fig. 7.33, $v_S(t) = 5u(t) - 5u(t-T)$. By linearity, the solution for the bit excitation is simply the sum $5v_C(t)$ $-5v_C(t-T)$ in Fig. 7.33 where $v_C(t)$ is the solution of Eq. (7.69) with $V_S = 1$ V.

7.5.5 Parallel Connected Second-Order RLC Circuit

Generic Representation of the Parallel RLC Circuit and Qualitative Operation

Consider a parallel connected LC block in the shaded box in Fig. 7.34. Using Norton's theorem, any network of resistors and power supplies connected to this block is represented as the parallel combination of a current source and a resistor R, as seen in Fig. 7.34. This is why we arrive at the very generic parallel RLC circuit shown in Fig. 7.34. The switch implies that the ideal current power supply is to be connected to the circuit at time instant $t = 0$. Qualitatively, just after closing the switch, the entire circuit current will flow through the capacitor. As time progresses, the capacitor current will decrease. Finally, the entire current will flow through the inductor, which becomes the short circuit under DC conditions.

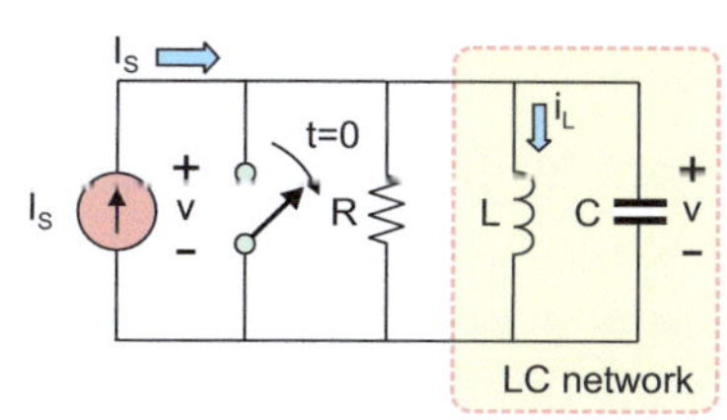

Fig. 7.34. Generic representation of any network of sources and resistances connected to the parallel LC circuit: the standard parallel RLC circuit.

Circuit Equation in Terms of Voltage

The same reasoning as mentioned for the series RLC circuit applies. There are two possible choices of the independent function for the RLC circuit in Fig. 7.34:

- Circuit (or capacitor) voltage, $v(t)$
- Inductor current $i_L(t)$

The capacitor voltage remains continuous over time (inductor "inertia") and so does the inductor current (inductor "inertia"). We use for derivation $v(t)$ first. KCL yields

$$C\frac{dv(t)}{dt}+\frac{v(t)}{R}+\frac{1}{L}\int_0^t v(t')dt' = I_S \tag{7.70}$$

at any *positive* time $t > 0$. Differentiation over time and division by C yield the expected *homogeneous* second-order ODE:

$$\frac{d^2v(t)}{dt^2}+\frac{1}{RC}\frac{dv(t)}{dt}+\frac{1}{LC}v(t)=0 \tag{7.71}$$

Eq. (7.71) is written in the form of Eq. (7.63):

$$\frac{d^2v(t)}{dt^2}+2\alpha\frac{dv(t)}{dt}+\omega_0^2 v(t)=0 \tag{7.72}$$

if we define the damping coefficient α and the undamped resonant frequency ω_0 as

$$\alpha=\frac{1}{2RC}, \quad \omega_0=\frac{1}{\sqrt{LC}} \tag{7.73}$$

Initial Conditions and Choice of Independent Function

The initial conditions for Eq. (7.72) are that the voltage across the capacitor and the inductor current must be continuous. Therefore, they must have the form

$$v(t=0)=0, \quad \frac{dv}{dt}(t=0)=\frac{I_S}{C} \tag{7.74}$$

The voltage derivative is not zero at the initial time moment. Eq. (7.74) is similar to Eqs. (7.65a, b). It has been stated that the step response of the second-order system is generally calculated with the homogeneous (zero) initial conditions. The second-order circuit ODE written in terms of the inductor current

$$\frac{d^2i_L(t)}{dt^2}+2\alpha\frac{di_L(t)}{dt}+\omega_0^2 v_L(t)=\omega_0^2 I_S \tag{7.75}$$

possesses zero initial conditions, i.e.,

$$i_L(t=0)=0, \quad \frac{di_L(t=0)}{dt}=0 \tag{7.76}$$

Those are the preferred conditions for the step response calculations. The derivation of Eq. (7.75) is similar to the derivation of Eq. (7.72); it is left as a homework problem.

Duality

Comparing Eqs. (7.66) and (7.67) for the series LCR circuit with Eqs. (7.76) and (7.75) for the parallel LCR circuit, we can establish the following substitutions:

$$V_S \leftrightarrow RI_S, \quad v_C \leftrightarrow Ri_L, \quad \frac{L}{R} \leftrightarrow RC \tag{7.77}$$

These substitutions make both sets of equations including all the constants mathematically *identical*. This fact reflects the *duality* of series/parallel RLC electric circuits. A similar duality is established for the steady-state RLC resonator circuits. Since the initial conditions are also the same, we conclude that the step response of the parallel RLC circuit is equivalent to the step response of the series RLC circuit.

Exercise 7.14: The damping coefficient of a second-order RLC circuit (i) does not depend on capacitance, (ii) decreases when resistance increases, and (iii) equals zero. Determine the circuit topology in every case.

Answer: (i) Series RLC circuit, (ii) parallel RLC circuit, (iii) series LC circuit.

Section 7.6 Step Response of the Series RLC Circuit

7.6.1 General Solution of the Second-order ODE

Solution for Step Response

The starting point is Eq. (7.69) of the previous section for the capacitor voltage of the series RLC circuit augmented with the homogeneous initial conditions, i.e.,

$$\frac{d^2 v_C(t)}{dt^2} + 2\alpha \frac{dv_C(t)}{dt} + \omega_0^2 v_C(t) = Au(t), \quad A = \omega_0^2 V_S \tag{7.78a}$$

$$v_C(t=0) = 0, \quad \frac{dv_C(t=0)}{dt} = 0 \tag{7.78b}$$

Similar to the first-order transient circuits with arbitrary sources, the general solution is also given by the sum of two parts: a *particular solution* of the inhomogeneous equation (7.78a), let us call it $x_p(t)$, and a *complementary solution*, let us call it $x_c(t)$, of the homogeneous equation (7.78a). Homogeneous implies that the right-hand side of the ODE equals zero. The particular solution is known as the *forced response* and the complementary solution is known as the *natural response*. As a result, the total solution is

$$v_C(t) = x_p(t) + x_c(t) \tag{7.79}$$

For the circuit with the DC voltage source and the switch acting as a step excitation, the particular solution is trivial. It is proved by direct substitution:

$$x_p(t) = V_S \Rightarrow v_C(t) = V_S + x_c(t) \tag{7.80}$$

The complementary solution carries information about the entire circuit and requires care.

Solution in Arbitrary Case

What if the right-hand side of Eq. (7.78a) is an arbitrary function of time? How is the solution obtained? We have already established that any such solution can be obtained on the basis of the step response. In general, the solution is expressed in terms of a *convolution integral*, which involves an arbitrary right-hand side of the second-order ODE and the time derivative of the step response. This interesting and fundamental question is studied further in signals and systems theory.

7.6.2 Derivation of the Complementary Solution: Method of Characteristic Equation

Similar to the first-order transient circuits, we seek a complementary solution (natural response) of the homogeneous Eq. (7.78a) in the most general exponential form

$$x_p(t) = K\exp(st) \tag{7.81}$$

where K and s are two arbitrary constants. The substitution yields

$$(s^2 + 2\alpha s + \omega_0^2)K\exp(st) = 0 \tag{7.82}$$

For a nontrivial solution, the *characteristic equation* $s^2 + 2\alpha s + \omega_0^2 = 0$ must be satisfied, that is,

$$s^2 + 2\alpha s + \omega_0^2 = 0 \Rightarrow s_{1,2} = \begin{cases} -\alpha + \sqrt{\alpha^2 - \omega_0^2} \\ -\alpha - \sqrt{\alpha^2 - \omega_0^2} \end{cases} \text{ or } s_{1,2} = \begin{cases} -\alpha\left(1 - \sqrt{\dfrac{\varsigma^2 - 1}{\varsigma^2}}\right) \\ -\alpha\left(1 + \sqrt{\dfrac{\varsigma^2 - 1}{\varsigma^2}}\right) \end{cases} \tag{7.83}$$

where the new constant $\zeta = \alpha/\omega_0$ is the *damping ratio* of the RLC circuit. Formally, this constant has units of 1/rad; it is often considered dimensionless. We must distinguish between three separate cases depending on the value of the damping ratio:

Case A This situation (*overdamping*) corresponds to $\varsigma > 1$. In this case, $s_{1,2}$ are both *real* and *negative*. Since the original ODE is linear, the general solution is simply the combination of two independent decaying exponential functions:

$$x_c(t) = K_1\exp(s_1 t) + K_2\exp(s_2 t) \tag{7.84a}$$

Case B This case (*critical damping*) corresponds to $\varsigma = 1$. Both roots $s_{1,2}$ become identical. Therefore, a solution in the form of Eq. (7.84a) with *two* independent constants can no longer be formed. Only one independent constant may be available. Fortunately, another solution in the form $t\exp(s_1 t)$ exists in this special case. This fact is proved by direct substitution. Thus, the general solution becomes

$$x_c(t) = K_1\exp(s_1 t) + K_2 t\exp(s_1 t) \tag{7.84b}$$

Case C This case (*underdamping*) corresponds to $\varsigma < 1$. Both roots $s_{1,2}$ become complex. This means that our initial simple guess Eq. (7.81) is no longer correct. One can prove by direct substitution that the general solution now has the oscillating form

$$x_c(t) = K_1\exp(-\alpha t)\cos\omega_n t + K_2\exp(-\alpha t)\sin\omega_n t \tag{7.84c}$$

where $\omega_n = \sqrt{\omega_0^2 - \alpha^2}$ is the (radian) *natural frequency* of the circuit. $1/\alpha$ is also called the *time constant* or *the time constant of the decay envelope*. The complementary solution in the form of Eqs. (7.84) always contains two independent integration constants. They should be used to satisfy the initial conditions, which complete the solution.

7.6.3 Finding Integration Constants

According to Eqs. (7.79) and (7.80), the capacitor voltage is $v_C(t) = x_c(t) + V_S$ where $x_c(t)$ is given by Eqs. (7.84). The integration constants may be found using Eq. (7.78b), which dictates that both the capacitor voltage and its derivative must vanish at the initial time $t = 0$. We then have from Eqs. (7.84)

$$\textbf{Case A.}\quad K_1 + K_2 + V_S = 0, \quad s_1 K_1 + s_2 K_2 = 0 \tag{7.85a}$$

$$\textbf{Case B.}\quad K_1 + V_S = 0, \quad s_1 K_1 + K_2 = 0 \tag{7.85b}$$

$$\textbf{Case C.}\quad K_1 + V_S = 0, \quad -\alpha K_1 + \omega_n K_2 = 0 \tag{7.85c}$$

The solution of Eqs. (7.85) has the form

$$\textbf{Case A.}\quad K_1 = \frac{s_2 V_S}{s_1 - s_2}, \quad K_2 = \frac{s_1 V_S}{s_2 - s_1} \tag{7.86a}$$

$$\textbf{Case B.}\quad K_1 = -V_S, \quad K_2 = s_1 V_S \tag{7.86b}$$

$$\textbf{Case C.}\quad K_1 = -V_S, \quad K_2 = -\frac{\alpha}{\omega_n} V_S \tag{7.86c}$$

Equations (7.83) through (7.86) complete the step response solution for the series RLC circuit. The circuit may behave quite differently depending on the value of the damping ratio ς.

7.6.4 Solution Behavior for Different Damping Ratios

We consider the series RLC circuit shown in Fig. 7.35. We will choose round numbers $L = 1\,\text{mH}$, $C = 1\,\text{nF}$. These values approximately correspond to an RLC transient circuit operating in the 100 kHz–1 MHz frequency band.

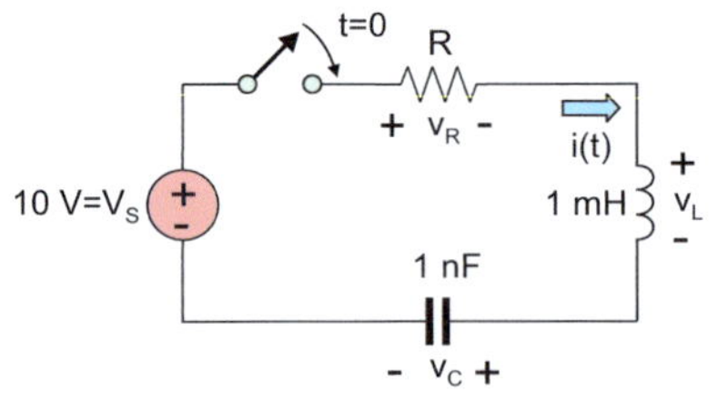

Fig. 7.35. RLC series circuit; the resistance value R may be varied.

Example 7.9: Determine the solution for the capacitor voltage for the circuit shown in Fig. 7.35 over a time interval from 0 to 25 μs for $R = 200\ \Omega,\ 1\ \text{k}\Omega,\ 2\ \text{k}\Omega$, and $20\ \text{k}\Omega$.

Solution: Equations (7.83) through (7.86) give

$R = 200\ \Omega \Rightarrow \alpha = 10^5, \omega_0 = 10^6 \Rightarrow \zeta = \mathbf{0.1}$ **(Case C—underdamped circuit)**

$$\Rightarrow v_C(t) = 10 - 10\exp(-10^5 t)\cos 9.95 \times 10^5 t - 1.005\exp(-10^5 t)\sin 9.95 \times 10^5 t \qquad (7.87\text{a})$$

$R = 1\ \text{k}\Omega \Rightarrow \alpha = 5 \times 10^5, \omega_0 = 10^6 \Rightarrow \zeta = \mathbf{0.5}$ **(Case C—underdamped circuit)**

$$\Rightarrow v_C(t) = 10 - 10\exp(-5 \times 10^5 t)\cos 8.66 \times 10^5 t - 5.77\exp(-5 \times 10^5 t)\sin 8.66 \times 10^5 t \qquad (7.87\text{b})$$

$R = 2\ \text{k}\Omega \Rightarrow \alpha = 10^6, \omega_0 = 10^6 \Rightarrow \zeta = \mathbf{1}$ **(Case B—critically damped circuit)**

$$\Rightarrow v_C(t) = 10 - 10\exp(-10^6 t) - 10^7 t\exp(-10^6 t) \qquad (7.87\text{c})$$

$R = 20\ \text{k}\Omega \Rightarrow \alpha = 10^7, \omega_0 = 10^6 \Rightarrow \zeta = \mathbf{10}$ **(Case A—overdamped circuit)**

$$\Rightarrow v_C(t) = 10 - 10.0252\exp(-5.013 \times 10^4 t) + 0.0252\exp(-1.995 \times 10^7 t) \qquad (7.87\text{d})$$

Equations (7.87) satisfy both the initial conditions to within numerical rounding error. Figure 7.36 shows the solution behavior for four distinct cases.

The first person who discovered and documented the oscillatory transient response of an electric circuit similar to that depicted in Fig. 7.36 was probably Félix Savary (1797–1841). A renowned astronomer and French academician, he worked with Ampère and discovered an oscillatory discharge of a Leyden jar (an early prototype of the battery) in 1823–1826. Some fifteen years later, the similar observation has been made by Joseph Henry.

7.6.5 Overshoot and Rise Time

One important result seen in Fig. 7.36a, b is the so-called dynamic overshoot caused by a sudden application of a voltage pulse and the associated voltage *ringing*. The dimensionless overshoot (overshoot percentage after multiplying by 100) M_p is the maximum voltage value minus the supply voltage divided by the supply voltage. For a slightly damped circuit, the overshoot may be quite large—see Fig. 7.36a, b. The *rise time* t_r (which is sometimes called the "transition time") in digital circuits is the time taken for the voltage to rise from $0.1V_S$ to $0.9V_S$ (T. L. Floyd, *Digital Fundamentals*, 9th, p. 8); see Figs. 8.14a–c. The circuit designer typically attempts to minimize both the rise time and the overshoot. An important example considered later in our text is a pulse train to be transmitted at a maximum speed (which requires minimum rise time) and with minimum distortion (which requires minimum overshoot). Figure 7.36 indicates that those goals are in fact conflicting. Decreasing the rise time increases the overshoot and vice versa. Designing the damping ratio close to unity, or *slightly below it*, is a reasonable compromise to quickly achieve the desired voltage level without a significant overshoot and ringing. The overshoot and rise time may be estimated analytically. We present here the estimates found in common control theory textbooks:

$$M_p = \frac{\exp(-\pi\zeta)}{\sqrt{1-\zeta^2}} \quad \text{for} \quad \zeta < 1, \quad M_p = 0 \quad \text{for} \quad \zeta \geq 1 \tag{7.88a}$$

$$t_r = \left(1 - 0.4167\zeta + 2.917\zeta^2\right)/\omega_n \quad \text{for} \quad \zeta < 1 \tag{7.88b}$$

Exercise 7.15: The damping coefficient of 15,000 neper/s and the natural frequency of 10 kHz are measured for an unknown series RLC circuit in laboratory via its step response. Given $R = 10\ \Omega$, determine L and C.

Answer: $L = 0.333$ mH, $C = 0.719$ μF.

As to mechanical engineering, it is interesting to note that an automotive suspension system is described by the same step response model and behaves quite similarly to the RLC circuit in Fig. 7.36. Well, driving a car everyday should certainly be a motivation for studying this topic.

7.6.6 Application Example: Nonideal Digital Waveform

Modeling Circuit

The series RLC block is an appropriate model to study the voltage pulse as it realistically occurs in digital circuits and in power electronic circuits involving pulse width modulation (PWM). The ideal square voltage pulse, shown in Fig. 7.33 of the previous section, is a crude approximation of reality. Parasitic capacitance, resistance, and inductance are always present in the circuit. As a result, the pulse form is distorted. To model the pulse form distortion, we consider the circuit shown in Fig. 7.37.

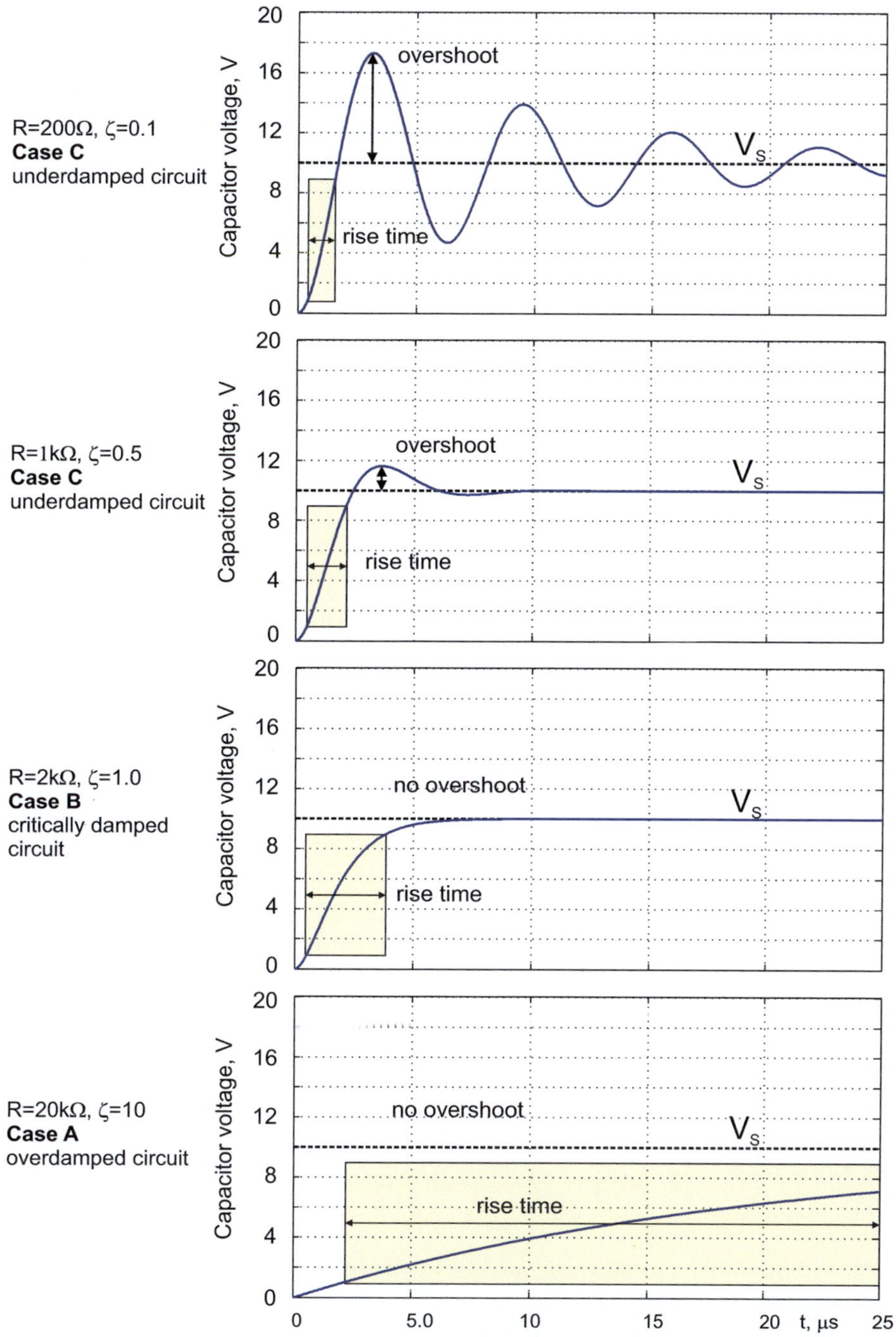

Fig. 7.36. Circuit responses in terms of capacitor voltages for different damping factors.

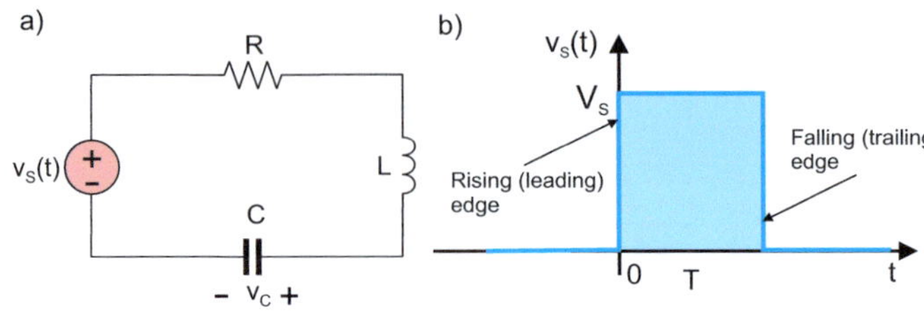

Fig. 7.37. The RLC circuit for studying the nonideal digital (pulse) waveform.

The switch is now removed and a time-varying voltage source $\upsilon_S(t)$ is introduced; it generates the voltage pulse (one bit) of duration T with amplitude V_S as seen in Fig. 7.37b. One may think of the source voltage as an "ideal" digital waveform and, for example, of the capacitor voltage as a nonideal (realistic) waveform influenced by parasitic capacitance, resistance, and inductance.

Solution

The solution to the pulse problem is derived as described at the end of the previous section. We know that Eqs. (7.83) through (7.86) determine the step response—the capacitor voltage $\upsilon_C(t)$ for the circuit with the DC voltage source shown in Fig. 7.35 after closing the switch. To obtain the solution $\upsilon_C^{\text{pulse}}(t)$ for the voltage pulse shown in Fig. 7.37, we simply combine two such step responses, i.e.,

$$\upsilon_C^{\text{pulse}}(t) = \upsilon_C(t) - \upsilon_C(t - T) \tag{7.89}$$

This operation again underscores the importance of the fundamental step response solution. Close inspection of Eq. (7.89) shows that the pulse will possess the dynamic overshoot and the nonzero rise time similar to the step response solution. This happens at the *rising (or leading) edge* of the pulse. At the same time, a *dynamic undershoot* and a nonzero *settling time* will happen at the *falling* or *trailing edge* as depicted in Fig. 7.38.

Example 7.10: Determine the solution for the capacitor voltage, $\upsilon_C^{\text{pulse}}(t)$, for the circuit shown in Fig. 7.37 with $L = 1\ \mu\text{H},\ C = 1\ \text{nF},\ V_S = 10\ \text{V},\ T = 0.5\ \mu\text{s}$ over the time interval from 0 to 1 μs for $R = 15\ \Omega,\ 30\ \Omega$, and $60\ \Omega$.

Solution: We find the step response $\upsilon_C(t)$ following Eqs. (7.83) through (7.86) first and then obtain the final solution using Eq. (7.89). For the step response, we obtain
$R = 15\ \Omega \Rightarrow \zeta = $ **0.24 (Case C—underdamped circuit)** $\Rightarrow$

$$\begin{aligned}\upsilon_C(t) = 10 - 10\exp\left(-7.5\times 10^6 t\right)\cos 3.07\times 10^7 t \\ - 2.44\exp\left(-7.5\times 10^6 t\right)\sin 3.07\times 10^7 t\end{aligned} \tag{7.90a}$$

Example 7.10 (cont.):

$R = 30\ \Omega \Rightarrow \zeta = \mathbf{0.47}$ **(Case C—underdamped circuit)** $\Rightarrow$

$$\begin{aligned} \upsilon_C(t) &= 10 - 10\exp\left(-1.5\times10^7 t\right)\cos 2.78\times10^7 t \\ &\quad - 5.39\exp\left(-1.5\times10^7 t\right)\sin 2.78\times10^7 t \end{aligned} \tag{7.90b}$$

$R = 60\ \Omega \Rightarrow \zeta = \mathbf{0.95}$ **(Case C—underdamped circuit)** $\Rightarrow$

$$\begin{aligned} \upsilon_C(t) &= 10 - 10\exp\left(-3.0\times10^7 t\right)\cos 1.0\times10^7 t \\ &\quad - 30\exp\left(-3.0\times10^7 t\right)\sin 1.0\times10^7 t \end{aligned} \tag{7.90c}$$

Figure 7.38a–c shows the distorted pulse forms for three particular cases. Figure 7.38a outlines the major pulse parameters: rise time, fall time, overshoot, undershoot, and pulse width. One can see that there is again a conflict between the desire to simultaneously decrease the rise time and the overshoot.

The overshoot and undershoot in Fig. 7.38 approximately coincide, and so do the rise time and the fall or settling time. Note that this is not always the case. The voltage pulse may be very significantly and *unsymmetrically* distorted when the initial pulse width, T, is comparable with the rise time. A good illustration is the previous example solved for $R = 15\,\Omega$ when $T = 0.25\,\mu\text{s}$ or less.

Exercise 7.16: Using a theoretical approximation, find the overshoot for the case of Fig. 7.38a and compare this value with value observed on the figure.

Answer: 48 % (theory) versus 50 % (observation).

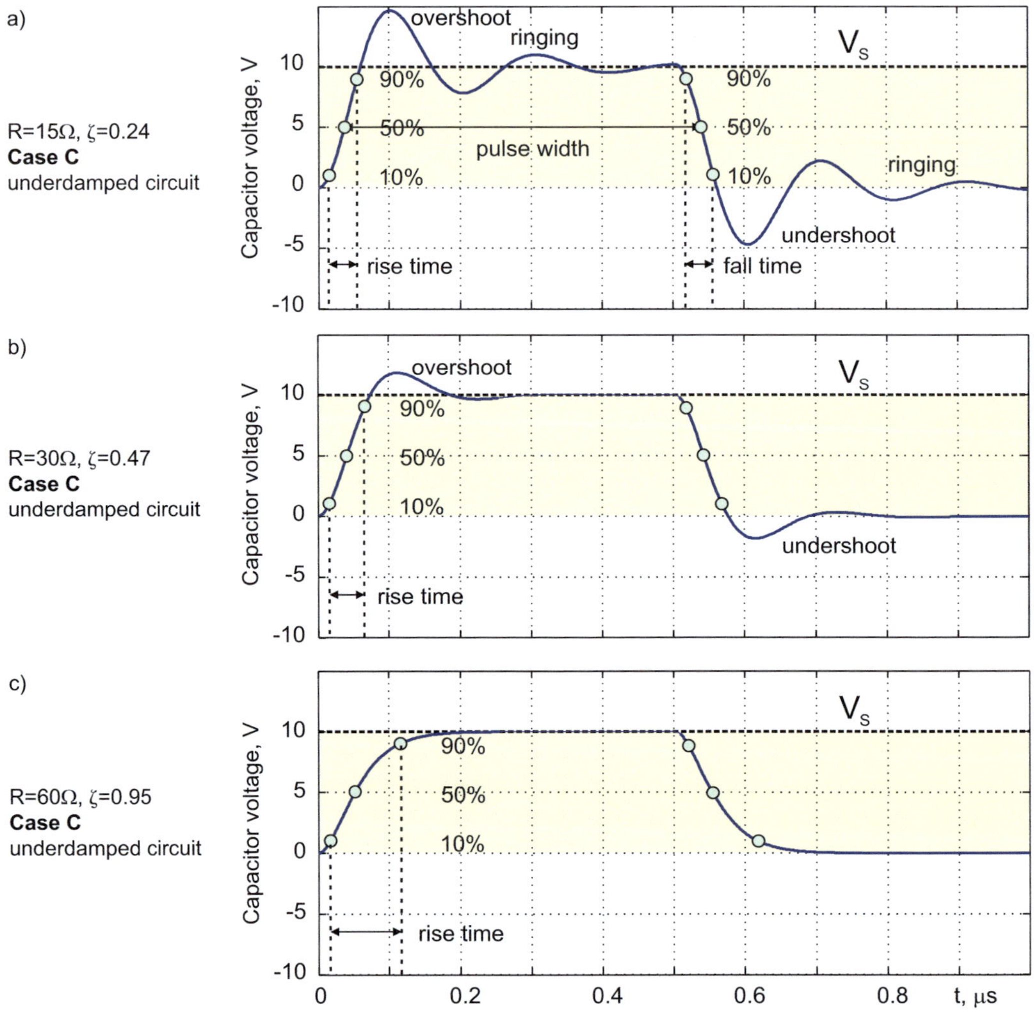

Fig. 7.38. Distorted pulse forms for three different values of the damping ratio. When the damping ratio increases, the overshoot decreases but the rise time increases.

The solution for the second-order circuits with arbitrary sources and arbitrary initial conditions can quite simply be obtained numerically. A straightforward finite-difference second-order method may be implemented in MATLAB or in other software packages with a few lines of the code. This method is the extension of the Euler method used for first-order transient circuits. Interestingly, the same method may be applied to radio-frequency pulse propagation in transmission lines and in free space, including problems such as signal penetration through walls.

Summary

General facts about transient circuits		
Voltage across the capacitor(s) remains continuous for all times—use this voltage as an unknown function		
Current through the inductor(s) remains continuous for all times—use this current as an unknown function		
When discharged through a small resistance, the (ideal) capacitor is able to deliver an extremely high power (and current) during a very short period of time		
When disconnected from the source, the (ideal) inductor in series with a large resistance is able to deliver an extremely high power (and voltage) during a very short period of time		
Transient circuit	**Generic circuit diagram**	**Solution plot**
Energy-release RC circuit $v_C(t) = V_0 \exp\left(-\frac{t}{\tau}\right)$ $\tau = RC$ ODE: $\frac{dv_C}{dt} + \frac{v_C}{\tau} = 0$	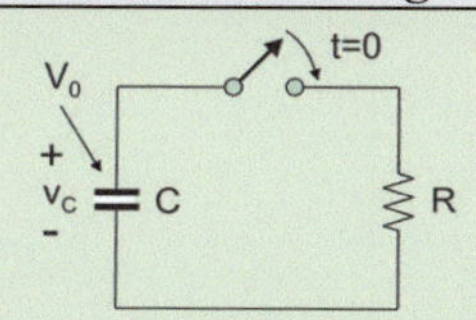	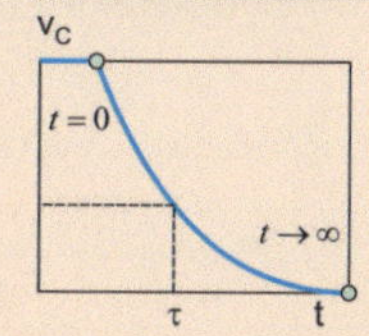
Energy-accumulating RC circuit $v_C(t) = V_S\left[1 - \exp\left(-\frac{t}{\tau}\right)\right]$ $\tau = RC$ ODE: $\frac{dv_C}{dt} + \frac{v_C}{\tau} = \frac{V_S}{\tau}$	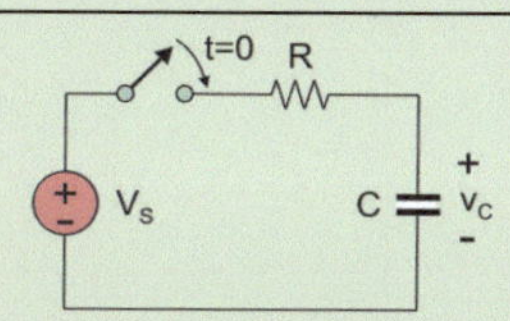	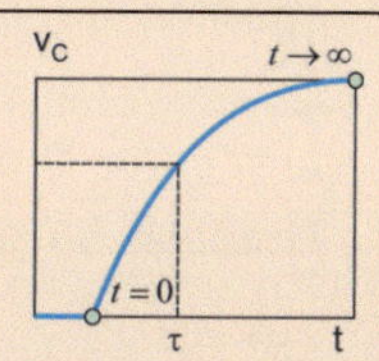
Energy-release RL circuit $i_L(t) = I_S \exp\left(-\frac{t}{\tau}\right)$ $\tau = L/R$ ODE: $\frac{di_L}{dt} + \frac{i_L}{\tau} = 0$	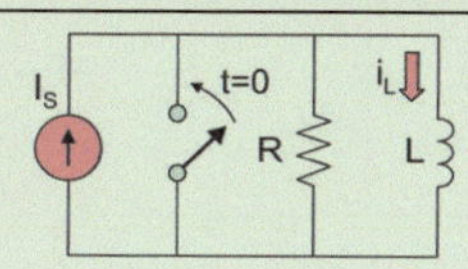	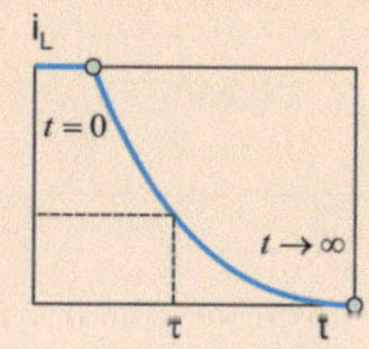
Energy-accumulating RL circuit $i_L(t) = I_S\left[1 - \exp\left(-\frac{t}{\tau}\right)\right]$ $\tau = L/R$ ODE: $\frac{di_L}{dt} + \frac{i_L}{\tau} = \frac{I_S}{\tau}$	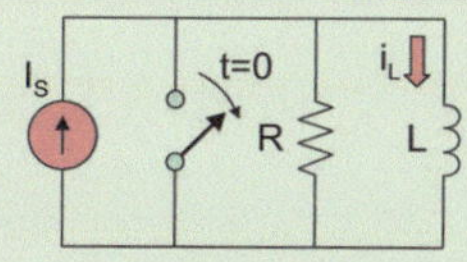	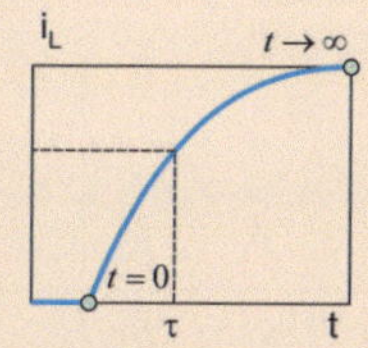
Energy-release RL circuit $i_L(t) = I_0 \exp\left(-\frac{t}{\tau}\right)$ $\tau = L/R,\ I_0 = V_S/R_0$ ODE: $\frac{di_L}{dt} + \frac{i_L}{\tau} = \frac{V_S}{R_0 \tau}$	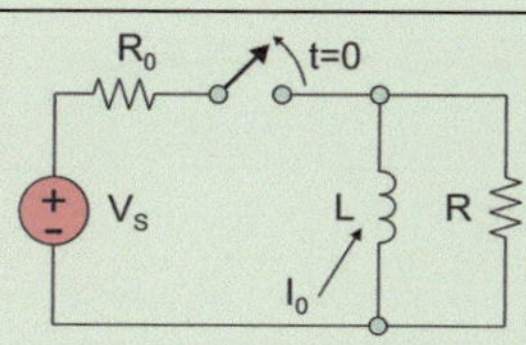	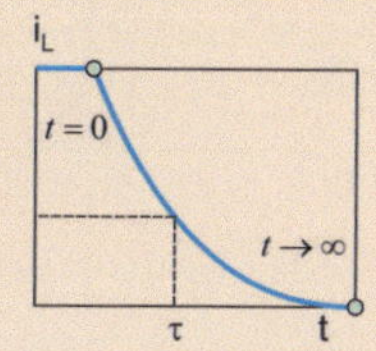

(continued)

	Voltage or current	Power
Generic energy-release curves for either dynamic element	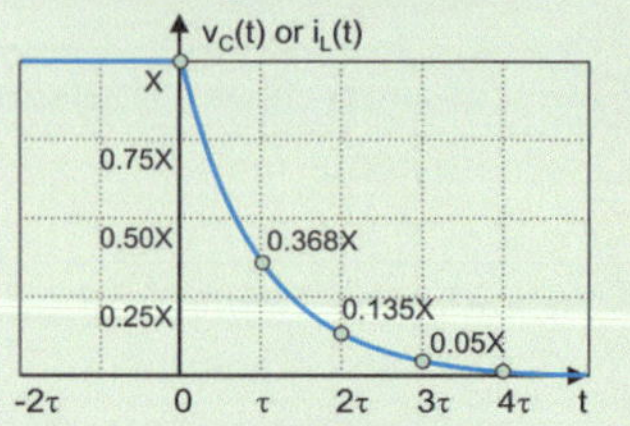	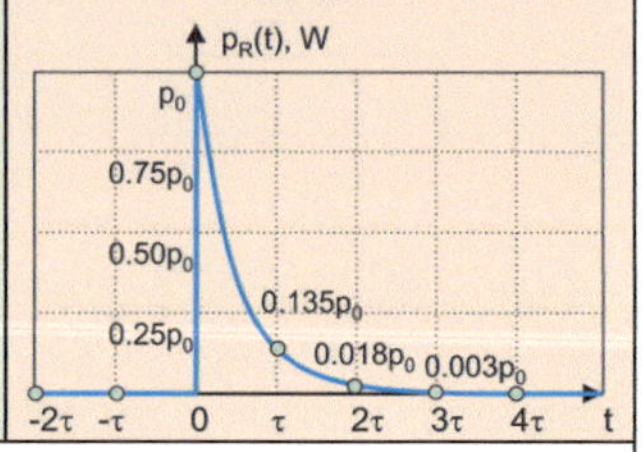
Switching RC oscillator (relaxation oscillator)—RC timer		
Bistable amplifier circuit with positive feedback. Two stable states: $v_{out} = +V_{CC}$ $v_{out} = -V_{CC}$ Threshold voltage(s): $v^* = \frac{R_1}{R_1 + R_2} v_{out}$	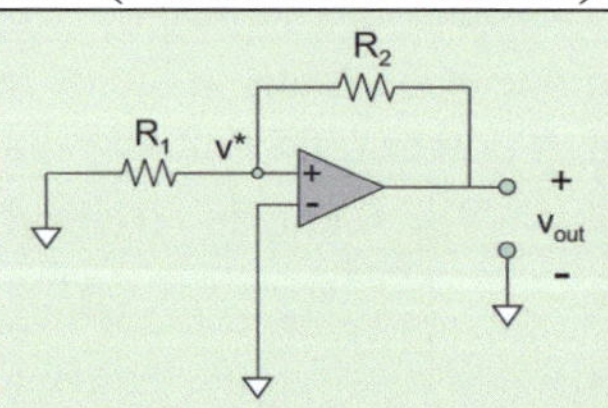	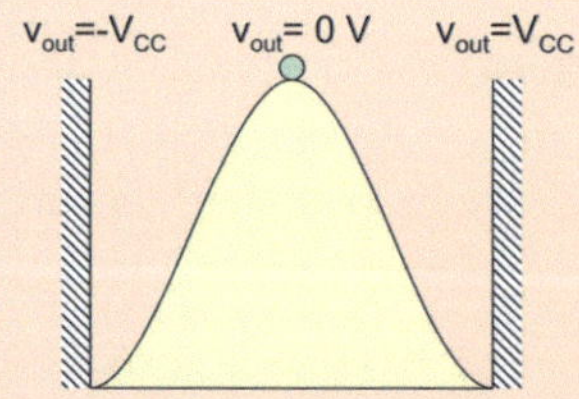
Relaxation oscillator with positive feedback: $v_{out} = +V_{CC}$ if $v^* > v_C$ $v_{out} = -V_{CC}$ if $v^* < v_C$ Threshold voltage(s): $v^* = \frac{R_1}{R_1 + R_2} v_{out}$	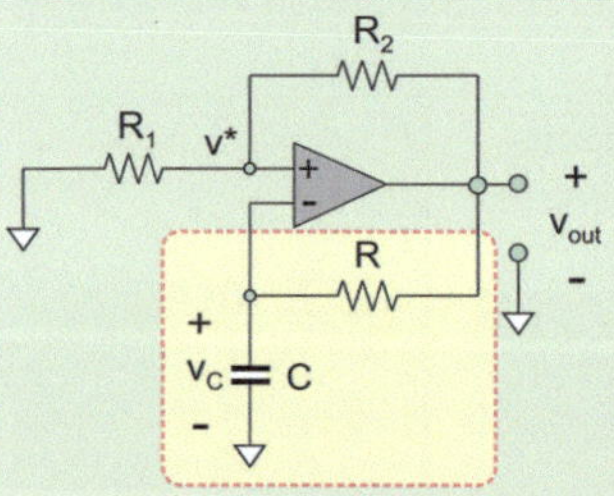	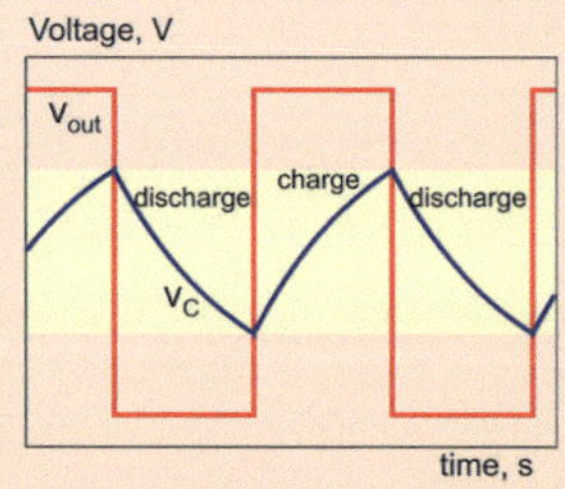
Period and frequency of the relaxation oscillator ($\beta = R_1/(R_1 + R_2)$)	$T = 2\tau \ln \frac{1+\beta}{1-\beta}$	$f = \frac{1}{T} = \frac{1}{2\tau}\left(\ln \frac{1+\beta}{1-\beta}\right)^{-1}$

Single-time-constant (STC) transient circuits		
$\tau = RC$ or $\tau = \frac{L}{R}$	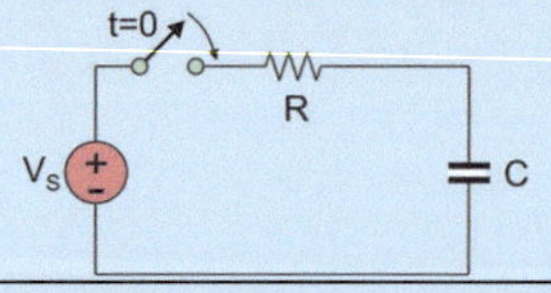	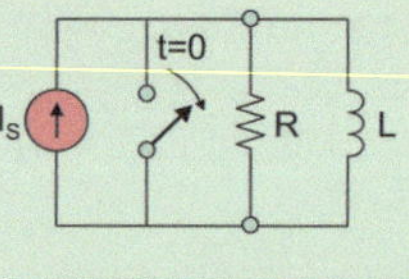
$\tau = (R_1 \| \| R_2)C$ or $\tau = \frac{L}{R_1 + R_2}$	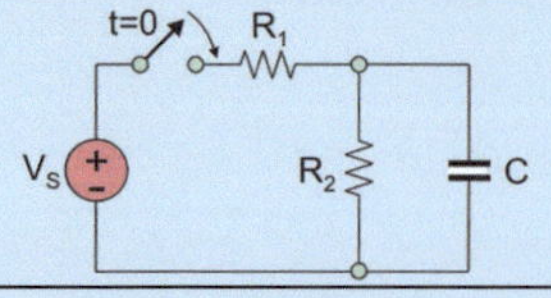	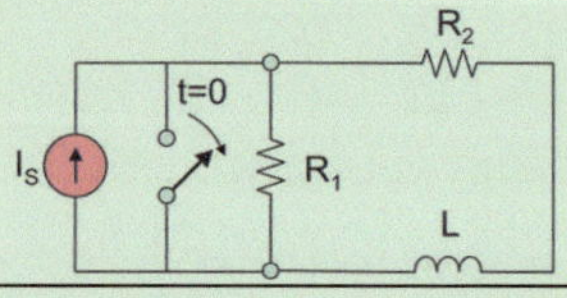
$\tau = R(C_1 + C_2)$ or $\tau = \frac{L_1 + L_2}{R}$	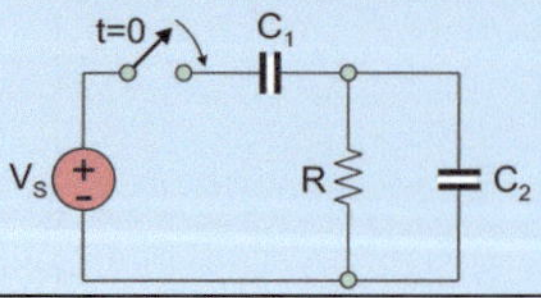	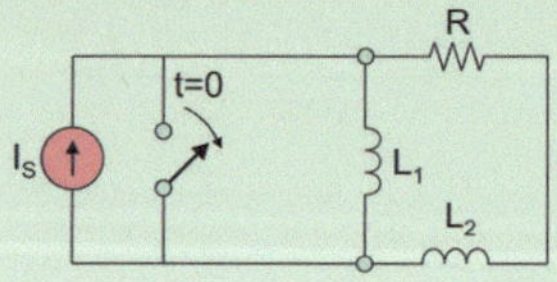

(continued)

$\tau = \dfrac{L_1 + L_2}{R_1 + R_2}$ $\tau = R_1 \| \| R_2 \times (C_1 + C_2)$	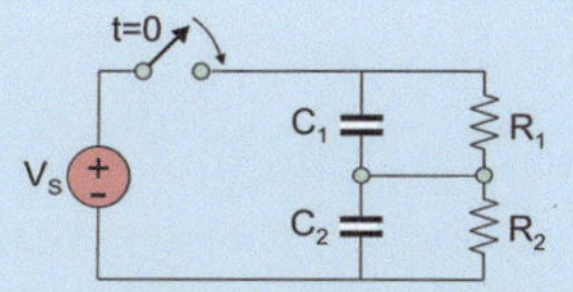 Arbtr. initial conditions are not allowed	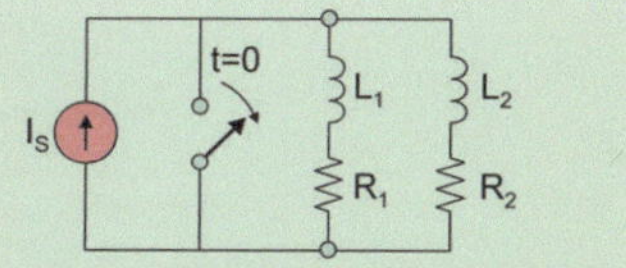 Arbtr. initial conditions are not allowed
STC circuits with general sources		
Bypass capacitor and decoupling inductor	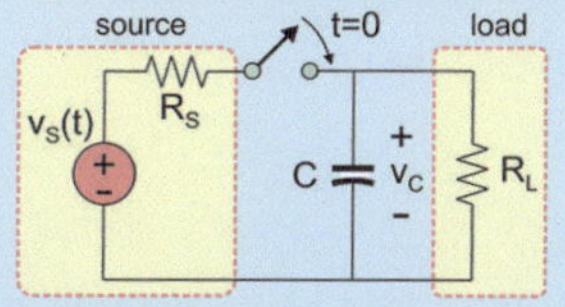 $v_S(t) = V_S + V_m \cos \omega t$, $\tau = (R_S \| \| R_L)C$	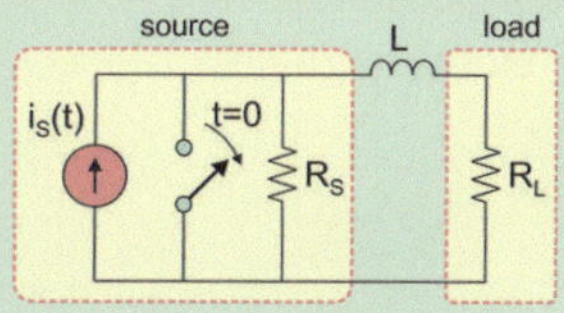 $i_S(t) = I_S + I_m \cos \omega t$, $\tau = L/(R_S + R_L)$
Solution for load voltage or load current	$v_L = \dfrac{R_L V_S}{R_S + R_L}\left(1 - \exp\left(-\dfrac{t}{\tau}\right)\right) + \dfrac{R_L V_m}{R_S + R_L} \times \dfrac{1}{1 + (\omega\tau)^2}\left[\cos\omega t + \omega\tau \sin\omega t - \exp\left(-\dfrac{t}{\tau}\right)\right]$	$i_L = \dfrac{R_S I_S}{R_S + R_L}\left(1 - \exp\left(-\dfrac{t}{\tau}\right)\right) + \dfrac{R_S I_m}{R_S + R_L} \times \dfrac{1}{1 + (\omega\tau)^2}\left[\cos\omega t + \omega\tau \sin\omega t - \exp\left(-\dfrac{t}{\tau}\right)\right]$
Second-order transient circuits		
With two identical dynamic elements	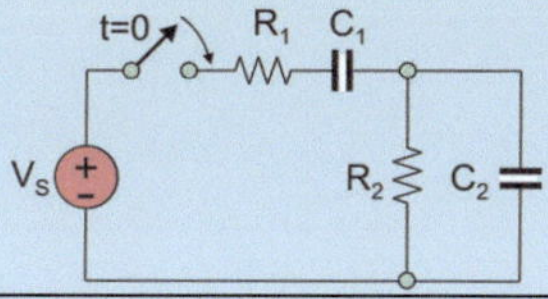	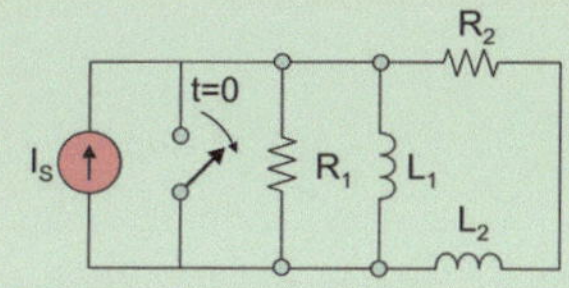
With series LC network	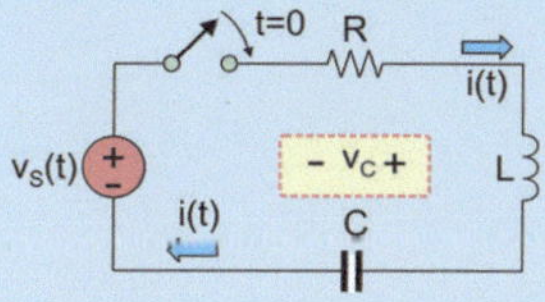 • Step response with zero initial conditions: use capacitor voltage $v_C(t)$	• Damping coefficient: $\alpha = R/(2L)$ • Undamped res. freq.: $\omega_0 = 1/\sqrt{LC}$ • Damping ratio: $\zeta = \alpha/\omega_0$ • Natural freq.: $\omega_n = \sqrt{\omega_0^2 - \alpha^2}$ Characteristic Eq.: $s^2 + 2\alpha s + \omega_0^2 = 0$
With parallel LC network	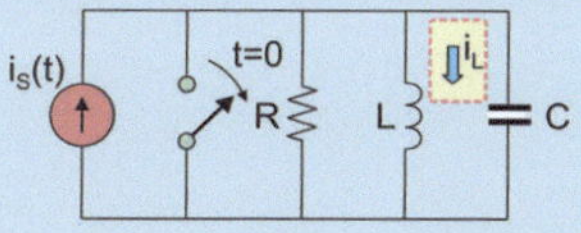 • Step response with zero initial conditions: use inductor current $i_L(t)$	• Damping coefficient: $\alpha = 1/(2RC)$ • Undamped res. freq.: $\omega_0 = 1/\sqrt{LC}$ • Damping ratio: $\zeta = \alpha/\omega_0$ • Natural freq.: $\omega_n = \sqrt{\omega_0^2 - \alpha^2}$ Characteristic Eq.: $s^2 + 2\alpha s + \omega_0^2 = 0$

(continued)

Overdamped, critically-damped, and underdamped RLC circuits	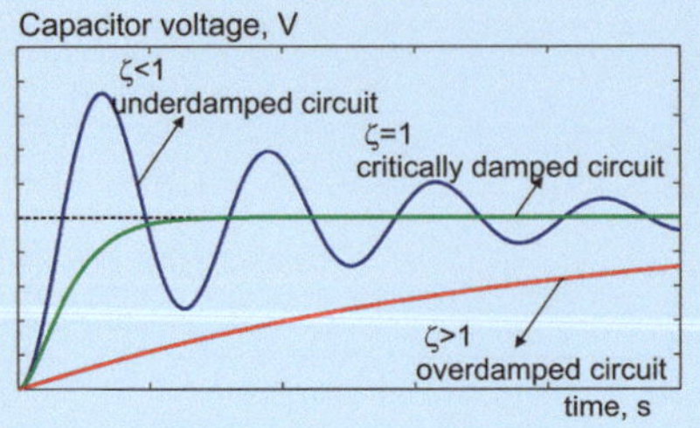	• *Overdamped* ($\zeta > 1$): $x_c(t) = K_1 \exp(s_1 t) + K_2 \exp(s_2 t)$ $K_1 = \frac{s_2 V_S}{s_1 - s_2},\ K_2 = \frac{s_1 V_S}{s_2 - s_1}$ • *Critically damped* ($\zeta = 1$): $x_c(t) = K_1 \exp(s_1 t) + K_2 t \exp(s_1 t)$ $K_1 = -V_S,\ K_2 = s_1 V_S$ • *Underdamped* ($\zeta < 1$): $x_c(t) = K_1 \exp(-\alpha t) \cos \omega_n t +$ $K_2 \exp(-\alpha t) \sin \omega_n t$ $K_1 = -V_S,\ K_2 = -\frac{\alpha}{\omega_n} V_S$
Non-ideal digital waveform: second-order circuit	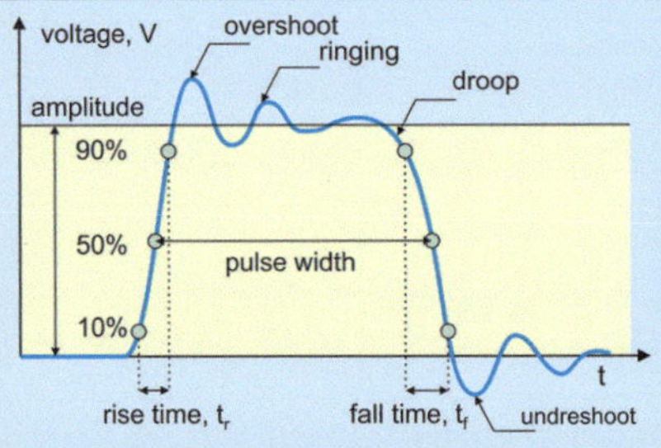	• *Overshoot* $M_p = \frac{\exp(-\pi\zeta)}{\sqrt{1-\zeta^2}}$ for $\zeta < 1$ • *Undershoot* is approximately overshoot for rise times small compared to pulse width • *Rise time* $t_r = (1 - 0.4167\zeta + 2.917\zeta^2)/\omega_n$ for $\zeta < 1$ *Fall time* is approximately rise time for rise times small compared to pulse width

Problems

7.1 RC Circuits

7.1.1 Energy-Release Capacitor Circuit

7.1.2 Time Constant of an RC Circuit and Its Meaning

7.1.3 Continuity of the Capacitor Voltage

Problem 7.1. For the capacitor as a dynamic circuit element, develop:

1. Equivalent circuit at DC
2. Relation between voltage and current
3. Expression for the time constant of a transient circuit that includes the dynamic element and a resistor *R*

Dynamic circuit element	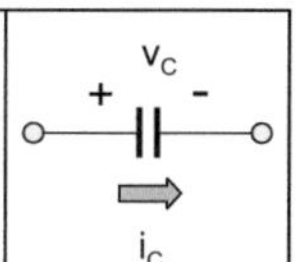
Equivalent circuit at DC (short or open)	
Relation between voltage and current (passive reference configuration)	
Expression for the time constant of a transient circuit that includes the dynamic element (C) and a resistor *R*.	$\tau =$

Problem 7.2. Using KCL and KVL, derive the differential equation for the circuit shown in the following figure, keeping the same labeling for the voltages and the currents.

A. Is the final result different from Eq. (7.5) of Section 7.1?

B. Could you give an example of a certain voltage and/or current labeling (by arbitrarily changing polarities and directions in the figure) that causes the differential equations to change?

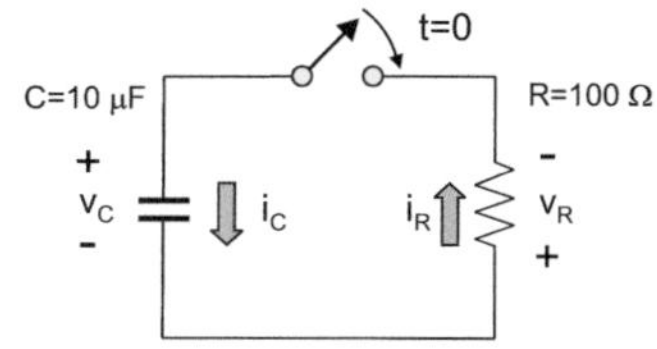

Problem 7.3. Prove that Eq. (7.6) is the solution to Eq. (7.5) (both from Section 7.1) using direct substitution and the differentiation that follows.

Problem 7.4.

A. Show that the time constant, τ, of an RC circuit has the units of seconds.

B. To obtain the slow discharge rate of lesser instantaneous power into the load, should the load resistance be small or large?

Problem 7.5. A 100-μF capacitor discharges into a load as shown in the following figure. The load resistance may have values of 100 Ω, 10 Ω, and 1 Ω. The capacitor is charged to 20 V prior to $t = 0$.

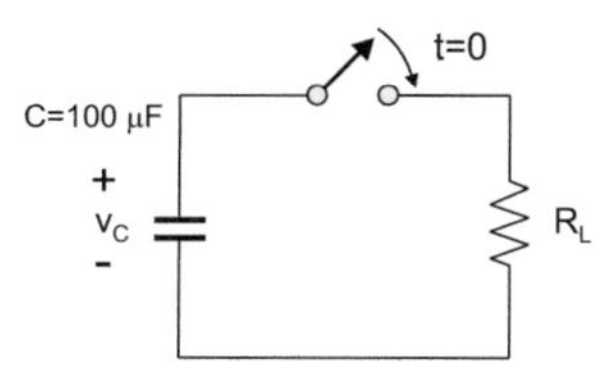

A. Find time constant τ and the maximum instantaneous power delivered to the load resistor in the very first moment for every resistor value—fill out the Table that follows.

Instantaneous load power right after the switch closes

R_L	τ, s	$p_L(t=+0)$, W
100 Ω		
10 Ω		
1 Ω		

B. Do the instantaneous power values from the Table depend on the capacitance value?

Problem 7.6. A 100-μF capacitor, shown in the following figure, discharges into a 10-Ω load resistor. The capacitor is charged to 15 V prior to $t = 0$.

A. Find the time constant of the circuit (show units).

B. Express the voltage across the capacitor as a function of time and sketch it to scale versus time over time interval from −2 ms to 5 ms.

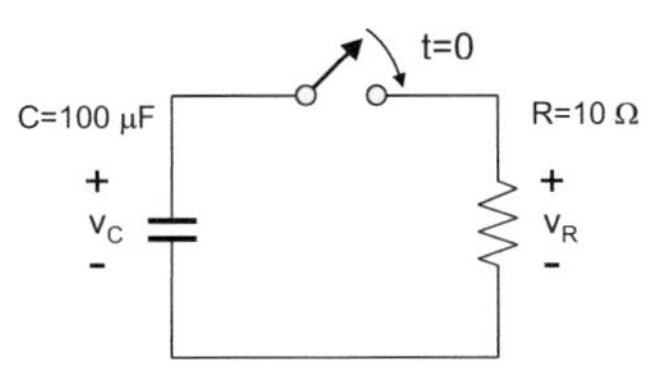

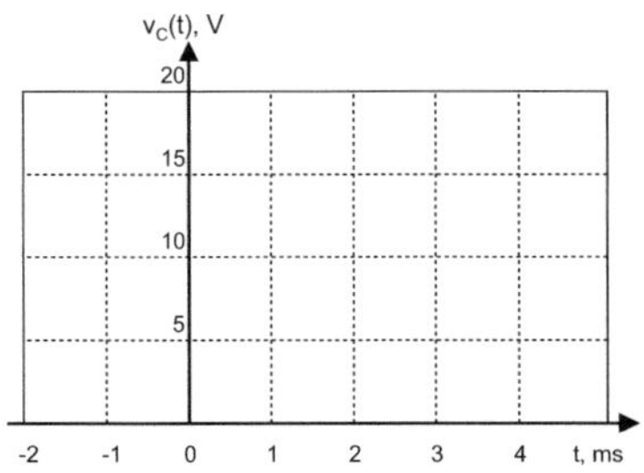

Problem 7.7. A 100-μF capacitor, shown in the following figure, discharges into a 5-Ω load resistor. The capacitor is charged to 20 V prior to $t = 0$.

A. Find an expression for the voltage across the capacitor as a function of time and sketch it to scale versus time over the interval from -2τ to 5τ.

B. Repeat the exercise for instantaneous power delivered to the resistor.

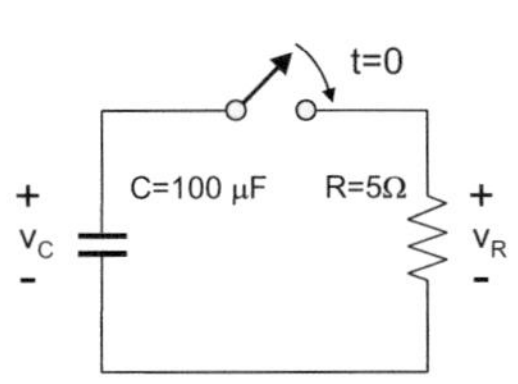

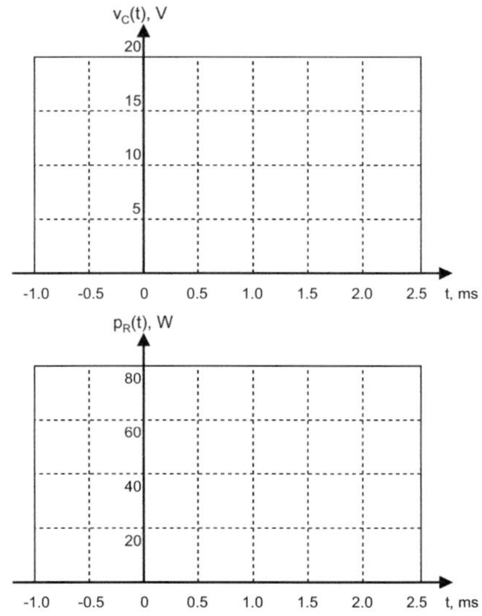

Problem 7.8. In the circuit shown in the following figure, the capacitor is charged to 20 V prior to $t = 0$.

A. Find an expression for the voltage across the capacitor as a function of time and sketch it to scale versus time over the interval from -2τ to 5τ.

B. Repeat for instantaneous power delivered to the rightmost resistor.

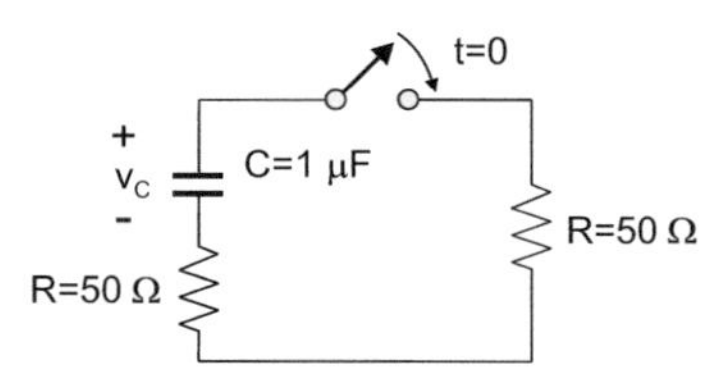

Problem 7.9. Present the text of a MATLAB script (or of any software of your choice) in order to generate Fig. 7.2d of Section 7.1. Attach the figure so generated to the homework report.

Problem 7.10. Prove that the integral of the load power in Fig. 7.2d given by Eq.(7.8c) is exactly equal to the energy stored in the charged capacitor, $E_C = \frac{1}{2}CV_0^2$ prior to $t = 0$.

Problem 7.11.

A. Create the generic capacitor voltage discharge curve similar to Fig. 7.2a but for an *arbitrary* capacitor powering an *arbitrary* load resistor over the time interval from -2τ to 5τ. The capacitor is charged to V_0 prior to $t = 0$. To do so, find the capacitor voltage as a fraction of V_0 for every unit of τ and fill out the Table that follows.

Capacitor voltage in terms of V_0

t	-2τ	$-\tau$	0	τ	2τ	3τ	4τ	5τ
$v_C(t)$								

B. Repeat the same task for Fig. 7.2d related to load power. Find the load power in terms of the maximum power just after closing the switch.

Problem 7.12. For an unknown energy-release RC circuit, capacitor voltage and capacitor current were measured in laboratory before and after closing the switch at $t = 0$ as shown in the figure that follows. Approximate R and C.

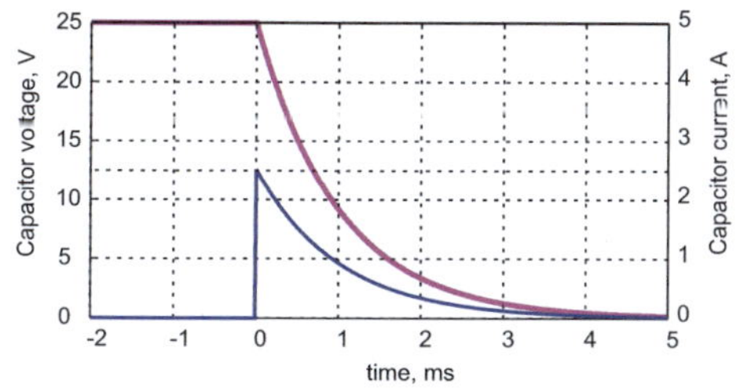

7.1.4 Application Example: Electromagnetic Railgun

7.1.5 Application Example: Electromagnetic Material Processing

Problem 7.13. An electromagnetic capacitor accelerator with permanent magnets has $B = 0.3$ T. The accelerating object has a length of 2 cm. Plot to scale the Lorentz force as a function of discharge current over the interval $0 < i_C < 1000$ A.

Problem 7.14. An electromagnetic capacitor accelerator needs to create an average force of 5 N over 2 ms on a moving object with length of 1 cm. The load (armature plus object) resistance is 1 Ω, and the external magnetic field is $B = 0.25$ T. Determine:

A. The required capacitor voltage prior to discharge

B. The required capacitance of the capacitor (bank of capacitors)

Hint: Assume that the average force acts over the time interval τ. Its value is approximately equal to 60 % of the initial force value.

Problem 7.15. Solve the previous problem when:

A. The average force increases to 50 N

B. The average force increases to 500 N

Problem 7.16. The world's largest capacitor bank is located in Dresden, Germany. The pulsed, capacitive power supply system was designed and installed for studying high magnetic fields by experts from Rheinmetall Waffe Munition. The bank delivers 200 kA of discharge current in the initial time moment (just after the switch closes). The time constant is 100 ms. Estimate the bank capacitance if the charging voltage is 200 kV.

7.1.7 Energy-Accumulating Capacitor Circuit

Problem 7.17. A 1-μF capacitor shown in the following figure is charged through the 1-kΩ load resistor. The initial capacitor voltage is zero.

A. Find the time constant of the circuit (show units).
B. Express the voltage across the capacitor as a function of time and sketch it to scale versus time over the interval from −2 ms to 5 ms.

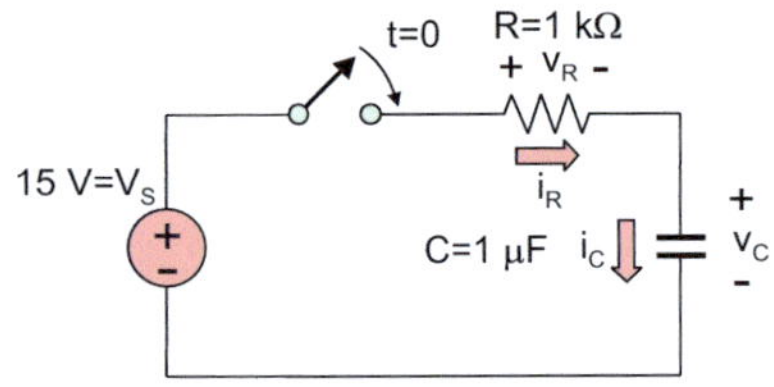

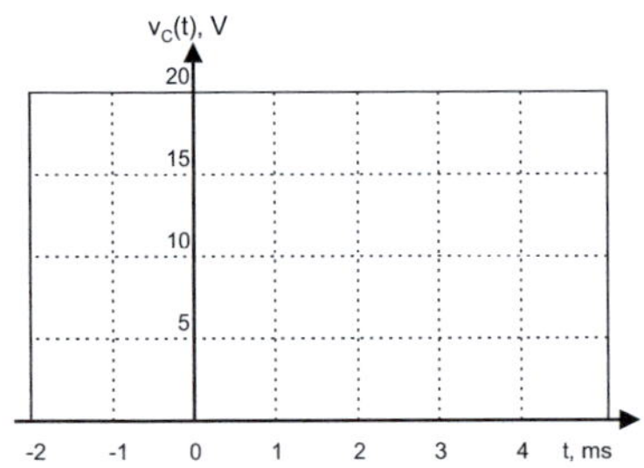

Problem 7.18. For the circuit shown in the following figure:

A. Find an expression for the capacitor voltage, v_C, and the capacitor current, i_C, including the value of time constant.
B. Sketch the capacitor voltage, v_C, and the capacitor current, i_C, to scale versus time over the interval from -2τ to 5τ.

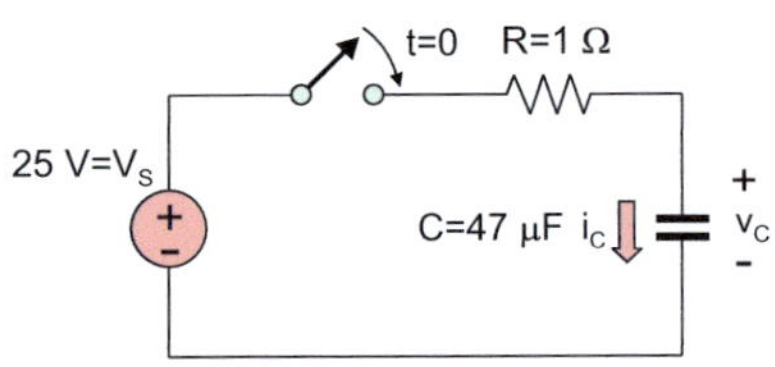

Problem 7.19. For the circuit shown in the following figure:

A. Find an expression for the capacitor voltage, v_C, and the capacitor current, i_C, including the value of time constant.
B. Sketch the capacitor voltage, v_C, and the capacitor current, i_C, to scale versus time over the time interval from -2τ to 5τ.

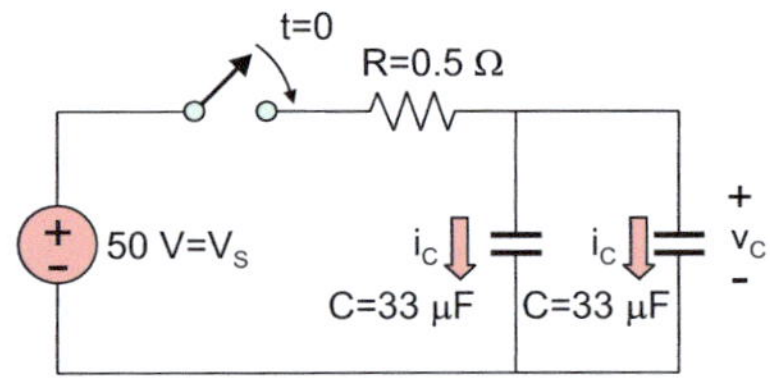

Problem 7.20. Sketch your own fluid-flow counterpart of the charging circuit shown in the figure and establish as many analogies between electrical (R, C, V_S) and mechanical parameters of your drawing as possible.

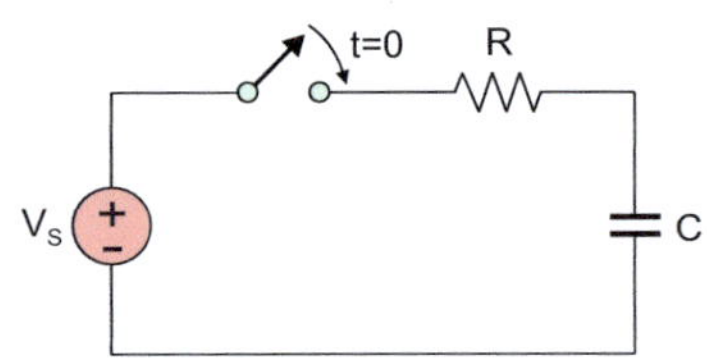

Problem 7.21. For the circuit shown in the following figure:

A. How much time does it take to charge the capacitor to 10 V?
B. To 25 V?

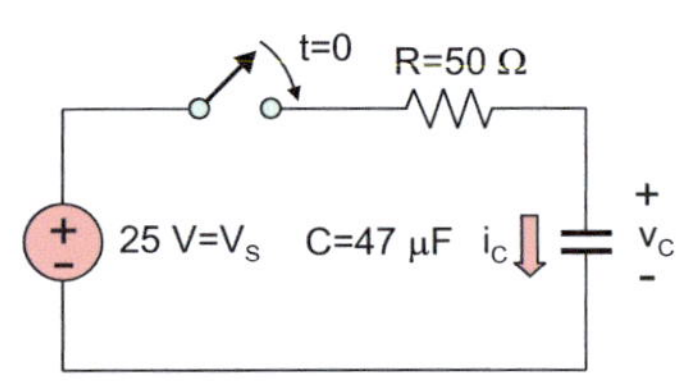

Problem 7.22. For the circuit shown in the figure, how much time does it take to charge

the capacitor, C_1, to 10 V? Assume that the initial voltages of both capacitors are zero.

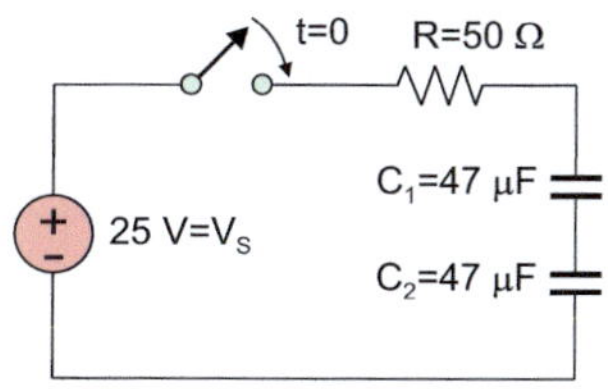

Problem 7.23. For an unknown energy-accumulating RC circuit, capacitor voltage and capacitor current were measured in laboratory before and after closing the switch at $t = 0$ as shown in the figure that follows. Approximate R and C.

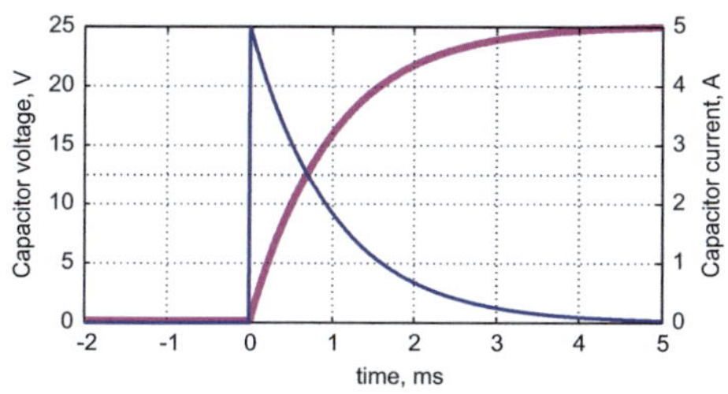

Problem 7.24.

A. Obtain an analytical solution for the capacitor voltage in the circuit shown in the following figure. When the switch is closed, the current source still generates current I_S at its terminals. However, the supply is shorted out – no current flows into the circuit. When the switch is open, the current flows into the circuit.

B. Could you convert this circuit to an equivalent RC transient circuit with the voltage source?

C. Plot the voltage across the resistor versus time over the time interval from -2τ to 5τ.

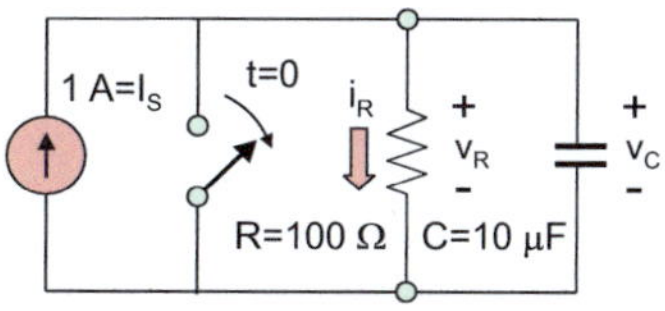

Problem 7.25. Repeat the previous problem for the circuit shown in the following figure.

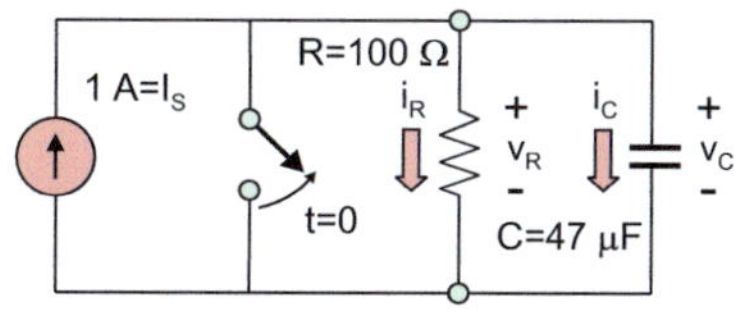

Problem 7.26. Obtain an analytical solution for the capacitor voltage in the circuit shown in the following figure at any time and express it in terms of I_s, R_1, R_2, C. Find the time constant of the circuit when $R_1 = R_2 = 100\ \Omega$, $C = 47\ \mu\text{F}$.

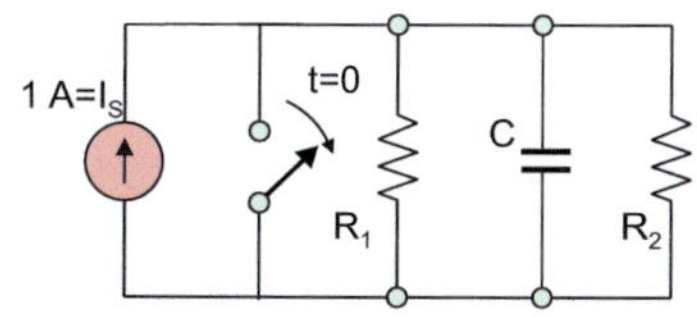

Problem 7.27. For the circuit shown in the figure:

A. Derive and solve the dynamic circuit equation after the switch opens. Assume the initial capacitor voltage equal to zero.

B. Plot the capacitor voltage to scale versus time over the time interval from -2τ to 5τ.

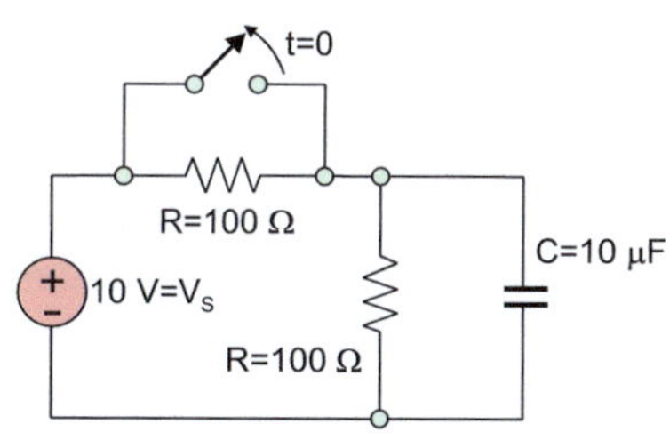

Problem 7.28. In the circuit that follows, the capacitor, C_1, is initially uncharged. The switch is closed at $t = 0$.

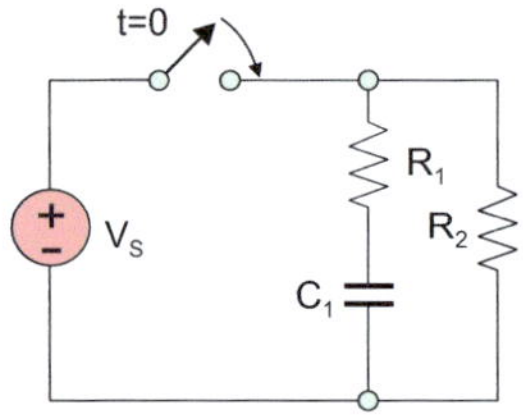

Give answers to the following questions based on known circuit parameters C_1, R_1, R_2, V_S:

A. What is the current through resistor R_2 as a function of time?
B. What is the maximum current through resistor R_1?
C. What is the current through resistor R_1 at a time long after the switch closes?
D. What is the charge, $Q^+(t)$, of the capacitor, C_1, as a function of time?

The switch is then opened a very long time after it has been closed – reset the time to $t = 0$.

A. What is the charge $Q^+(t)$ of the capacitor, C_1, as a function of time?
B. What is the current through resistor R_2 as a function of time? Specify the current direction in the figure.

7.2 RL Circuits

7.2.1 Energy-Release Inductor Circuit

7.2.2 Continuity of the Inductor Current

Problem 7.29. For the inductor as a dynamic circuit element, present:

1. Equivalent circuit at DC
2. Relation between voltage and current
3. Expression for the time constant of a transient circuit that includes the dynamic element and a resistor, R

Dynamic circuit element	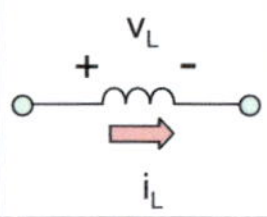
Equivalent circuit at DC (short or open)	
Relation between voltage and current (passive reference configuration)	
Expression for the time constant of a transient circuit that includes the dynamic element (L) and a resistor R	$\tau =$

Problem 7.30.

A. Using KCL and KVL, derive the differential equation for the inductor current in the circuit shown in the figure that follows, keeping the same labeling for the voltages and the currents.
B. Is the final result different from Eq. (7.19) of Section 7.2?

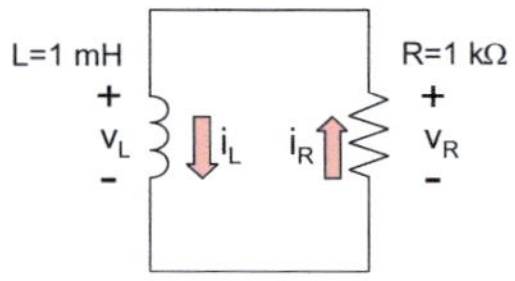

Problem 7.31. Prove that Eq. (7.20) is the solution to Eq. (7.19) using direct substitution and the corresponding differentiation.

Problem 7.32.

A. Show that the time constant τ has units of seconds for the *RL* circuit.
B. To ensure a slower energy release rate of the inductor, should the load resistance be small or large?
C. To ensure a faster energy release rate of the inductor, should the load resistance be small or large?

Problem 7.33. A 6.8-μH inductor releases its energy into a load resistor as shown in the following figure. The load resistance may have values of 10 Ω, 100 Ω, and 1 kΩ. The inductor current is 1 A prior to $t = 0$.

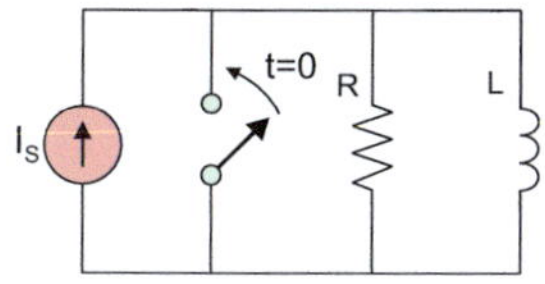

A. Find time constant τ and the maximum instantaneous power delivered to the load resistor in the very first moment for every resistor value—fill out the Table that follows.

Instantaneous load power just after the switch closes

R	τ, s	$p_R(t = +0)$, W
10 Ω		
100 Ω		
1 kΩ		

B. Do those instantaneous power values from the Table depend on the inductance value?

Problem 7.34. Prove that the integral from 0 to ∞ of the load power in Fig. 7.10d is exactly equal to the energy stored in the inductor $E_L = \frac{1}{2}LI_S^2$ prior to $t = 0$. *Hint:* The proof should include analytical integration of the instantaneous power in Eq. (7.21c).

Problem 7.35. A 2-mH inductor, shown in the following figure, releases its energy into the 2-kΩ load resistor. The supply current is 0.8 A.

A. Find the time constant of the RL circuit (show units).
B. Express the current through the inductor as a function of time and sketch it to scale versus time over time interval from −2 μs to 5 μs.

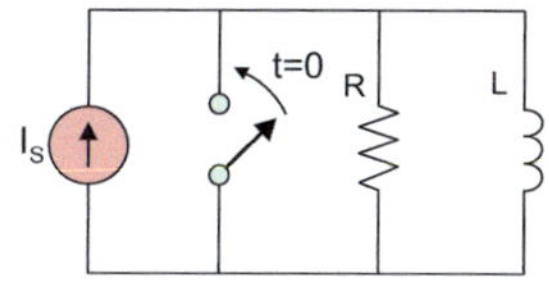

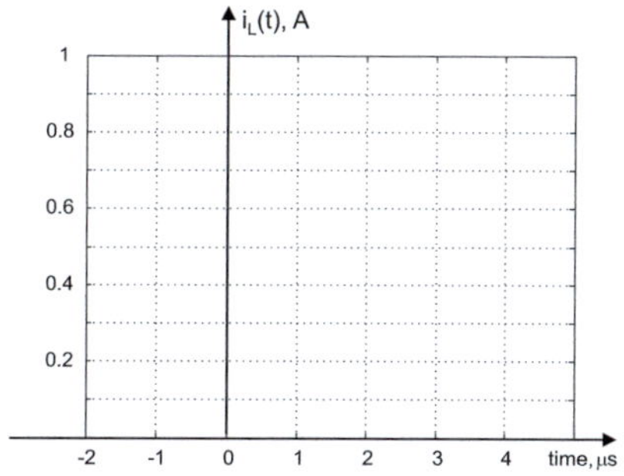

Problem 7.36. A 270-μH inductor, shown in the following figure, releases its energy into the 510-Ω load resistor. The supply current is 0.8 A.

A. Find the time constant of the RL circuit (show units).
B. Express the current through the inductor as a function of time and sketch it to scale versus time over time interval from −1 μs to 2.5 μs.
C. Express the resistor voltage as a function of time and sketch it to scale versus time over time interval from −1 μs to 2.5 μs.

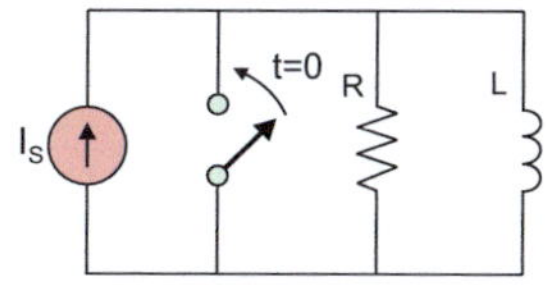

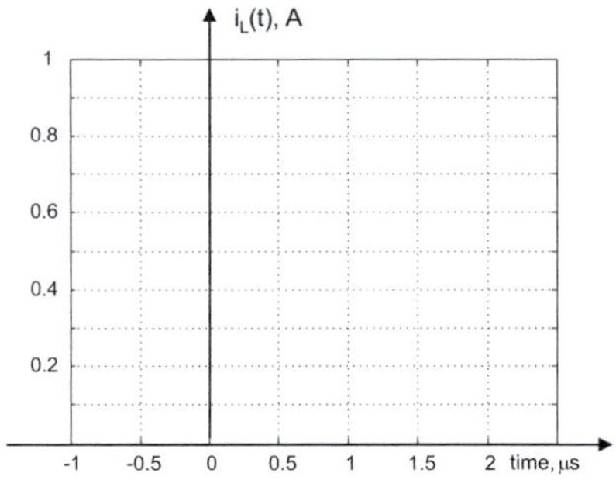

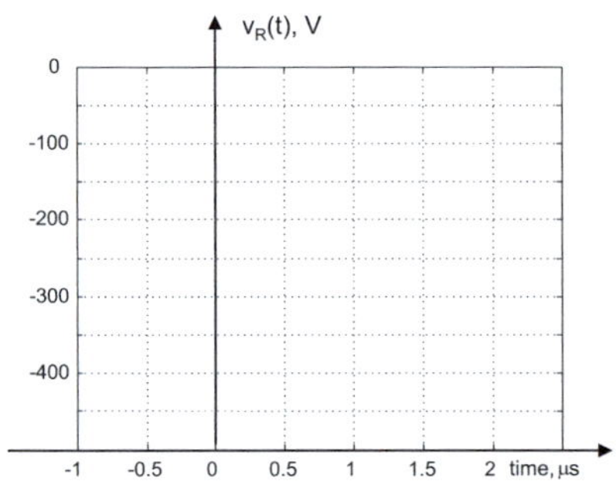

Problem 7.37. For an unknown energy-release RL circuit, inductor current and resistor voltage were measured before and after closing the switch at $t = 0$ as shown in the figure that follows. Approximate R and L.

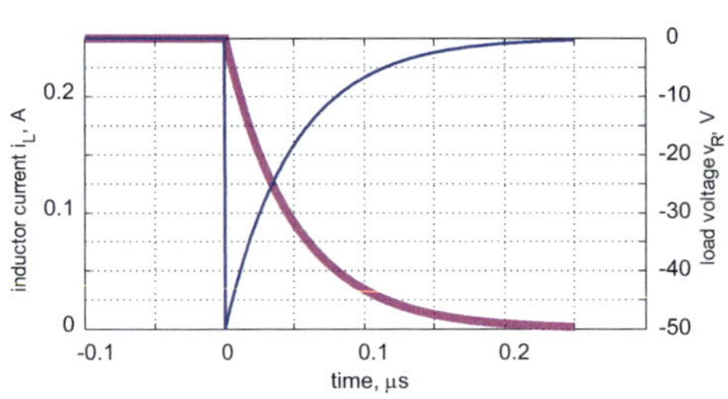

Problem 7.38. A 2-mH inductor, shown in the following figure, releases its energy into two 1-kΩ load resistors. The supply current is 100 mA.

A. Find the time constant of the RL circuit (show units).

B. Express the current through the inductor as a function of time and sketch it to scale versus time over time interval from −2 μs to 10 μs.

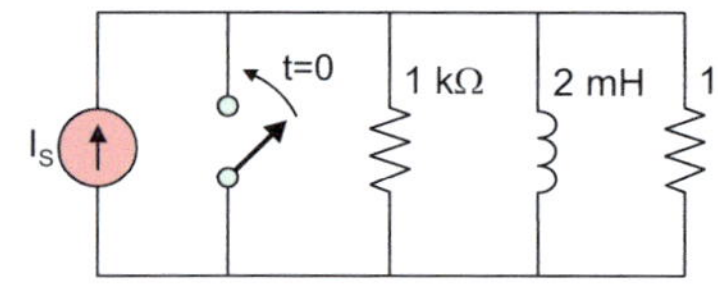

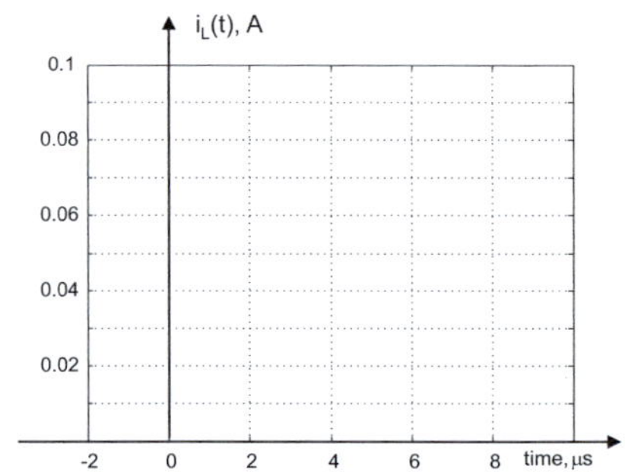

Problem 7.39.

A. Obtain the solution for the inductor current in the circuit shown in the figure at any time.

B. Plot to scale the current through the inductor versus time over the interval from -2τ to 10τ.

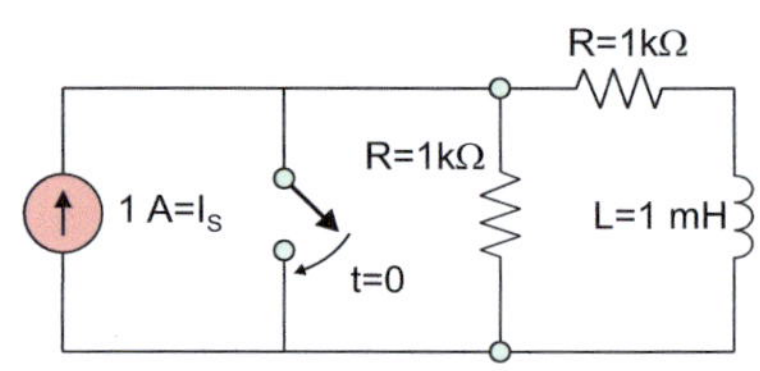

7.2.3 Energy-Accumulating Inductor Circuit

Problem 7.40. In the energy-accumulating RL circuit shown in the following figure, $R = 2$ kΩ and $L = 2$ mH. The supply current is 1 A.

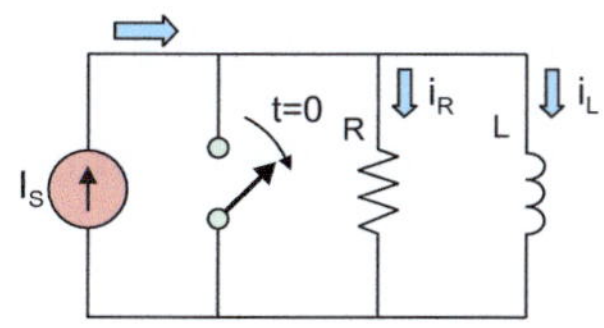

A. Find the time constant of the RL circuit (show units).
B. Express the current through the inductor as a function of time and sketch it to scale versus time over time interval from −2 μs to 5 μs.

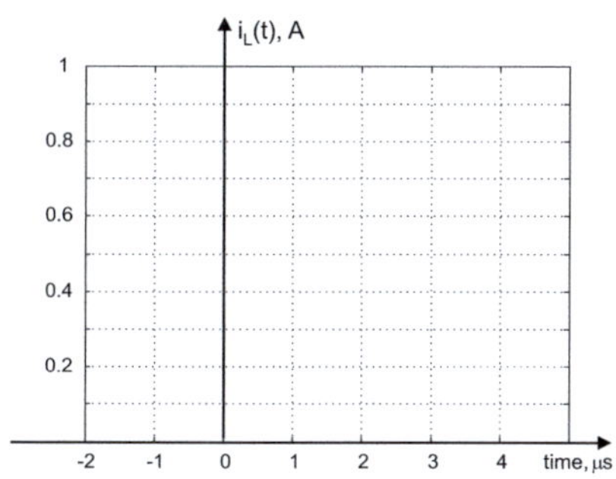

Problem 7.41. In the energy-accumulating RL circuit shown in the following figure, $R = 510$ Ω and $L = 270$ μH. The supply current is 1 A.

A. Find the time constant of the RL circuit (show units).
B. Express the current through the inductor as a function of time and sketch it to scale versus time over time interval from −1 μs to 2.5 μs.
C. Express the resistor voltage as a function of time and sketch it to scale versus time over time interval from −1 μs to 2.5 μs.

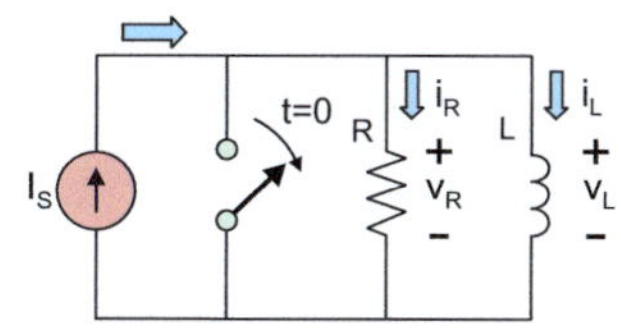

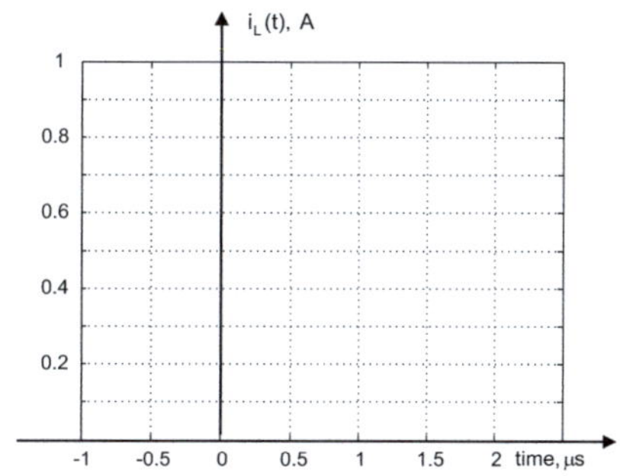

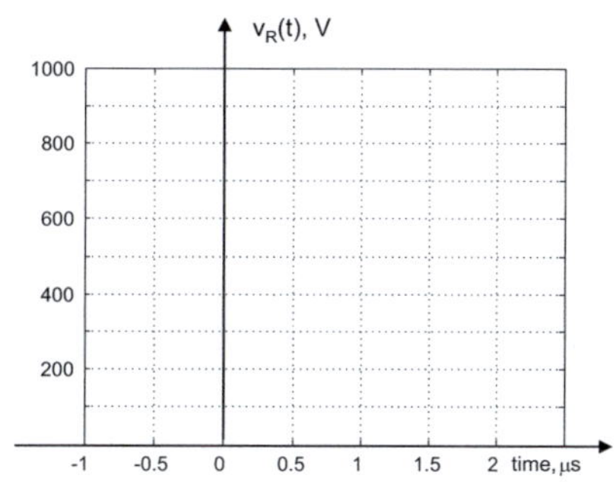

Problem 7.42.

A. Obtain the solution for the inductor current in the circuit shown in the figure at any time.
B. Plot the voltage across the rightmost resistor versus time over the interval from -2τ to 5τ.

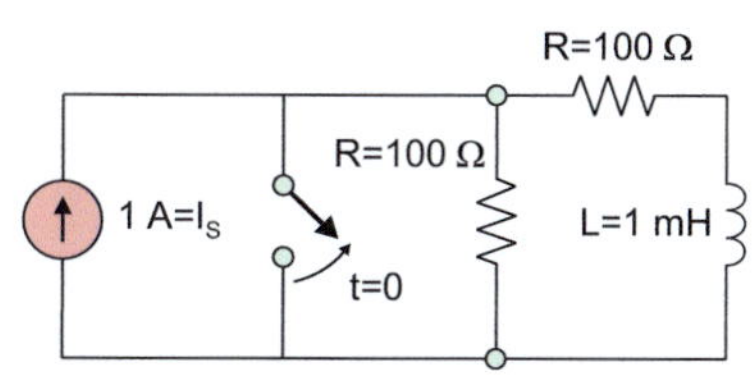

7.2.4 Energy-Release RL Circuit with the Voltage Supply

7.2.5 Application Example: Laboratory Ignition Circuit

Problem 7.43. A 270-μH inductor shown in the following figure releases its energy into the 510-Ω load resistor. The power supply voltage is 20 V. The switch opens at $t = 0$.

A. Present an expression for the inductor current as a function of time and sketch it to scale versus time over the interval from −1 μs to 2.5 μs.

B. Repeat the same task for the resistor voltage.

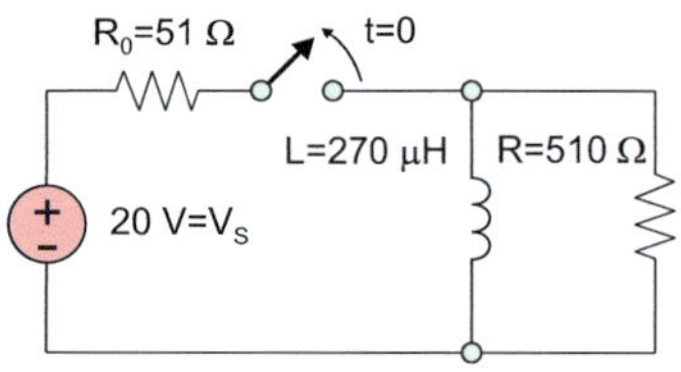

Problem 7.44. The circuit for the previous problem is converted to the energy-accumulating RL circuit by inversing the switch operation. Assume that the switch was open prior to $t = 0$. The switch closes at $t = 0$.

A. Derive an expression for the inductor current as a function of time.

B. Repeat the same task for the voltage across resistor R.

C. Could this circuit generate large voltage spikes, similar to the circuit from the previous problem?

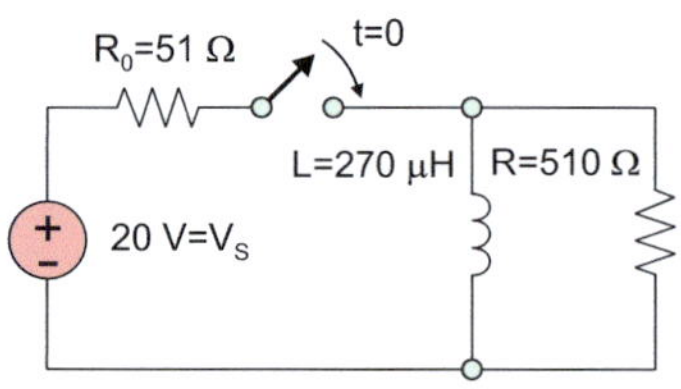

7.3 Switching RC Oscillator

7.3.2 Bistable Amplifier Circuit with the Positive Feedback

7.3.3 Triggering

Problem 7.45. The bistable amplifier circuit shown in the following figure (inverting Schmitt trigger) exists in the positive stable state. Amplifier's power supply rails are ±12 V. Determine output voltage when the applied trigger signal is

1. $v_{in} = 6$ V
2. $v_{in} = 2$ V
3. $v_{in} = -4$ V

Assume that the amplifier hits the power rails in saturation.

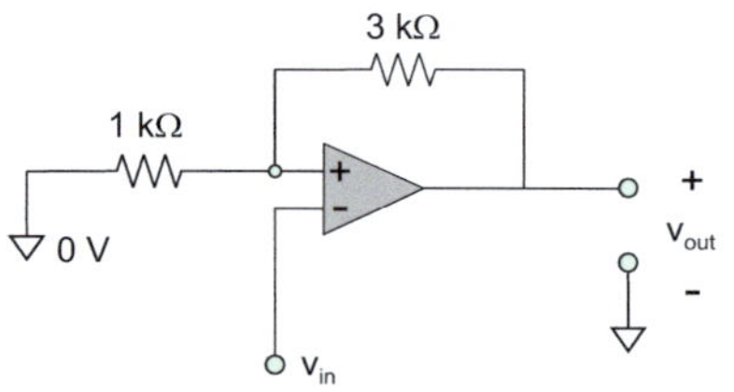

Problem 7.46. Repeat the previous problem when the initial stable state of the amplifier circuit is negative.

Problem 7.47. The bistable amplifier circuit shown in the following figure (non-inverting Schmitt trigger) exists in the positive stable state. Amplifier's power supply rails are ±15 V. Determine output voltage when the applied trigger signal is

A. $v_{in} = -1$ V

B. $v_{in} = -2$ V

C. $v_{in} = -4$ V

Assume that the amplifier hits the power rails in saturation.

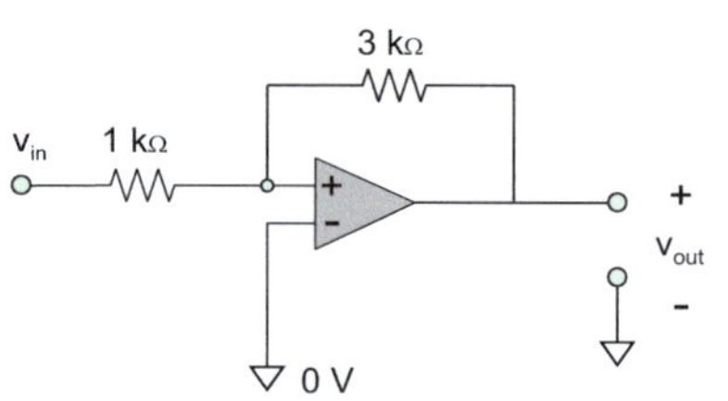

Problem 7.48. Repeat the previous problem when the initial stable state of the amplifier circuit is negative.

7.3.4 Switching RC Oscillator
7.3.5 Oscillation Frequency

Problem 7.49. A clock circuit (relaxation oscillator circuit) shown in the following figure is powered by a ±10-V power supply.

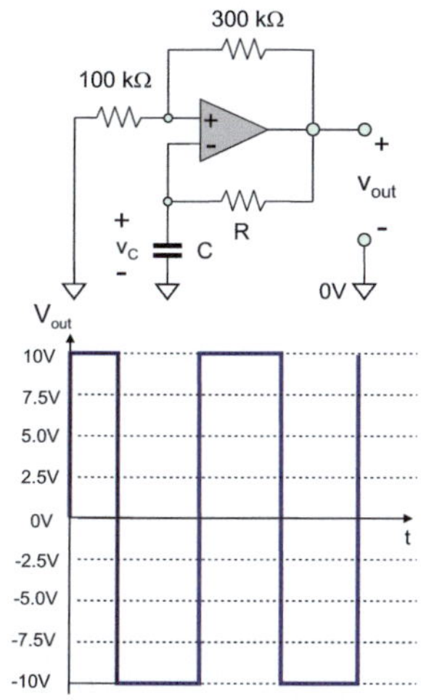

Sketch to scale the capacitor voltage, v_C, as a function of time. Assume that $v_C(t=0)=0$. Assume the ideal amplifier model. The specific values of R and C do not matter; they are already included in the time scale.

Problem 7.50. An RC clock circuit is needed with the oscillation frequency of 1 kHz and amplitude of the capacitor voltage of 4 V. Determine one possible set of circuit parameters R_1, R_2, R given that the capacitance of 100 nF is used. The power supply voltage of the amplifier is ±12 V. Assume that the amplifier hits the power rails in saturation.

Problem 7.51. A relaxation oscillator circuit may generate nearly triangular waveforms at the capacitor. Which values should the feedback factor $\beta = \frac{R_1}{R_1+R_2}$ attain to make it possible?

7.4 Single-Time-Constant (STC) Transient Circuits

7.4.1 Circuits with Resistances and Capacitances
7.4.2 Circuits with Resistances and Inductances

Problem 7.52. Determine whether or not the transient circuits shown in the following figure are the STC circuits. If this is the case, express the corresponding time constant in terms of the circuit parameters.

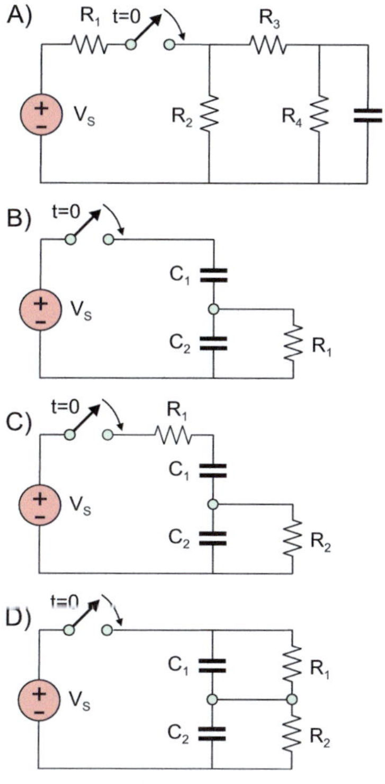

Problem 7.53. Repeat the previous problem for the circuits shown in the following figure.

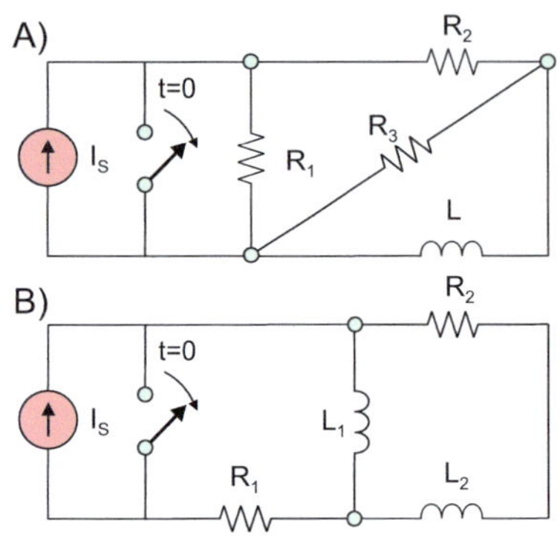

7.4.3 Example of a Non-STC Transient Circuit

Problem 7.54. Using software of your choice, generate the solution for the non-STC circuit in Fig. 7.25a over the time interval from 0 to $15\tau_0$ where $\tau_0 = RC$ with $C_1 = C_2 = C$, $R_1 = R_2 = R$. Use $R = 1$ kΩ, $C = 1$ μF, and $V_S = 10$ V. Plot two capacitor voltages and the circuit current as functions of time to scale.

Problem 7.55.

A. Derive the general ODE for the non-STC circuit in Fig. 7.25a in terms of v_1 for arbitrary circuit parameters.
B. Present its particular form when $C_1 = C_2 = C$ and $R_1 = 2R_2 = R$. Express all coefficients in terms of $\tau_0 = RC$.
C. Given that the solution for the homogeneous ODE has the form $\exp(-\alpha t/\tau_0)$, determine two possible solutions for the dimensionless coefficient α.
D. Using software of your choice, generate the circuit solution over the time interval from 0 to $15\tau_0$. Use $R = 1$ kΩ, $C = 1$ μF, and $V_S = 10$ V. Plot two capacitor voltages and the circuit current as functions of time to scale.

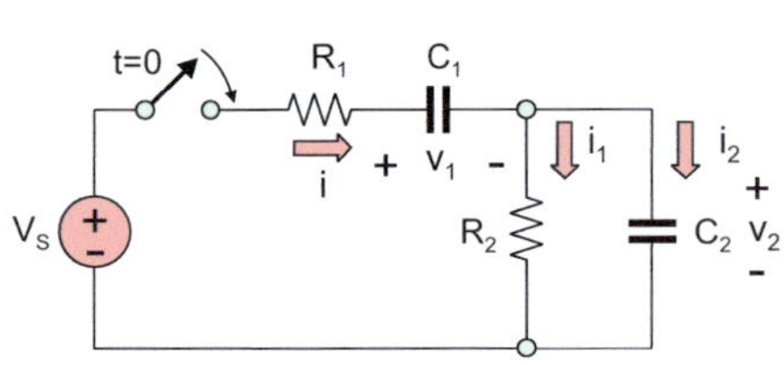

7.4.4 Example of a STC Circuit

Problem 7.56. For the circuit shown in Fig. 7.26, derive ODEs for inductor currents i_1, i_2.

7.4.5 Method of Thévenin Equivalent. Application Example: Circuit with a Bypass Capacitor

Problem 7.57. In the circuit from Fig. 7.27a, another resistance R_0 is present *in series* with the capacitance C. Determine the natural response of the circuit and find the corresponding time constant, τ.

Problem 7.58. A transient circuit with the DC voltage source V_S is shown in the following figure. Given that $V_S = 10$ V and $R_1 = 1$ kΩ, $R_2 = 1$ kΩ, and $C = 1$ μF:

A. Present the ODE for the capacitor voltage $v_C(t)$.
B. Determine the value of the time constant τ and the ODE right-hand side (the forcing function).
C. Present the solution for the capacitor voltage as a function of time assuming an initially uncharged capacitor

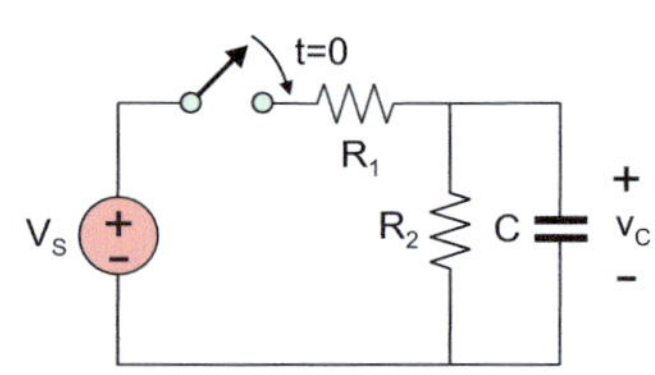

Problem 7.59. A transient circuit with the DC voltage source V_S is shown in the following figure. Given that $V_S = 10$ V and $R_1 = R_2 = 1$ kΩ, $R_3 = 2$ kΩ, and $C = 1$ μF:

A. Present the ODE for the capacitor voltage $v_C(t)$.
B. Determine the value of the time constant τ and the ODE right-hand side (the forcing function).
C. Present the solution for the capacitor voltage as a function of time assuming an initially uncharged capacitor.

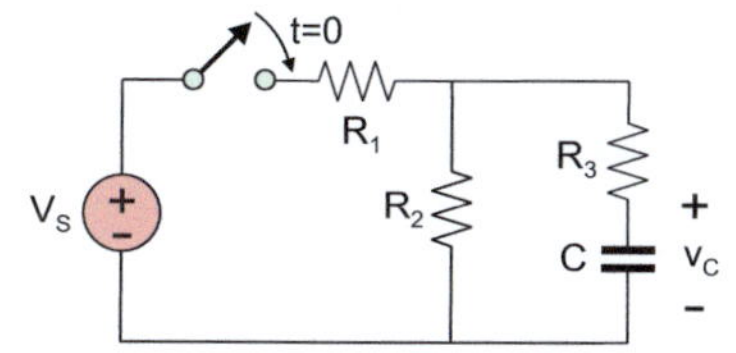

Problem 7.60. A transient circuit with the DC voltage source V_S is shown in the following figure. Given that $V_S = 10\,\text{V}$ and $R_1 = R_2 = 1\,\text{k}\Omega$, $R_3 = R_4 = 2\,\text{k}\Omega$, and $C = 1\,\mu\text{F}$:

A. Present the ODE for the capacitor voltage $v_C(t)$.
B. Determine the value of the time constant τ and the ODE right-hand side (the forcing function).
C. Present the solution for the capacitor voltage as a function of time assuming an initially uncharged capacitor.

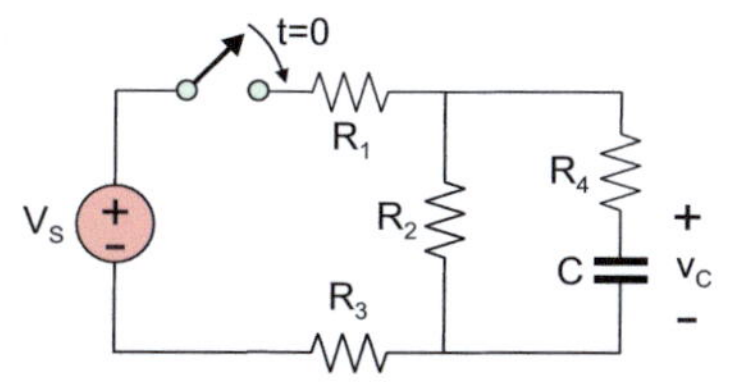

Problem 7.61. A transient circuit with the current source I_S is shown in the following figure. Given that $I_S = 10\ \text{mA}$ and $R_1 = R_2 = 1\ \text{k}\Omega$ and $L = 1\ \text{mH}$:

A. Present the ODE for the inductor current $i_L(t)$.
B. Determine the value of the time constant τ and the ODE right-hand side (the forcing function).
C. Present the solution for the inductor current as a function of time assuming the initial current equal to zero.

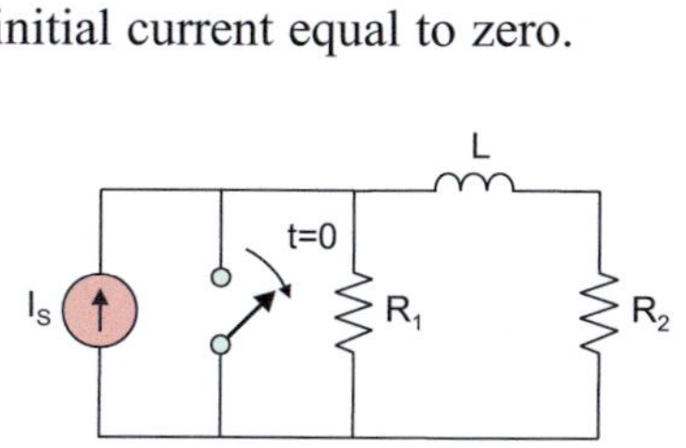

Problem 7.62. A transient circuit with the DC voltage source V_S is shown in the figure below. Given that $V_S = 10\ \text{V}$ and $R_1 = R_2 = 1\ \text{k}\Omega$, $L = 1\ \text{mH}$:

A. Present the ODE for the inductor current $i_L(t)$.
B. Determine the value of the time constant τ and the ODE right-hand side (the forcing function).
C. Present the solution for the inductor current as a function of time assuming the initial current equal to zero.

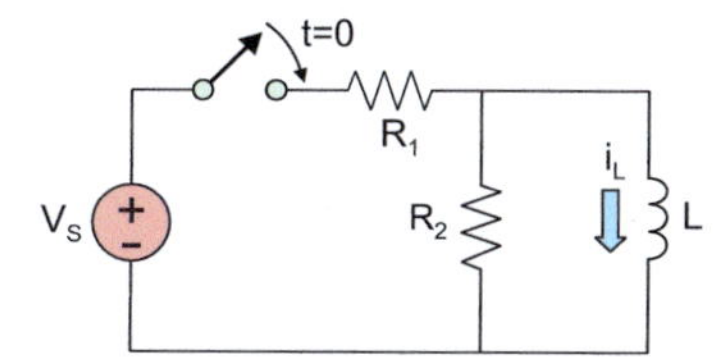

Problem 7.63. Consider the circuits in two previous problems at arbitrary values of R_1, R_2, L, I_S, V_S. What should be the relation between these parameters to guarantee the *same* solution for the inductor current in every case?

Problem 7.64. Describe the mathematical meaning of

A. Natural response
B. Forced response

for a first-order transient circuit in your own words. Do you think that this concept can be applied to any transient circuit?

Problem 7.65. If a transient circuit uses a DC supply (either voltage or current) and a switch, what is a general form of the forcing function?

Problem 7.66. If the forcing function of a first-order transient circuit is a combination of sine/cosine function and a constant, what is the general form of the forced response?

Problem 7.67. In the transient circuit shown in the figure below, $v_S(t) = V_S + V_m \sin \omega t$.

A. Write the solution for the voltage across the load resistor R_L in terms of the circuit parameters assuming an initially uncharged capacitor.

B. Write the solution for the voltage across the load resistor R_L when the bypass capacitor is absent.

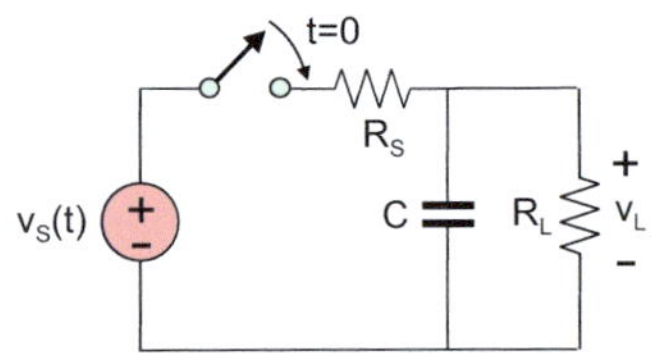

Problem 7.68. Plot to scale the load voltage for the circuit shown in Fig. 7.27a with and without the bypass capacitor over time interval from 0 to 50 ms. The circuit parameters are $R_S = 5\ \Omega$, $R_L = 1\ \text{k}\Omega$, $C = 500\ \mu\text{F}$. The source given by Eq. (7.50) is the superposition of the DC and AC components. The source parameters are $V_S = 10\ \text{V}$, $V_m = 1\text{V}$, $f = 250\ \text{Hz}$.

Problem 7.69. What is the asymptotic form of the solution given by Eqs. (7.57)–(7.58) when the source resistance, R_S, tends to zero?

Problem 7.70. In the transient circuit shown in the figure below, assume $i_S(t) = I_S + I_m \sin \omega t$.

A. Write the solution for the current through the load resistor R_L in terms of the circuit parameters assuming that the initial inductor current is equal to zero.

B. Write the solution for the current through the load resistor R_L when the decoupling inductor is absent

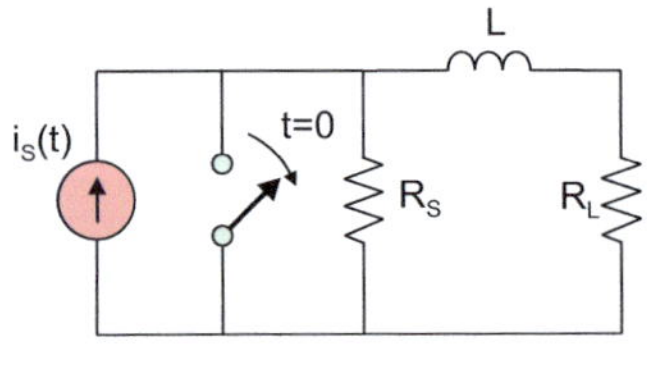

Problem 7.71. In the circuit for the previous problem, another resistor R_0 is present *in parallel* with the inductance L. Determine the natural response of the circuit and find the time constant, τ.

7.5 Description of the Second-Order Transient Circuits

7.5.1 First-order Transient Circuits Versus Second-order Transient Circuits

Problem 7.72. A transient circuit is shown in the following figure. Is it a first- or second-order transient circuit? Justify your answer.

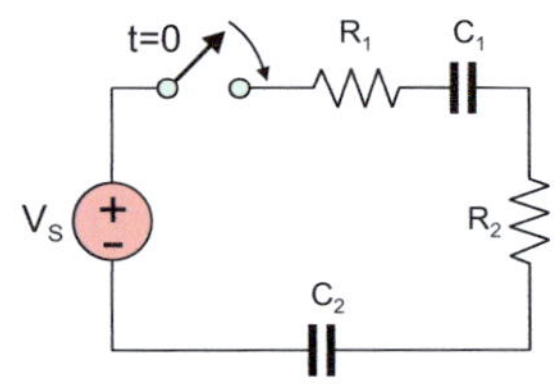

Problem 7.73. A transient circuit is shown in the figure below. Is it a first- or second-order transient circuit? Justify your answer.

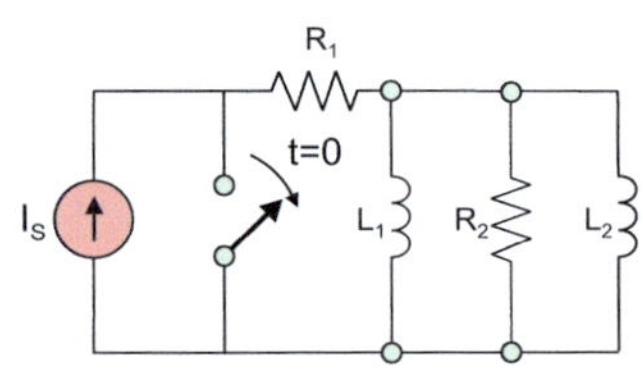

Problem 7.74. Establish the ODE for the transient circuit shown in the figure below. Both capacitors have zero voltage prior to closing the switch.

A. Assume $R_1 = R_2 = R$, $C_1 = C_2 = C$.

B. Assume arbitrary values of $R_{1,2}$, $C_{1,2}$.

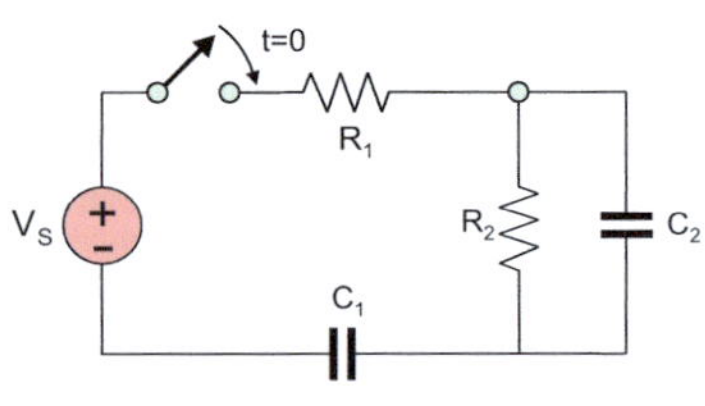

Problem 7.75. Establish the ODE for the transient circuit shown in the figure below. Inductor currents are zero prior opening the switch.

A. Assume $R_1 = R_2 = R$, $L_1 = L_2 = L$.

B. Assume arbitrary values of $R_{1,2}$, $L_{1,2}$.

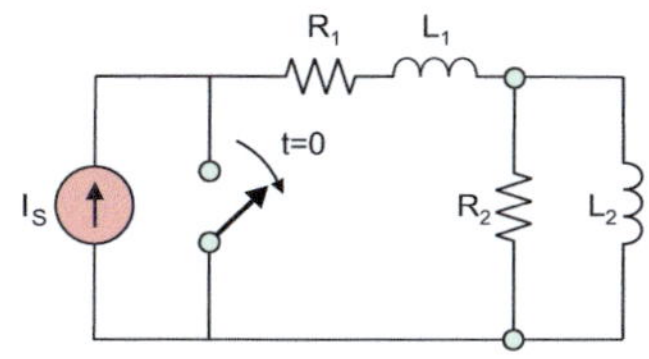

Problem 7.76. In the previous problem, add resistance R_3 as shown in the figure that follows and solve task B.

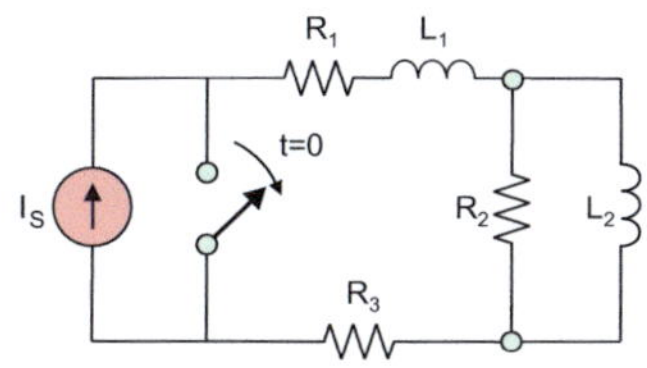

7.5.2 Series Connected Second-order RLC Circuit

Problem 7.77. Describe in your own words the mechanical counterpart of the series RLC circuit.

Problem 7.78. An RLC circuit in Fig. 7.30 has $R = 1\ \text{k}\Omega$, $C = 1\ \mu\text{F}$, $L = 1\ \text{mH}$.

A. Find the value of the damping coefficient, α (show units).
B. Find the value of undamped resonant frequency, ω_0.

Problem 7.79. How does the second-order ODE Eq. (7.63) for the circuit current in the series RLC circuit from Fig. 7.30 change if the capacitor was charged to $V_S/2$ prior to $t = 0$?

Problem 7.80. In the circuit shown in Fig. 7.30, the switch is replaced by a short circuit. The constant voltage source is replaced by an arbitrary time-varying voltage source $V_S \to v_S(t)$ as shown in the figure that follows. Derive the dynamic circuit equation for the circuit current similar to Eq. (7.63) of this section. Present your result in terms of damping coefficient α and undamped resonant frequency ω_0.

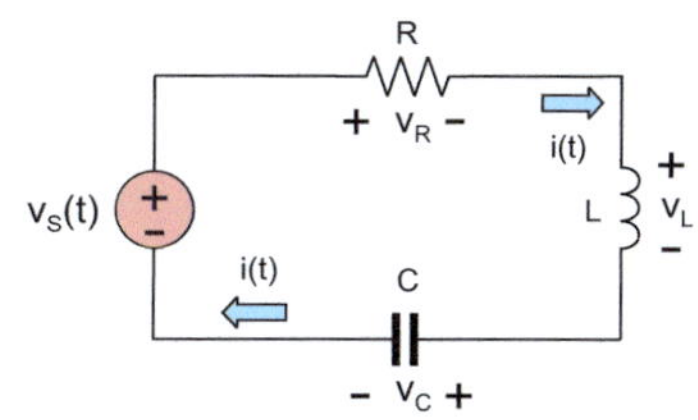

7.5.3 Choice of Independent Function: Initial Conditions

7.5.4 Step Response

Problem 7.81. The RLC circuit shown in Fig. 7.30 is described by dynamic equation (7.63) written in terms of the electric current. How do the initial conditions to this equation change if the capacitor was charged to $V_S/2$ prior to $t = 0$?

Problem 7.82. For the circuit shown in Fig. 7.30:

A. Derive the dynamic circuit equation (7.67) in terms of the capacitor voltage $v_C(t)$.
B. How does this equation change if the capacitor was charged to $V_S/2$ prior to $t = 0$?

Problem 7.83. For the series RLC circuit with the switch and the DC supply shown in Fig. 7.30, we know the solution, $v_C(t)$, of Eq. (7.67) or Eq. (7.69) for $V_S = 1$ V. The circuit shown in the following figure is now considered, with the voltage source in the form of a voltage pulse (one bit) centered about $t = 0$. Express the solution to the present problem in terms of $v_C(t)$.

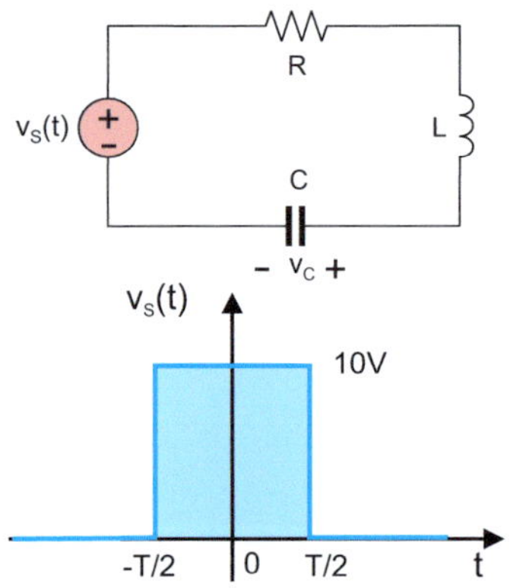

Problem 7.84. Repeat the previous problem for the voltage source shown in the figure that follows (a voltage pulse train of two bits).

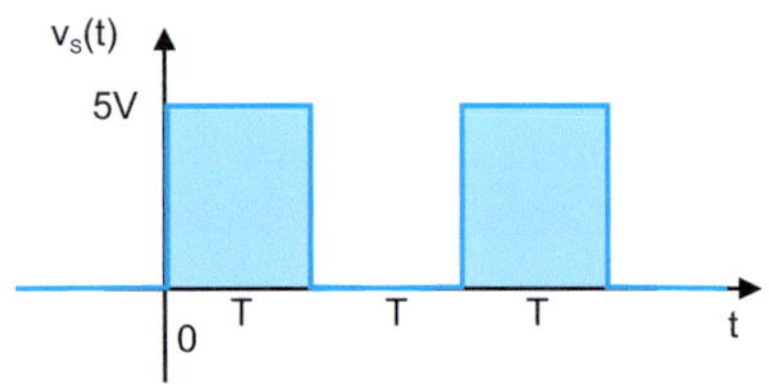

Problem 7.85. Repeat Problem 7.83 for the voltage source shown in the figure that follows (a bipolar voltage pulse).

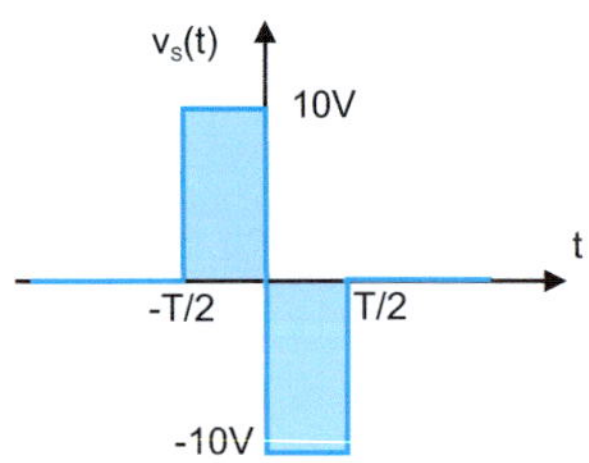

7.5.5 Parallel Connected Second-Order RLC Circuit

Problem 7.86. For the circuit shown in Fig. 7.34:

A. Derive the dynamic circuit equation (7.75) written in terms of the inductor current $i_L(t)$.
B. How does this equation change if the switch in Fig. 7.34 was open prior to $t = 0$ and closes at $t = 0$?
C. How do the initial conditions change in this case?

Problem 7.87. Describe the duality between series and parallel RLC circuits in your own words.

Problem 7.88. For the parallel RLC circuit with the switch and the DC current source shown in Fig. 7.34, we know the solution, $i_L(t)$, of Eq. (7.75) for $I_S = 1$ mA. The circuit shown in the following figure is now considered, with the current source in the form of a pulse (one bit) centered about $t = 0$. Express the solution to in terms of $i_L(t)$.

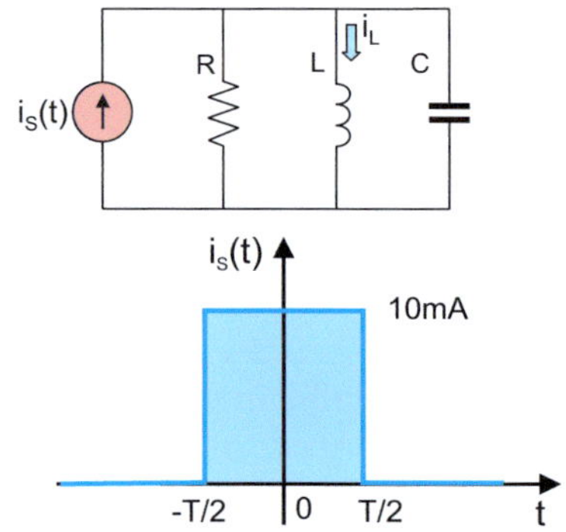

7.6. Step Response of the Series RLC Circuit

7.6.1 General Solution of the Second-order ODE

7.6.2 Derivation of Complementary Solution

Problem 7.89.

A. The complete solution to a second-order homogeneous ODE is a sum of two distinct components. Describe each of them.
B. Write three forms of the complementary solution (natural response) for the second-order homogeneous ODE.
C. What is a new parameter to be introduced for the underdamped circuit?

Problem 7.90. For the series RLC circuit shown in the following figure:

A. Find the value of the damping coefficient, α (show units).

B. Find the value of the undamped resonant frequency, ω_0 (show units).
C. Find the value of the damping ratio, ζ (show units).
D. Find the particular solution (forced response).
E. Outline the form of the complementary solution (natural response).
F. Which value should the circuit resistance have for a critically damped circuit?

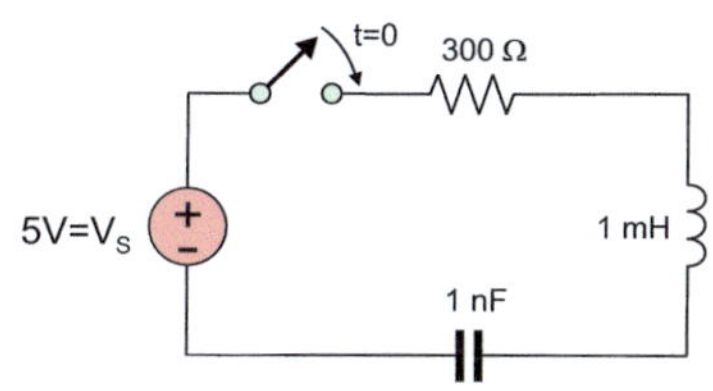

Problem 7.91. Repeat the previous problem for the series RLC circuit shown in the figure below.

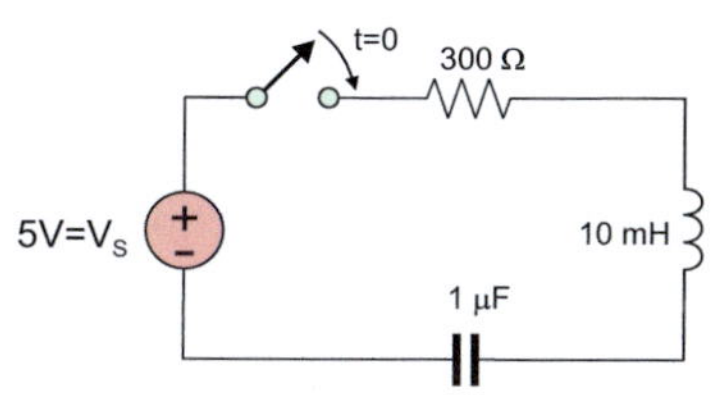

Problem 7.92. For the series RLC circuit shown in the following figure, fill out the table of circuit parameters.

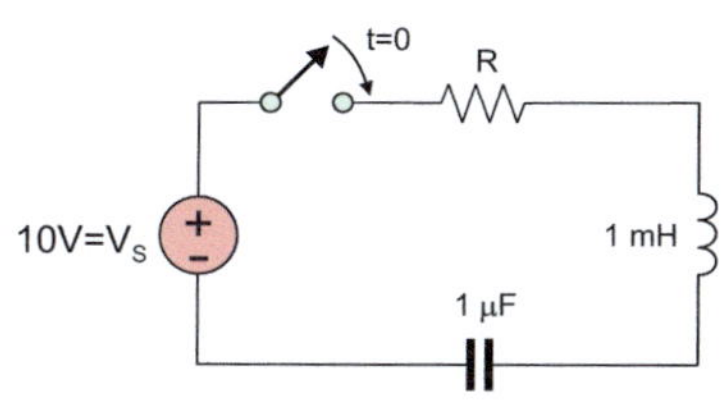

Table of circuit parameters

R, Ω	ζ	Circuit type (overdamped, critically damped, underdamped)
25		
50		
75		
100		

Given fixed L and C, which values of resistance (large or small) lead to the overdamped circuit?

Problem 7.93. For the series RLC circuit shown in the figure below, fill out the table of circuit parameters.

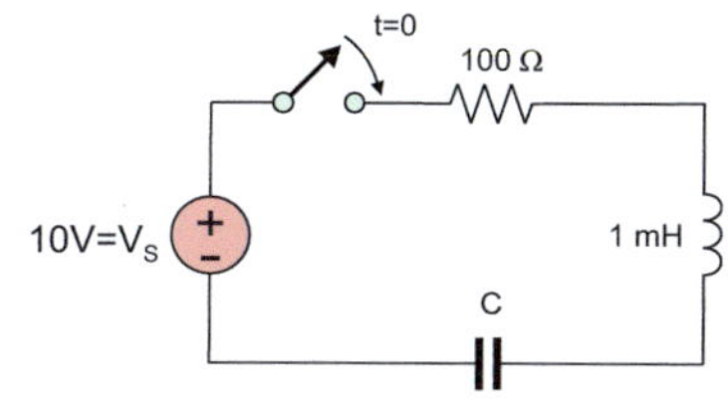

Table of circuit parameters

C, μF	ζ	Circuit type (overdamped, critically damped, underdamped)
0.01		
0.1		
0.4		
1.0		

Given fixed L and R, which values of capacitance (large or small) lead to the overdamped circuit?

Problem 7.94. For the series RLC circuit shown in the following figure, fill out the table of circuit parameters.

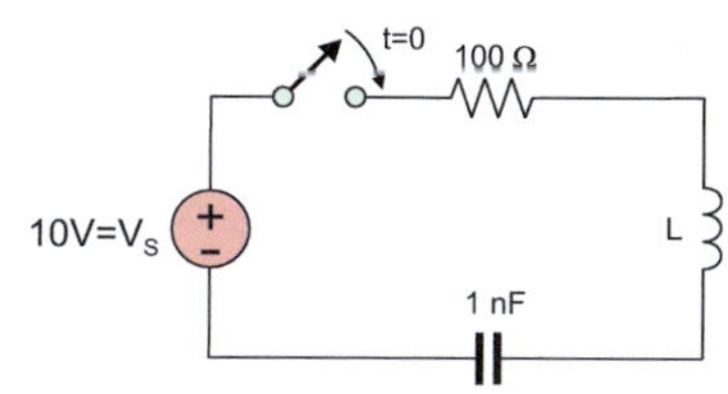

Table of circuit parameters

L, μH	ζ	Circuit type (overdamped, critically damped, underdamped)
0.1		
1		
2.5		
10		

Given fixed R and C, which values of inductance (large or small) lead to the overdamped circuit?

Problem 7.95. Show that underdamped solution and critically damped solutions coincide with each other when $\varsigma \to 1$.

7.6.3 Finding Integration Constants

7.6.4 Solution Behavior for Different Damping Ratios

Problem 7.96. For the circuit shown in the figure below:

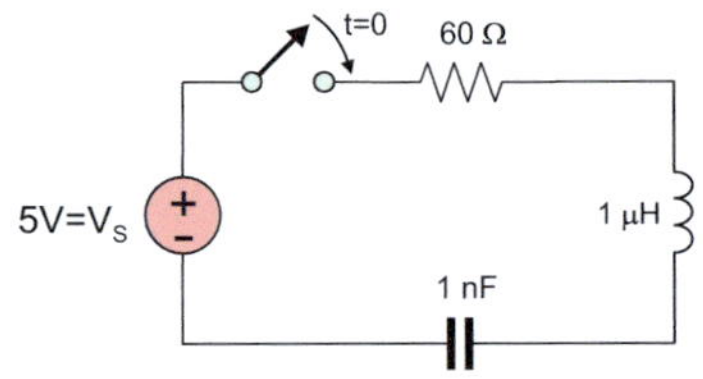

A. Determine damping coefficient α, undamped resonant frequency ω_0, and damping ratio ζ.
B. Determine constants K_1, K_2.
C. Write solution for the capacitor voltage with all constants defined.
D. Calculate and plot to scale capacitor voltage at 0, 0.05, 0.1, 0.2, and 0.3 μs.

Problem 7.97. For the circuit shown in the following figure:

A. Determine damping coefficient α, undamped resonant frequency ω_0, and damping ratio ζ.
B. Determine constants K_1, K_2.
C. Write the solution for the capacitor voltage with all constants defined.
D. Calculate capacitor voltage at 0, 1, 2, 3, 4, and 5 ms and plot it to scale versus time.

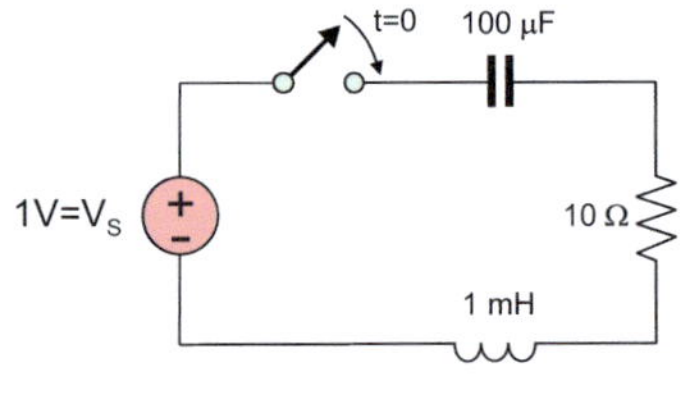

Problem 7.98. Repeat the previous problem for the circuit shown in the following figure.

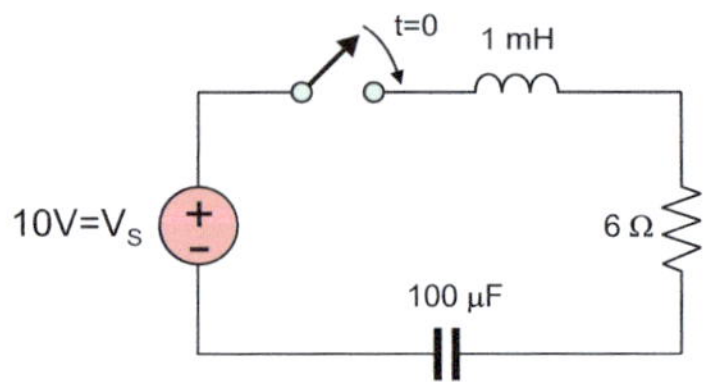

Problem 7.99. In the circuit shown in the figure below, the capacitor was charged to 10 V prior to closing the switch.

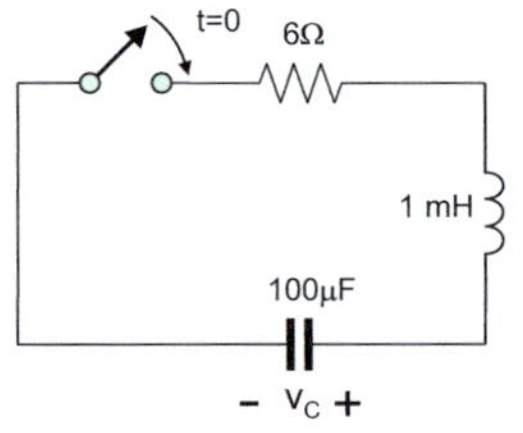

A. How are the circuit equation and initial conditions different from Eqs. (7.78a, b)?
B. Determine damping coefficient α, undamped resonant frequency ω_0, and damping ratio ζ.
C. Determine constants K_1, K_2.
D. Write the solution for capacitor voltage with all constants defined.
E. Calculate the capacitor voltage at 0, 1, 2, 3, 4, and 5 ms and plot it to scale versus time.

7.6.5 Overshoot and Rise Time

7.6.6 Application: Non-ideal Digital Waveform

Problem 7.100. The following figure shows the underdamped step response for a series RLC circuit. The DC source has the voltage of 10 V. Using the figure:

A. Estimate the overshoot percentage.
B. Estimate the rise time.
C. Do these estimates (approximately) agree with Eqs. (7.88a, b)?

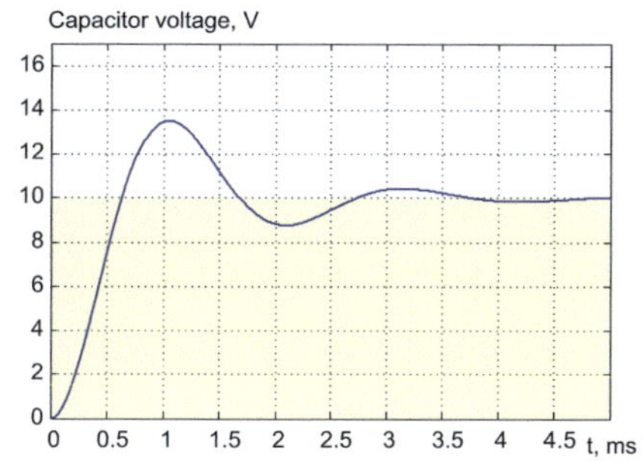

Problem 7.101. Capacitor voltage is measured in a series RLC circuit as shown in the figure to the previous problem. Given $R = 2\,\Omega$, estimate circuit inductance L and circuit capacitance C.

Problem 7.102. The figure that follows shows the distorted rectangular waveform (capacitor voltage) for the circuit shown in Fig. 7.33. The DC source has the voltage of 10 V.

A. Using the figure, estimate the overshoot and undershoot percentages.
B. Using the figure, estimate the rise time and the fall time.
C. Do these estimates (approximately) agree with Eqs. (7.88a, 7.88b)?

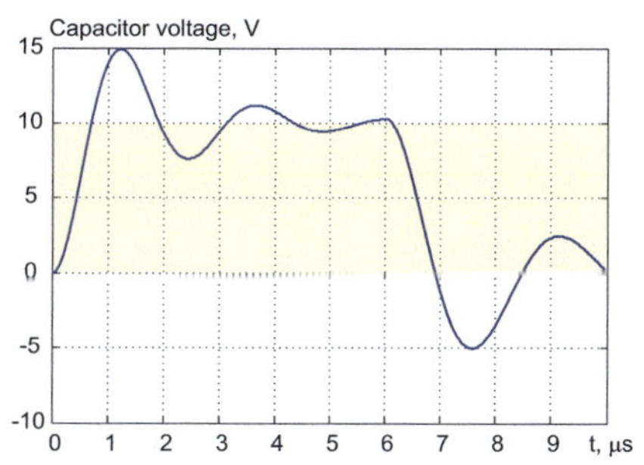

Problem 7.103. For the circuit shown in the following figure:

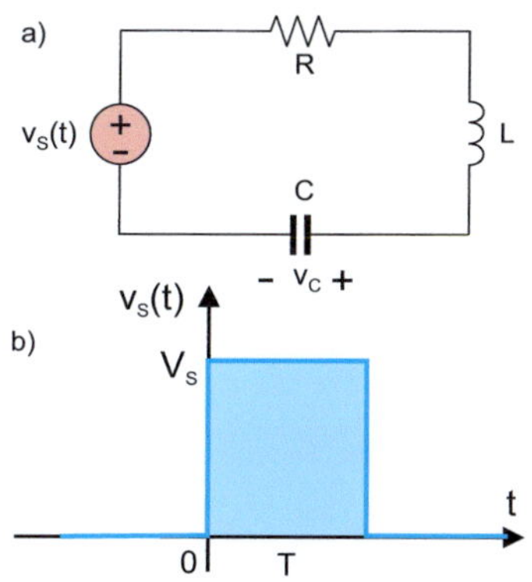

A. Determine the step response $v_C(t)$ for the circuit shown in figure (a) given that $L = 1\ \mu\text{H}$, $C = 1\ \text{nF}$, $V_S = 10\ \text{V}$, and $R = 75\ \Omega$.
B. Express the solution $v_C^{\text{pulse}}(t)$ for the voltage pulse shown in figure (b) in terms of the step response.
C. Given $T = 0.5\ \mu\text{s}$, calculate the solution for the voltage pulse over the time interval from 0 to 0.7 μs in steps of 0.1 μs and plot it to scale.

Problem 7.104. Repeat the previous problem assuming $T = 0.2\ \mu\text{s}$. Calculate the solution for the voltage pulse over the time interval from 0 to 0.5 μs in steps of 0.05 μs and plot it to scale versus time.

Part III

AC Circuits

Chapter 8: Steady-State AC Circuit Fundamentals

Overview

Prerequisites:
- Knowledge of DC circuit analysis (Chapters 2, 3, and 4)
- Knowledge of dynamic circuit elements (Chapter 6, optionally Chapter 7)
- Knowledge of complex arithmetic and calculus

Objectives of Section 8.1:
- Apply and work with the major parameters of the steady-state AC signals: amplitude, frequency, and phase
- Establish the concept of phase leading or lagging for AC voltages and currents
- Become familiar with the major function of the oscilloscope—measure periodic (AC) voltages in a circuit
- Understand the meaning of the phasor as a representation of the real signal
- Convert real signals to phasors and vice versa
- Perform basic operations with phasors
- Be able to construct the phasor diagram for real signals and restore the real signals from the phasor diagram
- Become familiar with the phasor (angle) notation

Objectives of Section 8.2:
- Provide the complete mathematical derivation of complex impedances
- Apply the impedance concept to resistor, capacitor, and inductor
- Understand the meaning of magnitude and phase of the complex impedance

Objectives of Section 8.3:
- Understand and apply the AC circuit analysis with phasors and impedances
- Appreciate the value of the phasor diagram as a tool for AC circuit analysis
- Transfer major circuit theorems to steady-state AC circuits
- Be able to solve multifrequency AC circuits using superposition principle

Application examples:
- Measurements of amplitude, frequency, and phase
- Impedance of a human body

S. N. Makarov et al., *Practical Electrical Engineering*, https://doi.org/10.1007/978-3-319-96692-2_8

Keywords:

Steady-state AC circuit, Signals, Harmonic, Amplitude, Pk-Pk value, Angular frequency, Phase, Period, Steady-state AC voltage, Steady-state alternating current, Leading signal, Lagging signal, Oscilloscope, Phasor, Phasor voltage, Phasor current, Phasor diagram, Phasor notation, Angle notation, Complex impedance, Impedance of the resistor, Impedance of the capacitor, Impedance of the inductor, Reactance, KVL in phasor form, KCL in phasor form, Source transformation in the frequency domain, Thévenin's theorem for steady-state AC circuits, Norton theorem for steady-state AC circuits, Superposition principle for multifrequency AC circuits

Section 8.1 Harmonic Voltage and Current: Phasor

Whether you use a kitchen appliance or operate a piece of industrial machinery, they are powered by alternating current (AC) power supplies. It may seem odd at first to create power sources with alternating polarities that result in fluctuating voltage and current directions through resistors, inductors, and capacitors. Consider the following situation: current flows through a coil from left to right with increasing and then decreasing magnitude, then changing direction, and again increasing and decreasing in magnitude before the process repeats itself. A benefit of using periodic, sinusoidal waveforms is related to the relative ease of stepping up or down AC voltages and currents with little, ideally no, power losses via a transformer. The economic importance of AC circuits can hardly be overstated; the bulk of the residential and industrial power demand is produced, transformed, and distributed via AC circuits. To understand the basics of AC circuit analysis, we start with the key characteristics of sinusoidal waveforms such as frequency, phase, and amplitude. Special attention is paid to the phase. We define the meaning of leading and lagging phase for steady-state AC voltages or currents. We note that in electrical engineering, AC voltages are often called *signals*.

8.1.1 Harmonic Voltages and Currents

In *steady-state AC circuits*, all voltages and currents measured across or through the elements are periodic and in the ideal case *harmonic* (i.e., *sine* or *cosine*) functions of time. These voltages and currents have the *same* frequency but different phases and amplitudes. Interestingly, the word *harmonic* originates from the reference to music sounds of pure, single tones, pitches, or *frequencies*. The word *steady-state* means that the circuit frequency, phases of all voltages and currents, and amplitudes of all voltages and currents do not change over time. Transient effects are entirely excluded from our consideration. Similarly, the AC voltage and currents are called *steady-state AC voltage* and *steady-state alternating current*. It is common to use terms *AC voltage* and *AC power*, but the term *AC current* does not make much sense, even though it might be a part of the electrical engineering jargon. Figure 8.1 shows an AC harmonic signal. All voltages and currents in an AC circuit will have exactly this form, regardless of whether they are measured over a resistor, capacitor, or inductor. For this figure we consider the voltage $v(t)$. The current $i(t)$ could be treated in an identical manner. As a harmonic function, the steady-state AC voltage can be written in the form

$$v(t) = V_{\mathrm{m}} \cos(\omega t + \varphi) \tag{8.1a}$$

where

V_{m} is the voltage *amplitude* (maximum *absolute* voltage), with the unit of volts.
ω is the *angular frequency*, with the unit of rad/s.
φ is the *phase*, with the unit of radians.

The use of a cosine function rather than a sine function, as a basis for any AC signal in Eq. (8.1a), is common. The angular frequency relates to the frequency *f* and period *T* by

$$\omega = 2\pi f, \quad T = 1/f \tag{8.1b}$$

where *f* is measured in hertz or Hz (1 Hz = 1 s^{-1}) and the period is recorded in seconds. As the name implies, the frequency *f* defines the number of cycles (positive-to-negative AC voltage transitions) per second. For example, the frequency of the AC wall plug in the USA is 60 cycles per second and is recorded as

$$f = 60\,\frac{1}{\text{s}} = 60 \ \text{Hz} \tag{8.1c}$$

The angular frequency ω is essentially a replica of the frequency *f*; its use is primarily a matter of convenience. Virtually all frequencies in electrical engineering are *measured* in terms of *f* and not in terms of angular frequency ω. For example, we say that the AC frequency in the USA is 60 Hz, or 50 Hz in Europe and other countries and not 377 rad/s or 314 rad/s. The frequency unit *hertz* honors Heinrich Hertz (1857–1894), a German scientist, who, at the age of 29, built the first radio-frequency transmitter based on a spark gap circuit and a receiver (a loop antenna). He thus confirmed Maxwell's theory of electromagnetic wave propagation.

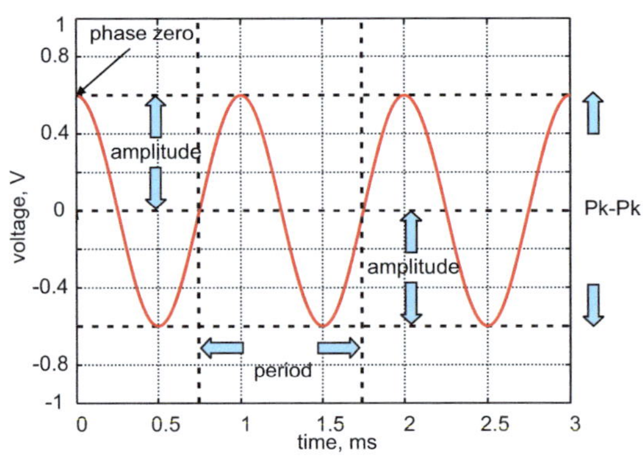

Fig. 8.1. Harmonic voltage signal of Eq. (8.1a) and its parameters. The phase is zero.

Example 8.1: Determine the frequency in Hz, the angular frequency in rad/s, and the amplitude of the harmonic voltage signal shown in Fig. 8.1.

Solution: The amplitude is the simplest parameter to find. By inspection, the maximum voltage value is 0.6 V. It is exactly the amplitude value, $V_m = 0.6$ V, and therefore is the absolute maximum voltage. The period is determined as the interval between two similar zero

Example 8.1 (cont.): crossings in Fig. 8.1; it is measured as one millisecond. Another (sometimes more accurate) way to measure the period is to observe that there are exactly three periods in the figure, recorded over the time interval from 0 to 3 ms. Dividing the time interval over the number of periods, we obtain the duration of one period. The frequency of the voltage is $f = 1/T = 1000\,\text{Hz} = 1\,\text{kHz}$. The angular frequency results in $\omega = 2\pi f = 6283.1$ rad/s, and the voltage in Fig. 8.1 is $v(t) = 0.6\cos(2\pi 1000t)$ [V]. In real circuits, you should *avoid* measuring the period by using two consecutive maxima or minima of the sinusoidal signal. Although the same result is obtained on paper, in reality the flat maximum plateau is frequently corrupted by noise. As a result, rather inaccurate maxima positions are acquired.

Note that the amplitude is well defined for a pure harmonic in Fig. 8.1. It is symmetric about the axis, i.e., does not have a DC offset. But what about other periodic signals that have a significant DC offset? How can we define their amplitudes? The key is the *peak-to-peak* (or in short: *Pk-Pk*) value shown in Fig. 8.1. For a harmonic AC voltage and current, the peak-to-peak value is simply twice the amplitude value. For non-sinusoidal periodic signals or sinusoidal signals with a DC offset, the peak-to-peak value is the *only* meaningful measure of the alternating signal strength. Therefore, an oscilloscope *always* uses the peak-to-peak value, instead of the amplitude.

8.1.2 Phase: Leading and Lagging

Perhaps the most confusing constant in Eq. (8.1a) is the phase φ. A nonzero phase means that the cosine function is shifted with respect to a zero-phase cosine in Fig. 8.1 either to the left or to the right. Let us ask the following question: if the phase in Eq. (8.1a) is positive, say $\varphi = +\pi/2$, is the cosine function shifted to the left or to the right versus the base cosine signal on the time axis? The answer might be somewhat unexpected: the positive phase means a shift to the *left*. Why? Because the given value of the cosine function occurs earlier in time than without the phase. Figure 8.2a shows a phase-shifted voltage signal $v(t) = 0.6\cos(2\pi 1000t + \pi/2)$ [V]; see the dashed curve; the base signal of zero phase is shown by the solid curve. At $t = 0$, the argument is already ahead by $+\pi/2$ and the cosine function is zero, whereas with zero phase the zero occurs later at $t = 0.5T$. A shift to the left means that all events (e.g., the peaks or zero crossings) happen *earlier* in time than for the base cosine with the phase zero. Therefore, the cosine with the positive phase always *leads* the cosine with the zero phase. In the present case, it leads the base cosine by 90°. In contrast, the cosine with the negative phase is shifted to the *right*. Figure 8.2b depicts the example of $\varphi = -\pi/2$. A shift to the right means that all events (e.g., the peaks or zero crossings) happen later in time than for the base cosine with zero phase. The voltage with the negative phase *lags* the cosine with the zero phase (in fact, it lags by 90° or $\pi/2$ in Fig. 8.2b). The concept of leading and lagging plays a prominent role in power electronics.

Exercise 8.1: Determine frequency in Hz, angular frequency in rad/s, and the amplitude of the harmonic voltage signals shown in Fig. 8.2.

Answer: All four signals in Fig. 8.2 have the same amplitude of 0.6 V, the same frequency of 1 kHz, and the same radian frequency of $\omega = 2\pi f = 6283.1 \ \ \mathrm{rad/s}$.

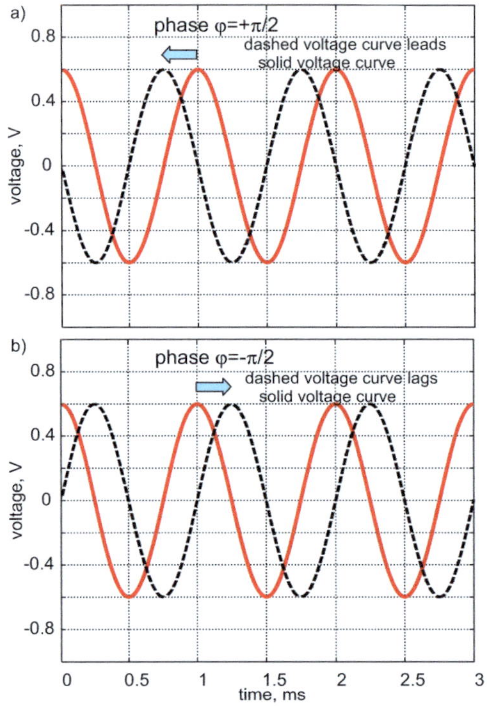

Fig. 8.2. Phase-shifted AC voltages $v(t) = 0.6\cos(2\pi 1000t \pm \pi/2)$ [V] versus the zero-phase voltage $v(t) = 0.6\cos(2\pi 1000t)$ [V] shown by a *solid curve*. In Fig. 8.2a, the *dashed* voltage leads the base signal, whereas in Fig. 8.2b it lags the base signal.

The most useful facts related to phase measurements are:

1. The phase in electrical engineering applications ranges from $-\pi$ to $+\pi$ radians; this corresponds in degrees to $-180°$ to $+180°$. The phase in degrees should be divided by 180 and multiplied by π to obtain the phase in radians. The phase in radians should be divided by π and then multiplied by 180 to obtain the phase in degrees.
2. The phase is a *relative* measure, with reference to a base signal. If the base signal is not present, the phase cannot be defined uniquely. By default the base signal is the cosine signal of the same frequency, that is, $\cos(\omega t)$.
3. If a measured phase for some reason exceeds π, we need to subtract its value from 2π. If it is less than $-\pi$, then we need to add 2π to its value. This will ensure that the phase will stay in the range from $-\pi$ to $+\pi$ radians.

Example 8.2: Determine the frequency, amplitude, and phase of the harmonic voltage signal shown in Fig. 8.3 versus the base cosine signal.

Solution: The amplitude is determined first: by inspection it is clear that $V_m = 1.0$V. Next the frequency is determined: the entire interval from 0 to 3 ms contains three full periods; hence $T = 1$ ms, and $f = 1/T = 1000$ Hz $= 1$ kHz. For the phase determination, we note that the first maximum in Fig. 8.3 occurs later in time than for the base cosine, which already peaks at $t = 0$. Therefore, the phase must be *negative*, that is, $\varphi < 0$. The absolute value of the phase in Fig. 8.3 is

$$|\varphi| = 2\pi\frac{\Delta T}{T} \tag{8.2}$$

which gives $\varphi = -\pi/3$ after measuring T and ΔT with a scale. A smarter way to obtain the same result is to note that the cosine function is equal to $0.5V_m$ at $t = 0$, so that $\varphi = -\cos^{-1}(0.5) = -\pi/3$. And an even "smarter" way is to use a calculator to plot the cosine function with all possible phases until it matches Fig. 8.3. This method only works well during an exam if enough time is available.

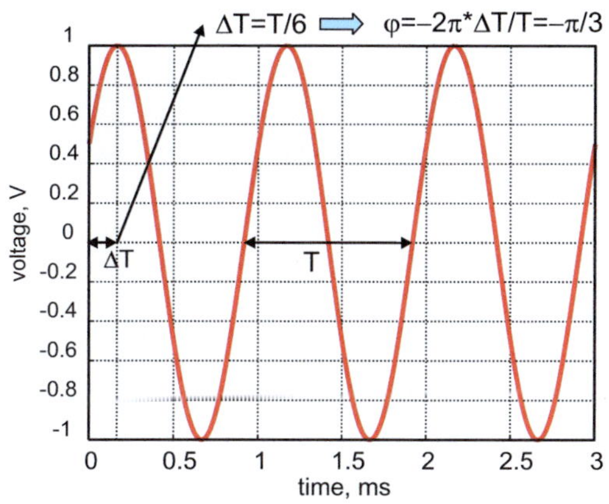

Fig. 8.3. A phase-shifted AC voltage $v(t) = 1\cos(2\pi 1000t - \pi/3)$ [V].

Example 8.3: Assume that an AC signal is given by $v(t) = V_m \sin(\omega t + \psi)$ [V]. How do we convert it to a cosine function?

Solution: The single most important trigonometric identity worth remembering is probably $\sin\alpha = \cos(\alpha - \pi/2)$. If you forget this identity, you may recall a visual picture: at $t = 0$ the sine with zero phase lags the cosine with zero phase by $\pi/2$ or by 90°. The result is then $v(t) = V_m\cos(\omega t + \psi - \pi/2)$ [V].

8.1.3 Application Example: Measurements of Amplitude, Frequency, and Phase

Different parameters of the steady-state AC voltage or of a general periodic voltage studied in this section are readily measured with the oscilloscope. The oscilloscope always measures the time-amplitude response of the voltage, *not* the current. Parameters include frequency, amplitude (or peak-to-peak value), the *rms* voltage, the mean voltage (voltage averaged over a period), etc. Figure 8.4 provides a practical measurement example. Several voltage signals can be measured simultaneously with several oscilloscope *channels*. The phase is not measured for a single signal, only the phase difference between two and more signals can be measured. The oscilloscope has two commonly used settings. In the first setting called *DC coupled*, any voltage supplied to the channel will cause a deflection of the trace. As a result, the *actual* value of the input voltage with respect to oscilloscope ground can be measured. In the second setting called *AC Coupled*, any DC offset to the periodic signal is eliminated via an internal coupling capacitor. The result, after a short transient period, is that the trace will settle at 0 V regardless of the magnitude of an applied DC voltage. This configuration is used whenever it is desired to ignore the constant DC component of an applied voltage waveform and only observe the AC component of the waveform with zero mean.

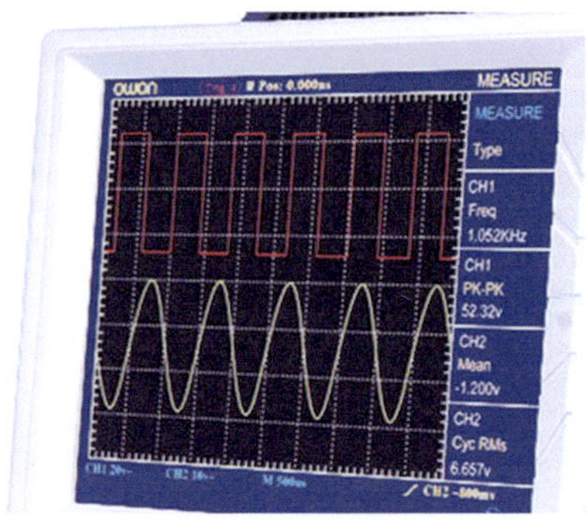

Fig. 8.4. Front panel of an inexpensive digital-storage oscilloscope. Note the measured signal parameters.

8.1.4 Definition of a Phasor

This topic is critical for the steady-state AC circuit analysis. We are about to introduce the method of solving AC circuits based on the use of so-called phasors or complex numbers. Working with phasors allows us to "cancel" out the frequency dependence and the time dependence. This is possible because in a linear system the harmonic behavior is the same for all circuit components, we only need to keep the amplitude and phase information for every sinusoid. The application of phasors will eventually allow us to reduce the AC circuit to an equivalent "DC" circuit that is solved using standard tools. However, there is no "free lunch"; the voltages and currents in the resulting "DC" circuit appear to be complex numbers. They carry information of two parameters: the phase and the amplitude of a sinusoid. The use of a complex number relies on two independent parameters, real and imaginary parts (or magnitude and phase); it is ideally suited to represent AC signals while entirely eliminating time-domain harmonics. The starting point of the phasor concept is rooted in Euler's formula in the form

$$e^{j\alpha} = \cos\alpha + j\sin\alpha \tag{8.3a}$$

The identity expresses the complex exponent in terms of the real-valued cosine and sine functions. Here, $j = \sqrt{-1}$ is the imaginary unit and α is an arbitrary, real number. Note that α can be equal to ωt or to $\omega t + \varphi$. In electrical engineering, the symbol $j = \sqrt{-1}$ is preferred over the mathematical symbol i which may be confused with the electric current. In terms of real and imaginary parts, we obtain

$$\cos\alpha = \mathrm{Re}\left(e^{j\alpha}\right), \quad \sin\alpha = \mathrm{Im}\left(e^{j\alpha}\right) \tag{8.3b}$$

Why do we need the complex exponent instead of cosine and sine? To answer this question, let us study the following identity:

$$e^{j(\omega t+\varphi)} = e^{j\omega t} \cdot e^{j\varphi} \tag{8.3c}$$

We will write the current or the voltage in the form of Eq. (8.3c) and use the multiplicative property of the exponent. Then, the factor $e^{j\omega t}$ can be exactly *canceled out* in *every* term in the underlying differential equation. This is a major simplification, because the ODE becomes an algebraic equation. The key is the function $\mathrm{Re}(\cdot)$ or the real part of a complex number:

$$\begin{aligned} v(t) &= V_m \cos(\omega t + \varphi) = \mathrm{Re}\left(V_m e^{j(\omega t+\varphi)}\right) \\ i(t) &= I_m \cos(\omega t + \psi) = \mathrm{Re}\left(I_m e^{j(\omega t+\psi)}\right) \end{aligned} \tag{8.4}$$

The remaining complex number available after cancellation of the time factor $e^{j\omega t}$ is called a *phasor*: *the phasor voltage* $\mathbf{V}$ and the *phasor current* $\mathbf{I}$:

$$\mathbf{V} = V_m e^{j\varphi}, \quad \mathbf{I} = I_m e^{j\psi} \tag{8.5}$$

Equation (8.5) as a definition tells us that the phasor is a complex number comprised of two parameters: amplitude and phase. For related operations with complex numbers, you can consult the chapter summary. You should notice that frequency is no longer present since it remains the same for all circuit elements, and it is equal to the known frequency of the voltage power supply. The phasor has *the same* units as the original quantity: the phasor voltage has units of volts, and the phasor current has units of amperes. The theory and application of the symbolic method of alternating currents by means of complex number algebra was started along with the famous book of C. P. Steinmetz, *Theory and Calculation of Alternating Current Phenomena*, New York, McGraw Hill Publishing Co., 1897. It happened only four years after the 1893 World's Fair, the World's Columbian Exposition in Chicago, where Nikola Tesla and George Westinghouse introduced visitors to AC power by using it to illuminate the exposition. Further, those methods received wide attention in control theory, in communications and signal processing, and in RF engineering.

8.1.5 From Real Signals to Phasors

Every AC voltage or current has its own phasor. The conversion from real-valued voltages and currents to phasors is performed exactly according to the phasor definition Eqs. (8.4) and (8.5).

Example 8.4: Determine the phasors for the real-valued AC voltages and currents:

$$\begin{aligned} v(t) &= 3\cos(\omega t - \pi/3)\ \text{[V]} \\ i(t) &= 1\cos(\omega t + \pi/6)\ \text{[A]} \end{aligned} \tag{8.6a}$$

Solution: To construct the phasor, we only need the amplitude (V_m or I_m) and the phase (φ or ψ) for every signal in Eq. (8.6a). The result then uses the phasor definition given by Eqs. (8.4) and (8.5):

$$\begin{aligned} \mathbf{V} &= 3e^{-j\pi/3} = 3\left(\cos\frac{\pi}{3} - j\sin\frac{\pi}{3}\right) = \frac{3}{2} - j\frac{3\sqrt{3}}{2}\ \text{[V]} \\ \mathbf{I} &= 1e^{j\pi/6} = 1\left(\cos\frac{\pi}{6} + j\sin\frac{\pi}{6}\right) = \frac{\sqrt{3}}{2} + j\frac{1}{2}\ \text{[A]} \end{aligned} \tag{8.6b}$$

All results are a direct consequence of using Euler's formula. The phasors are complex numbers that can be plotted in the complex plane as dots or vectors; we will provide an example of this later in this section. Moreover, we can add, subtract, multiply, and divide phasors, and the results are again phasors. However, the addition and subtraction of real signals is difficult, and the multiplication and division is practically impossible. Just imagine how you would divide a sine by a cosine and then convert the result into a cosine form.

To assure the uniqueness of phasor definition, the real signal must always be written in the standard form of the *cosine* function with the *positive* amplitude. For example, voltage $v(t) = -3\cos(\omega t)$ is to be converted to $v(t) = 3\cos(\omega t + \pi)$ and then to the phasor $\mathbf{V} = 3e^{j\pi}$.

Exercise 8.2: Determine the phasor for the real-valued alternating current $i(t) = -1\sin(\omega t - \pi/2)$ [A].

Answer: $\mathbf{I} = 1e^{j0} = 1$ [A].

8.1.6 From Phasors to Real Signals

If the phasors of the voltage and current are given, and the angular frequency of the AC source is known, the real-valued AC voltages and currents can be restored using the phasor definition given by Eqs. (8.4) and (8.5).

Example 8.5: The phasors of the AC voltage and current are given by

$$\begin{aligned} \mathbf{V} &= 12\angle\pi/3 = 12e^{j\pi/3} \text{ [V]} \\ \mathbf{I} &= 0.1\angle 60^\circ = 0.1e^{j60^\circ} \text{ [A]} \end{aligned} \tag{8.7a}$$

An AC source has the angular frequency ω. Restore the corresponding real-valued voltages and currents.

Solution: We construct the real signals in the form of Eq. (8.4) where the amplitude (V_m or I_m) and the phase (φ or ψ) of the corresponding sinusoidal function are extracted from the phasors in Eqs. (8.7a):

$$\begin{aligned} v(t) &= 12\cos(\omega t + \pi/3) \text{ [V]} \\ i(t) &= 0.1\cos(\omega t + \pi/3) \text{ [A]} \end{aligned} \tag{8.7b}$$

Exercise 8.3: The phasor voltage is given by $\mathbf{V} = -2\angle\pi/3$ [V]. Restore the corresponding real-valued voltage signal.

Answer: $v(t) = 2\cos(\omega t - 2\pi/3)$ [V].

8.1.7 Polar and Rectangular Forms: Phasor Magnitude

The phasor as a complex number has two forms: the *polar* form and the *rectangular* form. Equation (8.8a) includes both forms:

$$\mathbf{V} = \underbrace{V_m e^{j\varphi}}_{\text{polar form}} = \underbrace{V_m\cos\varphi + jV_m\sin\varphi}_{\text{rectangular form}} = \underbrace{x + jy}_{\text{rectangular form}} \tag{8.8a}$$

In general, the conversion from polar form to rectangular form uses Euler's identity. We emphasize that $V_m > 0$ is the *magnitude* of a complex number—the phasor. It is always positive and equals the amplitude of the corresponding real signal. As mentioned before, φ is the phase; it changes from $-\pi$ to $+\pi$ and equals the phase of the corresponding real signal. The phasor may be initially given in rectangular form, say as $\mathbf{V} = x + jy$. In that case, we may convert the phasor to polar form as follows:

$$\mathbf{V} = V_{\mathrm{m}} e^{j\varphi}, \quad V_{\mathrm{m}} = \sqrt{x^2 + y^2}, \quad \varphi = \tan^{-1}\left(\frac{y}{x}\right) \tag{8.8b}$$

Equation (8.8b) is *only* valid when x is positive. Otherwise, a factor of $\pm\pi$ must be added as explained in Fig. 8.5. The conversion to the polar form and vice versa is routinely done using a calculator, or MATLAB (function `angle`), or other software of your choice. For other operations with complex numbers, please refer to Tables 1 and 2 at the end of this section. Figure 8.5 plots a complex number (phasor) as given by Eqs. (8.8a,b) in the complex plane. The x-axis of the complex plane is the real part of the complex number, and the y-axis is the imaginary part. The complex number is either represented by a dot with the coordinates x, y or by a vector drawn from the origin to that dot. The *magnitude of the complex number* (phasor), $|\mathbf{V}| = V_{\mathrm{m}} = \sqrt{x^2 + y^2}$, is the length of this vector; the *phase* φ is the angle with the positive x-axis. It is straightforward to add two phasors; this operation corresponds to vector addition in Fig. 8.5.

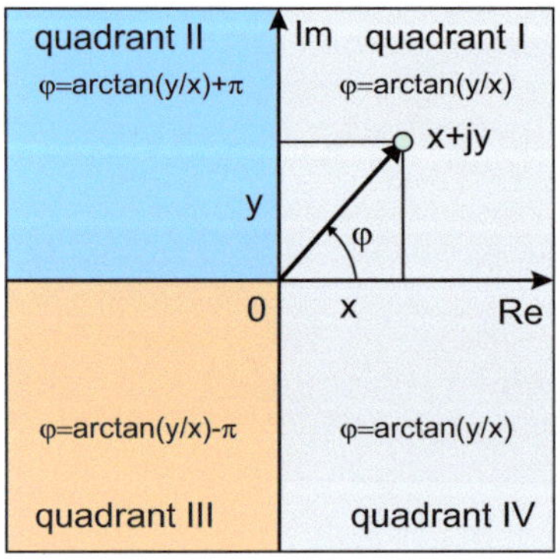

Fig. 8.5. A phasor in rectangular, or polar, form in the complex plane and the corresponding phase conversion. The phase of φ is given for different quadrants.

Example 8.6: The phasors of AC voltage and current are given in rectangular form

$$\begin{aligned} \mathbf{V} &= 5 + j5 \quad [\mathrm{V}] \\ \mathbf{I} &= -0.1 - j0.2 \quad [\mathrm{A}] \end{aligned} \tag{8.9}$$

Example 8.6 (cont.): An AC source has the angular frequency ω. Restore the corresponding real voltages and currents.

Solution: We convert the phasors to the polar form using Eq. (8.8b) and Fig. 8.5:

$$\begin{aligned} \mathbf{V} &= 7.07e^{j0.785} \quad [\mathrm{V}] \\ \mathbf{I} &= 0.224e^{-j2.03} \quad [\mathrm{A}] \end{aligned} \tag{8.10}$$

After that, we restore the real signals exactly following Example 8.5:

$$\begin{aligned} v(t) &= 7.07\cos(\omega t + 0.785) \quad [\mathrm{V}] \\ i(t) &= 0.224\cos(\omega t - 2.03) \quad [\mathrm{A}] \end{aligned} \tag{8.11}$$

The phase in Eqs. (8.11) is given in radians. Another equivalent form of the solution that should please your engineering professor implies replacing 0.785 by $\pi/4$ or 45° and 2.03 by $\pi/1.54$ or 116°.

8.1.8 Operations with Phasors and Phasor Diagram

Phasors allow us to perform a number of basic operations in AC circuits rather quickly. Let us consider a part of an AC circuit with two arbitrary series circuit elements, shown in Fig. 8.6. The AC voltages $v_1(t)$ and $v_2(t)$ across each element are known. How can we find the AC voltage $v(t)$ for the series combinations?

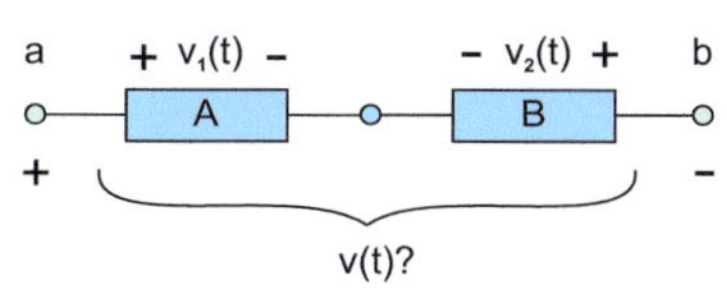

Fig. 8.6. A series combination of two AC circuit elements. The voltage across the series combination (voltage between terminals a and b) is desired. You should note the voltage polarities.

According to KVL we conclude $v(t) = v_1(t) - v_2(t)$, where the second voltage has opposite polarity. There are two ways to proceed from here. The first way in the time domain is to add two cosine functions directly and convert them into another cosine function. Unfortunately, if these cosine functions have different phases and amplitudes, the operation is not straightforward. It requires a search for the corresponding trigonometric formula and the accurate use of this formula, keeping in mind proper phase definition.

The second way is to use phasors. We will see that this approach has a clear intuitive background and is actually simpler to apply.

Example 8.7: Find the total voltage between terminals *a* and *b* in Fig. 8.6 if the element voltages are given by:

$$\begin{aligned} v_1(t) &= 6.08\cos(\omega t + 80.5^\circ) \quad [\mathrm{V}] \\ v_2(t) &= 5.00\cos(\omega t + 36.9^\circ) \quad [\mathrm{V}] \end{aligned} \tag{8.12a}$$

Solution: We construct the phasors first:

$$\begin{aligned} \mathbf{V}_1 &= 6.08\angle 80.5^\circ \quad [\mathrm{V}] \\ \mathbf{V}_2 &= 5.00\angle 36.9^\circ \quad [\mathrm{V}] \end{aligned} \tag{8.12b}$$

We convert them into rectangular form using Euler's formula next:

$$\begin{aligned} \mathbf{V}_1 &= 1.00 + j6.00 \quad [\mathrm{V}] \\ \mathbf{V}_2 &= 4.00 + j3.00 \quad [\mathrm{V}] \end{aligned} \tag{8.12c}$$

Both phasors are plotted in Fig. 8.7a. Such a plot is known as a *phasor diagram*. It is quite useful since it provides us with a visual picture of magnitude and phase. We subtract the phasors as complex numbers and obtain the resulting phasor for the desired voltage $v(t)$

$$v(t) = v_1(t) - v_2(t) \Rightarrow \mathbf{V} = \mathbf{V}_1 - \mathbf{V}_2 = 1 + j6 - 4 - j3 = -3 + j3 \quad [\mathrm{V}] \tag{8.13a}$$

The same subtraction operation is done in the vector form in the phasor diagram of Fig. 8.7a. The resulting phasor **V** is shown by a dashed arrow. We need to center it at the origin to obtain agreement with Eq. (8.13a). Next, we convert the phasor **V** into polar form using either Eq. (8.8b) or by just looking at the phasor diagram itself,

$$\mathbf{V} = -3 + j3 = 3\sqrt{2}e^{j135^\circ} = 4.23e^{j135^\circ} = 4.23\angle 135^\circ \quad [\mathrm{V}] \tag{8.13b}$$

Finally, we restore the voltage $v(t)$ from its phasor: $v(t) = 4.23\cos(\omega t + 135^\circ)$. Using the same method, we could find the terminal voltage for a series combination of any number of circuit elements with arbitrary polarities.

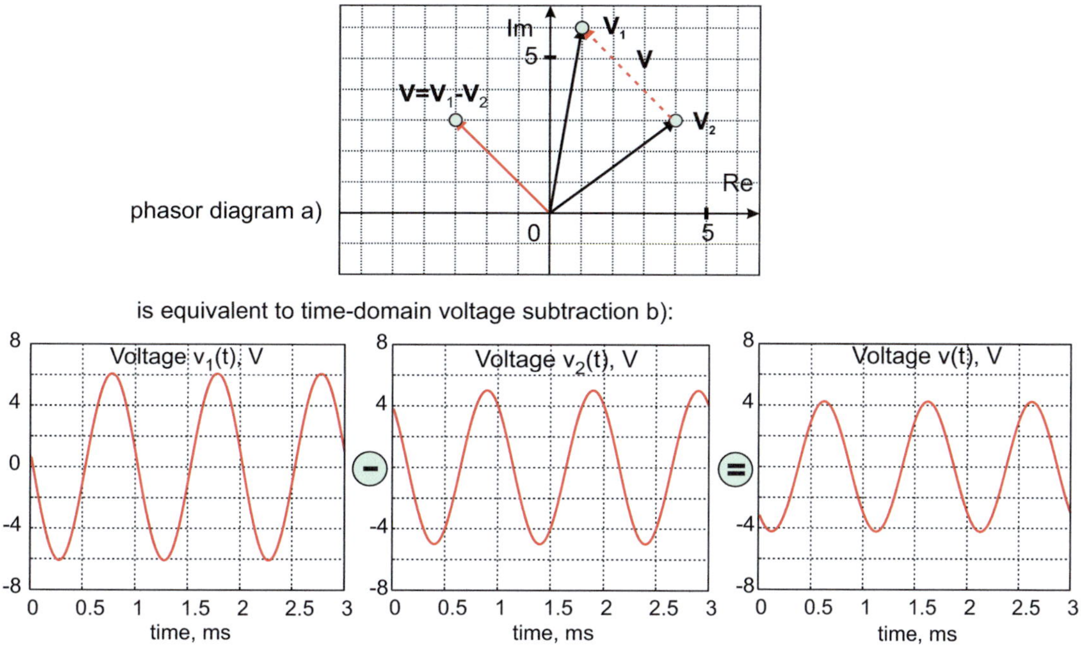

Fig. 8.7. (**a**) Finding series voltage for two circuit elements using phasors and (**b**) the time-domain representation of the same result.

The phasor diagram introduced in Fig. 8.7a also allows us to find the relative lag or lead for two or more AC voltages and/or currents, a topic you may recall from the previous section. As an example, in Fig. 8.8 we show the phasor diagram for three different time-domain voltages $v_1(t)$, $v_2(t)$, $v_3(t)$. Irrespective of the specific values of the amplitudes and phases, we may conclude that voltage $v_2(t)$ leads voltage $v_1(t)$ by 90°, but lags voltage $v_3(t)$.

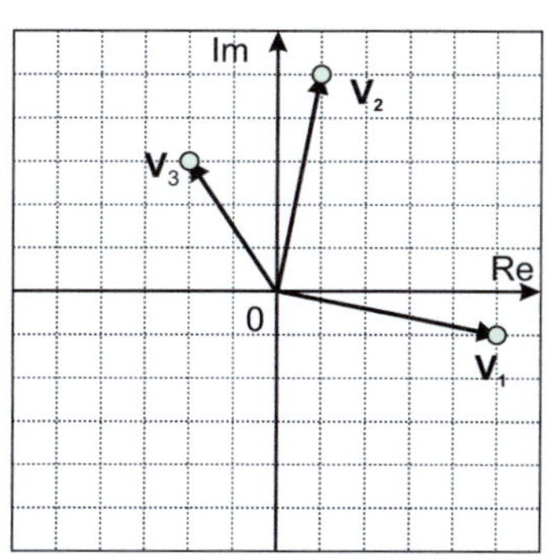

Fig. 8.8. Phasor diagram for three time-domain voltages $v_1(t)$, $v_2(t)$, $v_3(t)$. Voltage $v_2(t)$ leads voltage $v_1(t)$ by 90°, but lags voltage $v_3(t)$.

Exercise 8.4: Does the voltage signal $v(t) = 6.08\cos(\omega t + 80.5°)$ lead or lag the voltage signal $v(t) = 4.23\cos(\omega t + 135°)$?

Answer: It lags by 54.5°.

At this point a legitimate question arises: is adding sinusoids the only application of phasors? What will be the next step? Obviously, our goal now is to reduce an AC circuit to the equivalent "DC" circuit. This step requires a new concept known as *impedance*. It will be considered in the next section.

8.1.9 Shorthand Notation for the Complex Exponent

In electrical engineering and electronics, the shorthand notation (*phasor notation* or *angle notation*) $\mathrm{e}^{j\varphi} = \angle\varphi$ is commonly used to simplify the notation of the complex exponent. We will frequently use this shorthand notation in the following sections. All operations with phasors remain the same. For example, the multiplicative operations are written in the forms

$$V_1\mathrm{e}^{j\varphi}V_2\mathrm{e}^{j\psi} = V_1V_2\mathrm{e}^{j(\varphi+\psi)} \Rightarrow V_1\angle\varphi\; V_2\angle\psi = V_1V_2\angle(\varphi+\psi) \tag{8.14}$$

Section 8.2 Impedance

8.2.1 The Concept of Impedance

We can avoid solving ODEs for the AC circuits entirely if we establish a relation between the phasor voltage and the phasor current for the inductor and capacitor, similar to Ohm's law for the resistor. We consider three basic AC circuit elements shown in Fig. 8.9 in a passive reference configuration and determine their phasor voltages and phasor currents.

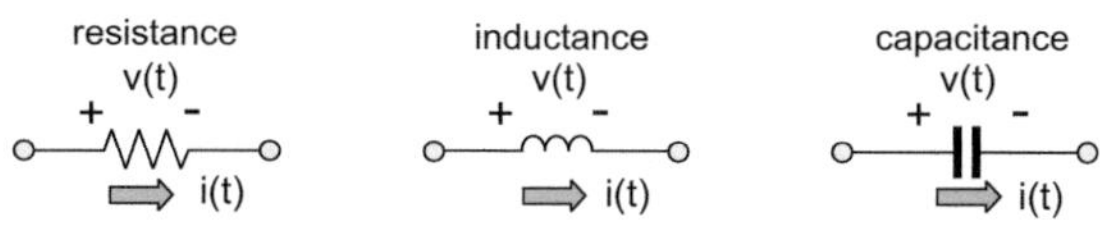

Fig. 8.9. Three AC circuit elements.

Resistance For the resistance, voltage will be given by a cosine function shown in Fig. 8.10a. One has

$$\begin{aligned} &v = Ri, \quad v(t) = V_{\mathrm{m}} \cos(\omega t + \varphi) \Rightarrow i(t) = \frac{V_{\mathrm{m}}}{R} \cos(\omega t + \varphi) \\ &\mathbf{V} = V_{\mathrm{m}} e^{j\varphi} \qquad \mathbf{I} = \frac{V_{\mathrm{m}}}{R} e^{j\varphi} \end{aligned} \tag{8.15}$$

Capacitance For the capacitance, voltage will again be given by a cosine function shown in Fig. 8.10b. One has

$$\begin{aligned} &i = C\frac{dv}{dt}, \quad v(t) = V_{\mathrm{m}} \cos(\omega t + \varphi) \\ &\Rightarrow i(t) = -\omega C V_{\mathrm{m}} \sin(\omega t + \varphi) = \omega C V_{\mathrm{m}} \cos(\omega t + \varphi + \pi/2) \\ &\mathbf{V} = V_{\mathrm{m}} e^{j\varphi} \qquad \mathbf{I} = \omega C V_{\mathrm{m}} e^{j\varphi + j\pi/2} \end{aligned} \tag{8.16}$$

Inductance For the inductance, current will be given by a cosine function shown in Fig. 8.10c. One has

$$\begin{aligned} &v = L\frac{di}{dt}, \quad i(t) = I_{\mathrm{m}} \cos(\omega t + \varphi) \\ &\Rightarrow v(t) = -\omega L I_{\mathrm{m}} \sin(\omega t + \varphi) = \omega L I_{\mathrm{m}} \cos(\omega t + \varphi + \pi/2) \\ &\mathbf{I} = I_{\mathrm{m}} e^{j\varphi} \qquad \mathbf{V} = \omega L I_{\mathrm{m}} e^{j\varphi + j\pi/2} \end{aligned} \tag{8.17}$$

Impedance **Z** of an *arbitrary linear circuit element* or of an arbitrary *one-port network comprised of such elements* is defined by

$$\mathbf{Z} \equiv \frac{\mathbf{V}}{\mathbf{I}} \tag{8.18}$$

Thus, the phasors for voltage and current are linked via a constant, which can be real or complex. In order to emphasize that this constant is not exactly a resistance, the constant is called the *impedance*. Substituting Eqs. (8.15), (8.16), and (8.17) into Eq. (8.18), we obtain:

$$\begin{aligned}
&\text{Resistance}: \quad \mathbf{Z}_{\mathrm{R}} \equiv \frac{\mathbf{V}}{\mathbf{I}} = R \quad [\Omega] \\
&\text{Capacitance}: \quad \mathbf{Z}_{\mathrm{C}} \equiv \frac{\mathbf{V}}{\mathbf{I}} = \frac{1}{\mathrm{e}^{j\pi/2}\omega C} = \frac{1}{\omega C}\angle -90^\circ = \frac{1}{j\omega C} \quad [\Omega] \\
&\text{Inductance}: \quad \mathbf{Z}_{\mathrm{L}} \equiv \frac{\mathbf{V}}{\mathbf{I}} = \mathrm{e}^{j\pi/2}\omega L = \omega L\angle 90^\circ = j\omega L \quad [\Omega]
\end{aligned} \tag{8.19}$$

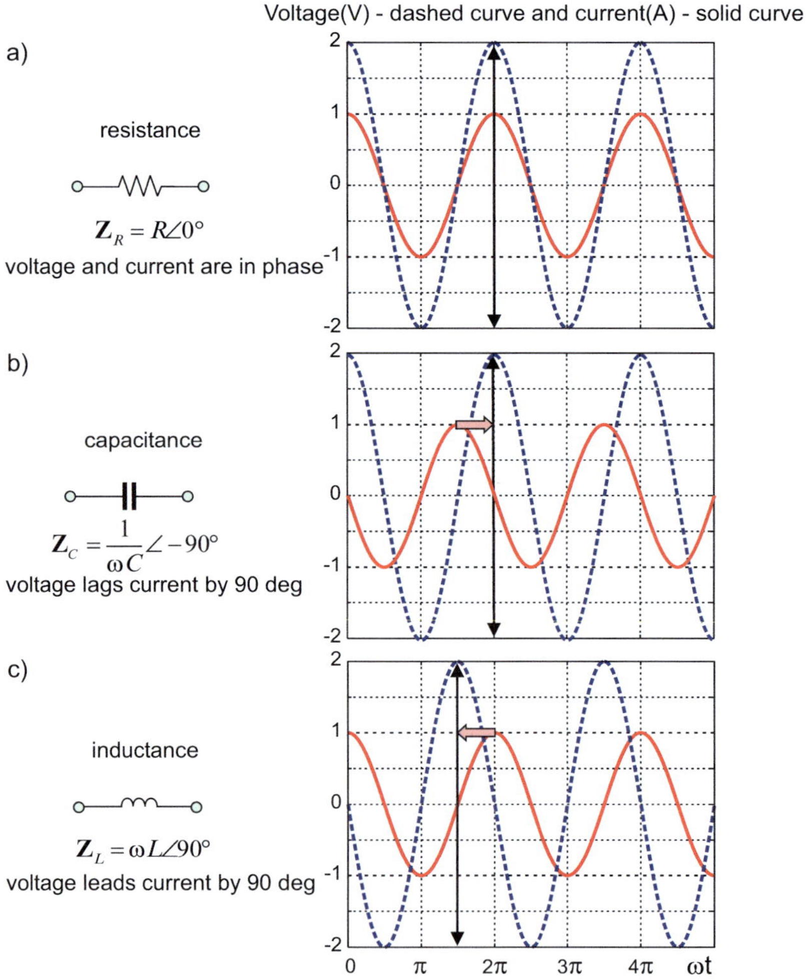

Fig. 8.10. Voltage and current sinusoids for resistor, inductor, and capacitor (at phase zero).

The impedance has units of ohms, exactly as the resistance does. The impedance is an extension of the familiar DC resistance concept for dynamic AC circuit elements or any combination of them. The term "impedance" (along with "inductance," "permeability," etc.) belongs to Oliver Heaviside (1850–1925), a brilliant self-taught English electrical engineer and mathematician. We note that the method of solving differential equations with phasors (or using Laplace transforms) originates from him. Heaviside invented and patented in England the first coaxial cable in 1880. His uncle was Sir Charles Wheatstone (1802–1875); do you remember the Wheatstone bridge?

8.2.2 Physical Meaning of Impedance

The problem with the impedance definition is that in general it is a complex number. What exactly is the physical meaning? For the resistor in Fig. 8.10a, we only need to stretch the current sinusoid in order to obtain the voltage sinusoid. Therefore, we only need *one* independent "stretching" parameter R to express voltage in terms of current. However, for the capacitor in Fig. 8.10b, we need not only stretch, but also shift the current sinusoid to the right in order to obtain the voltage sinusoid. To accomplish this, we need *two* independent parameters. The impedance as a complex number has exactly those *two independent parameters*. In polar form, it is the magnitude (responsible for stretching) and the phase (responsible for shifting). Thus, only a complex number can be used to express voltage in terms of current for the capacitor. Similarly, for the inductor in Fig. 8.10c, we need not only stretch, but also shift to the left the current sinusoid in order to obtain the voltage sinusoid. Therefore, we again need *two independent parameters* to express the voltage in terms of current; we need the complex impedance with two required independent parameters: the magnitude and the phase. In light of this reasoning, it is useful to express the impedances given by Eq. (8.20) in polar form. For the imaginary number, we use the equalities $j = \angle 90^\circ$ and $1/j = \angle -90^\circ$ and obtain

$$\mathbf{Z}_\mathrm{R} = R\angle 0^\circ, \quad \mathbf{Z}_\mathrm{C} = \frac{1}{\omega C}\angle -90^\circ, \quad \mathbf{Z}_\mathrm{L} = \omega L\angle 90^\circ \tag{8.20}$$

Example 8.8: Find impedances of resistor, capacitor, and inductor in Fig. 8.10.

Solution: In Fig. 8.10a,

$$\begin{aligned} i_\mathrm{R}(t) &= 1\cos(\omega t)\ [\mathrm{A}] \\ v_\mathrm{R}(t) &= 2\cos(\omega t)\ [\mathrm{V}] \end{aligned} \Rightarrow \mathbf{Z}_\mathrm{R} = R = 2\ [\Omega] \tag{8.21a}$$

The *resistor's impedance* is frequency independent. It is just the resistance R. In Fig. 8.10b,

Example 8.8 (cont.):

$$i_C(t) = 1\cos(\omega t + \pi/2)\ [\mathrm{A}] \quad \Rightarrow \mathbf{Z}_\mathrm{C} = 2e^{-j\pi/2} = -j2\ [\Omega] \quad (8.21b)$$
$$v_C(t) = 2\cos(\omega t)\ [\mathrm{V}]$$

The *capacitor's impedance* is thus an *imaginary negative number.* This is always true. In Fig. 8.10c,

$$i_L(t) = 1\cos(\omega t)\ [\mathrm{A}] \quad \Rightarrow \mathbf{Z}_\mathrm{L} = 2e^{+j\pi/2} = +j2\ [\Omega] \quad (8.21c)$$
$$v_L(t) = 2\cos(\omega t + \pi/2)\ [\mathrm{V}]$$

The *inductor's impedance* is thus an *imaginary positive number.* This is always true.

8.2.3 Magnitude and Phase of Complex Impedance

With reference to Fig. 8.10, the *magnitude of the impedance* is a factor by which we should multiply (or stretch) the current sinusoid in order to obtain the voltage sinusoid. The phase (or the polar angle) of the impedance is a factor by which we should shift *to the left* the current sinusoid in order to obtain the voltage sinusoid.

1. For the resistor, we multiply the current by $|\mathbf{Z}_\mathrm{R}| = R = 2\ \Omega$ and do not shift.
2. For the capacitor, we multiply the current by $|\mathbf{Z}_\mathrm{C}| = \frac{1}{\omega C} = 2\ \Omega$ and shift it by $-90°$ (or by a quarter of the period) to the *right* in Fig. 8.10b.
3. For the inductor, we multiply the current by $|\mathbf{Z}_\mathrm{L}| = \omega L = 2\ \Omega$ and shift by $+90°$ (or by a quarter of the period) to the *left* in Fig. 8.10c.

The above operations hold for any resistor, inductor, and capacitor. Thus, the impedance concept extends Ohm's law to dynamic circuit elements in steady-state AC circuits.

Exercise 8.5: Establish phase relationships for voltages and currents in Fig. 8.10.

Answer: For the resistor, current and voltage are in phase. For the capacitor, current *leads* voltage by 90° (or voltage *lags* current by 90°). For the inductor, voltage *leads* current by 90° (or current *lags* voltage by 90°).

Example 8.9: For two AC circuits shown in Fig. 8.11, find the impedance of the resistor, capacitor, and inductor.

Solution: The *resistor's impedance* is frequency independent. It is just the resistance R,

Example 8.9 (cont.):

$$\mathbf{Z}_{\mathrm{R}} = R = 2\ [\Omega] \tag{8.22a}$$

The *capacitor's impedance* does depend on frequency. In Fig. 8.11a, $\omega = 5000$ rad/s. Therefore,

$$\mathbf{Z}_{\mathrm{C}} = \frac{1}{j\omega C} = \frac{1}{j5000 \cdot 10^{-4}} = \frac{1}{j0.5} = \frac{2}{j} = -j2\ [\Omega] \tag{8.22b}$$

The *inductor's impedance* also depends on frequency. In Fig. 8.11b, $\omega = 20$ rad/s. Therefore,

$$\mathbf{Z}_{\mathrm{L}} = j\omega L = j20 \cdot 0.1 = j2\ [\Omega] \tag{8.22c}$$

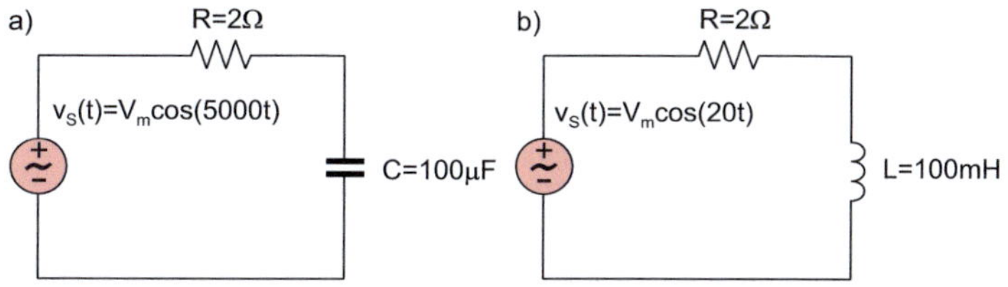

Fig. 8.11. Two types of a series AC circuit with a voltage supply.

An equivalent representation of impedances in Eqs. (8.22a, b, c) is the *impedance phasor diagram* shown in Fig. 8.12 where we plot the corresponding complex numbers.

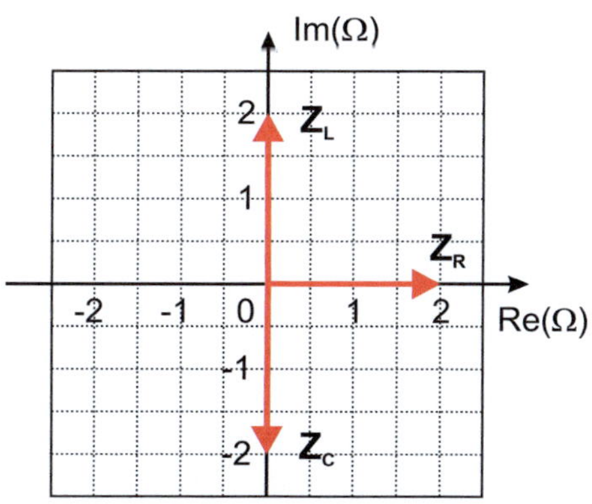

Fig. 8.12. Complex impedances displayed in the phasor diagram.

Note that the magnitudes of complex impedances in Fig. 8.12 are all equal to 2 Ω:

$$|\mathbf{Z}_R| = |\mathbf{Z}_R| = |\mathbf{Z}_R| = 2\ \Omega \tag{8.23}$$

8.2.4 Application Example: Impedance of a Human Body

The impedance is not only the characteristic of an electric circuit but also of an arbitrary conducting object, which may be modeled as a combination of ideal resistance, inductance, and capacitance. They can be connected in series, in parallel, or a combination of series and parallel. As an example, we consider here the impedance of a human body in the frequency range from 10 kHz to 3 MHz. The impedance magnitude is shown in Fig. 8.13. It was measured for about 400 human subjects of ages between 18 and 70 years and then averaged. The subject stood on a large aluminum sheet as a ground plane. The electrode was a cylindrical brass rod for making a grasping contact with the hand. In its simplest form, the concept of impedance measurement implies the simultaneous measurements of harmonic voltage and current and then the extraction of the amplitude ratio (magnitude of the impedance) and the phase difference (impedance phase).

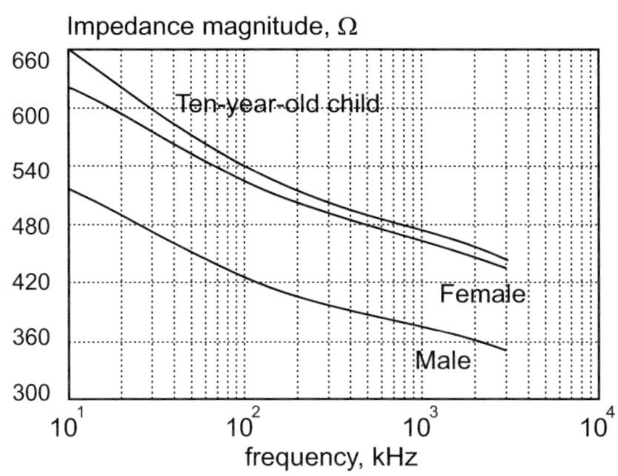

Fig. 8.13. Magnitude of the average impedance of the human body. From: I. Chatterjee et al., "Human Body Impedance and Threshold Currents for Perception and Pain for Contact Hazard Analysis in the VLF-MF Band," *IEEE Trans Biomedical Eng.*, May 1986.

Figure 8.13 indicates that the impedance magnitude decreases with frequency. Therefore, the human body impedance, at least at relatively low frequencies, behaves similarly to the impedance of a capacitor, where the magnitude also decreases with frequency. A purely resistive component is also present. The impedance measurements have been used for the extraction of various biomedical data such as assessment of a fat-free mass.

Section 8.3 Principles of AC Circuit Analysis

8.3.1 AC Circuit Analysis: KVL, KCL, and Equivalent Impedances

The generic process of conducting an AC circuit analysis is outlined in Fig. 8.14 for a steady-state circuit with a capacitor and a resistor. We replace the circuit components by their impedances and replace the voltages and currents by their phasors. Hence, the "real" time-domain AC circuit in Fig. 8.14a becomes an "imaginary" circuit with complex phasor voltages $\mathbf{V}_\mathrm{S}$, $\mathbf{V}_\mathrm{R}$, $\mathbf{V}_\mathrm{C}$ and a complex phasor current $\mathbf{I}$ "flowing" through the circuit; see Fig. 8.14b. However, the phasor $\mathbf{V}_\mathrm{S}$ is a constant and the resulting circuit is the "DC" circuit.

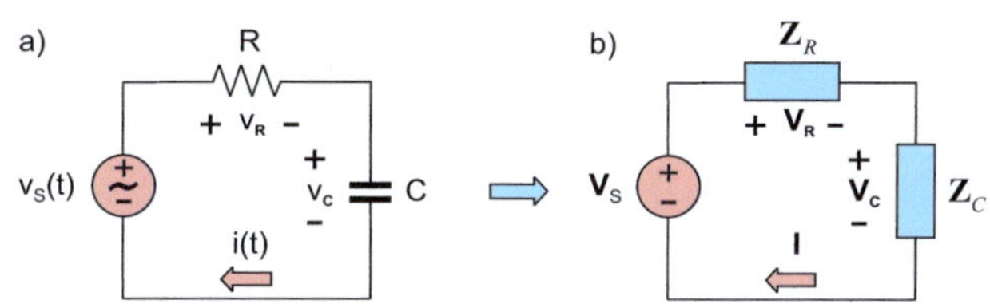

Fig. 8.14. Transformation of an AC circuit to phasor/impedance form.

KVL for the circuit in Fig. 8.14a in the time domain reads

$$-v_\mathrm{S}(t)+v_\mathrm{R}(t)+v_\mathrm{C}(t)=0 \tag{8.24}$$

Its phasor counterpart, Fig. 8.14b, in the frequency domain has exactly the same algebraic form

$$-\mathbf{V}_\mathrm{S}+\mathbf{V}_\mathrm{R}+\mathbf{V}_\mathrm{C}=0 \tag{8.25}$$

This result can be proven using the method described in the Chapter's summary. KCL is formulated in terms of phasors exactly in the same manner. According to KCL, the same phasor current $\mathbf{I}$ flows through all the elements in the circuit of Fig. 8.14b. This results in

$$-\mathbf{V}_\mathrm{S}+\mathbf{Z}_\mathrm{R}\mathbf{I}+\mathbf{Z}_\mathrm{C}\mathbf{I}=0\Rightarrow\mathbf{I}=\frac{\mathbf{V}_\mathrm{S}}{\mathbf{Z}_\mathrm{R}+\mathbf{Z}_\mathrm{C}}=\frac{\mathbf{V}_\mathrm{S}}{\mathbf{Z}_\mathrm{eq}} \tag{8.26}$$

where phasor voltages are related to phasor currents though the concept of impedances. Thus, once the AC circuit has been transformed to phasors and impedances (frequency domain), the impedances may be combined *as if they were simple resistors*. The series/parallel equivalents for the impedances are *equally applicable*. After plugging in numbers for ω, R, C, $\mathbf{V}_S$, the phasor current and phasor voltages can be found and converted back to time domain. This is the method of solving in AC circuits.

8.3.2 Complete Solution for an AC Circuit: KVL and KCL on Phasor Diagram

Let us assume that the power supply voltage in Fig. 8.14a has the form $v_S(t) = V_m \cos\omega t$, and its amplitude and frequency are given by $V_m = 5$ V, $\omega = 1000$ rad/s. Further, we know that $C = 1$ μF, $R = 1$ kΩ in Fig. 8.14a. The solution to this AC circuit includes several steps as discussed above. First, we convert the circuit to the phasor/impedance form as shown in Fig. 8.14b. All currents/voltages are replaced by their phasors, and all circuit elements are replaced by their impedances. Next, we solve the phasor circuit in Fig. 8.14b as if it were a "DC" circuit. The element impedances are

$$\mathbf{Z}_R = 1000\ [\Omega], \quad \mathbf{Z}_C = \frac{1}{j\omega C} = \frac{1}{j1000 \cdot 1 \times 10^{-6}} = -j1000\ [\Omega] \tag{8.27}$$

Note that the impedances here are written in rectangular instead of polar form. We must now find the equivalent impedance. From the series impedance combination:

$$\begin{aligned}\mathbf{Z}_{eq} &= \mathbf{Z}_R + \mathbf{Z}_C = 1000 - j1000 = \sqrt{1000^2 + 1000^2}\angle - \arctan(1)\\ &\approx 1414\angle - 45°\ [\Omega]\end{aligned} \tag{8.28}$$

The phasor current (circuit current) has the form

$$\mathbf{I} = \frac{\mathbf{V}_S}{\mathbf{Z}_{eq}} = \frac{5}{1414\angle - 45°} \approx 3.54\angle 45°\ [\text{mA}] \tag{8.29}$$

The phasor voltages across the resistor and the capacitor are found according to Ohm's law, that is:

$$\mathbf{V}_R = \mathbf{Z}_R\mathbf{I} = 1000 \times 0.00354\angle 45° = 3.54\angle 45°\ \ [\text{V}] \tag{8.30}$$

$$\mathbf{V}_C = \mathbf{Z}_C\mathbf{I} = -j1000 \times 0.00354\angle 45° = 3.54\angle - 90°\angle 45° = 3.54\angle - 45°\ [\text{V}] \tag{8.31}$$

The phasor voltages are plotted in the phasor diagram as depicted in Fig. 8.15.

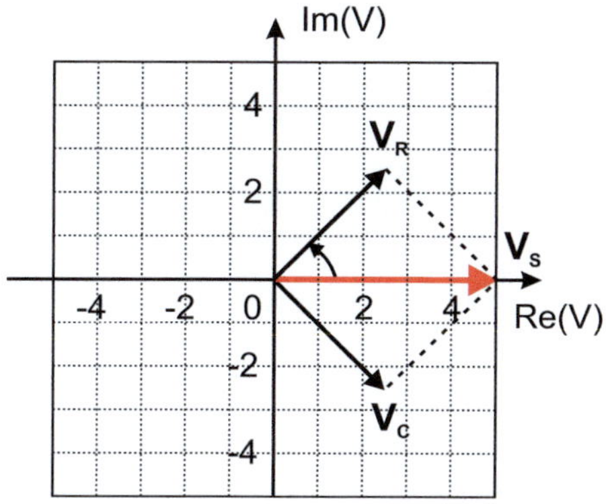

Fig. 8.15. Phasor diagram for the circuit in Fig. 8.14 and KVL in the vector form.

We may conclude that *KVL in phasor form* is equivalent to the addition of two vectors $\mathbf{V}_R$, $\mathbf{V}_C$, which equals the supply phasor voltage $\mathbf{V}_S$. For an AC circuit in the form of a current divider, *KCL in phasor form* will have exactly the same representation.

As a final step to arrive at the solution, we convert the phasors in Eqs. (8.30) and (8.31) to real-valued voltages:

$$\begin{aligned} v_R(t) &= 3.54\cos(\omega t + 45^\circ) \quad [\mathrm{V}] \\ v_C(t) &= 3.54\cos(\omega t - 45^\circ) \quad [\mathrm{V}] \end{aligned} \tag{8.32}$$

The AC circuit is thus solved. We note that the amplitudes of both voltages are the same; in this particular case the voltage divider splits the power supply voltage "equally." In fact, $\sqrt{3.54^2 + 3.54^2} = 5\mathrm{V}$, which is KVL in terms of the voltage amplitudes.

Exercise 8.6: How does the phasor diagram in Fig. 8.15 change if the voltage source in Fig. 8.14 is given by $v_S(t) = V_m \cos(\omega t - 45^\circ)$?

Answer: The entire phasor diagram rotates clockwise by 45°.

8.3.3 Source Transformation

After KVL and KCL in terms of phasors (in the frequency domain) have been established, the circuit laws and principles from Chapters 3 and 4 are straightforwardly extended to the steady-state AC *circuits operating at a single frequency.* The first example is the *source transformation in the frequency domain* shown in Fig. 8.16.

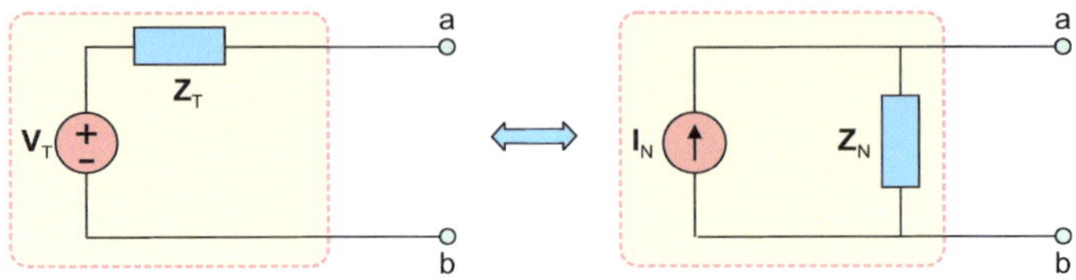

Fig. 8.16. Source transformation in the frequency domain.

A voltage source with the phasor voltage $\mathbf{V}_T$ and in series with a passive circuit element having impedance $\mathbf{Z}_T$ is equivalent to a current source with the phasor current $\mathbf{I}_N$ and in parallel with impedance $\mathbf{Z}_N$ given that

$$\mathbf{Z}_T = \mathbf{Z}_N, \quad \mathbf{V}_T = \mathbf{Z}_T\mathbf{I}_N \Leftrightarrow \mathbf{I}_N = \frac{\mathbf{V}_T}{\mathbf{Z}_T} \tag{8.33}$$

Equation (8.33) is the direct extension of the source transformation principle established in Chapter 4 for DC circuits.

Example 8.10: Determine phasor voltage $\mathbf{V}_1$ in the AC circuit shown in Fig. 8.17 using the method of source transformation. The impedance values are given at the frequency of interest. Note that, for every impedance box in Fig. 8.17, its physical counterpart is shown inside this box.

Solution: The source transformation follows Eq. (8.33) and leads to the AC circuit shown in Fig. 8.18. This circuit is easier to solve. We can see that the circuit in Fig. 8.18 becomes the current divider between the $10\Omega||10\Omega = 5\Omega$ impedance and the $j10\Omega - j5\Omega = j5\Omega$ impedance, respectively. Therefore, the phasor current through the impedance of interest is given by current division

$$\mathbf{I} = \frac{5}{5+j5} \times 1\angle 90^\circ = \frac{5}{5\sqrt{2}\angle 45^\circ} \times 1\angle 90^\circ = \frac{1}{\sqrt{2}}\angle 45^\circ \quad [\text{A}] \tag{8.34}$$

Thus, the resulting voltage is given by

$$\mathbf{V}_1 = -j5 \times \mathbf{I} = 5\angle -90^\circ \times \frac{1}{\sqrt{2}}\angle 45^\circ = \frac{5}{\sqrt{2}}\angle -45^\circ \quad [\text{V}] \tag{8.35}$$

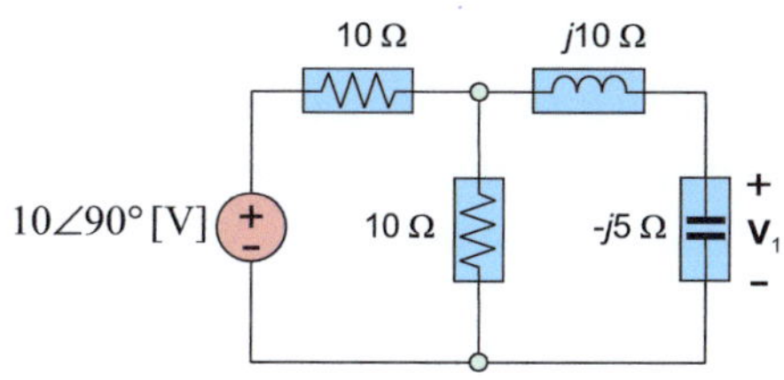

Fig. 8.17. An AC circuit solved with the help of source transformation. Note that every impedance box has a physical counterpart shown within this box.

Exercise 8.7: Three impedances $j5\Omega$, 10Ω, 10Ω are combined in parallel. What is the equivalent impedance?

Answer: $2.5 + j2.5\ \Omega$.

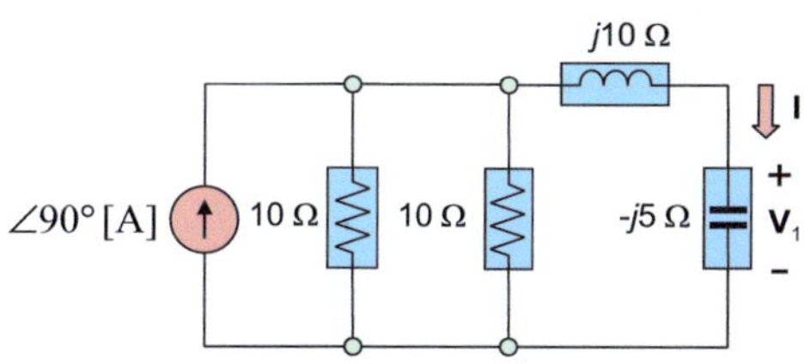

Fig. 8.18. Source transformation applied to the AC circuit from the previous figure.

Exercise 8.8: In Fig. 8.18, the impedance of the capacitor changes to $-j10\Omega$. What will be the phasor current, **I**, through the capacitor?

Answer: $\mathbf{I} = 1\angle 90^\circ$ [A]

8.3.4 Thévenin and Norton Equivalent Circuits

The *Thévenin's theorem for steady-state AC circuits* is formulated as follows. Any linear AC network of resistors/capacitors/inductors and voltage/current power sources *operating at the same frequency* can be represented in the form of a Thévenin equivalent network shown in Fig. 8.19b (phasor form). This result is a direct extension of Thévenin's theorem for DC circuits stated in Chapter 4. Thévenin's theorem allows us to reduce complicated AC circuits to a simple source and impedance configuration, with the same power output to a load. The AC frequency remains the same. Phasor voltage $\mathbf{V}_T$ is known as *Thévenin voltage* or simply the *source voltage*; impedance $\mathbf{Z}_T$ is called *Thé venin impedance* or *source impedance*. The phasor voltage may have amplitude and phase that are different from the original AC sources. *The Norton theorem for steady-state AC circuits* replaces the Thévenin equivalent circuit by the Norton equivalent circuit from Fig. 8.16. As in resistive circuits, the Thévenin phasor voltage is equal to the open-circuit phasor voltage of the original circuit in Fig. 8.19a:

$$\mathbf{V}_{\mathrm{T}} = \mathbf{V}_{\mathrm{ab}} = \mathbf{V}_{\mathrm{OC}} \tag{8.36}$$

To find the Thévenin impedance $\mathbf{Z}_{\mathrm{T}}$, we need to know the phasor, $\mathbf{I}_{\mathrm{SC}}$, for the short-circuit current of the network. The short-circuit current is obtained by shorting out the output terminals *a* and *b*. This results in the source impedance:

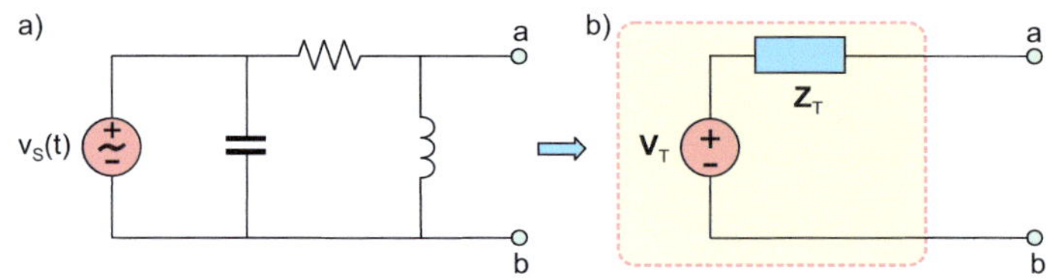

Fig. 8.19. Representation of a Thévenin equivalent circuit. The original AC circuit in (**a**) is transformed into its Thévenin equivalent circuit (**b**) by providing the same voltage and current to a load. Both circuits are indistinguishable when looking from the load.

$$\mathbf{Z}_{\mathrm{T}} = \frac{\mathbf{V}_{\mathrm{OC}}}{\mathbf{I}_{\mathrm{SC}}} = \frac{\mathbf{V}_{\mathrm{T}}}{\mathbf{I}_{\mathrm{SC}}} \tag{8.37}$$

To find $\mathbf{Z}_{\mathrm{T}}$ we often use another, somewhat simpler, approach. We short out the voltage source(s), i.e., replace them by wires. Then we zero the current sources (if present), i.e., replace them by an open circuit. This enables us to find the equivalent impedance of the resulting purely passive circuit, which is equal to $\mathbf{Z}_{\mathrm{T}}$.

Example 8.11: Find the Thévenin equivalent, i.e., $\mathbf{V}_{\mathrm{T}}$ and $\mathbf{Z}_{\mathrm{T}}$, for the circuit shown in Fig. 8.19a when $\omega = 377$ rad/s, $L = 26.5$ mH, $R = 10\ \Omega$, $C = 220\ \mu$F, and $v_S(t) = 10\cos\omega t$ [V].

Solution: We convert the circuit in Fig. 8.19a into phasor form first. $\mathbf{V}_{\mathrm{T}}$ is the phasor voltage between terminals *a* and *b*, i.e., the phasor voltage across the inductor. The resistor and the inductor form a voltage divider with regard to the supply phasor voltage of 10 V. According to the voltage division principle:

Example 8.11 (cont.):

$$\mathbf{V}_T = \mathbf{V}_L = \frac{\mathbf{Z}_L}{\mathbf{Z}_L + \mathbf{Z}_R}10 = \frac{j10}{10 + j10}10 = \frac{10\angle 90^\circ}{14.14\angle 45^\circ}10 = 7.07\angle 45^\circ \quad [\mathrm{V}] \tag{8.38}$$

To find the Thévenin resistance, we short out the voltage source (by using the second approach). This operation simultaneously shorts out the capacitor, with the result that the resistor and the inductor are now in parallel. This gives

$$\mathbf{Z}_T = \mathbf{Z}_L \| \mathbf{Z}_R = \frac{j10 \times 10}{10 + j10} = \frac{100\angle 90^\circ}{14.14\angle 45^\circ} = 7.07\angle 45^\circ \quad [\Omega] \tag{8.39}$$

8.3.5 Summary of AC Circuit Analysis at a Single Frequency

As long as an AC circuit includes a single AC source or sources which all operate at the same frequency, it can always be directly analyzed in terms of phasors/complex impedances. All DC analysis techniques:

- Series/parallel equivalents and voltage/current dividers
- Superposition principle
- Source transformation; Thévenin and Norton equivalent circuits
- Nodal and mesh analyses

are equally applicable to steady-state AC circuits. The specific frequency value does not matter. Indeed, we simply need to perform calculations with complex numbers instead of real ones.

8.3.6 Multifrequency AC Circuit Analysis: Superposition Theorem

Unfortunately, when an AC circuit includes sources operating at different frequencies and/or includes the DC sources, the situation is no longer straightforward. The key to the circuit analysis in this case is the general superposition theorem, which is applicable to all linear circuits with arbitrary independent and linear-dependent sources. The *superposition theorem for multifrequency AC circuits* is explained in Fig. 8.20. The original circuit with two AC voltage sources of different frequencies and with one DC source is replaced by three partial circuits, with two of three voltage sources shorted out at a time. Every partial circuit is solved *independently*, either using the phasor/impedance method (for single-frequency AC circuits) or the DC circuit analysis. Finally, the complete solution is obtained as a sum of three real-valued contributions.

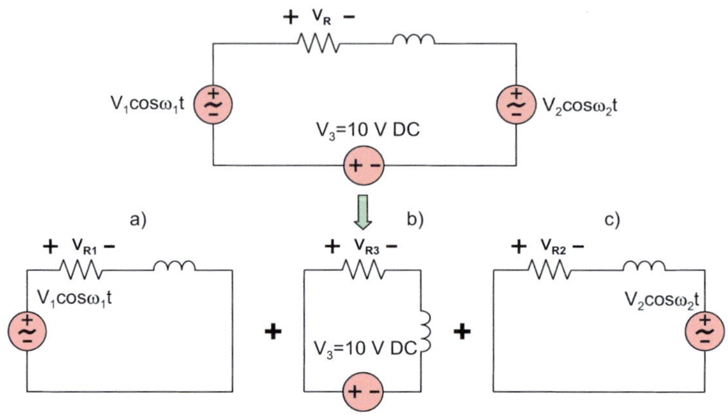

Fig. 8.20. Superposition theorem for a circuit with sources operating at different frequencies (including a DC source).

Example 8.12: Find real-valued voltage $v_R(t)$ across the resistor for the circuit in Fig. 8.20 using the superposition principle. You are given $R = 10\ \Omega$, $V_1 = 10$ V, $V_2 = 1$ V. The impedance of the inductor is $j2\ \Omega$ at frequency ω_1 and $j6\ \Omega$ at frequency ω_2, respectively.

Solution: Three partial equivalent circuits are shown in Fig. 8.20a,b,c. We solve the AC circuits in Fig. 8.20a,c using the phasor/impedance method. Applying voltage division for the circuits in frequency domain gives

$$\mathbf{V}_{R1} = +\frac{10}{10+j2}10 = 9.81\angle-11^\circ\ \text{V} \tag{8.40}$$

$$\mathbf{V}_{R2} = -\frac{10}{10+j6}1 = 0.86\angle+149^\circ\ \text{V} \tag{8.41}$$

The DC circuit in Fig. 8.20b gives $V_{R3} = 10$ V since the inductor is the short circuit in the DC steady state. Combining the solutions yields the resistor voltage,

$$v_R(t) = 10 + 9.81\cos(\omega_1 t - 11^\circ) + 0.86\cos(\omega_2 t + 149^\circ)\ \text{V} \tag{8.42}$$

Exercise 8.9: How does the solution of the previous example change when a 20 Ω resistor is placed in series with the DC source?

Answer: The solution becomes

$$v_R(t) = 3.33 + 3.33\cos(\omega_1 t - 4^\circ) + 0.33\cos(\omega_2 t + 169^\circ)\ \text{V}.$$

Summary

Term	Meaning/Figure
Steady-state AC voltage (steady-state alternating current)	$v(t) = V_m \cos(\omega t + \varphi)$ $V_m > 0$ is the voltage amplitude [V] $\omega = 2\pi f > 0$ is the angular frequency [rad/s] $f > 0$ is the frequency [Hz] $T = 1/f > 0$ is the period [s] $-\pi \le \varphi \le \pi$ is the phase [rad] or [deg] 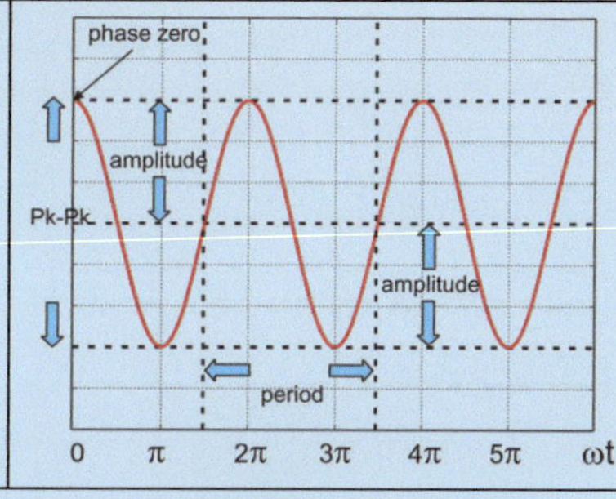
Leading/lagging	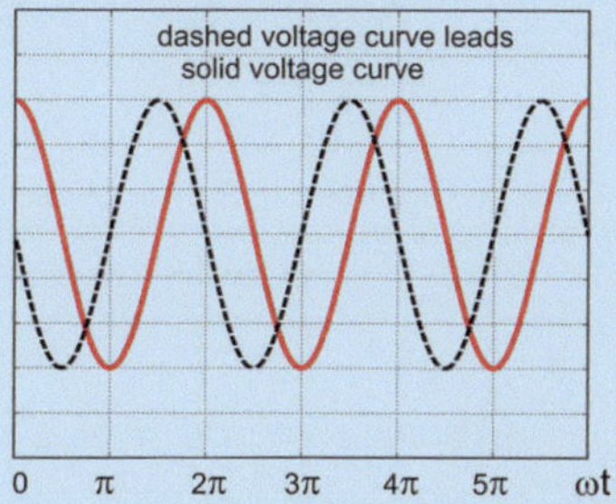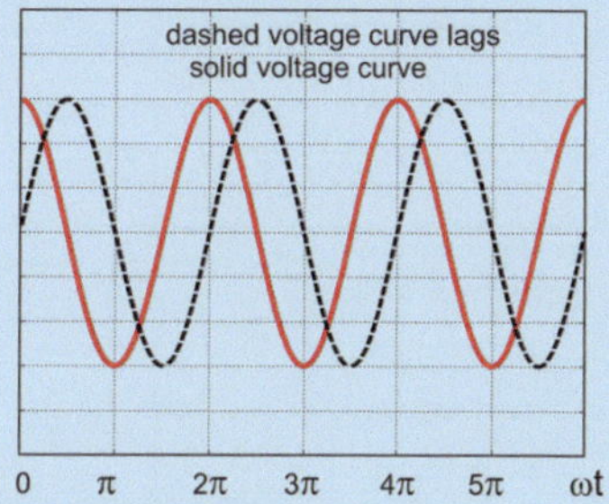
Euler's identity	$e^{j\alpha} = \cos\alpha + j\sin\alpha$, $e^{j\pi/2} = j$, $e^{-j\pi/2} = -j$
Time-domain signal $v(t)$ versus its phasor $\mathbf{V}$; phasor diagram	$v(t) = V_m \cos(\omega t + \varphi)$ $\mathbf{V} = V_m e^{j\varphi}$ 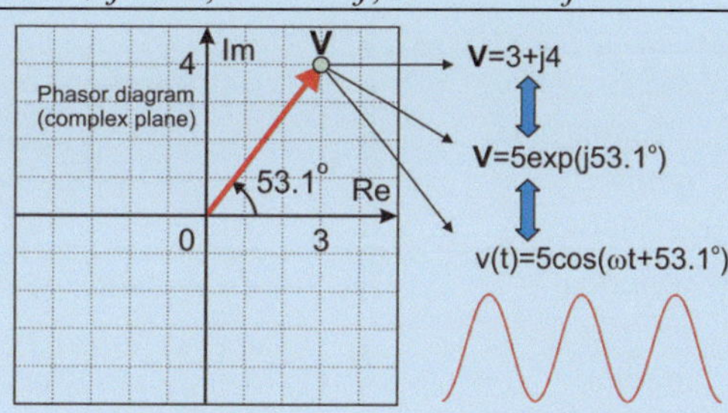
Complex phasors and impedances $\mathbf{Z}_R = R$ $\mathbf{Z}_C = \frac{1}{j\omega C}$ $\mathbf{Z}_L = j\omega L$	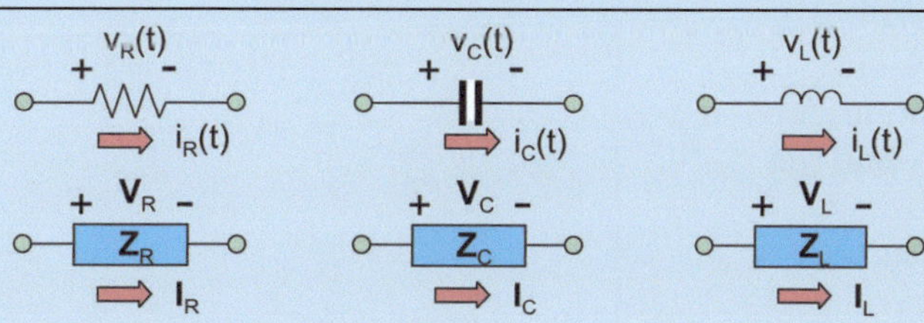

(continued)

Term	Meaning/Figure
Meaning of complex impedance	$\mathbf{Z}_R = R\angle 0°$ multiply current by R and shift by 0° $\mathbf{Z}_C = \frac{1}{\omega C}\angle -90°$ multiply current by 1/ωC and shift by -90° $\mathbf{Z}_L = \omega L\angle 90°$ multiply current by ωL and shift by 90° 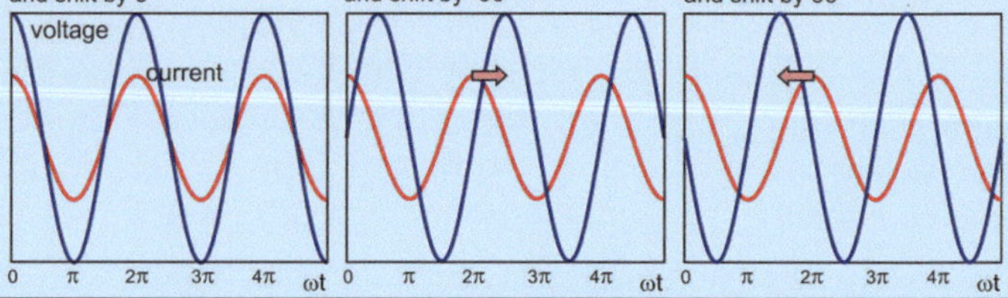
Solution for an AC circuit	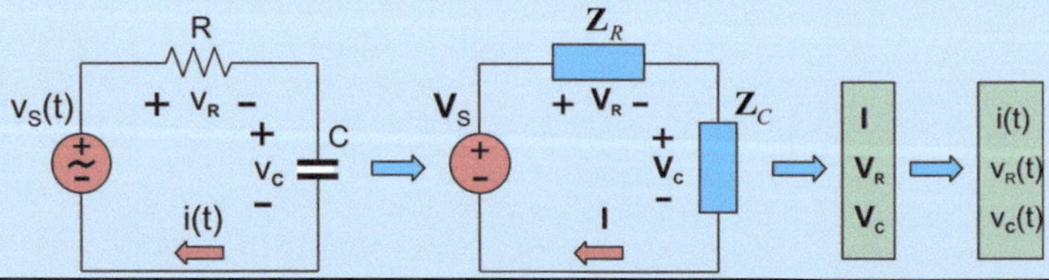
Proof of the conversion from time domain to frequency domain	$-v_S(t) + v_R(t) + v_C(t) = 0 \Rightarrow$ $-\mathrm{Re}(\mathbf{V}_S e^{j\omega t}) + \mathrm{Re}(\mathbf{V}_R e^{j\omega t}) + \mathrm{Re}(\mathbf{V}_C e^{j\omega t}) = 0 \Rightarrow$ $\mathrm{Re}(-\mathbf{V}_S e^{j\omega t} + \mathbf{V}_R e^{j\omega t} + \mathbf{V}_C e^{j\omega t}) = 0 \Rightarrow$ $-\mathbf{V}_S e^{j\omega t} + \mathbf{V}_R e^{j\omega t} + \mathbf{V}_C e^{j\omega t} = 0 \Rightarrow$ $-\mathbf{V}_S + \mathbf{V}_R + \mathbf{V}_C = 0$
Analytical solution method (calculator)	perform addition/subtraction in rectangular form: $\mathbf{Z}_{eq} = \mathbf{Z}_R + \mathbf{Z}_C$ perform multiplication/division in polar form: $\mathbf{I} = \frac{\mathbf{V}_S}{\mathbf{Z}_{eq}}$

Numerical solution method (MATLAB)	`%   Solution for the circuit in Fig. 8.14 of Sec. 4`
	 `omega= 1e3; R = 1e3; C=1e-6; ZC=-j/(omega*C);`
	 `VR=5*R/(R+ZC); vR.Vm=abs(VR); vR.phi=angle(VR)*180/pi`

Source transformation; Thévenin and Norton equivalent circuits	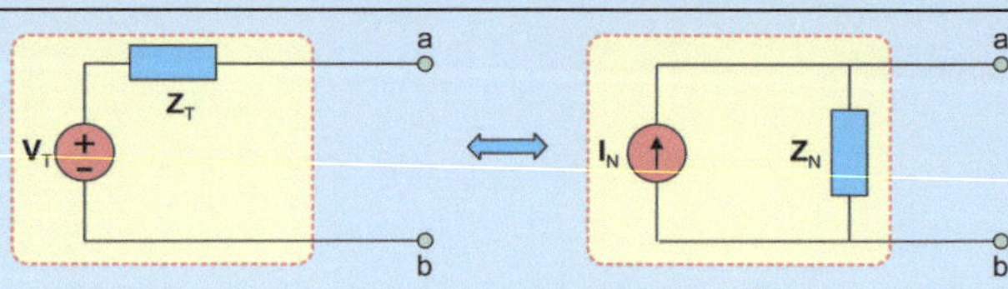 $\mathbf{Z}_T = \mathbf{Z}_N,\ \mathbf{V}_T = \mathbf{Z}_T\mathbf{I}_N,\ \mathbf{I}_N = \frac{\mathbf{V}_T}{\mathbf{Z}_T}$

Table 1. Some basic operations with complex numbers.

$e^{j0} = 1, \quad e^{j\pi/2} = j, \quad e^{-j\pi/2} = -j, \quad e^{j\pi} = -1, \quad e^{-j\pi} = -1, \quad \lvert e^{j\alpha}\rvert = 1$
$j = \angle 90^\circ, \quad \frac{1}{j} = \angle -90^\circ, \quad j^2 = -1, \quad \frac{1}{j} = -j, \quad \lvert j\rvert = 1$
$V_\mathrm{m}e^{j\varphi} = V_\mathrm{m}\angle\varphi = V_\mathrm{m}(\cos\varphi + j\sin\varphi), \quad V_\mathrm{m}e^{-j\varphi} = V_\mathrm{m}\angle -\phi = V_\mathrm{m}(\cos\varphi - j\sin\varphi)$ $\lvert V_\mathrm{m}e^{j\varphi}\rvert = \lvert V_\mathrm{m}e^{-j\varphi}\rvert = V_\mathrm{m}$
$x + jy = V_\mathrm{m}e^{j\varphi}, \quad V_\mathrm{m} = \sqrt{x^2 + y^2}, \quad \varphi = \arctan\left(\frac{y}{x}\right), \quad x \geq 0$
$\frac{1}{x+jy} = \frac{x}{x^2+y^2} - j\frac{y}{x^2+y^2}$
$\frac{1}{x+jy} = \frac{1}{V_\mathrm{m}}e^{-j\varphi}, \quad V_\mathrm{m} = \sqrt{x^2 + y^2}, \quad \varphi = \arctan\left(\frac{y}{x}\right), \quad x \geq 0$
$(x + jy)^* = x - jy, \quad (V_\mathrm{m}e^{j\varphi})^* = V_\mathrm{m}e^{-j\varphi}, \quad (V_\mathrm{m}\angle\varphi)^* = V_\mathrm{m}\angle -\varphi$
$V_\mathrm{m}\angle\varphi \cdot I_\mathrm{m}\angle\psi = (V_\mathrm{m}e^{j\varphi})(I_\mathrm{m}e^{j\psi}) = V_\mathrm{m}I_\mathrm{m}e^{j(\varphi+\psi)} = V_\mathrm{m}I_\mathrm{m}\angle(\varphi + \psi)$
$\frac{V_\mathrm{m}\angle\varphi}{I_\mathrm{m}\angle\psi} = \frac{V_\mathrm{m}e^{j\varphi}}{I_\mathrm{m}e^{j\psi}} = \frac{V_\mathrm{m}}{I_\mathrm{m}}e^{j(\varphi-\psi)} = \frac{V_\mathrm{m}}{I_\mathrm{m}}\angle(\varphi - \psi)$

Table 2. Selected trigonometric identities.

$\sin\alpha = \cos(\alpha - \pi/2), \quad -\sin\alpha = \cos(\alpha + \pi/2), \quad \cos 2\alpha = 2\cos^2\alpha - 1, \quad \sin 2\alpha = 2\sin\alpha\cos\alpha$
$\cos(\alpha + \beta) = \cos\alpha\cos\beta - \sin\alpha\sin\beta \quad \cos\alpha\cos\beta = 0.5(\cos(\alpha + \beta) + \cos(\alpha - \beta))$ $\sin(\alpha + \beta) = \sin\alpha\cos\beta + \cos\alpha\sin\beta \quad \sin\alpha\sin\beta = 0.5(\cos(\alpha - \beta) - \cos(\alpha + \beta))$ $\sin\alpha\cos\beta = 0.5(\sin(\alpha + \beta) + \sin(\alpha - \beta))$
$\cos\alpha + \cos\beta = 2\cos\left(\frac{\alpha+\beta}{2}\right)\cos\left(\frac{\alpha-\beta}{2}\right) \quad \sin\alpha + \sin\beta = 2\sin\left(\frac{\alpha+\beta}{2}\right)\cos\left(\frac{\alpha-\beta}{2}\right)$
$\arctan\varphi = \frac{\pi}{2} - \arctan\frac{1}{\varphi}, \quad \varphi > 0 \quad \arctan\varphi = -\frac{\pi}{2} - \arctan\frac{1}{\varphi}, \quad \varphi < 0$
$C_1\cos\omega t + C_2\sin\omega t = \sqrt{C_1^2 + C_2^2}\cos(\omega t + \varphi), \quad \varphi = -\arctan\left(\frac{C_2}{C_1}\right) \quad C_1, \; C_2 > 0$
$C_1\cos(\omega t + \varphi) \cdot C_2\cos(\omega t + \psi) = 0.5C_1C_2(\cos(\varphi - \psi) + \cos(2\omega t + \varphi + \psi))$
$C_1\cos(\omega t + \varphi) + C_2\cos(\omega t + \psi) = C\cos(\omega t + \varphi)$ $C = \sqrt{C_1^2 + C_2^2 + 2C_1C_2\cos(\varphi - \psi)}$ $\phi = \varphi - \arctan\left(\frac{C_2\sin(\varphi - \psi)}{C_1 + C_2\cos(\varphi - \psi)}\right) + \begin{cases} 0 & C_1 + C_2\cos(\varphi - \psi) > 0 \\ \pi & C_1 + C_2\cos(\varphi - \psi) < 0 \end{cases}$

Problems

8.1 Harmonic Voltage and Current: Phasor

8.1.1 Harmonic Voltages and Currents

8.1.2 Phase: Leading and Lagging

Problem 8.1

A. Write a general expression for the AC harmonic voltage signal (steady-state AC voltage) using the cosine function.
B. Identify amplitude, angular frequency, and phase.
C. Write relations between the angular frequency, frequency, and the period.

Problem 8.2

A. Determine frequency in Hz, angular frequency in rad/s, and amplitude of the harmonic voltage signal shown in the figure below (*show units for every quantity*).
B. Write the AC voltage in the form of a cosine function with the corresponding amplitude, frequency, and phase.

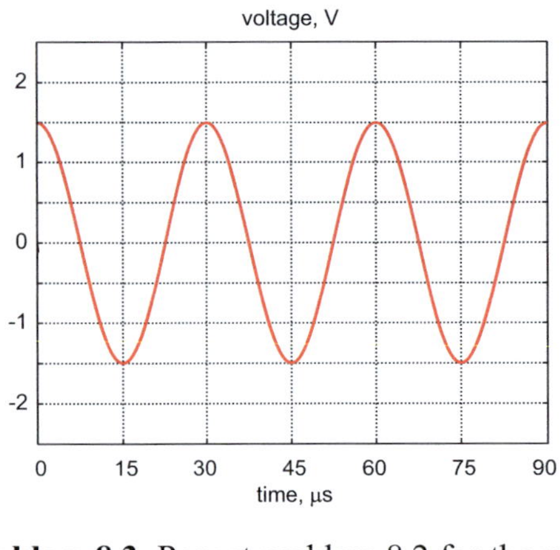

Problem 8.3. Repeat problem 8.2 for the voltage signal shown in the figure below.

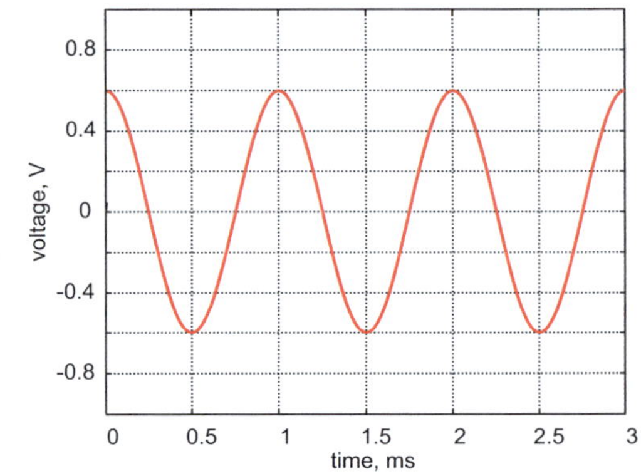

Problem 8.4. Repeat problem 8.2 for a harmonic voltage signal with a DC offset shown in the figure below.

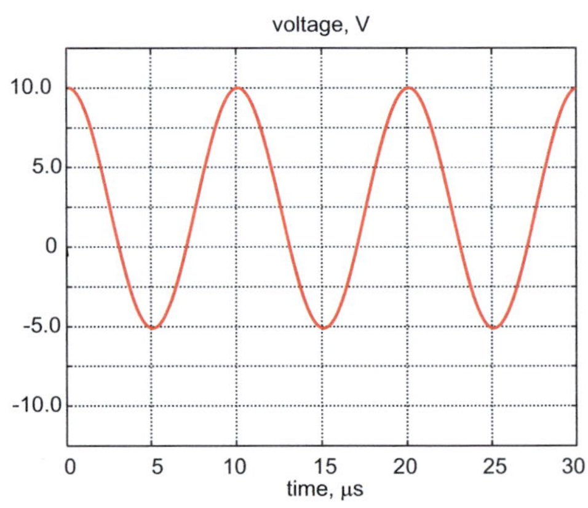

Problem 8.5

A. Determine frequency in Hz, angular frequency in rad/s, amplitude, and phase (versus the base $\cos \omega t$ signal) of the harmonic voltage shown in the figure below (*show units for every quantity*).
B. Write the AC voltage in the form of a cosine function with the corresponding amplitude, frequency, and phase.

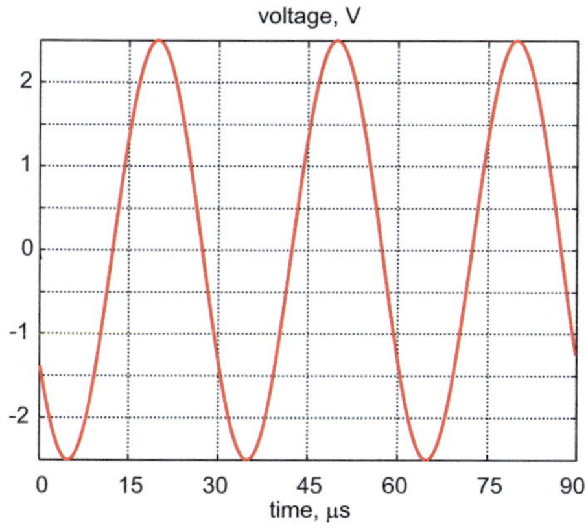

Problem 8.6. Repeat problem 8.5 for the voltage signal shown in the figure below.

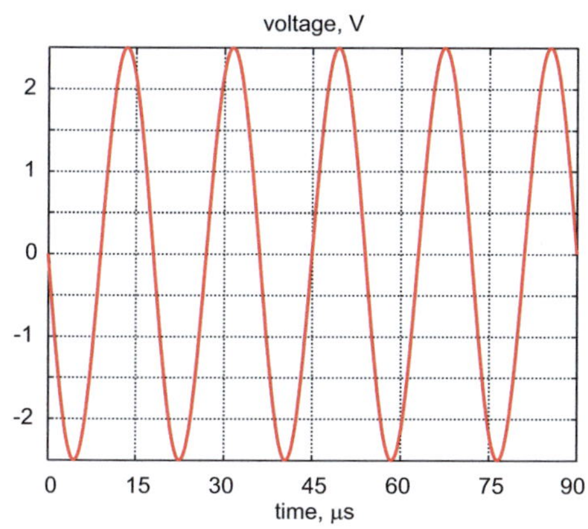

Problem 8.7. Repeat problem 8.5 for the voltage signal shown in the figure below.

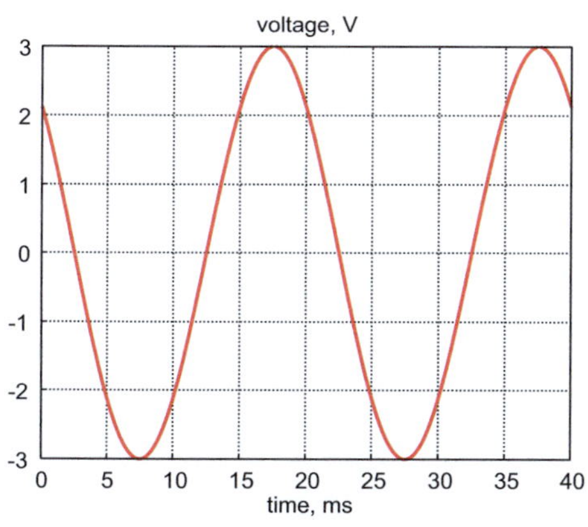

Problem 8.8. Repeat problem 8.5 for the voltage signal shown in the figure below.

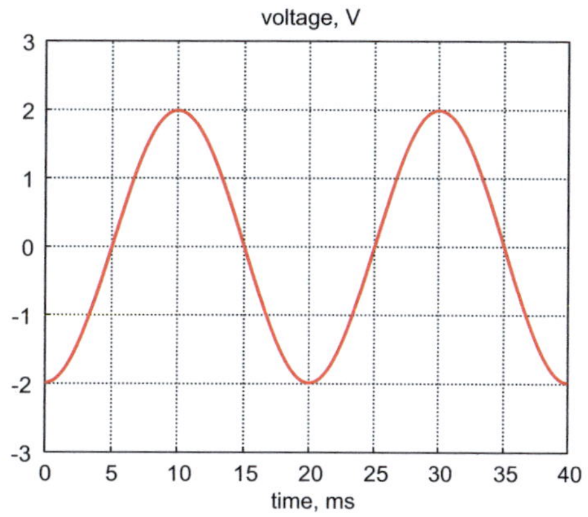

Problem 8.9. An AC voltage in a circuit is given by $v(t) = 10\cos(2\pi 50t)$ [V]. Using software of your choice, plot the voltage to scale over the time interval of two periods, i.e., $0 \le t \le 2T$. Label the axes.

Problem 8.10. An AC voltage in a circuit is given by the voltage expression $v(t) = 10\cos(1000t - \pi/3)$ [V]. Using software of your choice, plot the voltage to scale over the time interval of four periods, i.e., $0 \le t \le 4T$. Label the axes.

Problem 8.11. The reference voltage is shown by a solid curve in the figure; the AC voltage under study is shown by a dashed curve. Determine if the AC voltage under study leads or lags the reference voltage, and, if so, by how many degrees.

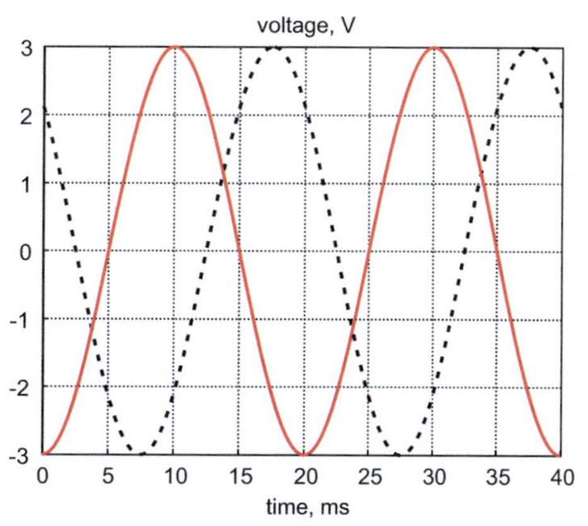

Problem 8.12. Repeat problem 8.11 for the voltage signal shown in the figure below.

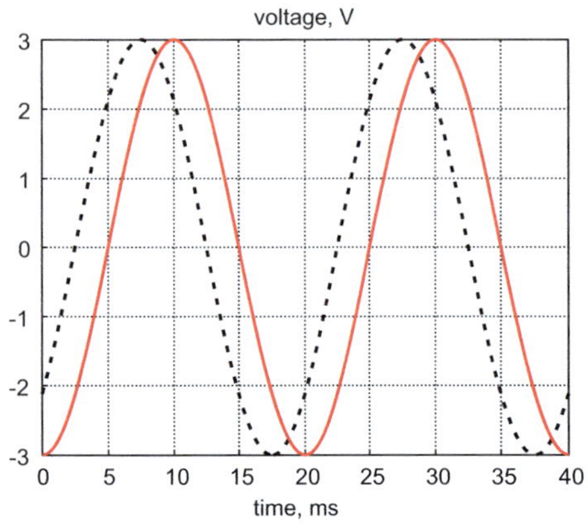

Problem 8.13. Repeat problem 8.11 for the voltage signal shown in the figure below.

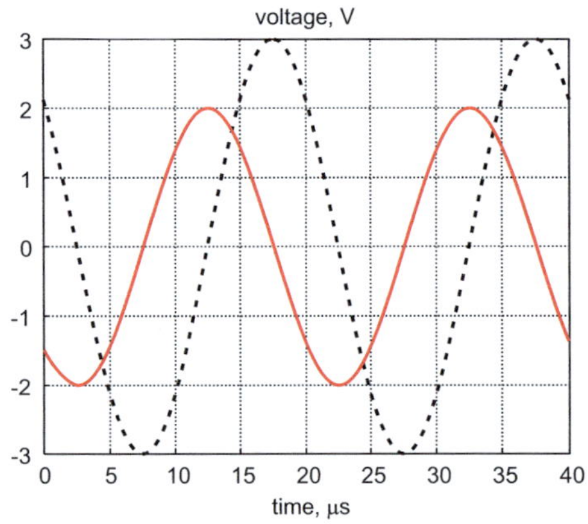

Problem 8.14. Determine the frequency in Hz, period in s, amplitude in V, and phase in degrees (versus the base cosine signal) of the voltage signal in the form $v(t) = 15\sin(100t + 45°)$ [V]. *Hint:* Convert the signal to the base cosine form first.

Problem 8.15. Determine the frequency in Hz, period in s, amplitude in V, and phase in radians (versus the base cosine signal) of the voltage signal in the form $v(t) = 15\sin(1000t - 35°)$ [V]. *Hint:* Convert the signal to the base cosine form first.

Problem 8.16. Determine the frequency in Hz, period in s, amplitude in V, peak-to-peak value in V, and phase in radians (versus the base cosine signal) of the voltage signal in the form $v(t) = 5\ \sin(100t + 225°)$ [V]. *Hint:* Convert the signal to the base cosine form first.

Problem 8.17. The AC voltage is given by a combination of two sinusoids:

A. $v(t) = 1\sin(\omega t + \pi/2) - 2\sin(\omega t - \pi/2)$
B. $v(t) = 1\ \sin(\omega t + \pi/2) - 2\ \sin(\omega t - \pi/3)$

Convert this voltage to the basic cosine form $v(t) = V_{\mathrm{m}}\cos(\omega t + \varphi)$ and determine the amplitude and the phase (versus the base cosine signal).
Hint: Trigonometric identities may be found in the summary to this chapter.

8.1.4 Definition of a Phasor
8.1.5 From Real Signals to Phasors
8.1.6 From Phasors to Real Signals

Problem 8.18. Determine the phasors for the real-valued AC voltages and currents. Show units. Express all phase angles in radians.

$$v(t) = 10\cos(\omega t + \pi/3)\quad [\mathrm{V}]$$
$$v(t) = 3\cos(\omega t - 30°)\quad [\mathrm{V}]$$
$$i(t) = 12\cos(\omega t + \pi/6)\quad [\mathrm{A}]$$
$$i(t) = -1\cos(\omega t - \pi/2)\quad [\mathrm{A}]$$

Problem 8.19. Determine the phasors for the real-valued AC voltages and currents. Use the shorthand notation $\angle$ for the complex exponent. Show units. Express all phase angles in degrees.

$$v(t) = 10\sin(\omega t + \pi/3)\quad [\mathrm{V}]$$
$$v(t) = 3\cos(100t - 30°)\quad [\mathrm{V}]$$
$$i(t) = -12\sin(\omega t + \pi/6)\quad [\mathrm{A}]$$
$$i(t) = \cos(\omega t) + \sin(\omega t)\quad [\mathrm{A}]$$

Problem 8.20. The phasors of the AC voltage and current are given by

$$\mathbf{V} = 5\angle\pi/3\quad [\mathrm{V}]$$
$$\mathbf{V} = 3\angle\pi\quad [\mathrm{V}]$$
$$\mathbf{I} = 2.1\angle 45°\quad [\mathrm{A}]$$
$$\mathbf{I} = 1\angle -180°\quad [\mathrm{A}]$$

The AC source has the angular frequency ω. Restore the corresponding real-valued voltages and currents. Show units; express all phase angles in radians.

Problem 8.21. The phasors of the AC voltage and current are given by

$$\mathbf{V} = 10\angle\pi/2 \quad [\mathrm{V}]$$
$$\mathbf{V} = 15\angle -\pi/3 \quad [\mathrm{V}]$$
$$\mathbf{I} = -20\angle -16^\circ \quad [\mathrm{A}]$$
$$\mathbf{I} = j\angle 45^\circ \quad [\mathrm{A}]$$

The AC source has the angular frequency ω. Restore the corresponding real-valued voltages and currents as functions of time. Show units. Express all phase angles in radians.

Problem 8.22. The phasors of the AC voltage and current are given by

$$\mathbf{V} = -1 \quad [\mathrm{V}]$$
$$\mathbf{V} = \angle\pi + \angle -\pi \quad [\mathrm{V}]$$
$$\mathbf{I} = -2\angle 45^\circ \quad [\mathrm{A}]$$
$$\mathbf{I} = \angle 45^\circ + \angle -45^\circ \quad [\mathrm{A}]$$

The AC source has the angular frequency ω. Restore the corresponding real-valued voltages and currents. Show units; express all phase angles in degrees.

8.1.7 Polar and Rectangular Forms: Phasor Magnitude

8.1.8 Operations with Phasors and Phasor Diagram

Problem 8.23. A complex number $\mathbf{V}$ is given by $\mathbf{V} = 4 + j2$.

A. Convert it into polar form; express the phase angle in degrees.
B. Plot the number on the phasor diagram.
C. If this number is a phasor of a voltage signal with the units of volts, what is the voltage signal in time domain? Assume the angular frequency ω.

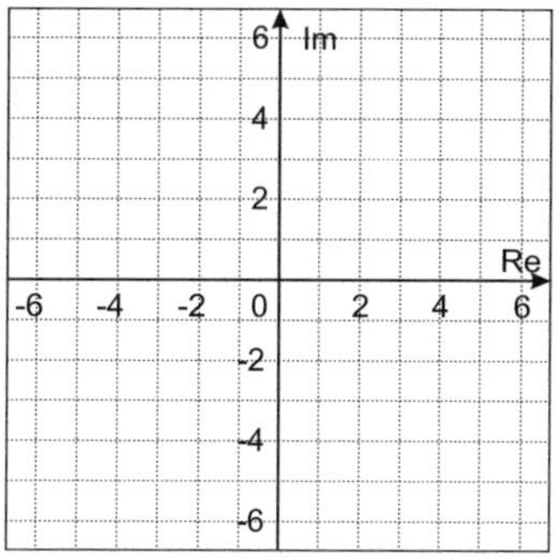

Problem 8.24. Repeat problem 8.23 for $\mathbf{V} = 4\exp(-j\pi/6)$, but convert this number to the rectangular form.

Problem 8.25. Repeat problem 8.23 for $\mathbf{V} = 25/(3 + j4)$.

Problem 8.26. Phasors of three AC voltage signals are shown in Fig. 8.8. Every division in the figure corresponds to 1 V. The AC source has the angular frequency ω. Restore the corresponding real-valued voltages in time domain. Show units. Express all phase angles in radians.

Problem 8.27. The phasors of the AC voltage and current are given in the rectangular form:

$$\mathbf{V} = 3 + j2 \quad [\mathrm{V}]$$
$$\mathbf{I} = -2 + j3 \quad [\mathrm{A}]$$

The AC source has the angular frequency ω. Restore the corresponding real-valued voltages and currents. Show units, express all phase angles in radians.

Problem 8.28*. Solve the previous problem using MATLAB. Present the corresponding MATLAB script.

Problem 8.29. The phasors of the AC voltage and current are given in the rectangular form:

$$\mathbf{V} = (3 + j2)^2 \quad [\mathrm{V}]$$
$$\mathbf{V} = (-2 + j3)(7 + j) \quad [\mathrm{V}]$$
$$\mathbf{V} = \frac{1}{-0.2 + j0.1} \quad [\mathrm{V}]$$

The AC source has the angular frequency ω. Restore the corresponding real-valued voltages and currents. Show units; express all phase angles in radians.

Problem 8.30. Solve the previous problem using MATLAB. Present the corresponding MATLAB script.

Problem 8.31. Two phasor voltages are shown in the phasor diagram. The AC source has the angular frequency ω. Restore the corresponding real-valued voltages. Show units. Express all phase angles in radians.

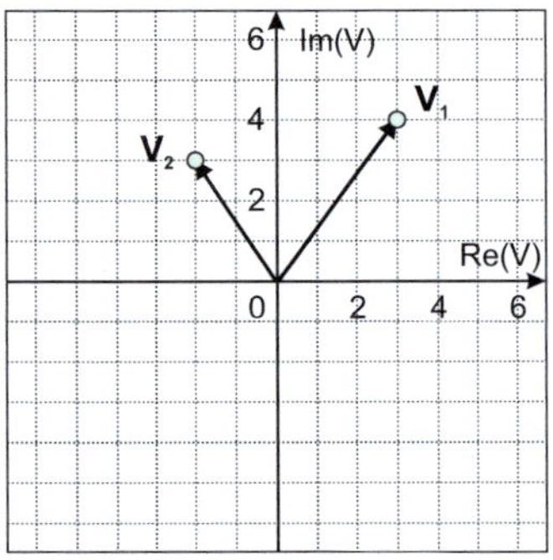

Problem 8.32. Phasors of two AC current signals are shown in the following figure. The AC source has the angular frequency ω. A sum of two current signals is desired. Restore the real-valued current corresponding to the sum of three signals in time domain. Express the phase angle in degrees.

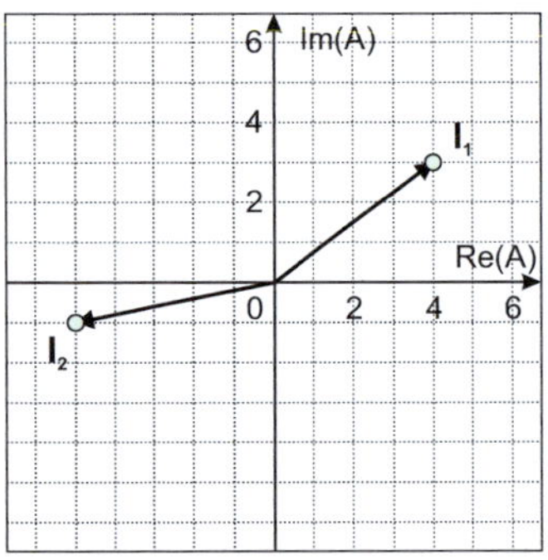

Problem 8.33. Phasors of three AC voltage signals are shown in Fig. 8.8. Every division in the figure corresponds to 1 V. The AC source has the angular frequency ω. A sum of three voltage signals is desired. Restore the real-valued voltage corresponding to the sum of three signals in time domain. Express the phase angle in degrees.

Problem 8.34. Solve the previous problem using MATLAB. Present the corresponding MATLAB script.

Problem 8.35. Voltages of two series elements, shown in the figure below,

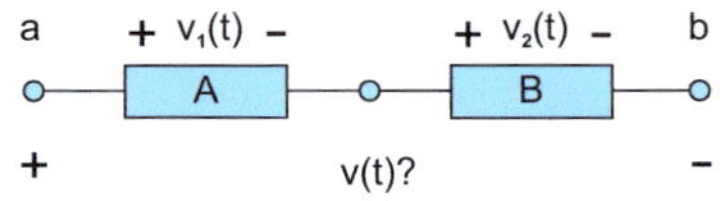

are given by

$$v_1(t) = 5\cos(\omega t + 45^\circ) \quad [\mathrm{V}]$$
$$v_2(t) = 5\cos(\omega t - 45^\circ) \quad [\mathrm{V}]$$

A. Draw the phasor diagram and show phasors $\mathbf{V}_1$, $\mathbf{V}_2$ to scale as two vectors in the complex plane.
B. Show the sum $\mathbf{V} = \mathbf{V}_1 + \mathbf{V}_2$ as a vector in the complex plane.
A. Find voltage $v(t)$ between terminals a and b using the phasor method. Express the phase angle in degrees.
C. Does voltage $v_1(t)$ lag or lead voltage $v_2(t)$?

Problem 8.36. Repeat the previous problem for:

$$v_1(t) = 1\cos(\omega t + 60^\circ) \quad [\mathrm{V}]$$
$$v_2(t) = 2\cos(\omega t - 30^\circ) \quad [\mathrm{V}]$$

Problem 8.37. Electric currents through two parallel circuit elements are shown in the following figure:

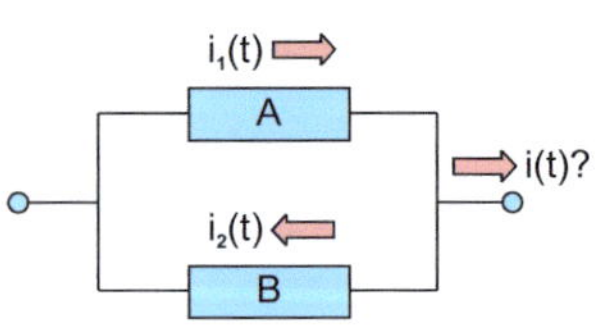

The current expressions are given by

$$i_1(t) = 4\cos(\omega t + 45°) \quad [\text{A}]$$
$$i_2(t) = 2\cos(\omega t - 60°) \quad [\text{A}]$$

A. Draw the phasor diagram and show phasors $\mathbf{I}_1$, $\mathbf{I}_2$ to scale as two vectors in the complex plane.
B. Show the difference $\mathbf{I} = \mathbf{I}_1 - \mathbf{I}_2$ as a vector in the complex plane.
C. Find the net current $i(t)$ of the parallel combination using the phasor method. Express the phase angle in degrees.
D. Does the net current $i(t)$ lag or lead current $i_1(t)$?

8.2 Impedance

8.2.1 The Concept of Impedance

8.2.2 Physical Meaning of the Impedance

Problem 8.38. Prove that the impedance of the inductor, $\mathbf{Z}_L = j\omega L$, has units of ohms.
Hint: The imaginary unit j is dimensionless. However, in the context of phasors, it may be assigned the units of rad^{-1}.

Problem 8.39. Prove that the impedance of the capacitor, $\mathbf{Z}_C = \frac{1}{j\omega C}$ has units of ohms.
Hint: The imaginary unit j is dimensionless. However, in the context of phasors, it may be assigned the units of rad^{-1}.

Problem 8.40. An ECE student discovers a "new" dynamic passive circuit element N (in addition to the inductor and the capacitor) that is described by the voltage-to-current relation $v_N = N\frac{d^2 i_N}{dt^2}$ where N is a *positive* constant:
A. Obtain the impedance for this circuit element.
B. Do you think such a circuit element may exist? Why or why not? *Hint:* The real part of the impedance is the element resistance. If the resistance is negative, the element delivers power instead of absorbing it.

Problem 8.41. Another ECE student discovers a "truly new" dynamic passive circuit element N (in addition to the inductor and the capacitor) that is described by the voltage-to-current relation $v_N = N\frac{d^3 i_N}{dt^3}$ where N is a constant:
A. Obtain the impedance for this circuit element.
B. Do you think such a circuit element may exist? Why or why not? *Hint:* The real part of the impedance is the element resistance. If the resistance is negative, the element delivers power instead of absorbing it.

Problem 8.42. For three AC circuits shown in the following figure, find the impedance of the resistor, inductor, and capacitor (when present). Show units. Express the result in rectangular form. Also express the result in polar form using the shorthand notation ∠. Determine the magnitude of the impedance. Does the strength of the power supply have an effect on the obtained impedance values?

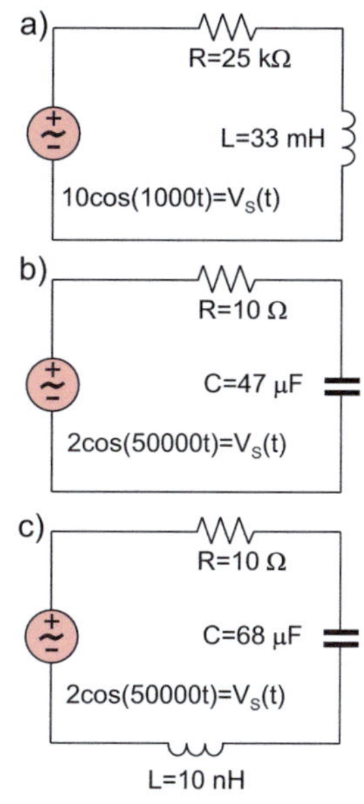

Problem 8.43. Repeat problem 8.42 if the AC voltage power supply in every figure is replaced by an alternating current power supply with the same frequency and amplitude of 0.5 A.

Problem 8.44. Voltages (dashed curves) and the corresponding currents (solid curves) for three unknown circuit elements are shown in the figure below.

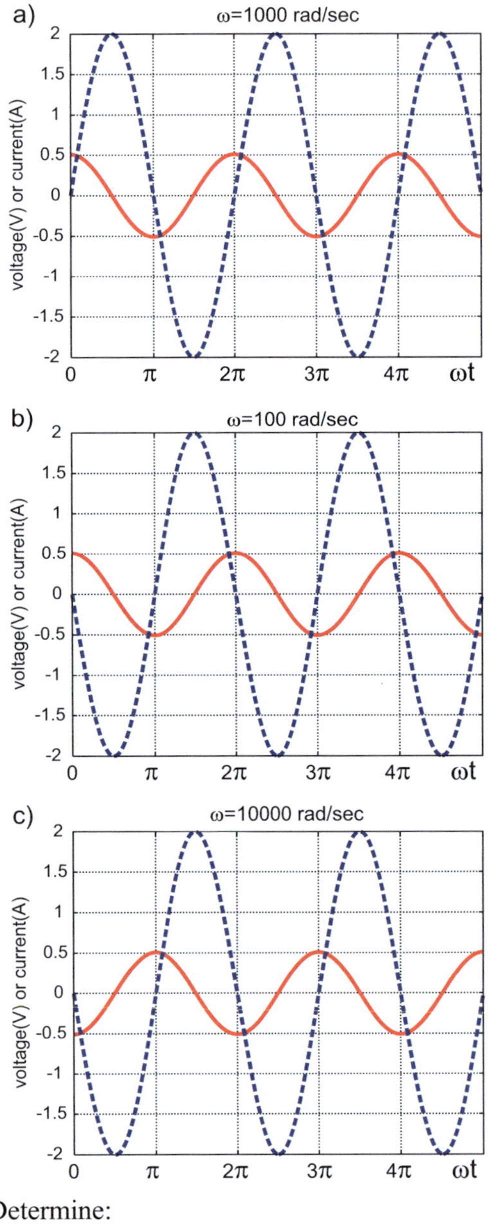

Determine:

A. The type of the element (resistor, capacitor, or inductor)
B. The value of the corresponding resistance, inductance, or capacitance

Note that the angular frequency is different in every case.

Problem 8.45. Voltages (dashed curves) and currents (solid curves) for three unknown circuit elements are shown in the figure below.

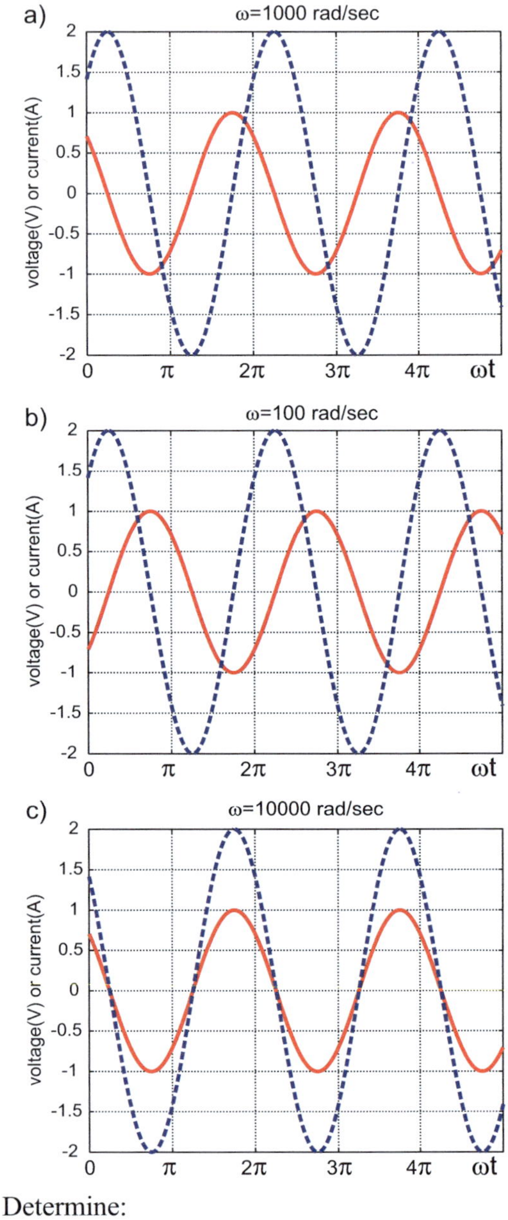

Determine:

A. The type of the element (resistor, capacitor, or inductor)
B. The value of the corresponding resistance, inductance, or capacitance

Problem 8.46. Phasor voltages and currents for three unknown circuit elements are shown in the figure below. Determine the type of the element (*R*, *L*, or *C*) and the value of *R*, *L*, or *C*.

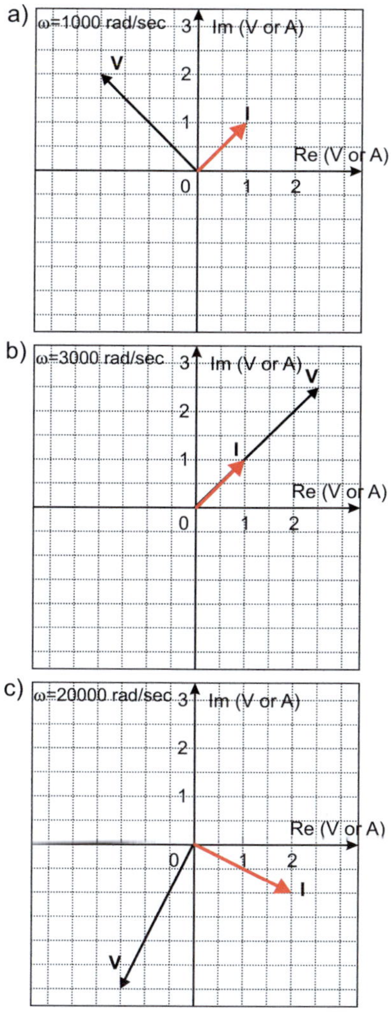

Problem 8.47. Phasor voltages and phasor currents for three unknown circuit elements are shown in the figure below. Determine the type of the element (*R*, *L*, or *C*) and the value of *R*, *L*, or *C* when appropriate.

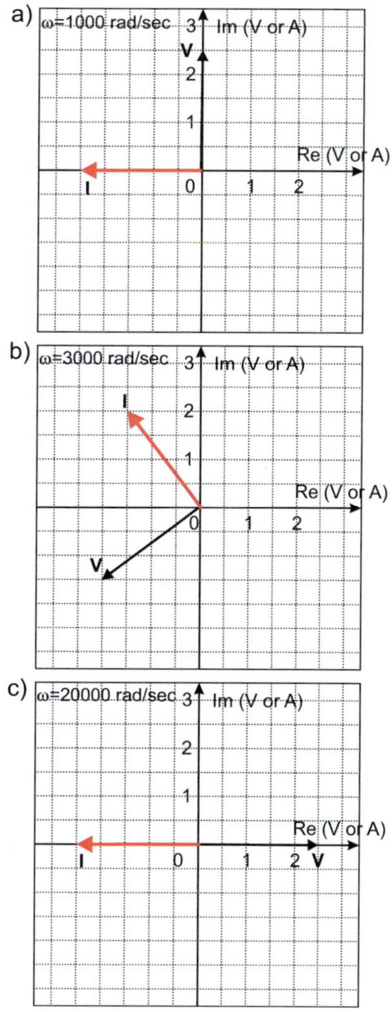

Problem 8.48*. The following MATLAB script plots the real-valued signals in time domain corresponding to the phasor voltage $\mathbf{V} = 5\angle 30°$ [V] and to the phasor current $\mathbf{I} = 2\angle -60°$ [A] for an inductor.

```
clear all
f  = 2e6;                    % frequency, Hz
T  = 1/f;                    % period, sec
dt = T/100;                  % sampling int.
t  = [0:dt:2.5*T];           % time vector
vL = 5*cos(2*pi*f*t+pi/6);   % voltage
iL = 2*cos(2*pi*f*t-pi/3);   % current
t  = t/T;                    % time in periods
```

```
plot(t, vL, 'b');
grid on; hold on;
plot(t, iL, 'r');
xlabel('t/T');
ylabel('voltage/current')
```

Modify the script and plot the real-valued voltages and currents corresponding to the phasors shown in the phasor diagrams for Problem 8.46.

Problem 8.49. Repeat the previous problem for the phasors shown in the phasor diagrams for Problem 8.47.

8.3 Principles of AC Circuit Analysis

8.3.1 AC Circuit Analysis: KVL, KCL, and Equivalent Impedances

8.3.2 Complete Solution for an AC Circuit: KVL and KCL on Phasor Diagram

Problem 8.50. For the AC circuit element combinations shown in the figure that follows,

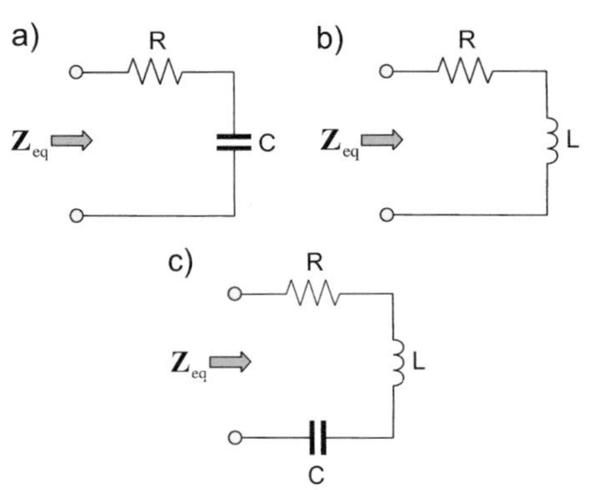

A. Find the equivalent impedance $\mathbf{Z}_{eq}$ in polar form given that $\omega = 10000$ rad/s, $C = 0.1$ μF, $L = 100$ mH, $R = 1$ kΩ.
B. Plot the result for the partial impedances and for $\mathbf{Z}_{eq}$ on the corresponding phasor diagram.

Problem 8.51. Find $\mathbf{Z}_{eq}$ in the polar form for the circuit element combination shown in the figure below when $\omega = 100,000$ rad/s, $C = 100$ nF, $L = 1$ mH, $R = 100$ Ω.

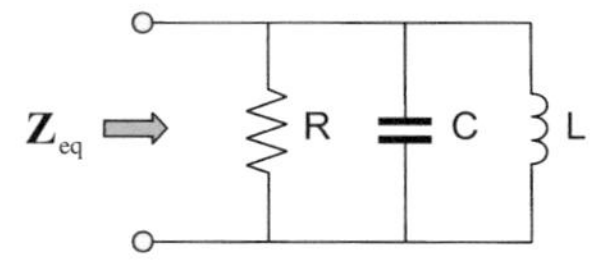

Problem 8.52. For three circuit element combinations shown in the figure below, find $\mathbf{Z}_{eq}$ given that $\omega = 2000$ rad/s, $C = 5$ μF, $L = 50$ mH, $R = 1$ kΩ.

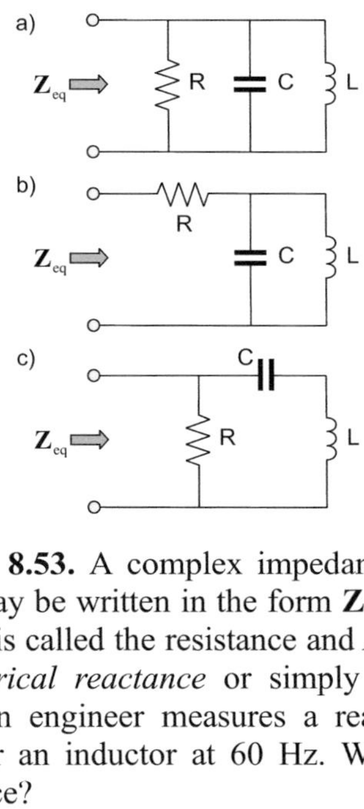

Problem 8.53. A complex impedance of any circuit may be written in the form $\mathbf{Z} = R + jX$ where R is called the resistance and X is called the *electrical reactance* or simply the *reactance*. An engineer measures a reactance of 2 Ω over an inductor at 60 Hz. What is the inductance?

Problem 8.54. The same engineer measures a reactance of −1 kΩ over a capacitor at 60 Hz. What is the capacitance?

Problem 8.55. For the circuit shown in the figure below, $v_S(t) = 10\cos\omega t$ [V], $\omega = 10,000$ rad/s, $C = 1$ μF, $R = 100$ Ω:

A. Find phasor current $\mathbf{I}$ and phasor voltages, $\mathbf{V}_R$, $\mathbf{V}_C$, and construct the *voltage* phasor diagram for phasors $\mathbf{V}_R$, $\mathbf{V}_C$, $\mathbf{V}_S$.

B. Find voltages across the resistor and capacitor, $v_R(t)$ and $v_C(t)$, as functions of time.

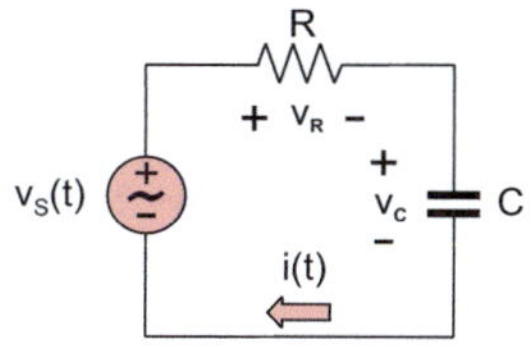

Problem 8.56. For the circuit shown in the figure below, $i_S(t) = 1\cos\omega t$ [A] and $\omega = 10,000$ rad/s, $C = 1$ µF, $R = 100\Omega$.

A. Find phasor voltages $\mathbf{V}_R$, $\mathbf{V}_C$, $\mathbf{V}$ and construct the *voltage* phasor diagram for phasors $\mathbf{V}$, $\mathbf{V}_R$, $\mathbf{V}_C$.
B. Find voltages across the resistor and capacitor, $v_R(t)$ and $v_C(t)$, as functions of time.

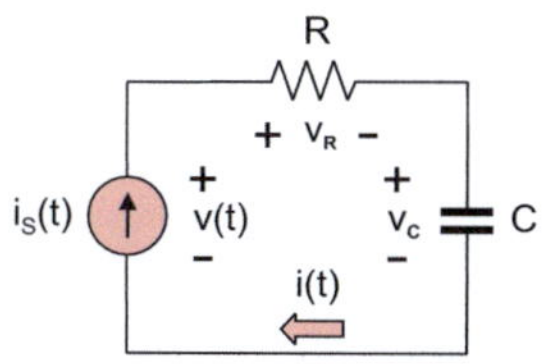

Problem 8.57. Repeat problem 8.55 when $C = 2.2$ µF. The rest of the parameters are the same.

Problem 8.58. Repeat problem 8.55 when $C = 2.2$ µF and $f = 500$ Hz. The rest of the parameters are the same.

Problem 8.59. In the circuit shown in the figure below, $v_S(t) = 12\cos\omega t$ [V], $\omega = 10,000$ rad/s, $L = 1$ mH, $R = 10$ Ω:

A. Find phasor current $\mathbf{I}$ and phasor voltages, $\mathbf{V}_R$, $\mathbf{V}_L$, and construct the *voltage* phasor diagram for phasors $\mathbf{V}_R$, $\mathbf{V}_C$, $\mathbf{V}_S$.
B. Find voltages across the resistor and inductor, $v_R(t)$ and $v_L(t)$, as functions of time.

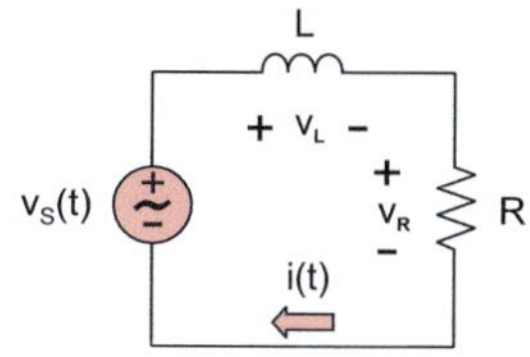

Problem 8.60. For the circuit shown in the figure, $i_S(t) = 1\cos\omega t$ [A] and $\omega = 10,000$ rad/s, $L = 10$ mH, $R = 100$ Ω:

A. Find phasor voltages, $\mathbf{V}_R$, $\mathbf{V}_L$, $\mathbf{V}$ and construct the *voltage* phasor diagram for phasors $\mathbf{V}$, $\mathbf{V}_R$, $\mathbf{V}_C$.
B. Find voltages across the resistor and inductor, $v_R(t)$ and $v_L(t)$, as functions of time.

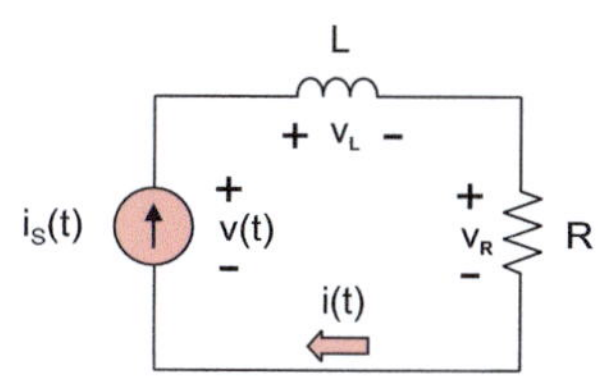

Problem 8.61. Repeat Problem 8.59 when $L = 1.9$ mH. Assume the other parameters to be the same.

Problem 8.62. Repeat Problem 8.59 when $L = 6.8$ mH and $f = 1000$ Hz. The rest of the parameters are the same.

8.3.3 Source Transformation

8.3.4 Thévenin and Norton Equivalent Circuits

8.3.5 Summary of AC Circuit Analysis at a Single Frequency

8.3.6 Multi-frequency AC Circuit Analysis: Superposition Principle

Problem 8.64. A current source with the phasor current $\mathbf{I}_N$ and in parallel with the impedance $\mathbf{Z}_N$ shown in the following figure is equivalent to a voltage source with the phasor voltage $\mathbf{V}_\mathrm{T}$ and in series with the impedance $\mathbf{Z}_\mathrm{T}$. Determine $\mathbf{V}_\mathrm{T}$ and $\mathbf{Z}_\mathrm{T}$ given that $\mathbf{I}_N = 2 + j3$ A, $\mathbf{Z}_N = 2 - j3$ Ω. Express your result in the polar form.

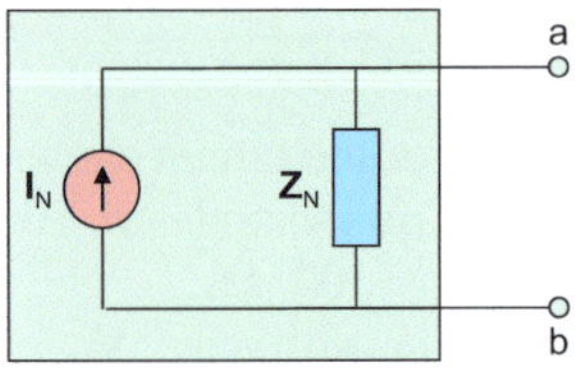

Problem 8.65. Determine phasor voltage $\mathbf{V}_1$ in the AC circuit shown in the following figure using the method of source transformation. The impedance values are given at the frequency of interest.

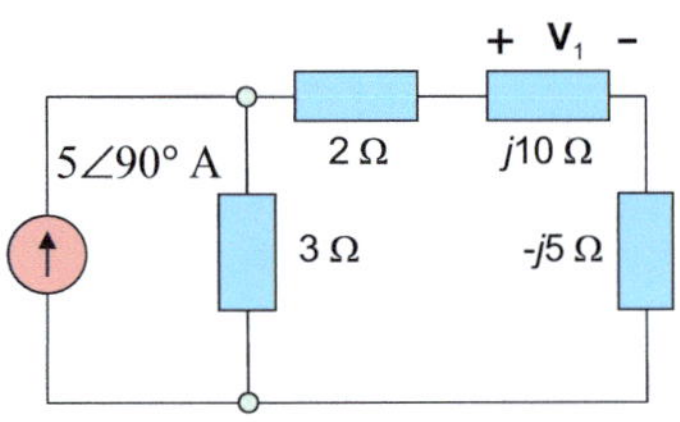

Problem 8.66. Find the Thévenin equivalent circuit, i.e., $\mathbf{V}_\mathrm{T}$ and $\mathbf{Z}_\mathrm{T}$, for the circuit shown in the figure below when $\omega = 377$ rad/s, $L = 26.5$ mH, $R = 10\ \Omega$, $C = 220\ \mu\mathrm{F}$, $v_\mathrm{S}(t) = 10\cos\omega t$ [V].

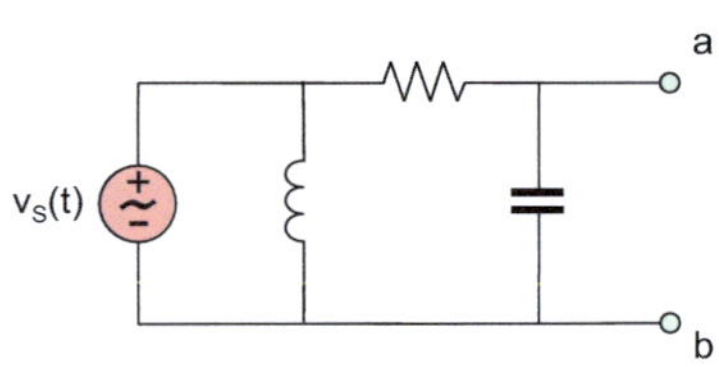

Problem 8.67. Find the Thévenin equivalent circuit, i.e., $\mathbf{V}_\mathrm{T}$ and $\mathbf{Z}_\mathrm{T}$, for the circuit shown in the figure below when $\omega = 377$ rad/s, $L = 26.5$ mH, $R = 10\ \Omega$, $C = 500\ \mu\mathrm{F}$, and $V_\mathrm{S}(t) = 10\cos\omega t$ [V].

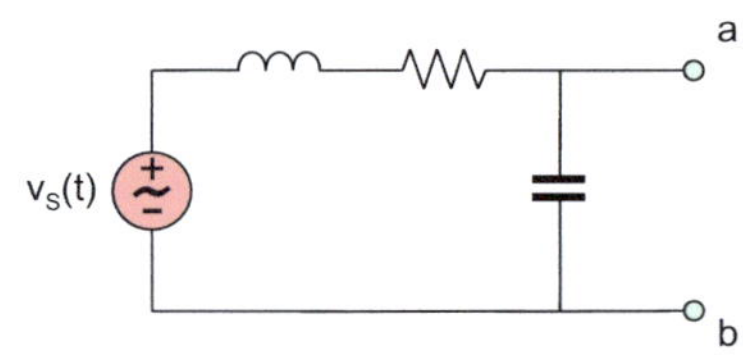

Problem 8.68. Describe the meaning of the superposition principle for multifrequency AC circuits in your own words.

Problem 8.69. Find real-valued voltage $v_\mathrm{R}(t)$ across the resistor for the circuit in the following figure using the superposition principle. You are given $V_1 = 10$ V, $V_2 = 1$ V. The AC frequencies are $\omega_1 = 377$ rad/s and $\omega_2 = 3\omega_1$, respectively.

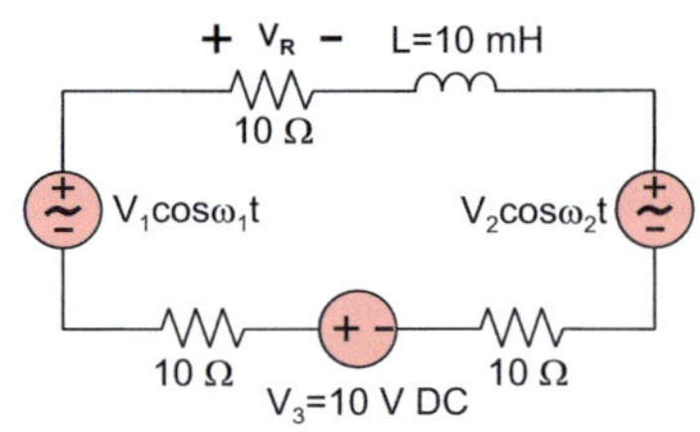

Chapter 9: Filter Circuits: Frequency Response, Bode Plots, and Fourier Transform

Overview

Prerequisites:
- Knowledge of complex arithmetic
- Knowledge of superposition principle for linear circuits (Chapter 3)
- Knowledge of harmonic voltage and current behavior (Chapter 8)
- Knowledge of phasor/impedance method for AC circuit analysis (Chapter 8)
- Knowledge of an operational amplifier with negative feedback (Chapter 5)

Objectives of Section 9.1:
- Establish the concept of a first-order analog filter as a two-port network
- Understand the difference between high-pass and low-pass filters
- Understand the effect of filter termination
- Become familiar with the fundamental filter characteristics including transfer function, break frequency, roll-off, and high-/low-frequency asymptotes
- Understand the construction of the Bode plot including decibels; become familiar with some of the jargon used by electrical engineers
- Establish the close agreement between first-order RC and RL filters; become familiar with the concept of cascaded filter networks

Objectives of Section 9.2:
- Establish the model for the open-loop gain of an operational amplifier as a function of frequency
- Understand the meaning of datasheet parameters such as unity-gain bandwidth and gain-bandwidth product
- Establish the model for the closed-loop gain of an operational amplifier as a function of frequency from first principles
- Find the frequency bandwidth for any practical operational amplifier circuit using the datasheet

S. N. Makarov et al., *Practical Electrical Engineering*, https://doi.org/10.1007/978-3-319-96692-2_9

Objectives of Section 9.3:

- Obtain an introductory exposure to the continuous Fourier transform and be able to compute the transform for simple examples including the meaning of a sinc function
- Be able to relate continuous and discrete Fourier transform via the Riemann sum approximation
- Be able to define sampling points of the DFT in both time and frequency domain
- Understand the structure and ordering of the DFT frequency spectrum including its relation to negative frequencies
- Apply the DFT to a filter with a given transfer function and generate the discrete frequency spectrum of the output signal
- Apply the DFT to filter operation with input pulse or nonperiodic signals
- Apply the DFT (FFT and IFFT) in MATLAB

Application examples:

- Effect of a load connected to the filter
- Effect of next-stage filter load
- Finding bandwidth of an amplifier circuit using the datasheet
- Selection of an amplifier IC for proper frequency bandwidth
- Numerical differentiation via the FFT
- Filter operation for an input pulse signal
- Converting computational electromagnetic solution from frequency domain to time domain

Keywords:

Analog filter, RC filter, RL filter, Port, Two-port network, First-order high-pass filter, First-order low-pass filter, Filter termination, Amplitude transfer function, Phase transfer function, Power transfer function, Complex transfer function, Frequency response, Break frequency, Half-power frequency, 3-dB frequency, Corner frequency, Bode plot, Decibel, Roll-off, High-frequency asymptote, Low-frequency asymptote, Frequency band, Passband, Stopband, Decade, Octave, Power gain, Open-loop amplifier gain, Unity-gain bandwidth, Gain-bandwidth product, Internal compensation, Open-loop AC gain, Closed-loop AC gain, Amplifier circuit bandwidth, **Fourier transform continuous** (direct inverse Fourier spectrum, direct inverse Fourier spectrum, bandlimited spectrum, reversal property, sinc function, mathematical properties, amplitude-modulated signal, Parseval's theorem, energy spectral density), **Fourier transform discrete** (Fast digital signal processing (DSP), sampling points, sampling interval, sampling frequency, sampling theorem, Riemann sum approximation, rectangle rule, fundamental frequency, direct (DFT), inverse (IDFT), standard form, reversal property, structure of discrete spectrum, numerical differentiation, filter operation for pulse signals)

Section 9.1 First-Order Filter Circuits and Their Combinations

AC voltage divider circuits (either RC or RL) generally operate as analog *filters*. They pass certain voltage signals but stop or cut out other signals, depending on the signal's frequency content. The analog filters studied in this section are *first-order* filters since they may be described by first-order differential equations—we discussed them in Chapter 7. The phasor/impedance method is applied to solve the AC circuit, both in analytical and in numerical form. Although over the years the value of the numerical analysis has greatly increased in engineering, the analytical method remains important if we are interested in a parametric study such as a rigorous filter analysis. The analytical method involves (multiple) conversions of complex numbers or expressions from the rectangular to polar form and vice versa. Generally, division and multiplication are better carried out in polar form, whereas addition and subtraction require a rectangular form.

9.1.1 RC Voltage Divider as an Analog Filter

The RC voltage divider circuit shown in Fig. 9.1a is perhaps the oldest and best-known version of an *analog filter*. In order to understand its operation, we must obtain a general solution to the RC circuit in Fig. 9.1a. Even though the solution has to work at any frequency f or angular frequency ω of interest, it is not difficult to find.

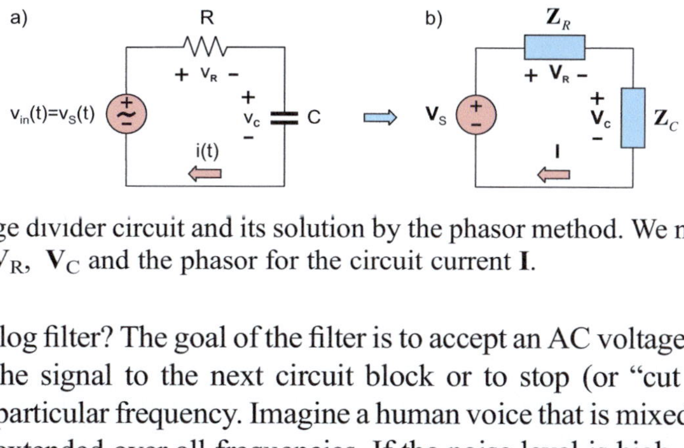

Fig. 9.1. RC voltage divider circuit and its solution by the phasor method. We note the phasors for the voltages $\mathbf{V}_S$, $\mathbf{V}_R$, $\mathbf{V}_C$ and the phasor for the circuit current **I**.

What is an analog filter? The goal of the filter is to accept an AC voltage signal at its input and either pass the signal to the next circuit block or to stop (or "cut out") the signal, depending on its particular frequency. Imagine a human voice that is mixed with white noise with a spectrum extended over all frequencies. If the noise level is high, we will probably only hear screaming at high frequencies. However, if we only pass the voltage signals with frequencies below 3 kHz, where most of the voice power is concentrated, the resulting total signal will be much clearer for listening. The analog filter is an electric circuit which, in its simplest form, is identical to the circuit in Fig. 9.1.

General Solution

Let us first convert the circuit in Fig. 9.1a to a phasor form as shown in Fig. 9.1b. We assume that $v_S(t) = V_m \cos \omega t$; therefore $\mathbf{V}_S = V_m$. Next, we solve the resulting "DC circuit" in the complex domain. The voltage division yields

$$\mathbf{V}_C = \frac{\mathbf{Z}_C}{\mathbf{Z}_R + \mathbf{Z}_C} V_m = \frac{\frac{1}{j\omega C}}{R + \frac{1}{j\omega C}} V_m = \frac{1}{1 + j\omega RC} V_m = \frac{1}{1 + j\omega \tau} V_m \ [\mathrm{V}] \tag{9.1a}$$

$$\mathbf{V}_R = \frac{\mathbf{Z}_R}{\mathbf{Z}_R + \mathbf{Z}_C} V_m = \frac{R}{R + \frac{1}{j\omega C}} V_m = \frac{j\omega RC}{1 + j\omega RC} V_m = \frac{j\omega \tau}{1 + j\omega \tau} V_m \ [\mathrm{V}] \tag{9.1b}$$

where $\tau = RC$ is exactly the *same* time constant that appears for transient circuits in Chapter 7. Converting Eq. (9.1a) and (9.1b) into polar form gives

$$\mathbf{V}_C = \frac{1}{\sqrt{1 + (\omega \tau)^2}} V_m \angle \varphi_C, \quad \varphi_C = -\tan^{-1}(\omega \tau) \tag{9.1c}$$

$$\mathbf{V}_R = \frac{\omega \tau}{\sqrt{1 + (\omega \tau)^2}} V_m \angle \varphi_R, \quad \varphi_R = \frac{\pi}{2} - \tan^{-1}(\omega \tau) \tag{9.1d}$$

After the polar form has been obtained, the real-valued voltages are found in the form

$$\begin{aligned} v_C(t) &= V_{mC} \cos(\omega t + \varphi_C) \ [\mathrm{V}], \quad V_{mC} = \frac{1}{\sqrt{1 + (\omega \tau)^2}} V_m \ [\mathrm{V}] \\ v_R(t) &= V_{mR} \cos(\omega t + \varphi_R) \ [\mathrm{V}], \quad V_{mR} = \frac{\omega \tau}{\sqrt{1 + (\omega \tau)^2}} V_m \ [\mathrm{V}] \end{aligned} \tag{9.1e}$$

The general solution of the RC circuit in Fig. 9.1a is now complete. The key observations are that the amplitudes of the resistor voltage and the capacitor voltage now become functions of frequency.

Qualitative Analysis

The circuit in Fig. 9.1 is a voltage divider. The supply voltage (or the input voltage to the filter) is divided between the capacitor and the resistor. Which voltage dominates at low frequencies and which at high frequencies? To answer those questions, we consider Eq. (9.1e). When $\omega \to 0$,

$$V_{\mathrm{mC}} = \frac{1}{\sqrt{1+(\omega\tau)^2}} V_{\mathrm{m}} \to V_{\mathrm{m}}$$
$$V_{\mathrm{mR}} = \frac{\omega\tau}{\sqrt{1+(\omega\tau)^2}} V_{\mathrm{m}} \to 0 \tag{9.2a}$$

Therefore, at *low frequencies*, the *capacitor voltage* dominates; it is approximately equal to the supply voltage. This fact is quite clear because the capacitor acts like an open circuit for DC, implying that the capacitor voltage "sees" nearly all the supply voltage. On the other hand, when $\omega \to \infty$,

$$V_{\mathrm{mC}} = \frac{1}{\sqrt{1+(\omega\tau)^2}} V_{\mathrm{m}} \to 0$$
$$V_{\mathrm{mR}} = \frac{\omega\tau}{\sqrt{1+(\omega\tau)^2}} V_{\mathrm{m}} \to V_{\mathrm{m}} \tag{9.2b}$$

Therefore, at *high frequencies*, the *resistor voltage* dominates; it approximately equals the supply voltage. This fact is also easy to understand because the capacitor acts like a *short circuit for a high-frequency AC*, $|\mathbf{Z}_{\mathrm{C}}| = 1/(\omega C) \to 0$, so that the capacitor voltage is nearly zero and all the supply voltage is "seen" by the resistor.

Filter Concept: Two-Port Network

Now, we will explore the concept of an *analog low-pass* RC filter. We consider the power supply AC voltage as the *input voltage* $v_{\mathrm{in}}(t)$ into the filter. We consider the capacitor voltage as the *output voltage* $v_{\mathrm{out}}(t)$ of the filter. According to Eq. (9.2a, 9.2b),

$v_{\mathrm{out}}(t) \approx v_{\mathrm{in}}(t)$ at low frequencies
$v_{\mathrm{out}}(t) \approx 0$ at high frequencies

The circuit so constructed passes voltage signals with lower frequencies (like the human voice) but cuts out voltage signals with higher frequencies (like noise). Figure 9.2 on the left depicts the corresponding circuit transformation. This transformation implies that the input voltage is acquired from another circuit block and the output voltage is passed to another circuit block. The qualitative filter description is complete. You should note that both circuits on the right of Fig. 9.2 are called *two-port networks*. A *port* is nothing else but a pair of voltage terminals, either related to the input voltage or to the output voltage, respectively. Can we construct an RC filter that passes high frequencies but cuts out low frequencies? In other words, can we create a so-called *high-pass* filter? The solution is simple and elegant; the output voltage is now the resistor voltage, not the capacitor voltage. Figure 9.2b shows the corresponding circuit transformation.

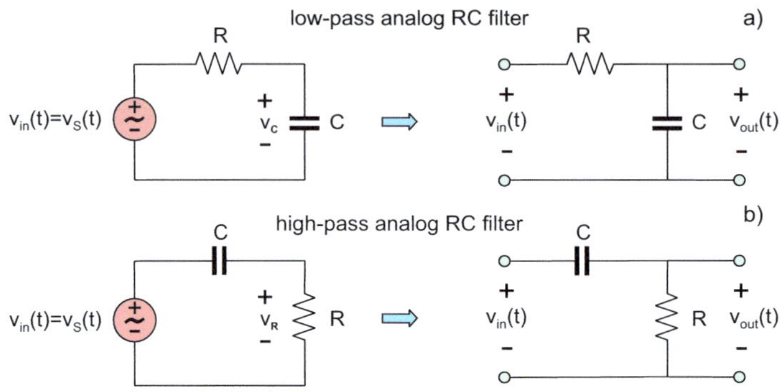

Fig. 9.2. (**a**) Transformation of the series RC circuit into the low-pass analog RC filter. (**b**) A similar transformation into the high-pass RC filter.

Example 9.1: The input voltage to the low-pass filter in Fig. 9.2a is a combination of three harmonics: $v_{in}(t) = 10\cos\omega_1 t + 10\cos\omega_2 t + 10\cos\omega_3 t$ [V], each with an amplitude of 10 V. The filter has the following parameters: $C = 530$ nF, $R = 100\ \Omega$. Determine the output voltage $v_{out}(t)$ given that:

1. $f_1 = 20$ Hz (lower frequency of the acoustic range)
2. $f_2 = 3000$ Hz (frequency below which most of the acoustic power is present)
3. $f_3 = 20,000$ Hz (higher frequency of the acoustic range)

Solution: The key is the *superposition principle*, which is based on circuit *linearity*. Using the superposition principle, we apply Eq. (9.1e) (and Eq. (9.1c) reporting the phases) for the capacitor voltage (the output voltage to the filter) to each harmonic separately and then find the sum of three partial solutions. This will be the filter output voltage, which is given by

$$\begin{aligned} v_{out}(t) = {} & 10.00\cos(\omega_1 t - 0.4°) + 7.07\cos(\omega_2 t - 45.0°) \\ & + 1.48\cos(\omega_3 t - 81.5°) \quad [\text{V}] \end{aligned} \tag{9.3}$$

The filter reduces the amplitudes of higher-frequency harmonics and simultaneously creates a certain phase shift. The high-pass filter operates in an opposite manner. The phase shift becomes positive.

Example 9.2: Solve Example 9.1 using MATLAB.

Solution: The text of the corresponding MATLAB is listed below. It is vectorized in the sense that *any* number of input harmonics may be taken into consideration:

Example 9.2 (cont.):

```
Vm      = [10 10 10];          %  input voltage amplitudes, V
f       = [20 3000 20000];     %  input voltage frequencies, Hz
omega   = 2*pi*f;              %  angular frequencies, rad/sec
R       = 100;                 %  resistance, Ohm
C       = 530e-9;              %  capacitance, F
tau     = R*C;
VmC     = 1./sqrt(1+(omega*tau).^2).*Vm      %  output voltage ampl., V
phiC    = - atan(omega*tau)*180/pi           %  output phases in deg
```

Exercise 9.1: The input voltage to a low-pass filter circuit is a combination of two harmonics, $v_{\text{in}}(t) = 2\cos\omega_1 t + 2\cos\omega_2 t$, with the amplitude of 2 V each. The filter has the following parameters: $R = 100\ \text{k}\Omega$ and $C = 1.59\ \text{nF}$. Determine the output voltage $v_{\text{out}}(t)$ to the filter given that $f_1 = 100\ \text{Hz}$ and $f_2 = 100\ \text{kHz}$.

Answer: $v_{\text{out}}(t) = 1.99\cos(\omega_1 t - 5.7°) + 0.02\cos(\omega_2 t - 89.4°)$ [V].

Application Example: Effect of a Load Connected to the Filter

The initial excitement about the simplicity of the theoretical filter model often fades quickly once we try to construct the filter circuit of Fig. 9.2a or Fig. 9.2b in the laboratory. And the circuit does not work. The major reason for this is the effect of a *load* connected to the filter. We consider the low-pass filter in Fig. 9.3.

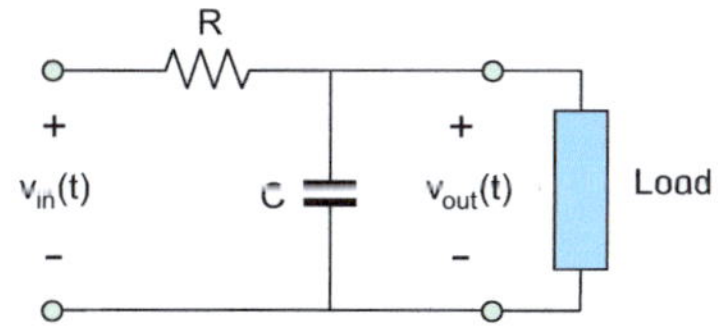

Fig. 9.3. A generic load connected to the low-pass RC filter.

To solve the circuit with the load, we need to apply the phasor method. The input voltage is now divided between the resistance R and the parallel combination of the capacitor impedance and the load resistance, R_{L}. Instead of Eq. (9.1c), we will have

$$\mathbf{V}_{\text{C}} = \frac{1}{\sqrt{(1 + R/R_{\text{L}})^2 + (\omega\tau)^2}} V_{\text{m}} \angle \varphi_{\text{C}}\ [\text{V}], \quad \varphi_{\text{C}} = -\tan^{-1}\left(\frac{\omega\tau}{1 + R/R_{\text{L}}}\right) \tag{9.4}$$

The proof of this result is suggested in Problems 9.5 and 9.6. The necessary condition for proper filter operation (both high pass or low pass) is that *the filter termination resistance* should be *much greater* than the filter's resistance R. Put in approximate mathematical terms: $R/R_L \ll 1$. The low-resistance load (e.g., a loudspeaker) would simply short out the capacitor! To avoid this effect, a buffer amplifier may have to be inserted between the load and the filter.

9.1.2 Half-Power Frequency and Amplitude Transfer Function

Low-Pass Filter

We are going to show how to construct a low-pass RC filter for a particular application. The design engineer needs to know at approximately which frequency the signal should be cut out. It is a common agreement to choose this frequency so that the amplitude of the output voltage is exactly $1/\sqrt{2} \approx 0.707$ of the input voltage amplitude V_m. In other words, the output filter power, which is proportional to the square of the output voltage, becomes exactly *half* of the input power. The corresponding frequency is called the *break frequency* or *half-power* frequency of the low-pass filter. According to Eq. (9.1e), the break frequency f_b is found using the amplitude of the output (capacitor) voltage in the following way:

$$\frac{1}{\sqrt{1+(\omega_b\tau)^2}} = \frac{1}{\sqrt{2}} \Rightarrow \omega_b\tau = 1 \Rightarrow \omega_b = \frac{1}{\tau} \Rightarrow f_b = \frac{\omega_b}{2\pi} = \frac{1}{2\pi\tau} = \frac{1}{2\pi RC} \quad [\mathrm{Hz}] \tag{9.5}$$

Expressed in terms of the break frequency, the amplitude of the output voltage or the voltage across the capacitor in Eq. (9.1e), has the form $V_m/\sqrt{1+(f/f_b)^2}$ since $\omega\tau = f/f_b$. With the input voltage amplitude to the filter being V_m, the ratio of the two amplitudes is the *amplitude transfer function* of the low-pass filter H_m. This transfer function is given by

$$H_m(f) = \frac{1}{\sqrt{1+(f/f_b)^2}} \le 1 \tag{9.6a}$$

We note that the transfer function is dimensionless (or has the units of V/V). For a given input voltage, the amplitude transfer function allows us to determine the output voltage amplitude. The behavior of Eq. (9.6a) is such that the amplitude transfer function is always less than one: the output voltage cannot exceed the input voltage.

High-Pass Filter

The break frequency, ω_b or f_b, of the high-pass filter has the meaning of reducing the voltage amplitude by a factor of $1/\sqrt{2}$ and reducing the signal power by the factor of ½. According to Eq. (9.1e) for the resistor voltage, it is found using the equality $\omega_b\tau/\sqrt{1+(\omega_b\tau)^2} = 1/\sqrt{2}$, which gives us exactly the *same* value as the break frequency for the low-pass filter; see Eq. (9.5). In other words, the definitions of the break frequency or the half-power frequency *coincide* for the first-order low-pass filter and the first-order high-pass filter, respectively. In terms of the break frequency, the amplitude of the output voltage to the high-pass filter, the voltage across the resistor in Eq. (9.1e), has the form $V_m f/f_b/\sqrt{1+(f/f_b)^2}$, whereas the input voltage amplitude to the filter is still V_m. The ratio of the two amplitudes is the *amplitude transfer function* of the high-pass filter, denoted here by the same letter H_m. This transfer function is given by

$$H_m(f) = \frac{f/f_b}{\sqrt{1+(f/f_b)^2}} \le 1 \tag{9.6b}$$

We note again that H_m cannot exceed 1. The implication is that the output voltage is always less than or equal to the input voltage; the filter cannot amplify the input.

Example 9.3: With the values of $C = 530\,\text{nF}$, $R = 100\,\Omega$, determine the break frequency of both the low-pass RC filter and the high-pass RC filter, respectively.

Solution: We utilize the definition of Eq. (9.5), $f_b = \frac{1}{2\pi RC}$ [Hz], and obtain $f_b = 3.00$ kHz in either case. This is exactly why the particular signal at 3 kHz in Eq. (9.3) of Example 9.1 (the example uses the same parameters) was reduced by a factor of 0.707 at the output of the low-pass filter. If a high-pass filter were used, the corresponding output signal would have exactly the same form but with the phase shift of +45° instead of −45°.

Exercise 9.2: The input signal to a high-pass RC filter includes a 180-Hz component. Its amplitude is to be reduced by a factor of 10. What break frequency should the filter have?

Answer: 1791 Hz.

9.1.3 Bode Plot, Decibel, and Roll-Off

The *Bode plot* displays the amplitude transfer function defined by Eq. (9.6a, 9.6b) as a function of frequency on a logarithmic scale. It was first suggested by an electrical engineer and mathematician Hendrik Wade Bode (1905–1982), Bell Labs, NJ, USA. The advantage of the logarithmic scale is the ability to simultaneously observe the (very large) function variations at small and large frequencies. Furthermore, you can more clearly see the *roll-off* of the transfer function as a straight line (*asymptote*). This is impossible to see when using the linear scale. As the x-variable, we will always choose frequency f (*and avoid the angular frequency* ω). As the y-variable, we plot the logarithmic function

$$H_{\mathrm{m}}(f)_{\mathrm{dB}} = 20\log_{10} H_{\mathrm{m}}(f) \quad [\mathrm{dB}] \tag{9.7}$$

The *dimensionless* units for the amplitude transfer function in Eq. (9.7) are *decibels* or dB. Figure 9.4 shows the Bode plot for transfer function Eq. (9.6a) with $f_{\mathrm{b}} = 100$ Hz. The selected values of the transfer function are given in Table 9.1 and where the last row is given in dB. The particular values of the resistance and capacitance are yet to be found; only their combination $\tau = RC = 1/(2\pi f_{\mathrm{b}}) = 1.6$ ms is really important for the Bode plot. Despite the apparent simplicity of this operation, the Bode plot for an RC filter is a very likely question on the entrance exam for an industrial position in electrical engineering.

Table 9.1. Values of amplitude transfer function for a low-pass filter with $f_{\mathrm{b}} = 100$ Hz.

f, Hz	1	10	**100**	1000	10^4	10^5	10^6	10^7
$H_{\mathrm{m}}(f)$	1.000	0.995	**0.707**	1.0×10^{-1}	1.0×10^{-2}	1.0×10^{-3}	1.0×10^{-4}	1.0×10^{-5}
$20\log_{10} H_{\mathrm{m}}(f)$	-0.0004	-0.0432	$\mathbf{-3.0103}$	-20.043	-40.000	-60.000	-80.000	-100.00

Historical: The decibel is named in honor of Alexander Graham Bell (1847–1922), a Scottish scientist and inventor who later became a professor at Boston University, MA. Bell invented the first practical telephone at the age of 28 (US Patent 174,465) and very quickly became a millionaire. His father-in-law Gardiner Greene Hubbard founded the Bell Telephone Company in 1878, which subsequently transformed into American Telephone & Telegraph Company (AT&T).

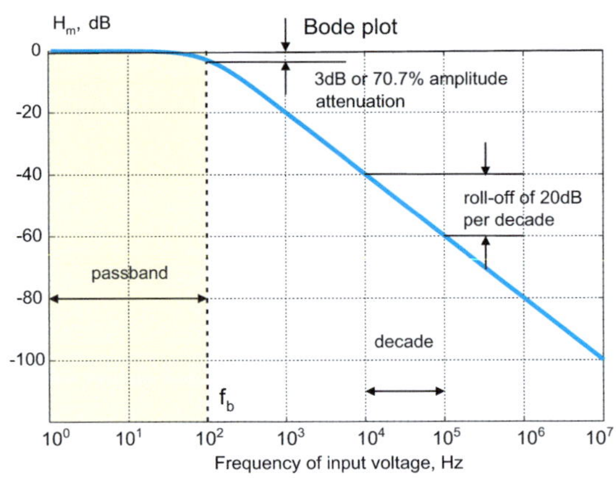

Fig. 9.4. Construction of a Bode plot for the amplitude transfer function of a low-pass RC filter with break frequency of $f_{\mathrm{b}} = 100$ Hz.

A legitimate question to ask is what is the meaning of the factor 20 in Eq. (9.7)? The answer is based on the equality $20\log_{10} H_{\mathrm{m}}(f) = 10\log_{10} H_{\mathrm{m}}^2(f)$ where $H_{\mathrm{m}}^2(f)$ is not the amplitude transfer function but rather attempts to represent *power*, which is proportional to voltage squared for a resistor. Therefore, Eq. (9.7) in fact attempts to plot the *power transfer function* even though the capacitor in Fig. 9.2a does not consume any power in the average sense, see Chapter 11. Also note that, when $f = f_b$, the transfer function in Table 9.1 is approximately -3 dB. Therefore, the break frequency is also called the *3-dB frequency* for obvious reasons. Another name, the *corner frequency*, will be explained shortly.

The interval on the Bode plot for which the frequencies differ by a factor of 10 is called a *decade*. Every division on the x-axis in Fig. 9.4 is *one decade*. The transfer function for *any* first-order low-pass filter decreases by 20 dB per decade or has the *20-dB-per-decade roll-off* as seen in Fig. 9.4. This not only occurs away from the break frequency, i.e., when $f \gg f_b$, but it is also approximately valid in the interval from f_b to $10f_b$; see Table 9.1. Note that an interval of frequencies is called the *frequency band*. The roll-off of 20 dB per decade (or equivalently, the slope of -20 dB per decade) means that the output amplitude of the filter decreases by a factor of 10 per decade (see Table 9.1), whereas the output power decreases by the factor of 100. Figure 9.4 shows a frequency band from 0 to f_b, which is the *passband* of the low-pass filter. The passband is the range of frequencies that are passed through a filter without being (significantly) attenuated. The opposite of the passband is the *stopband*. The required attenuation within the stopband may be specified between 20 and 120 dB as compared to the value of 0 dB, which means no attenuation. Besides the decade, the relative frequency interval of one *octave* is sometimes used. In this interval, the frequencies differ by the factor of 2, not 10. For example, a TV antenna that has the bandwidth of one octave (400–800 MHz) may be used to receive most of the (digital) commercial TV channels in the USA. It can be shown that the RC filter has a 6-dB-per-octave roll-off, away from the break frequency.

Historical: The career of Hendrik Wade Bode (1905–1982), a pioneer of modern control theory and electronic telecommunications, gives us an example of how important it is to have a comprehensive education in calculus and a solid background in electrical engineering. A graduate of the Ohio State University (BS in mathematics at the age of 19 and then MS in mathematics two years later), Hendrik Bode started his job at Bell Labs as a designer of electronic filters and invented the asymptotic plots we now call them Bode plots in 1938. These plots have proven to be extremely useful in feedback control theory. Today, any electrical engineer who works with amplifiers and their frequency responses is relying on Bode plots. Some consider Bode a pioneer of *robotics* as well, based on his invention of robotic antiaircraft artillery during WWII.

Exercise 9.3: The following values of the amplitude transfer function are given: $H_m(f) = 0.707,\ 0.0707,$ and 0.00707. Find the corresponding values of $H_m(f)_{dB}$.

Answer: -3.01 dB, -23.01 dB, and -43.01 dB.

Example 9.4: Design a medium-frequency-range RC low-pass filter (LPF) that has a break frequency of 1 kHz. The filter load has the resistance of $R = 100\ \text{k}\Omega$. Create the amplitude Bode plot in the range from 10 Hz to 100 kHz. Label the filter passband. Repeat the same task for the high-pass filter (HPF).

Solution: The condition $f_b = 1/(2\pi RC)$ yields $C = 1/(2\pi R f_b) = 1.6\,\text{nF}$. The Bode plot may be generated by finding transfer function values for (at least) every decade and filling out a table similar to Table 9.1. The result is shown in Fig. 9.5a. The passband is the frequency band from 0 to f_b. For the high-pass filter, we repeat the same steps but replace the transfer function given by Eq. (9.6a) by the transfer function given by Eq. (9.6b). The result is shown in Fig. 9.5b. The passband extends from f_b to infinity and is only limited by the upper frequency of the Bode plot. Note that the Bode plot for the high-pass filter has the same form, but it is *mirror reflected* about the break frequency. This is another advantage of the logarithmic scale.

Figure 9.5 indicates that the amplitude response of both the low-pass filter and the high-pass filter follows two straight lines, which are known as *high-frequency and low-frequency asymptotes*. The corner between them is the break frequency, also called the *corner frequency*. Note that, for the high-pass filter, the meaning of high-frequency and low-frequency asymptotes is *interchanged* in Fig. 9.5b.

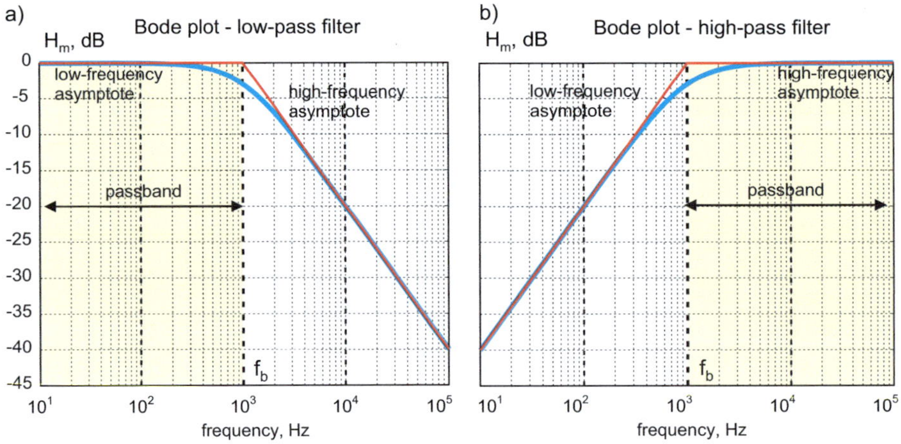

Fig. 9.5. (**a**) Bode plot for the amplitude transfer function of the low-pass RC filter with the break frequency $f_b = 1$ kHz. (**b**) The same Bode plot but for the high-pass RC filter. Note high-frequency and low-frequency asymptotes.

9.1.4 Phase Transfer Function and Its Bode Plot

According to Eq. (9.1e), it is not only the amplitude but also the phase of the input signal that undergoes a transformation when the signal is passed through the filter. The phase transformation is important since different frequencies (or harmonics) of the input signal may have a certain phase relation that is distorted by the filter. The *phase transfer function* is given by the phase variation of the filter's output voltage, which is either the capacitor voltage for the low-pass filter or the resistor voltage for the high-pass filter. From Eq. (9.1c) for the low-pass filter, the phase transfer function has the form

$$\varphi_{\mathrm{H}}(f) = -\tan^{-1}(\omega\tau) = -\tan^{-1}\left(\frac{f}{f_{\mathrm{b}}}\right) \quad \text{low-pass RC filter} \tag{9.8a}$$

From Eq. (9.1d) for the high-pass filter, the phase transfer function has the form

$$\varphi_{\mathrm{H}}(f) = \frac{\pi}{2} - \tan^{-1}(\omega\tau) = \frac{\pi}{2} - \tan^{-1}\left(\frac{f}{f_{\mathrm{b}}}\right) \quad \text{high-pass RC filter} \tag{9.8b}$$

where the break frequency is given by Eq. (9.5).

Example 9.5: Generate the phase Bode plots for the low-pass filter and the high-pass filter, respectively, with the same break frequency $f_{\mathrm{b}} = 1$ kHz. The frequency band is from 10 Hz to 100 kHz.

Solution: The phase Bode plots in Fig. 9.6 may be generated by calculating the phase transfer function according to Eq. (9.8a, 9.8b) for (at least) every decade. The result is shown in Fig. 9.6. You can see that the Bode plots only differ by a phase shift of 90°. Alternatively, a MATLAB script may be used:

```
f       = logspace(1, 5);          % frequency vector, Hz (from 10^1 to 10^5 Hz)
fb      = 1000;                    % break frequency, Hz
phiH1   = -atan(f/fb);             % low-pass filter phase transfer function
phiH2   = pi/2-atan(f/fb);         % high-pass filter phase transfer function
semilogx(f, phiH1/pi*180); grid on;
title('Bode plot'); ylabel('phase transfer function, deg'); xlabel('f, Hz')
```

9.1.5 Complex Transfer Function: Cascading Filter Circuits

The *complex transfer function* of the filter, $\mathbf{H}(f)$, is often called the *frequency response* of the filter. It describes not only the amplitude transformation but also the phase transformation. The transfer function now becomes a *complex expression*. It is equal to the ratio of two phasors; specifically, it denotes the ratio of the output phasor voltage to the input phasor voltage. The low-pass filter has the form of Fig. 9.1 with the input voltage equal to the supply voltage and the output voltage equal to the capacitor voltage. Its complex transfer function is given by Eq. (9.1c) divided by V_{m}. The high-pass filter also has the form of Fig. 9.1 with the input voltage equal to the supply voltage and the output voltage equal to the resistor voltage. Its complex transfer function is given by Eq. (9.1d) divided by V_{m}. Thus, we obtain

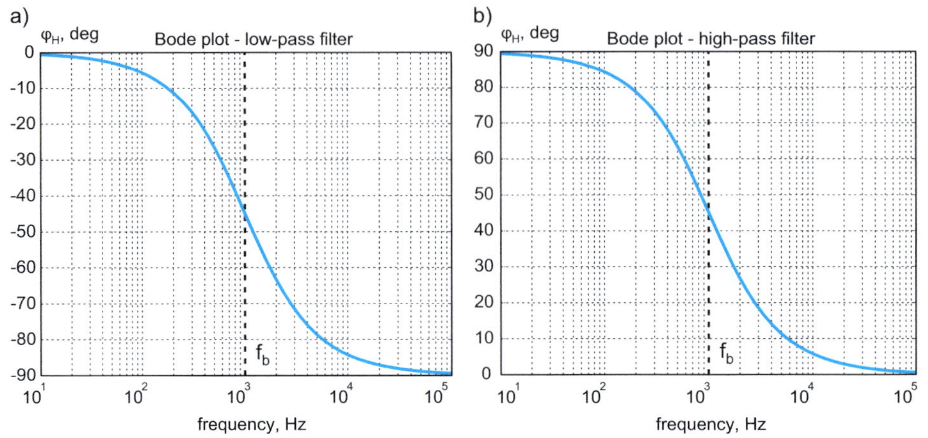

Fig. 9.6. Comparison of the phase Bode plots for (**a**) the low-pass-filter and (**b**) for the high-pass filter with the same break frequency $f_b = 1$ kHz. Both plots are identical to within a phase shift.

$$\mathbf{H}(f) \equiv H_m(f)\angle\varphi_H = \begin{cases} \dfrac{1}{1+j(f/f_b)} = \dfrac{1}{\sqrt{1+(f/f_b)^2}}\angle - \tan^{-1}\left(\dfrac{f}{f_b}\right); & \text{low-pass RC filter} \\ \dfrac{(f/f_b)}{1+j(f/f_b)} = \dfrac{f/f_b}{\sqrt{1+(f/f_b)^2}}\angle\dfrac{\pi}{2} - \tan^{-1}\left(\dfrac{f}{f_b}\right); & \text{high-pass RC filter} \end{cases} \tag{9.9a}$$

This is consistent with Eqs. (9.6a, b) and (9.8a, b), respectively. Given the phasor of input voltage $\mathbf{V}_{in}$, the phasor of the output voltage is simply expressed by

$$\mathbf{V}_{out} = \mathbf{H}(f)\mathbf{V}_{in} \tag{9.9b}$$

Equation (9.9b), which is valid for any linear electronic filter and other linear systems, fully describes the filter operation and has great practical value.

Example 9.6: For a low-pass RC filter with the values $C = 530\,\text{nF}$, $R = 100\,\Omega$, determine the output voltage in time domain when the input voltage is given by $v_{in}(t) = 1\cos(\omega t + 30°)$ [V] where $\omega = 2\pi \times 3000$ rad/s.

Solution: The break frequency of the low-pass filter is $f_b = 3.00$ kHz, which coincides with the signal frequency in this particular case. According to first Eq. (9.9a) and Fig. 9.6a, at that frequency, $\mathbf{H}(f) = \frac{1}{\sqrt{2}}\angle - 45°$; therefore, the output voltage has the form $\mathbf{V}_{out} = \frac{1}{\sqrt{2}}\angle - 15°$ or $v_{out}(t) = 0.71\cos(\omega t - 15°)$ [V]. The same analysis may be applied at any frequency and phase of the input harmonic voltage signal.

Another advantage of the complex transfer function lies in the fact that the *series* or *cascade combination* of any number of filters (or two-port networks) shown in Fig. 9.7 has a transfer function that is simply the product of the corresponding transfer functions:

$$\mathbf{H}(f) = \mathbf{H}_1(f)\mathbf{H}_2(f) \Rightarrow H_m(f) = H_{m1}(f)H_{m2}(f) \tag{9.10}$$

In this manner, a more advanced filter may be constructed from the individual filter blocks. To prove Eq. (9.10), we state that the phasor for the intermediate output voltage $v_{out1}(t)$ in Fig. 9.7 is given by $\mathbf{V}_{out1} = \mathbf{H}_1(f)\mathbf{V}_{in}$. Hence, the phasor for the output voltage $v_{out}(t)$ in Fig. 9.7 becomes $\mathbf{V}_{out} = \mathbf{H}_1(f)\mathbf{V}_{out1} = \mathbf{H}_1(f)\mathbf{H}_2(f)\mathbf{V}_{in}$ which is equivalent to Eq. (9.10). Due to the logarithmic scale of the Bode plot, the product in Eq. (9.10) is replaced by the sum of two contributions when the decibel scale is used:

$$H_m(f)_{dB} = H_{m1}(f)_{dB} + H_{m1}(f)_{dB} \tag{9.11}$$

Thus, we simply add up two magnitude transfer functions in dB and obtain the resulting magnitude transfer function also in dB.

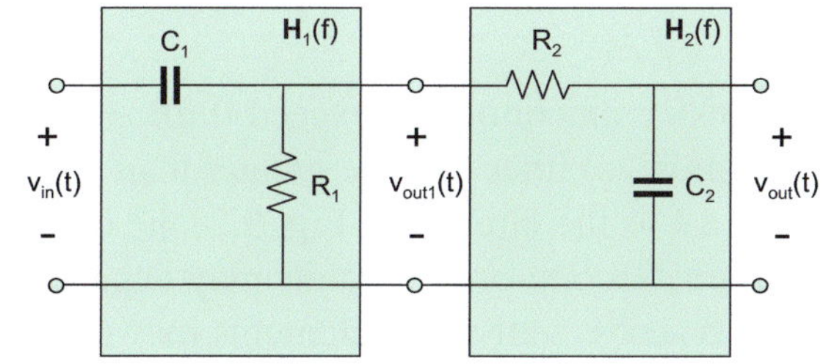

Fig. 9.7. Cascading a high-pass and a low-pass filter into a more complex filter structure.

Application Example: Effect of Next-Stage Filter Load

Equation (9.10) requires great care. For example, the equivalent impedance seen by the leftmost high-pass filter stage in Fig. 9.7 should be much greater than R_1; otherwise this stage will not operate as expected, and Eq. (9.10) will be inaccurate. In other words, a following filter stage should not appreciably *load* the previous one.

Example 9.7: For the combined circuit in Fig. 9.7, create the Bode plot for the transfer function of the cascade connection in the frequency band from 1 Hz to 1 MHz. You are given $R_1 = 159.1\ \Omega$, $C_1 = 10\ \mu F$ and $R_2 = 159.1\ \Omega$, $C_2 = 0.1\ \mu F$.

Solution: The break frequency of the high-pass filter is calculated as 100.0 Hz, and the break frequency of the low-pass filter is found to be 10.0 kHz. The combined Bode plot is generated using Eqs. (9.6a, 9.6b) and (9.10). Alternatively, the transfer functions in dB, specified by Eq. (9.7), may be added. The result is a *band-pass* filter as shown in Fig. 9.8 by the solid curve. This result is expected to be accurate only if $|R_2 + Z_{C_2}| \gg R_1$. Though valid at low frequencies below 1 kHz, this inequality is violated above 1 kHz. The exact transfer function is obtained by solving the complete AC circuit in Fig. 9.7 with the open-circuited capacitor C_2. It is plotted in Fig. 9.8 by a dashed curve. There is clearly a significant deviation from the solution given by Eq. (9.10) at higher frequencies above 1 kHz. To avoid the loading effect seen in Fig. 9.8, a buffer amplifier may be inserted between the filter stages shown in Fig. 9.7.

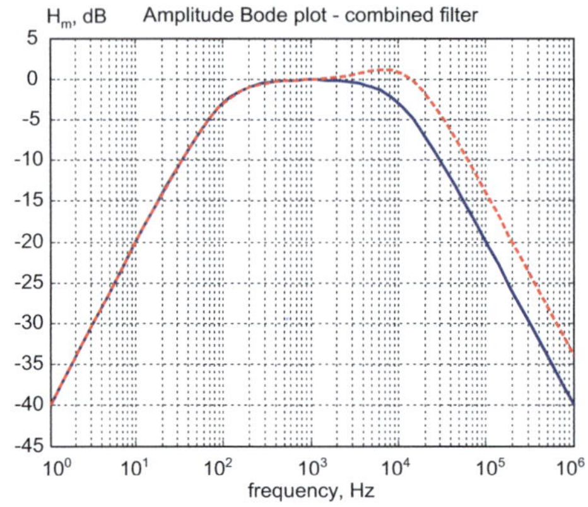

Fig. 9.8. *Solid curve*: Eq. (9.10) for the cascaded filters. *Dashed curve*: the exact solution with the open-circuited capacitor C_2.

In Fig. 9.8, the exact transfer function may exceed 0 dB. In other words, the voltage gain of the combined (still passive) filter may be greater than one. How is this possible? The answer is that, in contrast to the circuits in Fig. 9.2, the circuit in Fig. 9.7 is in fact already a *second-order* circuit. Second-order circuits may experience a resonance behavior where the circuit voltages across individual elements may (very considerably) exceed the original supply voltage. This effect, called *voltage multiplication*, is of great practical importance and will be considered in detail in Chapter 10 devoted to second-order AC circuits. Note that that the true *power gain* of a passive filter of any order and any topology is always less than one (less than 0 dB). Only electronic amplifiers may have a positive, and often high, power gain; this is discussed in the next section.

9.1.6 RL Filter Circuits

The RL circuits are used for the same filtering purposes as the RC circuits. Figure 9.9 depicts the concept. It may be demonstrated that the corresponding circuit theory and Eqs. (9.6a, 9.6b) for the transfer functions become equivalent to first-order RC filter circuits under the following conditions:

1. The time constant $\tau = RC$ is replaced by the time constant $\tau = L/R$, similar to the corresponding operation for the first-order transient circuits. The break frequency $f_b = 1/(2\pi\tau)$ remains the same.
2. The role of the capacitor and inductor are interchanged. For example, the RL circuit in Fig. 9.9a is a first-order high-pass filter because the inductor voltage, which is the output filter voltage, is exactly zero for a DC signal. However, it becomes a first-order low-pass filter if the inductor and resistor positions are interchanged, as shown in Fig. 9.9b.

3. Similarly, the RL circuit in Fig. 9.9b is a first-order low-pass filter simply because the inductor becomes a short circuit at DC and the DC signal will pass through. However, it becomes a first-order high-pass filter if the inductor and resistor positions are interchanged, as shown in Fig. 9.9a.

Furthermore, the filter specifications might require large inductance values, which lead to physically large inductor sizes.

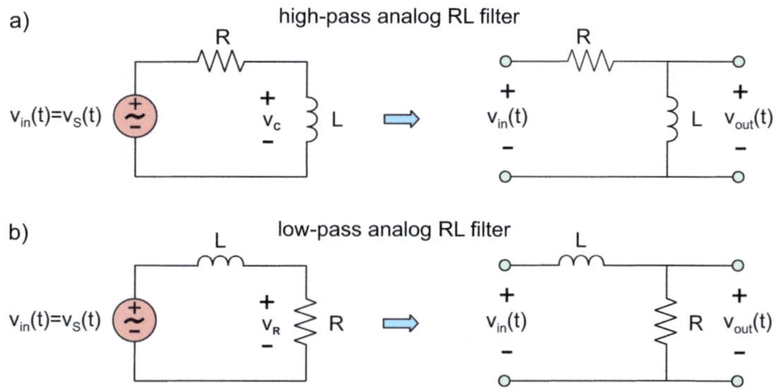

Fig. 9.9. (**a**) Transformation of a series RL circuit into a high-pass analog RL filter. (**b**) Similar transformation into the low-pass RL filter.

Example 9.8: For the two filter circuits in Fig. 9.9, create the amplitude Bode plots in the frequency band from 10 Hz to 100 kHz. You are given $R = 31.4\ \Omega,\ \ L = 5\,\mathrm{mH}$.

Solution: The break frequencies of the high-pass filter and the low-pass filter in Fig. 9.9 coincide. In either case, we obtain $f_b = 1/(2\pi\tau),\ \tau = L/R = 1.59\times\ 10^{-4}$ s. Thus, $f_b = 1.00$ kHz. The Bode plots may be generated by finding transfer function values based on Eqs. (9.6) for (at least) every decade and filling out a table similar to Table 9.1. The result is shown in Fig. 9.10 along with high- and low-frequency asymptotes. We again observe the 20-dB roll-off per decade. The Bode plots given in Fig. 9.10 coincide with the Bode plots for RC filters having the same break frequency, see Fig. 9.5. However, given an identical component topology, the filter function is interchanged.

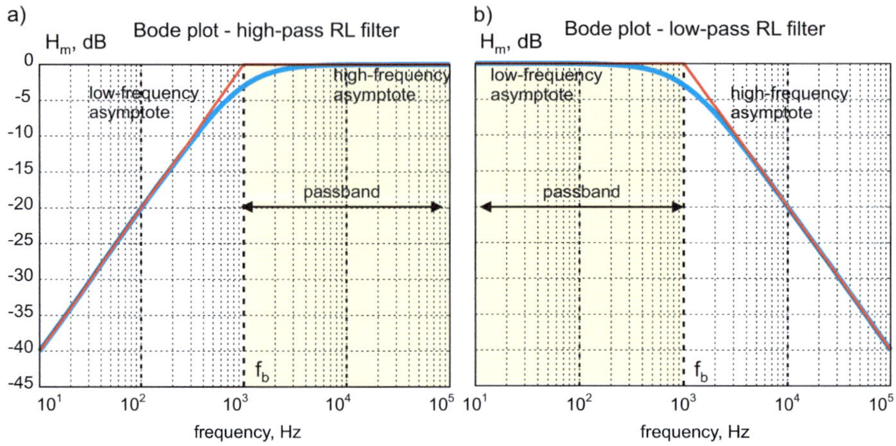

Fig. 9.10. (**a**) Bode plot for the amplitude transfer function of the high-pass RL filter with break frequency $f_b = 1\,\text{kHz}$. (**b**) The same Bode plot but for the low-pass RL filter.

Exercise 9.4: An RL filter circuit in Fig. 9.9a has $R = 100\,\Omega$ and $L = 1\,\text{mH}$. Establish the capacitance value of an equivalent RC filter, given that the resistances are the same in both cases.

Answer: 100 nF.

Section 9.2 Bandwidth of an Operational Amplifier

The operational amplifier circuits introduced earlier are implicitly assumed to operate equally well for any frequency of the input signal. In reality this is not true. An operational amplifier may operate only over a certain frequency band, and the associated *frequency bandwidth* is perhaps the most critical device parameter. Frequently we do not realize how severe this limitation can be and how difficult it is to build a high-frequency or radio-frequency amplifier. As an example, we should point out that none of the common amplifier ICs studied in introductory ECE classes can be used as a front-end amplifier for an AM radio receiver (520–1610 kHz), even if the noise levels were low. Indeed, high-frequency amplifiers with larger frequency bandwidths exist. A case in point is the accessible LM7171 chip. Key to understanding the amplifier frequency behavior is the theory of the first-order RC filters developed in the previous section.

9.2.1 Bode Plot of the Open-Loop Amplifier Gain

Open-Loop Amplifier Gain and Its Relation to the Previous Results

The (amplitude) *frequency response* of an operational amplifier is simply a plot of its gain magnitude versus frequency of the input AC voltage signal. This response is usually a Bode plot. The problem is that the gain of the amplifier (both open loop and closed loop) generally *decreases* with increasing frequency. We consider the *open-loop gain* (gain without the feedback loop) first. The open-loop gain magnitude will be denoted here by $A_{\mathrm{OL}} = A_{\mathrm{OL}}(f)$. Note that in Chapter 5 we have already introduced the *open-circuit gain*, A, of an amplifier at DC without the feedback loop. What is the relation between A_{OL} and A introduced previously? The answer is given by the equality $A = A_{\mathrm{OL}}(f = 0)$ as long as the amplifier is open circuited.

Open-Loop Gain Behavior

The open-loop gain decreases with increasing frequency of the input signal. Figure 9.11 shows the frequency response of an open-loop amplifier on a log-log scale. You may recall that the log-log scale used in this figure is simply the Bode plot introduced in the previous section. This figure is typical for the LM741 amplifier IC and similar general-purpose devices. Comparing the Bode plot in Fig. 9.11 with the Bode plot of the RC filter in Fig. 9.4 of the previous section, we discover that the amplifier's gain as a function of frequency is virtually identical to the transfer function of the RC filter for the same break frequency of 10 Hz, as seen in Fig. 9.11! In both cases, we have a roll-off of 20 dB per decade. Obviously, the scale is different. Why is this so? This occurs because the amplifier ICs are usually *internally compensated*, which means incorporating a simple RC filter network (in practice, it may be a single capacitor *C*) into the IC chip itself. This process is called *internal compensation* of the amplifier. The goal of such a modification is to ensure that the amplifier circuit will be stable. Stability refers to the amplifier's immunity to spontaneous oscillations. These undesired oscillations occur when the input frequency excites internal resonances, similar to a mechanical mass-spring system, that might continue ad infinitum.

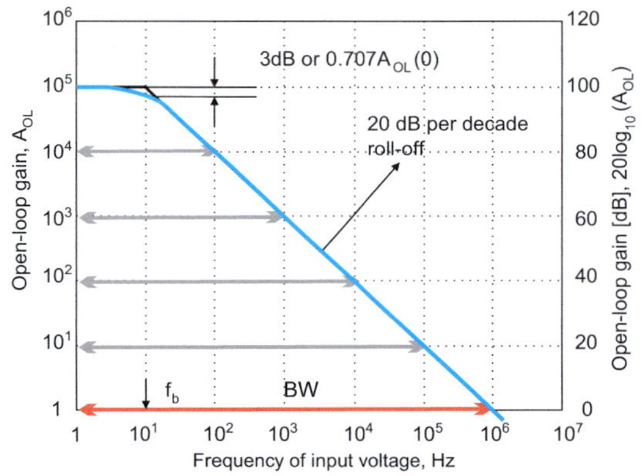

Fig. 9.11. Bode plot of the open-loop gain magnitude for the LM741-type amplifier IC. Note the logarithmic scale on the *left* and the corresponding scale in dB on the *right*. The frequency bandwidth given by the break frequency f_b is only 10 Hz.

9.2.2 Unity-Gain Bandwidth Versus Gain-Bandwidth Product

The amplifier gain in Fig. 9.11 decreases by a factor of 0.1 (or 20 dB) gain roll-off per frequency decade. The decay already starts at a relatively low break frequency of 10 Hz where the DC open-loop gain drops by the factor of 0.707 or $1/\sqrt{2}$. The corresponding value in dB is $20\log_{10}1/\sqrt{2} = -3$ dB. The gain continues to decrease further and reaches unity at the frequency of 1 MHz. This frequency is equal to the *unity-gain bandwidth* (BW) of the amplifier, i.e., for the amplifier IC depicted in Fig. 9.11:

$$\mathrm{BW} = 1\ \mathrm{MHz} \tag{9.12}$$

A remarkable observation from Fig. 9.11 is that the *gain-bandwidth product* (sometimes denoted by GBW or GB in datasheets) remains constant over the band for *every* particular gain value. The gain-bandwidth product is equal to the length of every single arrow (in Hz) in Fig. 9.11 times the corresponding gain value (dimensionless), that is,

$$\begin{aligned} f &= 10^2\mathrm{Hz} \Rightarrow \mathrm{GBW} = 10^2 \times 10^4 = 10^6\mathrm{Hz} = \mathrm{BW}, \\ f &= 10^3\mathrm{Hz} \Rightarrow \mathrm{GBW} = 10^3 \times 10^3 = 10^6\mathrm{Hz} = \mathrm{BW}, \\ f &= 10^4\mathrm{Hz} \Rightarrow \mathrm{GBW} = 10^4 \times 10^2 = 10^6\mathrm{Hz} = \mathrm{BW}, \end{aligned} \tag{9.13}$$

etc. Thus, the gain-bandwidth product is *exactly* equal to the unity-gain bandwidth BW; it is frequently specified in the manufacturer datasheet. In what follows, we will use the unity-gain bandwidth as the major parameter of interest. Note that instead of, or along with, the unity-gain bandwidth, the *rise time* of an amplifier may be specified in the datasheet. Approximately, we can state that $\mathrm{BW} = 0.35/\text{rise time}$ [Hz].

9.2.3 Model of the Open-Loop AC Gain

The open-loop gain dependence on the frequency has the form of a low-pass filter. We could therefore describe the open-loop gain in a *complex form* that is identical to the complex transfer function of the low-pass filter given, for example, by Eqs. (9.9a, b) of the previous section. The *open-loop AC gain* in complex phasor form states

$$\mathbf{A}_{\mathrm{OL}}(f) = \frac{A_{\mathrm{OL}}(0)}{1 + j(f/f_{\mathrm{b}})}, \quad A_{\mathrm{OL}}(0) \text{ is the open-loop DC gain} \tag{9.14}$$

For example, $A_{\mathrm{OL}}(0) = 10^5$ in Fig. 9.11. According to Eq. (9.14), the open-loop AC gain is a complex-valued frequency-dependent transfer function. This circumstance is reflected in a phase difference between the output and input voltages. To be consistent with Fig. 9.11 and with the previous DC amplifier analysis, the magnitude of the complex gain function in Eq. (9.14) is denoted by the *same* symbol, A_{OL}, i.e.,

$$|\mathbf{A}_{\mathrm{OL}}| = A_{\mathrm{OL}}(f) = \frac{A_{\mathrm{OL}}(0)}{\sqrt{1 + (f/f_{\mathrm{b}})^2}} \tag{9.15}$$

The Bode plot applied to Eq. (9.15) will give us exactly the dependence shown in Fig. 9.11. According to Eq. (9.15), the unity-gain bandwidth satisfies the equality

$$1 = \frac{A_{\mathrm{OL}}(0)}{\sqrt{1 + (\mathrm{BW}/f_{\mathrm{b}})^2}} \tag{9.16}$$

Since $\mathrm{BW}/f_{\mathrm{b}} \gg 1$, one has $\sqrt{1 + (\mathrm{BW}/f_{\mathrm{b}})^2} \approx \mathrm{BW}/f_{\mathrm{b}}$ with a high degree of accuracy. Therefore, according to Eq. (9.16),

$$\mathrm{BW} = A_{\mathrm{OL}}(0) f_{\mathrm{b}} \tag{9.17}$$

Looking at Fig. 9.11, we observe a very significant decrease of the open-loop gain, even in the audio frequency range. For example, the open-loop gain decreases by a factor of 1000 in the audio range from 10 Hz to 10 kHz. Does it mean that the LM741 or any general-purpose amplifier cannot be used in this range? The general answer is that the operational amplifier is mostly used with a negative feedback loop. When the *closed-loop DC gain* is not very high (say 10), the corresponding *closed-loop AC gain* appears to be nearly constant over a much wider bandwidth (say up to 100 kHz). This critical result will be proved mathematically shortly.

Example 9.9: The internally compensated LM148-series amplifiers (LM148/248/348) have a unity-gain bandwidth BW of 1 MHz. The typical large-signal voltage gain at room temperature reported in the datasheet is 160 V/mV.

A. Find the open-loop DC gain in dB and the open-loop break frequency f_b.
B. Find the open-loop gain at 100 Hz, 1 kHz, and 10 kHz.

Solution: The open-loop DC gain is $A_{OL}(0) = 160,000$ or $20\log_{10}(160,000) = 104$ dB. The break frequency may be found from Eq. (9.17):

$$f_b = \frac{BW}{160,000} = 6.25\ \text{Hz} \qquad (9.18)$$

According to Eq. (9.15), the open-loop gain at 100 Hz, 1 kHz, and 10 kHz becomes 10^4, 10^3, and 100, which corresponds to 80 dB, 60 dB, and 40 dB.

Exercise 9.5: For an internally compensated amplifier IC, the open-loop DC gain is 120 dB. The break frequency is 100 Hz. Determine the unity-gain bandwidth.

Answer: BW = 100 MHz.

9.2.4 Model of the Closed-Loop AC Gain

Consider a negative feedback amplifier in an inverting configuration, as shown in Fig. 9.12. Since the open-loop gain significantly decreases with frequency, we can no longer apply the second summing-point constraint (the differential input voltage is zero), which was justified based on the condition of the very high (ideally infinite) open-loop gain. However, the first summing-point constraint of no current into the amplifier is still valid. Therefore, a direct theoretical derivation of the closed-loop gain can be performed.

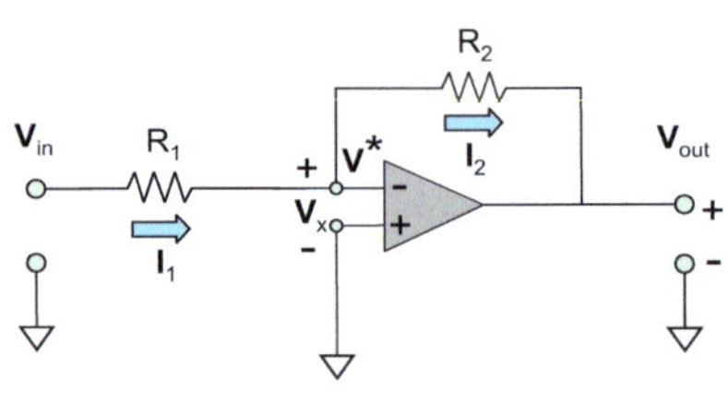

Fig. 9.12. Circuit configuration for deriving the frequency-dependent closed-loop gain.

We use the complex open-loop gain given by Eq. (9.14) and employ phasor voltages. Looking at Fig. 9.12, we conclude that $\mathbf{A}_{OL}(0\text{V} - \mathbf{V}^*) = \mathbf{V}_{out}$, based on the amplifier definition. This definition is valid for either real (time-dependent) voltages or complex phasors. By KCL at the node associated with **V***, we can develop

$$\begin{aligned}
&\frac{\mathbf{V}_{\text{in}}-\mathbf{V}^*}{R_1}=\frac{\mathbf{V}^*-\mathbf{V}_{\text{out}}}{R_2}\Rightarrow\frac{\mathbf{V}_{\text{in}}+\mathbf{V}_{\text{out}}/\mathbf{A}_{\text{OL}}}{R_1}=\frac{-\mathbf{V}_{\text{out}}/\mathbf{A}_{\text{OL}}-\mathbf{V}_{\text{out}}}{R_2}\Rightarrow\\
&\frac{\mathbf{V}_{\text{in}}}{R_1}=\left(-\frac{1}{\mathbf{A}_{\text{OL}}R_1}-\frac{1}{\mathbf{A}_{\text{OL}}R_2}-\frac{1}{R_2}\right)\mathbf{V}_{\text{out}}
\end{aligned}\tag{9.19}$$

It follows from Eq. (9.19) that the output phasor voltage to the amplifier and the closed-loop amplifier phasor gain $\mathbf{A}_{\text{CL}}$ become

$$\mathbf{V}_{\text{out}}=-\frac{R_2}{R_1}\frac{\mathbf{V}_{\text{in}}}{1+\frac{1}{\mathbf{A}_{\text{OL}}}\left(1+\frac{R_2}{R_1}\right)}\Rightarrow\mathbf{A}_{\text{CL}}\equiv\frac{\mathbf{V}_{\text{out}}}{\mathbf{V}_{\text{in}}}=-\frac{R_2}{R_1}\frac{1}{1+\frac{1}{\mathbf{A}_{\text{OL}}}\left(1+\frac{R_2}{R_1}\right)}\tag{9.20}$$

Next, we substitute Eq. (9.15) into Eq. (9.20) and rearrange terms to obtain the form

$$\mathbf{A}_{\text{CL}}(f)=-\frac{R_2}{R_1}\frac{1}{\left[1+\frac{1}{A_{\text{OL}}(0)}\left(1+\frac{R_2}{R_1}\right)\right]+\frac{1}{A_{\text{OL}}(0)}\left(1+\frac{R_2}{R_1}\right)j\frac{f}{f_{\text{b}}}}\tag{9.21}$$

The first term in the denominator on the right-hand side of Eq. (9.20) is one with a high degree of accuracy since $A_{\text{OL}}(0)\approx 10^5-10^8$. This approximation is valid for any realistic resistor values. Therefore, we again arrive at the first-order low-pass filter response:

$$\begin{aligned}
&\mathbf{A}_{\text{CL}}(f)=\frac{A_{\text{CL}}(0)}{1+j\left(f/f_{\text{b}}^{\text{closed loop}}\right)},\\
&A_{\text{CL}}(0)=-\frac{R_2}{R_1},\quad f_{\text{b}}^{\text{closed loop}}=\frac{A_{\text{OL}}(0)f_{\text{b}}}{1+R_2/R_1}=\frac{\text{BW}}{1+R_2/R_1}
\end{aligned}\tag{9.22}$$

but with a very different break frequency $f_{\text{b}}^{\text{closed loop}}$. A similar treatment holds for the *non-inverting amplifier configuration*. The result is *identical* to Eq. (9.22); however, the closed-loop DC gain $A_{\text{CL}}(0)$ is now given by

$$A_{\text{CL}}(0)=1+\frac{R_2}{R_1}\tag{9.23}$$

9.2.5 Application Example: Finding Bandwidth of an Amplifier Circuit

The relation reported in Eq. (9.22) is perhaps the single most important result with regard to the AC behavior of operational amplifiers. It reveals that the *closed-loop AC gain* has conceptually the *same* RC filter response as the open-loop gain; see Eq. (9.15). However, the corresponding break frequency $f_b^{\text{closed loop}}$ is *much larger*, namely, by a factor of $A_{OL}(0)/(1+R_2/R_1)$. This implies that the frequency response remains *flat* up to a very high frequency. The *amplifier bandwidth* in the closed-loop configuration *coincides* with the break frequency $f_b^{\text{closed loop}}$ determined by Eq. (9.22). Therefore, the bandwidth is directly proportional to the unity-gain bandwidth BW reported in the datasheet and inversely proportional to the factor $1+R_2/R_1$, which is straightforwardly calculated using the known values of the feedback resistances.

Example 9.10: An amplifier with the open-loop gain of Fig. 9.11 ($A_{OL}(0)=10^5$, $f_b=10\,\text{Hz}$) is used in the closed-loop inverting configuration with $R_2/R_1=9$ (the DC inverting gain is -9). Create the Bode plot for the gain magnitude $A_{CL}(f)$, compare this result with the open-loop gain, and determine the bandwidth of the amplifier.

Solution: According to Eq. (9.22), the gain magnitude is given by

$$A_{CL}(f)=\frac{R_2}{R_1}\frac{1}{\sqrt{1+\left(f/f_b^{\text{closed loop}}\right)^2}},\quad f_b^{\text{closed loop}}=\frac{10^5}{10}\times 10\ \text{Hz}=100\ \text{kHz} \tag{9.24}$$

In Fig. 9.13, we plot the closed-loop gain versus the open-loop gain given by Eq. (9.15). The amplifier bandwidth in the closed-loop configuration is now a respectable 100 kHz.

Exercise 9.6: The unity-gain bandwidth of an amplifier IC is 1 MHz. Determine the bandwidth of the non-inverting amplifier circuit with a gain of 200.

Answer: 5 kHz.

9.2.6 Application Example: Selection of an Amplifier IC for Proper Frequency Bandwidth

The required bandwidth and closed-loop gain usually are known to the circuit designer. Using Eq. (9.22), we can estimate whether or not a specific amplifier IC will meet those requirements. There is clearly a trade-off between the closed-loop gain and bandwidth according to Fig. 9.13 and Eq. (9.22). For a given amplifier IC, the lower the closed-loop gain requirement, the wider the achievable bandwidth.

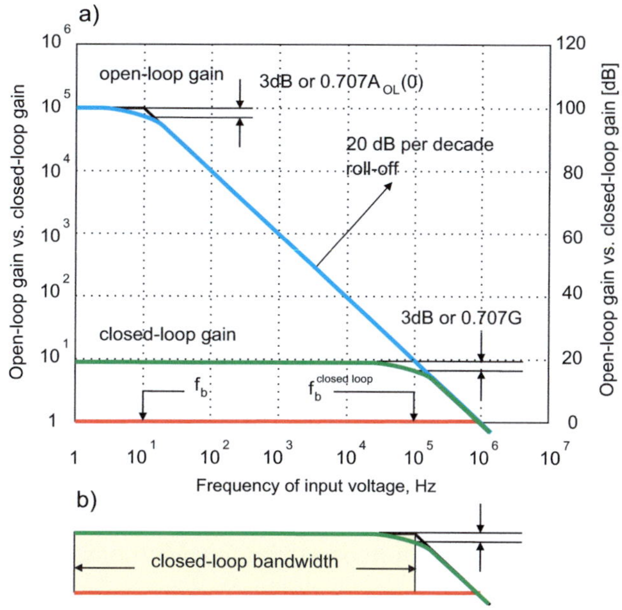

Fig. 9.13. Closed-loop AC gain $A_{CL}(f)$ (*lower curve*) versus open-loop AC gain $A_{OL}(f)$ (*upper curve*) for an inverting amplifier with $A_{OL}(0) = 10^5$ and $1 + R_2/R_1 = 10$ (the amplifier DC gain is -9).

Example 9.11: An inverting amplifier with a gain of -20 and bandwidth of at least 20 kHz is needed. Is the LM348 chip appropriate for this purpose?

Solution: From the LM348 datasheet, we obtain BW = 1 MHz. Because the inverting gain is -20, we should use a ratio of $R_2/R_1 = 20$. According to Eq. (9.22), this gives $f_b^{\text{closed loop}} = 47.6$ kHz. The closed-loop 3-dB bandwidth of the amplifier coincides with this value. Therefore, the LM348 chip is sufficient for our purposes. However, if its gain is forced to a higher value, say to 100, then the useful bandwidth reduces to 10 kHz.

Exercise 9.7: A non-inverting amplifier with a gain of 31 and a bandwidth of at least 90 kHz is needed. Is an LM741-based amplifier IC appropriate for this circuit?

Answer: No.

Section 9.3 Introduction to Continuous and Discrete Fourier Transform

9.3.1 Meaning and Definition of Fourier Transform

Consider a phasor as introduced in the previous chapter. This phasor is in fact a *transform*. It converts a harmonic sinusoidal time-domain signal into a complex number for easier, algebraic computation of circuit values. After determining the phasor value of a voltage or current signal, we transform it back to the time-domain expression. What if we do not have a pure sinusoidal tone, but an arbitrary voltage pulse $v(t)$ in the time domain? Another important example is a bit stream of arbitrary data, which can also be described by a certain voltage function $f(t)$. Could we still introduce a "phasor" for an arbitrary signal $f(t)$ in the time domain? The answer is yes; however, instead of a single complex number, we will have an entire complex function $F(\omega)$ of angular frequency ω. This function essentially consists of individual phasors, corresponding to all possible harmonic signals, which form the time-domain signal $f(t)$. Mathematically, the *direct Fourier transform* (from time domain to frequency domain) is given by

$$F(\omega) \equiv \int_{-\infty}^{\infty} f(t)\mathrm{e}^{-j\omega t}dt, \tag{9.25a}$$

whereas the *inverse Fourier transform* (from frequency domain to time domain) is given by a similar integral

$$f(t) \equiv \frac{1}{2\pi}\int_{-\infty}^{\infty} F(\omega)\mathrm{e}^{j\omega t}d\omega \tag{9.25b}$$

The pair of integrals in Eqs. (9.25a, 9.25b) completely describes the Fourier transform. Function $F(\omega)$ is called the *Fourier spectrum* (or simply the *spectrum*) of the signal $f(t)$. This function is generally complex; however, in contrast to the previous convention, we will not use boldface here in order to preserve the most common mathematical notations.

Exercise 9.8: Establish a relation between $F(-\omega)$ and $F(\omega)$ for a real signal $f(t)$, which is called a *reversal property* of the Fourier transform.

Answer:

$$F(-\omega) = F^*(\omega) \tag{9.26}$$

where the star denotes complex conjugate.

The spectrum is said to be *bandlimited* if $F(\omega)$ is zero above a certain angular frequency $\omega_{\max}$. According to Eq. (9.26), this simultaneously means that $F(\omega)$ is zero below $-\omega_{\max}$. Many useful signals are approximately bandlimited.

Example 9.12: Derive the Fourier transform of a rectangular pulse in the form of one bit of data shown in Fig. 9.14a.

Solution: The integral in Eq. (9.25a) is reduced to

$$F(\omega) = V_{\mathrm{m}} \int_{-T/2}^{T/2} \mathrm{e}^{-j\omega t} dt = \frac{V_{\mathrm{m}}}{-j\omega} \mathrm{e}^{-j\omega t} \Big|_{-T/2}^{T/2} = V_{\mathrm{m}} T \frac{\sin \omega T/2}{\omega T/2} \tag{9.27}$$

The function $\mathrm{sinc}(x) \equiv \sin \pi x/(\pi x)$ is called a *sinc function*. Using its definition, the final result for the spectrum has the form

$$F(\omega) = V_{\mathrm{m}} T \mathrm{sinc}\left(\frac{\omega T}{2\pi}\right) \tag{9.28}$$

and is plotted in Fig. 9.14b using a few lines of MATLAB code:

```
Vm      = 1;                            %  input voltage amplitude, V
T       = 1e-6;                         %  pulse duration, s
omega   = linspace(-12*pi/T, 12*pi/T);  %  angular frequency, rad/s
F       = Vm*T*sinc(omega*T/(2*pi));    %  inverse Fourier transform
plot(omega, F); grid on;
```

In contrast to the original signal, the pulse spectrum is not bounded and extends to infinity. This is due to the fact that the original pulse has sharp edges, which are described by higher-frequency harmonics. The pulse spectrum in the form of a sinc function is famous in communications theory. Figure 9.15 shows the shape of the sinc function depicted above the entrance of an electronics store in Silicon Valley, CA.

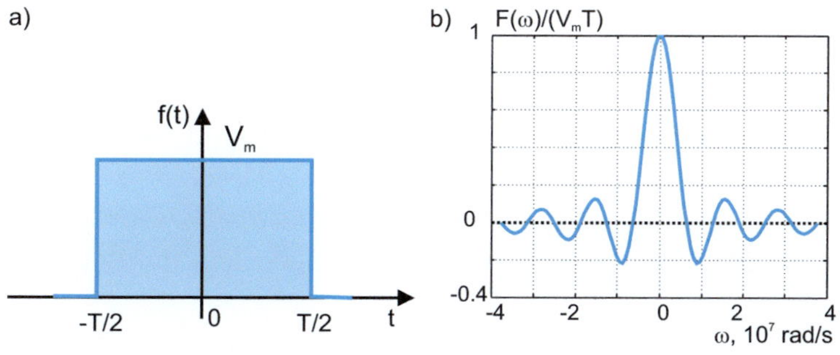

Fig. 9.14. (**a**) Rectangular pulse $f(t)$ of duration $T = 1$ μs and (**b**) its Fourier spectrum in the form of a sinc function.

Fig. 9.15. Fry's Electronics store in Sunnyvale, Silicon Valley, with an emblem depicting the sinc function.

9.3.2 Mathematical Properties of Fourier Transform

Major *mathematical properties* of the Fourier transform follow from its definition and are listed in Table 9.2.

Table 9.2. Major mathematical properties of Fourier transform.

$f(t)$	$F(\omega)$
$Kf(t)$	$KF(\omega)$
$f_1(t) - f_2(t) + f_3(t)$	$F_1(\omega) - F_2(\omega) + F_3(\omega)$
$d^n f(t)/dt^n$	$(j\omega)^n F(\omega)$
$\int_{-\infty}^{t} f(\tau)d\tau, \int_{-\infty}^{\infty} f(t)dt = 0$	$\frac{1}{j\omega}F(\omega)$
$f(at)$	$(1/a)F(\omega/a), \quad a > 0$
$f(t-a)$	$e^{-j\omega a}F(\omega)$

The first two properties follow from Fourier transform linearity. Multiplication of $f(t)$ by a constant corresponds to multiplying $F(\omega)$ by the same constant. Also, addition (subtraction) in the time domain corresponds to addition (subtraction) in the frequency domain. The next two properties (differentiation and integration) make the Fourier transform useful for solving ODEs since the time-domain derivatives and integrals will correspond to multiplication and division by $j\omega$ in the frequency domain. The two last properties (scaling and translation) directly follow from Eq. (9.25a).

Exercise 9.9: The Fourier transform of $f(t)$ is $F(\omega)$. What is the Fourier transform of $df(t)/dt + 5f(t)$?

Answer: $(5 + j\omega)F(\omega)$.

We emphasize that the properties listed in Table 9.2 also apply to the discrete Fourier transform studied below, but the corresponding indexing of discrete frequencies has to be carefully arranged.

9.3.3 Discrete Fourier Transform and Its Implementation

Direct Discrete Fourier Transform

Present and future demands are such that we must process continuous signals by discrete methods. Perhaps the most important method is the *discrete Fourier transform* (DFT) and its fast versions: *fast Fourier transform* (FFT) and *inverse fast Fourier transform* (IFFT). Let $f(t)$ be a continuous pulse signal which is the source of the data. We assume that $f(t)$ is zero outside of the interval $0 \le t < T$. Let $f(t_n), \quad n = 0, \ldots, N-1$ be its values at N uniformly distributed *sampling points* $t_n = \Delta T n, \quad n = 0, \ldots, N-1$ within the interval of interest. Here

$$\Delta T = \frac{T}{N} \tag{9.29}$$

is the *sampling interval*. Then, the integral of the direct Fourier transform in Eq. (9.25a) may be found using the *rectangle rule* (or the *Riemann sum approximation*)

$$F(\omega) = \Delta T \sum_{n=0}^{N-1} \mathrm{e}^{-j\omega n \Delta T} f(t_n) \tag{9.30}$$

We could in principle evaluate this expression at any value of ω. However, with *only* N data points to start with, *only* N final outputs will be significant. We choose those N uniformly distributed frequency sampling points as $\omega_m = \omega_0 m, \quad m = 0, \ldots, N-1$, where

$$\omega_0 = \frac{2\pi}{T} \tag{9.31}$$

is the *fundamental frequency* (with one period over the interval T). Let $F(\omega_m), m = 0, \ldots, N-1$ be the values of $F(\omega)$ at the frequency sampling points. Then, Eq. (9.30) gives

$$F(\omega_m) = \Delta T \sum_{n=0}^{N-1} \mathrm{e}^{-j\frac{2\pi}{N} mn} f(t_n), \quad m = 0, \ldots, N-1 \tag{9.32}$$

Inverse Discrete Fourier Transform (IDFT)

A very similar operation is applied to the integral of the inverse Fourier transform given by Eq. (9.25b). We first assume that $F(\omega)$ is zero outside of the interval $0 \le \omega < N\omega_0$; in other words, it is *bandlimited*. Then, the corresponding integral in Eq. (9.25b) is again approximated using the rectangle rule so that the final result has the form

$$f(t_n) = \frac{1}{N\Delta T}\sum_{m=0}^{N-1} e^{j\frac{2\pi}{N}mn} F(\omega_m), \quad n = 0, \ldots, N-1 \tag{9.33}$$

Definition of Discrete Fourier Transform

It is rather inconvenient to keep the factor ΔT in both Eqs. (9.32) and (9.33), respectively. Therefore, we may introduce the notation

$$f[n] \equiv \Delta T f(t_n), \quad F[m] \equiv F(\omega_m) \tag{9.34}$$

and obtain the *standard form* of the discrete Fourier transform

$$F[m] = \sum_{n=0}^{N-1} e^{-j\frac{2\pi}{N}mn} f[n], \quad m = 0, \ldots, N-1 \tag{9.35a}$$

$$f[n] = \frac{1}{N}\sum_{m=0}^{N-1} e^{j\frac{2\pi}{N}mn} F[m], \quad n = 0, \ldots, N-1 \tag{9.35b}$$

Here, $f[n]$ may be treated as an impulse having the area of $\Delta T f(t_n)$.

Exercise 9.10: Establish a relation between $F[N-m]$ and $F[m]$ for a real signal $f(t)$, which is called a *reversal property* of the discrete Fourier transform.

Answer:

$$F^*[N-m] = F[m] \tag{9.36}$$

where the star again denotes complex conjugate.

Example 9.13: It is possible to very significantly minimize the actual number of multiplications necessary to compute a given DFT in Eqs. (9.35a, b). The DFT so constructed is the *fast Fourier transform* (FFT) and *inverse fast Fourier transform* (IFFT). It works best when N is a power of two. For a pulse $f(t) = \exp\left(-2(t-5)^2\right)$, $0\ \text{s} \le t < 10\ \text{s}$, compute its FFT and then the IFFT and finally compare the end result with the original pulse form given that $N = 64$.

Solution: The solution is conveniently programmed using a few lines of a self-explanatory MATLAB code, which uses Eq. (9.29) and plots two final curves:

Example 9.13 (cont.):

```
T = 10; N = 64;
dT = T/N; t = dT*(0:N-1);
f0 = exp(-2*(t-5).^2);
F = fft(f0); f = ifft(F);
plot(t, f, t, f0, '*');
```

Both curves are virtually identical: the relative integral error (integral of signal difference magnitude over the integral of signal magnitude) does not exceed 10^{-16}.

Structure of Discrete Fourier Spectrum

The set of spectrum values $F[m]$, $m = 0, \ldots, N-1$, of the DFT has an important redundancy property illustrated in the following example.

Example 9.14: Express all discrete Fourier spectrum values $F[m]$ present in Eq. (9.35a) through $N/2$ first values of $F[m]$ only. *Hint:* Use Eq. (9.36).

Solution:

$$\begin{gathered} F[0], F[1], \ldots, F\left[\frac{N}{2}-1\right], F\left[\frac{N}{2}\right], F\left[\frac{N}{2}+1\right], \ldots, F[N-1] = \\ F[0], F[1], \ldots, F\left[\frac{N}{2}-1\right], F\left[\frac{N}{2}\right], F^*\left[\frac{N}{2}-1\right], \ldots, F^*[1] \end{gathered} \tag{9.37}$$

Equation (9.37) demonstrates how the output of the DFT (and of the FFT, in particular in MATLAB) is arranged in reality. It is a symmetric conjugate about $m = N/2$. Equation (9.37) is a key to finding derivatives and arbitrary filter transformations of the input signal with the FFT. Only a frequency with $m \le N/2$ is considered to be *valid*; its mirror reflection about $m = N/2$ is a higher "*aliasing frequency*." We emphasize that, according to Eq. (9.26), the complex conjugates may be replaced by spectrum values at a negative frequency, i.e., $F^*[1] = F[-1]$. Thus, the spectrum above $m = N/2$ corresponds to negative frequencies with $m > -N/2$.

9.3.4 Sampling Theorem

It follows from Example 9.14 that only frequency samples with $\omega_m \le \frac{N}{2}\omega_0$ are really needed. This fact is a consequence of the *sampling theorem*, which states that any signal bandlimited to $\omega_{\max}$ can be reproduced *exactly* using the discrete Fourier transform if

$$\omega_{\max} \le \frac{N}{2}\omega_0 \tag{9.38a}$$

Accordingly, the *maximum possible* sampling interval may be found from inequality

$$\Delta T \leq \frac{1}{2}\frac{1}{f_{\max}}, \quad f_{\max} = \frac{\omega_{\max}}{2\pi} \tag{9.38b}$$

Exercise 9.11: Examples of the maximum frequency of interest for some biomedical signals are:

1. Electrocardiogram (ECG) where $f_{\max} \approx 250$ Hz
2. Blood flow where $f_{\max} \approx 25$ Hz
3. Respiratory rate where $f_{\max} \approx 10$ Hz
4. Electromyogram where $f_{\max} \approx 10$ kHz

Establish the maximum possible sampling interval of the DFT and the minimum possible *sampling frequency*, which is equal to $1/\Delta T$.

Answer: (1) 2 ms and 500 Hz; (2) 20 ms and 50 Hz; (3) 50 ms and 20 Hz; (4) 50 μs and 20 kHz.

9.3.5 Applications of Discrete Fourier Transform

The DFT is one of the most important tools in *digital signal processing* (DSP). In particular, the DFT can calculate a signal's frequency spectrum. This is a direct examination of information encoded in the frequency, phase, and amplitude of the component sinusoids. For example, human speech and hearing use signals with this type of encoding. Second, the DFT or rather its variation, the *discrete cosine transform*, is used in sound compression; the MP3 format is one such example. The DFT is also an important image processing tool which is used to decompose an image into its sine and cosine components. The output of the transformation represents the image in the Fourier or frequency domain, while the input image is the spatial domain equivalent. In the Fourier domain image, each point represents a particular frequency contained in the spatial domain image. In particular, the JPEG format is using a modification of the DFT for image compression; the DFT is also used for image filtering and reconstruction. Along with this, the DFT is used widely in bioinformatics/computational biology to analyze DNA sequences. Last but not least, many computational modeling tools, such as antenna and high-speed circuit simulators, typically operate at one particular signal frequency (in the frequency domain). Collecting the solutions at many such frequencies makes it possible to establish evolution of an arbitrary signal or wave field in time.

9.3.6 Application Example: Numerical Differentiation via the FFT

We have established that a filter is characterized by its transfer function $\mathbf{H}(f)$ or $\mathbf{H}(\omega)$ and found this transfer function for simple cases. Given the input sinusoidal signal, we have also shown how to evaluate the filter's output when its transfer function is known. But what if the input signal is an arbitrary pulse? How could the solution for the output pulse be obtained? The answer relies upon an observation that the transfer function given by Eq. (9.9b) may be applied to every harmonic component of the input signal $f_{in}(t)$ *separately*. Those harmonics are all described by the Fourier spectrum of the pulse, $F(\omega)$. Therefore, the output Fourier pulse spectrum is given by

$$F_{out}(\omega) = \mathbf{H}(\omega)F_{in}(\omega) \tag{9.39}$$

The remaining part is to find the output pulse itself, which is clearly the inverse Fourier transform:

$$f_{out}(t) \equiv \frac{1}{2\pi}\int_{-\infty}^{\infty} F_{out}(\omega)e^{j\omega t}d\omega \tag{9.40}$$

When moving from continuous toward discrete Fourier transform and toward digital signal processing, Eq. (9.39) becomes a somewhat tricky operation. According to Eq. (9.37), the discrete version of Eq. (9.39) must have the form

$$\begin{aligned}\mathbf{H}F \to \mathbf{H}[0]F[0],\ \mathbf{H}[1]F[1], \ldots,\ \mathbf{H}\left[\frac{N}{2}-1\right]F\left[\frac{N}{2}-1\right],\ \mathbf{H}\left[\frac{N}{2}\right]F\left[\frac{N}{2}\right],\\ \mathbf{H}^*\left[\frac{N}{2}-1\right]F\left[\frac{N}{2}+1\right], \ldots\ \mathbf{H}^*[1]F[N-1]\end{aligned} \tag{9.41}$$

This version corresponds to the full list of monotonic frequency data $\omega_m = \omega_0 m, \quad m = 0, \ldots, N-1$. Also note that, in all realistic linear systems,

$$\mathbf{H}^*(m) = \mathbf{H}(-m) \tag{9.42}$$

Therefore, Eq. (9.41) simultaneously describes a set of data for the following non-monotonic frequency list 0, $\omega_m, \ldots, \frac{N}{2}\omega_0, \left(1-\frac{N}{2}\right)\omega_0, \left(2-\frac{N}{2}\right), \ldots, -\omega_0$, which also includes the negative frequencies.

Example 9.15 (numerical differentiation via the FFT): Prove Eq. (9.41) for a pulse $f(t) = \exp\left(-2(t-5)^2\right), \quad 0\ \text{s} \le t < 10\ \text{s}$ and for $\mathbf{H}(\omega) = j\omega$. Such a transfer function corresponds to numerical differentiation via the FFT. Use the FFT and IFFT with $N = 64$.

Solution: The solution is conveniently programmed in a self-explanatory MATLAB code, which uses Eq. (9.41) and plots two final results in Fig. 9.16, the numerical pulse derivative and the analytical derivative, respectively:

Example 9.15 (numerical differentiation via the FFT) (cont.):

```
T = 10; N = 64;
dT = T/N; t = dT*(0:N-1);
f = exp(-2*(t-5).^2);                          %   input pulse
omega = (2*pi/T)*[0:N/2];                      %   non-aliasing frequencies
H       = j*omega;                             %   H at non-aliasing frequencies
F       = fft(f);                              %   FFT spectrum
HF      = F.*[H, conj(H(end-1:-1:2))];         %   HF according to Eq. (9.40)
fder    = real(ifft(HF));                      %   numerical derivative
fder0   = -4*(t-5).*f;                         %   analytical derivative
plot(t, fder0, t, fder, 'd');                  %   compare both derivatives
```

Both curves are virtually identical: the relative integral error (integral of signal difference magnitude over the integral of analytical signal magnitude) does not exceed 1.3×10^{-15}.

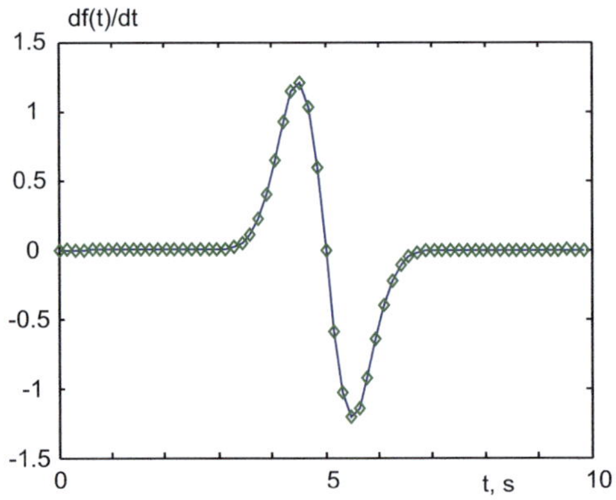

Fig. 9.16. Analytical (*solid curve*) and numerical (*diamonds*) differentiation of the original Gaussian pulse.

9.3.7 Application Example: Filter Operation for an Input Pulse Signal

The filter operation for an input pulse signal exactly follows Example 9.15 but with a different transfer function $\mathbf{H}(\omega)$.

Example 9.16: A pulse $f(t) = \exp\left(-2(t-5)^2\right)$, $0\ \text{s} \le t < 10\ \text{s}$ is an input to a first-order high-pass filter. Find the filter output when its (angular) break frequency is given by a) $\omega_0 = 1$ rad/s and b) $\omega_0 = 10$ rad/s. Use the FFT and IFFT with $N = 64$.

Solution: The solution is performed and programmed exactly as described in the previous example, but the transfer function is now given by Eq. (9.1b):

```
H = j*omega/omega0./(1+j*omega/omega0);
```

Example 9.16 (cont.):
Figure 9.17 plots two output pulse forms corresponding to two different values of the break frequency. Note that the value $\omega_0 = 10$ rad/s approximately corresponds to a mean value for non-aliasing frequencies of the FFT. Also note that when the break frequency becomes sufficiently high, the HPF behaves as an ideal differentiator but with a significant amplitude decay.

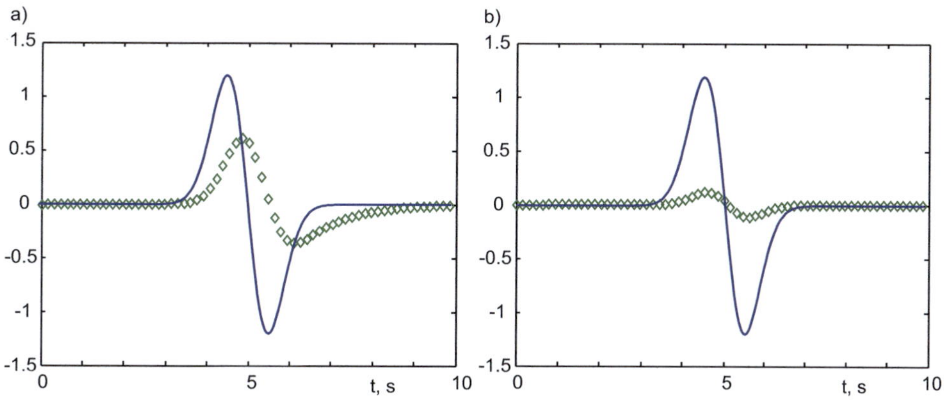

Fig. 9.17. *Diamonds*: HPF output for (**a**) $\omega_0 = 1$ rad/s and (**b**) $\omega_0 = 10$ rad/s, respectively. *Solid curve*: analytical derivative of the input Gaussian pulse.

9.3.8 Application Example: Converting Computational Electromagnetic Solution from Frequency Domain to Time Domain

Many computational modeling tools operate at one particular signal frequency or at a set of those (in frequency domain). To obtain the solution for an arbitrary pulse at an arbitrary point in space, we can again use the method of the transfer function and the FFT described previously. As an example, we consider a TMS (*transcranial magnetic stimulation*) coil above the head of a computational human phantom in Fig. 9.18a. Once a current pulse is applied to the coil, an electric field will be excited in the brain according to Faraday's law of induction. This field may help to reestablish some neuron connections lost, for example, in Parkinson's disease. For safety considerations, the field at arbitrary locations within the body needs to be evaluated, let's say at node 2 in Fig. 9.18a. In order to do so, the problem is first solved for about 40 single-frequency excitations, which will presumably cover the spectrum content of the desired TMS pulse in Fig. 9.18b well. The ratio of the electric field phasor at the observation point to the coil current phasor is the transfer function value at a desired frequency, $\mathbf{H}(\omega)$. This ratio does not depend on particular amplitude of the coil current. Next, we introduce the DFT of size N for the original pulse shown in Fig. 9.18b, interpolate the transfer function over $N/2 + 1$ required frequency points, and apply the method of Examples 91.5 and 9.16 with this new transfer function. The result is an electric field pulse at node 2 shown in Fig. 9.18c (the dominant z-component has been plotted), which is excited by the coil current shown in Fig. 9.18b.

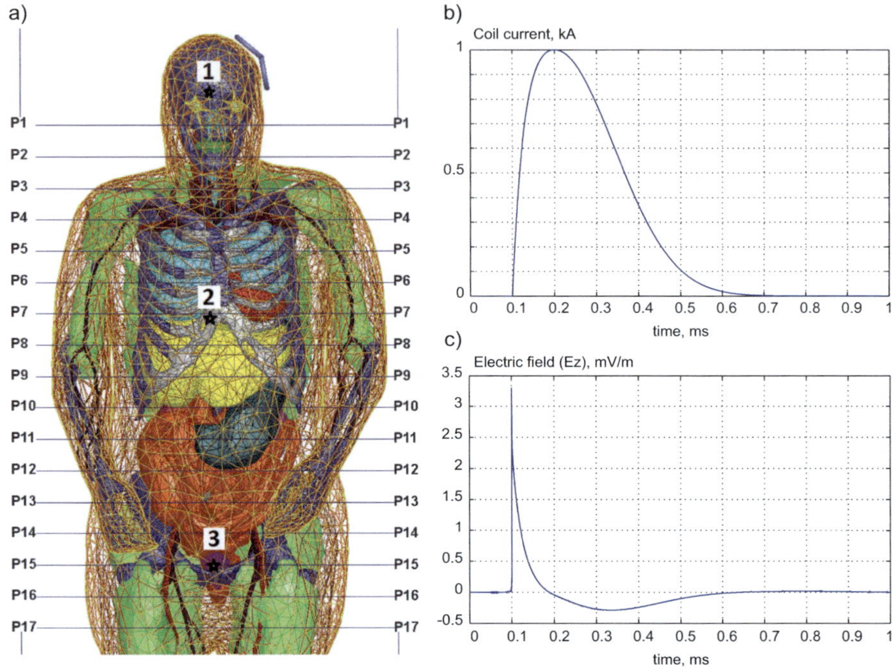

Fig. 9.18. Time-domain computational solution for the induced electric field within a human body obtained from the frequency-domain data via the FFT.

Summary

Property	First-order low-pass filter	First-order high-pass filter
Circuit schematic	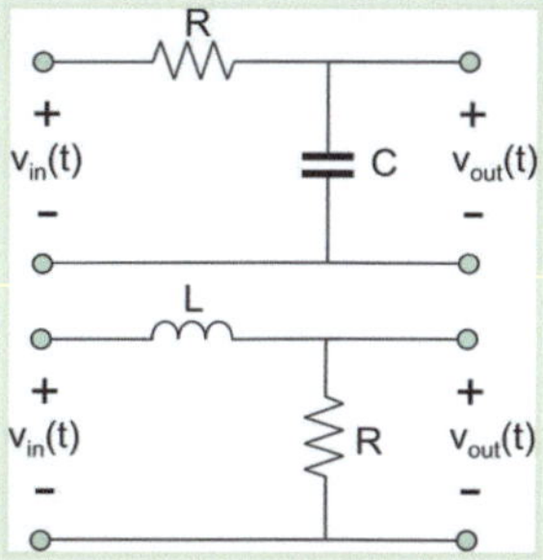	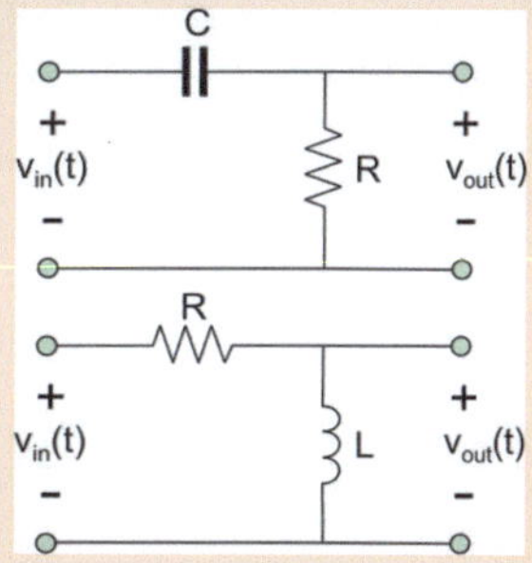
Transmission at $f = 0$ (DC)	**1** (DC path through the resistor)	**0** (No DC path)
Transmission at $f \to \infty$	**0** (Inductor is an open circuit at $f \to \infty$)	**1** (DC path through the resistor)
Transfer function $\mathbf{H}(f)$	$\dfrac{1}{1+j(f/f_b)}$	$\dfrac{(f/f_b)}{1+j(f/f_b)}$
Decibels of $H = \lvert\mathbf{H}\rvert$	$20\log_{10}H$ [dB]	$20\log_{10}H$ [dB]
Decibels of 1 and 0.1	0 dB and −20 dB	0 dB and −20 dB
Transfer function magnitude $H_m(f)$	$\dfrac{1}{\sqrt{1+(f/f_b)^2}}$	$\dfrac{f/f_b}{\sqrt{1+(f/f_b)^2}}$
Transfer function phase $\angle\varphi_H$	$\angle - \tan^{-1}\left(\dfrac{f}{f_b}\right)$	$\dfrac{\pi}{2} - \tan^{-1}\left(\dfrac{f}{f_b}\right)$
Break frequency, (half-power frequency, 3-dB frequency, corner frequency)	$f_b = \dfrac{1}{2\pi\tau}$ [Hz] $\tau = RC$ or $\dfrac{L}{R}$ [s]	$f_b = \dfrac{1}{2\pi\tau}$ [Hz] $\tau = RC$ or $\dfrac{L}{R}$ [s]
Passband (3 dB bandwidth), Hz	From 0 to f_b	From f_b to ∞
Filter with a resistive load R_L	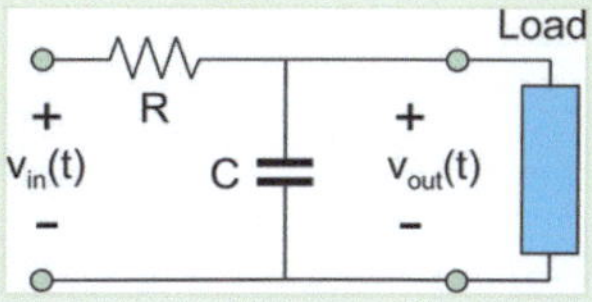	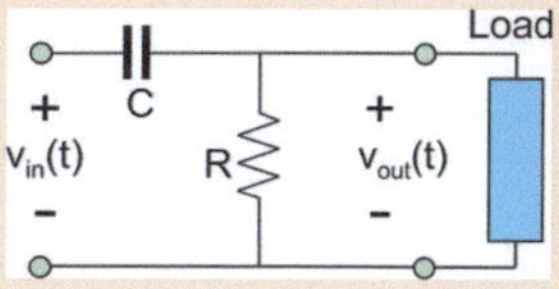

(continued)

<table>
<tr>
<td>Transfer function with the load $\mathbf{H}(f)$</td>
<td>$$\mathbf{H}(f)=\frac{1}{1+R/R_L+j(f/f_b)}$$ $$f_b=\frac{1}{2\pi\tau},\quad \tau=RC$$</td>
<td>$$\mathbf{H}(f)=\frac{(f/f_b)}{1+j(f/f_b)}$$ $$f_b=\frac{1}{2\pi\tau},\quad \tau=(R||R_L)C$$</td>
</tr>
<tr>
<td>Amplitude Bode plots</td>
<td>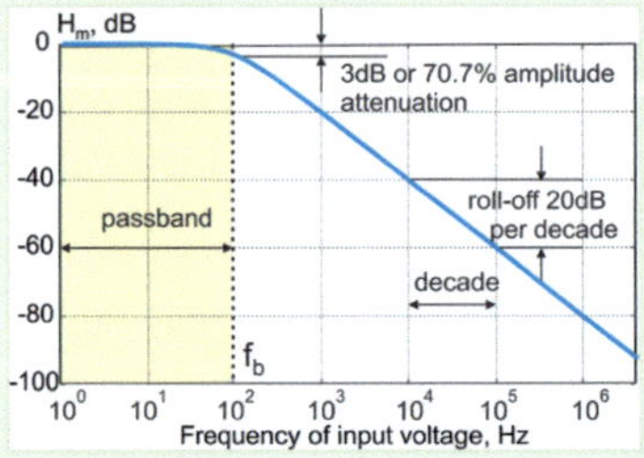
</td>
<td>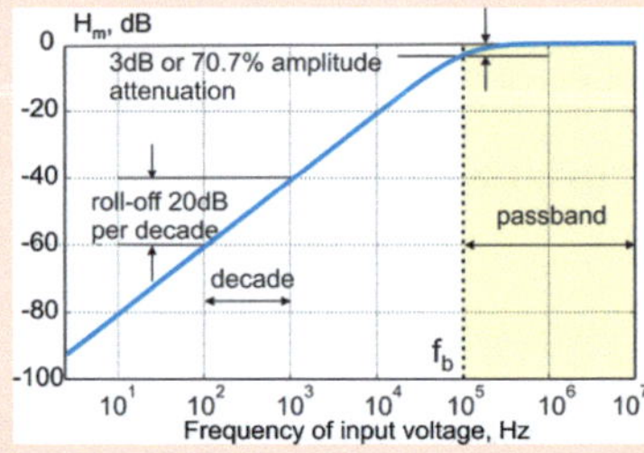
</td>
</tr>
<tr>
<td>Phase Bode plots</td>
<td>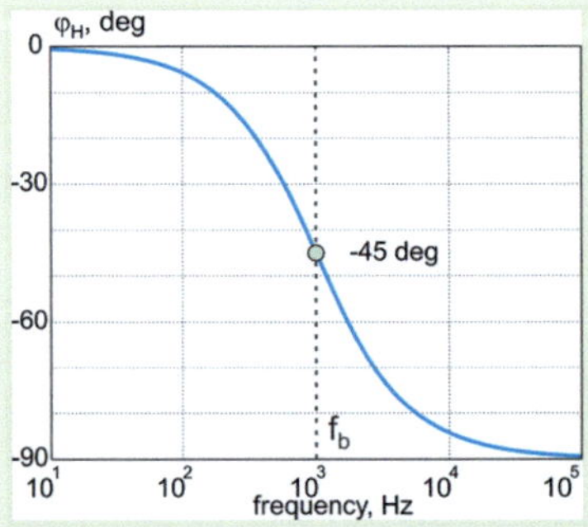
</td>
<td>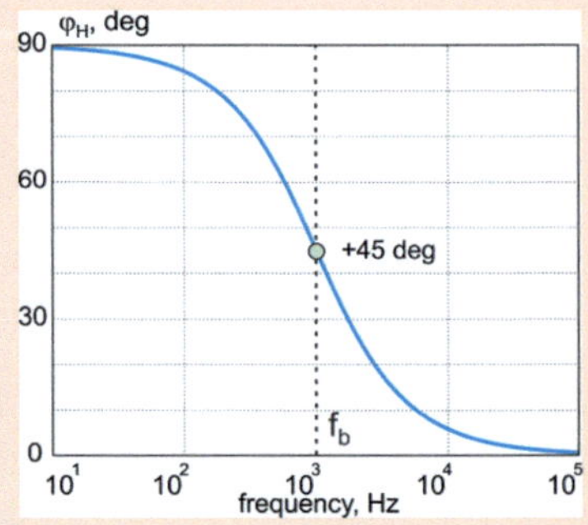
</td>
</tr>
<tr>
<td>Meaning of the transfer function for harmonic signals represented by phasors</td>
<td colspan="2">$$\mathbf{V}_{out}=\mathbf{H}(f)\mathbf{V}_{in}$$ where $\mathbf{V}_{in}$ is the input voltage phasor and $\mathbf{V}_{out}$ is the output voltage phasor</td>
</tr>
<tr>
<td>Cascading filters and linear systems (series combination)</td>
<td colspan="2">$$\mathbf{H}(f)=\mathbf{H}_1(f)\mathbf{H}_2(f)$$ only if the loading effect of individual blocks is minimized</td>
</tr>
<tr>
<td>3-dB bandwidth of an operational amplifier circuit (inverting or non-inverting amplifier configuration)</td>
<td colspan="2">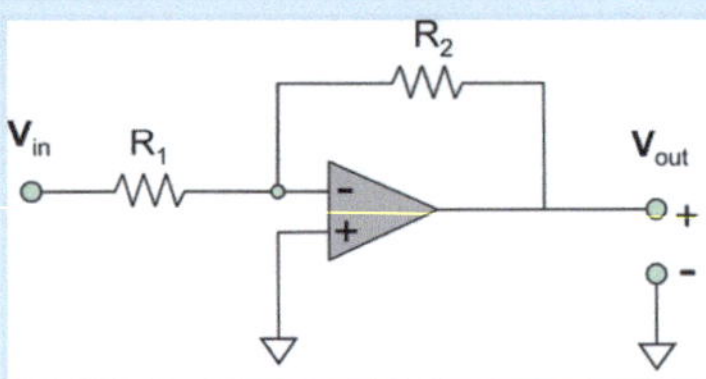

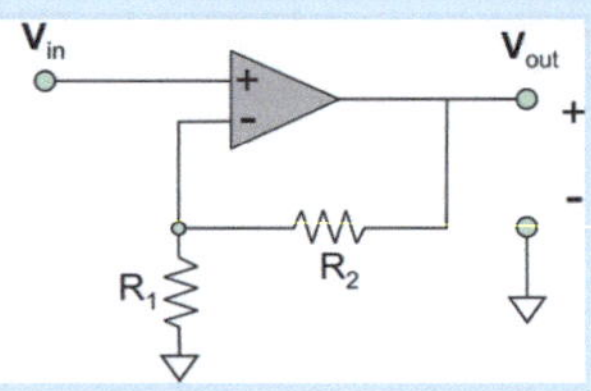

From 0 to $f_b^{\text{closed loop}}$ (closed-loop 3-dB frequency) where $$f_b^{\text{closed loop}}=\frac{\text{BW}}{1+R_2/R_1}$$ BW is the unity-gain bandwidth reported in the datasheet</td>
</tr>
</table>

(continued)

<table>
<tr><th>Property</th><th>Continuous and discrete Fourier transform</th></tr>
<tr><td>Fourier transform definition</td><td>

$$F(\omega) \equiv \int_{-\infty}^{\infty} f(t)e^{-j\omega t}dt \qquad f(t) \equiv \frac{1}{2\pi}\int_{-\infty}^{\infty} F(\omega)e^{j\omega t}d\omega$$

$F(-\omega) = F^*(\omega)$ if $f(t)$ is real

</td></tr>
<tr><td>Fourier transform of a rectangular pulse</td><td>

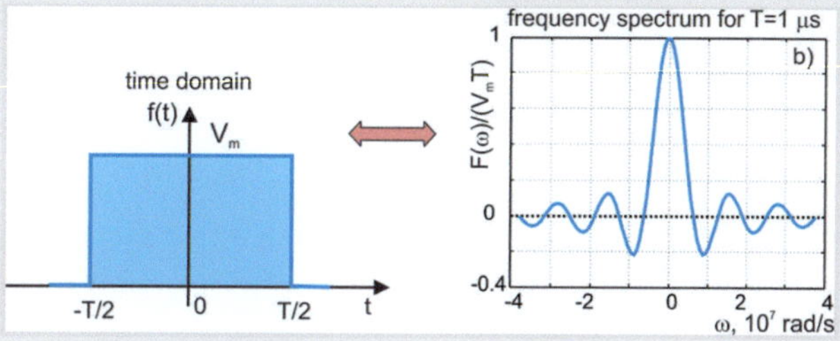

$$F(\omega) = V_m T \ \mathrm{sinc}\left(\frac{\omega T}{2\pi}\right)$$

</td></tr>
<tr><td>Fourier transform of a Gaussian pulse</td><td>

$$f(t) = e^{-at^2} \Leftrightarrow F(\omega) = \sqrt{\frac{\pi}{a}e^{-\omega^2/(4a)}}$$

</td></tr>
<tr><td>Major properties of Fourier transform</td><td>

<table>
<tr><td>$f(t)$</td><td>$F(\omega)$</td></tr>
<tr><td>$Kf(t)$</td><td>$KF(\omega)$</td></tr>
<tr><td>$f_1(t)-f_2(t)+f_3(t)$</td><td>$F_1(\omega)-F_2(\omega)+F_3(\omega)$</td></tr>
<tr><td>$d^n f(t)/dt^n$</td><td>$(j\omega)^n F(\omega)$</td></tr>
<tr><td>$\int_{-\infty}^{t} f(\tau)d\tau, \quad \int_{-\infty}^{\infty} f(t)dt = 0$</td><td>$\frac{1}{j\omega}F(\omega)$</td></tr>
<tr><td>$f(at)$</td><td>$(1/a)F(\omega/a), \quad a>0$</td></tr>
<tr><td>$f(t-a)$</td><td>$e^{-j\omega a}F(\omega)$</td></tr>
</table>

</td></tr>
<tr><td>Definition of sampling points: discrete Fourier transform</td><td>

$$t_n = \Delta T n, \quad n = 0, \ldots, N-1, \ T = N\Delta T$$
$$\omega_m = \omega_0 m, \quad m = 0, \ldots, N-1, \ \omega_0 = \frac{2\pi}{T}$$

</td></tr>
<tr><td>Definition of samples: discrete Fourier transform</td><td>

$$f[n] \equiv \Delta T f(t_n), \quad n = 0, \ldots, N-1$$
$$F[m] \equiv F(\omega_m), \quad m = 0, \ldots, N-1$$

</td></tr>
<tr><td>Discrete/fast Fourier transform</td><td>

$$F[m] = \sum_{n=0}^{N-1} e^{-j\frac{2\pi}{N}mn} f[n], \quad m = 0, \ldots, N-1$$
$$f[n] = \frac{1}{N}\sum_{m=0}^{N-1} e^{j\frac{2\pi}{N}mn} F[m], \quad n = 0, \ldots, N-1$$

</td></tr>
</table>

(continued)

Sampling theorem	1. Any signal bandlimited to $\omega_{\max}$ can be reproduced exactly using the discrete Fourier transform if $\omega_{\max} \le \frac{N}{2}\omega_0$ 2. Alternatively, the sampling interval must satisfy inequality $\Delta T \le \frac{1}{2}\frac{1}{f_{\max}}, \quad f_{\max} = \frac{\omega_{\max}}{2\pi}$
Structure of discrete Fourier spectrum	$F^*[N-m] = F[m]$ $\Downarrow$ $F[0], F[1], \ldots, F\left[\frac{N}{2}-1\right], F\left[\frac{N}{2}\right], F\left[\frac{N}{2}+1\right], \ldots, F[N-1] =$ $F[0], F[1], \ldots, F\left[\frac{N}{2}-1\right], F\left[\frac{N}{2}\right], F^*\left[\frac{N}{2}-1\right], \ldots, F^*[1]$
Equivalent frequency samples for negative frequencies	$0, \ \omega_m, \ \ldots, \ \frac{N}{2}\omega_0, \left(1-\frac{N}{2}\right)\omega_0, \left(2-\frac{N}{2}\right), \ \ldots, \ -\omega_0$
Transfer function multiplication	$\mathbf{H}F \to \mathbf{H}[0]F[0], \ \mathbf{H}[1]F[1], \ \ldots, \ \mathbf{H}\left[\frac{N}{2}-1\right]F\left[\frac{N}{2}-1\right], \ \mathbf{H}\left[\frac{N}{2}\right]F\left[\frac{N}{2}\right],$ $\mathbf{H}^*\left[\frac{N}{2}-1\right]F\left[\frac{N}{2}+1\right], \ \ldots, \ \mathbf{H}^*[1]F[N-1]$

Problems

9.1 First-Order Filter Circuits and Their Combinations

9.1.1 RC Voltage Divider as an Analog Filter

Problem 9.1

A. Explain the function of an analog RC filter.
B. Write the capacitor and resistor voltages $v_R(t)$ and $v_C(t)$ of the series RC circuits in the general form, as functions of the AC angular frequency.
C. Which circuit element (or which voltage) dominates at low frequencies? At high frequencies?

Problem 9.2

A. Draw a schematic of the low-pass analog RC filter. Show the input and output ports.
B. Repeat the same task for the high-pass analog RC filter.

Problem 9.3. The input voltage to the filter circuit shown in the following figure is a combination of two harmonics, $v_{in}(t) = 1\cos\omega_1 t + 1\cos\omega_2 t$, with the amplitude of 1 V each. The filter has the following parameters: $R = 100$ kΩ and $C = 1.59$ nF. Determine the output voltage $v_{out}(t)$ to the filter given that $f_1 = 100$ Hz and $f_2 = 100$ kHz. Express all phase angles in degrees.

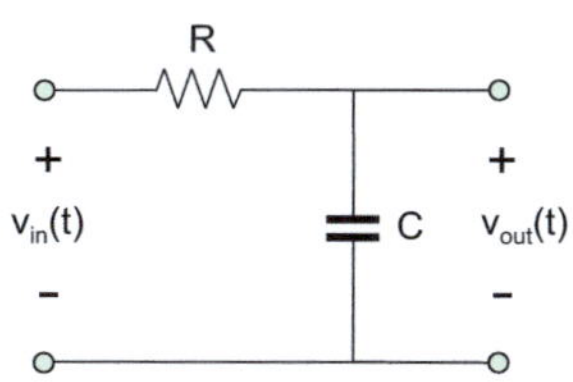

Problem 9.4. Repeat the previous problem for the filter circuit shown in the following figure. All other parameters remain the same.

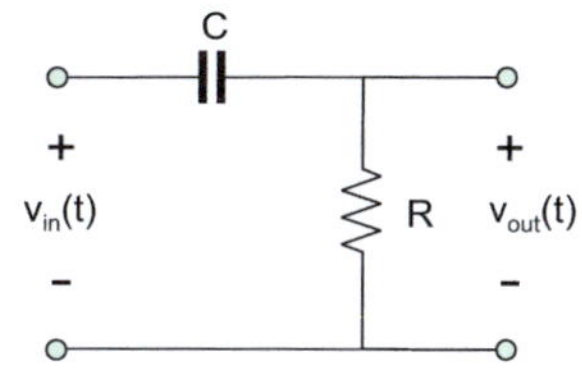

Problem 9.5. The input voltage to the RC filter circuit shown in the figure is $V_{in}(t) = 5\cos\omega t$ [V]. The filter has the following parameters: $C = 1$ μF and $R = 100$ Ω. The filter operates in the frequency band from 100 Hz to 50 kHz. The filter is connected to a load with the load resistance of 1 MΩ. By solving the corresponding AC circuit, determine the output voltage amplitude across the load (and its percentage versus the input voltage amplitude) *with and without the load* at $f = 100$ Hz, $f = 1592$ Hz, and $f = 50$ kHz.

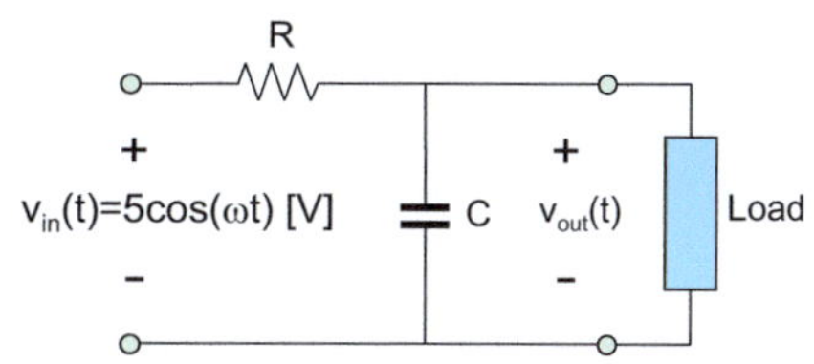

Problem 9.6. Repeat the previous problem when the load resistance changes from 1 MΩ to 100 Ω (decreases).

Problem 9.7. Repeat Problem 9.5 for the filter circuit shown in the following figure. Assume the load resistance of 100 Ω.

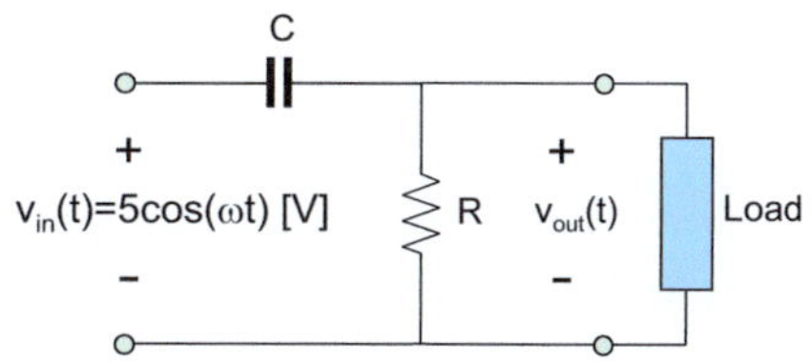

9.1.2 Half-Power Frequency and Amplitude Transfer Function

9.1.3 Bode Plot, Decibel, and Roll-off

Problem 9.8.

A. Describe the physical meaning of the (half-power) break frequency in your own words.
B. Give the expression for the break frequency in terms of circuit parameters of an RC filter. Is it different for low-pass and high-pass filters?

Problem 9.9. Given $R = 100\ \text{k}\Omega$ and $C = 1.59\ \text{nF}$, determine the break frequency of the low-pass RC filter and of the high-pass RC filter, respectively.

Problem 9.10. List all possible alternative names for the break frequency.

Problem 9.11. Write the amplitude transfer function for the low-pass RC filter. Repeat for the high-pass RC filter. Indicate units (if any).

Problem 9.12. The input signal to a high-pass RC filter includes a 60-Hz component. Its amplitude is to be reduced by a factor of 10. What break frequency should the filter have?

Problem 9.13. The input signal to a low-pass RC filter includes a 10-kHz component. Its amplitude is to be reduced by a factor of 5. What break frequency should the filter have?

Problem 9.14. Describe the meaning of the Bode plot in your own words.

Problem 9.15. It is known that $H_{\text{m}}(f)_{\text{dB}} = 0, \ -6, \ -20$ [dB]. Find the corresponding values of $H_{\text{m}}(f)$.

Problem 9.16. The following values are given $H_{\text{m}}(f) = 1, \ 0.707, \ 0.1,$ and 100. Find the corresponding values of $H_{\text{m}}(f)_{\text{dB}}$.

Problem 9.17

A. When the ratio of the amplitudes of two signals is $\sqrt{2}$, what is the difference between the two corresponding decibel measures in dB?
B. When the ratio of the amplitudes of two signals is $1/\sqrt{2}$, what is the difference between the two corresponding decibel measures in dB?
C. When the ratio of the amplitudes of two signals is $\sqrt{20}$, what is the difference between the two corresponding decibel measures in dB?
D. When the ratio of the powers of two signals is 1000, what is the difference between the two corresponding decibel measures in dB?

Problem 9.18. What do engineers mean by one decade? One octave?

Problem 9.19. For the filter circuit shown in the following figure, given that $R = 100\ \text{k}\Omega$ and $C = 159\ \text{pF}$:

A. Create the amplitude Bode plot by finding transfer function values for (at least) every decade.
B. Label the break frequency.
C. Label the filter passband.

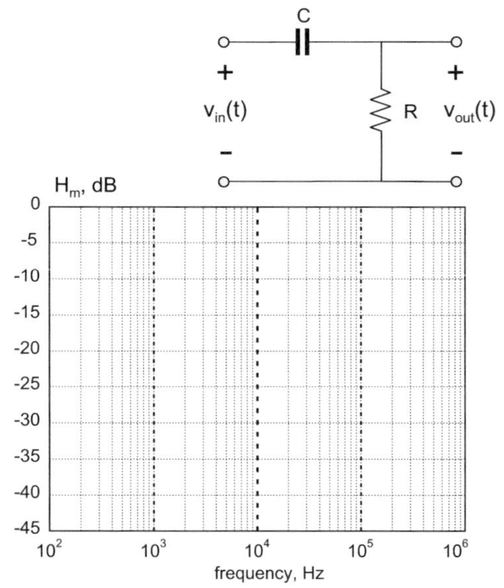

Problem 9.20. Repeat the previous problem with $R = 100\ \text{k}\Omega$ and $C = 53\ \text{pF}$.

Problem 9.21. For the filter circuit shown in the following figure, assume the values $R = 10\ \text{k}\Omega$ and $C = 1.59\ \text{nF}$.

A. Create the amplitude Bode plot by finding the transfer function values for (at least) every decade.
B. Label the break frequency.
C. Label the filter passband.

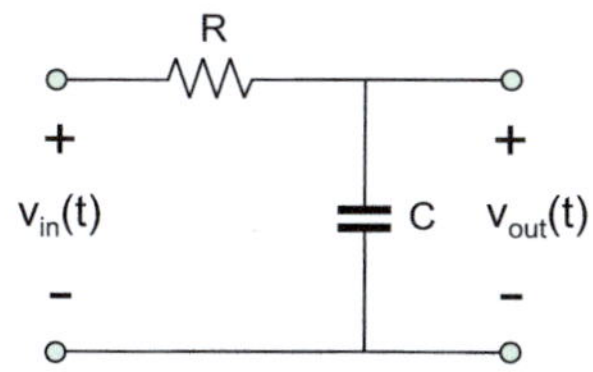

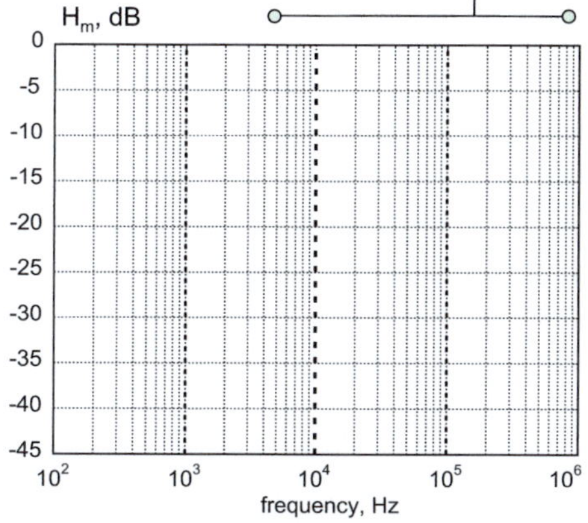

Problem 9.22. Repeat the previous problem with $R = 100\ \text{k}\Omega$ and $C = 53\ \text{pF}$.

Problem 9.23. An amplitude Bode plot for a certain RC filter is shown in the figure below.

A. Approximately determine the filter's resistance R if it is known that $C = 265\ \text{pF}$. Describe each step of your approach.
B. Suggest a way to verify your solution.

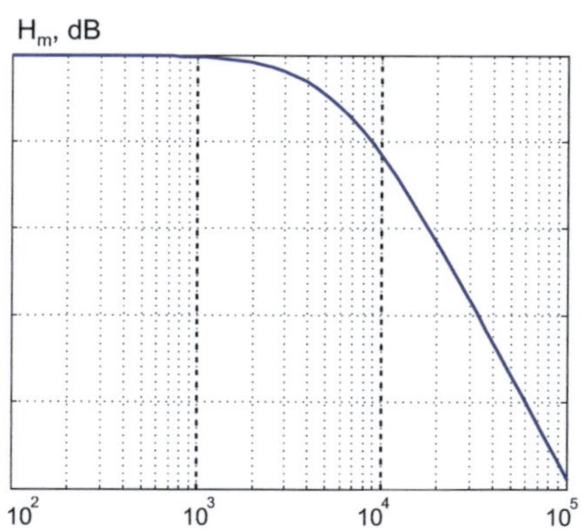

Problem 9.24. An amplitude Bode plot for a certain RC filter is shown in the figure below.

A. Approximately determine the filter's capacitance, C, for a given $R = 100\ \text{k}\Omega$. Describe each step of your approach.
B. Suggest a way to verify your solution.

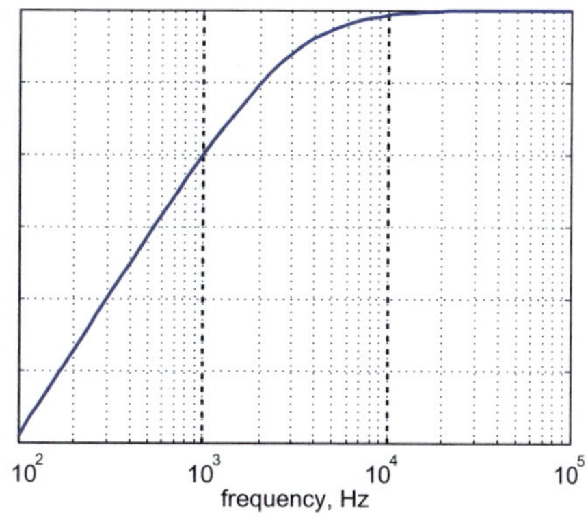

Problem 9.25. Prove analytically that the amplitude transfer functions of the low-pass filter and the high-pass filter are the mirror reflections of each other about the break frequency in the Bode plot.

9.1.4 Phase Transfer Function and Its Bode Plot

Problem 9.26. Write the phase transfer function for the low-pass RC filter. Repeat for the high-pass RC filter. Show units.

Problem 9.27. The input voltage to a low-pass RC filter has a zero phase. At what frequency in terms of the break frequency f_b is the phase shift at the output equal to $-1°$, $-45°$, and $-89°$?

Problem 9.28. The input voltage to a high-pass RC filter has a zero phase. At what frequency in terms of the break frequency f_b is the phase shift at the output equal to $5°$, $45°$, and $85°$?

Problem 9.29. A low-pass RC filter has the break frequency of 10 kHz. Create the phase Bode plot by finding the transfer function values for (at least) every decade.

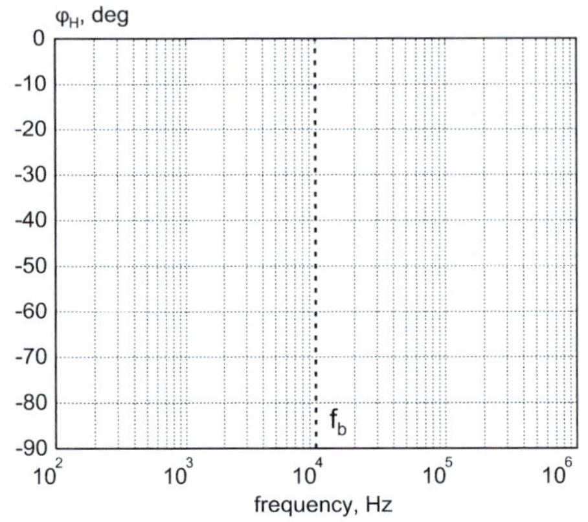

Problem 9.30. Repeat the previous problem for a high-pass RC filter with the same break frequency.

9.1.5 Complex Transfer Function: Cascading Filter Circuits

9.1.6 RL Filter Circuits

Problem 9.31. For the filter circuit shown in the figure below, create the amplitude response of the Bode plot by finding transfer function values for (at least) every decade. The two individual filter blocks both have the break frequency of 1 kHz. Assume that the loading effect of the filter stages is negligibly small; in practice, a buffer amplifier stage could be used.

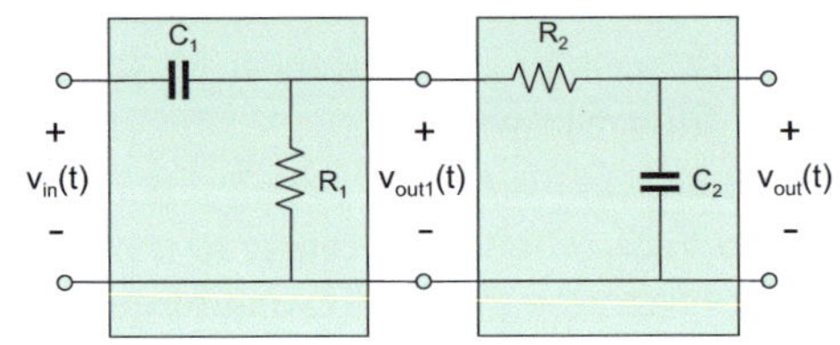

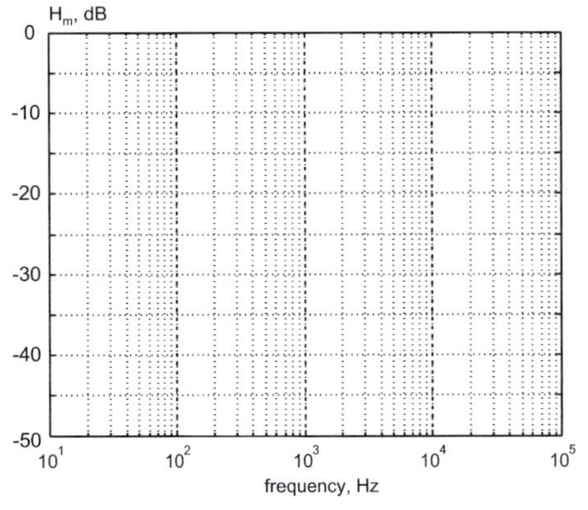

Problem 9.32

A. Repeat the previous problem for the filter circuit shown in the figure below.
B. Analytically determine the roll-off per decade in dB.

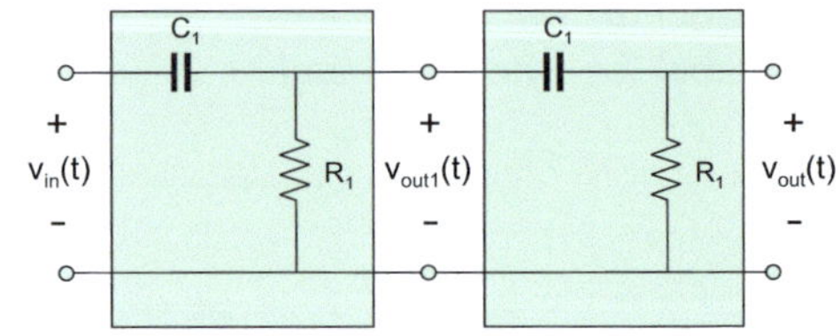

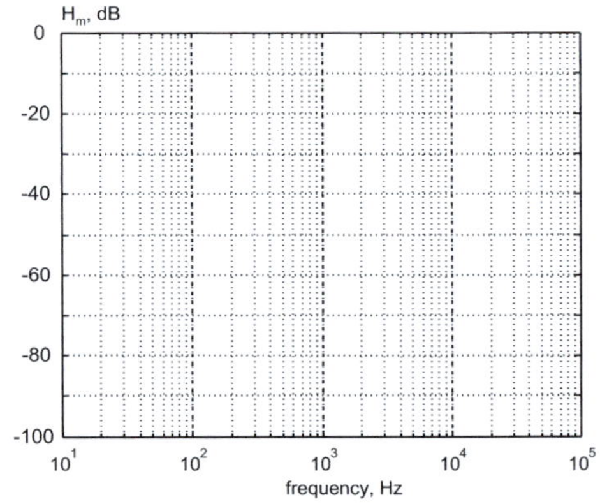

Problem 9.33. For two RL filter circuits with $R = 31.4\ \Omega$ and $L = 1\ \text{mH}$ shown in the figure below:

A. Determine the break frequency.
B. Draw the corresponding RC counterpart.
C. Establish the capacitance values of the RC filters, which assure the equivalent transfer functions, given that the resistances of the RC filters are 100 kΩ in both cases.

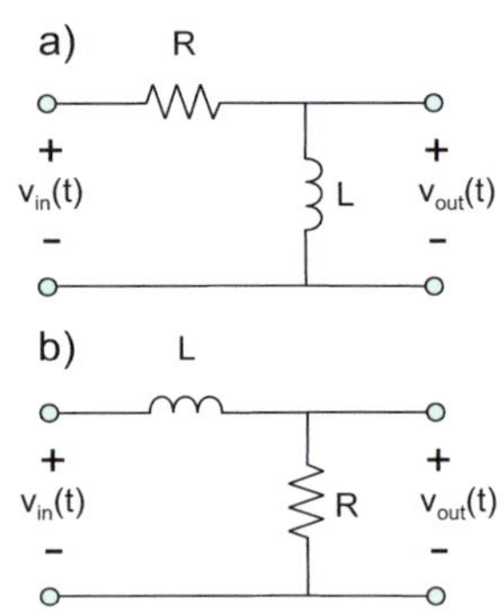

Problem 9.34. For the filter circuit shown in the figure below, assume the values $R_1 = 628$ Ω and $L_1 = 10$ mH.

A. Create the amplitude Bode plot by finding transfer function values for (at least) every decade.
B. Determine the roll-off per decade in dB.

Assume that the loading effect of the filter stages is negligibly small (e.g., a buffer amplifier stage is used).

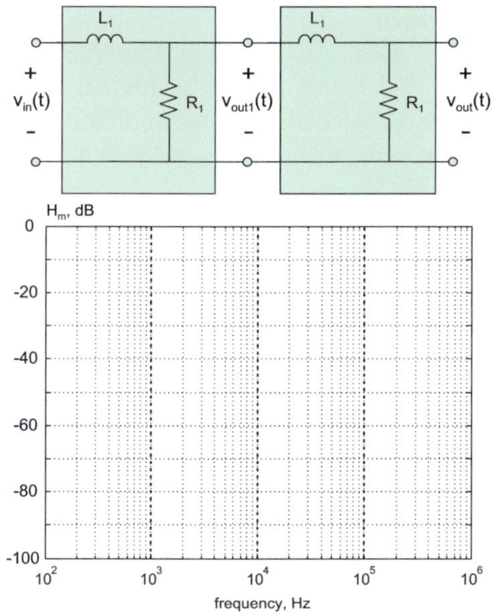

Problem 9.35. The transfer function of a filter circuit is given by $H(f) = \frac{1+j(f/1000)}{1+(f/1000)^2}$. Create its amplitude and phase Bode plots in the frequency band from 10 Hz to 100 kHz by finding transfer function values for (at least) every decade.

9.2 Bandwidth of an Operational Amplifier

9.2.1 Bode Plot of the Open-loop Amplifier Gain

9.2.2 Unity-gain Bandwidth Versus Gain-Bandwidth Product

Problem 9.36. An amplifier has the unity-gain bandwidth BW of 5 MHz. What exactly does this mean? Explain and provide equations.

Problem 9.37. Using a manufacturing company's website (usually it is a more accurate frequently updated source) or the corresponding datasheet, find the unity-gain bandwidth for the following amplifier ICs:

A. TL082
B. LM741
C. LM7171

Problem 9.38. Frequency response of an amplifier is characterized by the open-loop DC gain $A_{OL}(0) = 1.41 \times 10^6$ and the break frequency of $f_b = 20$ Hz. Numerically calculate the gain-bandwidth product for the amplifier at:

A. 20 Hz,
B. 2 kHz,
C. 2 MHz.

9.2.3 Model of the Open-Loop AC Gain

Problem 9.39. Frequency response of an amplifier is characterized by the open-loop DC gain $A_{OL}(0) = 10^6$ and the break frequency of $f_b = 20$ Hz. Plot the open-loop gain magnitude in dB over the range of frequencies (the frequency band) from 1 Hz to 10 MHz on the log-log scale (the Bode plot) and label the axes.

Problem 9.40. In the previous problem, find the unity-gain bandwidth BW of the amplifier.

Problem 9.41. Internally compensated LM358-series amplifiers have the unity-gain bandwidth (BW) of 1 MHz. The typical large-signal DC voltage gain at room temperature is 100 V/mV.

A. Find the open-loop DC gain in dB and the open-loop break frequency f_b.
B. Find the open-loop gain at 100 Hz, 1 kHz, and 10 kHz.

Problem 9.42. The open-loop gain magnitude of an internally compensated high-frequency amplifier has been given as

$$A_{OL}(100 \text{ Hz}) = 0.9 \times 10^6,$$
$$A_{OL}(1 \text{ MHz}) = 1.0 \times 10^2$$

at room temperature. Determine:

A. 3-dB break frequency,
B. DC open-loop gain,
C. Unity-gain bandwidth BW of the amplifier.

Problem 9.43. Repeat the previous problem for

$$A_{OL}(100\ \text{Hz}) = 0.5 \times 10^6,$$
$$A_{OL}(1\ \text{MHz}) = 1.0 \times 10^2.$$

9.2.4 Model of the Closed-loop AC Gain

9.2.5 Application Example: Finding Bandwidth of an Amplifier Circuit

Problem 9.44. An amplifier with the open-loop gain described by the first-order RC circuit response with $A_{OL}(0) = 10^5$ and $f_b = 20$ Hz is used in a closed-loop inverting configuration with $R_2/R_1 = 9$ and $R_2/R_1 = 99$, respectively.

A. Using the template that follows, create the Bode plots for the corresponding frequency response (closed-loop gain), $G(f)$, in the band from 10 Hz to 10 MHz on the same graph. Plot the gain values for (at least) every decade.
B. Also on the same graph, plot the open-loop gain as a function of frequency.
C. Determine the bandwidth of the closed-loop amplifier so constructed in every case.

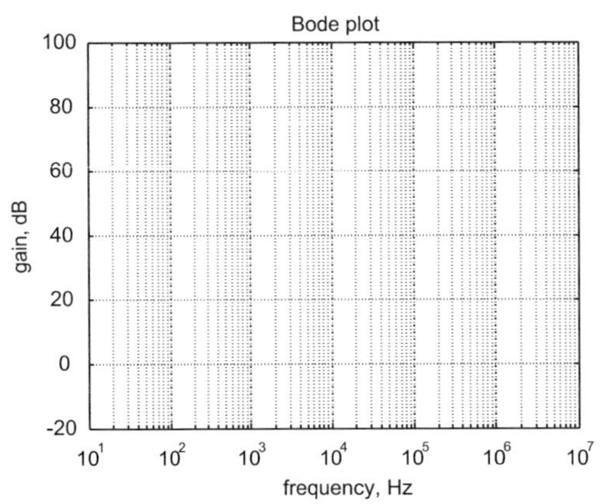

Problem 9.45. The unity-gain bandwidth of an amplifier IC is 1 MHz. Determine the bandwidth of the following amplifier circuits:

A. An inverting amplifier with the gain of −1,
B. An inverting amplifier with the gain of −10,
C. A non-inverting amplifier with the gain of 100,
D. A voltage follower (buffer amplifier).

constructed using the same IC.

9.2.6 Application Example: Selection of an Amplifier IC for Proper Frequency Bandwidth

Problem 9.46. An inverting amplifier with a gain of −20 and a bandwidth of at least 200 kHz is needed. Which amplifier chip is appropriate for this circuit (and which is not)?

A. LM358
B. TL082
C. LM741
D. LM7171
E. LM8272

Problem 9.47. A non-inverting amplifier with a gain of 31 and a bandwidth of at least 90 kHz is needed. Which amplifier chip is appropriate for this circuit (and which is not)?

A. TL082
B. LM7171
C. LM8272

9.3 Introduction to Continuous and Discrete Fourier Transform

9.3.1 Meaning and Definition of Fourier Transform

Problem 9.48. Establish all values of the angular frequency ω in Fig. 9.14b at which the Fourier spectrum $F(\omega)$ of a rectangular pulse crosses the frequency axis (becomes zero). Express your result in terms of pulse duration T.

Problem 9.49

A. Establish the value of the Fourier transform $F(\omega)$ for the pulse shown in the following figure at $\omega = 0$.

B. Establish the complete pulse spectrum $F(\omega)$ at all values of angular frequency ω.

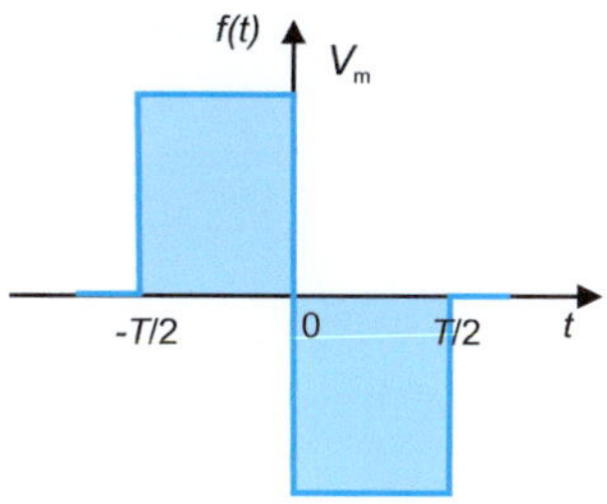

Problem 9.50
Establish the Fourier transform $F(\omega)$ for the following voltage signals in time domain:

A. $f(t) = A\sin\left(\frac{\pi}{2}t\right), \quad -2 \le t < 2$
$f(t) = 0, \quad$ otherwise

B. $f(t) = A\cos\left(\frac{\pi}{2}t\right), \quad -2 \le t < 2$
$f(t) = 0, \quad$ otherwise

Problem 9.51
Show that for an arbitrary real voltage signal $f(t)$:

A. The real part of $F(\omega)$ is an even function of angular frequency ω.
B. The imaginary part of $F(\omega)$ is an odd function of angular frequency ω.
C. The magnitude of $F(\omega)$ is an even function of angular frequency ω.
D. Replacing ω by $-\omega$ generates the complex conjugate of $F(\omega)$; in other words, $F(-\omega) = F^*(\omega)$.

9.3.2 Mathematical Properties of Fourier Transform

Problem 9.52. The Fourier transform of $f(t)$ is $F(\omega)$. What is the Fourier transform of $d^2f(t)/dt^2 - 2\int_{-\infty}^{t} f(\tau)d\tau$?

Problem 9.53. The Fourier transform of $f(t)$ is $F(\omega)$. What is the Fourier transform of $f(-t)$?

Problem 9.54. The function $f(t)\cos\omega_0 t$ is an *amplitude-modulated signal*: a high-frequency carrier $\cos\omega_0 t$, which is transmitted wirelessly, has a low-frequency envelope $f(t)$, which carries information and is being demodulated at the receiver. If the Fourier transform of $f(t)$ is $F(\omega)$, what is the Fourier transform of $f(t)\cos\omega_0 t$?

Problem 9.55. If $f(t)$ represents the voltage across a 1-Ω load, then $f^2(t)$ is the power delivered to the load and $\int_{-\infty}^{\infty} f^2(t)dt$ is the total energy delivered to the load. Prove *Parseval's theorem*,

$$\int_{-\infty}^{\infty} f^2(t)dt = \frac{1}{2\pi}\int_{-\infty}^{\infty} |F(\omega)|^2 d\omega,$$

which relates the total energy to an integral of the *energy spectral density*, $|F(\omega)|^2 = F(\omega)F^*(\omega)$, of the signal.
Hint: Use the reversal property of the Fourier transform given by Eq. (9.26).

Problem 9.56. Based on Parseval's theorem established in the previous problem, find the value of the integral $\int_{-\infty}^{\infty} \mathrm{sinc}^2(t)dt$.

9.3.3 Discrete Fourier Transform and Its Implementation

Problem 9.57. You are using the discrete Fourier transform of length 8 ($N = 8$) for a signal $f(t) = \sin t$ over a time interval from 0 to 2π s.

A. Compute all sampling points in the time domain.
B. Compute all sampling points in the frequency domain.

C. Compute equivalent frequency samples using negative frequencies.
D. Compute all discrete samples $f[n]$.
E. Compute all discrete samples $F[m]$ using the definition of the discrete Fourier transform. Explain the physical meaning of their values.
F. Repeat the previous step using function `fft` of MATLAB. Compare both sets of $F[m]$.
G. Restore all discrete samples $f[n]$ using the definition of the inverse discrete Fourier transform. Compare them with the exact function values.
H. Repeat the previous step using function `ifft` of MATLAB. Compare both sets of $f[n]$.

Problem 9.58. Repeat the previous problem for the signal $f(t) = \cos t$. All other parameters remain the same.

Problem 9.59. For Problem 9.57, establish and prove a discrete version of Parseval's theorem formulated in Problem 9.55.

Problem 9.60. An input signal to a filter has a discrete frequency spectrum $F[m],\quad m = 0, \ldots, N-1$ computed via the FFT. You are given filter transfer function $\mathbf{H}$ computed at $\frac{N}{2}+1$ frequency points of the FFT, $\mathbf{H}[m],\quad m = 0, \ldots, N/2$. Compute the discrete spectrum of the filter's output to be fed into the IFFT.

9.3.6 Application Example: Numerical Differentiation via the FFT

9.3.7 Application Example: Filter Operation for an Input Pulse Signal

Problem 9.61*. Present the text of a MATLAB script that numerically differentiates the input signal $f(t) = \sin t$ over the time interval from 0 to 4π s using the FFT with 4096 sampling points and plot the resulting signal derivative.

Problem 9.62. Repeat the previous problem for the signal $f(t) = \exp\left(-(t-2\pi)^2\right)$. All other parameters remain the same.

Problem 9.63. A monopolar pulse $f(t) = \exp\left(-2(t-5)^2\right),\quad 0 \le t < 10$ s is an input to a series combination of two identical first-order high-pass filters. Find the output of the filter combination when the (angular) break frequency is given by:
A. $\omega_0 = 0.5$ rad/s
B. $\omega_0 = 10$ rad/s

Use the FFT and IFFT with $N = 64$. Plot the filter output and explain the output signal behavior in every case.

Problem 9.64. A bipolar pulse $f(t) = (5-t)\exp\left(-2(t-5)^2\right),\ 0 \le t < 10$ s is an input to a first-order low-pass filter. Find the filter output when its (angular) break frequency is given by:
A. $\omega_0 = 0.5$ rad/s
B. $\omega_0 = 5$ rad/s

Use the FFT and IFFT with $N = 64$. Plot the filter output along with the input signal on the same graph and explain the output signal behavior in both cases.

Chapter 10: Second-Order RLC Circuits

Overview

Prerequisites:
- Knowledge of complex arithmetic
- Knowledge of phasor/impedance method for AC circuit analysis (Chapter 8)

Objectives of Section 10.1:
- Learn the concept of a resonant circuit and its relation to other engineering disciplines
- Understand the internal dynamics of the series/parallel RLC resonator including voltage and current behavior near the resonant frequency
- Establish the meaning and be able to calculate resonant frequency, quality factor, and bandwidth of the second-order resonant circuits
- Establish and quantify the duality between series and parallel RLC resonators

Objectives of Section 10.2:
- Construct four major types of the second-order RLC filters
- Relate all filter concepts to the corresponding circuit diagrams
- Specify two filter design parameters: the undamped resonant frequency and the quality factor
- Realize the advantages of the second-order filters versus the first-order filters

Objectives of Section 10.3:
- Become familiar with the concept of the near-field wireless link
- Apply the theory of the series resonant RLC circuit to the basic design of the near-field wireless transmitter and receiver
- Understand the operation of proximity sensors based on resonant RLC circuits

Application examples:
- Near-field wireless link in undergraduate laboratory
- Proximity sensors

S. N. Makarov et al., *Practical Electrical Engineering*, https://doi.org/10.1007/978-3-319-96692-2_10

Keywords:
Self-oscillating LC circuit, Series resonant RLC circuit, Parallel resonant RLC circuit, Series RLC tank circuit, Parallel RLC tank circuit, Undamped resonant frequency, Resonant frequency, Quality factor of the series resonant RLC circuit, Quality factor of the parallel resonant RLC circuit, Quality factor (general definition, interpretation, mechanical analogy, trade-off between Q-factor and inductance value), Bandwidth of the series resonant RLC circuit, Bandwidth of the parallel resonant RLC circuit, Half-power bandwidth, Upper half-power frequency, Lower half-power frequency, Duality of series/parallel RLC circuits, Ideal filter, Cutoff frequency, Second-order band-pass RLC filter, Second-order low-pass RLC filter, Second-order band-reject (or band-stop or notch) RLC filter, Second-order high-pass RLC filter, Quality factor of the filter circuit, Center frequency of the band-pass filter, Lower and upper half-power frequencies, Butterworth response, Quality factor of the nonideal inductor, Voltage multiplier circuit, Voltage multiplication, Near-field wireless link, Horseshoe coil

Section 10.1 Theory of Second-Order Resonant RLC Circuits

In this section, we study the last group of standard AC circuits—the resonators. They are second-order AC circuits in LC or RLC configuration. The term second order means that the circuits will be described by second-order differential equations if we work in the time domain. The value of a resonator circuit in electronics cannot be overstated. In order to proceed with any type of wireless communication, we first need to create a high-frequency AC signal as part of a resonator circuit. Beyond high-frequency circuits, resonators are often used in power electronics and as sensors. In this section, we apply the phasor/impedance method to analyze resonator circuits. We will discover that the most important characteristic is the *resonant frequency*. Another important parameter is the *quality factor*, which also determines the *resonator bandwidth*.

10.1.1 Self-Oscillating Ideal LC Circuit

The circuit shown in Fig. 10.1a includes an inductor and a capacitor and there is no power source connected to the circuit. The circuit is also *ideal*, which means that there is no resistance. In other words, the parasitic resistance of the inductor, parasitic resistance of the capacitor, and the wire resistance are all neglected. We assume that the power supply (voltage or current) was disconnected at $t = 0$, after the resonator was excited. The steady-state alternating current and the AC voltages across the circuit elements are sought once the oscillation process has been stabilized, i.e., at $t \to \infty$.

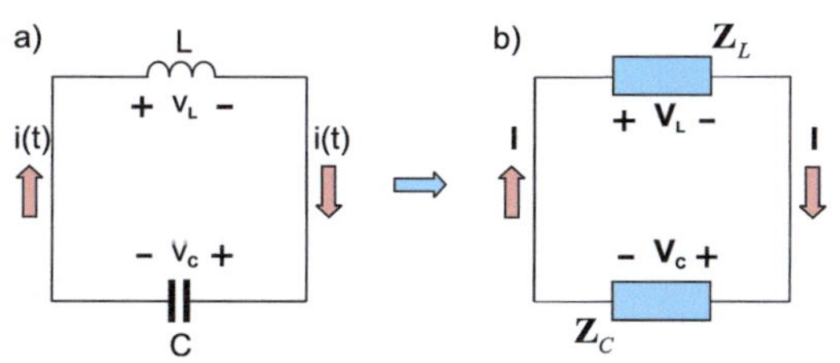

Fig. 10.1. Self-oscillating ideal LC circuit and its phasor representation.

When we apply the phasor/impedance method to the circuit in Fig. 10.1a, we obtain the circuit shown in Fig. 10.1b. KVL in phasor form yields (note the passive reference configuration)

$$\mathbf{V}_\mathrm{L} + \mathbf{V}_\mathrm{C} = 0 \Rightarrow \mathbf{Z}_\mathrm{L}\mathbf{I} + \mathbf{Z}_\mathrm{C}\mathbf{I} = 0 \Rightarrow (\mathbf{Z}_\mathrm{L} + \mathbf{Z}_\mathrm{C})\mathbf{I} = 0 \tag{10.1}$$

Generally, Eq. (10.1) requires the phasor current $\mathbf{I}$ to be zero. Obviously, if the phasor current is zero, then the real current is also zero and so are the voltages across the inductor and the capacitor. The circuit is not functioning. However, you should note that, if

$$\mathbf{Z}_\mathrm{L} + \mathbf{Z}_\mathrm{C} = 0 \tag{10.2}$$

in Eq. (10.1), the phasor current does not have to be zero and may have *any* value depending on the initial excitation. Equation (10.2) is satisfied at only *one* single frequency f_0:

$$\begin{aligned} &j\omega_0 L + \frac{1}{j\omega_0 C} = 0 \Rightarrow \text{(multiply by } j) \Rightarrow -\omega_0 L + \frac{1}{\omega_0 C} = 0 \Rightarrow \\ &\omega_0 = \frac{1}{\sqrt{LC}} \Rightarrow f_0 = \frac{1}{2\pi\sqrt{LC}} \end{aligned} \tag{10.3}$$

which is the *undamped resonant frequency* of the LC circuit. Equation (10.3) is perhaps the most important result of resonator theory. Once Eq. (10.3) is satisfied, the solution for the circuit current is obtained in the form:

$$\mathbf{I} = \mathbf{I}_0 \Rightarrow i(t) = I_\mathrm{m} \cos \omega_0 t \tag{10.4a}$$

The current amplitude I_m may be arbitrary; it is determined by the initial excitation. The voltages are found accordingly:

$$\begin{aligned} \mathbf{V}_\mathrm{L} &= \mathbf{Z}_\mathrm{L}\mathbf{I} \Rightarrow v_\mathrm{L}(t) = \omega_0 L I_\mathrm{m} \cos(\omega_0 t + 90^\circ) \\ \mathbf{V}_\mathrm{C} &= \mathbf{Z}_\mathrm{C}\mathbf{I} \Rightarrow v_\mathrm{C}(t) = 1/(\omega_0 C) I_\mathrm{m} \cos(\omega_0 t - 90^\circ) \end{aligned} \tag{10.4b}$$

The ideal *self-oscillating LC circuit* in Fig. 10.1 can oscillate indefinitely long. What is the physical basis of self-oscillations in an LC circuit? To answer this question, let us take a closer look at Eqs. (10.4). When the circuit current is at its maximum, the magnetic field energy stored in the inductor also has reached its maximum. Since the voltages are shifted by $\pm\pi/2$ versus the current, they are exactly zero at that time instance. The zero capacitor voltage means that no energy of the electric field is stored in the capacitor. All of the energy stored in the circuit is concentrated in the inductor. When the circuit current reaches zero, the situation becomes the opposite: the capacitor stores the entire circuit energy, and the inductor does not have any stored energy. As time progresses, the process continues so that the current flows back and forth in the circuit charging and discharging the capacitor (and in certain sense the inductor) periodically. Figure 10.2 shows the ideal mechanical counterpart of the circuit in Fig. 10.1. A massive wheel with a rotational inertia represents the inductor and the flexible membrane, the capacitor. The fluid flows back and forth either rotating the wheel (increasing its rotational energy) or bending the membrane (increasing its release energy).

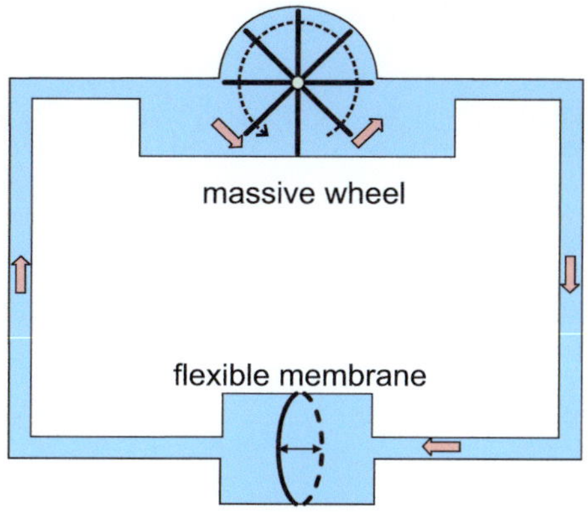

Fig. 10.2. Self-oscillating mechanical counterpart of the LC circuit shown in Fig. 10.1.

Example 10.1: An LC circuit in Fig. 10.1 has the circuit parameters $L = 1\ \mu\text{H}$, $C = 1\ \mu\text{F}$. Determine its resonant frequency, also known as the self-oscillation frequency.

Solution: Equation (10.3) is applied, which gives $\omega_0 = 10^6\ \text{rad/s} \Rightarrow f_0 = 159\ \text{kHz}$. For practical reasons, the resonant frequency is most often measured and reported in Hz, instead of rad/s.

10.1.2 Series Resonant Ideal LC Circuit

What if an alternating pressure pump is connected to the oscillating mechanical system in Fig. 10.2, which will add a small pressure "push" at every period of oscillation? Since there is no friction, the oscillations may grow up indefinitely. This phenomenon is known as *resonance*. The corresponding electrical counterpart is the circuit shown in Fig. 10.3a. The circuit in Fig 10.3a is solved using the phasor method, see Fig. 10.3b. We assume $v_S(t) = V_m \cos \omega t$. The equivalent impedance is given by,

$$\mathbf{Z}_{\text{eq}} = \mathbf{Z}_{\text{L}} + \mathbf{Z}_{\text{C}} = j\omega L + \frac{1}{j\omega C} = j\left(\omega L - \frac{1}{\omega C}\right) = -j\frac{1 - LC\omega^2}{\omega C} \quad [\Omega] \tag{10.5a}$$

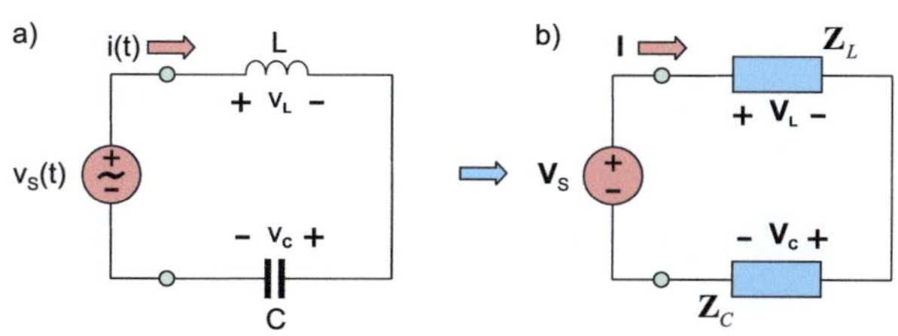

Fig. 10.3. Series resonant ideal LC circuit and its phasor representation.

The phasor voltages and phasor current become

$$\mathbf{I} = V_{\mathrm{m}}/\mathbf{Z}_{\mathrm{eq}} = \frac{V_{\mathrm{m}}\omega C}{1 - LC\omega^2}\angle 90^\circ, \quad \mathbf{V}_{\mathrm{L}} = \mathbf{Z}_{\mathrm{L}}\mathbf{I} = \frac{V_{\mathrm{m}}LC\omega^2}{1 - LC\omega^2}\angle 180^\circ,$$
$$\mathbf{V}_{\mathrm{C}} = \mathbf{Z}_{\mathrm{C}}\mathbf{I} = \frac{V_{\mathrm{m}}}{1 - LC\omega^2}\angle 0^\circ \quad (10.5\mathrm{b})$$

The real-valued circuit parameters are given by

$$i(t) = \frac{V_{\mathrm{m}}\omega C}{1 - LC\omega^2}\cos(\omega t + 90^\circ), \quad \upsilon_{\mathrm{L}}(t) = \frac{V_{\mathrm{m}}LC\omega^2}{1 - LC\omega^2}\cos(\omega t + 180^\circ),$$
$$\upsilon_{\mathrm{C}}(t) = \frac{V_{\mathrm{m}}}{1 - LC\omega^2}\cos(\omega t) \quad (10.5\mathrm{c})$$

The solution remains finite at any source frequency except the undamped resonant frequency $\omega_0 = 1/\sqrt{LC}$ or $f_0 = \omega_0/(2\pi)$. The closer the source frequency approaches the undamped resonant frequency, the higher the circuit current, capacitor voltage, and the inductor voltage become. Eventually, at the exact undamped resonant frequency, they all become infinitely high! The denominator in Eq. (10.5c) approaches zero and the circuit starts to "resonate." At the undamped resonant frequency, the impedances of the inductor and capacitor cancel out and their combination is a short circuit: an ideal wire of zero resistance. Moreover, the voltage source is shorted out. Note that the resonant frequency of an LC circuit was first derived by James Clerk Maxwell in 1868. A young man at this point, he spent a night working over this problem, which arose from an experiment of Sir William Grove, and wrote a report to him the next morning.

Example 10.2: Find the sum of the real-valued voltages $\upsilon_{\mathrm{L}}(t)$, $\upsilon_{\mathrm{C}}(t)$ in Fig. 10.3a.

Solution: The capacitor and inductor voltages are in *antiphase* (the phases differ by 180°). Therefore, they largely cancel out. According to Eq. (10.5c), the sum of the voltages is exactly the supply voltage $\upsilon_{\mathrm{S}}(t)$, irrespective of how high the individual voltages are.

Exercise 10.1: For the ideal series resonant ideal LC circuit in Fig. 10.3, determine the phasor voltages across the inductor and capacitor given that $V_{\mathrm{m}} = 1$ V and $\omega^2 = 0.9\omega_0^2$.

Answer: $\mathbf{V}_{\mathrm{L}} = 9\angle 180^\circ$ [V], $\mathbf{V}_{\mathrm{C}} = 10\angle 0^\circ$ [V]

10.1.3 Series Resonant RLC Circuit: Resonance Condition

The ideal LC circuit shown in Fig. 10.1a, or the series LC resonator shown in Fig. 10.3a, never exists in practice. Internal power supply resistance, wire resistance, or parasitic resistances of realistic capacitors and inductors lead to the realistic resonant RLC circuit model shown in Fig. 10.4a. Here, the resistance R models the combined resistances present in the circuit. The resistance reduces the resonant effect and leads to finite voltages/currents at the resonance.

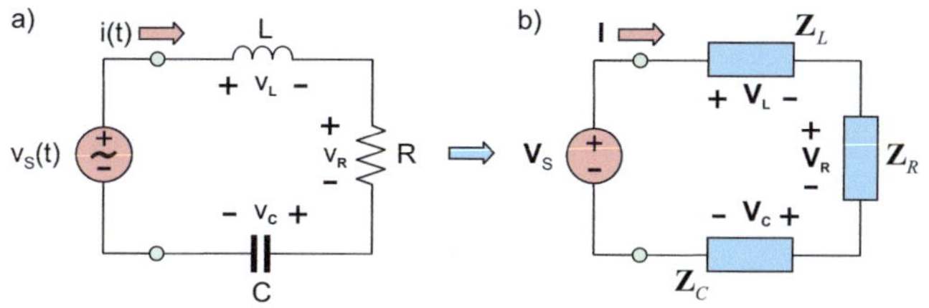

Fig. 10.4. Realistic series resonant RLC circuit and its phasor representation.

The circuit in Fig 10.4a is solved using the phasor method, see Fig. 10.4b. We again assume $v_S(t) = V_m \cos \omega t$. The equivalent impedance is given by

$$\mathbf{Z}_{eq} = \mathbf{Z}_R + \mathbf{Z}_L + \mathbf{Z}_C = R + j\omega L + \frac{1}{j\omega C} = R + j\left(\omega L - \frac{1}{\omega C}\right) \quad [\Omega] \tag{10.6a}$$

The *resonance condition* for *any* AC circuit, and not necessarily the circuit shown in Fig. 10.4, states that the imaginary part (the *reactance X*) of the equivalent circuit impedance *seen by the power source* must be equal to zero:

$$\mathrm{Im}(\mathbf{Z}_{eq}) = X = 0 \quad [\Omega] \tag{10.6b}$$

When applied to Eq. (10.6a), this condition defines the circuit's *resonant frequency* in the form $\omega_0 = 1/\sqrt{LC}$, $f_0 = 1/(2\pi\sqrt{LC})$. For the series RLC circuit, the resonant frequency clearly coincides with the undamped resonant frequency of the ideal LC circuit. Unfortunately, this is not always the case for general RLC circuits. This important question is addressed in the homework problems.

At the resonant frequency, the impedances of the inductor and capacitor in Eq. (10.6a) cancel out; their combination is a short circuit since only the resistance R remains. The real-valued circuit current and the real-valued voltages are given by

$$i(t) = \frac{V_m}{R}\cos(\omega_0 t), \quad v_L(t) = \frac{V_m}{R}\omega_0 L \cos(\omega_0 t + 90°), \quad v_C(t) = \frac{V_m}{R}\frac{1}{\omega_0 C}\cos(\omega_0 t - 90°) \tag{10.6c}$$

at resonance. Those are the *highest* amplitudes of the circuit current and the individual voltages that could be achieved in the series RLC circuit. If the frequency deviates from the resonant frequency, smaller amplitude values are obtained. When the circuit resistance is small, large circuit current and large capacitor and inductor voltages may be achieved at the resonance. You have to be aware of the fact that it is not uncommon to measure voltage amplitudes of 50–500 V across the individual elements in the laboratory, whereas the driving source voltage may only have an amplitude of 10 V. The circuit in Fig. 10.4 is also called *the series RLC tank circuit*.

Exercise 10.2: In the series resonant RLC circuit shown in Fig. 10.4, $V_m = 10$ V, $L = 50$ μH, $C = 0.5$ nF, $R = 50$ Ω. Determine the real-valued circuit current and the inductor/capacitor voltages at the resonance.

Answer:

$$i(t) = 0.2\cos(\omega t)\ [\mathrm{A}],\quad \upsilon_L(t) = 63.3\cos(\omega t + 90°)\ [\mathrm{V}],$$
$$\upsilon_C(t) = 63.3\cos(\omega t - 90°)\ [\mathrm{V}]. \tag{10.7}$$

Could we increase the resonant voltage amplitudes of the series RLC circuit in Fig. 10.4a [see Eq. (10.6c)] while keeping the voltage source and the circuit resistance unaltered? Yes we can. However, one more concept is required for this and similar problems: the concept of the *quality factor* of a resonator.

10.1.4 Quality Factor Q of the Series Resonant RLC Circuit

Multiple factors in front of resonant voltages and currents expressions can be reduced to one single factor. Using the definition of the resonant frequency ω_0, Eq. (10.6c) at the resonance may be rewritten in the simple form

$$i(t) = \frac{V_m}{R}\cos(\omega_0 t),\quad \upsilon_L(t) = QV_m\cos(\omega_0 t + 90°),\quad \upsilon_C(t) = QV_m\cos(\omega_0 t - 90°) \tag{10.8}$$

where the dimensionless constant

$$Q = \frac{\sqrt{L/R}}{\sqrt{RC}} = \frac{\sqrt{L/C}}{R} \tag{10.9}$$

is called the *quality factor of the series resonant RLC circuit*. The equivalent forms are

$$Q = \frac{1}{\omega_0 RC} = \omega_0\frac{L}{R} \tag{10.10}$$

Thus, in order to increase the resonant voltage amplitudes in Eq. (10.8), we should simply increase the quality factor of the resonator. Even if the circuit resistance remains the same, we can still improve Q by increasing the ratio of L/C in Eq. (10.9). This observation provides one physical interpretation of the quality factor: it determines the *maximum amplitude* of the resonant oscillations. A higher *Q-factor* results in larger amplitudes. Yet another, perhaps even more important, interpretation is related to the "sharpness" of the resonance at frequencies close to ω_0. What is the physical meaning of Eq. (10.9)? Why does the Q-factor increase with increasing the inductance but not the capacitance? To answer these questions, consider the fluid mechanics analogy in Fig. 10.2. The high Q implies a massive wheel (note: high inductance is equivalent to high wheel mass). Simultaneously, it implies a large membrane stiffness (the small capacitance, which is inversely proportional to the stiffness). The mechanical resonator so constructed will be less susceptible to losses at resonance but will not resonate at all if the driving force has a frequency far away from the resonance.

A *general definition* of the quality factor also applicable to mechanical engineering and physics is as follows. The quality factor is 2π times *the ratio per cycle of the energy stored in the resonator to the energy supplied by a source*, while keeping the signal amplitudes constant at the resonant frequency. According to Eq. (10.6c), the instantaneous energies stored in the inductor and capacitor are given by

$$\begin{aligned} E_{\mathrm{L}}(t) &= \frac{1}{2}Li^2(t) = \left(\frac{V_{\mathrm{m}}}{R}\right)^2 \frac{L}{2}\cos^2(\omega_0 t), \\ E_{\mathrm{C}}(t) &= \frac{1}{2}Cv_{\mathrm{C}}^2(t) = \left(\frac{V_{\mathrm{m}}}{R}\right)^2 \frac{1}{2\omega_0^2 C}\sin^2(\omega_0 t) \end{aligned} \tag{10.11}$$

Since $1/(\omega_0^2 C) = L$, the coefficients in front of the cosine squared and sine squared terms are equal. It means that even though both energies vary over time, their sum is a *constant*:

$$E_{\mathrm{L}}(t) + E_{\mathrm{C}}(t) = \left(\frac{V_{\mathrm{m}}}{R}\right)^2 \frac{L}{2} \tag{10.12a}$$

The energy dissipated in the resistance is the integral of the instantaneous absorbed power over the period; this integral is equal to

$$E_{\mathrm{diss}} = \int_0^{2\pi/\omega_0} \frac{V_{\mathrm{m}}^2}{R}\cos^2(\omega_0 t)dt = \frac{V_{\mathrm{m}}^2}{2R}\int_0^{2\pi/\omega_0} (1 + \cos(2\omega_0 t))dt = \frac{\pi V_{\mathrm{m}}^2}{\omega_0 R} \tag{10.12b}$$

The ratio of the two energies times 2π precisely equals Eq. (10.9).

Example 10.3: A series resonant LC circuit is driven by a laboratory AC voltage source with amplitude $V_m = 10$ V and an internal resistance of 50 Ω (a function generator). Which value should the ratio *L*/*C* have to obtain the amplitude of the capacitor voltage to be equal to 50 V at resonance?

Solution: We replace the realistic voltage source by its Thévenin equivalent: the ideal voltage source with the amplitude of 10 V and the series resistance of 50 Ω. Here, we again arrive at the standard RLC circuit shown in Fig. 10.4a. According to Eq. (10.8), the *Q*-factor of the RLC circuit should be equal to 5. From the definition of the *Q*-factor given by Eq. (10.9), one has

$$\sqrt{L/C} = RQ = 250 \Rightarrow L/C = 62,500\ \Omega^2 \tag{10.13}$$

This result is valid for *any* resonant frequency. For example, the set $L = 1$ mH, $C = 16$ nF will give us the desired amplitude value.

Note that large *Q*-factors usually imply large inductances which increase the series resistance of the inductor coil and thus increase the net circuit resistance (increase circuit loss). Therefore, there is a *trade-off* between the circuit *Q* and the inductance value.

Example 10.4: A series resonant RLC circuit is needed with the resonant frequency of 100 kHz and a *Q*-factor of 50. The circuit resistance is 10 Ω. Determine the necessary values of *L* and *C*.

Solution: From the definition of the resonant frequency and the *Q*-factor, we obtain

$$\begin{aligned} &1/\sqrt{LC} = 2\pi \times 10^5, \quad \sqrt{L/C} = RQ = 500 \Rightarrow \\ &\quad \frac{1}{C} = 2\pi \times 10^5 \times 500 \Rightarrow C = 3.2\ \text{nF} \end{aligned} \tag{10.14}$$

Consequently, $L = 0.80$ mH.

10.1.5 Bandwidth of the Series Resonant RLC Circuit

The *bandwidth of the series resonant RLC circuit* may be obtained by analyzing the behavior of the amplitude of the circuit current as a function of *source frequency*. An alternative definition implies analyzing the behavior of the amplitude of the resistor voltage; however, both quantities are equal to within a constant *R*. The phasor current at any frequency is obtained from Eq. (10.6a). It will be written here in terms of quality factor *Q*, frequency *f*, and resonant frequency f_0 in the form:

$$\mathbf{I} = \frac{V_m}{\mathbf{Z}_{eq}} = \frac{V_m}{R + j\left(\omega L - \frac{1}{\omega C}\right)} = I_m \frac{1}{1 + jQ\left(\frac{f}{f_0} - \frac{f_0}{f}\right)} \quad (10.15a)$$

Here, $I_m = V_m/R$ is the maximum (resonant) current amplitude in the series RLC circuit; $f_0 = 1/\left(2\pi\sqrt{LC}\right)$ is the resonant frequency in Hz. The real-valued circuit current and the real-valued resistor voltage are both found from the phasor current given by Eq. (10.15a):

$$\left.\begin{array}{l} i(t) = I_m H \cos(\omega t + \varphi) \ [\mathrm{A}] \\ v_R(t) = V_m H \cos(\omega t + \varphi) \ [\mathrm{V}] \end{array}\right\}, \quad H(f) = \frac{1}{\sqrt{1 + Q^2\left(\frac{f}{f_0} - \frac{f_0}{f}\right)^2}} \quad (10.15b)$$

Here, $H(f)$ is the dimensionless function of frequency that peaks at the resonant frequency $f = f_0, H(f_0) = 1$. One may treat $H(f)$ as an *amplitude transfer function* of an associated RLC filter with the input voltage being the source voltage and the output voltage being the resistor voltage. In this case, $H(f)$ is equal to the amplitude ratio of the two voltages. Simultaneously, $H(f)$ characterizes how fast the circuit current amplitude decays when the circuit frequency deviates from the resonant frequency f_0. To be specific, we assume $f_0 = 10$ kHz and $Q = 1,\ 2,\ 5$ in Eq. (10.15b). Figure 10.5 plots the function $H(f)$ in decibels, $H(f)_{dB} = 20\log_{10}(H(f))$ [dB], using a log-log scale, i.e., creates its *Bode plot*.

The *bandwidth B of the series resonant RLC circuit* is defined as the interval of frequencies over which the function $H(f)$ is greater than or equal to $1/\sqrt{2} = 0.707$. In other words, the signal power at the resistor (which is proportional to the square of $H(f)$) is no less than 50 % of the maximum power at the exact resonance. We call the bandwidth so defined the *half-power bandwidth*. In terms of the transfer function $H(f)_{dB}$ in decibels, this condition corresponds to the inequality $H(f)_{dB} \geq -3$ dB.

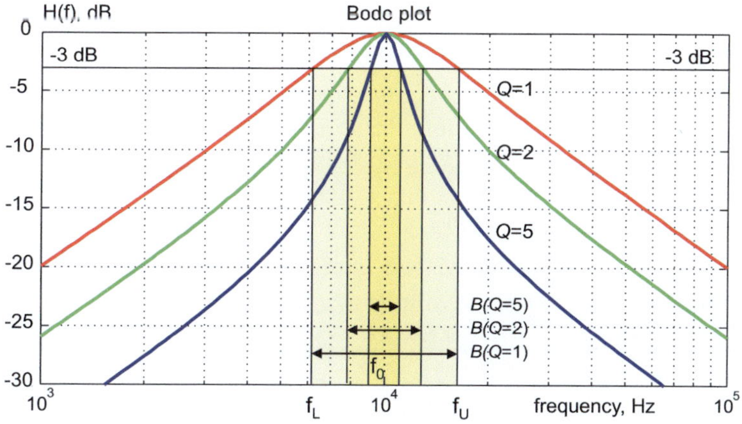

Fig. 10.5. Amplitude of the circuit current (or amplitude of the resistor voltage) normalized by its peak value at resonance. The resulting graph is the Bode plot.

The first resonant curve in the form of Fig. 10.5 was published by Heinrich Hertz in 1887 although he used a linear, not a logarithmic, frequency scale so that the curve did not look quite symmetric. Figure 10.5 indicates that the bandwidth increases when the quality factor decreases and vice versa. In other words, the low-Q resonant circuit has a large bandwidth; we may say it resonates "equally bad" over a wider band of frequencies. However, the high-Q circuit has a small bandwidth; it resonates well but only over a small band of frequencies. The *lower and upper half-power frequencies*, f_L, f_U, are obtained by setting $H(f) = 1/\sqrt{2}$ in Eq. (10.15b) and solving for f. The resulting quadratic equation has two roots:

$$\begin{aligned} f_L &= f_0\left(\sqrt{1 + 1/(2Q)^2} - 1/(2Q)\right), \\ f_U &= f_0\left(\sqrt{1 + 1/(2Q)^2} + 1/(2Q)\right) \quad \text{[Hz]} \end{aligned} \tag{10.16}$$

Despite the complexity of those expressions, the final result for the half-power bandwidth is surprisingly simple and understandable:

$$B \equiv f_U - f_L = \frac{f_0}{Q} = \frac{R}{2\pi L} \quad \text{[Hz]} \tag{10.17}$$

Exercise 10.3: Determine the bandwidth B of the series resonant RLC circuit with the resonant frequency of 1 MHz and a Q-factor of 10.

Answer: $B = 100$ kHz.

Example 10.5: A series resonant RLC circuit is needed with a resonant frequency of 500 kHz and a bandwidth of 20 kHz. Given the circuit resistance of 15 Ω, determine L and C.

Solution: From the bandwidth definition, the required Q-factor is equal to $500/20 = 25$. Further, we may follow the solution developed in Example 10.4. From the definition of the resonant frequency and the Q-factor, we subsequently obtain $1/\sqrt{LC} = 2\pi \times 5 \times 10^5$, $\sqrt{L/C} = RQ = 375 \Rightarrow C \approx 0.85$ nF. Next, we determine $L = C(RQ)^2 \approx 0.12$ mH. Alternatively, one could find the inductance L directly from Eq. (10.17), that is, $L = R/(2\pi B) = 0.12$ mH.

Example 10.6: Create Bode plots as seen in Fig. 10.5 using MATLAB.

Solution: We create only one bandwidth curve, for $Q = 2$. Other curves are obtained similarly, using the command `hold on`.

```
f        = logspace(3, 5, 101); % frequency vector, Hz(from 10^3 to 10^5 Hz)
f0       = 1e4;                 % resonant frequency, Hz
Q        = 2;                   % quality factor, dimensionless
H        = 1./sqrt(1+Q^2*(f/f0-f0./f).^2);
HdB      = 20*log10(H);
semilogx(f, HdB, 'r'); grid on;
title('Bode plot');
xlabel('frequency, Hz'); ylabel('H, dB')
axis([min(f) max(f) -30 0])
```

10.1.6 Parallel Resonant RLC Circuit: Duality

The parallel resonant RLC circuit is shown in Fig. 10.6. It is driven by an alternating current source $i_S(t) = I_m \cos \omega t$. The parallel RLC resonator is a current divider circuit, which is the dual to the series RLC resonator (which is a voltage divider) in Fig. 10.4. While the series RLC resonator is capable of creating large voltages (or "amplifying" the supply voltage), the parallel RLC resonator circuit is capable of producing large currents. The amplitudes of the currents through the inductor and the capacitor may be large, much larger than the supply current itself. The circuit in Fig. 10.6 is also called *the parallel RLC tank circuit.*

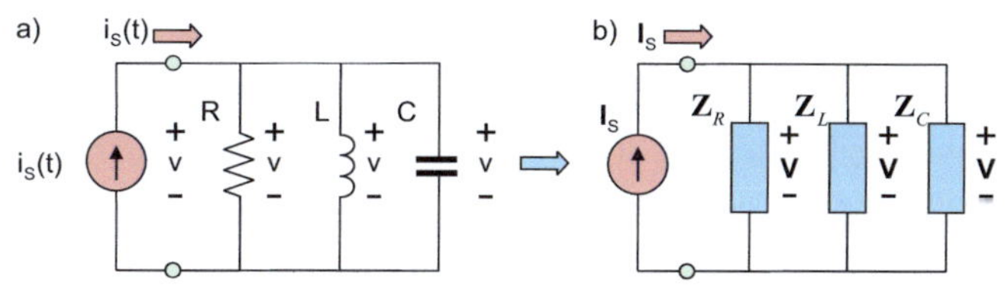

Fig. 10.6. Parallel resonant RLC circuit and its phasor representation.

The circuit in Fig 10.6a is solved by using the phasor method; see Fig. 10.6b. The equivalent impedance is given by

$$\frac{1}{\mathbf{Z}_{eq}} = \frac{1}{\mathbf{Z}_R} + \frac{1}{\mathbf{Z}_L} + \frac{1}{\mathbf{Z}_C} = \frac{1}{R} + \frac{1}{j\omega L} + j\omega C = \frac{1}{R} - j\frac{1 - LC\omega^2}{\omega L} \quad [\Omega] \tag{10.18}$$

The *resonance condition* for *any* AC circuit states that the impedance $\mathbf{Z}_{eq}$ must be a purely real number at resonance. If the impedance is real, its reciprocal, the *admittance*, is also real and vice versa. Therefore, from Eq. (10.18), we obtain the *resonant frequency*

$$\omega_0 = \frac{1}{\sqrt{LC}}, \; f_0 = \frac{1}{2\pi\sqrt{LC}} \tag{10.19}$$

which *coincides* with the resonant frequency of the series RLC tank circuit and with the undamped resonant frequency of the LC circuit. Thus, at resonance, $\mathbf{Z}_{\mathrm{eq}} = R$ and one has $\mathbf{V} = RI_{\mathrm{m}}$ for the phasor voltage in Fig. 10.6b. Knowing the phasor voltage, we can establish the phasor currents. The corresponding real-valued voltage and currents at resonance take on the forms

$$\begin{aligned} &v(t) = RI_{\mathrm{m}}\cos(\omega_0 t), \quad i_{\mathrm{L}}(t) = \frac{RI_{\mathrm{m}}}{\omega_0 L}\cos(\omega_0 t - 90^\circ), \\ &i_{\mathrm{C}}(t) = RI_{\mathrm{m}}(\omega_0 C)\cos(\omega_0 t + 90^\circ) \end{aligned} \tag{10.20}$$

The amplitude of the circuit voltage, along with the amplitudes of inductor and capacitor currents in Eq. (10.20), reaches a maximum at resonance. Next, we wish to introduce the Q-factor of the circuit, similar to Eq. (10.8) for the series resonator, that is,

$$\begin{aligned} &v(t) = RI_{\mathrm{m}}\cos(\omega_0 t), \quad i_{\mathrm{L}}(t) = I_{\mathrm{m}}Q\cos(\omega_0 t - 90^\circ), \\ &i_{\mathrm{C}}(t) = I_{\mathrm{m}}Q\cos(\omega_0 t + 90^\circ) \end{aligned} \tag{10.21}$$

Comparing Eq. (10.20) with Eq. (10.21), we obtain a different expression:

$$Q = \frac{\sqrt{RC}}{\sqrt{L/R}} = \omega_0 RC = \frac{R}{\omega_0 L} \tag{10.22}$$

which is exactly the reciprocal of the Q-factor of the series RLC circuit. This means that a high-Q parallel resonant circuit will require higher capacitances than inductances.

Fortunately, all the results related to the series resonant RLC circuit can directly be converted to the parallel RLC resonant circuit using the substitutions:

$$v(t) \to Ri(t), \quad i_{\mathrm{L}}(t) \to v_{\mathrm{C}}(t)/R, \quad i_{\mathrm{C}} \to v_{\mathrm{L}}(t)/R \tag{10.23a}$$

Here, the left-hand side corresponds to the parallel RLC circuit, whereas the right-hand side corresponds to the series RLC circuit. Furthermore, we need to replace V_{m} by RI_{m} and interchange the role of two partial time constants:

$$\frac{L}{R} \leftrightarrow RC \tag{10.23b}$$

in the original solution for the series RLC circuit; see Eq. (10.22). The solution so constructed will match exactly the solution of the parallel RLC circuit depicted in

Fig. 10.6. This fact is proved by direct substitution. Thus, Eqs. (10.23) reflect the *duality* of the series/parallel RLC AC steady-state electric circuits driven by voltage and current sources, respectively. It means that the results established for one circuit may be applied to the other circuit and vice versa. A similar duality is established for the transient RLC circuits. Consequently, we can concentrate our study on the series RLC circuit.

Exercise 10.4: For the parallel resonant RLC circuit in Fig. 10.6, determine resonant phasor currents $\mathbf{I}_R$, $\mathbf{I}_L$, and $\mathbf{I}_C$ if $I_m = 100$ mA, $L = 30$ μH, $C = 1$ μF, and $R = 100$ Ω.

Answer: $\mathbf{I}_R = \mathbf{I}_S = 0.1\angle 0°$, $\mathbf{I}_L = 1.83\angle -90°$, $\mathbf{I}_C = 1.83\angle +90°$ [A].

Example 10.7: The circuit voltage for the parallel RLC circuit in Fig. 10.6 at any frequency may be written in the form $v(t) = RI_m H(f)\cos(2\pi f t + \varphi)$ where $H(f)$ is a dimensionless amplitude transfer function, which peaks at the resonant frequency, $H(f_0) = 1$. Create the Bode plot of $H(f)$ at $f_0 = 10$ kHz and for $Q = 1, 2, 5$.

Solution: The amplitude transfer function $H(f)$ for the parallel RLC circuit coincides with the expression (10.15b) for the series RLC circuit. The Bode plot also *coincides* with the corresponding result for the series RLC circuit shown in Fig. 10.5. However, the Q-factor is now given by Eq. (10.22). The bandwidth of the parallel resonant circuit is still given by the expression $B = f_0/Q$ [Hz] but with the modified Q-factor. This results in $B = 1/(2\pi RC)$ [Hz].

Section 10.2 Construction of Second-Order RLC Filters

An immediate application of the RLC resonant circuits relates to the concept of filter design. We have already studied the first-order low-pass and high-pass filters on the basis of RC and RL steady-state AC circuits. In fact, the RLC circuits could also be used as low-pass and high-pass filters. They perform even better, i.e., more closely linked to the frequency response of the *ideal filter*, which implies passing all the frequency components below (or above) a certain *cutoff frequency* and rejecting all other frequency components. In other words, the frequency responses of the RLC low-pass and high-pass filters are "steeper" than first-order filters. Not only that, resonant RLC circuits can form *band-pass* and *band-reject* (or *band-stop* or *notch*) filters, a task which is impossible with first-order RC or RL circuits.

10.2.1 Second-Order Band-Pass RLC Filter

A series RLC circuit is shown in Fig. 10.7a. We consider the supply voltage as the *input voltage* $v_{\mathrm{in}}(t)$ into the filter and the resistor voltage $v_{\mathrm{R}}(t)$ as the *output voltage* $v_{\mathrm{out}}(t)$. Figure 10.7b depicts the corresponding circuit transformation. This transformation implies that the input voltage is provided by another circuit block and the output voltage is passed to another circuit block. The circuit in Fig. 10.7b is thus a *two-port network*. Qualitatively then, when the frequency of the input voltage is the resonant frequency of the RLC circuit, the LC block in Fig. 10.7 is replaced by a short circuit (a wire). The input voltage *passes* through unchanged. However, if the frequency differs from the resonant one, the LC block exhibits a large finite impedance that is added to the resistance *R*. As a result, the circuit current decreases in amplitude, as does the output voltage (voltage across the resistor), which is proportional to the current. Those frequencies are thus rejected. The filter so constructed is known as a *second-order band-pass RLC filter.*

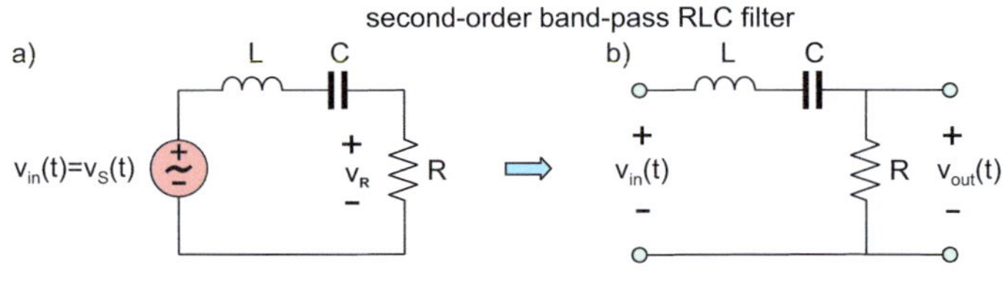

Fig. 10.7. Transformation of the series RLC circuit into the band-pass analog RLC filter.

We assume the source voltage (the input filter voltage) is given in the form $v_{\mathrm{S}}(t) = V_{\mathrm{m}} \cos \omega t$. The phasor current for the corresponding series RLC circuit was found in the previous section (see Eq. (10.15a)):

$$\mathbf{I} = \frac{V_m/R}{1 + jQ\left(\frac{f}{f_0} - \frac{f_0}{f}\right)} \tag{10.24a}$$

Here, $Q = \sqrt{L/C}/R$ is the corresponding quality factor of the series RLC circuit (*quality factor of the filter circuit*), and $f_0 = 1/\sqrt{LC}$ is its resonant frequency. The complex filter transfer function is defined by the ratio of the corresponding phasors:

$$\mathbf{H}(f) \equiv \frac{\mathbf{V}_R}{\mathbf{V}_S} = \frac{R\mathbf{I}}{V_m} \tag{10.24b}$$

Substitution of Eq. (10.24a) into Eq. (10.24b) gives the transfer function in the form:

$$\mathbf{H}(f) = \frac{1}{\sqrt{1 + Q^2\left(\frac{f}{f_0} - \frac{f_0}{f}\right)^2}} \angle - \tan^{-1}\left(Q\left(\frac{f}{f_0} - \frac{f_0}{f}\right)\right) \tag{10.24c}$$

which is equivalent to Eq. (10.15b) of the previous section. Therefore, the results derived for the series resonant RLC circuit are also valid for the band-pass RLC filter in Fig. 10.7. In particular, the *center frequency of the band-pass filter* is the circuit resonant frequency. The *half-power bandwidth of the filter* is given by Eq. (10.17), i.e., $B = f_0/Q = R/(2\pi L)$, and the *lower and upper half-power frequencies* are known from Eq. (10.16) of the previous section. Second-order filter circuits are designed by choosing the values of R, L, C in such a way as to obtain the required values of Q and f_0 (the filter center frequency and the required bandwidth). Thus, we have two equations for three unknowns. The remaining degree of freedom is used to match the filter impedances.

Example 10.8: A band-pass RLC filter is required with the center (resonant) frequency of 1 MHz and a half-power bandwidth B of 100 kHz. Create amplitude and phase Bode plots for the filter in the frequency band from 100 kHz to 10 MHz.

Solution: Clearly, $f_0 = 1$ MHz. The quality factor of the RLC circuit is found to be $Q = f_0/B = 10$. We plot the magnitude of the transfer function, $H(f)$, in decibels and its phase in degrees according to Eq. (10.24c). The result is shown in Fig. 10.8. You should note that far away from the passband, the filter follows a 20-dB-per-decade roll-off.

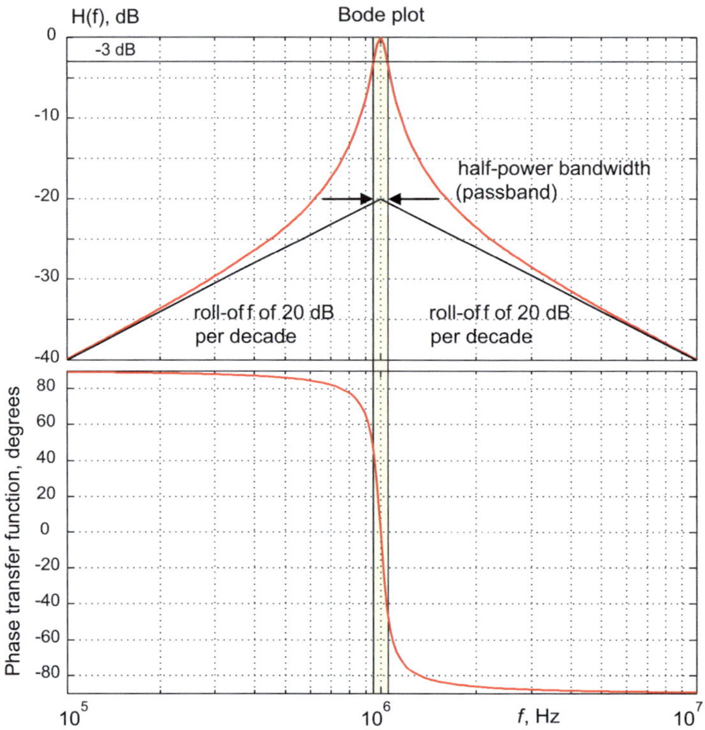

Fig. 10.8. Amplitude and phase Bode plot of a band-pass series RLC filter with $Q = 10$.

Exercise 10.5: In the band-pass filter circuit of Fig.10.7, $L = 100$ μH, $C = 15.9$ μF, and $R = 10\ \Omega$. What is the filter bandwidth?

Answer: 15.9 kHz.

Example 10.9: In the previous example, determine the necessary values of L and C given $R = 20\ \Omega$.

Solution: From the definition of the resonant frequency and the Q-factor, we obtain $1/\sqrt{LC} = 2\pi \times 10^6$, $\sqrt{L/C} = RQ = 200 \Rightarrow C \approx 796$ pF. Then, we find the required inductance, $L = C(RQ)^2 \approx 31.8$ μH. Alternatively, one could find the inductance L directly from the definition of the bandwidth for the series RLC resonator, that is, $L = R/(2\pi B) = 31.8$ μH.

10.2.2 Second-Order Low-Pass RLC Filter

A series RLC circuit is again shown in Fig. 10.9a. We consider the power supply AC voltage as the *input voltage* $v_{in}(t)$ into the filter. We monitor the capacitor voltage $v_C(t)$ as the *output voltage* $v_{out}(t)$ of the filter. Figure 10.9b depicts the corresponding circuit.

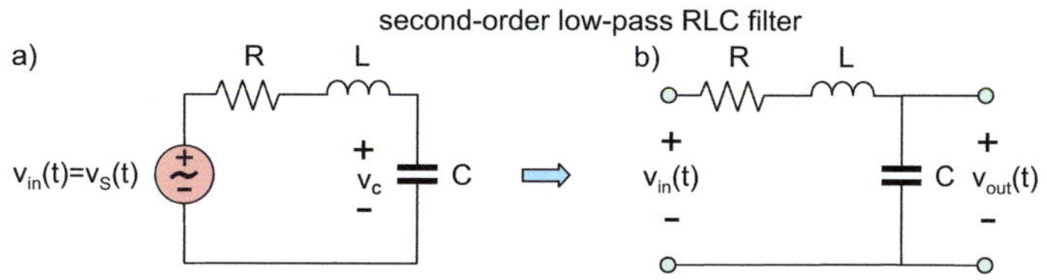

Fig. 10.9. Transformation of the series RLC circuit into the low-pass analog RLC filter.

The filter so constructed is a *second-order low-pass RLC filter*. Qualitatively then, when the frequency of the input voltage is low, the inductor behaves as a short circuit and the capacitor as an open circuit. The input voltage *passes* through unchanged. However, if the frequency is higher than the resonant frequency, both the inductor and the capacitor prevent transmission: the capacitor shorts out the output voltage, whereas the inductor reduces the circuit current. The complex filter transfer function is defined by the ratio of the corresponding phasors:

$$\mathbf{H}(f) \equiv \frac{\mathbf{V}_C}{\mathbf{V}_S} = \frac{\mathbf{I}}{j\omega C V_m} \tag{10.25a}$$

We substitute the expression for the phasor current of the series RLC circuit from Eq. (10.24a) and obtain the transfer function in the form:

$$\mathbf{H}(f) = Q\frac{f_0}{f}\frac{1}{\sqrt{1+Q^2\left(\frac{f}{f_0}-\frac{f_0}{f}\right)^2}} \angle -\frac{\pi}{2} - \tan^{-1}\left(Q\left(\frac{f}{f_0}-\frac{f_0}{f}\right)\right) \tag{10.25b}$$

Example 10.10: A low-pass RLC filter is required with a *passband* from 0 to 1 MHz. In other words, the low-pass *filter bandwidth*, which extends from zero hertz to the *half-power frequency*, should be 1 MHz. Create amplitude and phase Bode plots for the filter in the frequency band from 100 kHz to 10 MHz.

Solution: The critical point for the low-pass RLC filter design is the proper selection of the quality factor. The amplitude transfer function in Eq. (10.25b) can exhibit a sharp peak in the passband with its value higher than one. Such a peak (further investigated in the homework problems) occurs only for $Q = 1/\sqrt{2}$. The value $Q = 1/\sqrt{2}$ corresponds to the maximally flat but still steep transfer function (*maximally flat Butterworth*

Example 10.10 (cont.): *response*). We will use this value in Eq. (10.25b). Then, the half-power or 3-dB frequency of the filter will be *exactly* the resonant frequency f_0. The resulting Bode plots are shown in Fig. 10.10. The transfer function of the filter has 40-dB-per-decade roll-off.

Figure 10.10 shows three amplitude responses: for an ideal filter with the cutoff frequency of 1 MHz, for a second-order RLC filter with the 3-dB frequency which coincides with $f_0 = 1$ MHz, and for a first-order RC (or RL) filter with the break (half-power) frequency $f_b = f_0 = 1$ MHz. Clearly, the second-order filter better approaches the desired ideal response. This observation encourages us to consider filters of higher order.

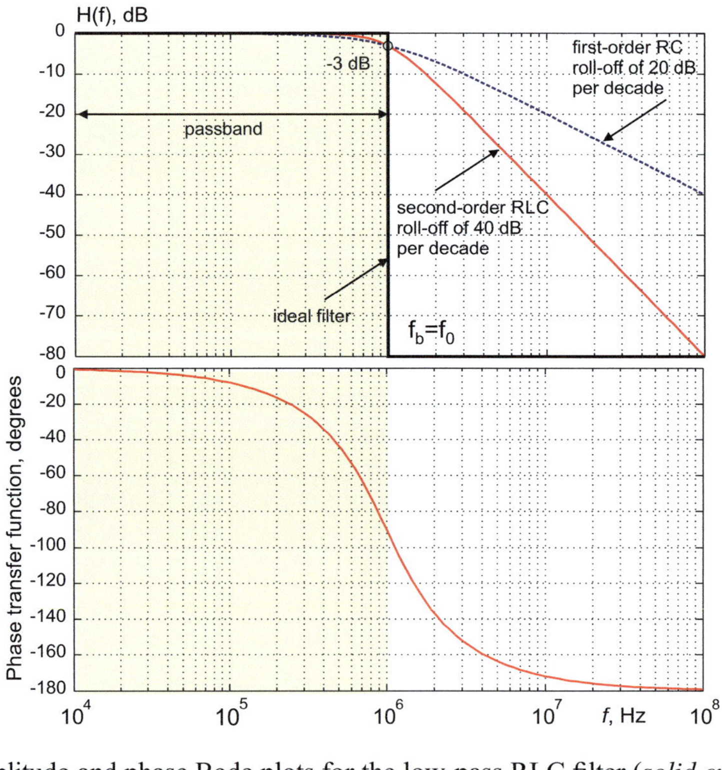

Fig. 10.10. Amplitude and phase Bode plots for the low-pass RLC filter (*solid curves*) compared with a first-order RC filter (*dotted curve*).

10.2.3 Second-Order High-Pass RLC Filter

A series RLC circuit is again shown in Fig. 10.11a. We consider the power supply AC voltage as the *input voltage* $\upsilon_{in}(t)$ into the filter. We next consider the inductor voltage $\upsilon_L(t)$ as the *output voltage* $\upsilon_{out}(t)$ of the filter, see Fig. 10.11b. The constructed circuit is a *second-order high-pass RLC filter*. Qualitatively, when the frequency of the input voltage is low, the capacitor behaves like an open circuit, while the inductor behaves like a short circuit. Both the inductor and the capacitor prevent transmission. However, if the frequency is higher than the resonant frequency, the capacitor becomes a short circuit and the inductor becomes an open circuit. The input voltage is *passed* through the filter nearly unchanged. The complex filter transfer function is defined by the ratio of the corresponding phasors:

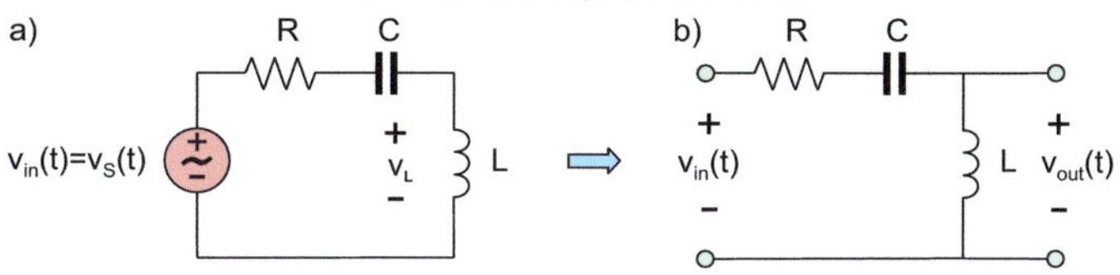

Fig. 10.11. Transformation of the series RLC circuit into the high-pass analog RLC filter.

$$\mathbf{H}(f) \equiv \frac{\mathbf{V}_L}{\mathbf{V}_S} = \frac{j\omega L\mathbf{I}}{V_m} \tag{10.26a}$$

We substitute the expression for the phasor current of the series RLC circuit from Eq. (10.24a) and obtain the transfer function in the form:

$$\mathbf{H}(f) = Q\frac{f}{f_0}\frac{1}{\sqrt{1+Q^2\left(\frac{f}{f_0}-\frac{f_0}{f}\right)^2}}\angle\frac{\pi}{2} - \tan^{-1}\left(Q\left(\frac{f}{f_0}-\frac{f_0}{f}\right)\right) \tag{10.26b}$$

The amplitude transfer function of the high-pass filter in Eq. (10.26b) is the mirror reflection of the amplitude transfer function for the low-pass filter in Eq. (10.25b) if a logarithmic frequency scale is used. This fact is seen by substituting $f \leftrightarrow 1/f, f_0 \leftrightarrow 1/f_0$, which makes both expressions identical.

Example 10.11: A high-pass RLC filter is required with the *passband* from 0 to 1 MHz. The high-pass filter *half-power frequency* should be 1 MHz. Create amplitude and phase Bode plots for the filter in the band from 100 kHz to 10 MHz.

Solution: The important point for the high-pass RLC filter design is again the proper selection of the quality factor. Similar to the low-pass filter, we choose the value $Q = 1/\sqrt{2}$, which corresponds to the maximally flat transfer function (*Butterworth response*). Then, the half-power or 3-dB frequency of the filter will be *exactly* the resonant frequency f_0. The resulting Bode plots are shown in Fig. 10.12 in comparison with the transfer function of the first-order high-pass filter. The amplitude transfer function of the filter again has the 40-dB-per-decade roll-off.

Exercise 10.6: A band-pass filter is as a series combination of the second-order low-pass RLC filter and the second-order high-pass RLC filter, respectively. Both filters have the same half-power frequency. What is the transfer function roll-off far away from the passband per one octave?

Answer: 12 dB.

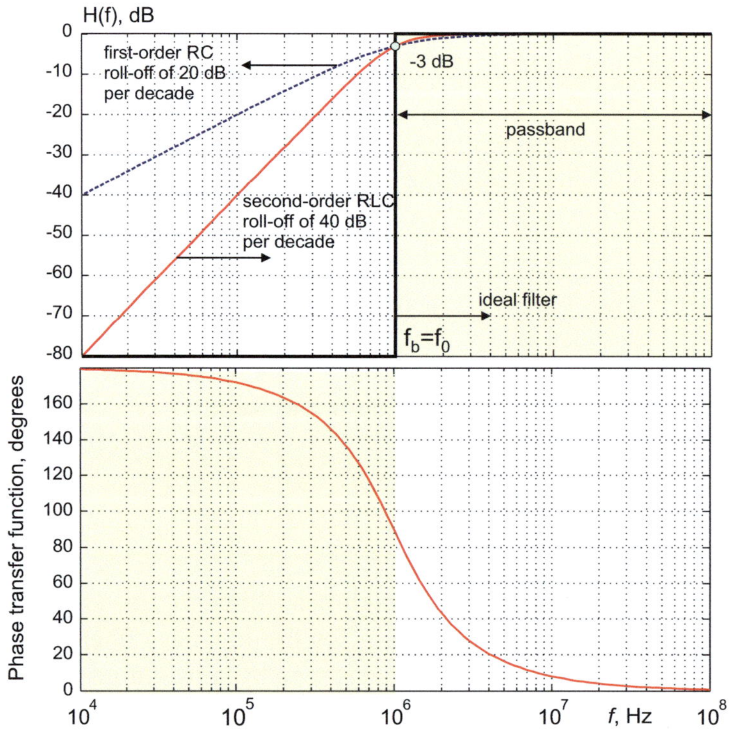

Fig. 10.12. Amplitude and phase Bode plots for the high-pass RLC filter (*solid curves*) and amplitude comparison with a first-order RC filter (*dotted curve*).

10.2.4 Second-Order Band-Reject RLC Filter

A series RLC circuit is again shown in Fig. 10.13a. We consider the power supply AC voltage as the *input voltage* $v_{in}(t)$ into the filter, and the voltage $v_{LC}(t)$ across the LC block is recorded as the *output voltage* $v_{out}(t)$. Figure 10.13b depicts the corresponding filter circuit.

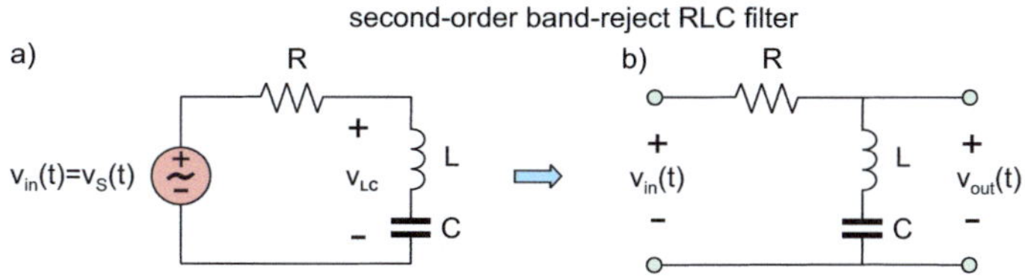

Fig. 10.13. Transformation of the series RLC circuit into the band-reject analog RLC filter.

The filter so constructed is a *second-order band-reject* (known as *band-stop* or *notch*) *RLC filter.* At resonance, the LC block forms a short circuit: the output filter voltage is thus shorted out. All other frequencies pass through. This filter is useful when a single tone (e.g., 60 Hz) needs to be rejected. By KVL, its transfer function is equal to one minus the transfer function of the band-pass filter in Eq. (10.24c), i.e.,

$$\mathbf{H}(f)=1-\frac{1}{1+jQ\left(\frac{f}{f_0}-\frac{f_0}{f}\right)}=Q\left|\frac{f}{f_0}-\frac{f_0}{f}\right|\frac{1}{\sqrt{1+Q^2\left(\frac{f}{f_0}-\frac{f_0}{f}\right)^2}}\angle\frac{\pi}{2}-\tan^{-1}\left(Q\left(\frac{f}{f_0}-\frac{f_0}{f}\right)\right) \tag{10.27}$$

Therefore, the filter behavior is the opposite of the band-pass filter previously analyzed.

Example 10.12: A band-reject RLC filter is required with the center frequency of 1 MHz and the half-power bandwidth, *B*, of 100 kHz. Create amplitude and phase Bode plots for the filter in the frequency band from 100 kHz to 10 MHz.

Solution: The quality factor of the RLC circuit is found to be $Q=f_0/B=10$. We plot the magnitude of the transfer function *H(f)* in decibels and its phase in degrees according to Eq. (10.27). The result is shown in Fig. 10.14. The filter response is very steep over the specified frequency range. We can lower the *Q*-factor, which will lead to a wider bandwidth. Note that the amplitude transfer function formally equals zero at the exact resonant frequency. This result is physically unrealizable since real inductors have a small parasitic series resistance.

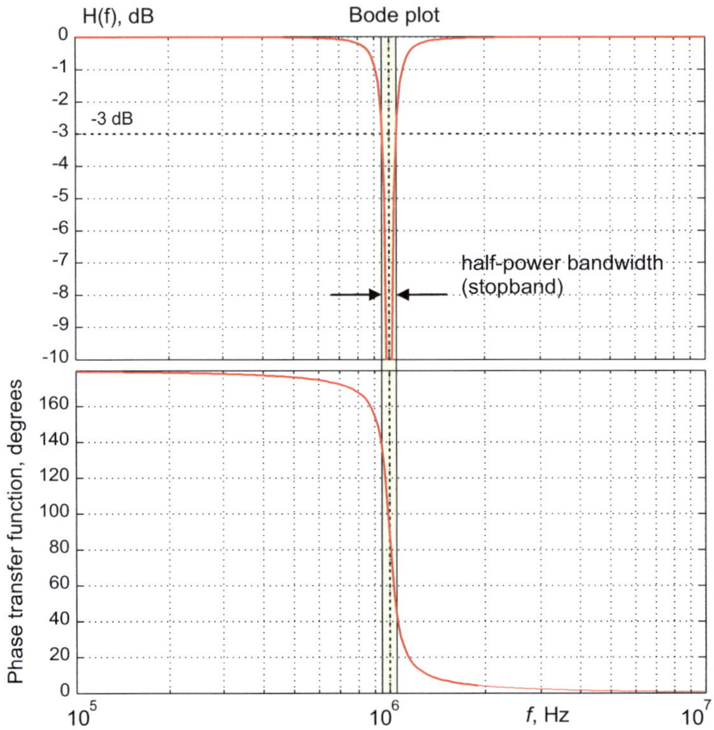

Fig. 10.14. Amplitude and phase Bode plots for the band-reject series RLC filter.

10.2.5 Second-Order RLC Filters Derived from the Parallel RLC Circuit

All second-order filters considered so far are derived from the *series* RLC circuit, with the same quality factor given by $Q = \omega_0 L/R$. The *natural* structure *after shorting out the input voltage source* is shown in Fig. 10.15a. A complementary group of these filter circuits exists; after shorting out the input voltage source, its natural structure is that of the parallel RLC circuit seen in Fig. 10.15b. These circuits operate quite similarly, but all of them have the quality factor of the parallel RLC resonator, that is, $Q = \omega_0 RC$.

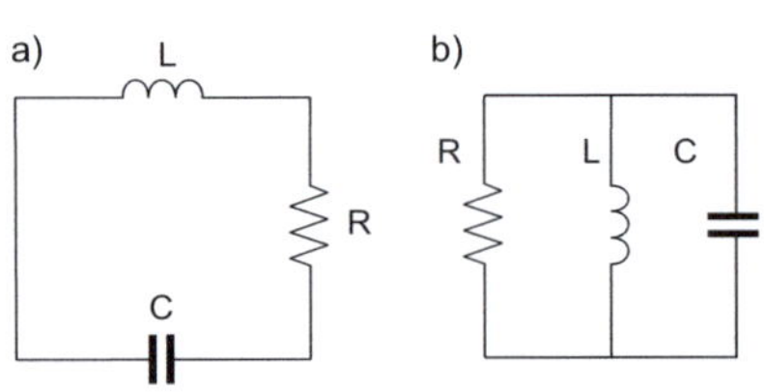

Fig. 10.15. (**a**) Series RLC circuit with no excitation and (**b**) parallel RLC circuit with no excitation.

For the filter circuits derived from the parallel RLC circuit, the resonant frequency still has to satisfy the condition that a *real* circuit impedance is "seen" by the voltage source. The resonant frequency found this way either does not equal the undamped resonant frequency $f_0 = 1/\left(2\pi\sqrt{LC}\right)$ or does not exist at all. However, the structure of the filter equations is *not* affected by this result. Only the undamped resonant frequency f_0 appears to be important for the voltage transfer function.

Section 10.3 RLC Circuits for Near-Field Communications and Proximity Sensors

10.3.1 Near-Field Wireless Link

Near-field wireless communication can transfer data, power, or both of them simultaneously. Common data-related applications include *radio-frequency identification* (RFID) systems of 125/134 kHz and 13.56 MHz, *electronic article surveillance* (EAS) for electronic anti-theft devices in shops, and mobile and other portable device *near-field communication* (NFC). Promising applications in biomedical engineering have also been explored. Figure 10.16 shows the key concept of a *near-field wireless link*. The transmitter and the receiver inductor coils share a common magnetic flux density $\vec{B}$ in the near field. The transmitter/receiver system in Fig. 10.16 is known as an *inductively coupled system*.

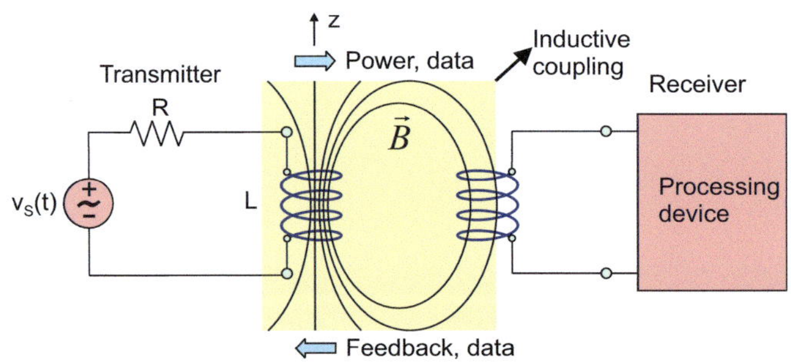

Fig. 10.16. The concept of the near-field wireless link; the magnetic flux density is shared between receiver and transmitter coils.

In contrast to the radio-frequency radiating fields, the near field $\vec{B}$ is very strong in the vicinity of the coil antenna. However, this field very rapidly decays at larger distances from the transmitter. For example, consider a transmitter coil with N loops of area A each. The corresponding near field of the transmitter coil in Fig. 10.16 with current $i(t)$ on the coil axis at the axial distances z much greater than the coil length (and the loop radius) may be found in the form:

$$B = N\left[\frac{\mu_0 A}{2\pi}\frac{i(t)}{z^3}\right]\ [\mathrm{T}],\quad \mu_0 = 4\pi\times 10^{-7}\ \mathrm{H/m} \tag{10.28}$$

where B is recorded in *tesla* T. The expression in square brackets is the contribution of a single loop. Thus, the near-field decay is inversely proportional to the *third* degree of the separation distance. This observation (obtained via an asymptotic analysis of the related magnetostatic expressions) is also valid for any quasi-static *magnetic* (and *electric*) *dipole*. Such a short-range wireless communication is both safe and effective.

10.3.2 Transmitter Circuit

In a transmitter circuit shown in Fig. 10.17a, the function generator is modeled by an ideal voltage source $v_S(t) = V_m \cos \omega t$ in series with resistance R. The function generator is connected to the transmitter coil modeled by the inductance L. The goal is to increase the magnetic flux density $\vec{B}$ of the transmitter. According to Eq. (10.28), the obvious choice is to increase the inductance of the transmitter. However, this operation would decrease the circuit current $i(t)$ due to an increase of the impedance magnitude. We will attempt to solve this problem with the series resonant RLC circuit shown in Fig. 10.17b.

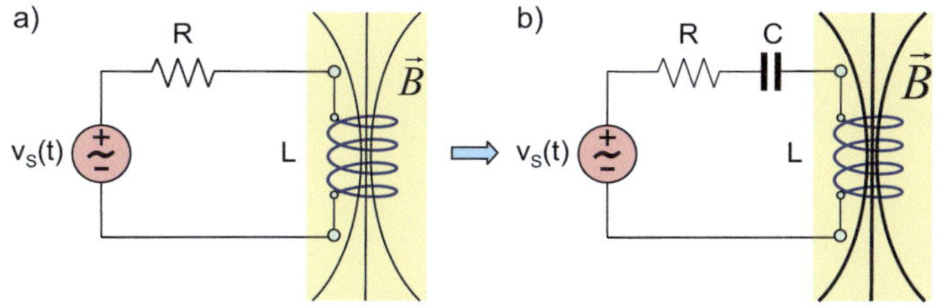

Fig. 10.17. Using a series capacitor in order to increase the circuit current.

The original and the modified circuits in Fig. 10.2 are both solved by using the phasor method. We denote the desired operating frequency by f_0. For the original circuit in Fig. 10.17a, the phasor current may then be written in the form:

$$\mathbf{I} = \frac{V_m}{\mathbf{Z}_{eq}} = \frac{V_m}{R + j\omega L} = \frac{V_m}{R} \frac{1}{\sqrt{1 + Q^2\left(\frac{f}{f_0}\right)^2}} \angle - \tan^{-1}\left(Q\frac{f}{f_0}\right), \quad Q = \omega_0(L/R) \tag{10.29}$$

For the series RLC circuit in Fig. 10.17b, the capacitance is chosen in such a way that the resonant frequency of the circuit coincides with the operation frequency f_0. The phasor current for the RLC circuit has been derived in the previous sections. It has the form:

$$\mathbf{I} = \frac{V_m}{\mathbf{Z}_{eq}} = \frac{V_m}{R + j\left(\omega L - \frac{1}{\omega C}\right)} = \frac{V_m}{R} \frac{1}{\sqrt{1 + Q^2\left(\frac{f}{f_0} - \frac{f_0}{f}\right)^2}} \angle - \tan^{-1}\left(Q\left(\frac{f}{f_0} - \frac{f_0}{f}\right)\right) \tag{10.30}$$

Note the presence of the quality factor, $Q = \omega_0(L/R)$, for the series RLC circuit in both Eqs. (10.29) and (10.30). At exactly the operation frequency, $f = f_0$, the ratio of current *magnitudes* (both phasors $\angle \cdot$ have the magnitude of one) becomes

$$\frac{I_{m \text{ circuit with series capacitor}}}{I_{m \text{ original circuit}}} = \sqrt{1 + Q^2} \approx Q \text{ for } Q \gg 1 \tag{10.31}$$

This ratio may significantly exceed one for high Q values. Thus, the series RLC circuit may considerably increase the circuit current and the associated magnetic field.

Exercise 10.7: It is suggested to increase the magnetic field for the circuit without the capacitor in Fig. 10.7a by simply doubling the number of coil turns and increasing the coil length by the factor of two. Given that:

1. $Q = \omega_0(L/R) \gg 1$ at the operation frequency
2. $Q = \omega_0(L/R) \ll 1$ at the operation frequency

how does the magnetic field B change?

Answer: (i) B remains the same. (ii) B doubles.

Example 10.13: Given the operation frequency (center band frequency) of $f_0 = 1$ MHz and $V_m = 10$ V, $L = 50$ μH, $R = 50$ Ω, plot the amplitude of the circuit current as a function of source frequency for the original (RL) and modified (resonant RLC) circuits in in Fig. 10.17 over the frequency band from 0.5 to 1.5 MHz.

Solution: The quality factor is found to be $Q = 6.283$. Next, we plot both current amplitudes based on Eqs. (10.29) and (10.30) using a linear scale. The result is shown in Fig. 10.18. The amplitude of the circuit current increases from 31.4 to 200 mA at the operation frequency f_0 (resonant frequency of the RLC circuit).

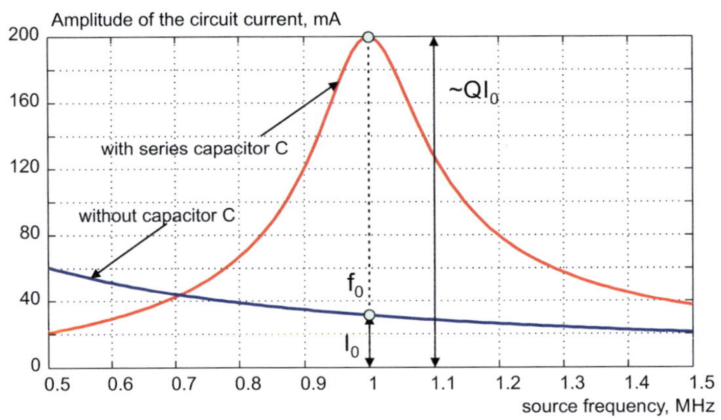

Fig. 10.18. Amplitudes of the circuit current for the original and modified circuits in Fig. 10.17.

10.3.3 Receiver Circuit

Consider the receiver coil in Fig. 10.19a. Its equivalent circuit includes the ideal inductor L in series with R, which is the resistance of the coil winding. It also includes an induced emf (electromotive force) voltage source $v_{emf}(t)$, which follows Faraday's law of induction:

$$\upsilon_{\text{emf}}(t) \equiv AN\frac{dB(t)}{dt} = V_{\text{m}}\cos(\omega t), \quad B(t) = B_{\text{m}}\sin(\omega t) \tag{10.32}$$

Here, $B(t)$ is the coaxial component of the external, time-varying magnetic flux density of the transmitter at the receiver location. The source voltage amplitude is given by $V_{\text{m}} = AN\omega B_{\text{m}}$ where A is the area of the receiver coil and N is the number of coil turns. The major parameter of interest is the (small) open-circuit voltage of the receiver coil, $\upsilon_{\text{out}}(t)$. It is desired to increase this voltage. For the circuit shown in Fig. 10.19a, $\upsilon_{\text{out}}(t)$ is always equal to $\upsilon_{\text{emf}}(t)$. However, the situation will change if we create a series RLC circuit as shown in Fig. 10.19b. The output voltage becomes the capacitor voltage. We will attempt to increase $\upsilon_{\text{out}}(t)$ by using the resonance condition.

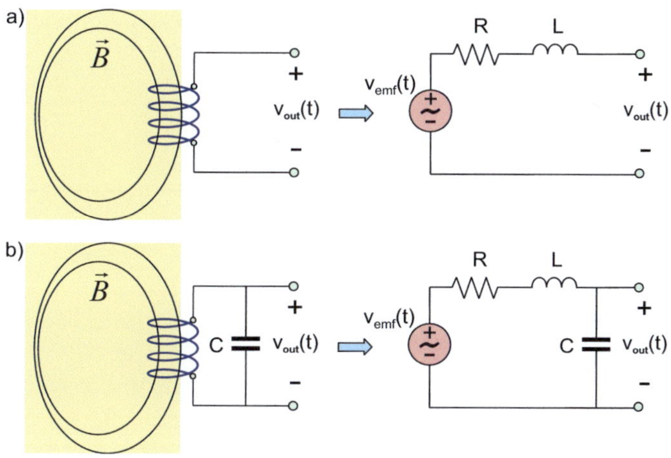

Fig. 10.19. (**a**) Receiver coil and (**b**) receiver coil with capacitance to increase the open-circuit voltage.

The circuit in Fig. 10.19b is solved using the phasor method. The desired resonant frequency (operating frequency) is f_0. The phasor for the output voltage has the form:

$$\mathbf{V}_{\text{out}} = \mathbf{V}_{\text{C}} = \frac{V_{\text{m}}/(j\omega C)}{R + j\omega L + \frac{1}{j\omega C}} = Q\frac{f_0}{f}\frac{V_{\text{m}}}{\sqrt{1 + Q^2\left(\frac{f}{f_0} - \frac{f_0}{f}\right)^2}}\left\{\angle -\frac{\pi}{2} - \tan^{-1}\left(Q\left(\frac{f}{f_0} - \frac{f_0}{f}\right)\right)\right\} \tag{10.33}$$

where $Q = \omega_0(L/R)$ is again the quality factor of the series RLC resonant circuit (and simultaneously the *quality factor of the nonideal inductor* with series resistance R). At the exact resonant frequency, the output voltage amplitude becomes

$$V_{\text{out}} = QV_{\text{m}} \tag{10.34}$$

This value may significantly exceed V_m given a high value of Q. Thus, the series RLC circuit formed with the help of the shunt capacitor C in Fig. 10.19b may considerably increase the received voltage.

Example 10.14: Given the operating frequency (center band frequency) of $f_0 = 1$ MHz and $V_m = 10$ mV, $L = 78$ μH, $R = 10$ Ω, plot the amplitude of the output voltage for the original and modified circuits in Fig. 10.19 as a function of source frequency over the band from 0.5 to 1.5 MHz.

Solution: We find the required capacitance value first. Specifically, $C = 1/\left(L(2\pi f_0)^2\right) \approx 325$ pF. The quality factor is given by $Q \approx 49.0$. Next, we plot both voltage amplitudes. The first one is simply V_m. The second plot is based on Eq. (10.33). The results are shown in Fig. 10.20. The amplitude of the output voltage increases from 10 to 490 mV at the operating frequency f_0 (resonant frequency of the series RLC circuit).

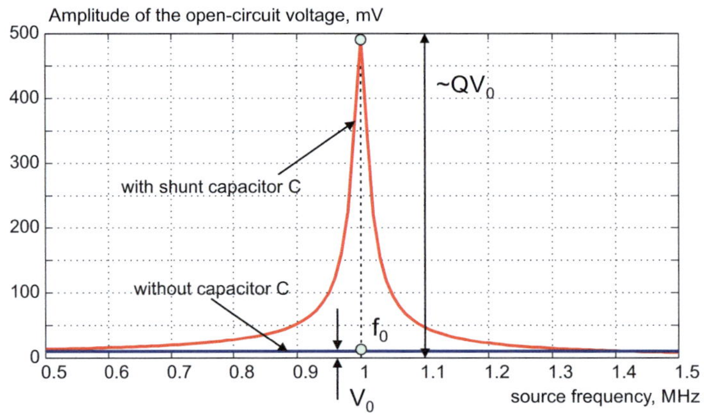

Fig. 10.20. Amplitudes of the output voltage for the original and modified circuits in Fig. 10.19.

The circuit in Fig. 10.19b is the low-pass second-order RLC filter studied in the previous section, right? Why is it boosting the source voltage instead of just passing it through? The key is the Q-factor. The present circuit operates as a filter at relatively small values of the quality factors, i.e., $Q \leq 1$. At higher Q values, the circuit generates a voltage spike close to the resonant frequency and operates as a *voltage multiplier.* This operation is similar to the operation of an electric transformer.

10.3.4 Application Example: Near-Field Wireless Link in Laboratory

Figure 10.21 shows a prototype of the near-field link implemented in an undergraduate laboratory. The operating frequency of the transmitter is tunable; it ranges from 400 kHz to 1.2 MHz. Despite this relatively high frequencies, the circuitry can still be implemented on standard protoboards. The key is the tunability of both the transmitter and the receiver, which simultaneously accounts for the parasitic capacitance of the board.

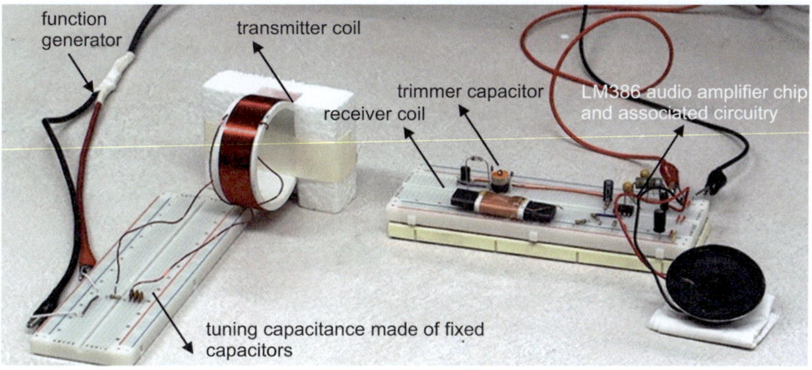

Fig. 10.21. Prototype of the near-field link implemented in laboratory.

The transmitter is driven by a function generator; the tuning is made by a bank of fixed capacitors from a laboratory kit. The receiver uses a single-ended magnetic-core coil (a *loopstick antenna*) with an inductance of approximately 1 mH and a series resistance of about 10 Ω. The RLC circuit at the receiver uses a trimmer capacitor of 10–180 pF range for tuning. When the transmitter is sending an amplitude-modulated signal from the function generator, the receiver operates as an AM radio given the subsequent rectifying circuit with a germanium diode and an audio amplifier IC. Frequency modulation is also possible; however, the receiver circuit has to be modified accordingly. The system operating range is up to two feet on average. When an external modulation input of the function generator is used, an audio clip may be transmitted.

10.3.5 Application Example: Proximity Sensors

The idea of the resonant RLC proximity sensor is quite simple. Assume that the inductance is a large coil or simply a loop of wire. When a metal object to be detected is brought in proximity to the loop, its (self) inductance changes. This causes a detectable change in the resonant frequency f_0 of an RLC tank circuit. After encoding, information may be extracted about the presence of the object and sometimes of its size. This is the well-known principle of a *metal detector*. A large variety of metal-detecting circuits already exist, and more are still awaiting discovery. Another idea is to change the capacitance by putting a dielectric object (such as a medical pill) inside the capacitor. A similar change in the resonant frequency may be observed and detected. Such a device may be used, for example, as an automatic pill counter.

The detector circuit itself can operate based on three different principles. First, a simple method is to use the series RLC tank circuit with the external AC power supply. The measured parameter is the amplitude of the circuit current (resistor voltage) at the frequency of the AC source. When the resonant frequency is close to the AC frequency, the circuit voltage is large. However, when the resonant frequency deviates from the source frequency, the circuit voltage becomes smaller. The change in the circuit voltage is detected. A second method is to make the tank circuit *self-resonant*, by using an amplifier with a *positive feedback*. A resonant circuit so built does not need an AC power supply. It oscillates *exactly* at f_0 when there is no object to be detected. When the object is present the oscillation frequency changes. The change in the AC frequency is encoded by another electronic circuit. Using self-resonant tank circuits is perhaps the most common method in practice. A third method is based on the effect of the resistance in the tank circuit. When a metal object is placed close to the coil, the coil's series resistance significantly increases, due to the so-called eddy current losses (for all metals) and, possibly, hysteresis losses (for magnetic metals such as iron, nickel, steel alloys, etc.). The increase in the resistance leads to smaller voltage oscillations in the self-resonant circuit. The circuit may be tuned in such a way as to stop oscillating at a given value of the extra resistance. Great sensitivity may be achieved with this method.

Figure 10.22 shows the inductor assembly in a resonant sensor for an *automatic traffic light*. The inductor now is a single-turn (or multi-turn) pavement loop. When a vehicle is located above the loop, its (self) inductance L decreases. This leads to an increase in the resonance frequency. The change in frequency, not the change in the amplitude, is typically detected and encoded. The latter is used to indicate the presence of a vehicle and to adjust the traffic light control. Most vehicle detectors based on loop inductors operate with frequencies from 10 to 100 kHz. A (simplified) equivalent tank circuit for the traffic light control is shown in Fig. 10.23b. We note the series resistance R, which is the parasitic resistance of the loop. The parasitic resistance includes both the effect of the passing vehicle and of the ground.

Fig. 10.22. Multiple vehicle detection loops after installation at an intersection. Courtesy of the US Traffic Corporation, Loop Application Note of 3/10/03.

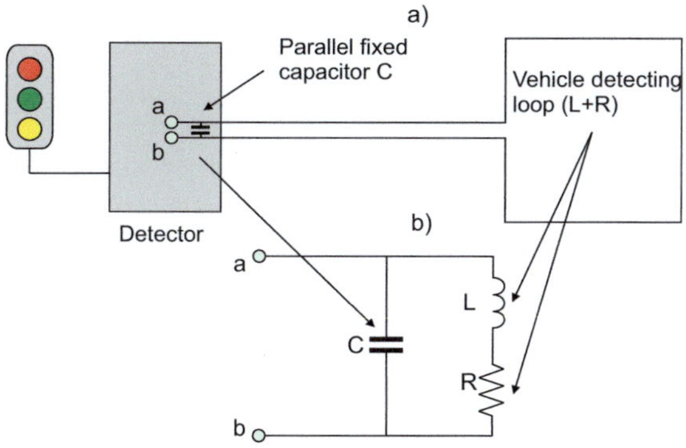

Fig. 10.23. (**a**) Simplified schematic of the vehicle detecting system and (**b**) equivalent resonant circuit.

The circuit in Fig. 10.23b may be analyzed exactly in the same way as the series/parallel RLC tank circuits in Section 10.1. The tricky part for the tank circuit block in Fig. 10.23b is that its resonant angular frequency is no longer $1/\sqrt{LC}$. However, it can still be found from the condition of a purely real equivalent impedance $\mathbf{Z}_{\mathrm{eq}}$, see the summary of this chapter.

Single coils of special shapes—the *horseshoe shape*—may be used to detect the level and the presence of molten metals through the walls of (large) casting molds and for other purposes. The equivalent circuit is the parallel RLC tank circuit. When properly tuned, the self-resonating circuit quantitatively detects variations in molten metal level through 4–5″ thick walls, see Fig. 10.24.

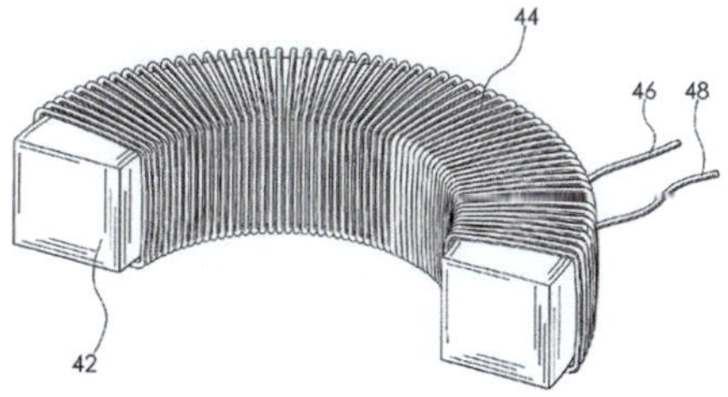

Fig. 10.24. A half-toroidal (*horseshoe*) coil used to concentrate the magnetic field between its tips in a molten metal detector (Foley, Biederman, Ludwig, and Makarov, US Patent 7,828,043 Nov. 9th 2010).

Summary

TERM	Series RLC circuit	Parallel RLC circuit
Series and parallel RLC resonators	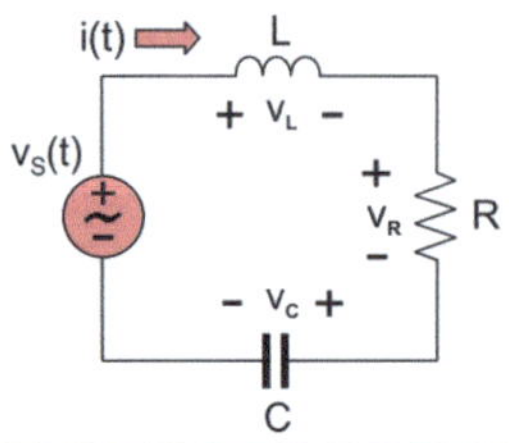	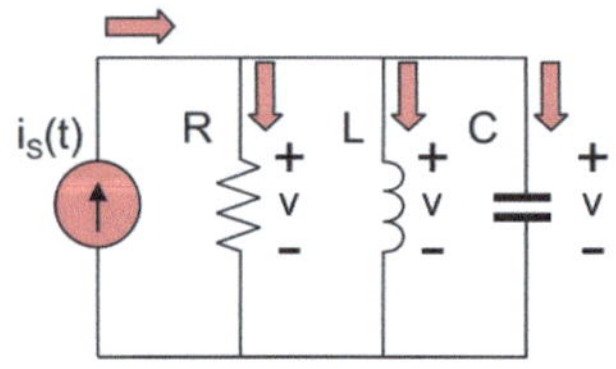
Resonant frequency	$\omega_0 = 1/\sqrt{LC},\ f_0 = 1/\left(2\pi\sqrt{LC}\right)$ Coincides with the undamped resonant frequency	$\omega_0 = 1/\sqrt{LC},\ f_0 = 1/\left(2\pi\sqrt{LC}\right)$ Coincides with the undamped resonant frequency
Quality factor of the resonant circuit	$Q = \frac{\sqrt{L/R}}{\sqrt{RC}} = \frac{1}{\omega_0 RC} = \omega_0(L/R)$ dimensionless	$Q = \frac{\sqrt{RC}}{\sqrt{L/R}} = \omega_0 RC = \frac{1}{\omega_0(L/R)}$ dimensionless
Bandwidth of the resonant circuit	$B \equiv f_U - f_L = \frac{f_0}{Q} = \frac{1}{2\pi(L/R)}$ [Hz]	$B \equiv f_U - f_L = \frac{f_0}{Q} = \frac{1}{2\pi RC}$ [Hz]
Half-power lower and upper frequencies	$f_{L,U} = f_0\left(\sqrt{1+\frac{1}{(2Q)^2}} \mp \frac{1}{2Q}\right)$ [Hz]	$f_{L,U} = f_0\left(\sqrt{1+\frac{1}{(2Q)^2}} \mp \frac{1}{2Q}\right)$ [Hz]
Other RLC resonators	Circuit diagram	Resonant frequency
L+R||C	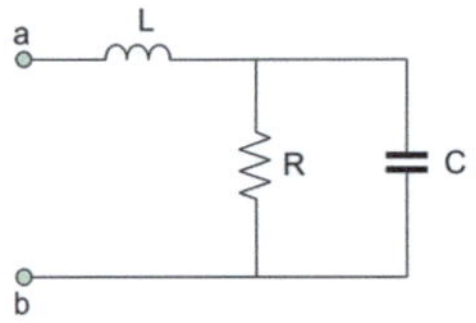	$\omega_0 = \frac{1}{(RC)}\sqrt{\frac{RC}{L/R}-1}$ Different from the undamped resonant frequency
C+L||R	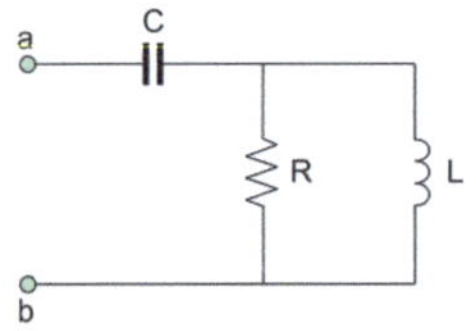	$\omega_0 = \frac{1}{(L/R)}\frac{1}{\sqrt{\frac{RC}{L/R}-1}}$ Different from the undamped resonant frequency
(R+L)||C	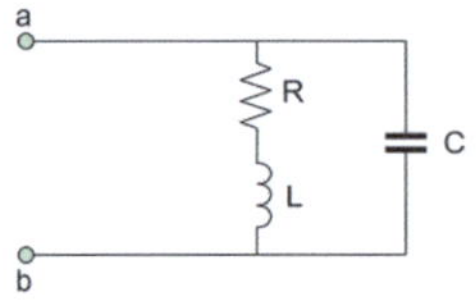	$\omega_0 = \frac{1}{(L/R)}\sqrt{\frac{L/R}{RC}-1}$ Different from the undamped resonant frequency

(continued)

RLC filter circuits derived from the *series* RLC circuit: $f_0 = 1/(2\pi\sqrt{LC})$, $Q = 1/(\omega_0 RC)$		
Band-pass filter	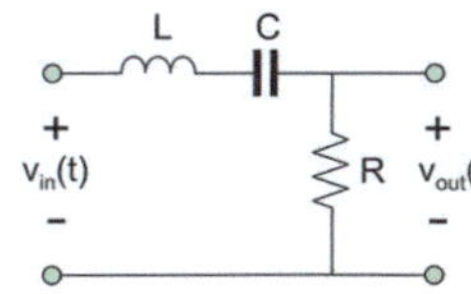	$\mathbf{H}_0(f) = \dfrac{1}{1 + jQ\left(\dfrac{f}{f_0} - \dfrac{f_0}{f}\right)}$
Low-pass filter	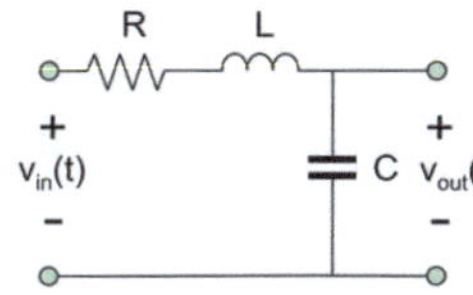	$\mathbf{H}(f) = Q\dfrac{f_0}{f}\mathbf{H}_0(f)$
High-pass filter	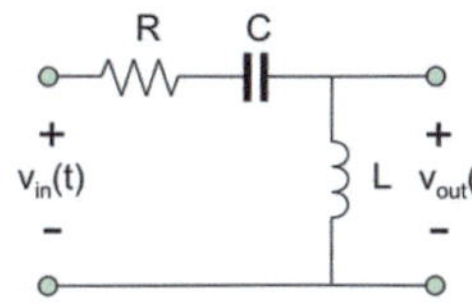	$\mathbf{H}(f) = Q\dfrac{f}{f_0}\mathbf{H}_0(f)$
Band-reject filter	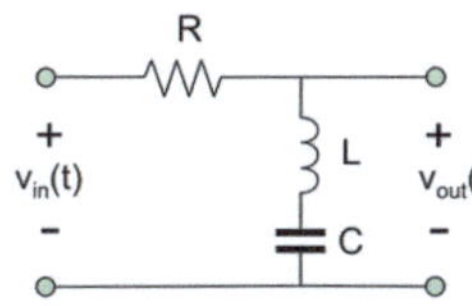	$\mathbf{H}(f) = jQ\left(\dfrac{f}{f_0} - \dfrac{f_0}{f}\right)\mathbf{H}_0(f)$
RLC filter circuits derived from the *parallel* RLC circuit: $f_0 = 1/(2\pi\sqrt{LC})$, $Q = \omega_0 RC$		
Band-pass filter based on parallel RLC circuit	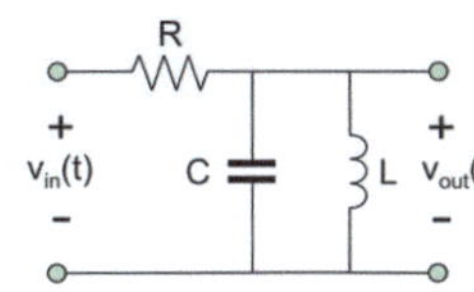	$\mathbf{H}_0(f) = \dfrac{1}{1 + jQ\left(\dfrac{f}{f_0} - \dfrac{f_0}{f}\right)}$
Low-pass filter based on parallel RLC circuit	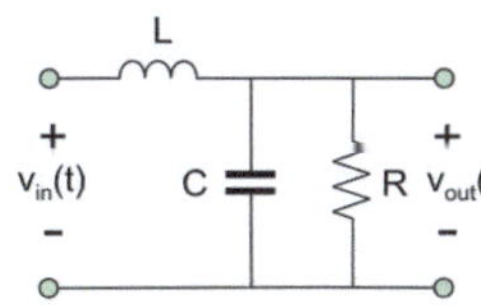	$\mathbf{H}(f) = Q\dfrac{f_0}{f}\mathbf{H}_0(f)$
High-pass filter based on parallel RLC circuit	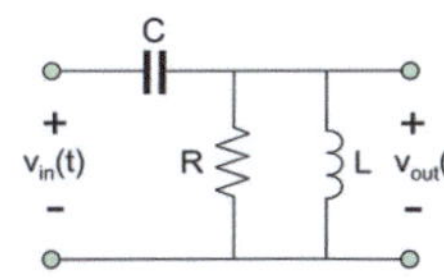	$\mathbf{H}(f) = Q\dfrac{f}{f_0}\mathbf{H}_0(f)$
Band-reject filter based on parallel RLC circuit	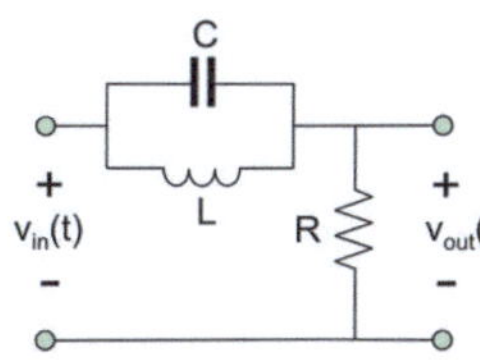	$\mathbf{H}(f) = jQ\left(\dfrac{f}{f_0} - \dfrac{f_0}{f}\right)\mathbf{H}_0(f)$

(continued)

<table>
<tr><th colspan="2">Near-field wireless transmitter/receiver</th></tr>
<tr><td>Resonant circuit at the transmitter (TX)</td><td>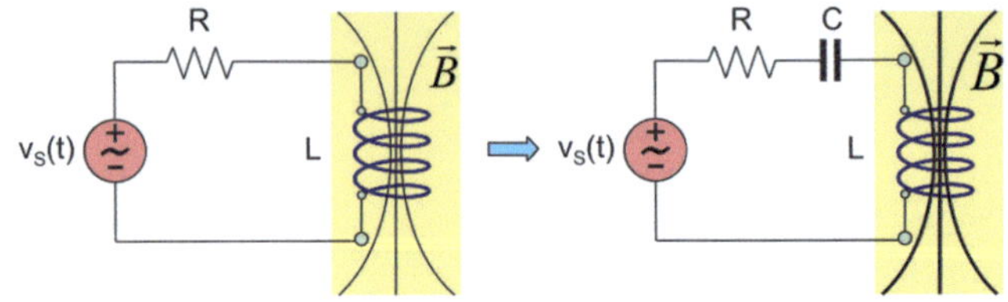

The series capacitor forms the series RLC circuit and increases the amplitude of the magnetic flux density anywhere in space by the factor $\sqrt{1+Q^2}$, $Q=\omega_0(L/R)$</td></tr>
<tr><td>Resonant circuit at the receiver (RX)</td><td>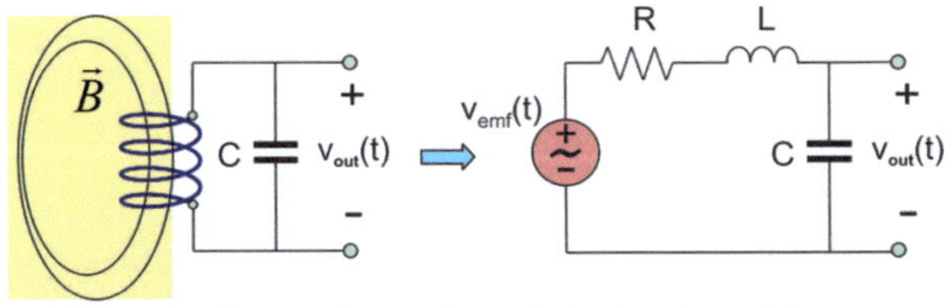

The shunt capacitor again forms the series RLC circuit and increases the amplitude of the output voltage by the factor Q, $Q=\omega_0(L/R)$</td></tr>
</table>

Problems

10.1. Theory of the Second-Order RLC Resonator

10.1.1 Self-Oscillating Ideal LC Circuit

10.1.2 Series Resonant Ideal LC Circuit

Problem 10.1. Give an example of a self-oscillating (resonant) mechanical system different from that in Fig. 10.2 of this section.

Problem 10.2. For the self-oscillating circuit shown in the figure below, the circuit current is specified by $i(t) = I_m \cos \omega t$. Given $I_m = 200$ mA, $L = 0.63$ mH, $C = 1$ μF:

A. Determine the undamped resonant frequency f_0.
B. Construct the phasor diagram for phasor voltages $\mathbf{V}_L$ and $\mathbf{V}_C$ and phasor current $\mathbf{I}$ on the same plot. Assume that every plot division corresponds to 1 V or to 100 mA.

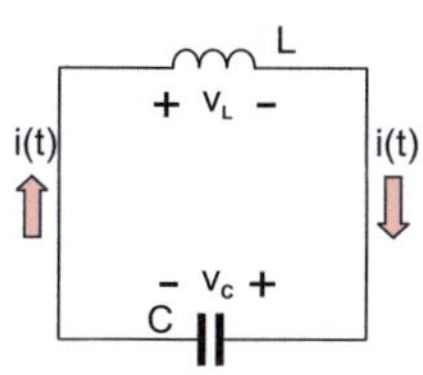

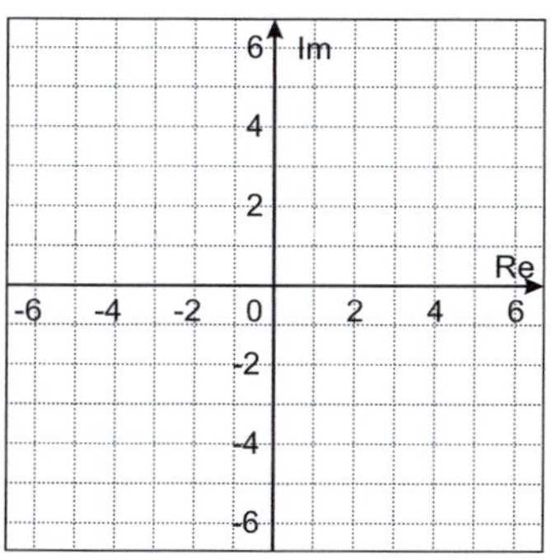

Problem 10.3. For a series ideal LC circuit shown in the figure below, the voltage source has the form $v_S(t) = V_m \cos \omega t$. Given $V_m = 5$ V, $L = 0.5$ mH, $C = 1$ μF:

A. Determine the undamped resonant frequency f_0 of the circuit.
B. Construct the phasor diagram for phasor voltages $\mathbf{V}_S$, $\mathbf{V}_L$, and $\mathbf{V}_C$ when the source frequency is 90 % of the resonant frequency.
C. Describe how your phasor diagram would change if the inductance becomes 1 mH instead of 0.5 mH.

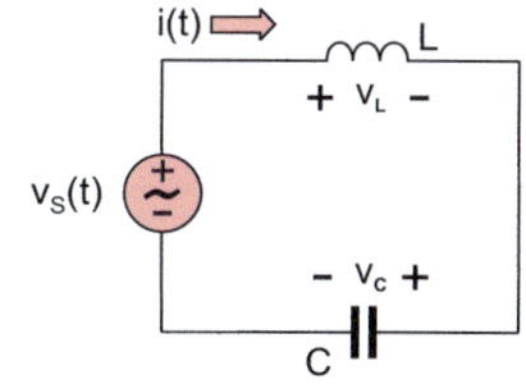

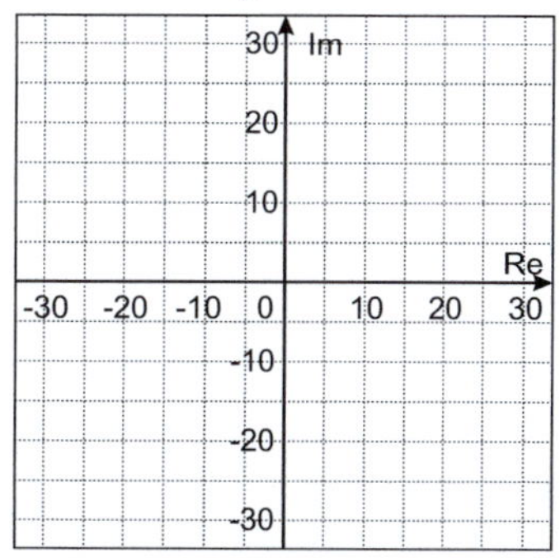

Problem 10.4. For a series ideal LC circuit shown in the figure below, the voltage source has the form $v_S(t) = V_m \cos \omega t$. Given $V_m = 5$ V, $L = 1$ mH, $C = 0.5$μF:

A. Determine the undamped resonant frequency f_0 of the circuit.
B. Construct the phasor diagram for phasor voltages $\mathbf{V}_S$, $\mathbf{V}_L$, and $\mathbf{V}_C$ when the source frequency is 111 % of the resonant frequency.
C. Describe how your phasor diagram would change if the capacitance becomes 1 μF instead of 0.5 μF.

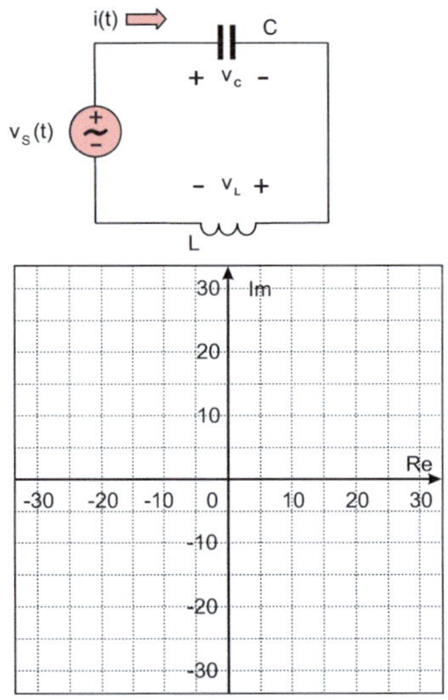

10.1.3 Series Resonant RLC Circuit: Resonance Condition

10.1.4 Quality Factor Q of the Series Resonant RLC Circuit

10.1.5 Bandwidth of the Series Resonant RLC Circuit

Problem 10.5. For a generic series resonant RLC circuit with the supply voltage $v_S(t) = V_m \cos \omega t$, resistance R, inductance L, and capacitance C, give the expressions (show units) for:

A. Equivalent circuit impedance at the resonance
B. Resonant frequency
C. Quality factor of the resonant circuit

Problem 10.6. Describe the physical meaning of the quality factor of the series RLC resonator circuit in your own words.

Problem 10.7. In the series resonant RLC circuit shown in the figure that follows, given $V_m = 1$ V, $L = 1$ mH, $C = 80$ pF, $R = 10\ \Omega$:

A. Determine resonant frequency and the Q-factor.
B. Determine resonant phasor current and phasor voltages $\mathbf{V}_R$, $\mathbf{V}_L$, and $\mathbf{V}_C$; construct the phasor diagram. Assume voltage scale in volts and current scale in milliamperes.
C. Determine the real-valued circuit current $i(t)$ and the inductor/capacitor voltages $v_L(t)$, $v_C(t)$ at resonance.

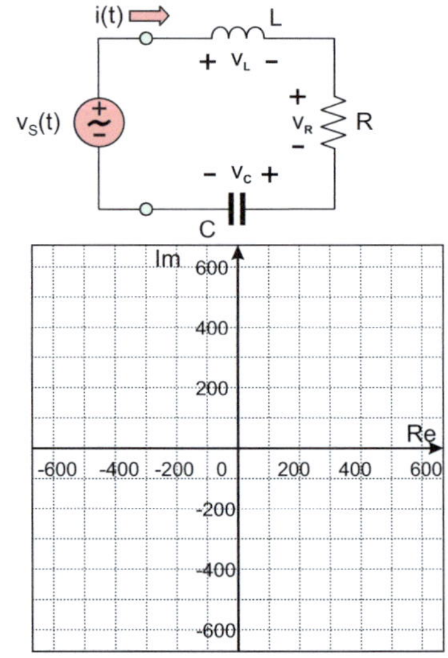

Problem 10.8. In the series resonant RLC circuit shown in the figure below, given $V_m = 5$ V, $L = 30$ μH, $C = 0.48$ nF, $R = 50\ \Omega$:

A. Determine resonant frequency and the Q-factor.
B. Determine resonant phasor voltages $\mathbf{V}_R$, $\mathbf{V}_L$, and $\mathbf{V}_C$; construct the phasor diagram.
C. Determine the real-valued resistor voltage $v_R(t)$ and the inductor/capacitor voltages $v_L(t)$, $v_C(t)$ at the resonance.

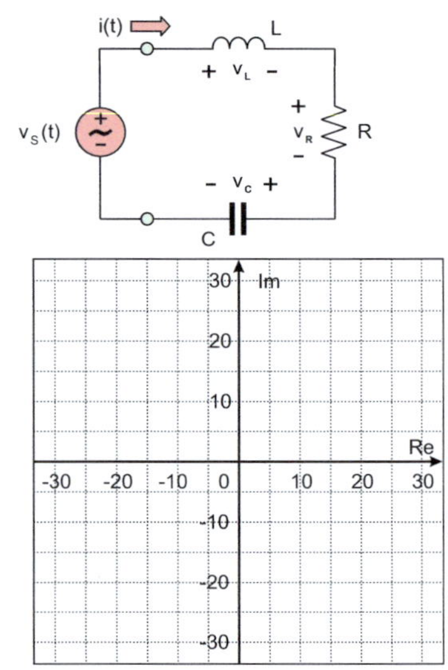

Problem 10.9. A series resonant LC circuit is driven by a laboratory AC voltage source with an amplitude $V_m = 12$ V and an internal resistance of 50 Ω (a function generator). Which value should the ratio L/C have to obtain the amplitude of the capacitor voltage equal to 200 V at the resonance?

Problem 10.10. A series resonant RLC circuit is needed with the resonant frequency of 1 MHz and a Q-factor of 100. The circuit resistance is 10 Ω. Determine the necessary values of L and C.

Problem 10.11. Describe the physical meaning of the resonance bandwidth of the series resonant RLC circuit in your own words.

Problem 10.12. A series resonant RLC circuit has the resonant frequency of 1 MHz and the quality factor of 10. Create the Bode plot for the amplitude of the circuit current normalized by its maximum value at the resonance over frequency band from 0.5 to 2 MHz.

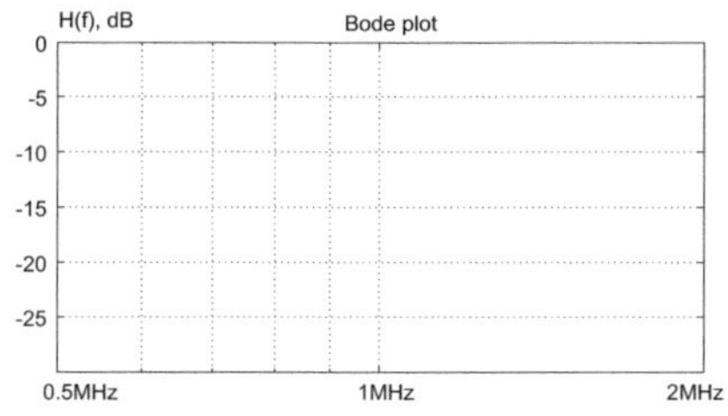

Problem 10.13. Determine the bandwidth, B, of the series resonant RLC circuit with the resonant frequency of 1 MHz and a Q-factor of 100.

Problem 10.14. A series resonant RLC circuit is needed with the resonant frequency of 1 MHz and the bandwidth of 10 kHz. Given the circuit resistance of 10 Ω, determine L and C.

Problem 10.15. For the RLC circuit block shown in the figure, establish the resonant frequency in terms of component values.

Hint: The resonance is defined by the condition of the purely real equivalent impedance between terminals a and b. In other words, Im $(\mathbf{Z}_{eq}) = 0$ at the resonance.

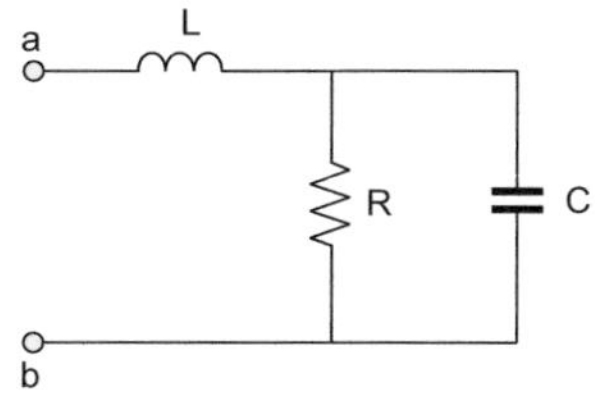

Problem 10.16. Repeat the previous problem for the circuit shown in the figure that follows.

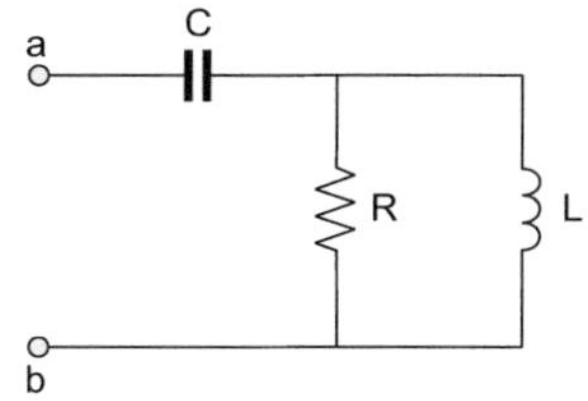

10.1.6 Parallel Resonant RLC Circuit: Duality

Problem 10.17. For a generic parallel RLC resonant circuit with the supply current $i_S(t) = I_m \cos \omega t$, resistance R, inductance L, and capacitance C, give the expressions (show units) for:

A. Equivalent circuit impedance at the resonance
B. Resonant frequency
C. Quality factor of the resonant circuit

Problem 10.18. In the parallel resonant RLC circuit shown in the figure that follows, given $I_m = 0.5$ A, $L = 30$ μH, $C = 0.43$ μF, $R = 50$ Ω:

A. Determine the resonant frequency and the Q-factor.
B. Determine resonant phasor currents $\mathbf{I}_R$, $\mathbf{I}_L$, and $\mathbf{I}_C$; construct the phasor diagram.
C. Determine the real-valued resistor current $i_R(t)$ and the inductor/capacitor currents $i_L(t)$, $i_C(t)$ at the resonance.

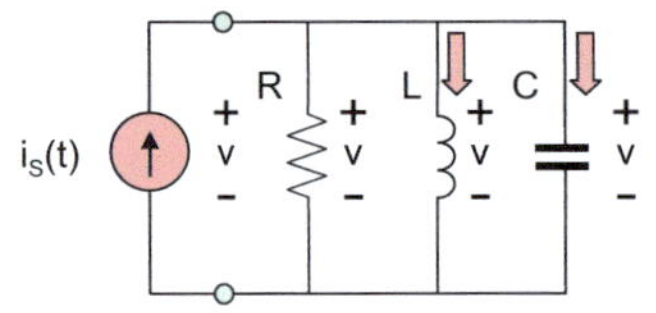

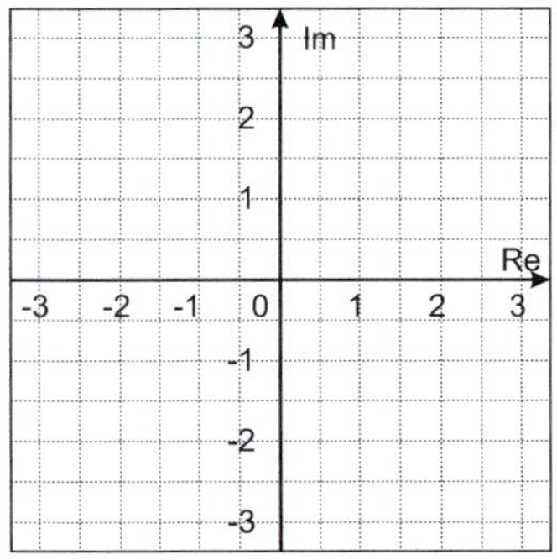

Problem 10.19. Determine the bandwidth, B, of the parallel resonant RLC circuit with the resonant frequency of 0.5 MHz and a Q-factor of 50.

Problem 10.20. A parallel resonant RLC circuit is needed with the resonant frequency of 1 MHz and the bandwidth of 10 kHz. Given the circuit resistance of 100 Ω, determine L and C.

10.2: Construction of Second-Order RLC Filters

10.2.1 Second-Order Band-Pass RLC Filter

10.2.2 Second-Order Low-Pass RLC Filter

10.2.3 Second-Order High-Pass RLC Filter

10.2.4 Second-Order Band-Reject RLC Filter

Problem 10.21

A. Draw the circuit diagram of the second-order RLC band-pass filter. Label R, L, and C.
B. Show the input and output ports (input and output voltages)
C. Define the resonant frequency and the Q-factor of the filter circuit.

Problem 10.22. A band-pass RLC filter is required with the center (resonant) frequency of 100 kHz and the half-power bandwidth, B, of 20 kHz.

A. Create its amplitude Bode plot in the frequency band from 10 kHz to 1 MHz.
B. Label the filter passband.
C. Determine the necessary values of L and C given $R = 20\ \Omega$.

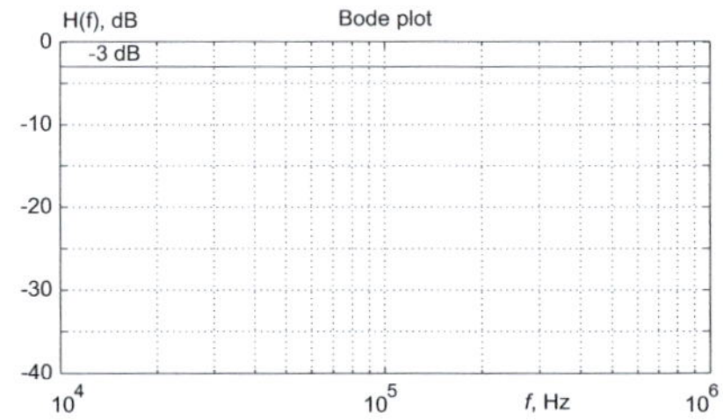

Problem 10.23

A. Draw the circuit diagram of the second-order RLC low-pass filter. Label R, L, and C.
B. Show the input and output ports (input and output voltages)
C. Define the resonant frequency and the Q-factor of the filter circuit.
D. Which Q-factor is required for the maximally flat response?
E. What is the filter's half-power frequency for the maximally flat response?

Problem 10.24. A low-pass RLC filter is required with the *passband* from 0 to 1 MHz. Create amplitude Bode plots for the filter in the frequency band from 100 kHz to 10 MHz given the resonant frequency of the filter circuit of 1 MHz and

A. $Q = 5$
B. $Q = 1/\sqrt{2}$
C. $Q = 0.2$

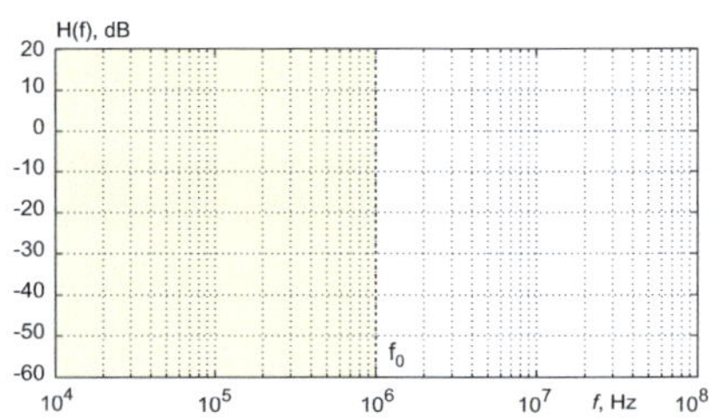

Problem 10.25*. Generate Fig. 10.4 of this section, the Bode plots for the low-pass filter using MATLAB.

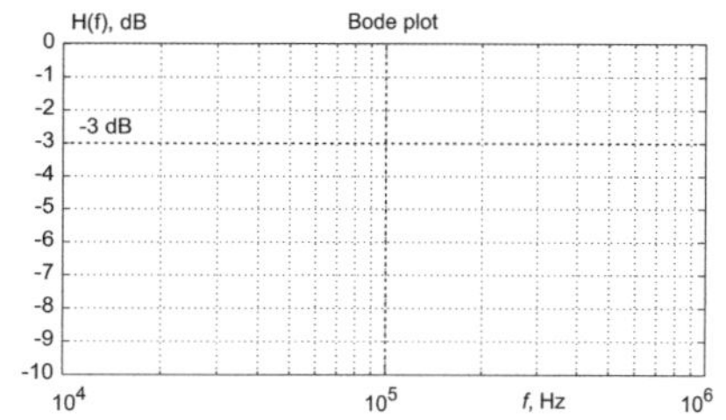

Problem 10.26

A. Draw the circuit diagram of the second-order RLC high-pass filter. Label *R*, *L*, and *C*.
B. Show the input and output ports (input and output voltages)
C. Define the resonant frequency and the *Q*-factor of the filter circuit.
D. Which *Q*-factor is required for the maximally flat response?
E. What is the filter's half-power frequency for the maximally flat response?

Problem 10.27. A high-pass RLC filter is required with the *passband* from 0 to 1 MHz. Create amplitude Bode plots for the filter in the frequency band from 100 kHz to 10 MHz given the resonant frequency of the filter circuit of 1 MHz and

A. $Q = 10$
B. $Q = 1/\sqrt{2}$
C. $Q = 0.1$

Problem 10.28

A. Draw the circuit diagram of the second-order RLC band-reject filter. Label *R*, *L*, and *C*.
B. Show the input and output ports (input and output voltages)
C. Define the resonant frequency and the *Q*-factor of the filter circuit.

Problem 10.29. A band-reject RLC filter is required with the center (resonant) frequency of 100 kHz and the half-power bandwidth, *B*, of 20 kHz.

A. Create its amplitude Bode plot in the frequency band from 10 kHz to 1 MHz.
B. Label the filter passband.
C. Determine the necessary values of *L* and *C* given $R = 20\ \Omega$.

10.2.6. Second-Order RLC Filters Derived from the Parallel RLC Circuit

Problem 10.30. Second-order RLC filters may be constructed either on the basis of the series RLC circuit or on the basis of the parallel RLC circuit. The undamped resonant frequency, $f_0 = 1/(2\pi\sqrt{LC})$, which is present in the filter equations, remains the same in either case. However, the quality factor does not. Three unknown second-order RLC filter circuits are shown in the figure that follows.

A. Determine the filter function (band-pass, low-pass, high-pass, or band-reject).
B. By analyzing filter's natural structure (after shorting out the input voltage source), determine the expression for the filter quality factor.

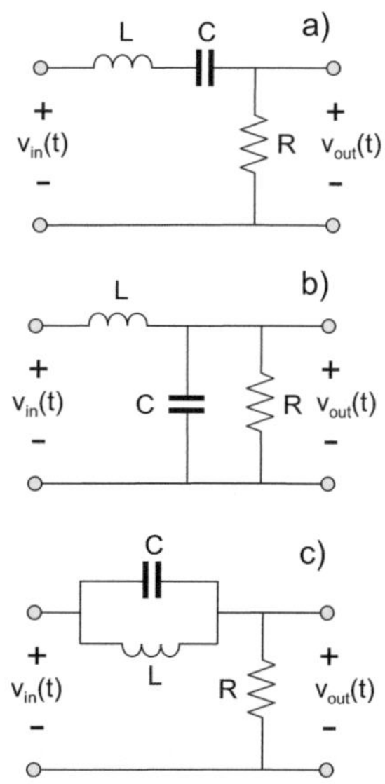

10.3. RLC Circuits for Near-Field Communications and Proximity Sensors

10.3.1 Near-Field Wireless Link

10.3.2 Transmitter Circuit

10.3.3 Receiver Circuit

10.3.4 Application Example: Near-Field Wireless Link in Laboratory

10.3.5 Application Example: Proximity Sensors

Problem 10.31. Describe the concept and purpose of the near-field wireless link in your own words. Think of an example where the link may be used solely for the power transfer.

Problem 10.32. In the circuit shown in the following figure, a capacitor C is introduced in series with an ideal coil having inductance L in order to set up a series resonant RLC circuit and increase the amplitude of the magnetic flux density $\vec{B}$ oscillating at 1 MHz. Determine the ratio of the magnetic flux amplitudes

$$\frac{B_{\mathrm{m}\ \text{circuit with series capacitor}}}{B_{\mathrm{m}\ \text{original circuit}}}$$

with and without the capacitor anywhere in space given that $L = 100\ \mu\text{H}$, $R = 25\ \Omega$.

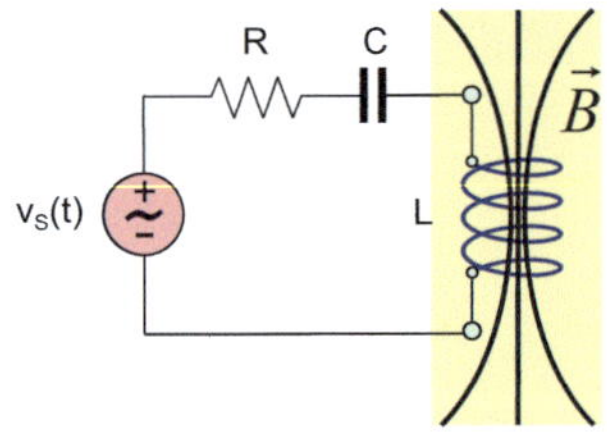

Problem 10.33. Given the operation frequency (center band frequency) of $f_0 = 1$ MHz and $V_\mathrm{m} = 1$ V, $L = 100\ \mu\text{H}$, $R = 20\ \Omega$, plot to scale the amplitude of the circuit current as a function of source frequency for the original (RL) and modified (resonant RLC) circuits in the figure to the previous problem over the frequency band from 0.5 to 1.5 MHz. Label the amplitude values at the operation frequency.

Problem 10.34. In the circuit shown in the figure that follows, a receiver coil antenna is subject to an external magnetic field oscillating at 1 MHz. A capacitor C is introduced in parallel with the coil having inductance L and series resistance R in order to set up a *series* resonant RLC circuit and increase the amplitude of the output voltage $v_{\mathrm{out}}(t)$. Determine the ratio of the output voltage amplitudes with and without the capacitor given that $L = 1000\ \mu\text{H}$, $R = 10\ \Omega$.

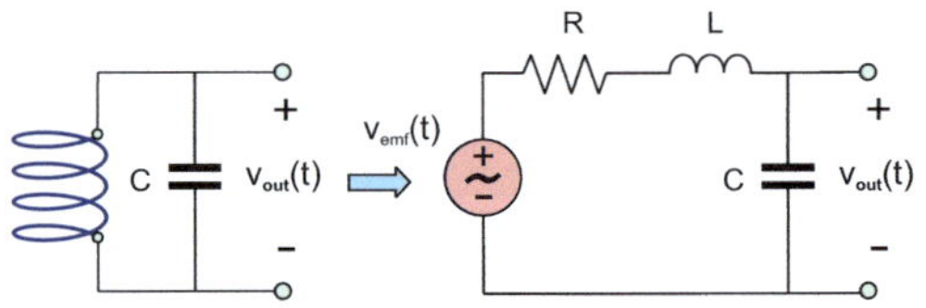

Problem 10.35. Given the operation frequency (center band frequency) of $f_0 = 1$ MHz and $v_{\mathrm{emf}}(t) = V_\mathrm{m}\cos(\omega t)$, $V_\mathrm{m} = 1$ mV, $L = 500\ \mu\text{H}$, $R = 50\ \Omega$, plot to scale the amplitude of the output voltage as a function of source frequency for the original (RL) and modified (resonant series RLC) circuits in the figure to the previous problem over the frequency band from 0.5 to 1.5 MHz. Label the amplitude values at the operation frequency.

Problem 10.36. In the circuit shown in the figure below, a receive coil antenna is subject to an external magnetic field oscillating at 1 MHz. A capacitor C is introduced in parallel with the coil having inductance L and series resistance R in order to set up a *series* resonant RLC circuit and increase the amplitude of the output voltage $v_{\mathrm{out}}(t)$. Determine the output voltage amplitudes with and without the capacitor given that $v_{\mathrm{emf}}(t) = 1\cos(\omega t)$ [mV] and that $L = 1000\ \mu\text{H}$, $R = 10\ \Omega$, $R_f = 100\ \Omega$.

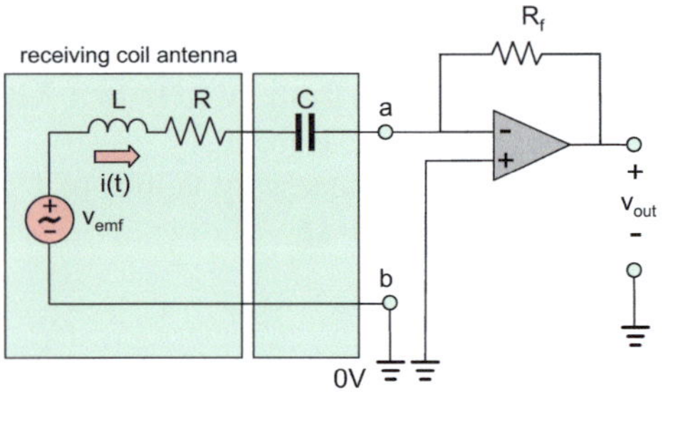

Chapter 11: AC Power and Power Distribution

Overview

Prerequisites:

- Knowledge of complex arithmetic
- Knowledge of basic circuit analysis (Chapters 3 and 4)
- Knowledge of phasor/impedance method for AC circuit analysis (Chapter 8)

Objectives of Section 11.1:

- Find average AC power for a resistive load and understand the rms values
- Express average power for any AC load in terms of power angle and power factor
- Express average power in terms of phasors/impedances
- Define major AC power types: average power, reactive power, complex power, and apparent power
- Be able to construct the power triangle and classify the load power factor

Objectives of Section 11.2:

- Be able to perform power factor correction of an inductive load (AC motor)
- Learn about maximum power efficiency technique in general
- Derive and test a simple condition for maximum power transfer to a load from an arbitrary AC source

Objectives of Section 11.3:

- Learn the structure of power distribution systems
- Establish the concept of the three-phase power transmission system
- Understand the meaning and realization of three-phase source and three-phase load
- Solve for phase and line voltages and line currents in the three-phase balanced wye-wye system
- Establish the meaning and the role of the neutral conductor in the wye-wye power distribution system

Objectives of Section 11.4:

- Establish that the instantaneous power in balanced three-phase systems is constant
- Extend the concepts of reactive power, complex power, and apparent power to the three-phase systems
- Compare conductor material consumption in single-phase and three-phase systems
- Become familiar with delta-connected three-phase sources and loads
- Establish equivalency between delta and wye topologies with no ground

S. N. Makarov et al., *Practical Electrical Engineering*, https://doi.org/10.1007/978-3-319-96692-2_11

Application Examples:
- rms voltages and AC frequencies around the world
- Wattmeter
- Automatic power factor correction system
- Examples of three-phase source and the load
- Conductor material consumption in three-phase systems

Keywords:
Time averaging, Average power, rms voltage, rms current, AC fuse, Root mean square, Sawtooth wave, Triangular wave, Noise signals, Power angle, Power factor, Reactance, Capacitive reactance, Inductive reactance, Active power, True power, Reactive power, Complex power, Apparent power, VAR (volt-amperes reactive), VA (volt-amperes), Power triangle, Lagging power factor, Leading power factor, Wattmeter, Wattmeter current coil, Wattmeter potential coil, AC power conservation laws, Power factor correction, Power factor correction capacitor, PFC capacitor, Principle of maximum power efficiency for AC circuits, Principle of maximum power transfer for AC circuits, Impedance matching, Single-phase two-wire power distribution system, Single-phase three-wire power distribution system, Neutral conductor, Neutral wire, Split-phase distribution system, Polyphase distribution systems, Three-phase four-wire power distribution system, Phase voltages, Line-to-neutral voltages, *abc* phase sequence, Positive phase sequence, *acb* phase sequence, Negative phase sequence, Balanced phase voltages, Wye (or Y) configuration, Balanced three-phase source, Wye-connected source, Wye-connected load, Wye-wye distribution system, Phase impedances, Load impedances per phase, Balanced three-phase load, Synchronous three-phase AC generator, Alternator, Rotor, Stator, Synchronous AC motor, Rotating magnetic field, Line-to-line voltages, Line voltages, Line currents, Superposition principle for three-phase circuits, Per-phase solution, Total instantaneous load power of the three-phase system, Average load power of the balanced three-phase system, Reactive load power of the balanced three-phase system, Complex load power of the balanced three-phase system, Balanced delta-connected load, Balanced delta-connected source, Delta-delta distribution system

Section 11.1 AC Power Types and Their Meaning

The present section studies the basics of AC power. We begin with the root-mean-square (rms) representation of AC voltages and currents. The rms concept enables us to develop a DC equivalent representation, which compares AC to DC conditions in terms of power delivered to the load. It is important to understand that the rms concept is a general power concept; it applies not only to periodic AC circuits but virtually to any circuit, even with nonperiodic power sources, like noise power sources. Further results presented in this section are primarily intended for power electronic circuits; they have an equal applicability to radio-frequency communication circuits.

11.1.1 Instantaneous AC Power

We consider an arbitrary load with resistance R, load current $i(t)$, and load voltage $v(t)$, in the passive reference configuration. The instantaneous power delivered into the load is given by

$$p(t) = v(t)i(t) = \frac{v^2(t)}{R} \tag{11.1}$$

according to Ohm's law. If we use the load voltage in the form $v(t) = V_{\rm m} \cos \omega t$ [V], then

$$p(t) = v(t)i(t) = \frac{V_{\rm m}^2}{R} \cos^2 \omega t = \frac{V_{\rm m}^2}{2R}(1 + \cos 2\omega t) \tag{11.2}$$

where we applied the trigonometric identity $\cos^2 \omega t = 0.5(1 + \cos 2\omega t)$. Interestingly, the load power is not constant; it varies in time, and the behavior is shown in Fig. 11.1 for a load voltage amplitude $V_{\rm m} = 3$V, frequency $f = 50$ Hz, and load resistance $R = 5\ \Omega$.

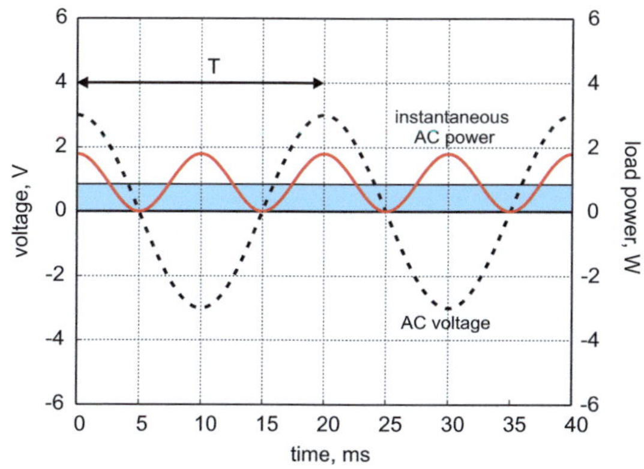

Fig. 11.1. Power (*solid line*) for a load voltage $v(t) = 3\cos(2\pi 50t)$ [V] (*dotted line*).

11.1.2 Time-averaged AC Power

An interesting question arises when we have to determine the bill for the variable AC power in Fig. 11.1. As far as the utility power is concerned, a consumer would prefer to pay for the minimum amount of power. The power minima in Fig. 11.1 occur at $t = 0.25T$, $0.75T$, $1.25T$, etc.; here T is the period of the AC voltage signal. Since the power is exactly zero at its minima, we would pay nothing. On the other hand, a utility would prefer to charge for the maxima of the power, which occur at $t = 0,\ 0.5T,\ 1.0T,\ 1.5T$, etc. A fair solution is clearly somewhere in the middle. It is based on *time averaging* the load power and then charging the consumer for the average (or mean) power as indicated in Fig. 11.1 by the shaded rectangle. Thus, we are interested in the averaged instantaneous power of Eq. (11.2). The time averaging is always done over a full period T of the AC voltage signal. The notation for the time-average value is often denoted by an overbar. Thus, the definition reads:

$$P = \overline{p(t)} \equiv \frac{1}{T}\int_0^T p(t)dt \tag{11.3}$$

where P is now the *average power* delivered to the load. We note that the average power times the period T gives us the energy E (in J or more often in W·h, 1 W·h = 3600 J) delivered to the load per period, i.e., $E = TP$.

rms Voltage and rms Current

Using Eq. (11.2) we obtain from Eq. (11.3)

$$P = \frac{1}{T}\int_0^T \frac{V_{\mathrm{m}}^2}{2R}(1 + \cos 2\omega\, t)\, dt = \frac{1}{T}\frac{V_{\mathrm{m}}^2}{2R}\left(\int_0^T 1 \cdot dt + \int_0^T \cos 2\omega\, t dt\right) \tag{11.4}$$

The first integral yields a nonzero contribution, whereas the second integral is exactly equal to zero, due to fact that the average of the sine or cosine functions over a period, or multiple periods, is zero. Thus,

$$\int_0^T 1 \cdot dt = T, \quad \int_0^T \cos 2\omega t dt = \frac{1}{2\omega}\sin 2\omega t|_0^T = \frac{1}{2\omega}\sin(4\pi t/T)|_0^T = 0 \tag{11.5}$$

Inserting these values into Eq. (11.4) results in

$$P = \frac{V_{\mathrm{m}}^2}{2R} = \frac{V_{\mathrm{rms}}^2}{R}, \quad V_{\mathrm{rms}} = \frac{V_{\mathrm{m}}}{\sqrt{2}} \tag{11.6}$$

where V_{rms} is the *rms* (*root-mean-square*) value for the load voltage $v(t) = V_{\mathrm{m}} \cos \omega t$ or simply the *rms voltage*. According to Eq. (11.6), the rms voltage is the *equivalent* DC voltage that provides the same power into the load. Once we know the rms voltage, the average power is given by the "DC" formula V_{rms}^2/R. The *rms* voltage is always less than the voltage amplitude by a factor of 0.707 (or 71 %). You should notice that Fig. 11.2 is a replica of Fig. 11.1; additionally, it shows the rms voltage and the averaged power for the load voltage $v(t) = 3\cos(2\pi 50t)$ [V]. If a nonzero phase is present in this expression, the result will not change. The signal will be shifted but all averages over the period will remain the same. The corresponding mathematical proof is suggested as one of the homework problems.

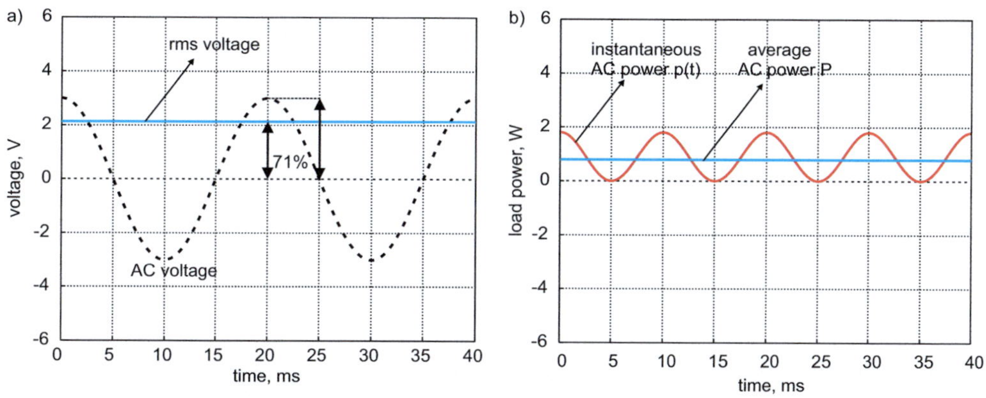

Fig. 11.2. (**a**) Load voltage $v(t) = 3\cos(2\pi 50t)$ [V] (*dotted line*) and its *rms* DC voltage, which delivers the same power into the load and (**b**) instantaneous power and the average power for the load voltage $v(t) = 3\cos(2\pi 50t)$ [V].

A similar expression is obtained for the alternating current $i(t) = I_{\mathrm{m}} \cos \omega t$ across the load. The *rms current* is given by

$$I_{\mathrm{rms}} = \frac{I_{\mathrm{m}}}{\sqrt{2}}, \quad V_{\mathrm{rms}} = RI_{\mathrm{rms}}, \quad P = R\frac{I_{\mathrm{m}}^2}{2} = RI_{\mathrm{rms}}^2 \tag{11.7}$$

Example 11.1: Determine average power delivered to a 10-Ω load when the applied AC voltage is given by $v(t) = 170\cos(2\pi 60t)$ [V] (US).

Solution: We find the rms voltage first:

$$V_{\mathrm{rms}} = 170/\sqrt{2} \approx 120.21 \text{ V} \tag{11.8}$$

Example 11.1 (cont.): The average power is then given by $P = \frac{V_{\text{rms}}^2}{R} = 1.445$ kW. Note that the rms current is equal to 12.02 A. Therefore, a 15-A or a 20-A AC *fuse* should be used. The fuse rating is based on the rms electric current value.

Exercise 11.1: Determine average power delivered to a 10-Ω load when the alternating load current is given by $i(t) = 5\cos(2\pi 60t)$ [A].

Answer: 125 W.

11.1.3 Application Example: rms Voltages and AC Frequencies Around the World

The AC voltage reported for the wall plug is an rms voltage. In the USA, the rms voltage typically ranges from 110 V to 127 V. Variations are caused by a specific type of a three-phase secondary distribution system studied later in this chapter. In this text we assume an average nominal value of $V_{\text{rms}} = 120$ V, perhaps a safe estimate. Using this value we obtain the voltage amplitude (the peak voltage value) of $V_{\text{m}} = \sqrt{2} \cdot 120 \text{ V} \approx 170$ V. This number is considerably greater than the reported 120 V AC. In other countries, the nominal rms wall plug voltage ranges from 220 V to 240 V, depending on the country. Using the nominal rms value of 220 V (People's Republic of China, Russia, France, Argentina, etc.), we obtain a voltage amplitude of $V_{\text{m}} = \sqrt{2} \cdot 220 \text{ V} = 311$ V. This number is again greater than the reported 220 V AC. India uses the nominal rms value of 230 V at 50 Hz, as do the European Union and Great Britain. This yields a voltage amplitude of $V_{\text{m}} = \sqrt{2} \cdot 230 \text{ V} = 325$ V. Depending on the country you live in, the AC frequency is either 50 Hz or 60 Hz; for instance, it is 60 Hz in the USA.

Historical: *From the IEEE Historical FAQ's and other sources*: The person responsible for adopting 60 Hz was probably Nikola Tesla who figured that for the Westinghouse-designed central stations for incandescent lamps, the efficient distribution was 59 Hz, and it was then rounded to 60 Hz. The German company AEG (Allgemeine Elektrizitäts-Gesellschaft), originally influenced by Thomas Edison, started using 50 Hz as a more "metric" number. Their standard spread to the rest of Europe and to other countries.

Figure 11.3 shows the rms voltage (and frequency) world map. Some countries have a dual distribution system that operates at 120 V and 220 V simultaneously. With the help of an electric transformer, studied in the following text, we can convert the voltages to higher or lower values. However, we cannot convert frequency with a linear circuit or a transformer. Some transformers are designed for both 50 Hz and 60 Hz but unfortunately not all. As time progresses, the frequency difference between the load and the source may have a severe effect on motorized applications and their transformers (power loss, overheating, and even eventual burnout).

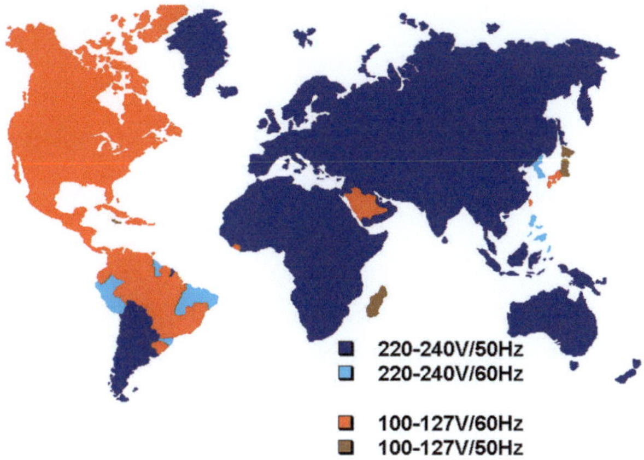

Fig. 11.3. The *rms* voltage world map, courtesy of Mr. Conrad H. McGregor, UK, and reproduced with the author's permission.

Example 11.2: Determine average power delivered to a 100-Ω load when the applied AC voltage has an *rms* value of 220 V (People's Republic of China).

Solution: The average power is given by

$$P = \frac{V_{\mathrm{rms}}^2}{R} = 484 \ \mathrm{W} \tag{11.9}$$

Note that the rms current is equal to 2.2 A. Therefore, a 5-A *AC fuse* (but not the 2-A fuse) is sufficient in the present case. The fuse is an overcurrent protective device; a soldered joint within the fuse is melted when the rms current exceeds a threshold.

Exercise 11.2: The load from Example 11.2 is connected to a wall plug in the USA. How would the average load power change?

Answer: The average load power will be 144 W.

11.1.4 rms Voltages for Arbitrary Periodic AC Signals

Analytical expressions given by Eqs. (11.6) and (11.7) are quite sufficient for finding the mean power but only for single-frequency AC signals. In certain cases, the signal may still be periodic with a period of T but may contain multiple frequency components. One such example is the clock signal for which we estimate the average electric power. In this case, we return to the definition Eq. (11.3) and rewrite it in the form, which literally explains the meaning of the *root-mean-square* value:

$$P = \frac{1}{T}\int_0^T p(t)\,dt = \frac{1}{R}\,\frac{1}{T}\int_0^T v^2(t)\,dt = \frac{\left(\sqrt{\frac{1}{T}\int_0^T v^2(t)\,dt}\right)^2}{R} \tag{11.10a}$$

We again wish to define the rms voltage as the DC voltage that gives the same power into the load resistance R. Therefore, it should be

$$V_{\text{rms}} = \sqrt{\frac{1}{T}\int_0^T v^2(t)\,dt} \Rightarrow P = \frac{V_{\text{rms}}^2}{R} \tag{11.10b}$$

For single-frequency voltage signals, this result reduces to Eq. (11.6). For more complicated voltages or for voltages measured directly, the calculation of the integral in Eq. (11.10b) may constitute some difficulties. At the end of this chapter, we provide a few homework problems tasking you to calculate the integral in Eq. (11.10b) directly. Once the rms voltage is found, the *rms* current through the resistive load is expressed by $I_{\text{rms}} = V_{\text{rms}}/R$, irrespective of the particular signal type.

Example 11.3: Determine the average power delivered to a 100-Ω load when the applied periodic voltage has the form $v(t) = 10t/T$ [V] over a period $T = 10$ ms. This signal is known as a *sawtooth* or a *triangular wave*.

Solution: We find the rms voltage first:

$$V_{\text{rms}} = \sqrt{\frac{1}{T}\int_0^T v^2(t)\,dt} = \sqrt{\frac{1}{T}\int_0^T 100t^2/T^2\,dt} = \sqrt{100/3} \approx 5.77 \text{ V} \tag{11.10c}$$

The average power is thus given by $P = V_{\text{rms}}^2/R \approx 333$ mW.

Frequently encountered voltage signals in microelectronic circuits are *noise signals*, which are neither sinusoidal nor periodic. In this case, Eq. (11.10b) applies again, but only in the limit as $T \to \infty$. Advanced analog electronics deals with certain electronic circuits where such noise sources become important and even critical.

11.1.5 Average AC Power in Terms of Phasors: Power Angle

For arbitrary dynamic circuit elements, the power analysis is carried out in terms of phasors. Consider element A in Fig. 11.4 which has real-valued voltage and current given by

$$v(t) = V_{\mathrm{m}} \cos(\omega t + \varphi)$$
$$i(t) = I_{\mathrm{m}} \cos(\omega t + \psi) \tag{11.11}$$

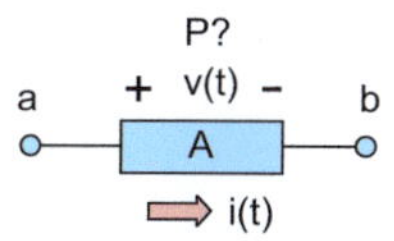

Fig. 11.4. Arbitrary circuit element in the passive reference configuration.

When element A is a resistor, the phases in Eq. (11.11) are the same, and finding the average power is straightforward. However, when element A is an inductor, capacitor, or a combination of resistor and inductor/capacitor, the situation becomes different. In this case, the phases of voltage and current in Eqs. (11.11) do not necessarily coincide. By definition:

$$P = \overline{p(t)} = \frac{1}{T}\int_0^T v(t)i(t)dt = \frac{1}{T}V_{\mathrm{m}}I_{\mathrm{m}}\int_0^T \cos(\omega t + \varphi)\cos(\omega t + \psi)dt \tag{11.12}$$

To manipulate the cosine expression in Eq. (11.12), we can use the trigonometric identity $\cos(\omega t + \varphi)\cos(\omega t + \psi) = 0.5\cos(\varphi - \psi) + 0.5\cos(2\omega t + \varphi + \psi)$. The integral of the second term in Eq. (11.12) will be equal to zero since it is the integral of the plain cosine function over two periods. The result then has the form:

$$P = \overline{p(t)} = \frac{1}{T}\int_0^T v(t)i(t)dt = \frac{V_{\mathrm{m}}I_{\mathrm{m}}}{2}\cos(\varphi - \psi) = \frac{V_{\mathrm{m}}I_{\mathrm{m}}}{2}\cos\theta = V_{\mathrm{rms}}I_{\mathrm{rms}}\cos\theta \tag{11.13}$$

Equation (11.13) is of great importance for power electronics since it introduces the so-called power angle θ

$$\theta = \varphi - \psi, \quad -90^{\circ} \le \theta \le +90^{\circ} \tag{11.14}$$

and the power factor

$$PF = \cos(\varphi - \psi) = \cos\theta \tag{11.15}$$

Both of these expressions determine the average power delivered to the circuit element. For any passive load, the power angle must be between −90° and +90°; i.e., the average power delivered to the element must be *nonnegative*! However, for an active load (an amplifier), it is possible that the power angle is no longer within those limits. Then, the element actually delivers power to the circuit rather than absorbing it. Equation (11.13) can now be expressed in terms of phasor voltage $\mathbf{V} = V_m \angle \varphi$ and phasor current $\mathbf{I} = I_m \angle \psi$. The result is simple and elegant:

$$P = \frac{\mathrm{Re}\left(\mathbf{V} \cdot \mathbf{I}^*\right)}{2} = \frac{\mathrm{Re}\left(\mathbf{V}^* \cdot \mathbf{I}\right)}{2} \tag{11.16}$$

where the star denotes the *complex conjugate*, $\left(\mathrm{e}^{j\alpha}\right)^* = \mathrm{e}^{-j\alpha}$, and Re is the real part of a complex number. The proof is based on the phasor substitution, that is,

$$\frac{\mathrm{Re}\left(\mathbf{V} \cdot \mathbf{I}^*\right)}{2} = \frac{\mathrm{Re}(V_m \angle \varphi \cdot I_m \angle -\psi)}{2} = \frac{V_m I_m}{2} \mathrm{Re}(\angle \varphi - \psi) = \frac{V_m I_m}{2} \cos(\varphi - \psi) \tag{11.17}$$

Example 11.4: The phasor voltage across a purely resistive load with a resistance $R = 10\ \Omega$ is given by $\mathbf{V} = -3 + j3$ [V]. Find the average power delivered to the load.

Solution: According to Eq. (11.16),

$$P = \frac{\mathrm{Re}\left(\mathbf{V} \cdot \mathbf{I}^*\right)}{2} = \frac{\mathrm{Re}\left(\mathbf{V} \cdot \mathbf{V}^*\right)}{2R} = \frac{|\mathbf{V}|^2}{2R} = \frac{3^2 + 3^2}{20} = \frac{18}{20} = 900\ \mathrm{mW}$$

Exercise 11.3: The phasor voltage and phasor current for an AC load are given by $\mathbf{V} = -3 + j3$ [V], $\mathbf{I} = +2 + j3$ [A]. Find the average power delivered to the load.

Answer: $P = 1.5$ W.

11.1.6 Average Power for Resistor, Capacitor, and Inductor

For arbitrary passive circuit elements, the phasor voltage **V** and the phasor current **I** are related by Ohm's law in the impedance form, $\mathbf{V} = \mathbf{Z}\mathbf{I}$, where **Z** is the element impedance (or the equivalent impedance of a circuit block). We can substitute this result into Eq. (11.16) and obtain

$$P = \frac{\mathrm{Re}(\mathbf{V}\cdot\mathbf{I}^*)}{2} = \frac{\mathrm{Re}(\mathbf{Z}\cdot\mathbf{I}\cdot\mathbf{I}^*)}{2} = \frac{\mathrm{Re}\left(\mathbf{Z}\cdot|\mathbf{I}|^2\right)}{2} = \frac{\mathrm{Re}(\mathbf{Z})|\mathbf{I}|^2}{2} \tag{11.18}$$

since for any complex number **I**, the following equality holds: $\mathbf{I}\cdot\mathbf{I}^* = I_{\mathrm{m}}\angle\psi\cdot I_{\mathrm{m}}\angle-\psi = I_{\mathrm{m}}^2 = |\mathbf{I}|^2 > 0$. Equation (11.18) is a remarkable result: if the impedance of an element is purely imaginary, the average power delivered to the circuit element must be zero. Indeed, so are the impedances for the inductor and the capacitor. Therefore, the average power delivered to either the inductor or the capacitor must be zero! The same result may be explained using Eq. (11.13). We put the phase of the current, ψ, equal to zero for simplicity. The voltage phase φ will then be $+90°$ for the inductor and $-90°$ for the capacitor. As the cosine of $\pm 90°$ is zero, Eq. (11.13) will also give zero average power. An additional explanation is related to the phasor diagrams for voltages and currents shown in Fig. 11.5. The average power is half of the dot products of two vectors (phasor voltage and phasor current) in the complex plane. The dot product of two perpendicular vectors is exactly zero.

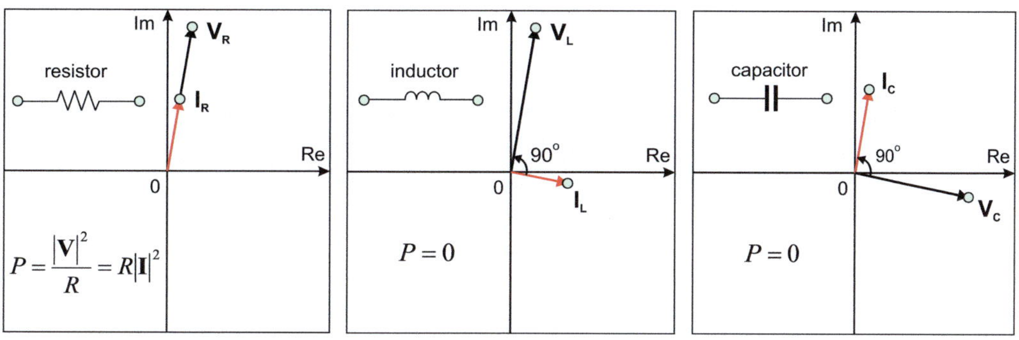

Fig. 11.5. Average power for a resistor, inductor, and capacitor and the related phasor diagrams.

11.1.7 Average Power, Reactive Power, and Apparent Power

We have just seen that the impedance of a load, **Z**, is most important for the average AC power delivered to the load. If the impedance is a pure resistance, there is no problem. Otherwise, almost no power may be delivered to the load even though large currents can flow in the circuit and large AC voltages are observed. For example, if the load is a pure inductance or capacitance, then no average power will be delivered to the load, no matter which voltages and currents we use. Instead, we will only heat up wires and other circuit components. We write the impedance for an arbitrary load both in rectangular and in polar form:

$$\mathbf{Z} = R + jX\ [\Omega], \quad \mathbf{Z} = |\mathbf{Z}|\angle\theta\ [\Omega], \quad R = |\mathbf{Z}|\cos\theta, \quad X = |\mathbf{Z}|\sin\theta \tag{11.19a}$$

$$|\mathbf{Z}| = \sqrt{R^2 + X^2}\ [\Omega], \quad \theta = \tan^{-1}\left(\frac{X}{R}\right) \tag{11.19b}$$

The real part of the impedance, R, is the *resistance* of the load, and the imaginary part, X, is the load *reactance*. For example, the inductor and the capacitor are purely reactive loads (have only X but not R), whereas the resistor is purely "resistive". The angle θ is the power angle of the load; the power angle has already been introduced in Eqs. (11.13), (11.14), and (11.15). Thus, the power factor PF is simply the cosine of the angle of the load impedance.

Example 11.5: Determine the resistance and the reactance of an RLC series load shown in Fig. 11.6. The AC angular frequency is 1000 rad/s.

Solution: The three impedances are combined in series (added to each other),

$$\mathbf{Z} = 100 + j\omega L - j\frac{1}{\omega C} = 100 + j1 - j100 = 100 - j99\ [\Omega] \tag{11.20}$$

The resistance is 100 Ω and the reactance is equal to −99 Ω. The reactance is negative, i.e., *capacitive*. In other words, the capacitive reactance dominates.

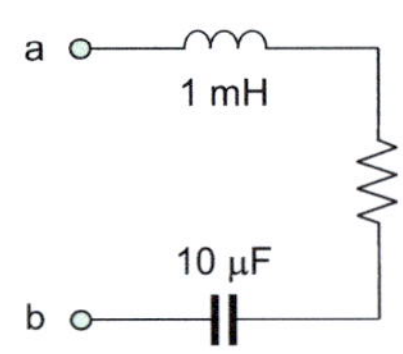

Fig. 11.6. A RLC series load.

Using resistance R and reactance X of the load, we now introduce three *different* AC power types for that load. The first type is the average or *active* power P studied before in this section. The active power is expressed by

$$P = \frac{\mathrm{Re}\left(\mathbf{V}\cdot\mathbf{I}^*\right)}{2} = \frac{R|\mathbf{I}|^2}{2} = \frac{|\mathbf{Z}||\mathbf{I}|^2}{2}\cos\theta = \frac{|\mathbf{V}||\mathbf{I}|}{2}\cos\theta = V_{\mathrm{rms}} I_{\mathrm{rms}}\cos\theta \quad [\mathrm{W}] \tag{11.21a}$$

since $|\mathbf{V}| = V_{\mathrm{m}} = \sqrt{2}V_{\mathrm{rms}}, \quad |\mathbf{I}| = I_{\mathrm{m}} = \sqrt{2}I_{\mathrm{rms}}$. This is *the true or useful power* delivered to the load, with the units of watts. Note the operations with complex magnitudes: $|\mathbf{ab}| = |\mathbf{a}||\mathbf{b}|, \quad |\mathbf{a}/\mathbf{b}| = |\mathbf{a}|/|\mathbf{b}|, \quad |\mathbf{a}^*| = |\mathbf{a}|$, which directly follow from the complex number definition in polar form. The second power type is the *reactive* power Q that is

$$Q = \frac{\mathrm{Im}(\mathbf{V}\cdot\mathbf{I}^*)}{2} = \frac{X|\mathbf{I}|^2}{2} = \frac{|\mathbf{Z}||\mathbf{I}|^2}{2}\sin\theta = \frac{|\mathbf{V}||\mathbf{I}|}{2}\sin\theta = V_{\mathrm{rms}}I_{\mathrm{rms}}\sin\theta \quad [\mathrm{VAR}] \tag{11.21b}$$

The physical units of the reactive power are also watts. However, to underscore the fact that this power is not an active useful power, the units of VAR *(volt-amperes reactive*) are used. The reactive power flows back and forth from the source to the load, through an electric line but does not do real work. The last power type is the *complex power* **S** that is simply

$$\mathbf{S} = \frac{\mathbf{V}\cdot\mathbf{I}^*}{2} = P + jQ \quad [\mathrm{VA}] \tag{11.21c}$$

The complex power is measured in *volt-amperes* (VA). The magnitude of the complex power $S = |\mathbf{S}|$ is called the *apparent power*. We can see that the apparent power is given by

$$S = |\mathbf{S}| = \frac{|\mathbf{Z}||\mathbf{I}|^2}{2} = \frac{|\mathbf{V}||\mathbf{I}|}{2} = V_{\mathrm{rms}}I_{\mathrm{rms}} \quad [\mathrm{VA}] \tag{11.21d}$$

The apparent power is the "best possible" load power that can be obtained if one measures current and voltage and ignores the phase shift between them. Equations (11.21) and (11.22) raise the obvious question: why do we need so many AC power types? The answer is that a purely resistive load (the power angle θ equals zero) is merely a dream and not realistic. Any AC load generally has a significant reactive impedance part. So does an electric motor, a small antenna in your cellphone, and even a household electric heater whose heating spiral is a series combination of a resistance and a small, but often visible, inductance. Therefore, we always deal with active and reactive power; the sum of their squares is the square of the magnitude of the apparent power. The reactive power increases the electric current flowing in the circuit and thus increases the unrecoverable losses in (sometimes very long) power lines, which have a finite resistance. Therefore, our goal is to decrease the percentage of the reactive power and thus decrease the net power loss. The three power definitions show us how to accomplish this task. In the next section, we will need to decrease the power angle θ by modifying the load through adding other circuit components; in other words, we are attempting to *load match* the circuit.

11.1.8 Power Triangle

Since $\cos^2\theta + \sin^2\theta = 1$, the three powers (average, reactive, and apparent) are interconnected by the relation

$$S^2 = P_{\mathrm{avg}}^2 + Q^2 \quad [\mathrm{W}] \tag{11.22}$$

This relation is also called the *power triangle*.

Example 11.6: Determine the average (or true) power and the reactive power for the inductive load shown in Fig 11.7a. Construct the corresponding power triangle. The circuit parameters are as follows:

$V_m = 170$ V, $\omega = 377$ rad/s, $L = 25.7$ mH, $R = 9.7$ Ω.

Solution: We need to find active and reactive powers according to Eq. (11.21). To do so, we need the load impedance and the load or circuit current. We convert the circuit to the phasor/impedance form. The equivalent impedance is given by

$$\mathbf{Z} = \mathbf{Z}_R + \mathbf{Z}_L = R + j\omega L = 9.7 + j9.7 = 9.7\sqrt{2}\angle 45° \ [\Omega] \quad (11.23)$$

The power angle (the phase of the complex impedance) is 45°. The phasor voltage **V** across the load is equal to $V_m = 170$V. The load phasor current is given by $\mathbf{I} = V_m/\mathbf{Z} = 170/(9.7\sqrt{2}\angle 45°) = 12.39\angle -45° \ [A]$. According to Eq. (11.21),

$$\begin{aligned} P &= 0.5 \times 170 \times 12.39 \times \cos 45° = 745 \ [W] \\ Q &= 0.5 \times 170 \times 12.39 \times \sin 45° = 745 \ [VAR] \end{aligned} \quad (11.24)$$

The corresponding power triangle is plotted in Fig. 11.8a.

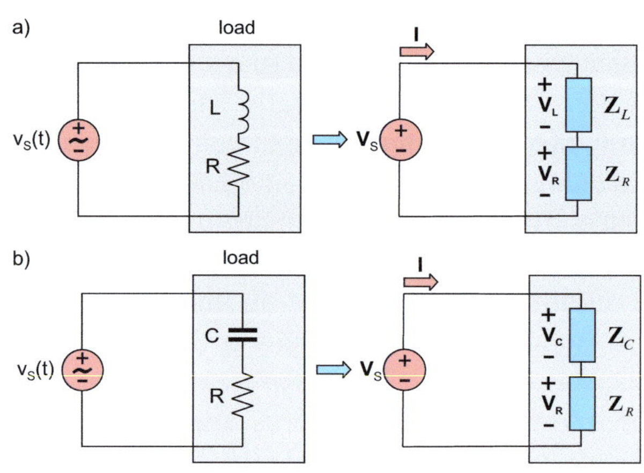

Fig. 11.7. Two circuits for power calculation of (**a**) inductive load and (**b**) capacitive load.

The power angle (the phase of the load impedance) of the power triangle in Fig. 11.8a is +45°. When the power angle θ is positive, as in the present case, the corresponding power factor is said to be *lagging*. The lagging power factor means that the load current lags the load voltage. Thus, the power factor in Fig. 11.8a is 0.707 *lagging* or, which is the same, is 70.7 % *lagging*. Similarly, the power factor in Fig. 11.8b is 44.8 % lagging. However, the power angle (the phase of the load impedance) in Fig. 11.8c is −45°.

When the power angle θ is negative, as is the case here, the corresponding power factor is said to be *leading*. The leading power factor means that the load current leads the load voltage. The power factor in Fig. 11.8c is 70.7 % *leading*. The lagging occurs for a predominantly inductive load, whereas the leading occurs for a predominantly capacitive load; see Fig. 11.7.

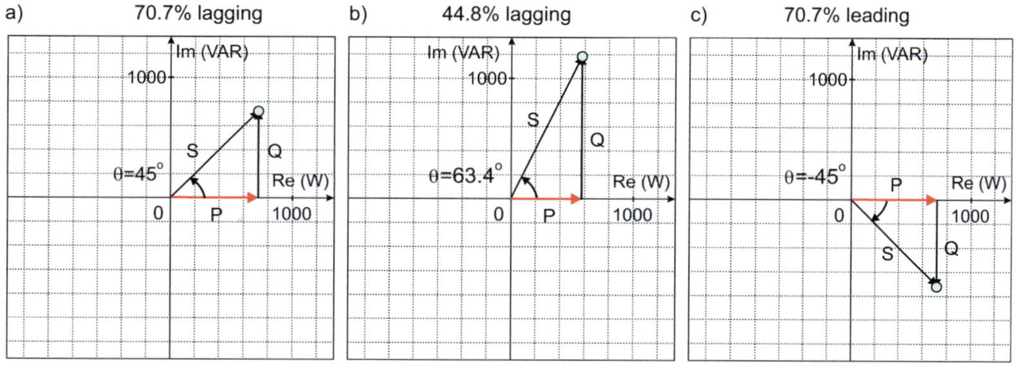

Fig. 11.8. Power triangles for inductive loads (**a**, **b**), and for a capacitive load (**c**). The real axis corresponds to the average (true) power; the imaginary axis is the reactive power.

Exercise 11.4: Determine the average (or true) power and the reactive power for the inductive load shown in Fig 11.7a. Construct the corresponding power triangle. You are given $V_m = 170$ V, $\omega = 377$ rad/s, $L = 26.5$ mH, $R = 5\ \Omega$.

Solution: $P = 578.9$ [W], $Q = 1156.7$ [VAR], $\theta = 63.4°$. The power triangle is plotted in Fig. 11.8b. The power factor is 44.8 % lagging.

Exercise 11.5: Determine the average (or true) power and the reactive power for the capacitive load shown in Fig 11.7b. Construct the corresponding power triangle. The circuit parameters are $V_m = 170$ V, $\omega = 377$ rad/s, $C = 265\ \mu$F, $R = 10\ \Omega$.

Answer: $P = 722.5$ [W], $Q = -722.5$ [VAR], $\theta = -45°$. Note that the reactive power becomes negative for the capacitive load. The corresponding power triangle is plotted in Fig. 11.8c. The power factor is 70.7 % leading.

It might be interesting to mention that the circuit in Fig. 11.7b is an equivalent circuit model of a short dipole or monopole *antenna*, the so-called whip antenna; it represents predominantly a capacitive load. Indeed, much higher frequencies are employed, but the concept remains the same. Whip antennas are common on ships, trucks, and other vehicles. Many radio amateurs use whip antennas as well.

11.1.9 Application Example: Wattmeter

AC power is measured with a *wattmeter.* The idea of an analog wattmeter operation is schematically illustrated in Fig. 11.9a. The wattmeter includes at least two coils: the massive immovable *current coil* and the lighter suspended, or pivoted, *voltage* (or *potential*) *coil*. The current coil has a *very low impedance*; it is connected in series with the load in Fig. 11.9b. The voltage coil has a *very high impedance*; it is connected in parallel with the load in Fig. 11.9b. The voltage coil typically has a high-value resistor connected in series to increase the impedance. The voltage coil is constructed of a fine wire, whereas the fixed (current) coil uses a thicker wire to carry the load current. When the current and voltage are in phase, the magnetic fluxes in both coils attempt to align with each other so that the arrow in Fig. 11.9a will move to the right. When the current and voltage are out of phase (phase difference of 180°; the load is in fact an AC source), the arrow in Fig. 11.9a would move to the left. When the phase difference is 90°, the arrow stays at the center. Thus, the power angle could be measured, and the average and reactive powers could be reported in an analog or digital way.

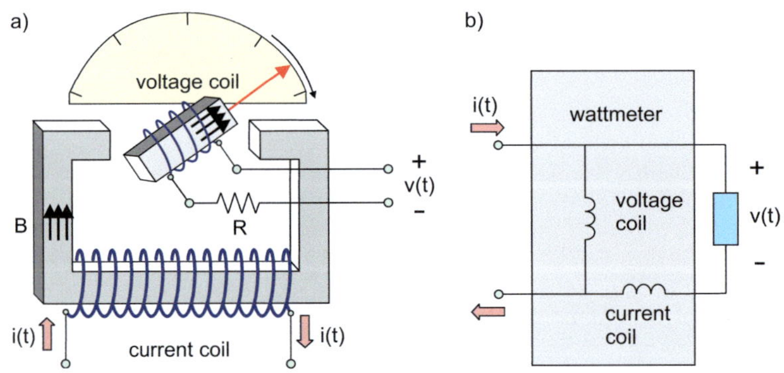

Fig. 11.9. (**a**) Wattmeter concept and (**b**) wattmeter coil connection to the load.

Section 11.2 Power Factor Correction: Maximum Power Efficiency and Maximum Power Transfer

11.2.1 Power Factor Correction

We are about to proceed with the *correction of the power factor* $PF = \cos\theta$ for an arbitrary AC load. The correction has several equivalent approaches. One can make:

1. the power angle exactly equal to zero.
2. the power factor equal to one.
3. the reactive power equal to zero.
4. the load impedance purely resistive.
5. the imaginary part of the load impedance (reactance) equal to zero.

The last approach is perhaps most useful from a practical point of view. We attempt to modify the load by adding extra circuit elements so that the impedance of the *modified* load becomes purely resistive (the imaginary part of the impedance is zero). It is worth noting that the condition of zero reactance is simultaneously the resonance condition for the various RLC tank circuits studied in the previous chapter. It means that we need to make the load "resonant" in order to correct the power factor! This is often achieved by converting the load to an RLC circuit: adding a capacitor to the inductive load or an inductor to a capacitive load. Most residential loads (washer, dryer, air conditioner, refrigerator, etc.) and industrial loads are powered by an induction motor. A simplified equivalent circuit of it is an inductive load shown in Fig. 11.10a. We intend to add a capacitor in parallel with the load as in Fig. 11.10b. We attempt to choose the capacitance value in such a fashion as to make the power factor of the modified load equal to one. The capacitor in Fig. 11.10b is called the *power factor correction capacitor* or the *PFC capacitor*.

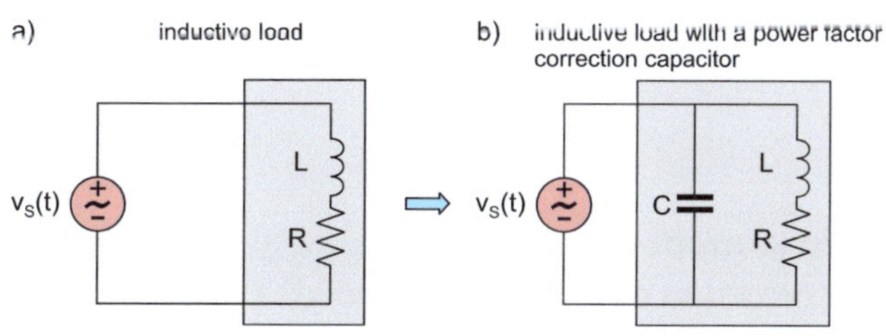

Fig. 11.10. Power correction for an inductive load with the shunt capacitor (capacitor in parallel).

We solve both circuits in Fig. 11.10 in the phasor form. The equivalent impedance (or better its reciprocal, the *admittance*) for the modified load in Fig. 11.10b is found first, that is,

$$\frac{1}{\mathbf{Z}}=\frac{1}{\mathbf{Z}_C}+\frac{1}{\mathbf{Z}_L+\mathbf{Z}_R}=j\omega C+\frac{R-j\omega L}{R^2+(\omega L)^2} \tag{11.25}$$

If the impedance is a real number, then the admittance is a real number and vice versa. Therefore, the condition of a real impedance is equivalent to the condition of a real admittance. From Eq. (11.25), one has

$$j\omega C+\frac{-j\omega L}{R^2+(\omega L)^2}=0 \Rightarrow R^2+(\omega L)^2=\frac{L}{C} \Rightarrow C=\frac{L}{\omega^2 L^2+R^2} \tag{11.26}$$

The capacitance value is thus found from the equation

$$C=\frac{L}{\omega^2 L^2+R^2} \tag{11.27}$$

Equation (11.27) is a mathematical statement for the power factor correction capacitor. Its practical value will become apparent from the example that follows. The equivalent impedance of the load with the matching capacitor is then found using the real part of Eq. (11.25), i.e.,

$$\frac{1}{\mathbf{Z}}=\frac{R}{R^2+(\omega L)^2} \Rightarrow \mathbf{Z}=R+\frac{(\omega L)^2}{R} \tag{11.28}$$

We are interested in the phasor circuit current **I** with and without the PFC capacitor. Given the voltage source $v_S(t)=V_m\cos\omega t$, one obtains for the circuits in Fig. 11.10:

$$\mathbf{I}=\overbrace{\underbrace{\frac{RV_m}{R^2+(\omega L)^2}}_{\text{with capacitor}}-\frac{j\omega L V_m}{R^2+(\omega L)^2}}^{\text{without capacitor}} \tag{11.29}$$

As you can see, the two terms of the expression without capacitor are reduced to the first term when the power correction capacitor is included. This completes the analysis of the circuits in Fig. 11.10.

Example 11.7: For the circuit in Fig. 11.10, find the average (or true) load power and the reactive load power with and without the power correction factor capacitor. You are given $V_m=170\ \text{V},\quad \omega=377\ \text{rad/s},\quad L=25.7\ \text{mH}, R=9.7\ \Omega$.

Example 11.7 (cont.):

Solution: To find the power expressions, we need the phasor voltage across the load, which is simply V_m. We also need the phasor current, which is given by Eq. (11.29) in either case. Plugging in the numbers we obtain,

$$\mathbf{I} = 8.77\ \text{A} \quad \text{or} \quad \mathbf{I} = 8.77 - j8.77\ \text{A} \tag{11.30}$$

with and without the PFC capacitor, respectively. Now, we use the power definitions:

$$P = \frac{\text{Re}(\mathbf{V} \cdot \mathbf{I}^*)}{2} \Rightarrow P = 745\ \text{W} \quad \text{or} \quad P = 745\ \text{W} \tag{11.31}$$

$$Q = \frac{\text{Im}(\mathbf{V} \cdot \mathbf{I}^*)}{2} \Rightarrow Q = 0\ \text{VAR} \quad \text{or} \quad Q = 745\ \text{VAR} \tag{11.32}$$

with and without the PFC capacitor, respectively.

In summary, we have found the following information from this example:

No power factor correction:

- Average (active or true) power: $P = 745\ \text{W}$
- Reactive power: $Q = 745\ \text{VAR}$
- Amplitude of the circuit current: 12.41 A

Power factor correction:

- Average (active or true) power: $P = 745\ \text{W}$
- Reactive power: 0
- Amplitude of the circuit current: 8.77 A

By correcting the load power factor with the capacitor in parallel, we did *not* change the average power (power delivered to the load), but we eliminated the reactive power and decreased the amplitude of the circuit current by 70.7 % ($1/\sqrt{2}$). This means that the ohmic losses in the electric line connecting load and generator will decrease by 50 %, since these losses are proportional to the square of the current amplitude. In other examples, the loss reduction factor may be even more significant. Is it worth doing a power factor correction? Well, if the electric power line is long enough or the initial power factor is not high enough, it is definitely a very useful and professional task. To support this conclusion, we mention a citation from *IEEE Transactions on Power Electronics*: "Everyone knows that correcting power factor is the easiest and fastest way to save energy dollars."

Exercise 11.6: Find the value of the power factor correction capacitor in Example 11.7.

Answer: $C = 136.73\ \mu\text{F}$.

Exercise 11.7: Find the value of the load impedance in Example 11.7 with and without the power factor correction capacitor. Express your result in polar form.

Answer: $\mathbf{Z} = 13.7\angle 45°\ \Omega$ and $\mathbf{Z} = 19.4\angle 0°\ \Omega$, respectively.

11.2.2 Application Example: Automatic Power Factor Correction System

The power factor correction capacitors are frequently seen on residential power poles in the form of pole-mounted capacitor banks. Figure 11.11 shows an automatically switched power factor correction system that measures all three power types (active, reactive, and apparent power) using the same wattmeter principle described in the previous section. Based on the recorded measurements, the required capacitor value is selected, which assures the targeted power factor.

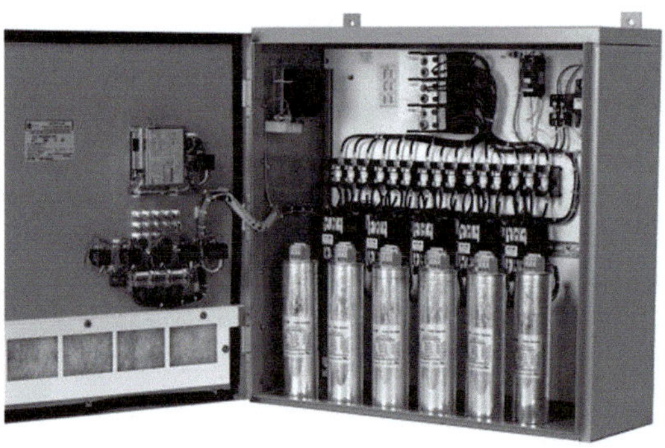

Fig. 11.11. Automatically switched power factor correction systems for low-voltage applications. Six capacitor cells are seen on the bottom. Technical Data TD02607001E Cutler-Hammer.

11.2.3 Principle of Maximum Power Efficiency for AC Circuits

Why is the power correction capacitor placed in parallel, not in series with the load? To answer this question, we should establish and understand the *principle of maximum power efficiency for AC circuits*. Consider a generic source-load AC circuit depicted in Fig. 11.12a in phasor form. It is based on a Thévenin equivalent circuit for an AC source with the source impedance $\mathbf{Z}_T$ connected to the load impedance $\mathbf{Z}_L$. The source impedance will also include the loss resistance of power lines. Figure 11.12b shows the corresponding DC counterpart, which is useful for the subsequent analysis. The source resistance R_T will include the loss resistance of power lines as well.

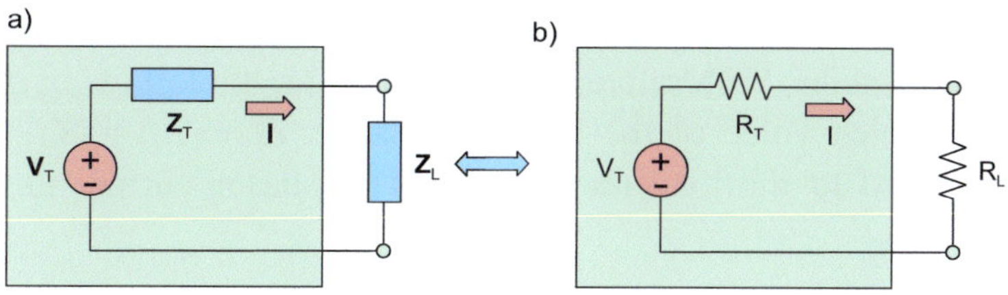

Fig. 11.12. Thévenin equivalent circuit for an AC source with the load impedance $\mathbf{Z}_L$ and its DC counterpart with the load resistance R_L.

In the DC case, the source-load configuration of Fig. 11.12b is *maximally efficient* when the load resistance is much greater than the source (loss) resistance. This fact has been established in Chapter 4. When $R_L \gg R_T$, the useful power delivered to the load resistance R_L is much larger than the power loss in R_T. We emphasize that the load power itself could be relatively small in this case as compared to the maximum available load power at $R_L = R_T$. The same situation occurs for the AC system shown in Fig. 11.12a. Do we wish to deliver the maximum available power of a remote megawatt AC source to the household? No, this is not our goal. We would rather deliver a reasonable amount of power but in a most efficient way. It means that not only do we need to make $\mathbf{Z}_L$ real but also *as high as possible.* This operation would further reduce the circuit current and the associated loss. Exactly this goal is accomplished by the shunt PFC capacitor in Fig. 11.10. If we consider the series-connected PFC capacitor as an alternative, we will obtain

$$\mathbf{Z}_L = R \quad \text{for series connection} \quad \text{versus} \quad \mathbf{Z}_L = R + \frac{(\omega L)^2}{R} \quad \text{for shunt connection.} \tag{11.33}$$

Both impedances in Eq. (11.33) are real; there is no reactive power in either case. However, the second impedance is considerably greater than the first one for poor power factors, i.e., for $\omega L \gg R$! Hence, considerably smaller circuit currents and considerably better efficiencies are achieved. Moreover, the parallel connection is easier to accomplish in practice—we remember how easy it is to connect a voltmeter as opposed to an ammeter

Exercise 11.8: Find the amplitude of the circuit current in Example 11.7 if the power correction capacitor were in series.

Answer: $I_m = 17.53$ A.

11.2.4 Principle of Maximum Power Transfer for AC Circuits

The *principle of maximum power transfer* is perhaps less important for residential power distribution systems where efficiency counts. However, it is critical for radio-frequency and communication circuits, which are conceptually the same AC circuits but operating at much higher frequencies. With reference to Fig. 11.12a, the following question should now be asked: at which value of the load impedance $\mathbf{Z}_L = R_L + jX_L$ is the average (true) power delivered to the load maximized? The phasor current in Fig. 11.12a is given by

$$\mathbf{I} = \frac{\mathbf{V}_T}{\mathbf{Z}_L + \mathbf{Z}_T} \quad [\mathrm{A}] \tag{11.34}$$

so that the average power delivered to the load becomes

$$P = \frac{R_L|\mathbf{I}|^2}{2} = \frac{R_L|\mathbf{V}_T|^2}{2|\mathbf{Z}_L + \mathbf{Z}_T|^2} = \frac{0.5R_L|\mathbf{V}_T|^2}{(R_L + R_T)^2 + (X_L + X_T)^2} \quad [\mathrm{W}] \tag{11.35}$$

Let us take a closer look at Eq. (11.35); in order to reach the maximum true power, the load reactance X_L should be equal to the generator reactance X_T taken with the opposite sign so that $X_L + X_T = 0$. This yields for the average load power

$$P = \frac{0.5R_L|\mathbf{V}_T|^2}{(R_L + R_T)^2} \quad [\mathrm{W}] \tag{11.36}$$

Consequently, the problem reduces to the maximum power transfer of a DC circuit as studied in Chapter 2. The corresponding condition for the maximum load power is

$$R_L = R_T \tag{11.37}$$

This condition, augmented by the equality for the reactances

$$X_L = -X_T \tag{11.38}$$

leads to a simple, yet very useful result for the maximum power transfer to the load:

$$\mathbf{Z}_L = \mathbf{Z}_T^* \Rightarrow P_{\max} = \frac{1}{8}\frac{|\mathbf{V}_T|^2}{R_T} \quad [\mathrm{W}] \tag{11.39}$$

We note that the load impedance should be the complex conjugate of the generator impedance. Along with the maximum power transfer, Eq. (11.39) assures that there is no reflection of radio-frequency waves propagating along the circuit transmission lines from the source to the load, which may even be a more important factor. A process of modifying the load impedance in order to satisfy Eq. (11.39) is called *impedance matching*.

Example 11.8: A generator impedance is 50 Ω. The load impedance is $10+j100$ Ω. What percentage of maximum available power (at a load impedance of 50 Ω) is transferred to the load?

Solution: According to Eq. (11.36), when the load impedance is exactly 50 Ω,

$$P = \frac{0.5 \times 50|\mathbf{V}_\mathrm{T}|^2}{(50+50)^2} = \frac{|\mathbf{V}_\mathrm{T}|^2}{8 \times 50} = 0.0025|\mathbf{V}_\mathrm{T}|^2 \quad [\mathrm{W}] \tag{11.40a}$$

When the load impedance is $10+j100\Omega$, Eq. (11.35) gives

$$P = \frac{0.5 \times 10|\mathbf{V}_\mathrm{T}|^2}{(50+10)^2 + 100^2} = \frac{|\mathbf{V}_\mathrm{T}|^2}{2720} = 0.00037|\mathbf{V}_\mathrm{T}|^2 \quad [\mathrm{W}] \tag{11.40b}$$

The ratio of the two power expressions is 0.147, or 14.7%. In other words, 85.3% of available power is lost!

Exercise 11.9: A generator's impedance is $50-j100$ Ω. What should the load impedance be for maximum power transfer?

Answer: $50+j100$ Ω.

Exercise 11.10: Solve Example 11.8 when the load impedance is $10-j100$ Ω.

Answer: The same result of 14.7 % is obtained.

Section 11.3 AC Power Distribution: Balanced Three-Phase Power Distribution System

11.3.1 AC Power Distribution Systems

Representative AC power distribution systems are shown in Fig. 11.13. A *single-phase two-wire power distribution system* is depicted in Fig. 11.13a. It consists of a generator with a voltage amplitude of V_m, an rms value of $V_{rms} = V_m/\sqrt{2}$, and a phase φ connected through two conductors to a load with impedance $\mathbf{Z}$. The previous analysis of AC power was solely restricted to this configuration. An extension is the *single-phase three-wire power distribution system* shown in Fig. 11.13b. Such a system contains two identical AC sources of the same amplitude and phase connected to two ($\mathbf{Z}_1$, $\mathbf{Z}_2$) loads or to one ($\mathbf{Z}$) load through two outer conductors and the *neutral conductor* (or *neutral wire*). This system is the common household distribution system. It allows us to connect both 120-V and 240-V appliances as shown in Fig. 11.13b; we sometimes called it the *split-phase distribution system*. The neutral wire is usually physically grounded. In contrast to those two cases, the power distribution systems shown in Fig. 11.13c, d are the *polyphase distribution systems* in the sense that they use AC sources with *different* phases. For example, Fig. 11.13c illustrates a *two-phase three-wire distribution system* with two voltage sources; the second one lags the former by 90°. Finally, Fig. 11.13d shows the most important and practical *three-phase four-wire power distribution system* with three sources and three load impedances $\mathbf{Z}_1$, $\mathbf{Z}_2$, and $\mathbf{Z}_3$. Generally, the three-phase system also uses a (grounded) neutral wire. We will show that this wire may be omitted for balanced power distribution circuits, with the earth itself acting as the neutral conductor. This is important for long-distance, high-power transmissions. Power systems designed in this way are grounded at critical points to ensure safety.

Today, a vast majority of electric power is generated and distributed via the three-phase power systems. Why is this so? You will soon learn that in contrast to the single-phase systems, the instantaneous power in balanced three-phase systems is constant or independent of time rather than pulsating. This circumstance results in more uniform power transmission and less vibration of electric machines. Furthermore, three-phase AC motors have a nonzero starting torque in contrast to the single-phase motors. Last but not least, it will be shown that the three-phase system surprisingly requires a *lesser* amount of wire compared to the single-phase system.

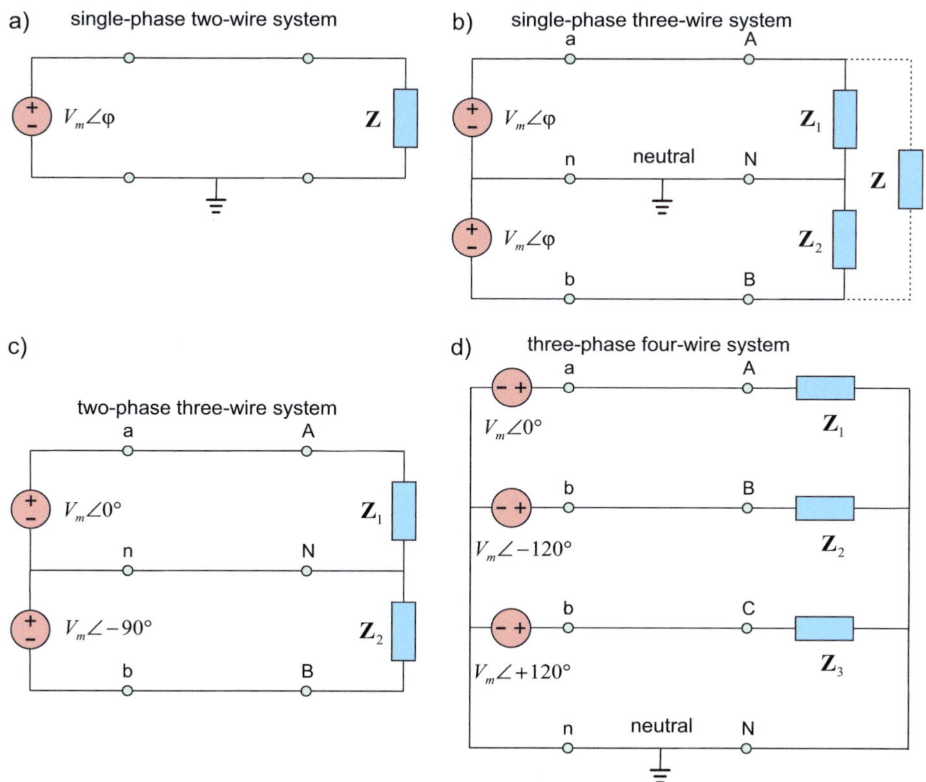

Fig. 11.13. Various AC power distribution systems. *N* or *n* indicates the neutral line.

11.3.2 Phase Voltages: Phase Sequence

The voltage sources in the three-phase system in Fig. 11.13d are set between lines *a*, *b*, *c* and the neutral line *n*. Those voltages are called *phase voltages* or *line-to-neutral voltages*. The phase voltages are 120° out of phase. One possible scenario for the real-valued phase voltages is

$$v_{\mathrm{an}}(t) = V_{\mathrm{m}} \cos(\omega t), \quad v_{\mathrm{bn}}(t) = V_{\mathrm{m}} \cos(\omega t - 120°), \\ v_{\mathrm{cn}}(t) = V_{\mathrm{m}} \cos(\omega t + 120°) \tag{11.41a}$$

$$\mathbf{V}_{\mathrm{an}} = V_{\mathrm{m}}, \quad \mathbf{V}_{\mathrm{bn}} = V_{\mathrm{m}}\angle - 120°, \quad \mathbf{V}_{\mathrm{cn}} = V_{\mathrm{m}}\angle + 120° \tag{11.41b}$$

Phase voltage v_{an} leads phase voltage v_{bn}, which in turn leads v_{cn}. This set of voltages is shown in Fig. 11.14. It has a *positive* or *abc phase sequence* since the voltages reach their peak values in the order *abc* as seen in Fig. 11.14. Simultaneously, the phasor voltages are obtained from each other by clockwise rotation in the phasor diagram. This is shown in Fig. 11.15a.

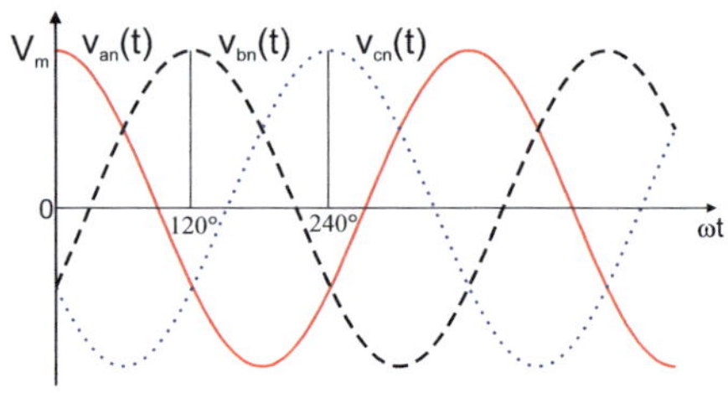

Fig. 11.14. Balanced phase voltages in positive phase sequence.

An alternative is the *negative* or *acb* phase sequence, which corresponds to 0, $+120°$, $-120°$ phases in Eqs. (11.41a, b). In this case, the phasor voltages are obtained from each other by counterclockwise rotation in the phasor diagram of Fig. 11.15a. Thus, the *balanced phase voltages* are those which have *equal* amplitudes and are out of phase with each other by 120° (either in positive 0, $-120°$, $+120°$ or in negative 0, $+120°$, $-120°$ phase sequence). An example of balanced voltages is given by Eqs. (11.41a, b). The concept of balanced phase voltages is critical for the subsequent analysis.

Example 11.9: Determine whether the phase voltages

$$v_{\mathrm{an}}(t) = 3\cos(\omega t - 90°), \quad v_{\mathrm{bn}}(t) = 3\cos(\omega t + 150°), \quad v_{\mathrm{cn}}(t) = 3\cos(\omega t + 30°) \qquad (11.42)$$

of a three-phase system are balanced or not. If yes, determine the corresponding phase sequence.

Solution: The amplitudes of the phase voltages are equal, which is the first necessary condition of the balanced sources. To analyze the phases, we plot the voltages in the phasor diagram and obtain Fig. 11.15b. Despite the common phase shift of $-90°$ as compared to Eqs. (11.41a, b), the phase voltages are still out of phase with each other by 120° and form the same positive phase sequence; see Fig. 11.15b.

Exercise 11.11: The phase voltage $\mathbf{V}_{\mathrm{bn}}$ is given by $V_{\mathrm{m}}\angle+45°$. Determine the remaining phase voltages $\mathbf{V}_{\mathrm{an}}$, $\mathbf{V}_{\mathrm{cn}}$ of the balanced three-phase system for the positive phase sequence. Express your result in phasor form.

Answer: $\mathbf{V}_{\mathrm{an}} = V_{\mathrm{m}}\angle 165°$, $\mathbf{V}_{\mathrm{cn}} = V_{\mathrm{m}}\angle -75°$.

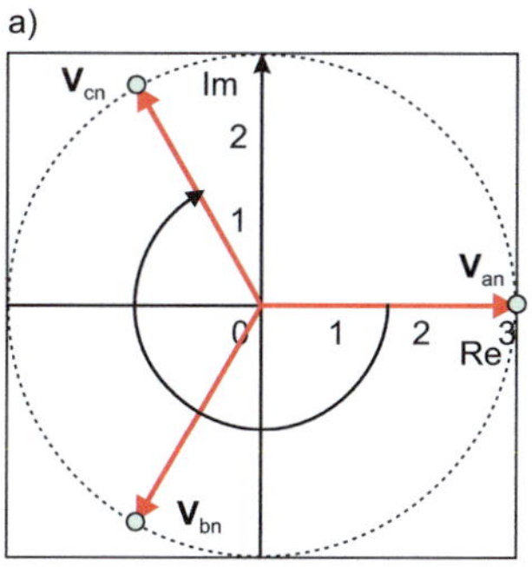

Fig. 11.15. (**a**) Phasor diagram for the phase sequence 0, $+120°$, $-120°$ and (**b**) phasor diagram for individual phase voltages from Eq. (11.42). Both three-phase sources are equivalent.

11.3.3 Wye (Y) Source and Load Configurations for Three-Phase Circuits

The voltage sources in the three-phase system in Fig. 11.13d are now rearranged as shown in Fig. 11.16a. This configuration is indeed equivalent to the original one; it is known as the *wye* (or *Y*) *configuration*. Accordingly, the *balanced three-phase source* in Fig. 11.16a is the *wye-connected source,* and the load in Fig. 11.16b is the *wye-connected load*.

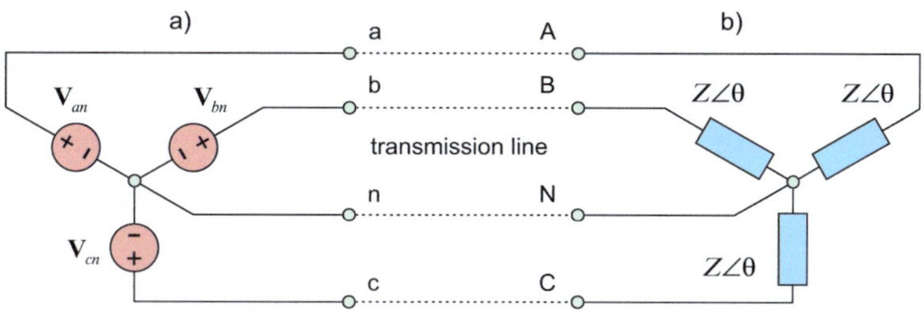

Fig. 11.16. Wye configuration for the three-phase source connected to a three-phase load.

The AC voltage source in Fig. 11.16a has four terminals. The corresponding load should also have four terminals. The concept is shown in Fig. 11.16b. This load assembly is also identical with the topology of Fig. 11.13d. The load includes three impedance elements *(phase impedances* or *load impedances per phase)* $\mathbf{Z} = Z\angle\theta$ with impedance magnitude Z and phasor angle (power angle) θ each. The load so assembled is the *balanced three-phase load*. In the balanced load, the phase impedances are equal in magnitude and phase. The source and the load are typically connected by (long) wire transmission lines. When necessary, the wire resistance may be added to each individual load impedance.

11.3.4 Application: Examples of Three-Phase Source and the Load

Synchronous Three-Phase AC Generator

Despite the apparent complexity, the three-phase source and the three-phase loads are relatively simple to realize in practice. Figure 11.17 shows *a synchronous three-phase AC generator* (or *alternator*), which is equivalent to the three-phase source in Fig. 11.16a.

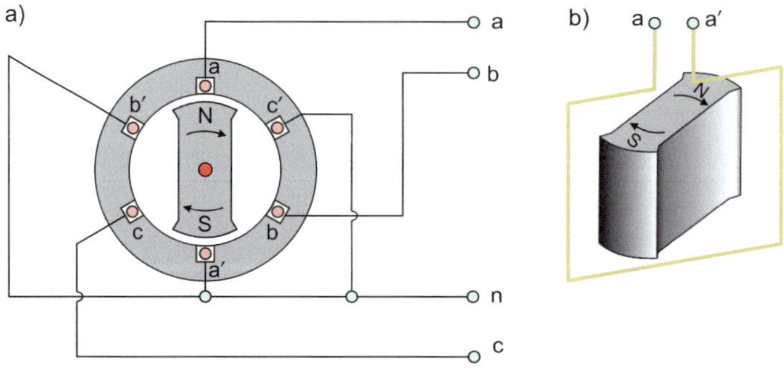

Fig. 11.17. Structure of the three-phase AC generator. (**a**) Cross-section view. (**b**) Simplified outline of one of the three windings.

Consider first the generator cross section shown in Fig. 11.17a. The generator's *rotor* is a permanent magnet (small scale) or an electromagnet (industrial scale) rotated by a mechanical torque (a turbine). Three individual coil windings aa', bb', and cc' in the *stator* are spaced exactly 120° apart around the stator. When the rotor moves, an induced emf (induced voltage) will be created in every individual winding according to Faraday's law of induction—see Fig. 11.17b. From the geometry considerations, the induced voltages are equal in magnitude and out of phase by 120°. When the coil terminals a', b', and c' are all connected to the neutral wire, see Fig. 11.17a, we obtain exactly the three-phase source with the neutral wire in Fig. 11.16a.

Automotive Alternator

The automotive alternator operates based on the same principle. However, the resulting three-phase voltage is further converted to the DC voltage (rectified, see Chapter 16).

Synchronous Three-Phase AC Motor

The counterpart of the three-phase generator is the three-phase AC motor (*the synchronous AC motor*). The stator, which is subject to the three-phase voltage source, creates a *rotating magnetic field*; the rotor magnet is aligned with this field at every time moment and rotates accordingly. The stator's circuit model is similar to the three-phase load model in Fig. 11.16b where each load impedance **Z** includes resistance and inductance of the individual (identical) coil windings. The induced emf should be included into our consideration as well. Why is the phase sequence important for power distribution? There is a simple answer to this question. Assume that the machine in Fig. 11.17 operates in the motor mode. Changing the phase sequence from *abc* to *acb* will reverse the direction of the magnetic field rotation and thus reverse the direction of the motor rotation! This method is used in practice since it requires interchanging only two connections.

Residential Household

Another example of the load impedance per phase is related to a typical residential household in the USA. A single phase of a three-phase residential distribution system is normally used to power them up; see Fig. 11.18. This single phase still has a high rms voltage (4800 V or 7200 V). A step-down center-tap transformer is used to decrease this voltage level to the desired level of 120–240 V and provide the neutral contact necessary for the three-wire single-phase residential system shown in Fig. 11.13b. This transformer case is also seen in Fig. 11.18. In the USA, a pole-mounted transformer in a suburban setting may supply one to three houses.

Fig. 11.18. Three-phase to three-wire residential system connected via a step-down transformer. From the pole transformer, the residential power system serving two houses is run down the pole underground. Cape Cod, MA.

11.3.5 Solution for the Balanced Three-Phase Wye-Wye Circuit

Phase Voltages and Line Voltages

A three-phase balanced circuit (*wye-wye configuration*) which includes the source and the load is shown in Fig. 11.19. We place the nodes *n* and *N* at the originally anticipated center positions. The positive phase sequence of 0, $-120°$, $+120°$ is assumed. The sum of phase voltages is to be found first. In the phasor form,

$$\begin{aligned}\mathbf{V}_{an}+\mathbf{V}_{bn}+\mathbf{V}_{cn} &= V_m(1+1\angle{-120°}+1\angle{+120°}) = V_m(1+2\cos 120°)\\ &= V_m(1-1)=0\end{aligned} \tag{11.43}$$

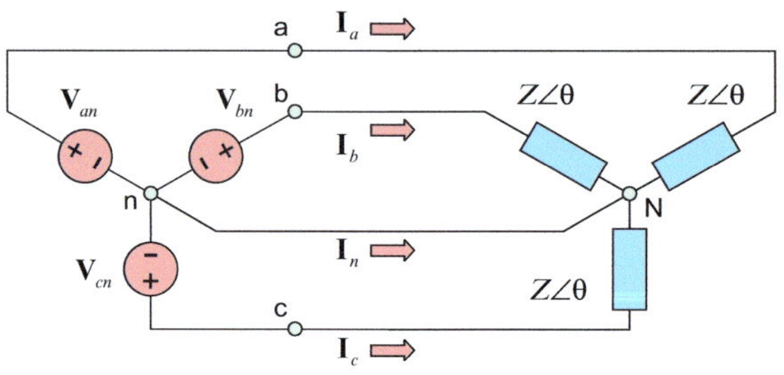

Fig. 11.19. Three-phase, four-wire balanced *wye-wye distribution system*. Ground connection is implied for the neutral wire.

Thus, the sum of the balanced phase voltages is exactly *zero*, either in the phasor form or in the time domain. Now, along with the phase (line-to-neutral) voltages, we define *line-to-line voltages* (or just *line voltages*) $\mathbf{V}_{ab}$, $\mathbf{V}_{bc}$, $\mathbf{V}_{ca}$ between nodes *a–b*, *b–c*, and *c–a*, as indicated in Fig. 11.19. These voltages are expressed through the phase voltages using KVL. Using the trigonometric identity $1 - 1\angle{-120^\circ} = \sqrt{3}\angle 30^\circ$ three times, it can be shown that

$$\begin{aligned}\mathbf{V}_{ab} &= \mathbf{V}_{an} - \mathbf{V}_{bn} = \sqrt{3}\mathbf{V}_{an}\angle 30^\circ,\\ \mathbf{V}_{bc} &= \mathbf{V}_{bn} - \mathbf{V}_{cn} = \sqrt{3}\mathbf{V}_{bn}\angle 30^\circ,\\ \mathbf{V}_{ca} &= \mathbf{V}_{cn} - \mathbf{V}_{an} = \sqrt{3}\mathbf{V}_{cn}\angle 30^\circ\end{aligned} \tag{11.44}$$

It is seen that the line voltages are higher in amplitude than the phasor voltages by a factor of $\sqrt{3} \approx 1.73$. Furthermore, they lead their corresponding phase voltages by 30°. According to Eqs. (11.43) and (11.44), the sum of the line voltages is also equal to zero. Both the phase voltages and the line voltages may be used in the three-phase system.

Exercise 11.12: Is Eq. (11.44) also valid for the negative phase sequence?

Answer: Not exactly. A substitution $30^\circ \to -30^\circ$ has to be made.

Example 11.10: The electric service for commercial buildings (university campus buildings) in the USA is a three-phase, four-wire wye system schematically shown in Fig. 11.19. Determine rms phase voltages if the line voltages are all equal to 208 V rms.

Solution: According to Eq. (11.44), we should divide the line voltage of 208 V rms by $\sqrt{3}$. This gives us exactly 120 V rms voltage per phase. Thus, the present wye system is also powering common 120 V wall plugs with any of the line-to-neutral voltages. Note that the source in Fig. 11.19 typically models an output of a three-phase transformer.

Line Currents: Per-Phase Solution

The currents $\mathbf{I}_{a,b,c}$ in Fig. 11.19 are called *line currents*. To find the line currents, the circuit may be solved separately for every phase using the *superposition principle*. The superposition principle implies shorting out two of the three voltage sources at a time. This method applies to both balanced and unbalanced circuits. Shorting out voltage sources $\mathbf{V}_{bn}$ and $\mathbf{V}_{cn}$ leads to a single-phase equivalent circuit, shown in Fig. 11.20, since the two remaining source impedances will be shorted out by the neutral wire. As long as the system is balanced, the same equivalent circuit will be derived for every other phase.

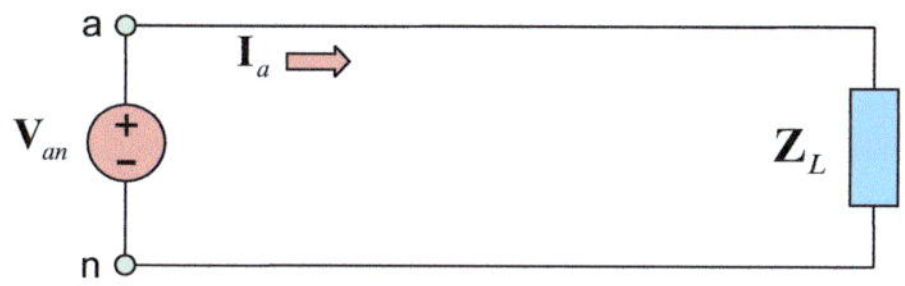

Fig. 11.20. A single-phase equivalent circuit by shorting out $\mathbf{V}_{bn}$ and $\mathbf{V}_{cn}$.

Applying this method to every phase, we obtain

$$\mathbf{I}_a = \mathbf{V}_{an}/\mathbf{Z} = I_m\angle-\theta, \quad \mathbf{I}_b = \mathbf{V}_{bn}/\mathbf{Z} = I_m\angle-120^\circ-\theta, \\ \mathbf{I}_c = \mathbf{V}_{cn}/\mathbf{Z} = I_m\angle120^\circ-\theta \tag{11.45}$$

where $I_m = V_m/Z$. The sum of the line currents is given by

$$\mathbf{I}_a + \mathbf{I}_b + \mathbf{I}_c = (\mathbf{V}_{an} + \mathbf{V}_{bn} + \mathbf{V}_{cn})/\mathbf{Z} = 0 \tag{11.46}$$

according to Eq. (11.43). Thus, the sum of the balanced line currents is also exactly *zero*, either in phasor form or in the time domain. Equations (11.43), (11.44), (11.45), and (11.46) hold for any phase sequence, with or without the common phase shift.

11.3.6 Removing the Neutral Wire in Long-Distance Power Transmission

Equation (11.46) for the line currents has an important implication. By taking into account Eq. (11.46), KCL for node *n* in Fig. 11.19 yields

$$\mathbf{I}_n = -(\mathbf{I}_a + \mathbf{I}_b + \mathbf{I}_c) = 0 \tag{11.47}$$

Equation (11.47) states that the neutral conductor in the *balanced circuit* carries *no* current. Such a wire could in principle be *removed* from the balanced circuit *without* affecting the rest of it. Removing the neutral conductor is economically beneficial in long-distance high-voltage power transmission, which utilizes the balanced circuits. In high-voltage power lines, the conductors in multiples of three are used; see Fig. 11.21.

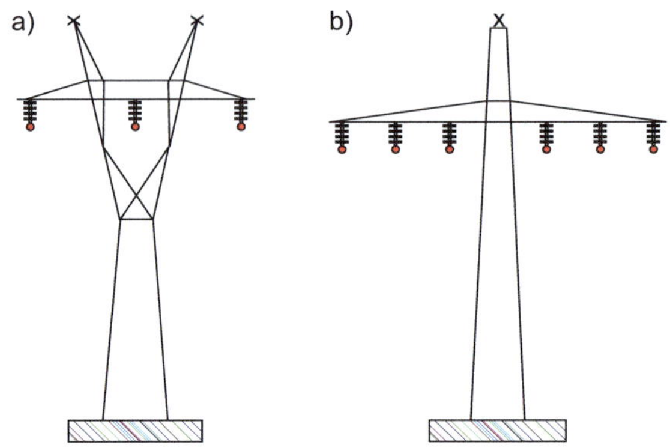

Fig. 11.21. (**a**) Three-phase single-circuit high-voltage overhead power transmission line and (**b**) three-phase double-circuit, high-voltage overhead line. Both lines include (thinner) shielding wires on top of it to protect against lightning strikes (F. Kiessling, et al., "*Overhead Power Lines: Planning, Design, Construction*," Springer 2003).

In fact, the neutral wire is not removed entirely since the earth ground itself plays the role of the neutral conductor. We will not draw the neutral wire in the three-phase *balanced* wye-wye circuit; in Fig. 11.22 only three wires are drawn. However, the meaning of the phase voltage or the phase-to-neutral voltage still remains unchanged—this voltage is simply defined with regard to the reference node n in Fig. 11.22.

If a three-phase circuit is unbalanced, like having different individual loads in Fig. 11.19, then a significant current may flow in the neutral wire. The neutral wire is thus meant to carry unbalanced currents in the electrical system. It should be kept in place for potentially unbalanced systems.

Example 11.11: Determine line currents in the balanced three-phase wye-wye circuit shown in Fig. 11.22 given the *acb* sequence of phase voltages $\mathbf{V}_{an} = 325\angle 0°$, $\mathbf{V}_{bn} = 325\angle 120°$, $\mathbf{V}_{cn} = 325\angle -120°$ [V] and load impedance per phase $\mathbf{Z} = 8.333 + j14.434\ \Omega$.

Solution: The three-phase circuit in Fig. 11.22 is balanced; hence, the single-phase circuit in Fig. 11.20 applies to *every* phase (to visualize the per-phase method, we can still imagine the neural wire present). We convert the load impedance to polar form first, i.e., $\mathbf{Z} = 16.667\angle 60°\ \Omega$. Then, we solve the circuit in Fig. 11.20 for every phase and obtain $\mathbf{I}_a = 19.5\angle -60°$, $\mathbf{I}_b = 19.5\angle 60°$, $\mathbf{I}_c = 19.5\angle 180°$ [A]. The solution is shown in the phasor diagram in Fig. 11.23. Note that the phasor voltages/currents are obtained from each other by counter clockwise rotation in the phasor diagram, which corresponds to the negative or *acb* phase sequence. Also note that the rms values for the phase voltages in this example are 230 V, which corresponds to the European residential power distribution system.

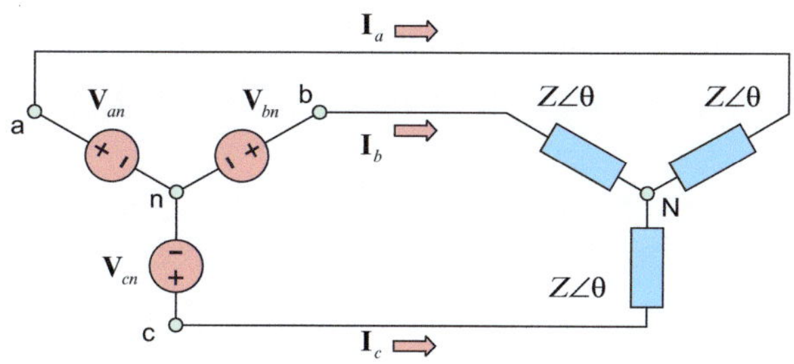

Fig. 11.22. Three-phase balanced wye-wye system with the neutral conductor removed. The neutral conductor may still be implied for the solution using the per-phase method.

The wye-wye circuit in Fig. 11.22, along with the similar circuits in Figs. 11.13 and 11.19, may contain extra impedances. Those are *line impedance*, which characterizes transmission line loss and inductance, and *source impedances*, which are present for nonideal voltage sources. Fortunately, all those (equal) impedances are combined in series along the line into one impedance **Z** which is called the *total load impedance per phase*. In this sense, Fig. 11.22 represents this general case as well.

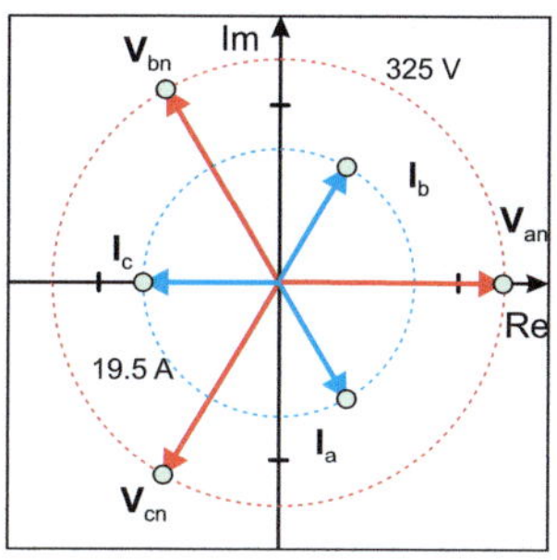

Fig. 11.23. Phasor diagram for the three-phase circuit of Example 11.11. Note the separate scales for the phasors of voltages and phasors of currents.

Section 11.4 Power in Balanced Three-Phase Systems: Delta-connected Three-Phase Circuits

11.4.1 Instantaneous Power

The analysis of the instantaneous power requires a source-load circuit in terms of real-value expressions of voltages and currents. This is shown in Fig. 11.24 for the wye-wye configuration. We assume a balanced source and a balanced load. This means that individual loads *a*, *b*, and *c* in Fig. 11.24 are identical. Each of them can be a mixed RLC load, with an arbitrary impedance $\mathbf{Z} = Z\angle\theta$.

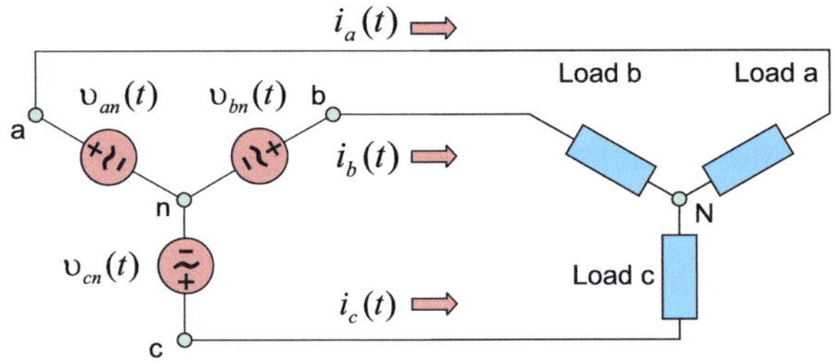

Fig. 11.24. Three-phase balanced wye-wye circuit in the time domain. Three individual loads are identical. Each of them is characterized by the impedance $\mathbf{Z} = Z\angle\theta$ in the frequency domain.

We consider the positive phase sequence. According to Eqs. (11.41) and (11.45) of the previous section, the phase voltages and line currents in Fig. 11.24 are given by

$$\upsilon_{\rm an}(t) = V_{\rm m}\cos(\omega t), \quad \upsilon_{\rm bn}(t) = V_{\rm m}\cos(\omega t - 120^\circ), \quad \upsilon_{\rm cn}(t) = V_{\rm m}\cos(\omega t + 120^\circ) \tag{11.48a}$$

$$\begin{aligned} &i_{\rm a}(t) = I_{\rm m}\cos(\omega t - \theta), \quad i_{\rm b}(t) = I_{\rm m}\cos(\omega t - 120^\circ - \theta), \\ &i_{\rm c}(t) = I_{\rm m}\cos(\omega t + 120^\circ - \theta) \end{aligned} \tag{11.48b}$$

The *total instantaneous load power of the three-phase system* is the sum of the three power contributions for each phase voltage, that is,

$$p(t) = \upsilon_{\rm an}(t)i_{\rm a}(t) + \upsilon_{\rm bn}(t)i_{\rm b}(t) + \upsilon_{\rm cn}(t)i_{\rm c}(t) \tag{11.49}$$

Every summand on the right-hand side of Eq. (11.49) is the product of two cosines. To transform this product back to cosines, we use the trigonometric identity $\cos\alpha\cos\beta = \frac{1}{2}(\cos(\alpha+\beta) + \cos(\alpha-\beta))$ and obtain

$$p(t) = \frac{3}{2}V_{\rm m}I_{\rm m}\cos\theta + \frac{1}{2}V_{\rm m}I_{\rm m}\left[\cos(2\omega t - \theta) + \underbrace{\cos(2\omega t - \theta + 120^\circ) + \cos(2\omega t - \theta - 120^\circ)}\right] \tag{11.50}$$

We then use the above trigonometric identity again to convert the underlined term to $2\cos(2\omega t - \theta)\cos(120^\circ) = -\cos(2\omega t - \theta)$. Consequently, the entire term in the square brackets in Eq. (11.50) is equal to zero, and the final result for the instantaneous power is

$$p(t) = \frac{3}{2}V_{\mathrm{m}}I_{\mathrm{m}}\cos\theta = 3V_{\mathrm{rms}}I_{\mathrm{rms}}\cos\theta = \text{const}\,! \tag{11.51}$$

where the rms values of phase voltages and line currents are indeed related to the amplitudes by $V_{\mathrm{m}} = \sqrt{2}V_{\mathrm{rms}}, \quad I_{\mathrm{m}} = \sqrt{2}I_{\mathrm{rms}}$.

Example 11.12: A balanced wye-wye three-phase system in Fig. 11.24 operates at 60 Hz. The line-to-neutral voltages have the amplitudes of 170 V, i.e. $V_{\mathrm{m}} = 170$ V. Every phase impedance is a 77.2-mH inductance in series with a 29.1-Ω resistance. Find the instantaneous load power.

Solution: The first step is to find the impedance for every phase of the load. We have

$$\mathbf{Z} = R + j\omega L = 29.1 + j29.1\ \Omega = 29.1\sqrt{2}\angle 45^\circ\ \Omega \tag{11.52}$$

Next, we find the line currents. Since the circuit is balanced, the per-phase solution applies, with the equivalent circuit shown in Fig. 11.20 of the previous section. It yields

$$I_{\mathrm{m}} = \frac{V_{\mathrm{m}}}{|\mathbf{Z}|} = \frac{V_{\mathrm{m}}}{Z} = \frac{170}{41.154} = 4.1309\ \mathrm{A} \tag{11.53}$$

The instantaneous load power follows Eq. (11.51) with the power angle, $\theta = 45^\circ$. Therefore, we obtain

$$p(t) = \frac{3}{2}V_{\mathrm{m}}I_{\mathrm{m}}\cos\theta = 745\ \mathrm{W} = \text{const} \tag{11.54}$$

Equation (11.51) is critical for three-phase systems. It tells us that the total instantaneous power delivered to the load remains constant at any instance in time. This is in contrast to the instantaneous power of every individual single phase, which is still pulsating in time. Equation (11.51) implies that a three-phase load (e.g., an induction motor) as well as the three-phase generator introduced in the previous section should generate or require a *constant* torque. Thus, they undergo less vibration since the net power transfer is uniform.

11.4.2 Average Power, Reactive Power, and Apparent Power

The AC power types defined for the single-phase power distribution in Section 11.1 of this chapter also apply for the three-phase circuits. The *average* or active load power P, the *reactive* load power Q, and the *complex* load power $\mathbf{S}$ *of the balanced three-phase system* are given by

$$P = 3V_{\text{rms}}I_{\text{rms}}\cos\theta \ \ [\text{W}], \quad Q = 3V_{\text{rms}}I_{\text{rms}}\sin\theta \ \ [\text{VAR}], \quad \mathbf{S} = 3V_{\text{rms}}I_{\text{rms}}\angle\theta \ \ [\text{VA}] \tag{11.55a}$$

While the instantaneous powers *per phase* are pulsating, their average values labeled with indexes a, b, c are exactly one third of the load powers. One has

$$\begin{aligned} &P_{\text{a,b,c}} = V_{\text{rms}}I_{\text{rms}}\cos\theta \ \ [\text{W}], \quad Q_{\text{a,b,c}} = V_{\text{rms}}I_{\text{rms}}\sin\theta \ \ [\text{VAR}], \\ &\mathbf{S}_{\text{a,b,c}} = V_{\text{rms}}I_{\text{rms}}\angle\theta \ \ [\text{VA}] \end{aligned} \tag{11.55b}$$

per phase. Equation (11.55) uses the rms values of phase voltages and line currents.

Example 11.13: For the previous example, determine the load average power, reactive power, and the apparent power. Do these powers coincide with the corresponding source measures?

Solution: The average power is simply the load instantaneous power, $P = 745$ W. The reactive power is $Q = \frac{3}{2}V_m I_m \sin\theta = 745$ VAR. The apparent power is $S = |\mathbf{S}| = \frac{3}{2}V_m I_m = 1053$ VA. And the apparent power can be also found from the power triangle. All load powers coincide with the corresponding source powers since the transmission lines in Fig. 11.24 are assumed to be ideal conductors.

Exercise 11.13: A three-phase induction motor is modeled by a balanced wye load in Fig. 11.24. The motor (active) power is 6 kW; the line current is 20 A rms, and the line voltage is 208 V rms. Determine the power factor of the motor, which is the ratio of the active load power P to the magnitude of the total apparent power, $|\mathbf{S}|$.

Answer: $PF \approx 5/6 = 0.833$.

11.4.3 Application Example: Material Consumption in Three-Phase Systems

A comparison is made between the conductor material consumption in a single-phase two-wire transmission system (shown in Fig. 11.25a) and the three-phase, three-wire transmission system shown in Fig. 11.25b. Both systems have the identical distance from the source to the load, the same average power P distributed to a purely resistive load, and the same rms line voltages *close* to the load. They all use the same conductor material. The distributed resistance per wire is modeled by a lumped resistor R for the single-phase line and by a lumped resistor R' for the three-phase line. Given the same load power and line voltage, the rms line currents are expressed as

$$I_{\text{single phase}} = \frac{P}{V}, \quad I_{\text{three phase}} = \frac{P}{\sqrt{3}\,V} \tag{11.56}$$

Equating power loss in the wire conductors, we obtain

$$2RI^2_{\text{single phase}} = 3R^{'}I^2_{\text{three phase}} \Rightarrow 2R\left(\frac{P}{V}\right)^2 = 3R^{'}\left(\frac{P}{\sqrt{3}\,V}\right)^2 \Rightarrow R^{'} = 2R \tag{11.57}$$

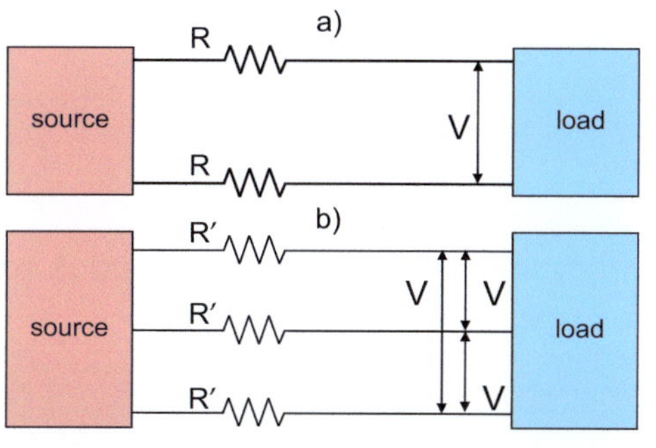

Fig. 11.25. Comparison of single-phase and three-phase transmission systems.

Since resistance R is twice as large as resistance R', the cross section of the corresponding cylindrical conductor is smaller by a factor of two in the three-phase configuration. Hence, its radius is $1/\sqrt{2}$ times less than the radius, r, of the single-phase line. Figure 11.26 depicts the radii of the equivalent conductors.

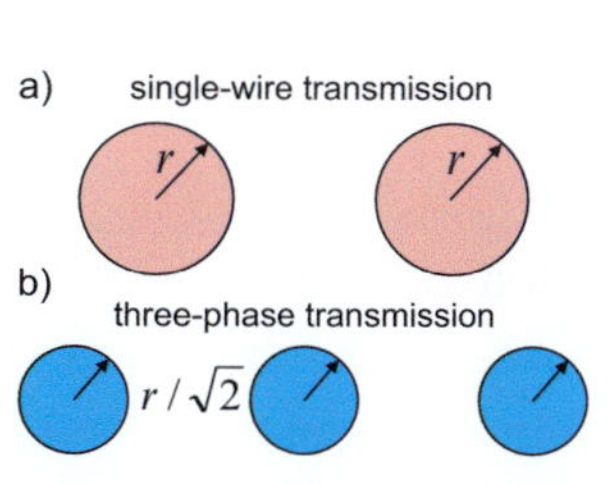

Fig. 11.26. Radii of equivalent conductors for the two systems in Fig. 11.25.

For the single-phase line, the total conductor cross section is $2\pi r^2$; for the three-phase line, the total cross section is $1.5\pi r^2$. Given the same length, the ratio of conductor material required is exactly the cross-section ratio, that is, $1.5/2 = 0.75$. In other words, the three-phase system consumes 25 % less conductor material compared to the single-phase system. The key is the absence of the neutral wire (or, possibly, using a much thinner neutral wire). Other examples for particular loads might result in even more dramatic savings.

11.4.4 Balanced Delta-Connected Load

Along with the wye-connected load, an important example of the three-phase load is the *delta-connected load*, which is shown in Fig. 11.27a in the *balanced configuration*. The balanced delta-connected load is common, along with a balanced wye-connected load. The delta-connected load inherently does not have a neutral port. This load may be converted to the wye-connected load shown in Fig. 11.27b by using the Y-Δ transformation algorithm established in Chapter 3. This algorithm equally applies to the impedance circuits. The algorithm considerably simplifies when the loads are balanced (load resistances or impedances are equal). With reference to Fig. 11.27, one has

$$\mathbf{Z}_{\mathrm{Y}} = \frac{1}{3}\mathbf{Z}_{\Delta} \leftrightarrow \mathbf{Z}_{\Delta} = 3\mathbf{Z}_{\mathrm{Y}} \tag{11.58}$$

for phase impedance transformation. Here, indexes Y and Δ refer to the wye-connected and delta-connected loads, respectively.

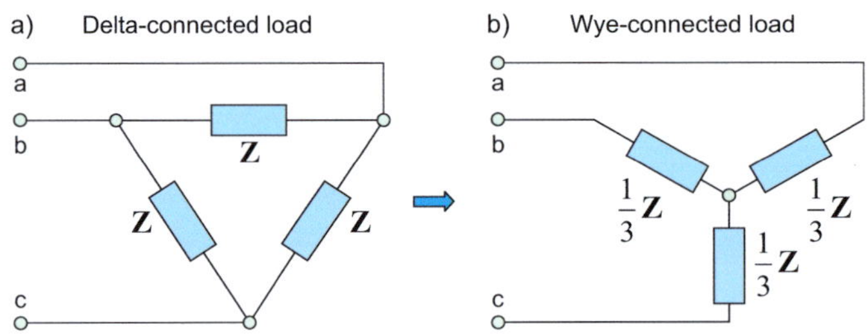

Fig. 11.27. Delta-connected load versus ad.

11.4.5 Balanced Delta-Connected Source

The *balanced delta-connected source* is shown in Fig. 11.28b. In its original configuration, it is not using the ground terminal or a neutral conductor. The delta-connected source so wound is generally less common and less safe than the wye-connected source. It may be created by the three-phase generator shown in Fig. 11.17 of the previous section, assuming the three individual coil windings aa', bb', and cc' are interconnected in a closed loop.

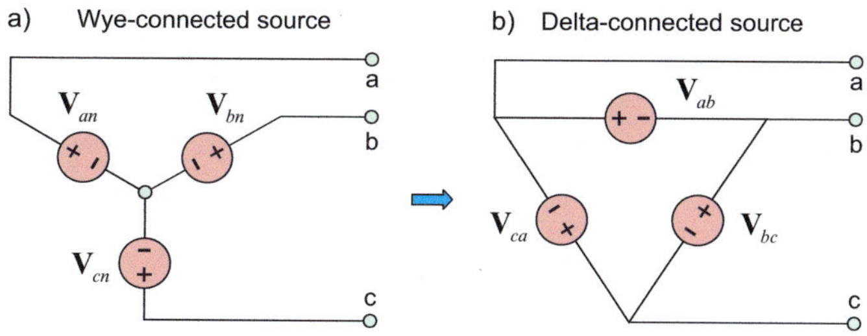

Fig. 11.28. Wye-connected source versus delta-connected source.

The balanced wye-connected source *without* a neutral or ground conductor can be easily converted to the balanced delta source and vice versa. The concept is shown in Fig. 11.28. The line voltages $\mathbf{V}_{\mathrm{ab}}$, $\mathbf{V}_{\mathrm{bc}}$, $\mathbf{V}_{\mathrm{ca}}$ between nodes *a–b*, *b–c*, and *c–a* of the wye source become the phase voltages of the delta source. The relation between two voltage types is given by Eq. (11.44) of the previous section, that is (positive phase sequence),

$$\mathbf{V}_{\mathrm{ab}} = \sqrt{3}\mathbf{V}_{\mathrm{an}}\angle 30^\circ, \quad \mathbf{V}_{\mathrm{bc}} = \sqrt{3}\mathbf{V}_{\mathrm{bn}}\angle 30^\circ, \quad \mathbf{V}_{\mathrm{ca}} = \sqrt{3}\mathbf{V}_{\mathrm{cn}}\angle 30^\circ \tag{11.59}$$

Thus, according to Eq. (11.59) and Fig. 11.28, the phase voltages of the *equivalent* delta-connected source $\mathbf{V}_{\mathrm{ab}}$, $\mathbf{V}_{\mathrm{bc}}$, $\mathbf{V}_{\mathrm{ca}}$ are greater in amplitude by a factor of $\sqrt{3} \approx 1.73$ as compared to the phase voltages $\mathbf{V}_{\mathrm{an}}$, $\mathbf{V}_{\mathrm{bn}}$, $\mathbf{V}_{\mathrm{cn}}$ of the equivalent wye-connected source in Fig. 11.28. The line voltages of the delta-connected source *coincide* with its phase voltages given lossless conductors and *coincide* with the line voltages of the wye-wye source; all of them are simply $\mathbf{V}_{\mathrm{ab}}$, $\mathbf{V}_{\mathrm{bc}}$, $\mathbf{V}_{\mathrm{ca}}$. Indeed, the sum of the phase voltages for the delta-connected source is still equal to zero according to Eq. (11.43) of the previous section. Hence, there is no current circulation in the (ideal) delta loop in Fig. 11.28b. Transformations given by Eqs. (11.58) and (11.59) allow us to consider four distinct source-load configurations in the three-phase systems: wye-wye, wye-delta, delta-wye, and delta-delta. All of them may be reduced to the wye-wye circuit or solved independently. Figure 11.29 shows one such configuration: a balanced *delta-delta distribution system*. In the delta-delta system, the line voltages coincide with the phase voltages, whereas the line currents $\mathbf{I}_{\mathrm{a}}$, $\mathbf{I}_{\mathrm{b}}$, and $\mathbf{I}_{\mathrm{c}}$ are different from the load (or phase) currents $\mathbf{I}_{\mathrm{AB}}$, $\mathbf{I}_{\mathrm{BC}}$, and $\mathbf{I}_{\mathrm{CA}}$. This is in contrast to the wye-wye system where the line and phase voltages are different, but the line and load currents remain the same.

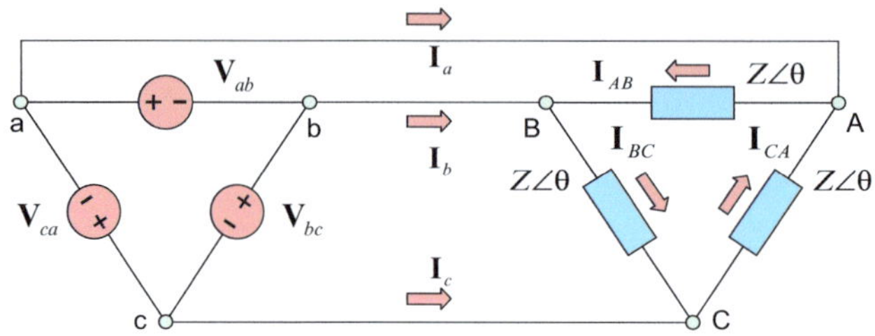

Fig. 11.29. Three-phase balanced *delta-delta distribution system*. Note the load (phase) currents circulating in the delta loop.

Example 11.14: A balanced delta-delta system in Fig. 11.29 operates at 60 Hz. The phase voltages of the delta source, $\mathbf{V}_{ab}$, $\mathbf{V}_{bc}$, $\mathbf{V}_{ca}$, have amplitudes of $V_m = 294.5$ V each. Moreover, each phase impedance is a 0.2315 H inductance in series with a 87.3 Ω resistance. Find the average load power.

Solution: We find the impedance for each phase of the load first. One has

$$\mathbf{Z} = R + j\omega L = 87.3 + j87.3\ \Omega = 87.3\sqrt{2}\angle 45^\circ\ \Omega \tag{11.60}$$

The power angle is thus given by $\theta = 45°$. Next, we find the load (phase) currents $\mathbf{I}_{AB}$, $\mathbf{I}_{BC}$, $\mathbf{I}_{CA}$ circulating in the delta-connected load. Since the individual voltage sources in Fig. 11.6 are now directly connected to the individual load phases, one has for the amplitude of the phase current $\mathbf{I}_{AB}$:

$$\mathbf{I}_{AB} = \frac{\mathbf{V}_{ab}}{\mathbf{Z}} \Rightarrow I_m = \frac{V_m}{|\mathbf{Z}|} = \frac{294.5}{Z} = \frac{294.5}{123.46} = 2.385\ \text{A} \tag{11.61}$$

The remaining phases have the same amplitudes: the per-phase method is used again. Both the average load power and the instantaneous load powers are the sum of three individual contributions, that is,

$$P = p(t) = 3 \times \left(\frac{1}{2}V_m I_m \cos\theta\right) = \frac{3}{2}294.5 \times 2.385 \times 0.707 = 745\ \text{W} \tag{11.62}$$

The instantaneous power may be calculated; it is constant and equals 745 W. Note that the rms line voltages in this example are 208 V.

Example 11.15: Solve the previous example by converting the delta-delta system to the equivalent wye-wye system.

Solution: First, the phase impedance of the wye load should be three times less than the phase impedance of the delta load, that is, $\mathbf{Z} = 29.1\sqrt{2}\angle 45^\circ\ \Omega$. Second, the amplitude of the wye phase source should be $1/\sqrt{3}$ times less than the amplitude of the delta phase source, that is, $V_m = 170\ \text{V}$. These numbers have been used in Examples 11.12 and 11.13 for the wye-wye system, which gave us exactly the same value of 745 W (one horsepower) for the average and instantaneous powers, respectively.

Apart from the circuit equivalence, one may look at Fig. 11.28 from a slightly different perspective. What if the voltage sources in Fig. 11.28 are all the same (the same windings of the three-phase generator just connected differently)? In this case, the wye connection gives us a line voltage $\sqrt{3}$ times greater than the delta connection. Hence, the line current, which is required for the same power transfer, will be $1/\sqrt{3}$ times *less*. Reducing line currents reduces line losses. This explains why the wye source connection is preferable for long-distance power transmission.

Summary

rms Voltages and currents in terms of sine/cosine amplitudes and in the general case
For sinusoidal signals: $V_{\mathrm{rms}} = \dfrac{V_m}{\sqrt{2}}, I_{\mathrm{rms}} = \dfrac{I_m}{\sqrt{2}}$ General periodic case: $V_{\mathrm{rms}} = \sqrt{\dfrac{1}{T}\int_0^T v^2(t)\,dt}$
Average power for resistive load: $P = \dfrac{1}{2}V_m I_m = V_{\mathrm{rms}} I_{\mathrm{rms}}, V_{\mathrm{rms}} = R I_{\mathrm{rms}}$
Power angle θ and power factor PF
$\begin{matrix} v(t) = V_m \cos(\omega t + \varphi) \\ i(t) = I_m \cos(\omega t + \psi) \end{matrix} \Rightarrow \theta = \varphi - \psi, \quad -90^\circ \le \theta \le +90^\circ$ $PF = \cos(\varphi - \psi) = \cos\theta$
Average power for arbitrary load: $P = \dfrac{V_m I_m}{2}\cos\theta = V_{\mathrm{rms}} I_{\mathrm{rms}} \cos\theta$ (zero for L and C)
Average power and power angle in terms of phasors: $P = \dfrac{\mathrm{Re}(\mathbf{V}\cdot\mathbf{I}^*)}{2}, \mathbf{Z} = \lvert\mathbf{Z}\rvert\angle\theta$
Average power P, reactive power Q, complex power $\mathbf{S}$, and apparent power S
$P = \dfrac{\mathrm{Re}(\mathbf{V}\cdot\mathbf{I}^*)}{2} = \dfrac{R\lvert\mathbf{I}\rvert^2}{2} = \dfrac{\lvert\mathbf{Z}\rvert\lvert\mathbf{I}\rvert^2}{2}\cos\theta = \dfrac{\lvert\mathbf{V}\rvert\lvert\mathbf{I}\rvert}{2}\cos\theta = V_{\mathrm{rms}} I_{\mathrm{rms}}\cos\theta\,[\mathrm{W}]$ $Q = \dfrac{\mathrm{Im}(\mathbf{V}\cdot\mathbf{I}^*)}{2} = \dfrac{X\lvert\mathbf{I}\rvert^2}{2} = \dfrac{\lvert\mathbf{Z}\rvert\lvert\mathbf{I}\rvert^2}{2}\sin\theta = \dfrac{\lvert\mathbf{V}\rvert\lvert\mathbf{I}\rvert}{2}\sin\theta = V_{\mathrm{rms}} I_{\mathrm{rms}}\sin\theta\,[\mathrm{VAR}]$ $\mathbf{S} = \dfrac{\mathbf{V}\cdot\mathbf{I}^*}{2} = P + jQ\,[\mathrm{VA}]$ $S = \lvert\mathbf{S}\rvert = \dfrac{\lvert\mathbf{Z}\rvert\lvert\mathbf{I}\rvert^2}{2} = \dfrac{\lvert\mathbf{V}\rvert\lvert\mathbf{I}\rvert}{2} = V_{\mathrm{rms}} I_{\mathrm{rms}}\,[\mathrm{VA}]$
Power triangle (lagging/leading power factor)
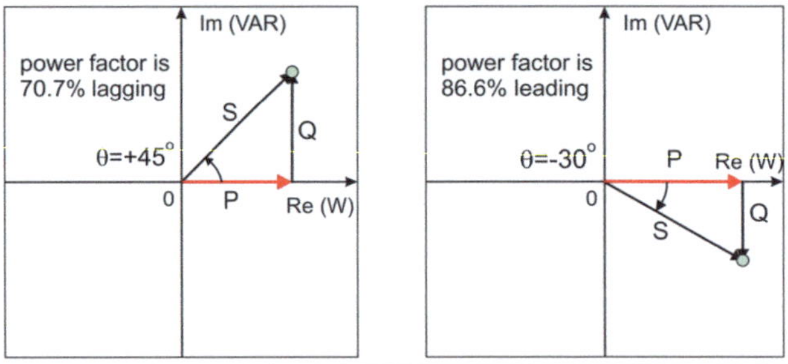
AC power conservation laws   For any network of N loads connected in series, parallel, or in general: $\mathbf{S} = \mathbf{S}_1 + \mathbf{S}_2 + \ldots \mathbf{S}_N, P = P_1 + P_2 + \ldots P_N, Q = Q_1 + Q_2 + \ldots Q_N$

(continued)

Power factor correction

for most common inductive (motor) load

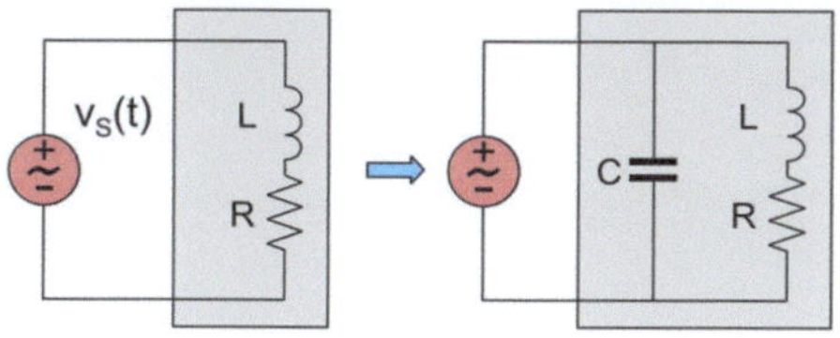

for capacitive load

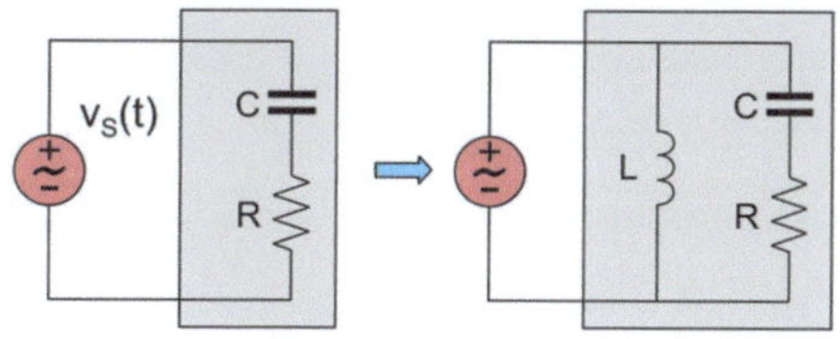

$$C = \frac{L}{\omega^2 L^2 + R^2} \Rightarrow \mathbf{Z} = R + \frac{(\omega L)^2}{R} \Rightarrow$$

$$\mathbf{I} = \overbrace{\underbrace{\frac{RV_m}{R^2 + (\omega L)^2}}_{\text{with capacitor}} - \frac{j\omega L V_m}{R^2 + (\omega L)^2}}^{\text{without capacitor}}$$

$$L = \frac{1 + \omega^2 R^2 C^2}{\omega^2 C} \Rightarrow \mathbf{Z} = R + \frac{1}{R(\omega C)^2} \Rightarrow$$

$$\mathbf{I} = \overbrace{\underbrace{\frac{R(\omega C)^2 V_m}{1 + (\omega RC)^2}}_{\text{with inductor}} + \frac{j\omega C V_m}{1 + (\omega RC)^2}}^{\text{without inductor}}$$

P remains exactly the same, Q becomes zero, PF becomes 100 %

Maximum power transfer

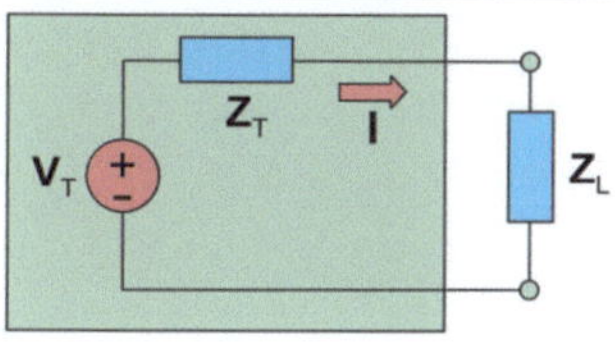

$$P = \frac{0.5 R_L |\mathbf{V}_T|^2}{(R_L + R_T)^2 + (X_L + X_T)^2} \text{ [W]}$$

$$P_{max} = \frac{1}{8} \frac{|\mathbf{V}_T|^2}{R_T} \text{ [W]}$$

at $\mathbf{Z}_L = \mathbf{Z}_T^*$

Some equivalent drawings of the *same* balanced *three-phase four-wire wye–wye* power distribution system

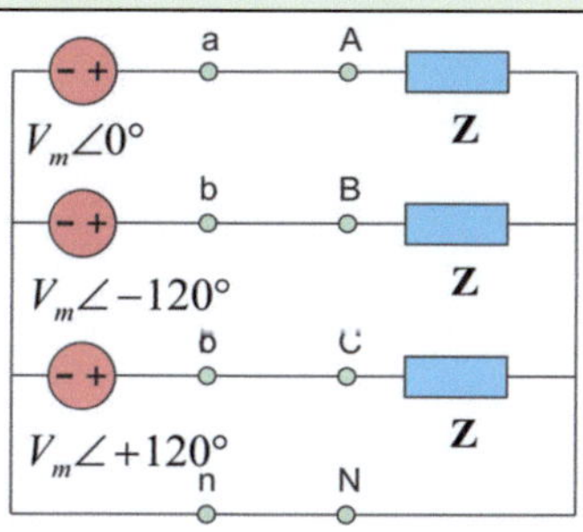

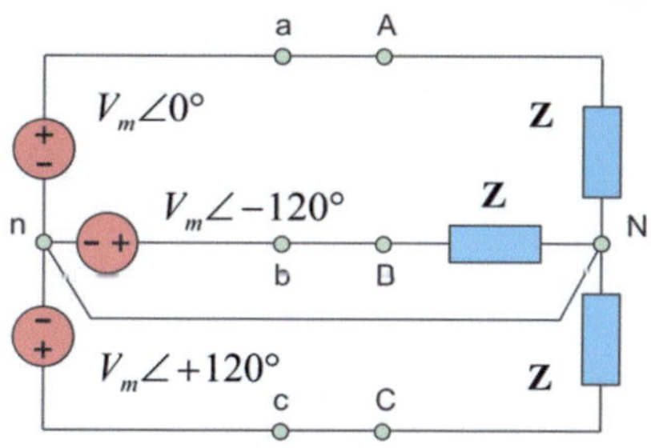

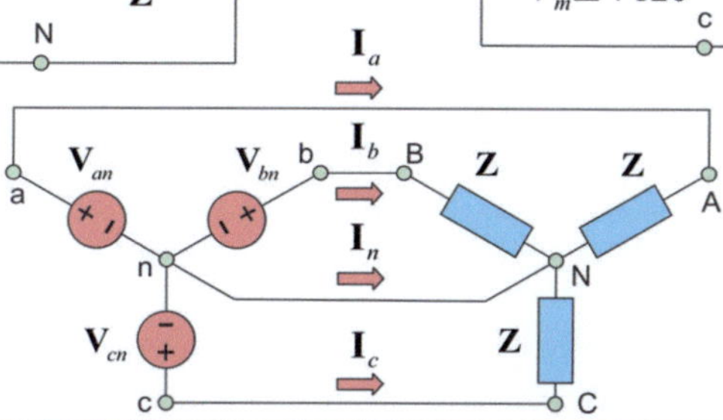

(continued)

Major parameters of the balanced three-phase four-wire wye–wye power distribution system
Positive phase sequence $\mathbf{V}_{an} = V_m\angle 0^\circ$, $\mathbf{V}_{bn} = V_m\angle -120^\circ$, $\mathbf{V}_{cn} = V_m\angle +120^\circ$
Negative phase sequence $\mathbf{V}_{an} = V_m\angle 0^\circ$, $\mathbf{V}_{bn} = V_m\angle +120^\circ$, $\mathbf{V}_{cn} = V_m\angle -120^\circ$
Current in the neutral wire: $\mathbf{I}_n = 0$
Per phase solution: $\mathbf{I}_a = I_m\angle -\theta$, $\mathbf{I}_b = I_m\angle -120^\circ - \theta$, $\mathbf{I}_c = I_m\angle 120^\circ - \theta$
$I_m = V_m/Z$, $\mathbf{Z} = Z\angle\theta$
Line voltages (positive phase sequence): $\mathbf{V}_{ab} = \mathbf{V}_{an} - \mathbf{V}_{bn} = \sqrt{3}\mathbf{V}_{an}\angle 30^\circ$, $\mathbf{V}_{bc} = \mathbf{V}_{bn} - \mathbf{V}_{cn} = \sqrt{3}\mathbf{V}_{bn}\angle 30^\circ$, $\mathbf{V}_{ca} = \mathbf{V}_{cn} - \mathbf{V}_{an} = \sqrt{3}\mathbf{V}_{cn}\angle 30^\circ$
Instantaneous/average load power: $p(t) = P = \frac{3}{2}V_m I_m\cos\theta = 3V_{\text{rms}}I_{\text{rms}}\cos\theta = \text{const}$
Apparent load power: $S = \frac{3}{2}V_m I_m = 3V_{\text{rms}}I_{\text{rms}}$

Some common wye distribution systems		
3-Phase, 4-Wire 208Y/120 V (US)	Line :	$V_{\text{rms}} = 208$ V, $V_{\text{m}} = 294$ V
	Phase :	$V_{\text{rms}} = 120$ V, $V_{\text{m}} = 170$ V
3-Phase, 4-Wire 400Y/230 V (EU, Others)	Line :	$V_{\text{rms}} = 400$ V, $V_{\text{m}} = 566$ V
	Phase :	$V_{\text{rms}} = 230$ V, $V_{\text{m}} = 325$ V

Wye load to delta load conversion

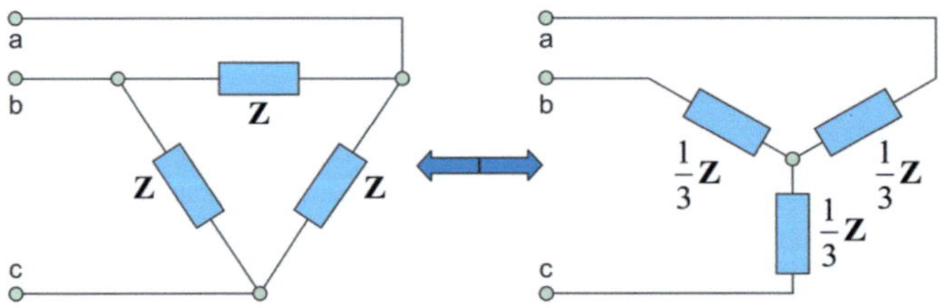

Wye source to delta source conversion

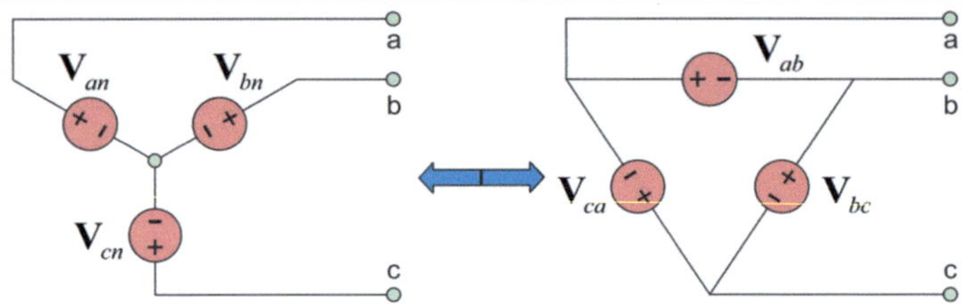

$\mathbf{V}_{ab} = \sqrt{3}\mathbf{V}_{an}\angle 30^\circ$, $\mathbf{V}_{bc} = \sqrt{3}\mathbf{V}_{bn}\angle 30^\circ$, $\mathbf{V}_{ca} = \sqrt{3}\mathbf{V}_{cn}\angle 30^\circ$ (positive phase sequence)

Problems

11.1 AC Power Types and Their Meaning

11.1.1 Instantaneous AC Power

11.1.2 Time-Averaged AC Power

Problem 11.1. An AC voltage signal across a resistive load with $R = 10\ \Omega$ is given by:

A. $v(t) = V_m \cos 1000t$ [V]
B. $v(t) = V_m \sin 60t$ [V]
C. $v(t) = V_m \cos(60t + 45°)$ [V]

where $V_m = 10$ V. Determine the average AC power into the load in every case.

Problem 11.2. An alternating current through a resistive load with $R = 100\ \Omega$ is given by:

A. $i(t) = I_m \cos 10^6 t$ [V]
B. $i(t) = I_m \cos 37t$ [V]
C. $i(t) = I_m \sin(2011t + 45°)$ [V]

where $I_m = 1$ A. Determine the average AC power into the load in every case.

Problem 11.3. An rms voltage across a resistive load with $R = 100\ \Omega$ is given by:

A. $V_{rms} = 5$ V
B. $V_{rms} = 100$ V
C. $V_{rms} = 0$ V

Determine the average power into the load in every case.

Problem 11.4. An rms current through a resistive load with $R = 1$ kΩ is given by:

A. $I_{rms} = 1$ A
B. $I_{rms} = 100$ μA
C. $I_{rms} = 0$ A

Determine the average power into the load in every case.

Problem 11.5. An AC voltage signal is given by:

A. $v(t) = V_m \cos(\omega t + \varphi)$ [V]
B. $v(t) = 1\text{V} + V_m \cos(\omega t + \varphi)$ [V]
C. $v(t) = 1\ \text{V} - V_m \sin(\omega t + \varphi)$ [V]

where $V_m = 1$ V, $\omega = 100$ rad/s, and $\varphi = \pi/2$ rad. Find the time-average voltage $\overline{v(t)}$ in every case.

Problem 11.6. Present a mathematical proof of the fact that the expression for the average power, $P = \frac{V_m^2}{2R}$, holds for an AC voltage signal given by $v(t) = V_m \cos(\omega t + \varphi)$ [V] where φ is an arbitrary phase.

11.1.3 Application Example: rms Voltages and AC Frequencies Around the World

Problem 11.7. A 100 Ω resistive load is connected to an AC wall plug in:

A. Peoples Republic of China
B. India
C. USA
D. Germany

Determine the average power delivered to the load in every case. Also determine the rms load current in every case.

Problem 11.8. What do you think is a major

A. Advantage
B. Disadvantage

of having a higher AC voltage?

11.1.4 rms Voltages for Arbitrary Periodic AC Signals

Problem 11.9. Determine the average power delivered to a 100 Ω resistive load when the applied periodic voltage signal has the form $v(t) = (5t + 0.01)/T$ [V] over one period $T = 0.01$ s. This signal is known as the sawtooth or the triangular wave:

A. Use the analytical calculation of the rms voltage.
B. Use the rms voltage found numerically, based on a MATLAB script or any software of your choice.

Problem 11.10. Determine the average power delivered to a 100 Ω resistive load when the applied periodic voltage has the form $v(t) = \sqrt{t}/T$ [V] over one period $T = 0.01$ s:

A. Use the analytical calculation of the rms voltage.
B. Use the rms voltage found numerically, based on a MATLAB script or any software of your choice.

Problem 11.11. Of the two periodic voltage signals shown in the figures below,

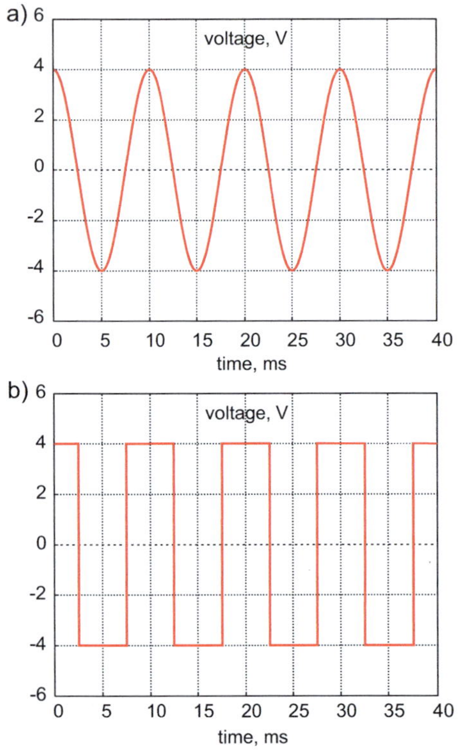

which signal delivers more average power into a resistive load? The periodic voltage on the top graph is the cosine function. Explain your answer and provide an analytical proof (find the *rms* voltages and the average power in every case).

Problem 11.12. Of the two periodic signals shown in the figures that follow, which signal delivers more average power into a resistive load? Explain your answer and provide:

A. An analytical proof—find the *rms* voltage and the average power in every case

B. A numerical proof (use MATLAB or any software of your choice).

The periodic voltage on the top graph is the cosine function. The periodic voltage on the bottom graph is given by $v(t) = 3.2 \times 10^5(t - 0.005)^2 - 4$ [V] over the time interval from 0 to T.

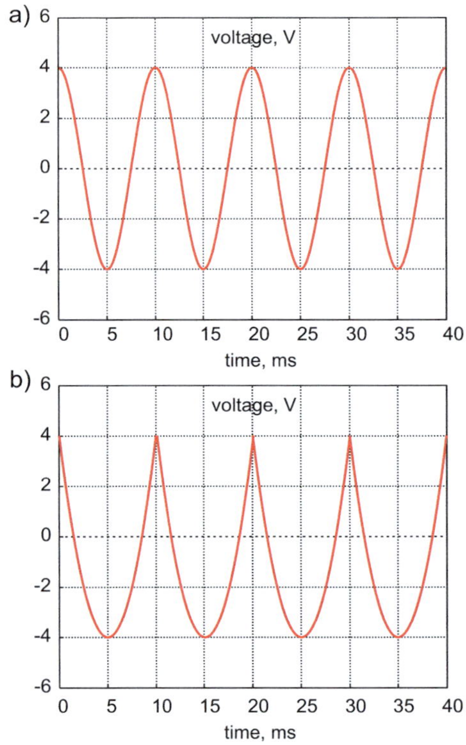

Problem 11.13. Of the two periodic signals shown in the figure that follows, which signal delivers more average power into a resistive load? The periodic voltage on the top graph is the cosine function. Explain your answer and provide:

A. An analytical proof—find the *rms* voltage and the average power in every case

B. A numerical proof (use MATLAB or any software of your choice).

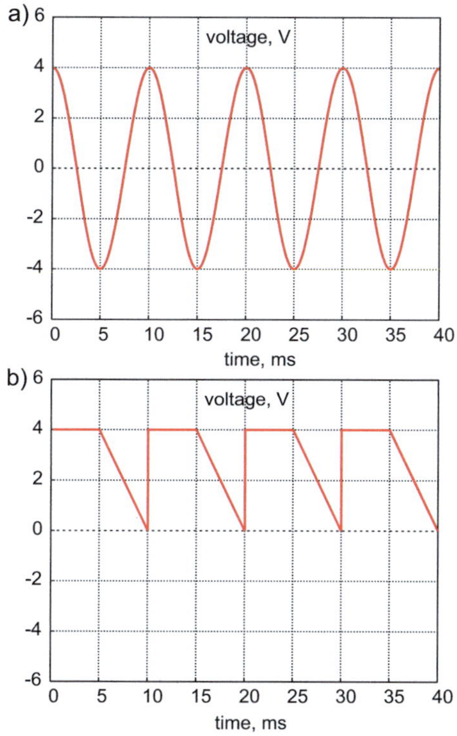

11.1.5 Average AC Power in Terms of Phasors: Power Angle

11.1.6 Average Power for the Resistor, Capacitor, and Inductor

Problem 11.14. The phasor voltage across a purely resistive load with the resistance $R = 100\ \Omega$ is given by $\mathbf{V} = -2 - j1.5$ [V]. Find the average power delivered to the load.

Problem 11.15. The phasor current through a purely resistive load with the resistance $R = 100\ \Omega$ is given by $\mathbf{I} = -1 - j0.5$ [A]. Find the average power delivered to the load.

Problem 11.16. The phasor voltage across an AC load and the phasor current through the same AC load are given by:

$$\mathbf{V} = -3 + j3 \quad [\mathrm{V}]$$
$$\mathbf{I} = \ +j0.1 \quad [\mathrm{A}]$$

A. Find the average power delivered to the load analytically.
B. Find the average power delivered to the load numerically using MATLAB or any software of your choice.

Problem 11.17. Repeat the previous problem for phasor voltage and phasor current in the form:

$$\mathbf{V} = 2 + j2 \quad [\mathrm{V}]$$
$$\mathbf{I} = \ 1 - j1 \quad [\mathrm{A}]$$

Problem 11.18. Express the average power given by $P = \frac{\mathrm{Re}(\mathbf{V}\cdot\mathbf{I}^*)}{2}$ in terms of the following three quantities: magnitude of the phasor voltage, $|\mathbf{V}|$; the impedance magnitude, $|\mathbf{Z}|$; and the real part of the impedance, $\mathrm{Re}(\mathbf{Z})$.

Problem 11.19. Determine the average power delivered to the load circuit between terminals *a* and *b* shown in the figure that follows. The AC angular frequency is 100 rad/s.

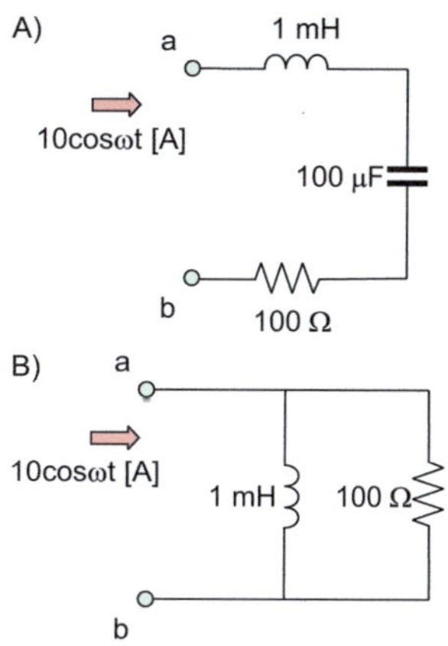

Problem 11.20. Determine the average power delivered to the load circuit between terminals *a* and *b* shown in the figure that follows. The AC angular frequency is 1000 rad/s.

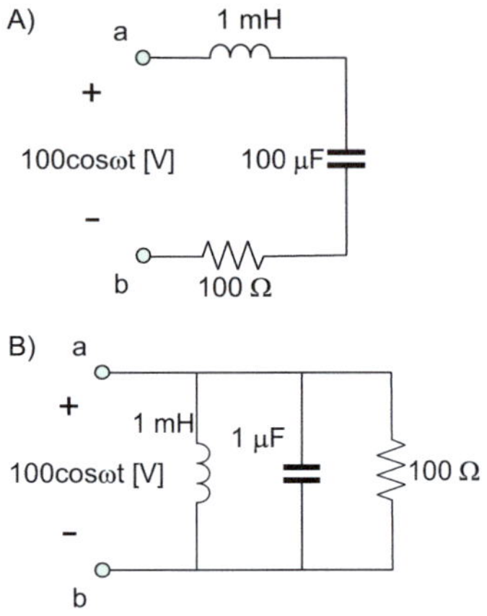

Problem 11.21. Determine the average power delivered to the load circuit shown in the figure below. The AC signal frequency is 10^6 Hz.

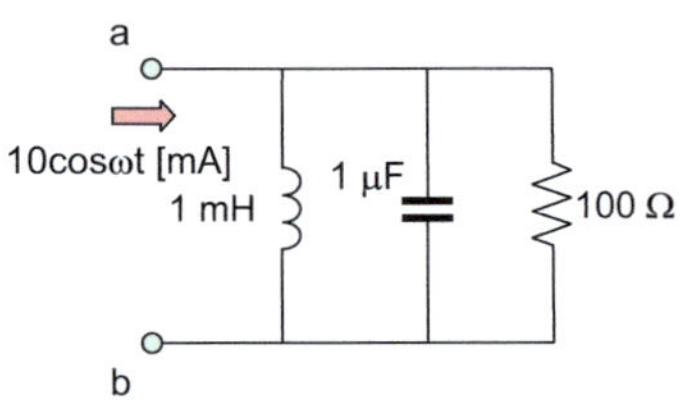

11.1.7. Average Power, Reactive Power, and Apparent Power

11.1.8. Power Triangle

Problem 11.22. Determine the resistance and the reactance of the circuit blocks (the load) shown in the figure. The AC angular frequency is 1000 rad/s.

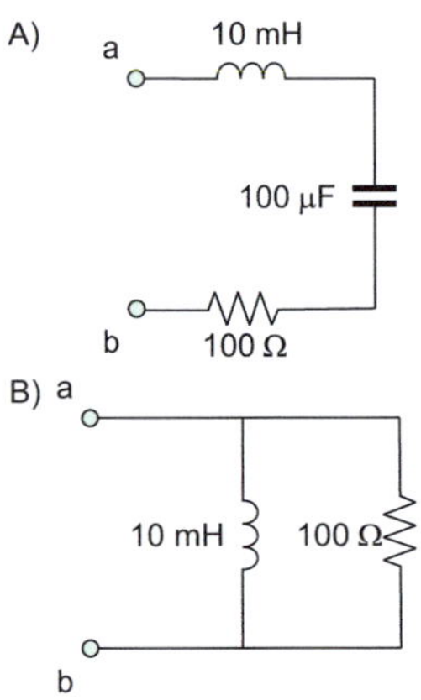

Problem 11.23. Determine the resistance and the reactance of the circuit block (the load) shown in the figure in terms of R, L, and C in a general form. The AC angular frequency is ω.

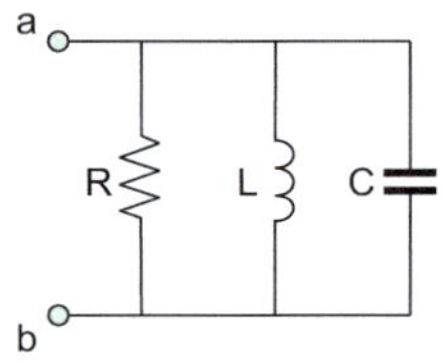

Problem 11.24. Write the expressions (and show units) for the average power P and the reactive power Q in terms of:

A. Phasor current **I** through the load and the load resistance R and the reactance X

B. Phasor voltage **V** across the load, the load impedance magnitude $|\mathbf{Z}|$, and the impedance phase (or the power angle) θ

Problem 11.25. For the circuit shown in the figure with the parameters $V_m = 170$ V, $\omega = 377$ rad/s, $L = 26.5$ mH, $R = 25$ Ω:

A. Determine the power angle and the power factor.

B. Determine the average (or true) power and the reactive power for the inductive load shown in the figure.

C. Construct the corresponding power triangle.

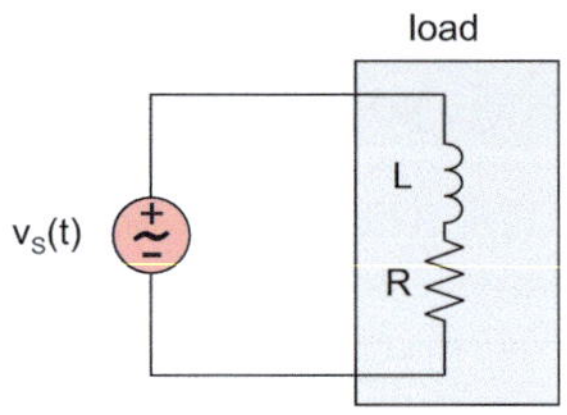

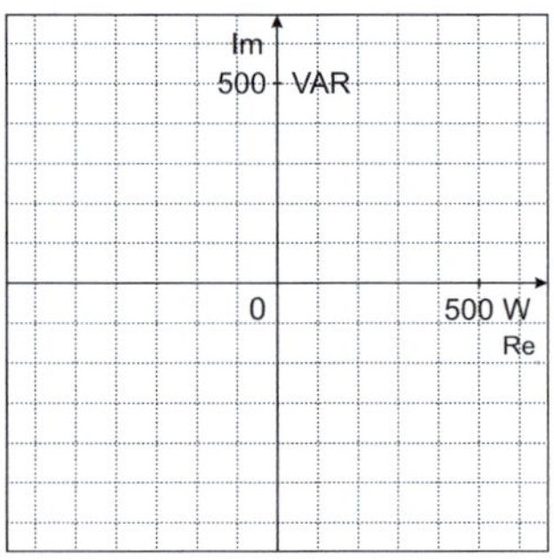

Problem 11.26. For the circuit shown in the figure with the parameters $V_{\mathrm{m}} = 170$ V, $\omega = 377$ rad/s, $C = 500$ µF, $R = 10$ Ω:

A. Determine the power angle and the power factor.
B. Determine the average (or true) power and the reactive power for the capacitive load shown in the figure.
C. Construct the corresponding power triangle.

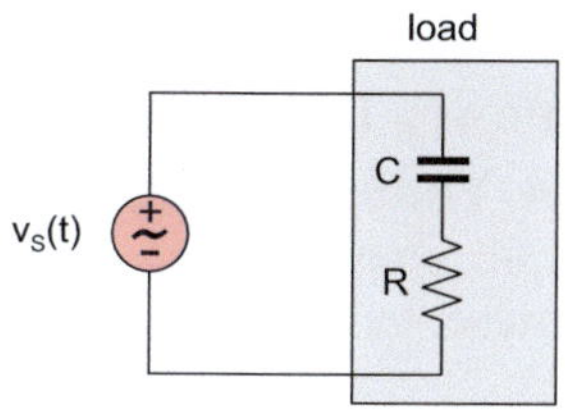

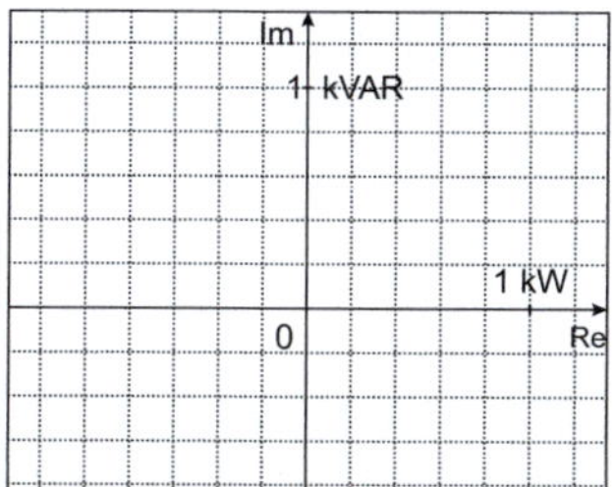

Problem 11.27. For the circuit shown in the figure with the parameters $V_{\mathrm{m}} = 170$ V, $\omega = 377$ rad/s, $L = 14.07$ mH, $C = 500$ µF, $R - 10$ Ω:

A. Determine the power angle and the power factor.
B. Determine the average (or true) power and the reactive power for the complex load shown in the figure.
C. Construct the corresponding power triangle.

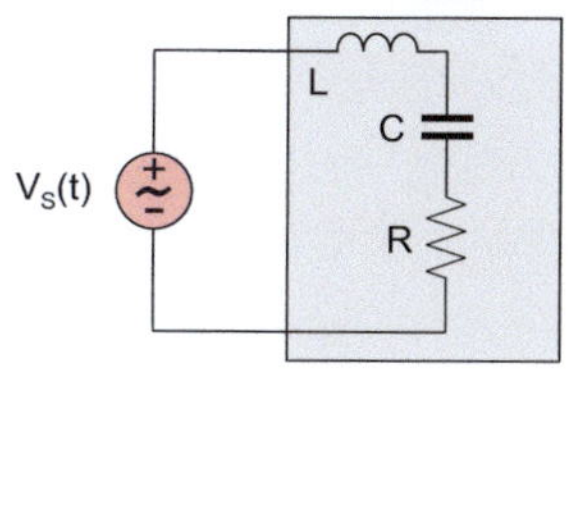

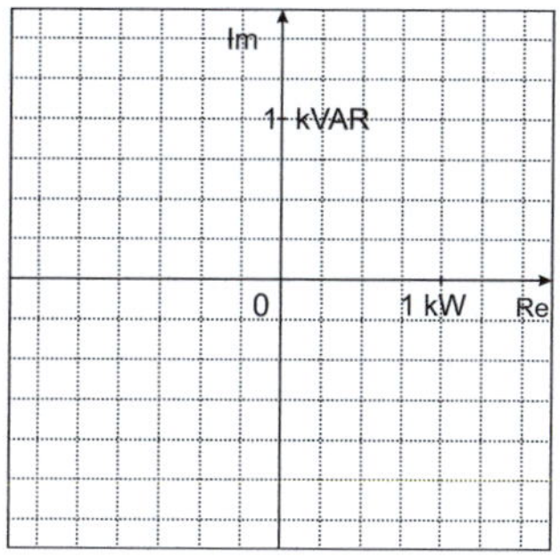

11.2 Power Factor Correction: Maximum Power Efficiency and Maximum Power Transfer

11.2.1 Power Factor Correction

11.2.3 Principle of Maximum Power Efficiency for AC Circuits

Problem 11.28. Correct the power factor for the inductive load shown in the figure below. The circuit parameters are $V_{\mathrm{m}} = 170$ V, $\omega = 377$ rad/s and $L = 53$ mH, $R = 10$ Ω:

A. Present the circuit diagram of the modified load and determine the required capacitance.
B. Determine average (true) power, reactive power, power factor, and amplitude of the circuit current before the power factor correction.
C. Determine average (true) power, reactive power, power factor, and amplitude of the circuit current after the power factor correction.

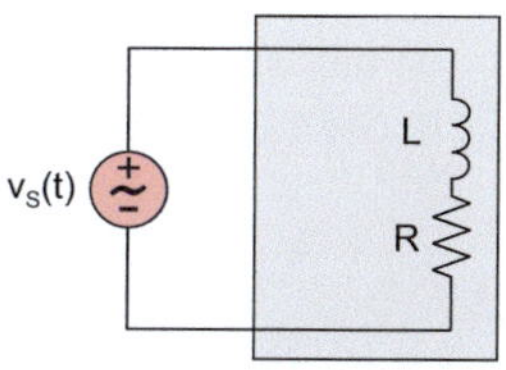

Problem 11.29. Correct the power factor for the capacitive load shown in the figure that

follows. The circuit parameters are $V_m = 170$ V, $\omega = 377$ rad/s and $C = 265$ μF, $R = 10$ Ω:

A. Present the circuit diagram of the modified load and determine the required inductance.
B. Determine average (true) power, reactive power, power factor, and amplitude of the circuit current before the power factor correction.

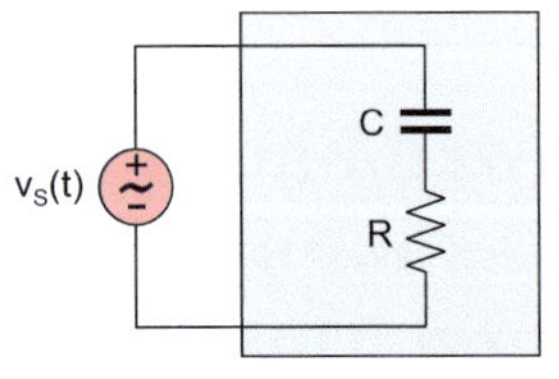

Problem 11.30. A whip monopole antenna used in US Coast Guard ships has an equivalent electric circuit shown in the figure of the previous problem. Its (radiation) resistance is 1 Ω, and the reactance is $-j1000$ Ω. By modifying the antenna circuit with a lumped inductor, it is required to make the antenna impedance real and as large as possible:

A. Present the circuit diagram of the modified load
B. Determine the required impedance of the inductor.

11.2.4 Principle of Maximum Power Transfer for AC Circuits

Problem 11.31. Describe in your own words the difference between the concepts of maximum power efficiency and maximum power transfer for AC circuits.

Problem 11.32. A generator's impedance is $50\angle 30^\circ$ [Ω]. What should the load impedance be to allow the maximum power transfer to the load?

Problem 11.33

A. A generator's impedance is 50 Ω. The load impedance is $1+j50$ [Ω]. What percentage of the maximum available power (at the load impedance of 50 Ω) is transferred to the load?
B. Repeat the same task for the load impedance of $1-j50$ [Ω].
C. Repeat the same task for the load impedance of $5+j50$ [Ω].

Hint: Derive the general expression for the power ratio first and then plug in the numbers.

11.3 AC Power Distribution: Balanced Three-Phase Power Distribution System

11.3.1 AC Power Distribution Systems

11.3.2 Phase Voltages: Phase Sequence

Problem 11.34. Draw generic circuits for the following representative AC power distribution systems:

A. Single-phase two-wire system
B. Single-phase three-wire system
C. Two-phase three-wire system
D. Three-phase four-wire system

Show loads and phasor voltages with the corresponding phases.

Problem 11.35. Determine the phase sequence for the phase voltages given by:

$$v_{an}(t) = 240\cos(314t + 75^\circ)\ [\text{V}],$$
$$v_{bn}(t) = 240\cos(314t - 165^\circ)\ [\text{V}],$$
$$v_{cn}(t) = 240\cos(314t - 45^\circ)\ [\text{V}].$$

To simplify the solution, construct the corresponding phasor diagram in the figure below:

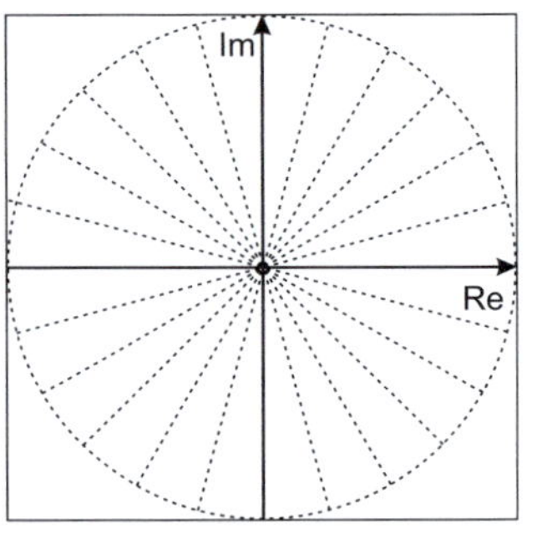

Problem 11.36. Given $\mathbf{V}_{bn} = 120\ \angle 45°[\mathrm{V}]$, find $\mathbf{V}_{an}$ and $\mathbf{V}_{cn}$ assuming:

A. The positive *abc* phase sequence
B. The negative *acb* phase sequence

Express your result in phasor form. Make sure that the phase ranges from −180° to +180°. To check the solution, you may want to use the corresponding phasor diagram shown in the figure for the previous problem.

11.3.3 Wye (Y) Source and Load Configurations for Three-phase Circuits
11.3.4 Application: Examples of Three-phase Source and the Load
11.3.5 Solution for the Balanced Three-Phase Wye-Wye Circuit
11.3.6 Removing the Neutral Wire in Long-Distance Balanced High-power Transmission

Problem 11.37

A. Draw the circuit diagram for a generic three-phase four-wire balanced wye-wye power distribution system.
B. Label phase voltages and phase impedances (load impedances per phase).
C. Label line currents.

Problem 11.38. A three-phase circuit is shown in the figure that follows:

A. Is it a balanced wye-wye circuit?
B. If not, show your corrections on the figure.

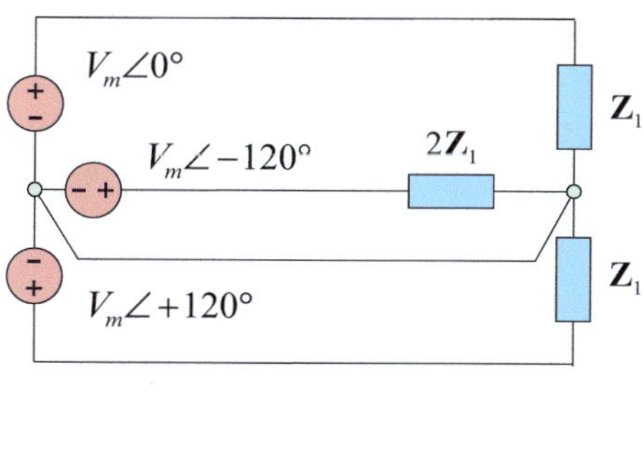

Problem 11.39. Repeat the previous problem for the circuit shown in the figure below:

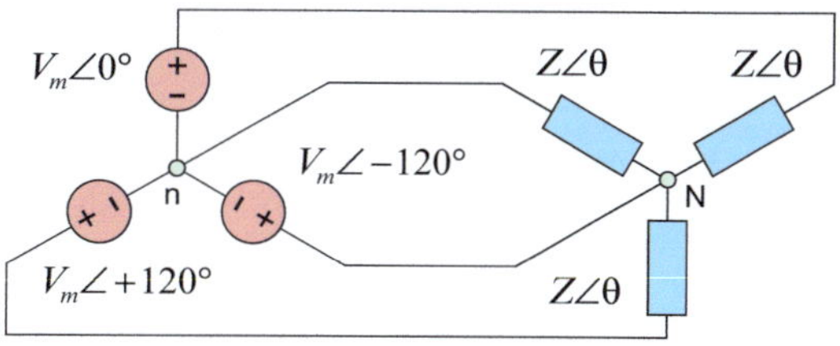

Problem 11.40. Repeat Problem 11.38 for the circuits shown in the figure that follows:

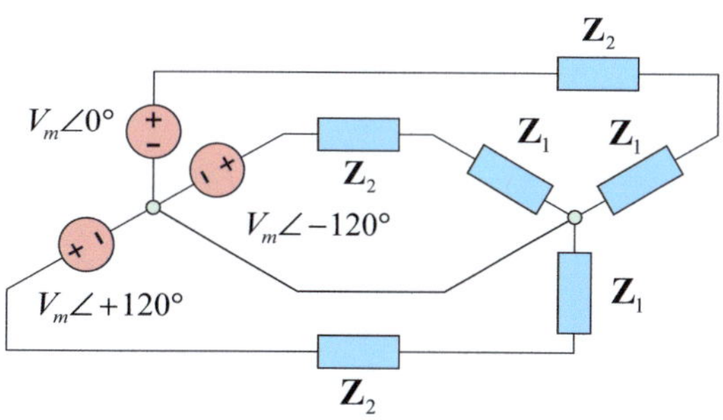

Problem 11.41. Prove that Eq. (11.28) of this chapter for line voltages also holds for the negative phase sequence to within the substitution 30° → −30°.

Problem 11.42. The local electric service in the European Union is provided by a three-phase four-wire *abcn* wye system with the line voltages equal to 400 V rms each (so-called Niederspannungsnetz):

A. Determine the rms phase voltages.
B. By connecting terminals *abcn* in any sequence of your choice, could you in principle obtain the rms voltages higher than 400 V?

Problem 11.43. Determine line currents in the balanced three-phase wye-wye circuit shown in the figure that follows. You are given the *acb* sequence of phase voltages $\mathbf{V}_{an} = 170\angle 0°\ [\mathrm{V}]$,

$\mathbf{V}_{\mathrm{bn}} = 170\angle 120°$ [V], $\mathbf{V}_{\mathrm{bn}} = 170\angle{-120°}$ [V], and load impedance per phase, $\mathbf{Z} = 8 + j30\ \Omega$.
Plot phasor currents on the phasor diagram that follows.

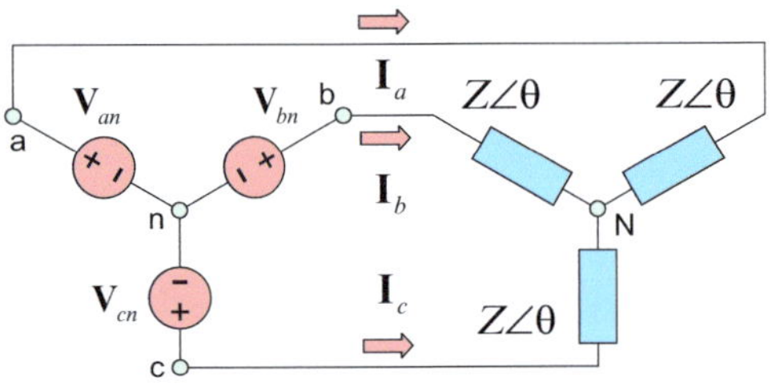

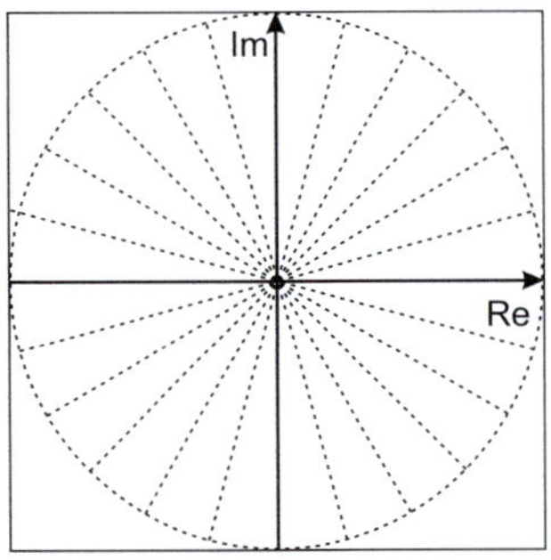

Problem 11.44. In the balanced three-phase wye-wye circuit shown in the figure that follows, the power line resistance and inductance are additionally included into consideration. The three-phase source operates at 60 Hz; $R = 2\ \Omega$, $L = 9.6$ mH. You are given the *abc* sequence of phase voltages $\mathbf{V}_{\mathrm{an}} = 170\angle 0°$ [V], $\mathbf{V}_{\mathrm{bn}} = 170\angle - 120°$ [V], $\mathbf{V}_{\mathrm{bn}} = 170\angle 120°$ [V], and load impedance per phase, $\mathbf{Z} = 7 + j30\ \Omega$.

A. Determine line currents.
B. Plot phasor line currents on the phasor diagram to the previous problem.

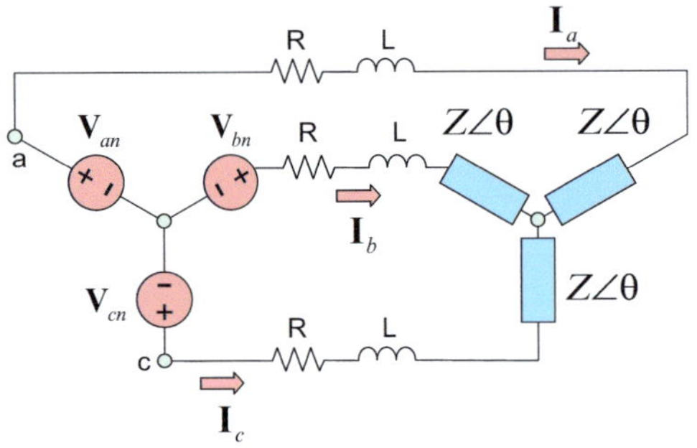

11.4 Power in Balanced Three-Phase Systems: Delta-connected Three-Phase Circuits

11.4.1 Intantaneous Power

11.4.2 Average Power, Reactive Power, and Apparent Power

Problem 11.45. In a three-phase balanced wye-wye system, the rms phase voltages are 120 V, and the rms line currents are 10 A. The impedance has the power angle of $\theta = 75°$. Find:

1. The instantaneous load power
2. The average load power

Problem 11.46. In a three-phase balanced wye-wye system, the rms line voltages are 400 V, and the rms line currents are 10 A. The impedance has the power angle of $\theta = 60°$. Find:

1. The instantaneous load power
2. The average load power

Problem 11.47. In the three-phase system shown in the figure that follows, $\mathbf{Z} = 40\angle 60°$. The sources have the relative phases $0, \ -120°, \ +120°$. The rms *line* voltages are 208 V. Determine:

A. The type of the three-phase system
B. Instantaneous power delivered to the three-phase load
C. Average power delivered to the three-phase load

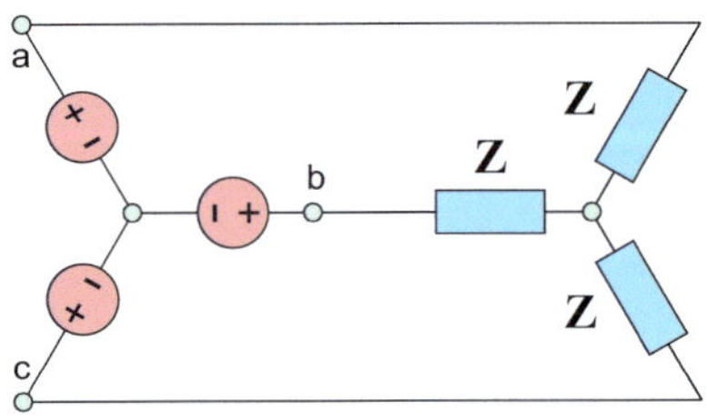

Problem 11.48. A balanced wye-wye three-phase system in the figure that follows uses lossless transmission lines and operates at 60 Hz.

The line-to-neutral voltages have the amplitudes of 170 V, $V_m = 170$ V. Every phase impedance is a 92-mH inductance in series with a 20 Ω resistance. Find the instantaneous load power.

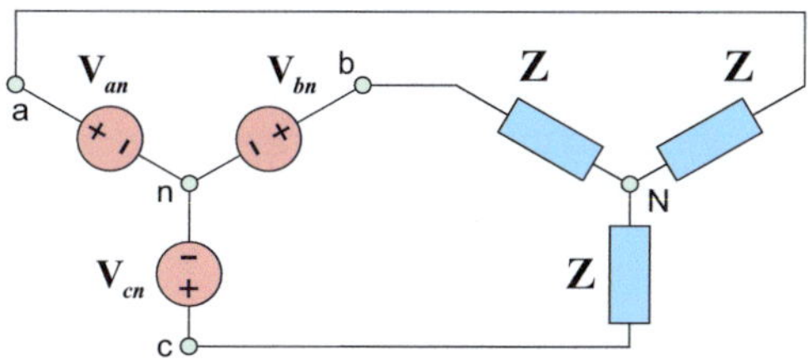

Problem 11.49. In the previous problem:

A. Determine the load average power, reactive power, and the apparent power.
B. Do these powers coincide with the corresponding source measures?

Problem 11.50. A three-phase induction motor is modeled by a balanced wye load. The motor (active) power is 2.5 kW; the line current is 10 A rms, and the phase voltage of a three-phase wye source is 120 V rms. Determine the power factor of the motor.

Problem 11.51. In the previous problem, the motor (active) power is 9 kW; the line current is 15 A rms, and the line voltage of a three-phase wye source is 400 V rms. Determine the power factor of the motor.

11.4.4 Balanced Delta-Connected Load

11.4.5 Balanced Delta-Connected Source

Problem 11.52. A three-phase balanced wye-wye system is shown in the figure that follows. Its delta-delta equivalent is sought, which is shown in the same figure. For the delta-delta system, write the corresponding voltage and impedance values in the phasor form close to every circuit element.

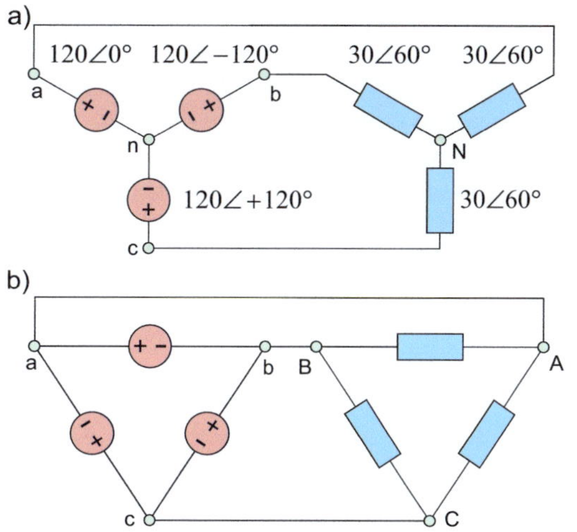

Problem 11.53. A three-phase balanced delta-delta system is shown in the figure that follows. Its wye-wye equivalent is sought, which is shown in the same figure. For the wye-wye system, write the corresponding voltage and impedance values in the phasor form close to every circuit element.

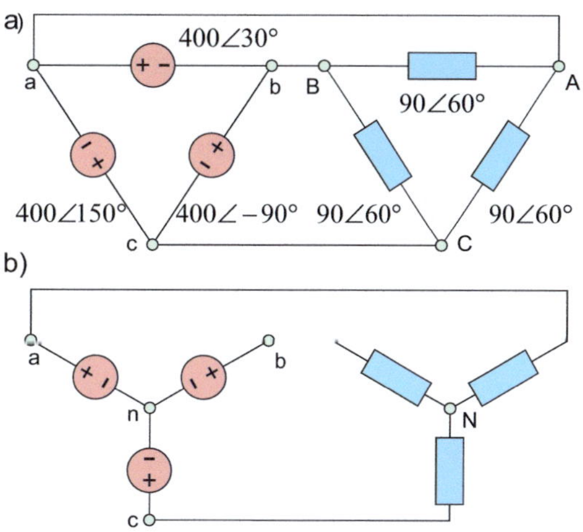

Problem 11.54. A balanced delta-delta system shown in the figure below operates at 50 Hz. The phase voltages of the delta source, $\mathbf{V}_{ab}$, $\mathbf{V}_{bc}$, $\mathbf{V}_{ca}$, have the amplitudes of $V_m = 563.4$ V each. Every phase impedance is a 0.21 H inductance in series with a 38-Ω resistance:

A. Find the average load power,
B. Find the instantaneous load power,
C. Find the apparent load power.

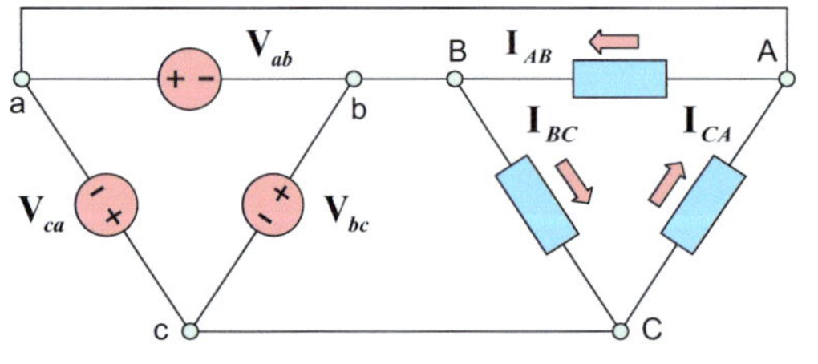

Problem 11.55. A balanced wye-delta system shown in the figure below operates at 60 Hz. The phase voltages of the wye source, $\mathbf{V}_{an}$, $\mathbf{V}_{bn}$, $\mathbf{V}_{cn}$, have the amplitudes of $V_m = 170$ V each. Each phase impedance is a 0.18 H inductance in series with a 90 Ω resistance:

A. Find the average load power,
B. Find the instantaneous load power,
C. Find the apparent load power.

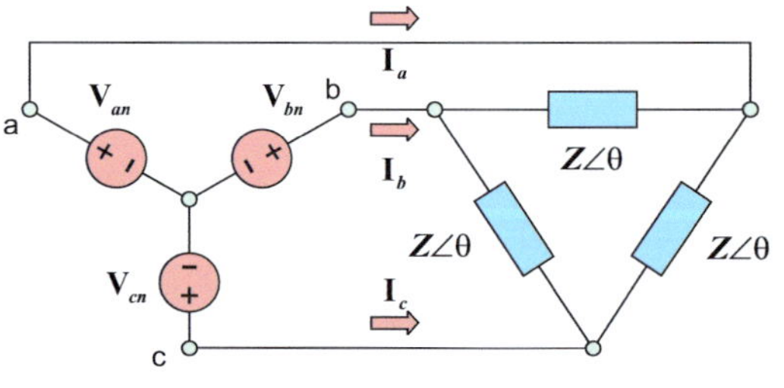

Chapter 12: Electric Transformer and Coupled Inductors

Overview

Prerequisites:
- Knowledge of complex arithmetic
- Knowledge of basic circuit analysis (Chapters 3 and 4)
- Knowledge of self- and mutual inductances (Chapter 6)
- Knowledge of phasor/impedance method for AC circuit analysis (Chapter 8) and of basic AC power analysis (Chapter 11)

Objectives of Section 12.1:
- Derive the ideal transformer model from the first principles
- Understand the role of the ideal magnetic core
- Understand the role of Faraday's law and Ampere's law
- Prepare the background for introducing magnetic circuits
- Understand and apply the dot convention
- Relate ideal transformer model to a model with dependent sources

Objectives of Section 12.2:
- Be able to analyze electric circuits with ideal transformer
- Learn about load and source reflections
- Learn about impedance matching via transformers
- Learn about electric power transfer via transformers

Objectives of Section 12.3:
- Derive equations for useful transformer types—autotransformer, multiwinding transformer, and center-tapped transformer—from the first principles
- Understand the role of the center-tapped transformer for single-ended to differential transformation and for power division

Objectives of Section 12.4:
- Understand the physical background of the Steinmetz model and relate the model parameters to real transformers
- Be able to analyze the nonideal transformer model
- Define transformer voltage regulation and power efficiency
- Briefly discuss the high-frequency transformer model

S. N. Makarov et al., *Practical Electrical Engineering*, https://doi.org/10.1007/978-3-319-96692-2_12

Objectives of Section 12.5:
- Introduce the model of two coupled inductors from the first principles
- Learn how to analyze electric circuits with coupled inductors
- Learn about the useful conversion to the T-network of uncoupled inductances
- Obtain the basic exposure to wireless inductive power transfer including its major features and challenges

Application examples:
- Electric power transfer via transformers
- Wireless inductive power transfer
- Coupling of nearby magnetic radiators

Keywords:
Electric transformer (primary winding, secondary winding, circuit symbol, isolation transformer, instrumentation transformer, current transformer, clamp on ammeter, potential transformer, exciting current, magnetizing current, magnetizing inductance, magnetizing reactance, power conservation, stored energy, turns ratio, step-up transformer, step-down transformer, high-voltage side, low-voltage side, transformer rating, dot convention, dotted terminals, voltage polarity, current reference directions, summary of reference directions, mechanical analogies), **Ideal transformer model** (ideal magnetic core, ideal open-circuited transformer, ideal transformer equations, ideal transformer equations in phasor form, power conservation, stored energy, model in terms of dependent sources), **Ampere's law** (linked current, for ideal magnetic core, for multiwinding transformer), Referred (reflected) source network in the secondary, Referred (reflected) load impedance in the primary, Load reflection, Source reflection, Reflected resistance, Reflected inductance, Reflected capacitance, Transformer as a matching circuit, Matching real-valued impedances, Matching arbitrary complex impedances, Partial matching condition, **Power transfer via transformers** (for fixed load voltage, sending-end voltages, for fixed source voltage), **Autotransformer** (step down, step up, circuit symbol, ideal transformer equations), **Multiwinding transformer** (ideal circuit equations, Ampere's law, telephone hybrid circuit), **Center-tapped transformer** (ideal transformer equations, single-ended to differential transformation, 180° power divider, 180° power splitter), **Real transformer** (nonideal low-frequency model, Steinmetz model, **Steinmetz parameters** (magnetizing reactance, core loss resistance, primary leakage reactance, secondary leakage reactance, primary ohmic resistance, secondary ohmic resistance), nonideal transformer model terminology/analysis, voltage regulation, transformer power efficiency, nonideal high-frequency model), **Model of two coupled inductors** (equations, circuit symbol, circuit analysis, solution for N coupled inductors, stored energy, conversion to T-network), **Mutual inductance** (definition, of ideal transformer, of two coaxial coils), **Coupling coefficient** (definition, largest possible value, trend), **Wireless inductive power transfer** (application, basic model, examples of), **Mutual coupling for nearby magnetic radiators** (arrays of magnetic radiators), Lenz's law

Section 12.1 Ideal Transformer as a Linear Passive Circuit Element

12.1.1 Electric Transformer

General

An *electric transformer* is a simple and versatile device, which targets AC power transfer from one electric circuit to another. These circuits are coupled via a time-varying magnetic flux linking two or more coils. There is no direct electric connection between the coils. The transformer cannot transfer power between the DC circuits. Analysis of transformers involves many principles that are basic to the understanding of electric machines. Transformers are primarily used to:

1. Change the voltage level in power electronics AC circuits.
2. Insulate one AC (or RF) circuit from another (*isolation transformers*).
3. Match the impedance of the source and the load in electronic circuits.
4. Measure AC voltages and currents (*instrumentation transformers*).

As another everyday application example, we mention various DC power supplies (AC to DC *converters* or *adapters*), both switching and linear. These DC supplies power PCs, printers, modems, cordless phones, video game consoles, etc. at your home. Low-frequency (bulky) or high-frequency (smaller) transformers are very important parts of these supplies, irrespective of their particular construction.

Function

Although the transformer typically consists of two coupled inductors—see Fig. 12.1—its function is *principally different* from that of the familiar inductance. While the inductance is an energy-storage (and energy-release) circuit element, the *ideal transformer*, as a new circuit element, never stores any instantaneous energy. It does not possess any inductance (or impedance in general) either.

Approach

The model a two-winding electric transformer introduced in this section does yet not use the concept of a *magnetic circuit*. Instead, we accurately formulate and employ Faraday's law and Ampere's law directly. The same transformer model in the framework of magnetic circuits is revisited in the next chapter. Through the text, we use opposite reference directions for the transformer currents; the equal directions are also discussed in the text.

12.1.2 Ideal Open-Circuited Transformer: Faraday's Law of Induction

We will perform transformer analysis in several steps illustrated in Fig. 12.1.

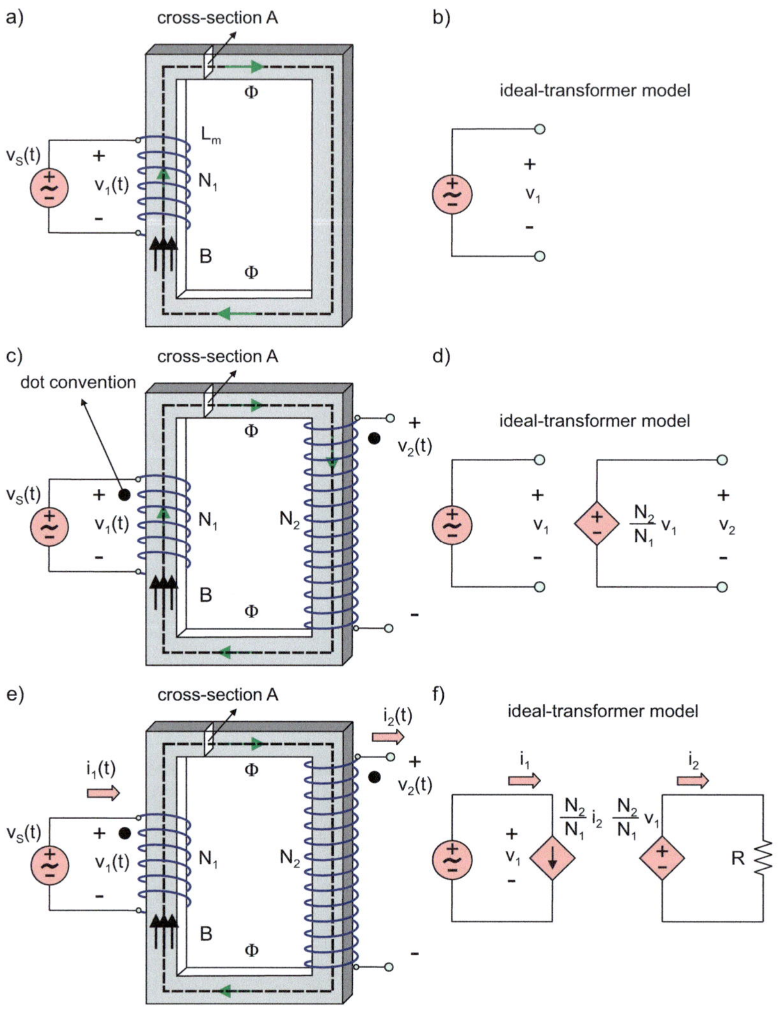

Fig. 12.1. Construction of an ideal electric transformer. The negligibly small exciting current is not shown. In all three cases, the magnetic flux $\Phi(t)$ in the core remains the *same*.

Using Faraday's Law of Induction

Transformer construction can start with a *single* lossless inductor with a closed magnetic core directly connected to an AC voltage source, see Fig. 12.1a. The inductor is a coil with N_1 turns—the *primary winding* of the transformer. Its inductance neglecting the flux leakage effect will be denoted here by L_m. By KVL,

$$v_1(t) = v_S(t) \tag{12.1}$$

The primary winding establishes a *finite* time-varying magnetic flux $\Phi(t) = AB(t)$ in the core. Here, B is the flux density and A is the core cross section. According to Faraday's law of induction—see Eq. (6.19) and (6.33)—this flux is uniquely determined by the winding voltage (which is equal to the source voltage) in the form

$$v_1(t) = N_1 \frac{d\Phi(t)}{dt} \tag{12.2}$$

Exciting Current and Inductance for the Ideal Magnetic Core

An inductor current $i_\Phi(t)$ which would be present in Fig. 12.1a is called the *exciting current* (or *magnetizing current*). Accordingly, inductance L_m is called the *magnetizing inductance*. The magnetizing inductance is found using Eq. (6.23). The exciting current is then found using Eq. (6.19) of the same chapter, which is valid in both static and dynamic cases. This gives

$$L_m = \frac{\mu_0 \mu_r A N_1^2}{l}, \quad i_\Phi(t) = N_1 \frac{\Phi(t)}{L_m} \tag{12.3}$$

where A is the core cross section shown in Fig. 12.1. When the relative magnetic permeability μ_r of the core is very high, the coil inductance L_m is very large. Therefore, the corresponding inductor current $i_\Phi(t)$ is quite small. An *ideal magnetic core* assumes that $\mu_r \to \infty$. Therefore, according to Eq. (12.3),

$$L_m \to \infty, \quad i_\Phi(t) = 0, \quad \mu_r \to \infty \tag{12.4}$$

Equation (12.4) corresponds to the *ideal transformer model*. The primary winding becomes an open circuit of infinite inductance as shown in Fig. 12.1b. However, the finite magnetic flux $\Phi(t)$ is still established in the core. There is no contradiction here since a negligible exciting current $i_\Phi(t)$ is necessary to establish the finite flux $\Phi(t)$ in a core with the infinitely high permeability (infinitely high inductance). The situation is somewhat similar to an operational amplifier with the negative feedback where the negligible input different voltage controls the large-signal amplifier operation.

Using Faraday's Law a Second Time: Relation Between Transformer Voltages

As a next step, another coil with N_2 turns—the *secondary winding* of the transformer—can be added as shown in Fig. 12.1c. The core flux $\Phi(t)$ also links the secondary winding and creates an open-circuit voltage at its terminals. No flux leakage in air is permitted in the ideal model. Faraday's law is applied a second time, which yields

$$v_2(t) = N_2 \frac{d\Phi(t)}{dt} \tag{12.5}$$

The plus sign implies the dot convention to be discussed shortly. From Eqs. (12.2) and (12.5), the voltage ratio becomes

$$\frac{v_2}{v_1} = \frac{N_2}{N_1} \tag{12.6}$$

at any time instant. Thus, the voltage ratio is the ratio of respective turns. The equivalent circuit of the open-circuited transformer is that of Fig. 12.1d. The secondary winding operates as a dependent voltage-controlled voltage source. Thus far, it has zero effect on the circuit connected to the primary winding.

Exercise 12.1: You are given the source voltage in the form $v_S(t) = 170\cos(2\pi 60t)$ [V], the number of turns of the primary winding $N_1 = 200$, the finite relative permeability of the magnetic core, $\mu_r = 5000$, the coil length of the primary winding of 0.1 m, and the core cross-section of $A = 0.001\ \mathrm{m}^2$.

A. Find the exciting current (no-load current) in the primary winding of this non-ideal transformer. Justify the ideal-transformer approximation Eq. (12.4).
B. Reduce N_1 from 200 to 20 and repeat the solution.

Answers:

A. $i_\Phi(t) = 179\sin(2\pi 60t)$ [mA]. The current amplitude and its rms value are much less than typical current amplitudes of 5 A (3.54 A rms) observed in residential AC circuits. The ideal-transformer approximation is justified.
B. $i_\Phi(t) = 17.9\sin(2\pi 60t)$ [A]. The current amplitude is quite high; the ideal-transformer approximation is severely violated.

Dot Convention

Windings of transformers are marked to indicate *the relative voltage polarities* of voltages v_1 and v_2. We indicate the relative polarities by the *dot convention*. Namely, voltages v_1 and v_2 with the *positive polarity* at the *dotted terminals* will be strictly *in phase*, see Fig. 12.1c. It does not matter where the dots are exactly located. Yet another meaningful definition of the dotted terminals is as follows. Currents *entering* the dotted terminals (which means the passive reference configuration not to be confused with Fig. 12.1e) would produce fluxes in the *same* direction in the core that forms the common magnetic path. If the windings are visually seen, the polarities are determined by examining clockwise or counterclockwise coil winding directions. Otherwise, an experiment could be employed, with a function generator and an oscilloscope. Emphasize that the dot convention only determines the voltage polarity; it has nothing in common with large currents which can flow in a loaded transformer to be discussed shortly.

Exercise 12.2: In Fig. 12.1c, the winding direction of the secondary winding is reversed. Where should the dot be placed?

Answer: To the lower end of the secondary winding.

12.1.3 Appearance of Transformer Currents

Finally, a load resistance R is connected across the terminals of the secondary winding, this resistance should be present in Fig. 12.1e, f. As a result, a large current $i_2(t)$ will flow in the secondary winding, which is given by Ohm's law:

$$i_2(t) = \frac{v_2(t)}{R} \tag{12.7}$$

In this case, the transformer performs its major function of power transfer from the source to the load. When a large current $i_2(t)$ starts to flow, it will drastically change the magnetic field in the core, core flux $\Phi(t)$, and consequently voltage $v_1(t)$ according to Eq. (12.2). However, any extra change in $v_1(t)$ is in contradiction with KVL stated by Eq. (12.1). Therefore, another large current $i_1(t)$ in the primary winding will immediately start to flow too, in order to undo this change. In other words, current i_2 "calls current i_1 into existence." *Ampere's law* is used to establish the quantitative relation between transformer currents i_1 and i_2 at *any* time instant and in *any* circuit, not necessarily the circuit from Fig. 12.1.

12.1.4 Ampere's Law

Ampere's law does not operate in terms of the magnetic flux Φ (or flux density $\vec{B}$), but rather in terms of the magnetic field $\vec{H}$ defined in Section 6.1.7. Consider a closed contour *abcd* in Fig. 12.2. Ampere's law expresses the field created by *a linked current i* via a contour integral:

$$\oint_{abcd} \vec{H} \cdot d\vec{l} = i \tag{12.8a}$$

It does not matter whether the magnetic core is present or not. The "linked current" i is the *total current* that passes through a surface bounded by the contour in Fig. 12.2. The direction in which the current traverses the contour and reference direction of the loop *abcd* (clockwise in Fig. 12.2) are related by the *right-hand rule*. For the contour *abcd* in Fig. 12.2, the field in the magnetic core is a constant and is parallel to the contour. In this case, Eq. (12.8a) is simplified as

$$lH = i \tag{12.8b}$$

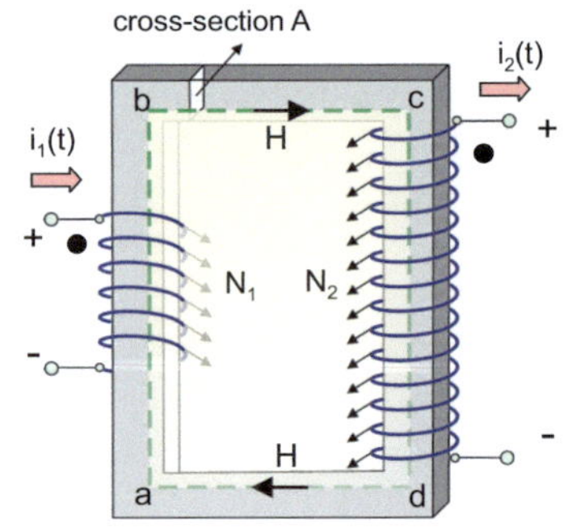

Fig. 12.2. Ampere's law for a transformer.

where $l = l_{ab} + l_{bc} + l_{cd} + l_{da}$ is the total contour length. Ampere's law is further simplified for time-varying currents linking the *ideal* magnetic core with $\mu_r \to \infty$. Note the magnetic flux in the core is given by $\Phi(t) = AB(t) = A\mu_0\mu_r H(t)$. When $\mu_r \to \infty$, any finite magnetic field $H(t)$ would create an infinite flux in the core and thus infinite voltages across the coils, which is impossible. Therefore, we must set $H = 0$ in Eq. (12.8b) and obtain

$$i = 0 \quad \text{for} \quad \mu_r \to \infty \tag{12.8c}$$

In order words, for the ideal magnetic core, the time-varying linked current must be *zero*.

12.1.5 Ideal Loaded Transformer

For the transformer in Fig. 12.2, the linked current is given by

$$i = N_1 i_1 - N_2 i_2 \tag{12.9}$$

since current i_1 is inflowing and current i_2 is outflowing through the surface bounded by the contour. Therefore, according to Eq. (12.8c)

$$N_1 i_1 - N_2 i_2 = 0 \Rightarrow \frac{i_1}{i_2} = \frac{N_2}{N_1} \tag{12.10}$$

at any time instant. Equation (12.9) completes the analysis of the ideal transformer. It states that the current ratio is the *inverse* turns ratio. This result does not depend on the particular value of the load resistance in Fig. 12.1f. It is perhaps even more important that Eq. (12.10) does not require the source to be directly connected to the primary. Thus, the ideal transformer as a new passive linear circuit element is *completely* described by Eqs. (12.6) and (12.10) (*ideal transformer equations*), respectively. The corresponding circuit model may be expressed in terms of two dependent sources—the voltage-controlled voltage source in the secondary and the current-controlled current source in the primary—see Fig. 12.1f.

Typical Transformer Circuit

Figure 12.3a shows a typical single-phase transformer circuit along with the ideal transformer *circuit symbol* (with a magnetic core). The transformer acts as *a linear* interface between an arbitrary AC source circuit and an arbitrary AC load circuit. Both the source and the load do not have to be linear. Figure 12.3b shows the corresponding dependent-source model. Such a model may be quite useful in transformer circuits where extra coupling between the source and the load exists, for example, via a capacitor. This may occur in high-frequency transformers.

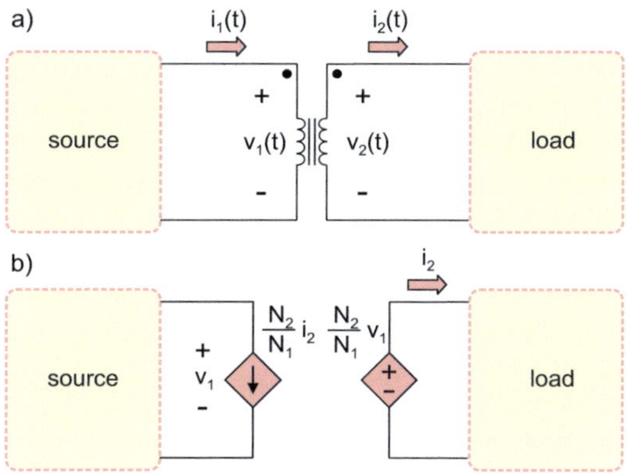

Fig. 12.3. Transformer interfacing a source and a load and its model in terms of dependent sources.

Power Conservation

Combining Eqs. (12.6) and (12.10), we obtain for the instantaneous power delivered to the load

$$p_2 = v_2 i_2 = \frac{N_2}{N_1}\frac{N_1}{N_2} v_1 i_1 = v_1 i_1 = p_1 \tag{12.11}$$

Equation (12.11) states that, in the absence of losses, instantaneous (and indeed average) power is conserved during voltage/current transformation; there is no net gain or loss of power for an ideal transformer. As soon as power is consumed by the load, the same power is drawn from the source. The transformer, therefore, provides a physical isolation between the load and the source while maintaining electric power transfer. Note that Fig. 12.3b has similarities with the equivalent amplifier circuit model. If the transformer may be used as a voltage booster, why do we need the semiconductor amplifier then? The answer is that it is the power that counts. When the transformed voltage increases, the transformed current decreases so that the output power remains the same. An amplifier, on the other hand, boosts the total power, which is true amplification of the source signal. Still, the transformer may be used at the amplifier output to provide a significant extra voltage (but not power) gain.

Stored Energy

The ideal loaded transformer in Fig. 12.1e, f does not possess any stored magnetic-field energy. The instantaneous magnetic-field energy stored in a magnetic core of volume V is given by Eq. (6.18):

$$E(t) = \frac{1}{\mu_0\mu_r}\int_V \vec{B}(t)\cdot\vec{B}(t)dv = \frac{Al}{\mu_0\mu_r}B^2(t) = \frac{l}{A\mu_0\mu_r}\Phi^2(t) \tag{12.12}$$

The energy becomes exactly zero when $\mu_r \to \infty$. Here, l is the length of the centerline of the magnetic core and A is its cross section. The same proof is valid for the open-circuited ideal transformer.

Exercise 12.3: Under conditions of Exercise 12.1, determine the instantaneous stored energy of a non-ideal transformer given that the core centerline length is 30 cm.

Answers:

A. $E(t) = 0.24\sin^2(2\pi 60t)$ [J], which is a small value.
B. $E(t) = 24\sin^2(2\pi 60t)$ [J], which is a quite significant number.

Exercise 12.4: In the circuit in Fig. 12.1f, $v_S(t) = 325\cos(2\pi 50t)$ [V]. For a 400 Ω resistive load, determine load voltage, load current, and average power delivered to the load when the turns ratio is equal to 10:1, 1:1, and 1:10.

Answers:

For 1:10 turns ratio:
$v_R(t) = 32.5\cos(2\pi 50t)$ [V], $i_R(t) = 0.08125\cos(2\pi 50t)$ [A], $P = 1.320$ W

For 1:1 turns ratio:
$v_R(t) = 325\cos(2\pi 50t)$ [V], $i_R(t) = 0.8125\cos(2\pi 50t)$ [A], $P = 132.0$ W

For 10:1 turns ratio:
$v_R(t) = 3250\cos(2\pi 50t)$ [V], $i_R(t) = 8.125\cos(2\pi 50t)$ [A], $P = 13.2$ kW

12.1.6 Ideal Transformer Versus Real Transformer: Transformer Terminology

Ideal Transformer Versus Real Transformer

The major assumptions of the ideal transformer discussed above are:

1. Infinite permeability of the magnetic core and no magnetic flux leakage from the core in air.
2. No coil resistance loss (through using copper wires of a very small resistance).

3. No other loss in the core called the *iron loss*. The iron losses would include *hysteresis loss* and *eddy current loss*.

Real transformers studied in Section 12.4 deviate from this ideal circuit model often very significantly. The ratio of output power to input power is called the efficiency of the transformer. For large power transformers, the efficiency can be in excess of 98 %. For RF (radio-frequency) transformers, the efficiency is typically much lower. Two methods of analysis can be used to study realistic transformers:

1. An extended equivalent circuit model that includes the present ideal transformer model plus extra inductances and resistances, see Section 12.4.
2. A different mathematical model of magnetically coupled circuits with self- and mutual inductances, see Section 12.5.

Terminology

Engineers have adopted a special terminology when dealing with transformers:

A. The ratio $N_1 : N_2$ is the *turns ratio* of the transformer. A transformer with a primary winding of 100 turns and a secondary winding of 200 turns has a turns ratio of 1:2. A transformer with a primary winding of 200 turns and a secondary winding of 150 turns has a turns ratio of 4:3.
B. When $N_2 > N_1$, the transformer increases the input AC voltage; it is called a *step-up transformer.*
C. When $N_2 < N_1$, the transformer decreases the input AC voltage; it is called a *step-down transformer.*
D. The winding with a higher number of turns is the *high-voltage* (HV) *side* of the transformer.
E. The winding with a smaller number of turns is the *low-voltage* (HV) *side* of the transformer.

In Figs. 12.1, 12.2, and 12.3, we have used the opposite *current reference directions* for the two dotted terminals. Quite often, the *same* reference directions are employed. This is to underscore the fact that either winding may serve as the input of the transformer. Sign minus should then be inserted into Eq. (12.10) which relates i_1 and i_2.

Transformer Rating

Power transformers seldom drive purely resistive loads. Therefore, their power rating is given in VA (*volt-amperes*) or kVA instead of watts, identical to the complex power defined in Section 11.1. More precisely, this is the apparent load power defined by Eq. (11.21d). Consider a popular example of a transformer that carries the following information on a nameplate or in a reference manual: 10 kVA, 1100:110 V. The voltage rating means the one transformer winding (high-voltage side) is rated for 1100 V, whereas another (low-voltage side) for 110 V. The turns ratio is the voltage ratio, $N_1 : N_2 = 10$. The corresponding current ratings are 9.09 A rms and 90.9A rms, respectively.

Exercise 12.4: A power transformer is rated as 1 kVA, 120:480 V. Determine transformer type, turns ratio, and the rated current on the low-voltage side.

Answers: This is a step-up transformer with turns ratio 1:4 and a current on the low-voltage side of 8.333 A rms.

Example 12.1: In the circuit shown in Fig. 12.4, $v_S(t) = 170\cos(2\pi\omega t)$ [V]. The ideal-transformer model is used. Determine all circuit currents and voltages when:

A. The switch is open.
B. The switch is closed.
C. The switch is closed and $\omega \to 0$.

Solution A: The current through the secondary winding is zero, so is the current through the primary. The open-circuit condition thus *passes through* the transformer. The physical background for this is an infinite permeability of the magnetic core and hence an infinite inductance of the primary coil. You should know that the infinite inductance is an open circuit for an AC signal. By KVL, the voltage of the primary winding $v_1(t)$ is equal to the source voltage. Open-circuit voltage of the secondary winding $v_2(t)$ is exactly five times smaller.

Solution B: Both winding voltages remain identical to the voltages in the previous case. The load voltage is now the voltage of the secondary winding. The load current, by Ohm's law, is $i_R = i_2 = v_R/R_L = 0.850\cos\omega t$ [A]. The primary-winding current is exactly five times smaller than the load current.

Solution C: When the frequency approaches zero, the transformer loses its functionality. The induced voltages v_1 and v_2 tend to zero. Moreover, the primary winding of the transformer becomes a short circuit, which prohibits using transformers loaded with DC power sources.

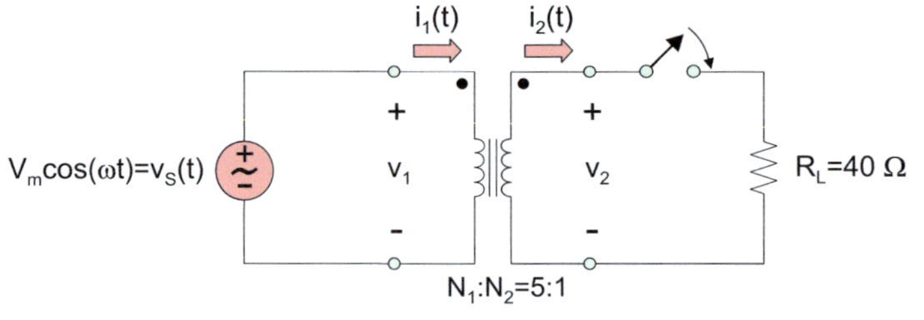

Fig. 12.4. A transformer circuit with a load and a switch.

Exercise 12.5: In the circuit shown in Fig. 12.4, the current through resistance is given by $i_2(t) = 1\cos(2\pi 60t)$ [A]. With the switch closed, find the source voltage.

Answer: $v_S(t) = 200\cos(2\pi 60t)$ [V].

12.1.7 Mechanical Analogies of a Transformer

An electric transformer operates with alternating currents. One mechanical analogy of the transformer is a gear transmission or gearbox, see Fig. 12.5a.

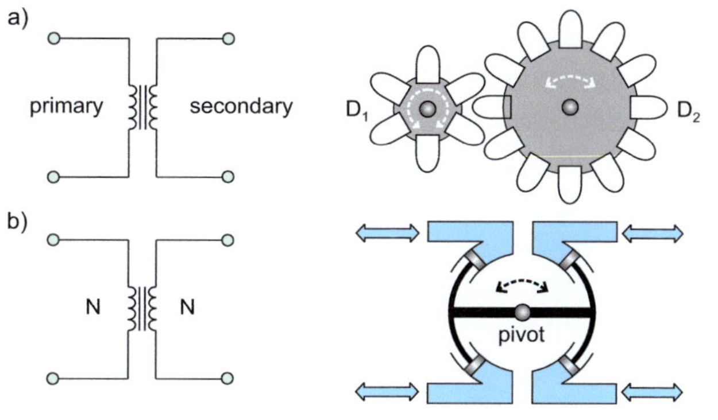

Fig. 12.5. Analogies for a transformer.

In terms of angular speed ω [rad/s] and developed torque T $[\mathrm{N}\cdot\mathrm{m}]$, one has

$$T_2 = \frac{D_2}{D_1} T_1, \quad \omega_2 = \frac{D_1}{D_2}\omega_1 \tag{12.13}$$

where $D_{1,2}$ are pitch diameters of gear wheels. Here, torque is the voltage and speed is the current. $D_{1,2}$ are similar to the number of turns, $N_{1,2}$, of the primary and secondary windings of the transformer, respectively. Emphasize that the present transformer analogy still ignores the field effect—magnetic coupling between the coils. Therefore, it will fail in the DC case. A more realistic hydraulic transformer analogy is shown in Fig. 12.5b. The model with four pistons transforms power from one circuit to another in the AC case only. It is drawn for a transformer whose turns ratio is 1:1. When, for example, a transformer with a turns ratio of 2:1 is required, the area of output pistons is doubled. This doubles the output current, but the output voltage (the force) will be halved.

Historical: The transformer action was independently invented and utilized by many researchers, starting with Michael Faraday, Joseph Henry, Nicholas Callan of Ireland, and Pavel Yablochkov of Russia. Hungarian engineer Ottó Titusz Bláthy introduced the name "transformer." A power transformer developed by Lucien Gaulard of France and John Dixon Gibbs of England was demonstrated in London in 1881 and then sold to the American company Westinghouse. In 1886 William Stanley, working for Westinghouse, built the refined, commercially used AC transformer used in the electrification of Great Barrington, Massachusetts.

Section 12.2 Analysis of Ideal Transformer Circuits

12.2.1 Circuit with a Transformer in the Phasor Form

Consider a generic transformer circuit shown in Fig. 12.6a in the frequency domain. The circuit is given in the phasor form assuming a harmonic signal source. A *source circuit* with the phasor voltage $\mathbf{V}_\mathrm{S}$ and impedance $\mathbf{Z}_\mathrm{S}$ is connected to a *load* with impedance $\mathbf{Z}_\mathrm{L}$ via an ideal transformer studied in the previous Section. Since any linear AC source network can be represented in the form of its Thévenin equivalent, and any linear passive source can be replaced by the equivalent impedance, Fig. 12.6a is a rather general interpretation of the transformer setup with linear networks in Fig. 12.4. When written in the phasor form, the ideal transformer model given by Eqs. (12.6) and (12.10) does not need a special treatment. We simply replace the real-valued voltages and currents by phasors:

$$v_2 = \frac{N_2}{N_1} v_1 \Rightarrow \mathbf{V}_2 = \frac{N_2}{N_1} \mathbf{V}_1 \tag{12.14a}$$

$$i_2 = \frac{N_1}{N_2} i_1 \Rightarrow \mathbf{I}_2 = \frac{N_1}{N_2} \mathbf{I}_1 \tag{12.14b}$$

In power electronics, phasor voltage and phasor current in Fig. 12.6a are often expressed in terms of *rms values* times the phasor (the complex exponent). This is in contrast to the previous analysis where we have used the amplitude of a sinusoidal function times the phasor. The circuit analysis remains the same, but the factor of 2 in the expressions for the power disappears. We will mention this convention *every time* when required.

12.2.2 Referred (Or Reflected) Source Network in the Secondary Side

What voltage and impedance does the load see in Fig. 12.6a? In other words, what is the Thévenin equivalent circuit of the source and the transformer combined? To answer this question, we find Thévenin equivalent voltage $\mathbf{V}_\mathrm{T}$ of the one-port network with terminals a and b in Fig. 12.6a as its open-circuit voltage. Using Eqs. (12.14a, b) and setting $\mathbf{Z}_\mathrm{L} = \infty$ in Fig. 12.6a yields

$$\mathbf{I}_2 = 0 \Rightarrow \mathbf{I}_1 = 0 \Rightarrow \mathbf{V}_\mathrm{T} \equiv \mathbf{V}_2 = \frac{N_2}{N_1} \mathbf{V}_1 = \frac{N_2}{N_1} \mathbf{V}_\mathrm{S} \tag{12.15}$$

The Thévenin equivalent impedance $\mathbf{Z}_\mathrm{T}$ is found by dividing the open-circuit voltage by the short-circuit current $\mathbf{I}_\mathrm{SC} = \mathbf{I}_2$. Setting $\mathbf{Z}_\mathrm{L} = 0$ in Fig. 12.6a, one finds the short-circuit current

$$\mathbf{Z}_\mathrm{L} = 0 \Rightarrow \mathbf{V}_2 = \mathbf{V}_1 = 0 \Rightarrow \mathbf{I}_\mathrm{SC} = \mathbf{I}_2 = \frac{N_1}{N_2} \mathbf{I}_1 = \frac{N_1}{N_2} \frac{\mathbf{V}_\mathrm{S}}{\mathbf{Z}_\mathrm{S}} \tag{12.16}$$

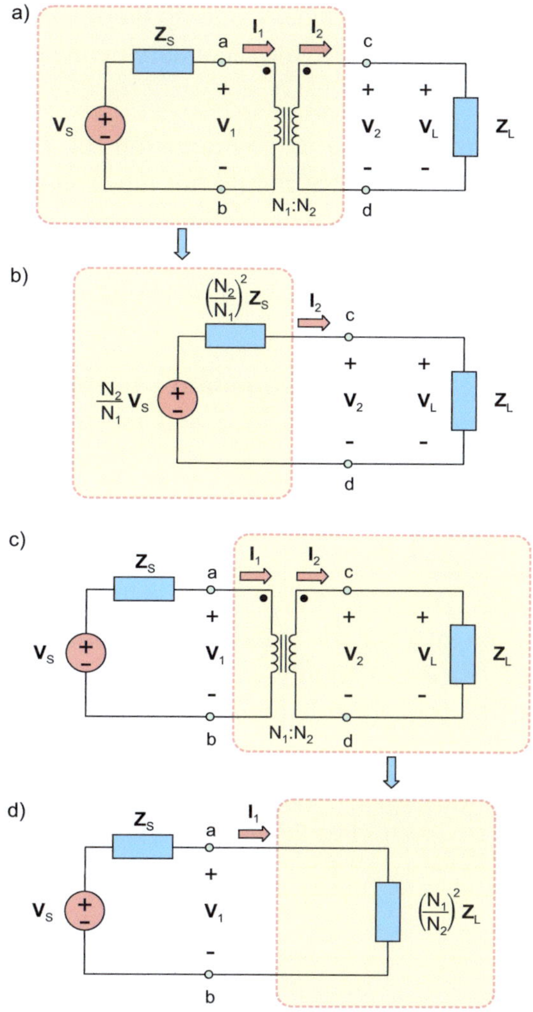

Fig. 12.6. (**a**) A source circuit and a load circuit connected to a transformer. (**b**) Equivalent circuit of the source with the transformer *when looking from the load*. (**c**) The same as (**a**). (**d**) Equivalent circuit of the load with the transformer *when looking from the source*.

In summary, the equivalent circuit of the source and the transformer combined is shown in Fig. 12.6b. It is characterized by

$$\mathbf{V}_{\mathrm{T}} = \frac{N_2}{N_1}\mathbf{V}_{\mathrm{S}}, \quad \mathbf{Z}_{\mathrm{T}} = \left(\frac{N_2}{N_1}\right)^2 \mathbf{Z}_{\mathrm{S}} \tag{12.17}$$

We conclude that, from the viewpoint of the load, the ideal transformer boosts the source voltage by a factor of N_2/N_1 and multiplies the source impedance by a factor of $(N_2/N_1)^2$. The meaning of Eq. (12.17) is commonly expressed in the following form:

1. The source is *reflected* to the secondary side by the inverse turns ratio.
2. The source impedance is *reflected* to the secondary by the square of the inverse turns ratio.

An inspective student may notice a contradiction in Eq. (12.16): as long as $\mathbf{V}_2 = \mathbf{V}_1 = 0$, the transformer apparently should not function at all. In fact, the equality $\mathbf{V}_2 = \mathbf{V}_1 = 0$ is never exactly satisfied; a (infinitesimally small) voltage across the transformer is assumed that supports its operation. Mathematically, one considers a finite but small load impedance and then obtains Eq. (12.16) in the limit $\mathbf{Z}_L \to 0$.

12.2.3 Referred (Or Reflected) Load Impedance to the Primary Side

What is the equivalent impedance of the load and the transformer combined? With reference to Fig. 12.6c, one has

$$\mathbf{Z}_T \equiv \frac{\mathbf{V}_1}{\mathbf{I}_1} = \left(\frac{N_1}{N_2}\right)^2 \frac{\mathbf{V}_2}{\mathbf{I}_2} = \left(\frac{N_1}{N_2}\right)^2 \mathbf{Z}_L \tag{12.18}$$

The equivalent circuit of the load and the transformer combined are shown in Fig. 12.6d. The meaning of Eq. (12.18) is commonly expressed in the following form: the load impedance is *reflected* to the primary side by the square of the turns ratio.

Exercise 12.6: Load impedance is 12 Ω. Find the equivalent impedance of the load combined with a 1:2 step-up transformer to the primary side.

Answer: $R_T = 3\ \Omega$.

Exercise 12.7: Source voltage is given by $v_S(t) = 1\cos\omega t$ [V] and the source impedance is 12 Ω. Find the equivalent circuit of the source combined with a 1:2 step-up transformer in the secondary side. Express your result both in frequency domain and in time domain.

Answer: $R_T = 48\ \Omega$, $\mathbf{V}_T = 2\angle 0°$ [V] or $v_T(t) = 2\cos\omega t$ [V].

Both source and load reflections are of great practical value since they *eliminate* the transformer from the circuit analysis and thereby simplify the overall circuit analysis and design. The analysis with Norton equivalent circuits is developed in a similar fashion.

Exercise 12.8: To bring out an important application, let the load be a series RLC network with resistance R, inductance L, and capacitance C. An equivalent impedance of the load combined with a $N_1 : N_2$ step-up transformer at the primary is again a series RLC network. Find the *reflected* (or "transformed") values of the resistance, inductance, and capacitance.

Answer: $R \to \left(\frac{N_1}{N_2}\right)^2 R$, $L \to \left(\frac{N_1}{N_2}\right)^2 L$, $C \to \left(\frac{N_2}{N_1}\right)^2 C$.

12.2.4 Transformer as a Matching Circuit

Matching Real-Valued Impedances

Consider a transformer circuit with real-valued impedances shown in Fig. 12.7. A *given* practical voltage source with resistance R_S is to be connected to a *given* load with resistance $R_L \neq R_S$. The principle of maximum power transfer states that there should be $R_L = R_S$ for the maximum power transfer from the source to the load. Could we still achieve the maximum available power from the source without changing the source and the load? The answer is yes; such a procedure is known as *impedance matching with a transformer*. A transformer used for this purpose is known as a *matching transformer*. The idea is to use a transformer with the turns ratio:

$$N_1 : N_2 = \sqrt{R_S / R_L} \tag{12.19}$$

To prove this fact we consider a transformer with an arbitrary turns ratio in the circuit of Fig. 12.7a. The load resistance is reflected to the primary side by the square of the turns ratio. The resulting equivalent circuit is a voltage divider. It includes the voltage source and two resistances in series. Instantaneous power delivered to the load is given by

$$p(t) = \left\{ \frac{xR_L}{(R_S + xR_L)^2} \right\} v_S^2(t), \quad x = \left(\frac{N_1}{N_2} \right)^2 \tag{12.20}$$

The expression in curly brackets is maximized with regard to *parameter x*, which may attain any positive values. Its maximum (and the maximum power) is achieved when

$$x = R_S / R_L \tag{12.21}$$

or when the reflected load resistance is exactly R_S, see Fig. 12.7b. The same circuit may be equally well solved by reflecting the source to the primary side; the related homework problems provide several examples.

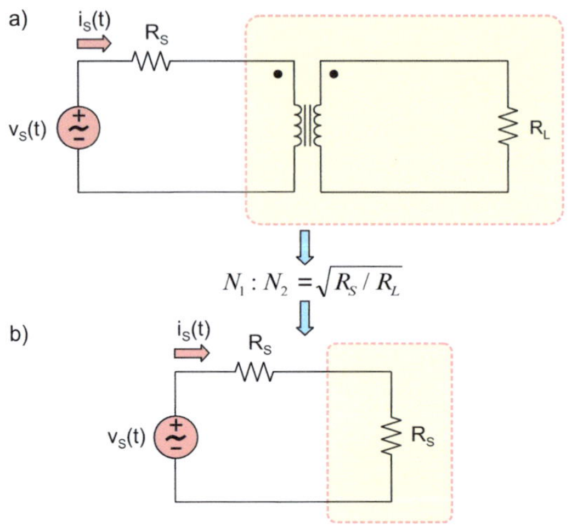

Fig. 12.7. (**a**) A matching transformer for real impedances. (**b**) Equivalent circuit of the load with the transformer *when looking from the source.*

Example 12.2: A passive RFID tag is modeled by a load resistance of 9 Ω. The tag is wirelessly powered; it is augmented with a small collecting antenna, which has a radiation resistance of 1 Ω. Find the ratio of the average received powers from the antenna with and without the matching transformer.

Solution: We use Fig. 12.7a to visualize the circuit. The antenna is the source, the tag is the load. For the received power without the transformer, one has

$$P_\mathrm{L} = \frac{1}{2}\frac{R_\mathrm{L}|\mathbf{V}_\mathrm{S}|^2}{(R_\mathrm{S}+R_\mathrm{L})^2} = 0.045|\mathbf{V}_\mathrm{S}|^2 \quad [\mathrm{W}] \tag{12.22}$$

where $\mathbf{V}_\mathrm{S}$ is a phasor voltage at the antenna terminals. With the matching transformer (1:3 turns ratio), the result has the form (after reflecting the source to the secondary side):

$$P_\mathrm{L} = \frac{1}{2}\frac{R_\mathrm{L}/9|\mathbf{V}_\mathrm{S}|^2}{(R_\mathrm{S}+R_\mathrm{L}/9)^2} = 0.125|\mathbf{V}_\mathrm{S}|^2 \quad [\mathrm{W}] \tag{12.23}$$

The ratio of the two powers is 2.78 or $10\log_{10}(2.78) = 4.4$ dB. This is an improvement of the tag performance which results in a greater reading range.

Exercise 12.9: A 4-Ω speaker is connected to an AC voltage source with the voltage amplitude of 10 V. The source's impedance is 1 Ω.

A. Determine the average power delivered to the speaker.
B. Given a 1:2 matching transformer, determine power delivered to the speaker.

Answer: A. 8 W; B. 12.5 W.

Matching Arbitrary Complex Impedances

Can a transformer match two arbitrary complex impedances? Unfortunately, it cannot. The transformer operates as an impedance multiplier; it multiplies (or divides) by a real number. On the other hand, the complex impedance match requires two complex conjugate impedances. A transformer could often provide a "better" match (see the summary to this chapter) but cannot perform impedance matching in full. Other circuit elements (capacitance or inductance) may be necessary to complete the task.

12.2.5 Application Example: Electric Power Transfer via Transformers

Circuit with a Fixed Load Voltage

Figure 12.8a shows a circuit for transmitting electric power over a long transmission line with the total resistance R and the total inductance L. The circuit in Fig. 12.8a is converted to a phasor form first. We consider the phasors in terms of *rms values*. Two competing schemes are studied: transmission without transformers (see Fig. 12.8b) and a transmission scheme with two 1:20 and 20:1 ideal transformers—see Fig. 12.8c. In order to compare the performance of two circuits (with and without transformers), it is assumed that the load phasor voltage $\mathbf{V}_{\mathrm{L}}$ (V rms) and the load phasor current $\mathbf{I}_{\mathrm{L}}$ (V rms) have the *same* values in both cases. This guarantees us the same average power delivered to the load. The power loss (ohmic loss) in the line resistance in Fig. 12.8b is

$$P_{\mathrm{loss}} = R|\mathbf{I}_{\mathrm{L}}|^2 \tag{12.24}$$

The power loss in the line resistance in Fig. 12.8c is decreased by a factor of 400:

$$P_{\mathrm{loss}} = \frac{1}{400} R|\mathbf{I}_{\mathrm{L}}|^2 \tag{12.25}$$

since the line current is exactly 20 times less than in the first case. This result is independent of the particular values of R and L. Simultaneously, the line voltages increase by a factor of 20, but the load voltage still remains the same due to the step-down transformer. Thus, using a pair of transformers allows us to choose an economically optimum voltage for transmitting a given amount of power. The line sees a high voltage of the secondary of the first transformer while the load essentially sees the source voltage. Not only does the use of transformers greatly decrease the line loss, but it also potentially allows us to use *smaller* source voltages (*sending-end voltages*). The required source power also decreases.

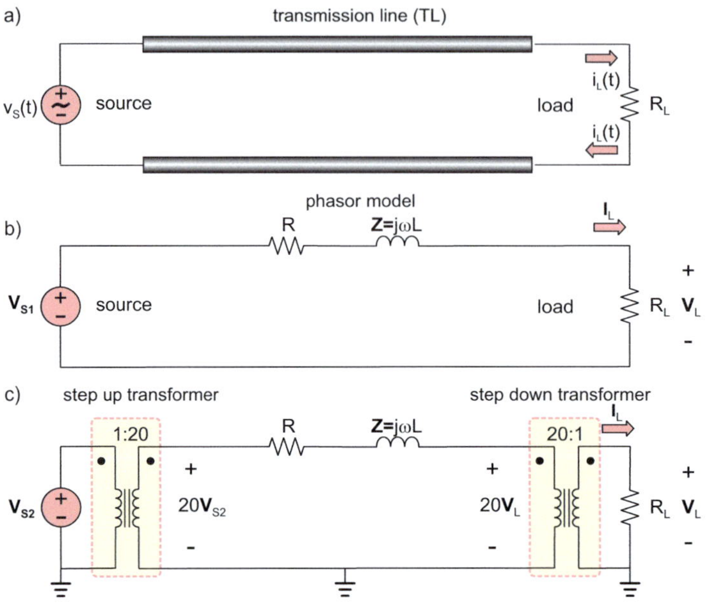

Fig. 12.8. (**a**) Schematic of a transmission line. (**b**) Transmission line without a transformer (phasor form). (**c**) Transmission line with step-up and step-down transformers (phasor form).

Circuit with a Fixed Source Voltage

Next, the performance of the circuit in Fig. 12.9 with a fixed source voltage is to be analyzed, with and without 1:10 step-up transformer and 10:1 step-down transformer, respectively. All phasors are given in terms of the rms values. We are interested in source current $\mathbf{I}_\mathrm{S}$ and active source power:

$$P_\mathrm{S} = \mathrm{Re}\left(\mathbf{V}_\mathrm{S} \cdot \mathbf{I}_\mathrm{S}^*\right) \tag{12.26}$$

We are also interested in load parameters $\mathbf{V}_\mathrm{L}, \mathbf{I}_\mathrm{L}$ and active load power:

$$P_\mathrm{L} = \mathrm{Re}\left(\mathbf{V}_\mathrm{L} \cdot \mathbf{I}_\mathrm{L}^*\right) \tag{12.27}$$

and, finally, in line power loss:

$$P_\mathrm{loss} = R|\mathbf{I}_\mathrm{line}|^2 \tag{12.28}$$

The circuit in Fig. 12.9a without the transformers is solved by finding the equivalent impedance and then the source current. All circuit currents will coincide, $\mathbf{I}_\mathrm{S} = \mathbf{I}_\mathrm{line} = \mathbf{I}_\mathrm{L}$. The circuit with the transformers is solved by reflecting the load impedance to the primary side twice, see Fig. 12.9b and c, respectively.

The simplified circuit in Fig. 12.9c allows us to find the source current. Then, we return to the circuit in Fig. 12.9b and find the line current following the ideal transformer model Eq. (12.14b). Finally, we return to the circuit in Fig. 12.9a and restore the load current

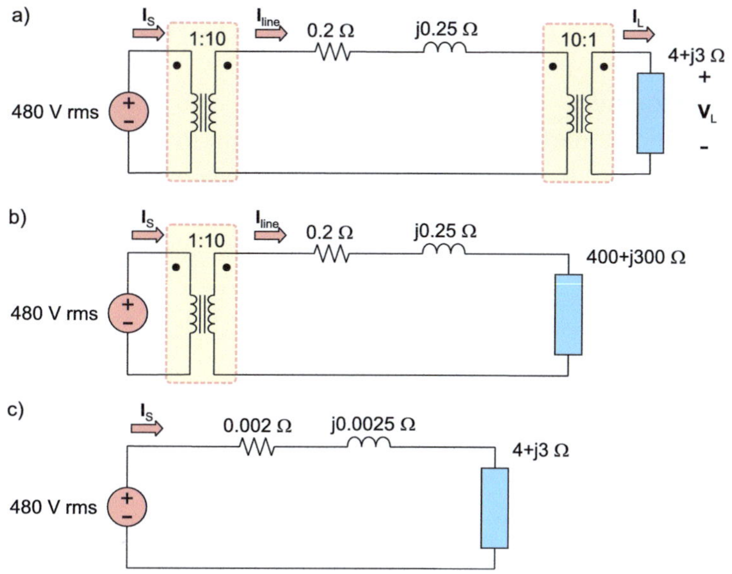

Fig. 12.9. Analysis of a particular power-transfer circuit.

from the line current. Table 12.1 summarizes the corresponding numerical values. It illustrates the reduction in the line loss, which is again approximately proportional to the *square* of the turns ratio.

Table 12.1. Circuit parameters (V rms, A rms, and W) for the power-transfer circuit in Fig. 12.9.

Circuit	$\mathbf{V}_S$	$\mathbf{I}_S$	P_S	$\mathbf{I}_{line}$	P_{loss}	$\mathbf{V}_L$	$\mathbf{I}_L$	P_L
No tr.	$480\angle 0°$	$90\angle -38°$	34,312	$90\angle -38°$	1634	$452\angle -1°$	$90\angle -38°$	32,678
W tr.	$480\angle 0°$	$96\angle -37°$	36,837	$9.6\angle -37°$	18.4	$480\angle 0°$	$96\angle -37°$	36,818

Historical: The use of AC in favor of DC is based on the efficiency of distributing electrical power. In DC systems, heavy conductors must be used to deliver electrical power over long distances. However, in AC systems, transformers boost voltages to high levels so that electricity can be distributed across country on relatively thin wires (high voltage, low current), and then converted back to low voltage, high current where needed. This is not possible with DC.

In the late 1880s, USA experienced what many citizens termed the "Battle of the Currents" where George Westinghouse and Nikola Tesla prevailed with their AC power distribution system over world-famous Thomas Edison who advocated DC power and tried to discredit AC by saying it was more dangerous. To sway the public opinion, Edison went to the extreme of secretly acquiring an AC generator and designing the first electric chair with it. The first public execution of a prisoner in this electric chair was William Kemmler in 1890 using 2000V AC. The electrocution required two attempts. This incident supposedly prompted Westinghouse to state that an axe would have been a more humane execution.

Section 12.3 Some Useful Transformers

12.3.1 Autotransformer

A transformer in which a part of the winding is common to both primary and secondary circuits—see Fig. 12.10—is known as an *autotransformer*. This common part may be controlled by a sliding contact as shown in the same figure.

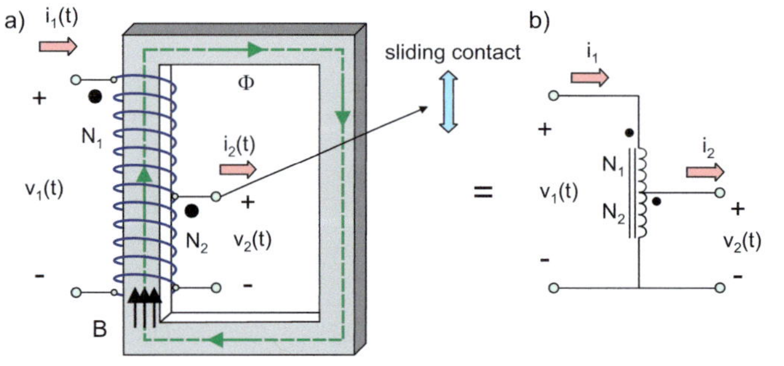

Fig. 12.10. Step-down autotransformer along with the corresponding circuit symbol. Circuit symbols for the autotransformer may vary.

An autotransformer is typically a step-down transformer. Autotransformers are used widely and come in many sizes. The autotransformer is also known as a *Variac* (*variable AC transformer*). Such autotransformers are used in the laboratory to obtain a variable output voltage by means of a movable contact in Fig. 12.10 that changes N_2 and consequently the turns ratio $N_1 : N_2$. An autotransformer will again be analyzed using the ideal transformer model. Primary winding N_1 and secondary winding N_2 in Fig. 12.10 share the same flux $\Phi(t)$. Both induced voltages $v_1(t)$ and $v_2(t)$ in Fig. 12.10 obey Faraday's law given by Eqs. (12.2) and (12.5), respectively. One has

$$v_1(t) = (N_1 + N_2)\frac{d\Phi}{dt}, \qquad v_2(t) = N_2\frac{d\Phi(t)}{dt} \tag{12.29}$$

Therefore,

$$v_2 = \frac{N_2}{N_1 + N_2} v_1 \tag{12.30}$$

Using Ampere's law for the ideal magnetic core Eq. (12.8c), we obtain an expression that is very similar to Eq. (12.10) of Section 12.1,

$$(N_1 + N_2)i_1 - N_2 i_2 = 0 \Rightarrow i_2 = \frac{N_1 + N_2}{N_2} i_1 \tag{12.31}$$

Thus, the model of the ideal autotransformer is indeed indistinguishable from the ideal transformer model. Advantages of the autotransformer are generally lower losses and variable output voltage. However, there is now a direct electric connection between the primary and secondary sides, which potentially enables direct current flow.

Exercise 12.9: A voltage source connected to the primary winding of an autotransformer has the form $v_S = 10\cos\omega t + 1$ [V]. Ohmic resistance of the primary winding is 1 Ω. Magnetic losses are ignored. Turns ratio established by a sliding contact is given by $N_1 + N_2 : N_2 = 5$. Determine the voltage in the secondary.

Answer: Using the superposition theorem for DC and AC voltages, one has $v_S = 2\cos\omega t + 0.2$ [V].

12.3.2 Multiwinding Transformer

A *multiwinding transformer* shown in Fig. 12.11 is one in which two or more secondary windings are placed on the same core.

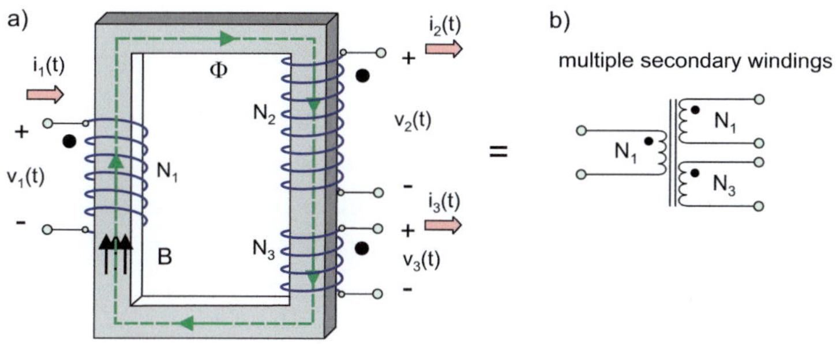

Fig. 12.11. A multiwinding transformer along with the corresponding circuit symbol. Circuit symbols for the multiwinding transformer may vary.

For example, a three-winding transformer shown in Fig. 12.11 has a primary winding N_1 and two secondary windings, N_2 and N_3, respectively. It is useful when lossless power division is required. The multiwinding transformer is also analyzed using the ideal transformer model. All induced voltages $v_1(t)$, $v_2(t)$, and $v_3(t)$ in Fig. 12.11 obey Faraday's law of induction. Therefore,

$$\frac{v_1}{N_1} = \frac{v_2}{N_2} = \frac{v_3}{N_3} \tag{12.32}$$

Using Ampere's law for the ideal magnetic core Eq. (12.8c), we obtain

$$N_1 i_1 - N_2 i_2 - N_3 i_3 = 0 \tag{12.33}$$

Example 12.3: In the circuit shown in Fig. 12.11, $N_1 = N_2 = N_3$. How is the instantaneous power partitioned between the two secondary windings?

Solution: According to Eqs. (12.32) and (12.33),

$$v_1 = v_2 = v_3, \quad i_1 = i_2 + i_3 \tag{12.34}$$

If two secondary windings are terminated into the same load resistances, then $i_2 = i_3 = 0.5i_1$. Therefore, the input power is divided equally. However, if one of the loads is an open circuit, all the input power is transferred to another load.

A multiwinding transformer may be used, for example, in *a telephone hybrid circuit* which is designed to convert a two-wire interface to a four-wire interface. A telephone hybrid is the circuit which separates the transmitted and received audio which are initially sent both at the same wire pair in two-wire normal telephone interface.

12.3.3 Center-Tapped Transformer: Single-Ended to Differential Transformation

Figure 12.12 shows a particular, yet widely used modification of the multiwinding transformer in Fig. 12.11. Two equal secondary windings, with $N_2/2$ turns each, are connected to each other and then to ground. The center tap of the secondary coil may be grounded. The output voltages of the two secondary windings versus the dotted terminals are still equal to each other and equal to voltage v_2. However, since the center tap is grounded, the absolute voltages at the output to the transformer are

$$+v_2(t), \ 0 \text{ V}, \text{ and } -v_2(t) \quad \text{where} \quad v_2 = \frac{N_2}{2N_1}v_1 \tag{12.35a}$$

versus ground. The output voltage is thus identical to a dual-polarity AC voltage source. More generally, this design is intended to convert a *single-ended* (*two-wire*) voltage signal v_1, i.e., a signal that is initially referenced to ground, to a balanced *differential* (*three-wire*) voltage signal with equal voltages with respect to ground but of opposite polarity.

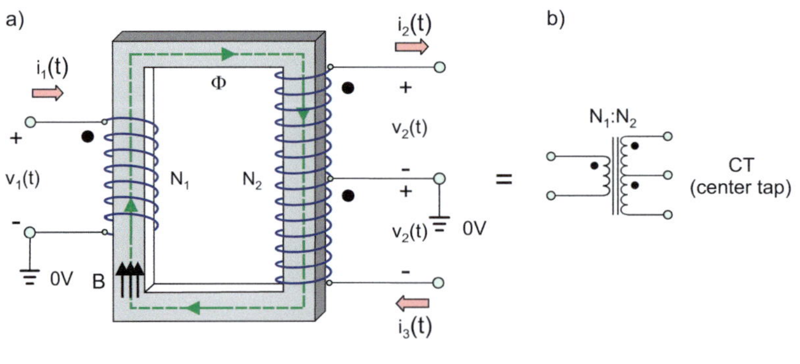

Fig. 12.12. Center-tapped transformer intended for single-ended to differential signal transformation (differential transformer) and the corresponding circuit symbol.

The differential signal has significant advantages; it can be used in different applications including power electronics, instrumentation, communication circuits, and high-speed digital circuits. If, and only if, the center-tapped transformer is connected to the two *identical* loads, the center tap carries *zero* current and with reference to Fig. 12.12,

$$i_2(t) = i_3(t) \quad \text{where} \quad i_1 = \frac{N_2}{N_1} i_2 \tag{12.35b}$$

The three-wire model described by Eq. (12.35a) is often called a *three-wire single-phase system*. Note that *all* of the AC power that is supplied to your house by a residential AC distribution network is a 120/240 V rms three-wire single-phase system, see Fig. 11.13.

Example 12.4: Common home appliances use a single-phase 120-V rms AC line (a refrigerator, TV, microwave, etc.) and a 240-V rms three-wire differential AC line (electric range, dryer, etc.). A house located in a remote area is powered from a local micro hydropower generator located a couple of miles away. Suggest a way to power all appliances when the input power is a not a residential AC distribution network, but a single-phase two-wire 240 V rms AC line from the generator.

Solution: A solution is shown in Fig. 12.13. The center-tapped transformer is used to accomplish the task. All loads are to be connected in parallel.

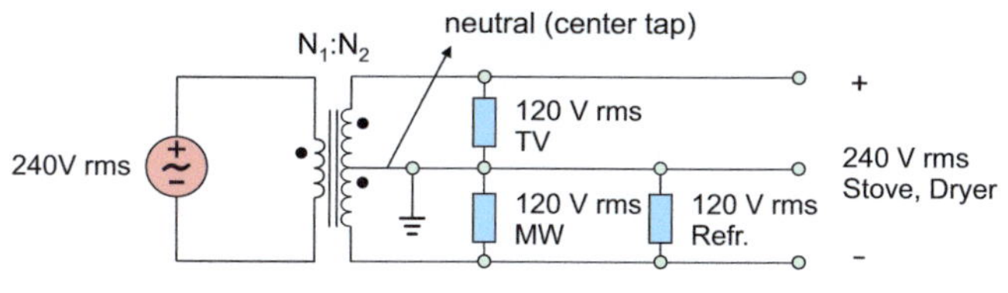

Fig. 12.13. Powering home appliances in remote areas using a single-phase two-wire 240 V rms AC line as an input. Note that the dot locations are less important for this case.

Example 12.5: A dipole antenna with two wings is to be fed as follows: one wing is fed with a voltage signal $+v(t)$ versus ground plane and another wing is fed with a voltage signal $-v_1(t)$ versus ground plane (180° phase shift). At the same time, the input from a transmitter is a single-phase coaxial line. Suggest a way of connecting the antenna to a transmitter.

Solution: A solution is shown in Fig. 12.14. The center-tapped transformer from Fig. 12.12 is again used to accomplish the task. From the viewpoint of an RF circuit design, the transformer in Fig. 12.14 is sometimes called a *balun transformer*; the word *balun* is an acronym for balanced-to-unbalanced converter. These balun structures are employed in such common RF and microwave components as mixers, antenna-feed networks, and frequency multipliers.

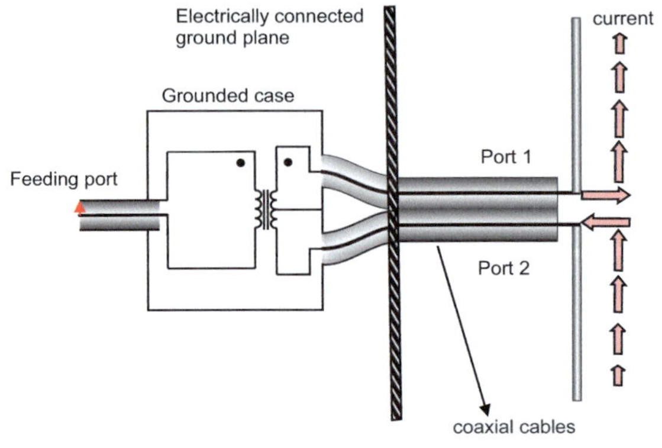

Fig. 12.14. A balun (center-tapped) transformer attached to a dipole antenna. The metal case of the transformer is grounded.

Note that the transformer in Fig. 12.12 is simultaneously a *power divider*, namely, the so-called 180° power divider or 180° power splitter. It is amazing to see how many different applications rely on the "good old" transformer.

12.3.4 Current Transformer

There is a device that has been around for a long time and which allows the engineer to monitor the electrical power system. It is called *the current transformer* and is shown in Fig. 12.15. The current transformer is a main measuring tool to determine current flow in a power system. The corresponding hardware device is known as a *clamp on ammeter.*

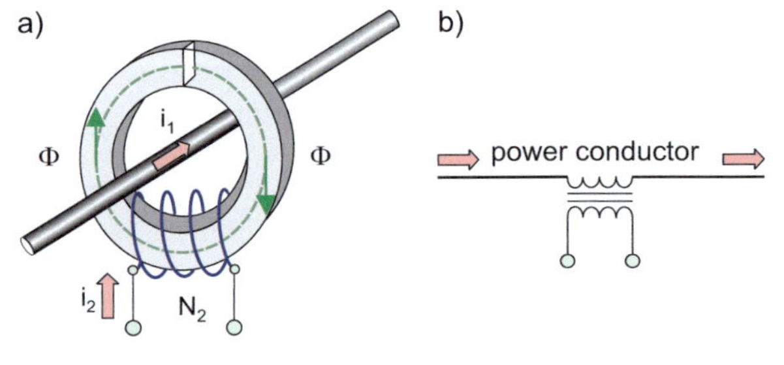

Fig. 12.15. A current transformer and the corresponding circuit symbol. Circuit symbols for the current transformer may vary.

Consider a straight conductor in Fig. 12.15 that carries current i_1 to be measured. This conductor effectively forms one turn of the primary coil. The secondary coil has N_2 turns. We again assume the ideal transformer model and apply Ampere's law for the ideal magnetic core Eq. (12.8c). The current through the secondary coil then follows the equality (note current directions in Fig. 12.15)

$$i_1 - N_2 i_2 = 0 \Rightarrow i_2 = \frac{1}{N_2} i_1 \tag{12.36}$$

which is a particular case of the ideal transformer equation with $N_1 = 1$. Thus, by measuring current i_2 or the associated voltage, the unknown current i_1 can be established. Note that Fig. 12.15 and Eq. (12.36) only demonstrate the very basic concept; the practical current transformer design is significantly more complicated. Another type of instrumentation transformers—*potential transformers*—is used for accurate AC voltage measurements.

Section 12.4 Real-Transformer Model

12.4.1 Model of a Nonideal Low-Frequency Transformer

Figure 12.16 shows a linear circuit model of a practical low-frequency (60 or 50 Hz) two-winding transformer. This circuit is also known as the *Steinmetz model* and its parameters as *Steinmetz parameters*. The complete model will be explained in several steps:

1. Consider the open-circuited transformer first. The primary winding is characterized by a large but finite magnetizing inductance, L_m, which was defined in Eq. (12.3). This is the standard inductance expression for a long solenoid with the magnetic core. It is shown in Fig. 12.16a.
2. A small amount of magnetic flux is still situated outside the core so that Eq. (12.3) needs to be refined. The total inductance as in Fig. 12.16 of the primary winding is therefore somewhat larger:

$$L_1 = L_m + L_{l1} \tag{12.36}$$

   where a small addition L_{l1} is called the *leakage inductance*, see Fig. 12.16a.
3. A practical primary winding has a certain ohmic resistance, R_1, which is placed in series with L_1. Simultaneously, the core loss (hysteresis and eddy current loss) in the magnetic material consumes some extra current. It is modeled by an equivalent resistance R_c, which is placed in parallel with L_m. R_c is often called the *core loss resistance*. The resulting equivalent circuit in Fig. 12.16b is also an equivalent circuit of a *nonideal inductor with the magnetic core.*
4. Finally, the secondary winding is added and a load is connected to the transformer. This results in the complete equivalent circuit model of Fig. 12.16c. Two new circuit parameters L_{l2}, and R_2 are the leakage resistance and the ohmic resistance, respectively, of the secondary winding.

In general, all model parameters, and especially L_m and R_c, are frequency dependent.

12.4.2 Model Parameters and Their Extraction

Table 12.2 lists typical equivalent circuit values for three distinct power transformers of different power ratings and compares them with the ideal transformer model. Not the inductances themselves but the rather reactance values are given at angular frequency $\omega = 2\pi f$ where $f = 50$ or 60 Hz.

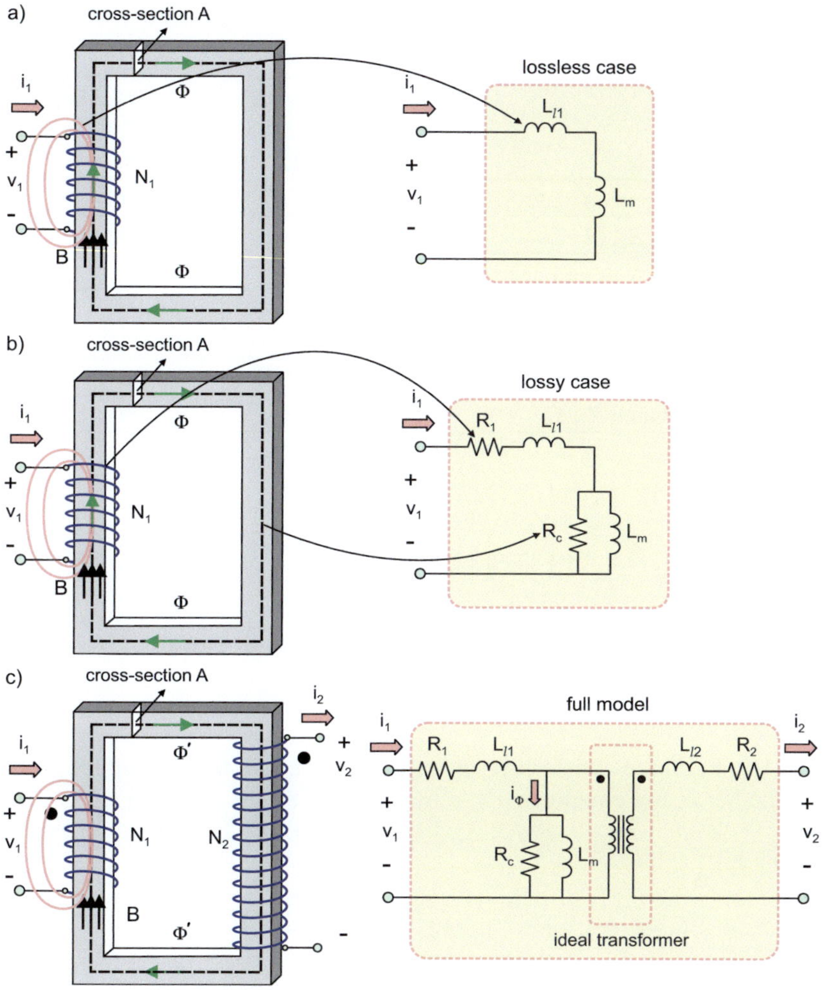

Fig. 12.16. Circuit model of a low-frequency transformer—the Steinmetz model.

It follows from Table 12.2 that the real-transformer data is reasonably well approximated by the ideal model. In particular, we may observe that

$$R_{1,2} \ll R_c, \quad X_{l1,2} \ll X_m \tag{12.37}$$

in all three cases considered. In a *well-designed transformer*, the following equalities should be satisfied:

Table. 12.2. Equivalent circuit values for three different practical power transformers as compared to the ideal transformer model.

Element nameplate	Ideal model	#1—2 kVA 230:115 V 50 Hz	#2—10 kVA 2300:230 V 60 Hz	#3—100 kVA 11,000:2200 V 60 Hz
Magnetizing reactance $X_m = \omega L_m$, Ω	∞	1437.5	69,400	57,300
Core loss resistance, R_c, Ω	∞	294.2	75,600	124,000
Primary leakage reactance $X_{l1} = \omega L_{l1}$, Ω	0	0.430	12	31.2
Secondary leakage reactance, $X_{l2} = \omega L_{l2}$, Ω	0	0.006	0.12	1.25
Primary ohmic resistance R_1, Ω	0	0.428	5.80	6.1
Secondary ohmic resistance R_2, Ω	0	0.123	0.0605	0.29

$$R_1 \approx \left(\frac{N_1}{N_2}\right)^2 R_2 = R_2', \quad X_{l1} \approx \left(\frac{N_1}{N_2}\right)^2 X_{l2} = X_2', \tag{12.38}$$

along with Eq. (12.37). Here, the prime denotes circuit parameters of the secondary winding *referred to the primary side*.

Exercise 12.10: For three transformers from Table 12.2, find the corresponding turns ratio and prove whether or not equalities given by Eq. (12.38) are approximately satisfied.

Answer:

- Model #1—turns ratio 2:1, first Eq. (12.38) is satisfied, but the second is not.
- Model #2—turns ratio 10:1, both Eq. (12.38) are satisfied.
- Model #3—turns ratio 5:1, both Eq. (12.38) are satisfied.

If the complete geometry and design characteristics of a transformer are available, all parameters of the equivalent circuit model can be calculated theoretically. Most complicated is the calculation of leakage resistances. At the same time, these parameters are directly and more easily found using measurements. *Open-circuit transformer test* and *short-circuit transformer test* (and often *the DC bridge test*) are performed to find all parameters of interest.

12.4.3 Analysis of Nonideal Transformer Model

When a harmonic voltage source is applied to a nonideal transformer and a linear load is connected to it, the corresponding equivalent circuit follows Fig. 12.17a. Although only a resistive load is indicated, the same circuit applies to an arbitrary RLC linear load.

The corresponding circuit in frequency domain is shown in Fig. 12.17b. Despite its apparently complicated nature, the circuit analysis is quite straightforward.

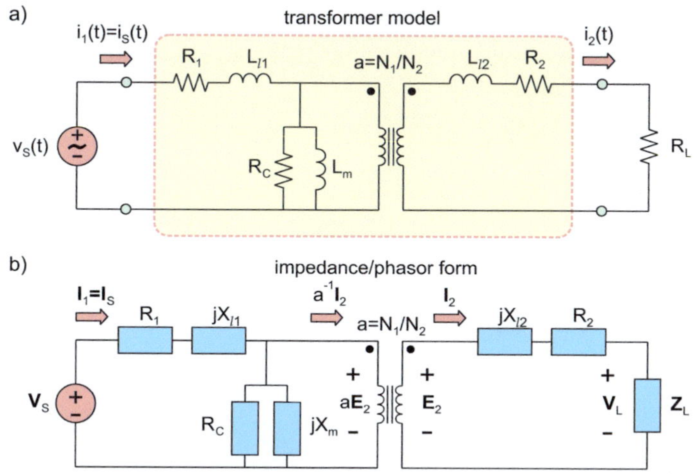

Fig. 12.17. Equivalent circuits for real transformers with harmonic source and linear load.

Terminology

1. In power electronics, phasor voltage and phasor current in Fig. 12.17b are expressed in terms of *rms values* times the phasor (the complex exponent). This is in contrast to the previously analyzed circuits where we have mostly used the amplitude of the sinusoidal function times the phasor. The analysis remains the same, but the factor of 2 in the expressions for the power disappears. For example, the input (source) and output (load) active powers are given by

$$P_{\mathrm{L}} = \mathrm{Re}\left(\mathbf{V}_{\mathrm{L}} \cdot \mathbf{I}_{\mathrm{L}}^{*}\right), \quad P_{\mathrm{S}} = \mathrm{Re}\left(\mathbf{V}_{\mathrm{S}} \cdot \mathbf{I}_{\mathrm{S}}^{*}\right) \tag{12.39}$$

2. The *rated load* is that which has the rms voltage corresponding to the transformer rating. For example, the rated load for transformer #1 in Table 12.2 has

$$\mathbf{V}_{\mathrm{L}} = 115\angle 0^{\circ} \;\; [\mathrm{V\ rms}] \tag{12.40}$$

It is customary to choose the phase of the load voltage equal to zero.

3. An impedance load has a power factor *PF*, lagging or leading. It means that the rated-load current is defined by

$$\mathbf{I}_{\mathrm{L}} = \frac{S}{\mathbf{V}_{\mathrm{L}}} \angle \mp \arccos(PF) \quad [\mathrm{A\ rms}] \tag{12.41}$$

where S is transformer power rating (the apparent load power). For example, for transformer #1 in Table 12.2, $S = 2$ kVA. In Eq. (12.41), the minus sign corresponds to the lagging power factor and the plus sign to the leading power factor, see Section 11.1.

Example 12.6: In the circuit in Fig. 12.17b, the load voltage $\mathbf{V}_L$ and the load current $\mathbf{I}_L$ are both known. Determine the source phasor voltage $\mathbf{V}_S$ and the source phasor current $\mathbf{I}_S$ in the general form. Use the notation $a = N_1/N_2$.

Solution: Applying KVL to the circuit connected to the secondary winding of the ideal transformer yields

$$\mathbf{E}_2 = \mathbf{V}_L + (R_2 + jX_{l2})\mathbf{I}_L \tag{12.42}$$

Next, we apply KCL to the circuit connected to the primary winding of the ideal transformer. This gives ($\mathbf{I}_2 = \mathbf{I}_L$)

$$\mathbf{I}_S = \mathbf{I}_1 = \frac{\mathbf{I}_L}{a} + \left(\frac{1}{R_C} + \frac{1}{jX_m}\right)a\mathbf{E}_2 \tag{12.43}$$

Finally, KVL for the primary circuit branch leads to

$$\mathbf{V}_S = a\mathbf{E}_2 + (R_1 + jX_{l1})\mathbf{I}_S \tag{12.44}$$

Example 12.7: Given the rated load with the power factor of 1, determine the source phasor voltage $\mathbf{V}_S$ and the source phasor current $\mathbf{I}_S$ for transformer #1 in Table. 12.2.

Solution: For the rated load, $\mathbf{V}_L = 115\angle 0°$ [V rms]. The load phasor current is found from Eq. (12.41), $\mathbf{I}_L = 17.39\angle 0°$ [A rms]. Further, we use the solution of the previous example with $a = 2$, plug in the model parameters from Table 12.2, and obtain

$$\mathbf{V}_S = 238.5\angle + 1.0°, \quad \mathbf{I}_S = 9.49\angle - 1.0° \tag{12.45}$$

Emphasize that the source voltage of 238.5 V rms is *higher* than the rated or nameplate value of 230 V rms. The additional potential of 8.5 V is needed to *overcome* the finite impedance of the non-ideal transformer. Indeed, for the ideal-transformer model, the agreement would be perfect.

Exercise 12.11: Repeat the previous example for transformer #2 and transformer #3 from Table. 12.2.

Answer:

- Model #2—$\mathbf{V}_S = 2354\angle + 2.5°$, $\mathbf{I}_S = 4.38\angle - 0.4°$.
- Model #3—$\mathbf{V}_S = 11142\angle + 2.9°$, $\mathbf{I}_S = 9.19\angle - 1.2°$.

Exercise 12.12: Repeat the previous example for every transformer model from Table 12.2 given the rated load with the power factor of 0.9 lagging (power angle of 25.8°). Does the deviation from the rated input voltage increase?

Answer:

- Model #1—$\mathbf{V}_S = 239.3\angle+0.1°$, $\mathbf{I}_S = 9.49\angle-24.7°$.
- Model #2—$\mathbf{V}_S = 2394\angle+1.7°$, $\mathbf{I}_S = 4.39\angle-26.1°$.
- Model #3—$\mathbf{V}_S = 11373\angle+2.3°$, $\mathbf{I}_S = 9.26\angle-26.7°$.

In all three cases, the deviation from the rated input voltage slightly increases when compared to the case of the rated resistive load.

12.4.4 Voltage Regulation and Transformer Efficiency

Voltage Regulation

Since the transformer model is nonideal, the voltage delivered to the load does depend on the load current. This means, for example, that the voltage delivered to lights will somewhat decrease when another parallel high-current load (a dryer) is turned on. This undesired effect is characterized by the *regulation of a transformer* (or *percentage regulation*):

$$\text{Regulation} = \frac{|\mathbf{V}_{\text{no-load}}| - |\mathbf{V}_L|}{|\mathbf{V}_L|} \times 100\,\% \tag{12.46}$$

Here, $|\mathbf{V}_{\text{no-load}}|$ is the rms voltage for the open-circuited load, and $|\mathbf{V}_L|$ is the rms voltage of the rated load. The regulation is determined by calculating $\mathbf{V}_S$ for the rated load and then using *the same* $\mathbf{V}_S$ to find the voltage across the open-circuited load. With reference to Fig. 12.17b, a simple yet accurate approximation for the no-load voltage is given by

$$\mathbf{V}_{\text{no-load}} \approx \frac{\mathbf{V}_S}{a} \tag{12.47}$$

To prove this fact we observe that $\mathbf{I}_2 = 0$ for the open-circuited load. Furthermore, almost all of $\mathbf{V}_S$ is applied to the primary.

Exercise 12.13: Given the rated load with the power factor of 1.0, determine the percentage regulation for three transformer models in Table. 12.2.

Answer:

- Model #1—Regulation = 3.8 %.
- Model #2—Regulation = 2.4 %.
- Model #3—Regulation = 1.3 %.

Exercise 12.14: Repeat the previous exercise for the power factor of 0.9 lagging.

Answer:

- Model #1—Regulation = 4.1 %.
- Model #2—Regulation = 4.1 %.
- Model #3—Regulation = 3.4 %.

Transformer Efficiency

The transformer equivalent circuit in Fig. 12.17 includes parasitic resistances and thus implies power loss in the transformer itself. This power loss is characterized by *transformer power efficiency* (*percentage efficiency*), η, which is defined by

$$\eta = \frac{P_L}{P_S} \times 100\ \% \tag{12.48}$$

where the input active power P_S and the output active power P_L are given by Eq. (12.39). In well-designed power transformers, the efficiency may approach 99 %.

Exercise 12.15: Given the rated load with the power factor of 1.0, determine the percentage efficiency for three transformer models in Table. 12.2.

Answer:

- Model #1—$\eta = 88.4$ %.
- Model #2—$\eta = 97.1$ %.
- Model #3—$\eta = 97.9$ %.

Exercise 12.16: Repeat the previous exercise for the power factor of 0.9 lagging.

Answer:

- Model #1—$\eta = 87.3$ %.
- Model #2—$\eta = 96.8$ %.
- Model #3—$\eta = 97.7$ %.

12.4.5 About High-Frequency Transformer Model

At medium (in the audio range) and higher frequencies, the effect of *winding capacitances* and other capacitances shown in Fig. 12.18 becomes important. When frequency increases, the transformer, as an RLC circuit, may ultimately exhibit a *resonance* that will cause the effective parameter $a = N_1/N_2$ to change. A number of capacitances may be added to the equivalent circuit model to model this effect.

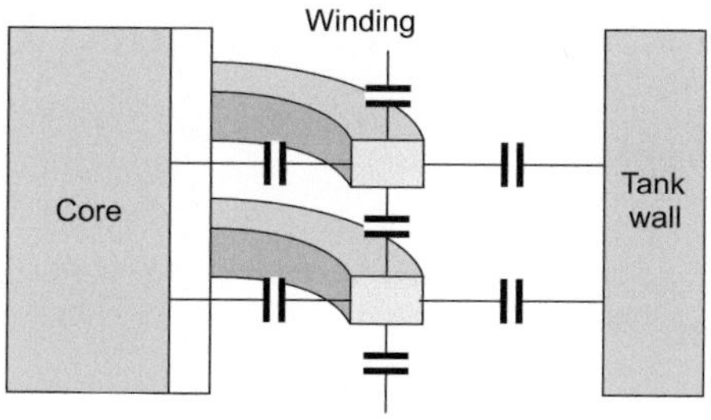

Fig. 12.18. Internal winding capacitances of a realistic high-frequency transformer.

Section 12.5 Model of Coupled Inductors

12.5.1 Model of Two Coupled Inductors

Definitions

Now it is time to examine a *model of two coupled inductors* L_1 and L_2. This model is perhaps more general than the transformer model. What does *coupled* mean? For simplicity, we start with a "broken" transformer model with a large air gap shown in Fig. 12.19. Primary and secondary windings have currents i_1 and i_2, respectively. However, in contrast to the previous study, *both* currents will now enter the dotted terminals. This means that the primary winding, inductor #1, as well as the secondary winding, inductor #2, will both satisfy the passive v-i reference configuration for the inductance. Therefore, when the air gap is very large, we will have the familiar inductor laws for both voltages v_1 and v_2 in Fig. 12.19, respectively:

$$v_1 = +L_1 \frac{di_1}{dt}, \quad v_2 = +L_2 \frac{di_2}{dt} \tag{12.49}$$

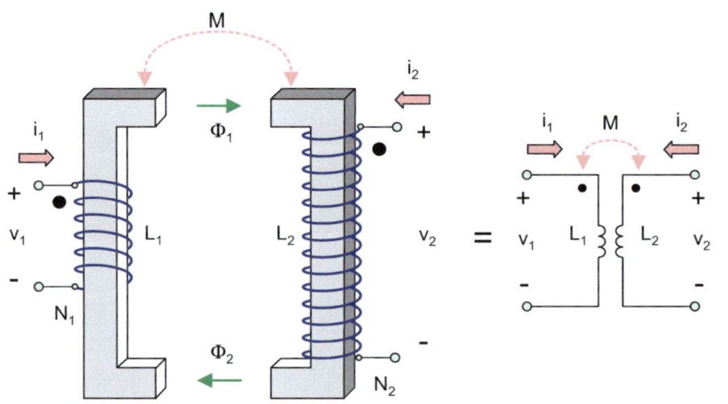

Fig. 12.19. Two coupled inductors. Note that we no longer use the transformer symbol. Also note the passive reference configuration for each inductance.

When the air gap is not very large, inductor #1 will be subject to a certain time-varying magnetic flux $\Phi_2(t)$ created by current $i_2(t)$ in the second inductor. According to Faraday's law, voltage $v_1(t)$ will therefore additionally depend on current $i_2(t)$, more precisely, on its time derivative. Quite similarly, inductor #2 will be subject to a certain time-varying magnetic flux $\Phi_1(t)$ created by current $i_1(t)$ in the first inductor. According to Faraday's law, voltage $v_1(t)$ will also depend on time derivative of current $i_1(t)$. As a result, instead of the inductance laws given by Eq. (12.49) for two uncoupled (literally noninteracting) inductors, we have the *model*

$$v_1 = +L_1 \frac{di_1}{dt} + M \frac{di_2}{dt}$$
$$v_2 = +M \frac{di_1}{dt} + L_2 \frac{di_2}{dt} \tag{12.50}$$

of two coupled inductors. Here the coefficient $M > 0$ is the *mutual inductance* of two coils, which is also measured in henries. You can see from the second expression in Eqs. (12.50) that the mutual inductance determines the voltage induced in inductor #2 due to changes of the electric current in inductor #1. Alternatively, from the first expression in Eqs. (12.50), the same mutual inductance determines the voltage induced in inductor #1 due to changes of the electric current in inductor #2. The signs in Eq. (12.50) are important. They follow a few rules:

1. Eqs. (12.50) corresponds to the dot convention and voltage polarities/current directions shown in Fig. 12.19.
2. If one of the current reference directions, say the direction of i_2, is selected oppositely, we will have to use the minus sign in the respective terms in Eqs. (12.50).
3. If one of the voltage polarities changes, we have will have to use the minus sign where required in Eqs. (12.50).

Figure 12.19 also shows the *circuit symbol for two coupled inductors*. This symbol does not include the magnetic core; event if it is present in reality. Also, the mutual inductance is shown by an arrow.

12.5.2 Analysis of Circuits with Coupled Inductors

Solving circuits with the coupled inductors requires care. Consider, for example, a simple circuit shown in Fig. 12.20 in frequency domain. We cannot apply the impedance relations following from Eq. (12.49). For two coupled inductors, $\mathbf{V}_1 \neq +j\omega L_1 \mathbf{I}_1$ and $\mathbf{V}_2 \neq j\omega L_2 \mathbf{I}_2$. Instead, we should convert Eqs. (12.50) to the phasor form first and obtain

$$\mathbf{V}_1 = j\omega L_1 \mathbf{I}_1 + j\omega M \mathbf{I}_2$$
$$\mathbf{V}_2 = j\omega M \mathbf{I}_1 + j\omega L_2 \mathbf{I}_2 \tag{12.51}$$

These are *two* equations for *four* unknowns $\mathbf{V}_1, \mathbf{I}_1, \mathbf{V}_2, \mathbf{I}_2$. The *two* remaining equations are KVL for the left part of the circuit and KVL for the right part of the circuit, i.e.,

$$\mathbf{V}_1 = +\mathbf{V}_S \quad \text{and} \quad \mathbf{V}_2 = -R_L \mathbf{I}_2 \tag{12.52}$$

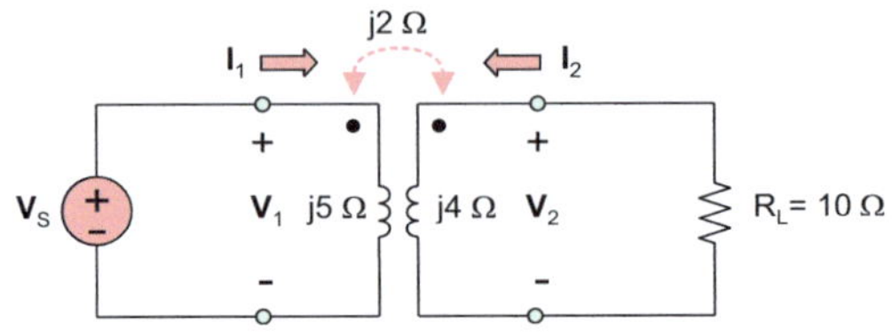

Fig. 12.20. Circuit with two coupled inductors in the phasor form.

Example 12.8: Solve the circuit with two coupled inductors in Fig. 12.20—determine the phasor load current given that $\mathbf{V}_S = 10\angle 0°$ [V].

Solution: First, we substitute into Eqs. (12.51) the expressions for the reactances $j\omega L_1 = j5\ \Omega, j\omega L_2 = j4\ \Omega, j\omega M = j2\ \Omega$ and obtain

$$\begin{aligned}\mathbf{V}_1 &= j5\mathbf{I}_1 + j2\mathbf{I}_2 \text{ [V]}\\ \mathbf{V}_2 &= j2\mathbf{I}_1 + j4\mathbf{I}_2 \text{ [V]}\end{aligned} \tag{12.53}$$

Since, by KVL, $\mathbf{V}_2 = -R_L\mathbf{I}_2 = -10\mathbf{I}_2$, from the second Eq. (12.53), one has

$$\mathbf{I}_1 = (-2 + j5)\mathbf{I}_2 \text{ [A]} \tag{12.54}$$

Since by KVL, $\mathbf{V}_1 = \mathbf{V}_S = 10$ [V], from the first Eq. (12.53), one has

$$10 = (-25 - j8)\mathbf{I}_2 \Rightarrow \mathbf{I}_2 = 0.381\angle 162° \text{ [A]} \tag{12.55}$$

The phasor source current $\mathbf{I}_1$ is given by $\mathbf{I}_1 = 2.052\angle -86°$ [A].

A general solution of the four Eqs. (12.51) and (12.52) is of practical interest. Following the method of Example 12.8, it is obtained in the form:

$$\mathbf{I}_2 = \frac{\mathbf{V}_S}{j\omega\dfrac{M^2 - L_1L_2}{M} - \dfrac{L_1R_L}{M}} \text{ [A]}, \quad \mathbf{I}_1 = -\left(\frac{L_2}{M} + \frac{R_L}{j\omega M}\right)\mathbf{I}_2 \text{ [A]} \tag{12.56}$$

Exercise 12.17: Using Eqs. (12.56) solve the previous example given that $j\omega L_1 = j4\ \Omega$, $j\omega L_2 = j1\ \Omega$, and the mutual inductance is given by $M = \sqrt{L_1L_2}$.

Answer: $\mathbf{I}_1 = 2.513\angle -84°$ [A], $\mathbf{I}_2 = 0.500\angle 180°$ [A], $\mathbf{V}_1 = 10.000\angle 0°$ [V], $\mathbf{V}_2 = 5.000\angle 0°$ [V].

Solution for N Coupled Inductors

The solution method for two coupled inductors may be straightforwardly extended to the case of N coupled inductors. In this case, Eqs. (12.51) will involve a *matrix* of self- and mutual inductances (the *inductance matrix* L) on the size $N \times N$. KVL applied to every individual inductor circuit gives N remaining algebraic equations. The coupled system of $2N$ algebraic equations is then solved for unknowns $\mathbf{I}_1, \ldots, \mathbf{I}_N$ and $\mathbf{V}_1, \ldots, \mathbf{V}_N$. The systems of coupled inductors are used in many applications, mostly for (multiple output) *switched-mode power conversion*. They could also be used for sensor and other purposes in bioelectromagnetics and other disciplines—see Fig. 12.21.

Fig. 12.21. An array of coupled inductors (small coils) located on top of a human-head phantom. This hypothetical setup was tested for applications related to brain stimulation.

Finding Mutual Inductance(s)

Despite the dynamic nature of Eqs. (12.50), the mutual inductance defined previously by Eq. (6.16) is inherently a *static quantity*. It may be computed from the corresponding 3D magnetostatic analysis (often quite complicated). We will briefly review this question at the end of this section. Right now, however, we will establish the *highest* possible value of M, which is achieved for the *ideal transformer*.

Conversion to T-Network

It may be convenient to replace a circuit with two coupled inductors by a circuit *without* magnetic coupling. This can be done using either a T-network or a Π-network of three inductances—see Section 3.3 of Chapter 3. Figure 12.22 illustrates a conversion to the T-network for the circuit from Fig. 12.20. For leftmost inductance L_a, rightmost inductance L_b, and shunt inductance L_c of the T-network, one has

$$L_a = L_1 - M, \quad L_b = L_2 - M, \quad L_c = M \tag{12.57}$$

The proof is based on establishing v–i relationships for both two-port networks in Fig. 12.22 and demonstrating their identity.

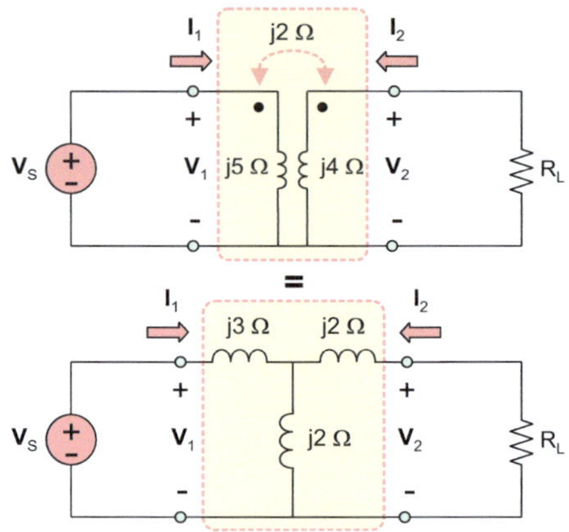

Fig. 12.22. Conversion of two coupled inductors to the T-network of three inductances.

12.5.3 Coupling Coefficient

Energy in a Coupled Circuit

Multiplying Eqs. (12.50) by i_1 and i_2, respectively, and combining them together, one can find expressions for the instantaneous power $p(t)$ and instantaneous energy $E(t)$ stored in the system of two coupled inductors.

$$\left.\begin{aligned} v_1 i_1 &= L_1 i_1 \frac{di_1}{dt} + M i_1 \frac{di_2}{dt} \\ v_2 i_2 &= M i_2 \frac{di_1}{dt} + L_2 i_2 \frac{di_2}{dt} \end{aligned}\right\} \Rightarrow p(t) = v_1 i_1 + v_2 i_2 = \frac{d}{dt}\left\{\frac{1}{2}L_1 i_1^2 + \frac{1}{2}L_2 i_2^2 + M i_1 i_2\right\} \Rightarrow$$
$$E(t) = \int_0^t p\left(t'\right)dt' = \frac{1}{2}L_1 i_1^2 + \frac{1}{2}L_2 i_2^2 + M i_1 i_2 \geq 0 \tag{12.58}$$

To complete the square in the expression for the energy, we both add and subtract the term $\sqrt{L_1 L_2} i_1 i_2$, which gives

$$E(t) = \frac{1}{2}\left(\sqrt{L_1} i_1 + \sqrt{L_2} i_2\right)^2 + \left(M - \sqrt{L_1 L_2}\right) i_1 i_2 \geq 0 \tag{12.59}$$

Mutual Inductance of the Ideal Transformer

For the ideal transformer, the energy stored in the transformer is zero, for *any* values of i_1, i_2. This fact has been proved in Section 12.1. Given arbitrary i_1, i_2, the condition $E(t) = 0$ in Eq. (12.58) is satisfied if and only if,

$$M = \sqrt{L_1 L_2} \quad [\mathrm{H}] \tag{12.60}$$

and

$$\frac{i_1}{i_2} = -\frac{\sqrt{L_2}}{\sqrt{L_1}} = -\frac{N_2}{N_1} \tag{12.61}$$

Equation (12.60) is the definition of the mutual inductance for the ideal transformer. It gives the *largest possible value* of the mutual inductance. Equation (12.61) uses Eq. (6.23) for the coil inductance. We note that this is indeed the ideal-transformer law (12.10), but with the minus sign due to the opposite current direction in Fig. 12.19.

Example 12.9: A system of two coupled inductors is characterized by Eq. (12.60). Is this system equivalent to an ideal transformer?

Solution: We check the voltage relation first. From Eq. (12.50),

$$\frac{v_2}{v_1} = \frac{\sqrt{L_2}\left(\sqrt{L_1}\frac{di_1}{dt} + \sqrt{L_2}\frac{di_2}{dt}\right)}{\sqrt{L_1}\left(\sqrt{L_1}\frac{di_1}{dt} + \sqrt{L_2}\frac{di_2}{dt}\right)} = \sqrt{\frac{L_2}{L_1}} = \frac{N_2}{N_1} \tag{12.62}$$

Thus, the ideal-transformer law for voltages is satisfied. However, the ideal-transformer law for currents is *not satisfied*. An example is given by Eqs. (12.56). In order to obtain the ideal-transformer model, we must additionally assume that

$$L_1 \to \infty, \quad L_2 \to \infty, \quad M \to \infty \tag{12.63}$$

Otherwise, the coupled-inductor model will additionally take into account the magnetizing inductance, which is already a part of the nonideal transformer model studied in the previous section.

Coupling Coefficient

The mutual inductance in the general case is expressed in terms of the two coils' self-inductances L_1 and L_2, in the form

$$M = k\sqrt{L_1 L_2} \quad [\mathrm{H}] \tag{12.64}$$

where k is a so-called *coupling coefficient*, $0 < k \leq 1$. The coupling coefficient determines the amount of total magnetic flux linkage from the first coil shared by the second coil and vice versa. For the ideal transformer, the flux is entirely concentrated within the common magnetic core so that $k = 1$. Figure 12.23 illustrates different values of the coupling coefficient.

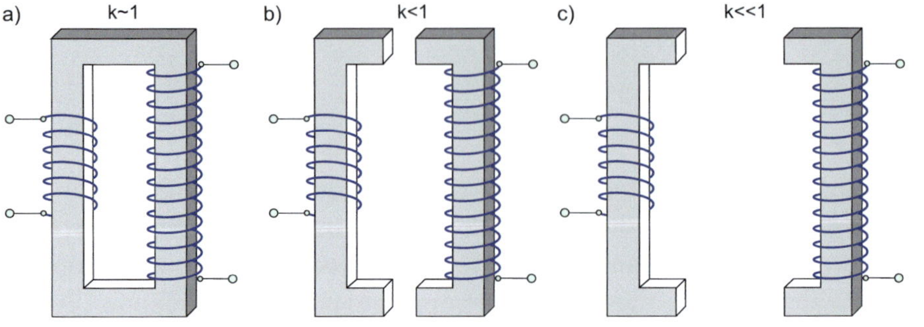

Fig. 12.23. Different configurations and coupling coefficient values for two coupled inductors.

12.5.4 Application Example: Wireless Inductive Power Transfer

Next, we shall consider two arbitrary coils in spatial proximity to each other. Equations (12.50) will always hold for any such coil configuration and for any distance, *d,* between the coils. However, the coupling coefficient, *k*, in Eq. (12.64) has to be determined separately for every particular geometry. The coupling coefficient may now be much less than one. Nonetheless, for any non-zero *k*, the variable current in coil #1 will create a variable voltage in coil #2. In other words, we could create *wireless power or data transfer* between two, or even more than two, inductors. This is due to the very distinct property of an inductor's magnetic near field. We are going to briefly discuss two examples of wireless data and/or power transfer with two coupled inductors. One approach is *radio-frequency identification* (RFID) which employs the use of a patch, known as an RFID tag, that is attached to the person or machinery to be monitored. The RFID technology has created a rapidly growing industry following the first patent issued in 1983 to Charles Walton (US Patent 4,284,288).

Example 12.10: The mutual inductance and the associated magnetic coupling provide a fundamental example of *wireless data transfer* between two physically distant circuits containing two coupled inductors. We should keep in mind that even though this is considered *wireless* coupling, it is quite different from true radio-frequency (RF) wireless data transmission over large distances. Still, many of the modern circuits, including the first two RFID standards in Table 12.3, use magnetic coupling and mutual inductance to enable wireless power transfer or data transfer from the tag to the reader and vice versa.

Example 12.10 (cont.):

Table 12.3. Standard frequencies for RFID tags (RFID TX/RX systems).

Low frequency (LF)	125–134 kHz
High frequency (HF)	13.56 MHz
Ultra high frequency (UHF)	868–930 MHz
Microwave	2.45 GHz
Microwave	5.80 GHz

Circuits with coupled inductors are highlighted.

Example 12.11: The mutual inductance and the associated magnetic coupling also provide a basic example of a *wireless power (not data) transfer* between two physically separated circuits. Let us assume that a power transfer is needed between two circuits, but without mechanical contact. For example, one circuit may be fixed, and another contained in a rotating or indexing machine. Clearly, we cannot use a wired wall plug for such a situation. An important example of near field wireless power transfer is a battery of an implanted device in a human body, which needs to be charged from time to time. A solution to the problem involves a pair of coupled inductors. Figure 12.24 depicts a 120-W inductive power transfer system from Mesa Systems Co., Medfield, MA. This system can be powered by any 12 V battery and also includes DC to AC converters.

Fig. 12.24. A 120-W inductive power-transfer system powered by a 12-V battery from Mesa Systems Co., Medfield, MA.

Basic Model of Inductive Power Transfer at Large Separation Distances

Two coaxial coupled inductors L_1 and L_2 in the wireless link configuration are shown in Fig. 12.25. One of them is the transmitter (TX) and the second is the receiver (RX). The separation distance between the inductors is d. The coupling coefficient k is usually much less than one, and the calculation of the mutual inductance requires care. It is prudent to calculate the mutual inductance of two single *coaxial* loops of current separated by d first. The result has the form

$$M_{\text{loop}} \approx \frac{\pi\mu_0}{2}\frac{r_1^2 r_2^2}{d^3} \quad [\text{H}], \quad d \gg r_1, r_2 \tag{12.65}$$

where r_1, r_2 are the loop radii. Note that d^3 is in the denominator indicating that the mutual inductance decreases very rapidly when the separation distance d increases. This is somewhat discouraging, but an unavoidable result for *magnetic near-field calculations*. The mutual inductance between the two air-core coils having radii r_1, r_2, and N_1 and N_2 turns, respectively, is obtained from Eq. (12.65) in the form

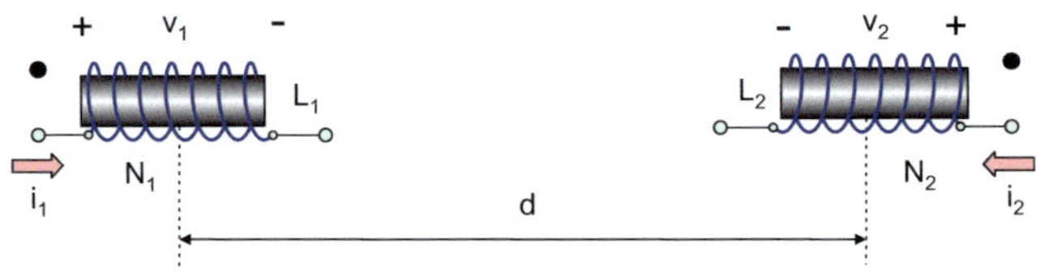

Fig. 12.25. Coupling between two coils in a typical near-field wireless link configuration.

$$M = N_1 N_2 M_{\text{loop}} \approx \frac{N_1 N_2 \pi\mu_0}{2}\frac{r_1^2 r_2^2}{d^3} \tag{12.66}$$

Knowing the mutual inductance, we can find the received current or voltage (and the received power) in the second coil once the transmitted voltage or the current is known in the first coil.

Example 12.12: Two coaxial air-core coils with $r_1 = r_2 = 1.0$ cm and with $N_1 = N_2 = 100$ are separated by 1 m. What is the voltage signal induced in the second coil (RX) if the current in the first coil (TX) is given by $i_1(t) = 100$ mA $\times$ sin (ωt)?

Solution: We find the mutual inductance first, using Eq. (12.67):

$$M = \frac{100 \times 100 \times 4\pi^2 \times 10^{-7}}{2}\frac{10^{-8}}{1} \approx 2.0 \times 10^{-10}\ [\text{H}] \tag{12.67}$$

A voltage induced in coil #2 according to the second equation in Eq. (12.50) is given by

$$v_2 = M\frac{di_1}{dt} = 2.0 \times 10^{-10} \times 10^{-1}\omega \cos(\omega t) \tag{12.68}$$

One can see that the higher the transmission frequency, the greater response we will obtain! Let us assume that the transmission frequency is given by $f = 120$ kHz. Then, $\omega = 2\pi f \approx 0.75 \times 10^6$ rad/s and Eq. (12.68) yields

Example 12.12 (cont.):

$$v_2 = 2.0 \times 0.75 \times 10^{-11} \times 10^6 \cos(\omega t)\ [\mathrm{V}] = 15\ \cos(\omega t)\ [\mu\mathrm{V}] \tag{12.69}$$

Such a weak voltage signal can hardly be observed on an oscilloscope, extracted from noise, and then amplified in a class laboratory setup. The most critical parameter is the coil radius; if it increases by the factor of 2, the received voltage will increase by a factor of 16. We see from this example that the near-field wireless link is tricky: the weak received signal must be carefully managed before conducting any experimentation.

The above analysis implies that the separation distance between the two coils is much greater than any of the coil dimensions. A finite magnetic core is not yet taken into account in Eq. (12.67). The equation is only valid for ceramic-core coils. Also note that the presence of steel conductors (armatures) nearby may greatly *increase* the efficiency of the wireless link. Such an effect is frequently observed in the laboratory. If the receive coil is terminated into a very small resistance, infinite received current and power may be obtained, even if the received voltage is small. Where is the contradiction? The point is that the received current creates its own magnetic field that opposes the changes in the transmitted magnetic field. As long as the received current is small, this is not an issue. The rule of thumb is to request the magnitude of the magnetic field created by the received current to be at *most* 10 % of the transmitted one at the receiver location.

Example 12.13: Compile a MATLAB script that will estimate the voltage signal (voltage amplitude) induced in the second coaxial coil (RX) if the periodic current (current amplitude) in the first coaxial coil (TX) is known. The signal strength (amplitude of the receiver voltage) is to be plotted as a function of the distance between two coils, *d*. We know coil radius, *r*, and number of turns, *N* (the same for both coils).

Solution: The text of the corresponding MATLAB script is given below. It uses Eq. (12.67). Next, it plots the voltage as a function of the separation distance. The plot for the previous example at 1 MHz is shown in Fig. 12.26.

```
mu0     = 4*pi*1e-7;         %   permeability of vacuum (air)
omega   = 2*pi*1e6;          %   angular frequency, rad/s
i1      = 0.1;               %   amplitude of exciting current i1, A
r       = 1e-2;              %   coil radius, m
l       = 0.1;               %   coil length, m
N       = 100;               %   number of turns
d       = [0.1:0.01:2];      %   separation distance, m
M0      = pi*mu0*r^4*N^2./(2*d.^3);  %   mutual inductance
v2      = M0*omega*i1;               %   received voltage, V
semilogy(d, v2*1000); grid on;
xlabel('distance d, m'); ylabel('received voltage, mV')
```

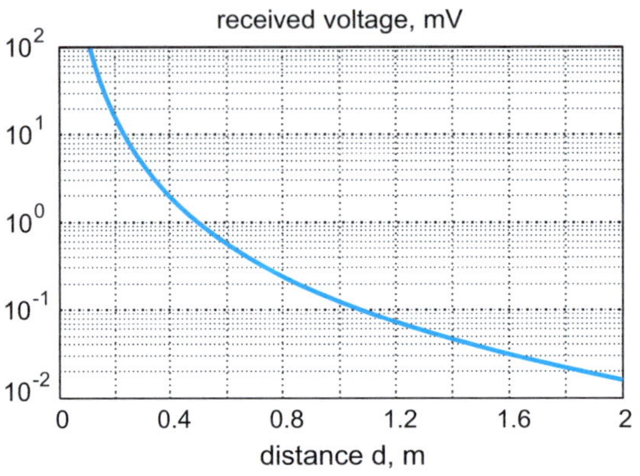

Fig. 12.26. Voltage amplitude in coil #2 when the current amplitude in coil #1 is 100 mA.

12.5.5 Application Example: Coupling of Nearby Magnetic Radiators

The mutual coupling between the two inductors (a transmitter and a receiver) is key for the wireless inductive power transfer. At the same time, the mutual coupling may have a *negative* effect when the transmitter includes two or more independently driven coils assembled in a *coil array*. The mutual coupling between the individual transmit coils may reduce the individual coil current, i.e., reduce the resulting total magnetic field.

Circuit with Two Identical Radiators

The circuit shown in Fig. 12.27 formalizes the problem. It models the coupling effect between two nearby identical magnetic radiators. This circuit is important for near-field wireless power transfer with coil arrays including medical applications. Our goal is to express the source phasor current $\mathbf{I}_S$ through the circuit parameters for two distinct cases:

A. Mutual coupling is absent.
B. Mutual coupling is present; the mutual reactance is $X_M > 0$.

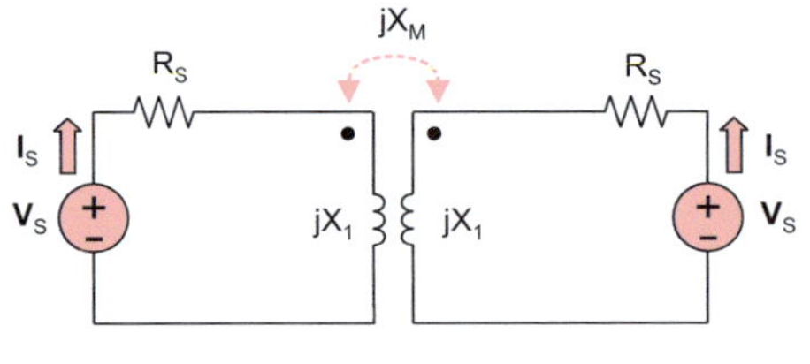

Fig. 12.27. Modeling mutual coupling of two transmit coils.

To solve the circuit, we convert the two coupled inductors to a T-network. The resulting circuit is shown in Fig. 12.28.

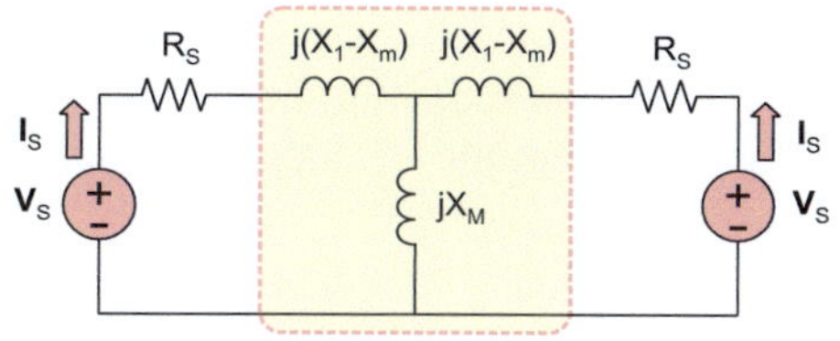

Fig. 12.28. Equivalent circuit with uncoupled inductors.

When the mutual coupling is absent, the central inductor becomes a wire; both currents are given by

$$\mathbf{I}_S = \frac{\mathbf{V}_S}{R_S + jX_1} \tag{12.70}$$

When the mutual coupling is present, we can use KVL for the either circuit loop, which gives

$$-\mathbf{V}_S + R_S\mathbf{I}_S + j(X_1 - X_M)\mathbf{I}_S + 2jX_M\mathbf{I}_S = 0 \tag{12.71}$$

Therefore,

$$\mathbf{I}_S = \frac{\mathbf{V}_S}{R_S + j(X_1 + X_M)} \tag{12.72}$$

Note that the sign in front of X_M may vary depending on coil orientation. For example, it is positive for two coaxial coils (or loops) with in-phase currents (or in-phase fluxes) and is negative otherwise.

Tuned Radiators

The individual radiator circuit should be tuned to the operating frequency to maximize the circuit current/magnetic field. The tuning is typically achieved by a series capacitor with reactance $-X_1$, which exactly cancels the inductor's reactance $+X_1$. If this is the case, Eq. (12.72) is transformed to

$$\mathbf{I}_S = \frac{\mathbf{V}_S}{R_S + jX_M} \Rightarrow |\mathbf{I}_S| = \frac{|\mathbf{V}_S|}{\sqrt{R_S^2 + (X_M)^2}} \tag{12.73}$$

Thus, for the tuned radiators, the mutual coupling reduces the circuit current and the magnetic field, *irrespective of* the coil orientation. When X_M is small compared to the source resistance, this effect is of little value. Altogether, it can be eliminated by adjusting the value of the tuning capacitors.

Exercise 12.18: Determine the relative reduction in the emanating magnetic field at any point in space for two tuned identical coupled radiators when $R_S = 10\ \Omega$ and $X_M = 1\ \Omega$.

Answer: 0.5 %.

Exercise 12.19: Which reactance should the tuning capacitors have in order to eliminate the effect of the mutual inductance in Fig. 12.28?

Answer: $-X_1 - X_M$.

Summary

Ideal-transformer model		
Concept	Circuit	Formulas/Meaning
Faraday's law 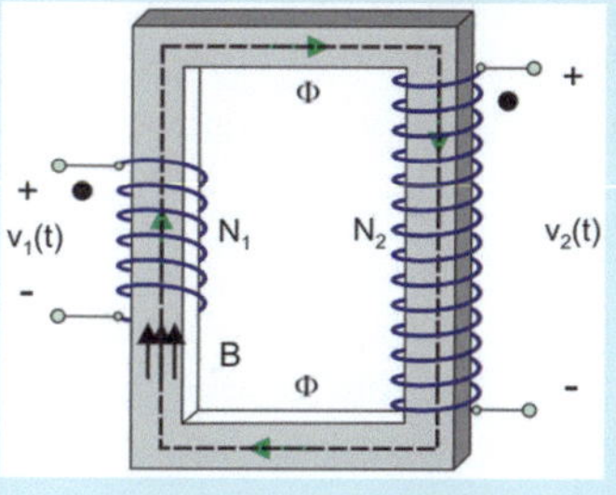	Dependent voltage source in the secondary winding 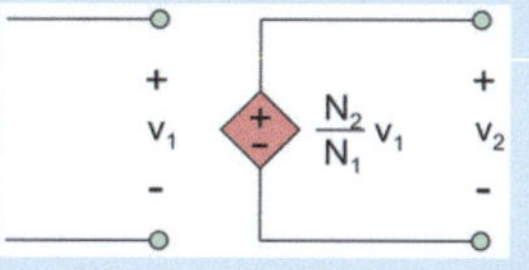	Primary winding: $v_1(t) = N_1 \frac{d\Phi(t)}{dt}$ Secondary winding: $v_2(t) = N_2 \frac{d\Phi(t)}{dt}$ Conclusion: $\frac{v_2}{v_1} = \frac{N_2}{N_1}$
Ampere's law 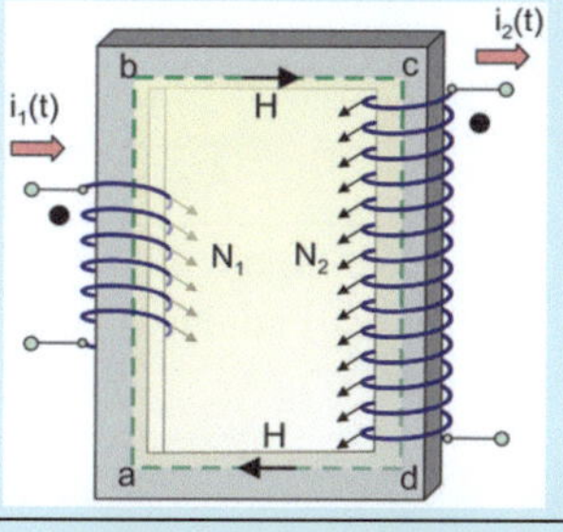	Dependent current source in the primary winding 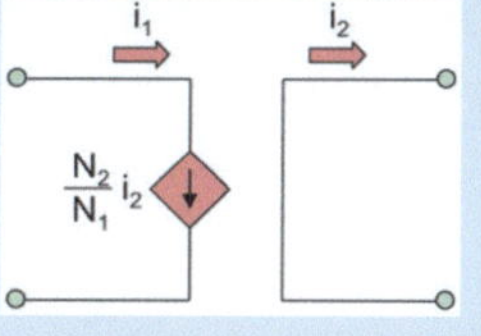	$l = l_{ab} + l_{bc} + l_{cd} + l_{da}$ General form: $lH = i$ Ideal core: $0 = i = N_1 i_1 - N_2 i_2$ Conclusion: $\frac{i_1}{i_2} = \frac{N_2}{N_1}$
Ideal-transformer model with two dependent sources 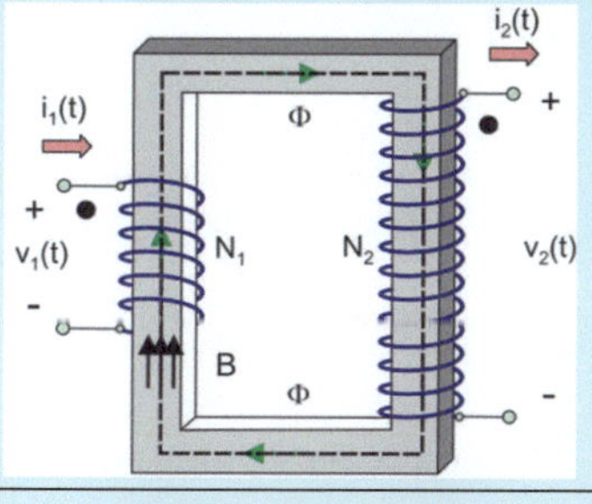	Equivalent circuit model of the ideal transformer 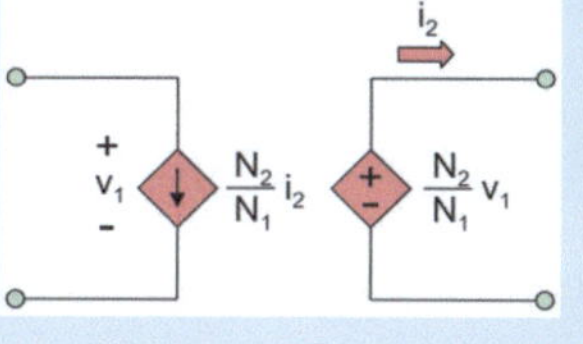	Ideal-transformer equations: $\frac{v_2}{v_1} = \frac{N_2}{N_1}$ $\frac{i_1}{i_2} = \frac{N_2}{N_1}$ Power conservation: $v_1 i_1 = v_2 i_2$ Stored energy: 0
Simple transformer circuit 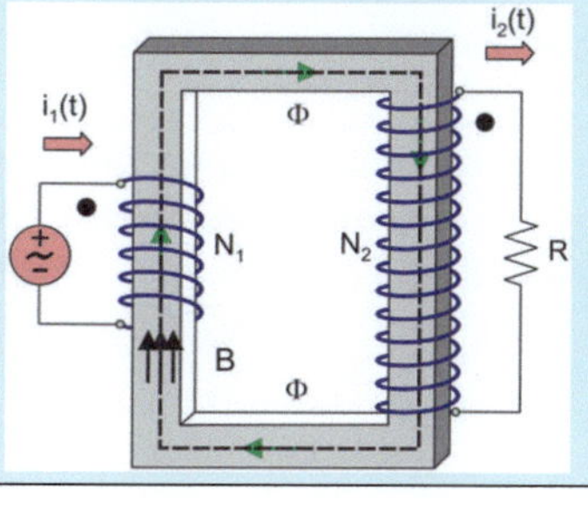	Circuit with transformer symbol 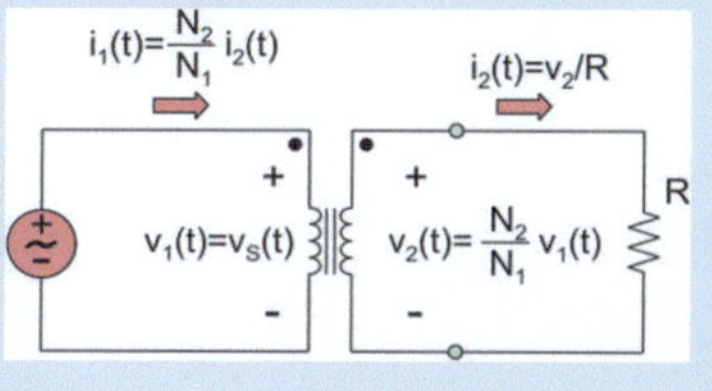	Load voltage: $v_R = \frac{N_2}{N_1} v_S$ Load current: $i_R = \frac{v_R}{R}$ Source power: $p_S(t) = \left(\frac{N_2}{N_1}\right)^2 \frac{v_S^2(t)}{R}$

(continued)

Terminology	
Turns ratio	$N_1 : N_2 = a$
Step up transformer	$N_2 > N_1$, $a < 1$ Secondary winding is the high-voltage side
Step down transformer	$N_2 < N_1$, $a > 1$ Primary winding is the high-voltage side
Ratings	Apparent load power and Primary voltage (rms) **:** Secondary voltage (rms)

<table>
<tr><th colspan="3">Some useful transformers</th></tr>
<tr>
<td>

Autotransformer

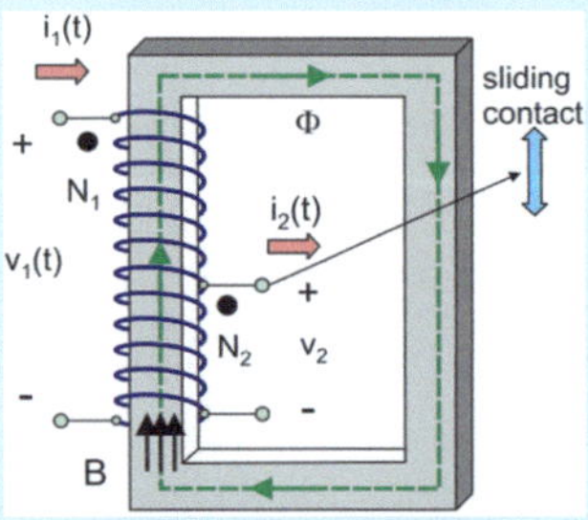

</td>
<td>

Circuit symbol

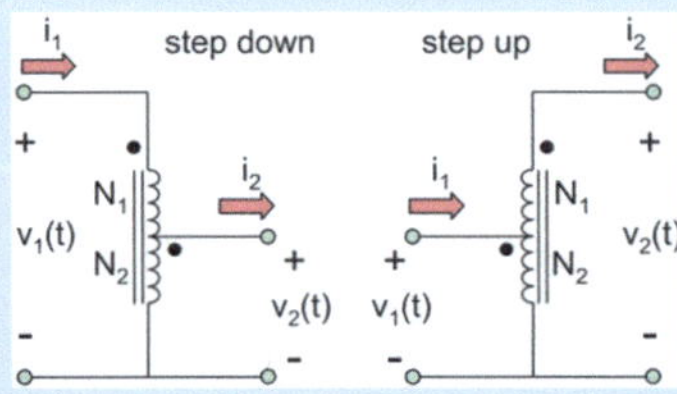

</td>
<td>

Step down setup:
Voltage relation:
(from Faraday's law)

$$v_2 = \frac{N_2}{N_1 + N_2} v_1$$

Current relation:
(from Ampere's law)

$$i_2 = \frac{N_1 + N_2}{N_2} i_1$$

</td>
</tr>
<tr>
<td>

Multiwinding transformer

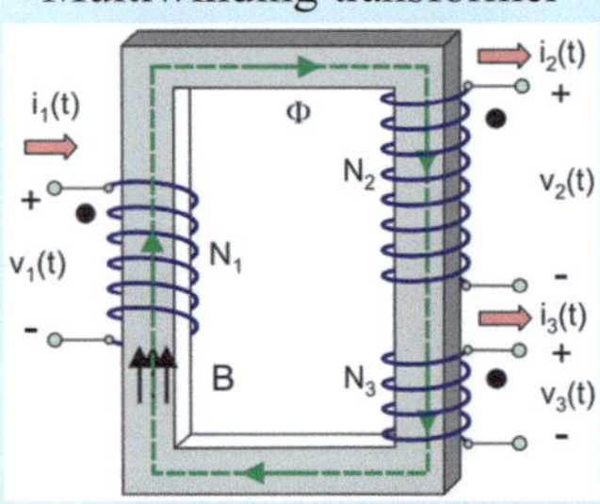

</td>
<td>

Circuit symbol

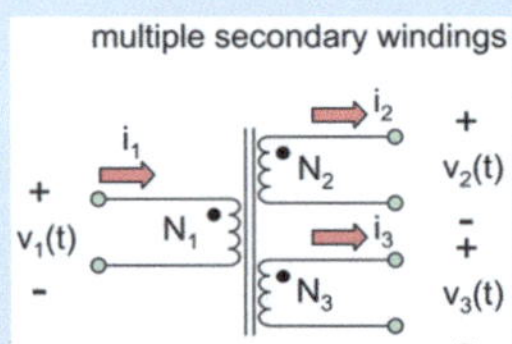

</td>
<td>

Voltage relation:
(from Faraday's law)

$$\frac{v_1}{N_1} = \frac{v_2}{N_2} = \frac{v_3}{N_3}$$

Current relation:
(from Ampere's law)

$$N_1 i_1 = N_2 i_2 + N_3 i_3$$

</td>
</tr>
<tr>
<td>

Center-tapped transformer

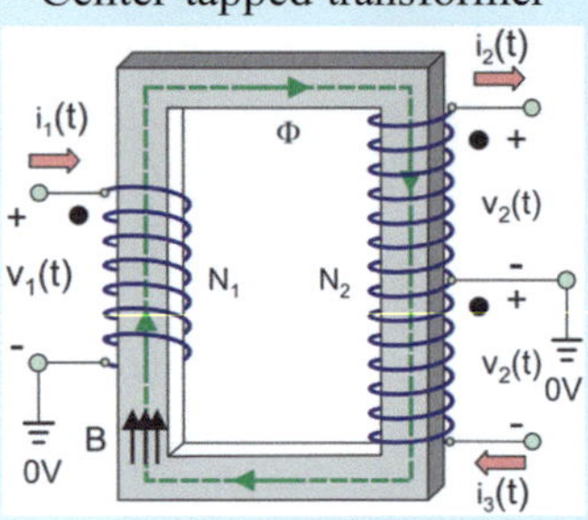

</td>
<td>

Circuit symbol

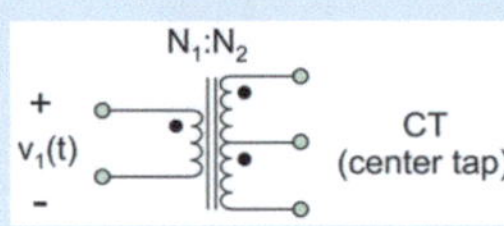

</td>
<td>

Voltage relation:
(from Faraday's law)

$$v_2 = \frac{N_2}{2N_1} v_1$$

Current relation:
(from Ampere's law for two identical loads)

$$\frac{N_1}{N_2} i_1 = i_2 = i_3$$

</td>
</tr>
<tr>
<td>

Current transformer

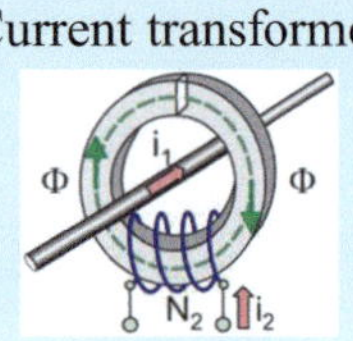

</td>
<td>

Circuit symbol

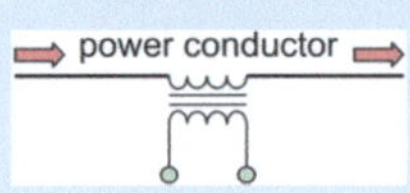

</td>
<td>

Instrumentation transformer
(to measure current)

$$i_2 = \frac{1}{N_2} i_1$$

</td>
</tr>
</table>

(continued)

Phasor form of a transformer circuit	
Particular types of dot convention	
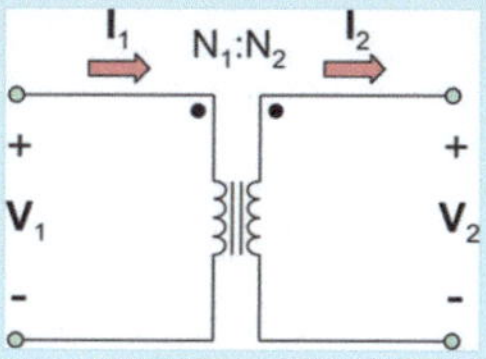	$\frac{\mathbf{V}_2}{\mathbf{V}_1}=+\frac{N_2}{N_1},\quad \frac{\mathbf{I}_2}{\mathbf{I}_1}=+\frac{N_1}{N_2}$
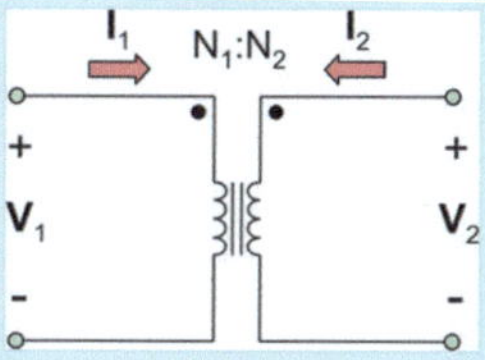	$\frac{\mathbf{V}_2}{\mathbf{V}_1}=+\frac{N_2}{N_1},\quad \frac{\mathbf{I}_2}{\mathbf{I}_1}=-\frac{N_1}{N_2}$
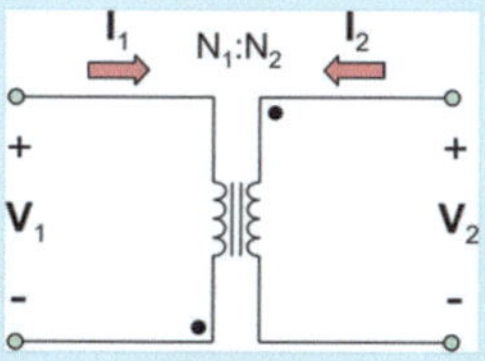	$\frac{\mathbf{V}_2}{\mathbf{V}_1}=-\frac{N_2}{N_1},\quad \frac{\mathbf{I}_2}{\mathbf{I}_1}=+\frac{N_1}{N_2}$
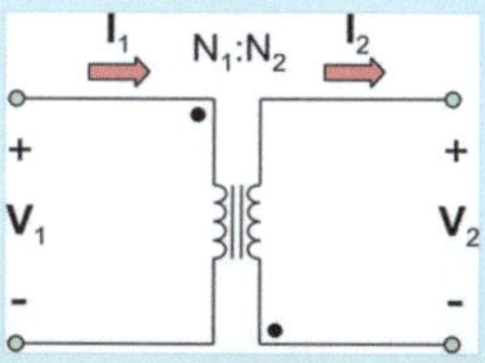	$\frac{\mathbf{V}_2}{\mathbf{V}_1}=-\frac{N_2}{N_1},\quad \frac{\mathbf{I}_2}{\mathbf{I}_1}=-\frac{N_1}{N_2}$
Reflecting the source to the secondary side	
Transformer circuit	Circuit equivalent (load voltage is the same)
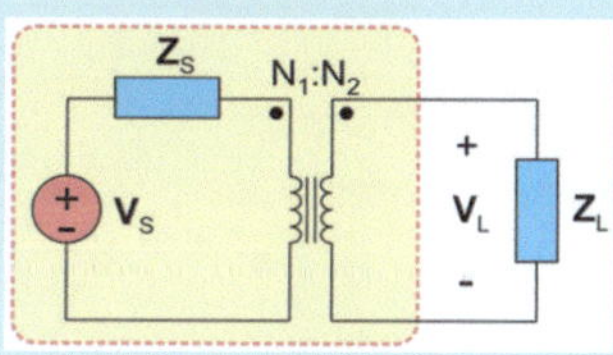	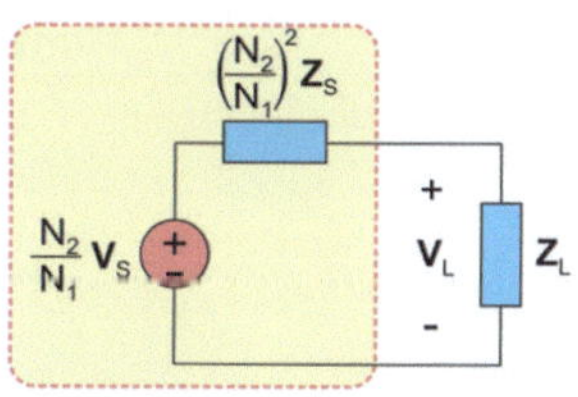
Reflecting the load to the primary side	
Transformer circuit	Circuit equivalent (source current is the same)
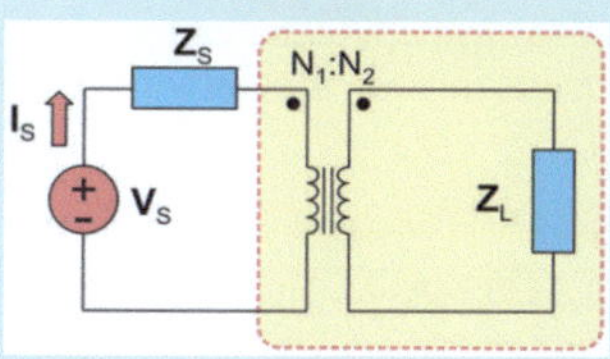	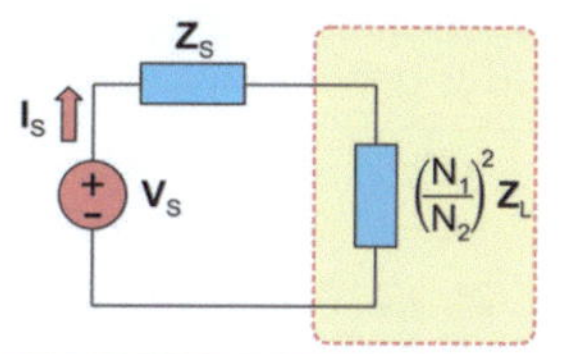

(continued)

Transformer-based impedance matching	
Exact matching with real impedances	
Transformer circuit 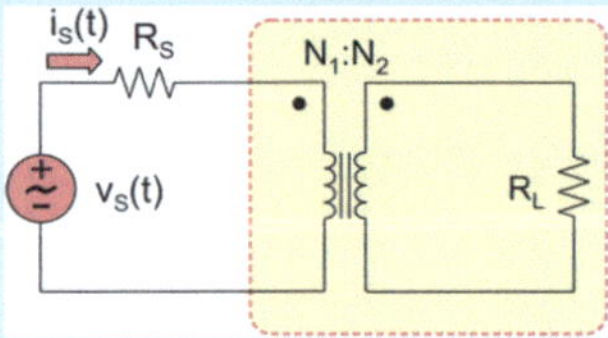	Circuit equivalent with $N_1 : N_2 = \sqrt{R_S/R_L}$ 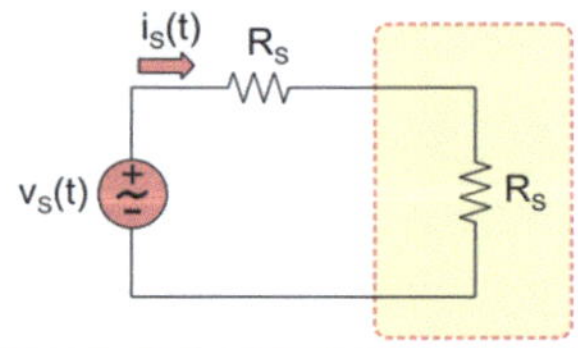
Approximate matching for maximum power transfer from source to load	
Transformer circuit 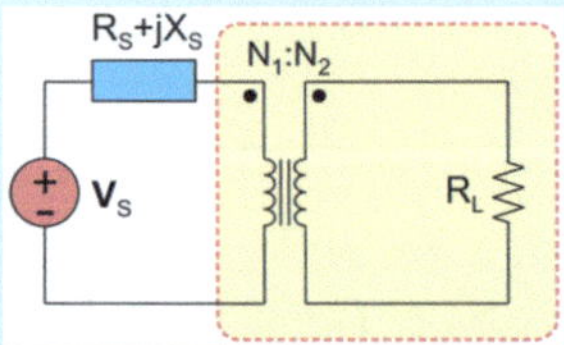	$N_1 : N_2 = \sqrt[4]{(R_S^2 + X_S^2)/R_L^2}$ 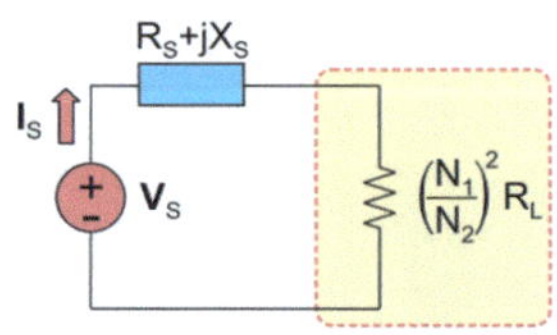
Non-ideal low-frequency transformer model (Steinmetz model)	
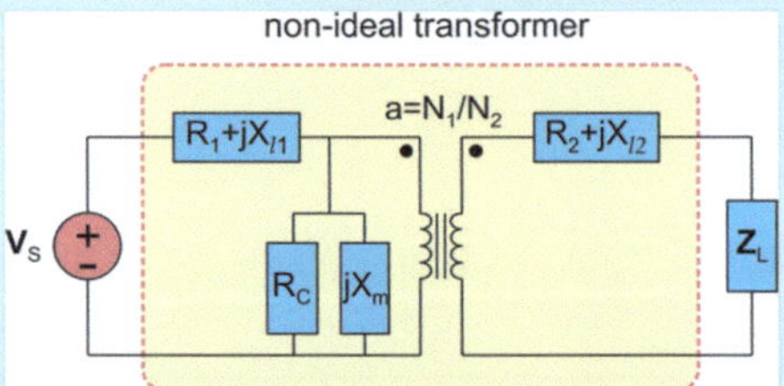	*Primary winding and magnetic core:* Magnetizing inductance L_m ($X_m = \omega L_m$); Core loss equivalent resistanceR_c; Leakage inductance L_{l1} ($X_{l1} = \omega L_{l1}$); Ohmic resistanceR_1; *Secondary winding*: Leakage inductance L_{l2} ($X_{l2} = \omega L_{l2}$); Ohmic resistance R_1
Model of coupled inductors	
Circuit symbol 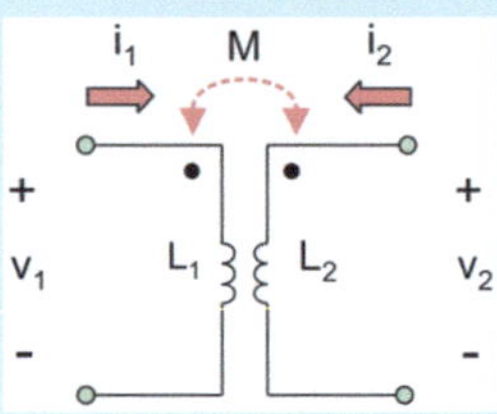	Physical counterpart 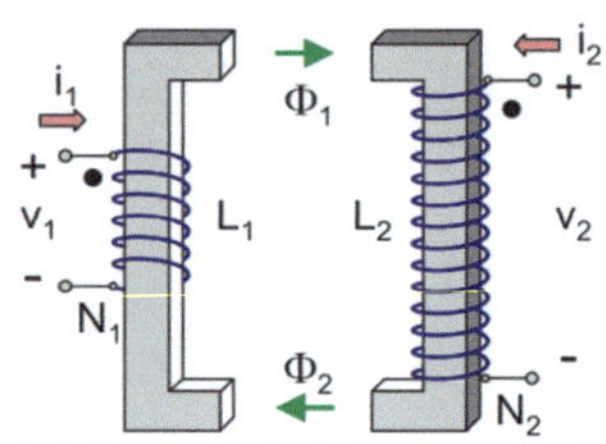
Reduction to ideal-transformer model 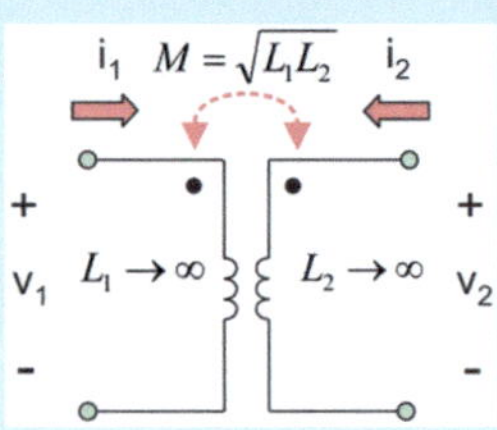	Ideal-transformer model 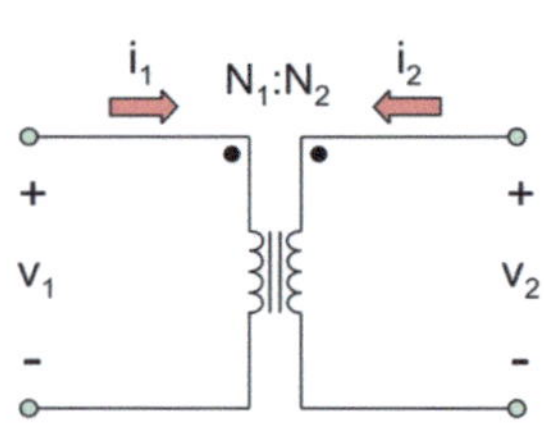

(continued)

<table>
<tr><td>

Phasor form

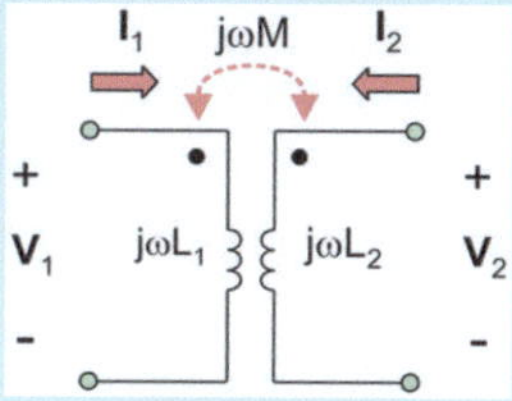

</td><td>

Constitutive relations:

$$\mathbf{V}_1 = j\omega L_1 \mathbf{I}_1 + j\omega M \mathbf{I}_2$$
$$\mathbf{V}_2 = j\omega M \mathbf{I}_1 + j\omega L_2 \mathbf{I}_2$$

Mutual inductance:

$$M = k\sqrt{L_1 L_2}$$

Dimensionless coupling coefficient:

$$0 \le k \le 1$$

</td></tr>
<tr><td>

Simple circuit

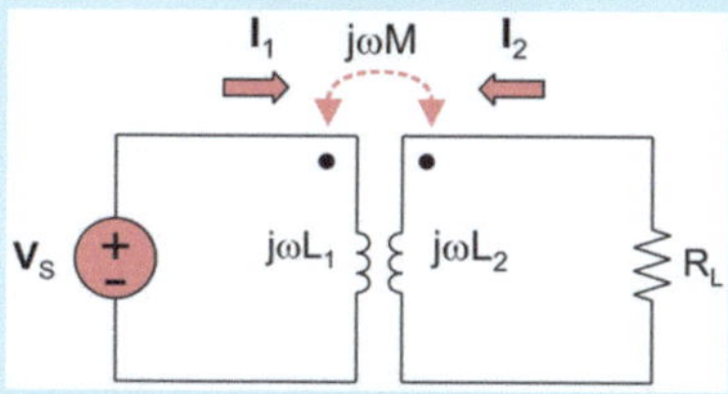

</td><td>

To solve the circuit use the constitutive relations and KVL for each loop. Solution:

$$\mathbf{I}_2 = \frac{\mathbf{V}_S}{j\omega \dfrac{M^2 - L_1 L_2}{M} - \dfrac{L_1 R_L}{M}} \text{ [A]}$$

$$\mathbf{I}_1 = -\left(\frac{L_2}{M} + \frac{R_L}{j\omega M}\right) \mathbf{I}_2 \text{ [A]}$$

</td></tr>
<tr><td>

Conversion to a T-network without coupling

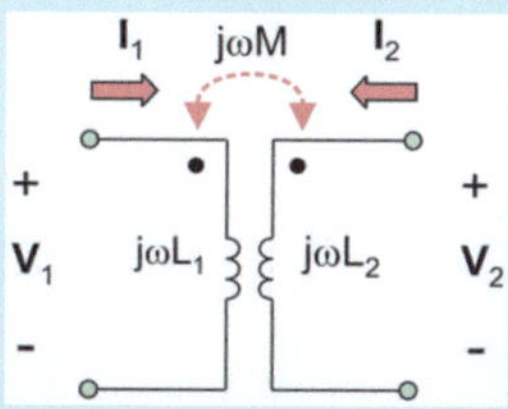

</td><td>

Equivalent T-network of uncoupled inductances

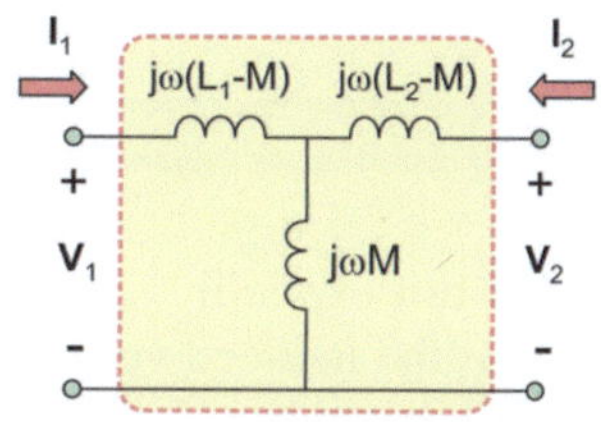

</td></tr>
<tr><td>

Mutual inductance for two short ceramic coaxial coils

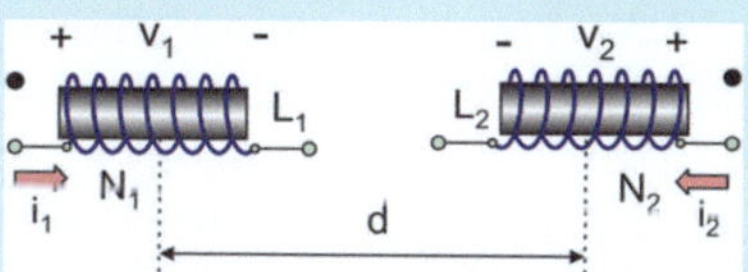

</td><td>

$$M \approx \frac{N_1 N_2 \pi \mu_0}{2} \frac{r_1^2 r_2^2}{d^3}$$

r_1, r_2 are the coil radii;

$d >> r_1, r_2$ is the coil separation distance

</td></tr>
<tr><td>

Coupling between two nearby near-field magnetic radiators

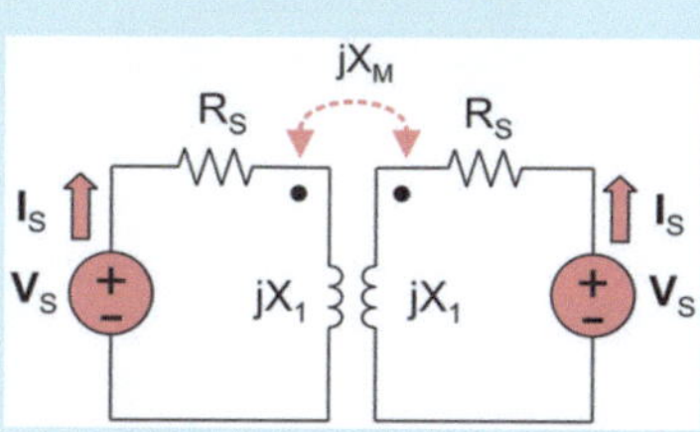

</td><td>

Mutual coupling between two radiators generally reduces the amplitude of the source current in every radiator: $\mathbf{I}_S = \dfrac{\mathbf{V}_S}{R_S + j(X_1 + X_M)}$ and the associated magnetic field for both of them

</td></tr>
</table>

Problems

12.1 Ideal Transformer as a Linear Passive Circuit Element

12.1.2 Ideal Open-Circuited Transformer: Faraday's Law of Induction

Problem 12.1. The following figure shows two open-circuited coils subject to a time-varying flux, $\Phi(t)$, in the core.

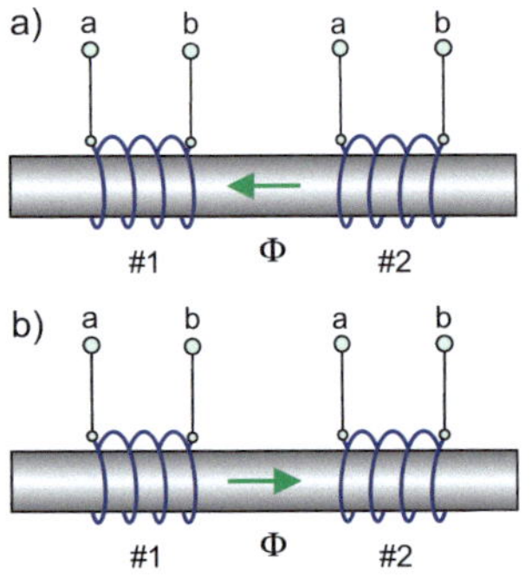

Faraday's law does not explicitly say anything about the polarity of the induced voltage in the coils. Nevertheless, the induced voltage does have a preferred polarity determined by *Lenz's law* (Heinrich Friedrich Emil Lenz (1804–1865), a Russian physicist of German ethnicity, taught at the University of St. Petersburg). Lenz's Law states that the circuit current *due to induced voltage* v produces a flux in such a direction as to *oppose* the change of the flux. In other words, the induced current always seeks to maintain the *status quo* of the magnetic field. To find the direction of the anticipated circuit current, you may imagine a resistance connected across the coil terminals. Now, assume that the magnetic flux in the core, Φ, is increasing in time. Label the polarity of the induced voltages for coils #1 and #2, respectively, by $\pm$ and show the direction of the anticipated circuit current.

Problem 12.2. Are the following figures correct?

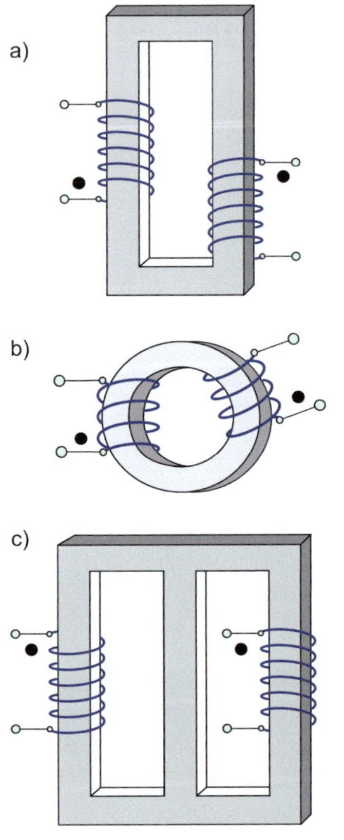

Problem 12.3. For the circuit shown in the following figure, determine voltage $v_2(t)$ if $v(t) = 100\cos\omega t$ [V] using Faraday's law of induction. Count the number of turns. Assume no flux leakage.

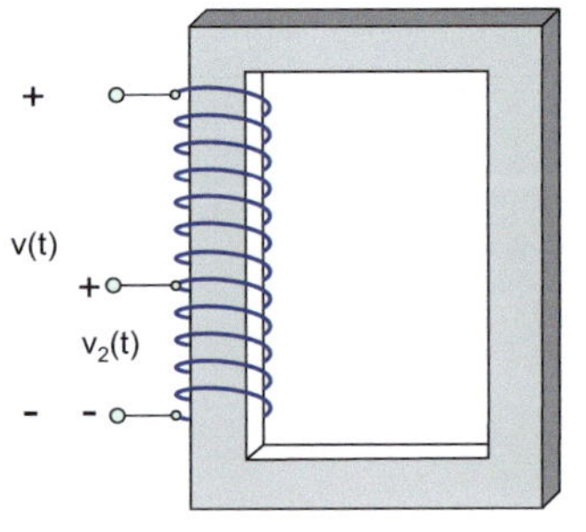

Problem 12.4. For the circuit shown in the figure below, determine voltage $v_2(t)$ if $v(t) = 70\cos\omega t$ [V] using Faraday's law of induction. Count the number of turns. Assume no flux leakage.

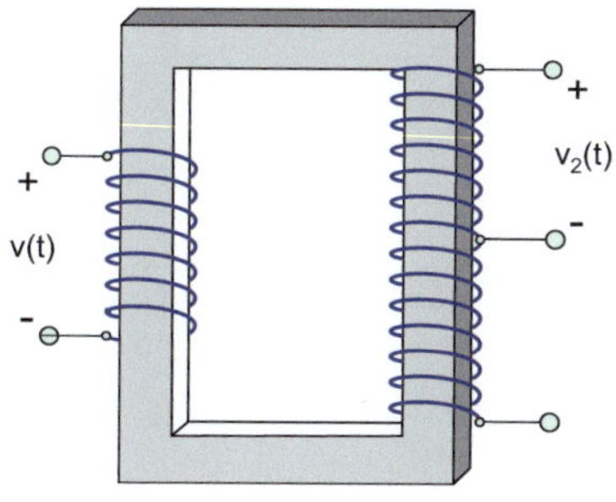

Problem 12.5. Explain the meaning of the ideal magnetic core. What is the inductance value for an inductor which uses the closed-loop ideal magnetic core?

Problem 12.6. You are given the source voltage in the form $v_S(t) = 340\cos(2\pi 60t)$ [V], the number of turns of primary winding $N_1 = 250$, the finite relative permeability of the magnetic core, $\mu_r = 7000$, the coil length of the primary winding of 9 cm, and the core cross section of $A = 0.0015$ m^2.

A. Find the exciting current $i_\Phi(t)$ (no-load current) in the primary winding of this nonideal transformer.
B. Reduce N_1 by the factor of 10 and repeat the solution.

12.1.4 Ampere's Law

12.1.5 Ideal Loaded Transformer

12.1.6 Ideal Transformer Versus Realistic Transformer: Transformer Terminology

Problem 12.7. For the circuit shown in the following figure, establish the relation between currents $i_1(t)$ and $i_2(t)$. Assume the ideal magnetic core. Count the number of turns.

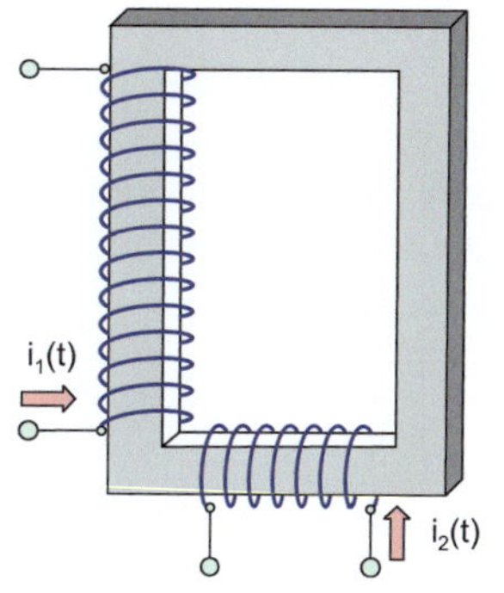

Problem 12.8. For the circuit shown in the following figure, establish the relation between currents $i_1(t)$, $i_2(t)$, and $i_3(t)$. Assume an ideal magnetic core. Count the number of turns.

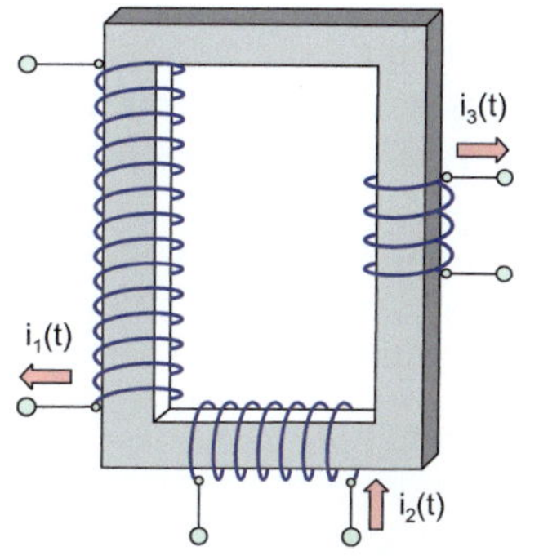

Problem 12.9. For the circuit shown in the figure below, determine current $i_2(t)$ given that $i_1(t) = 1\cos\omega t$ [A]. Count the number of turns.

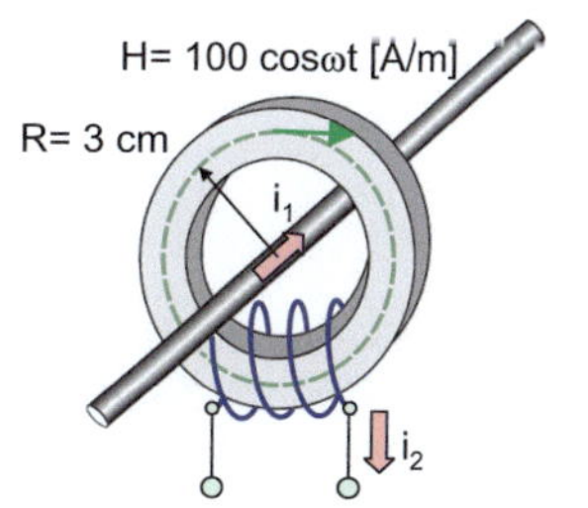

Problem 12.10. For the circuit shown in the figure below, establish the relation between currents $i_1(t)$, $i_2(t)$, and $i_3(t)$. Assume an ideal magnetic core. Count the number of turns. *Hint*: Ampere's law for the ideal core applies to every closed path.

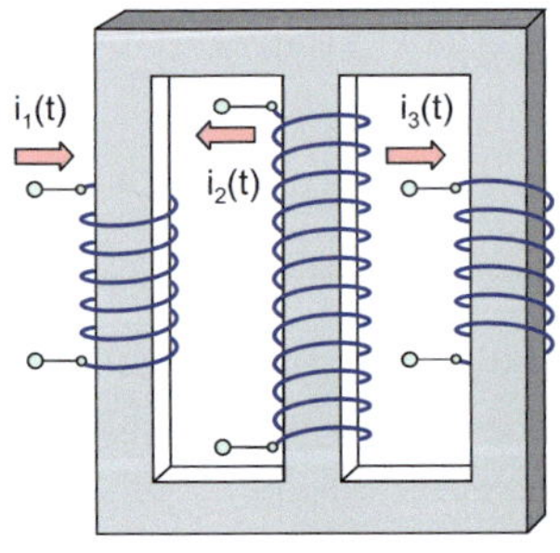

Problem 12.11. In Problem 12.6, determine the instantaneous stored energy of a nonideal transformer given a core centerline length of 30 cm.

Problem 12.12. In the circuit shown in the figure below, $v_S(t) = 170\cos(2\pi 60t)$ [V]. For a 100 Ω resistive load, determine:

- Source current $i_S(t)$
- Load voltage $v_R(t)$
- Load current $i_R(t)$
- Average power delivered to the load P

when the turns ratio is equal to 10:1, 1:1, and 1:10.

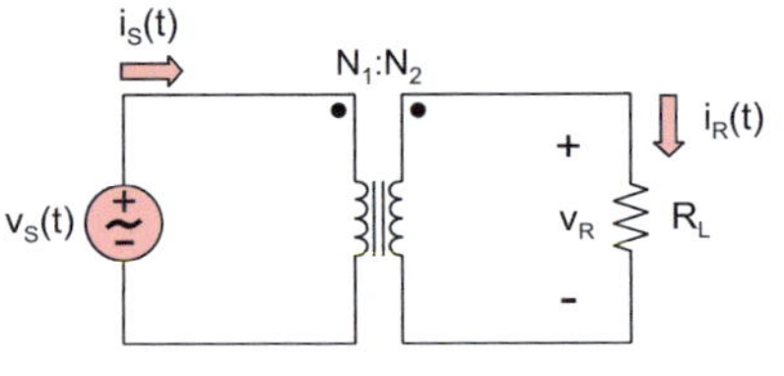

Problem 12.13. In the circuit shown in the figure below, $v_S(t) = 325\cos(2\pi\omega t)$ [V]. The ideal transformer model is used. Determine source current $i_S(t)$, load voltage $v_R(t)$, and load current $i_R(t)$ when:

A. The switch is open.
B. The switch is closed.
C. The switch is closed and $\omega \to 0$.

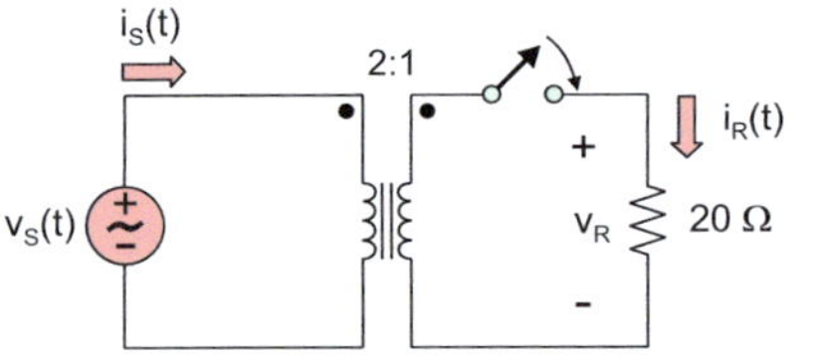

Problem. 12.14. In the circuit shown in the figure below, the current through resistance is given by $i_R(t) = 50\cos(2\pi 60t)$ [A]. Determine:

- Source voltage $v_S(t)$
- Source current $i_S(t)$

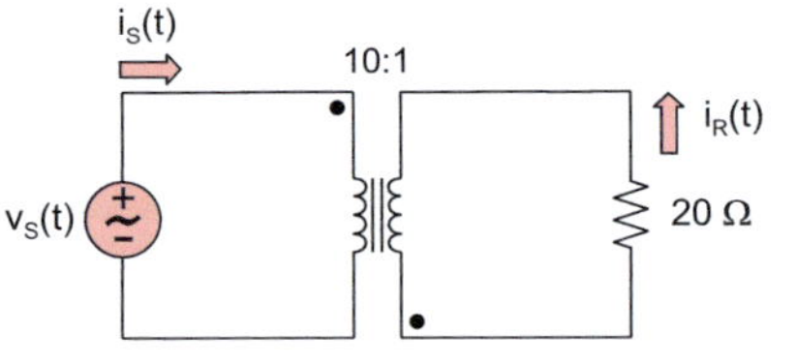

Problem 12.15. A power transformer is rated as 100 kVA, 11,000:2200 V, 60 Hz. Determine:

- Transformer type
- Turns ratio
- The rated current on the low-voltage side

Problem 12.16. Repeat the previous problem for a transformer rated as 10 kVA, 2300:230 V, 60 Hz.

Problem 12.17. Repeat Problem 12.15 for a transformer rated as 2 kVA, 230:115 V, 50 Hz.

12.2 Analysis of Ideal-Transformer Circuits

12.2.1 Circuit with a Transformer in the Phasor Form

12.2.2 Referred (or Reflected) Source Network in the Secondary Side

12.2.3 Referred (or Reflected) Load Impedance in the Primary Side

Problem 12.18. The load impedance is $12.5 - j2.5\ \Omega$. Find the equivalent impedance of the load $\mathbf{Z}_T$ combined with a 2:1 step-down transformer in the primary side.

Problem 12.19. Source voltage is given by $v_S(t) = 2.5\cos\omega t + 45°$ [V] and the source impedance is $5 - j0.5\ \Omega$. Find the equivalent circuit of the source (find $\mathbf{V}_T$ and $\mathbf{Z}_T$) combined with a 1:4 step-up transformer in the secondary side. Express your result both in frequency domain and in time domain.

Problem 12.20. A large capacitance value of 1.6 mF is required. The available physical component is a 100-μF capacitor. You are given an ideal transformer with an arbitrary turns ratio. Design the equivalent circuit for the 1.6 mF capacitance, specify the transformer turns ratio, and draw the corresponding circuit diagram.

Problem 12.21. For the circuit shown in the figure below, determine transformer's turns ratio.

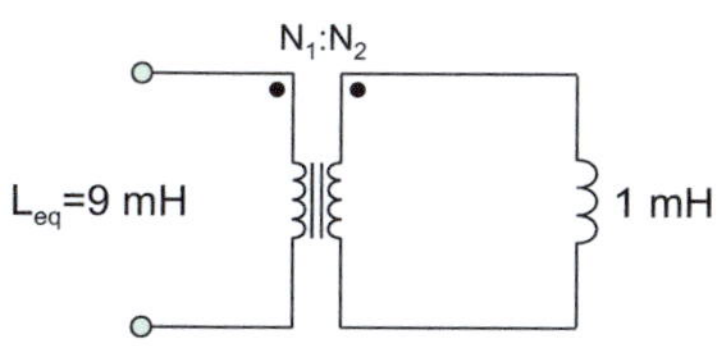

Problem 12.22. For the circuit shown in the following figure, $\mathbf{V}_S = 10\angle 45°$ [V], $\mathbf{Z}_S = 5 + j5\ \Omega$, $\mathbf{Z}_L = 5 + j5\ \Omega$, and $N_1 : N_2 = 1 : 2$. Find phasor voltage across the load, $\mathbf{V}_L$. Express your result in polar form. Assume the ideal transformer.

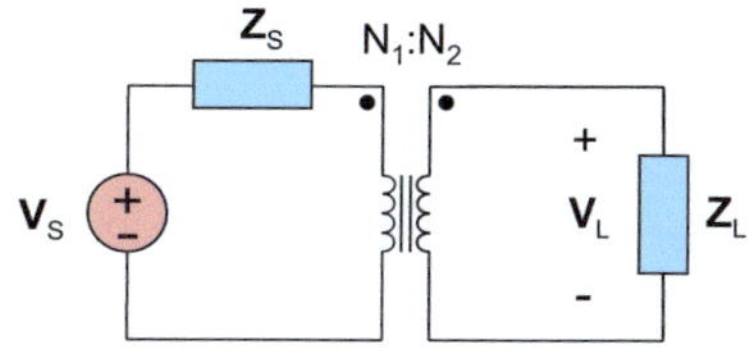

Problem 12.23. In the circuit shown in the figure below, $v_S(t) = 2.5\cos(\omega t + 45°)$ [V] where $\omega = 1000$ rad/s, $C = 10$ μF, $R_1 = 25\ \Omega$, $R_L = 200\ \Omega$. Furthermore, $N_1 : N_2 = 1 : 2$. Find voltage across the load $v_L(t)$ in time domain. Assume the ideal transformer.

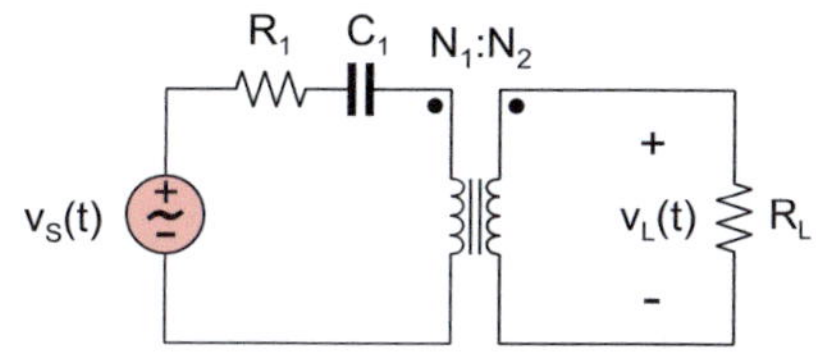

Problem 12.24. For the circuit shown in the following figure, $\mathbf{V}_S = 10\angle 45°$ [V], $\mathbf{Z}_S = 5 + j5\ \Omega$, $\mathbf{Z}_L = 5 + j5\ \Omega$, and $N_1 : N_2 = 1 : 2$. Find phasor current of the source, $\mathbf{I}_S$. Express your result in polar form. Assume the ideal transformer.

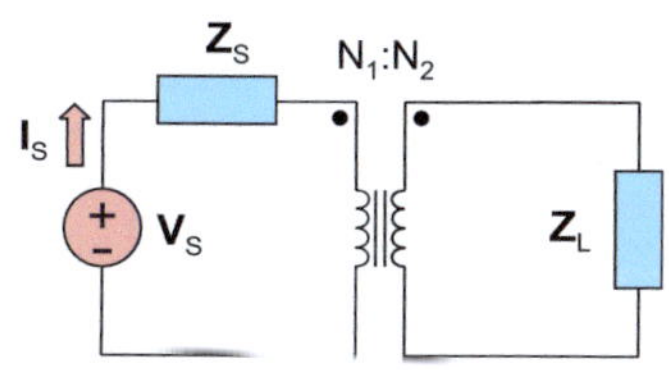

Problem 12.25. In the circuit shown in the following figure, $v_S(t) = 2.5\cos(\omega t + 45°)$ [V] where $\omega = 2000$ rad/s, $C = 10$ μF, $R_1 = 50\ \Omega$, $R_L = 20\ \Omega$. Furthermore, $N_1 : N_2 = 2 : 1$. Find source current $i_S(t)$ in time domain. Assume the ideal transformer.

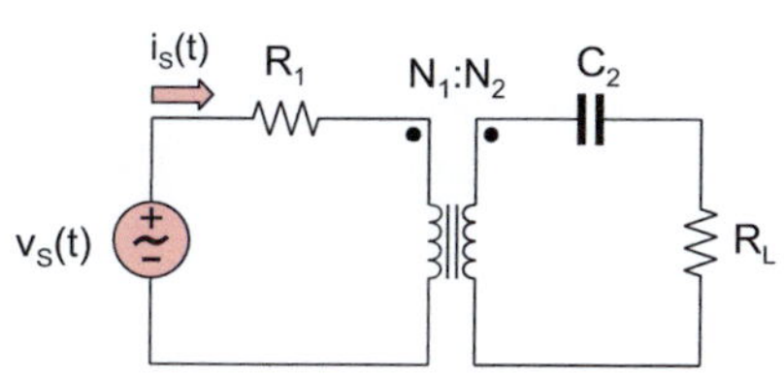

12.2.4 Transformer as a Matching Circuit

Problem 12.26. A 16-Ω load is connected to an AC voltage source with a voltage amplitude of 10 V and series resistance of 1 Ω. A matching transformer is used with the turns ratio $N_1 : N_2$. Determine the average power delivered to the load when

A. $N_1 : N_2 = 1 : 1$

B. $N_1 : N_2 = 1 : 2$

C. $N_1 : N_2 = 1 : 4$

D. $N_1 : N_2 = 1 : 5$

Problem 12.27. A passive RFID tag circuit measuring temperature is modeled by a load resistance of 16 Ω. The tag is wirelessly powered; it is augmented with a small collecting antenna, which has the radiation resistance of $1 - j2$ Ω. The negative antenna reactance is an inherent part of the small dipole antenna design. Find the ratio of the average received powers from the antenna with and without the matching transformer. Assume the load matching condition for the real part of the impedances.

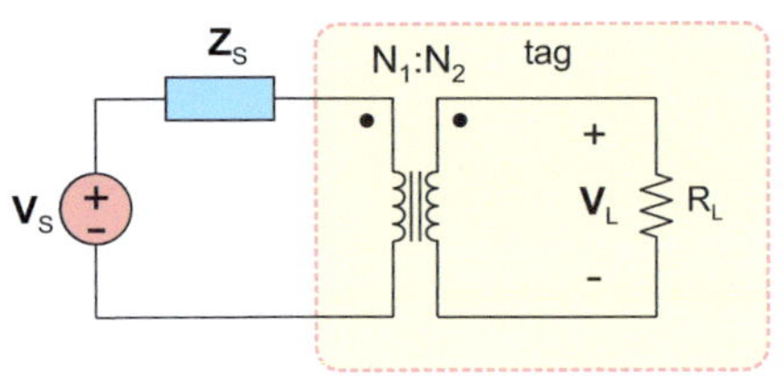

Problem 12.28. The previous problem may be generalized as follows. An antenna operates as an energy-harvesting source. The antenna is modeled as an impedance $\mathbf{Z}_S = R_S + jX_S$ in series with a fixed ideal voltage source. The nonzero antenna reactance is an inherent part of the small antenna design. The load is modeled as a resistance R_L. Express the turns ratio of a matching transformer, which is necessary for maximum average power transfer from the antenna to the load, in terms of three given problem parameters. Note that this problem has an elegant analytical solution.

12.2.5 Application Example: Electric Power Transfer via Transformers

Problem 12.29. The performance of a transmission line circuit in the following figure is to be analyzed, with and without 1:10 step-up transformer and 10:1 step-down transformer, respectively. All phasors are given in terms of the rms values. Solve the circuit with and without transformers and fill out the table that follows including active load power $P_L = \mathrm{Re}(\mathbf{V}_L \cdot \mathbf{I}_L{}^*)$, active source power, $P_S = \mathrm{Re}(\mathbf{V}_S \cdot \mathbf{I}_S{}^*)$, and the line power loss $P_{loss} = R|\mathbf{I}_{line}|^2$. All powers are to be reported in watts. You are given $\mathbf{Z}_L = 4 + j2$ Ω.

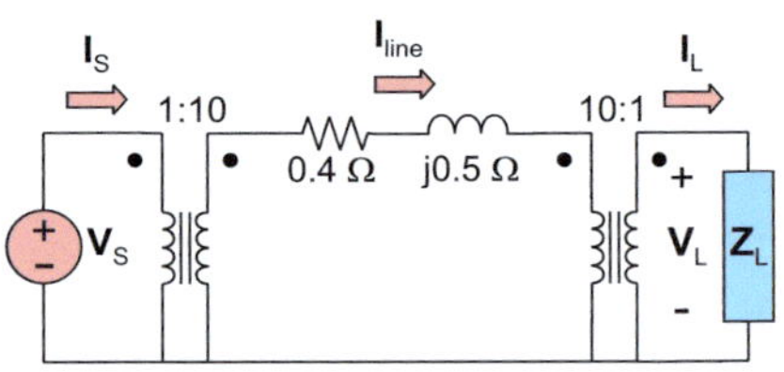

Param.	$\mathbf{V}_S$	$\mathbf{I}_S$	P_S	$\mathbf{I}_{line}$
No tr.	$240\angle 0°$			
W tr.	$240\angle 0°$			
Param.	P_{loss}	$\mathbf{V}_L$	$\mathbf{I}_L$	P_L
No tr.				
W tr.				

Problem 12.30. Repeat the previous problem when the transformer setup changes to a 1:5 step-up transformer and a 5:1 step-down transformer.

Problem 12.31. An AC-direct micro-hydro-power system is illustrated in the following figure.

Reprinted from *Micro-Hydropower Systems* Canada 2004, ISBN 0-662-35880-5

The system uses a single phase induction generator with the rms voltage of 240 V. The system serves four small houses, each connected to the generator via a *separate* transmission line with the same length of 3000 m. Each line uses AWG#10 solid aluminum wire with a diameter of 2.59 mm. The house load in every house is an electric range with the resistance of 20 Ω. Determine total active power delivered by the generator, P_S, total power loss in the transmission lines, P_{loss}, and total active useful power, P_L:

1. When no transformers are used;
2. When a 1:5 step-up transformer is used in powerhouse and a 5:1 step-down transformer is used at home.

Problem 12.32. Solve the previous problem when the distributed line inductance is additionally taken into account. The inductance per unit length of a two-wire line is given by $\frac{\mu_0}{\pi}\left(\ln\frac{d}{a}+\frac{1}{4}\right)$ where a is the wire radius and d is the separation distance. Assume the separation distance of 1 m. The operation frequency is 50 Hz.

12.3 Some useful transformers

12.3.1 Autotransformer

Problem 12.33. A voltage source connected to the primary winding of an ideal step-up autotransformer shown in the following figure has the form $v_S(t) = 5\cos\omega t$ [V]. The source current into the dotted terminal is $i_S(t) = 10\cos\omega t$ [A]. You are given $N_1 = 200$, $N_2 = 50$. Determine voltage $v_2(t)$ and current $i_2(t)$ in the secondary.

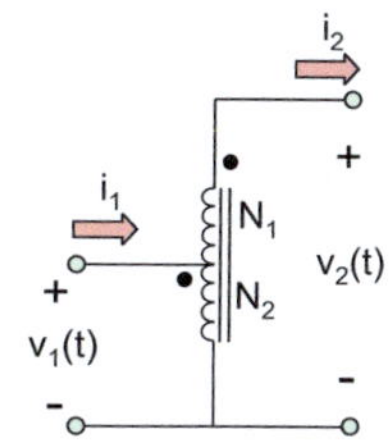

Problem 12.34. Solve the previous problem for the circuit shown in the following figure.

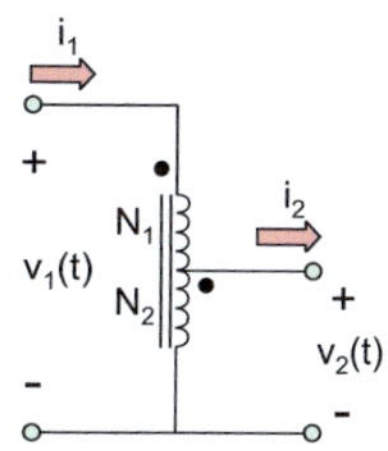

Problem 12.35. For the circuit shown in the following figure, $\mathbf{V}_S = 100\angle 45°$ [V], $\mathbf{Z}_S = 5 + j5\ \Omega$, $\mathbf{Z}_L = 5 + j5\ \Omega$, and $N_1 : N_2 = 4 : 1$. Find phasor current of the source, $\mathbf{I}_S$. Express your result in polar form. Assume the ideal autotransformer.

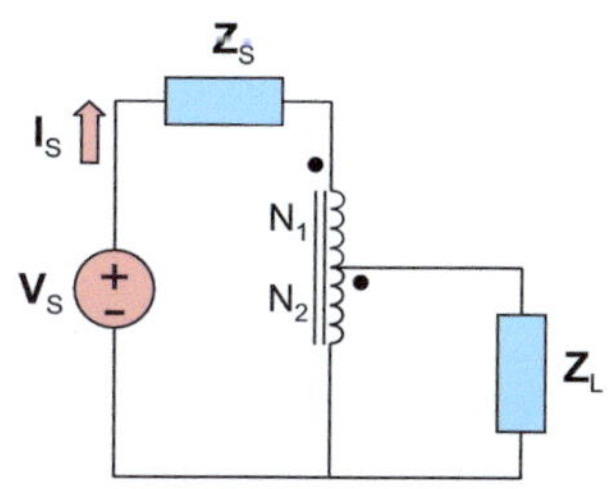

Problem 12.36. Find the equivalent input impedance, $\mathbf{Z}_{in}$, for two autotransformer circuits shown in the following figure.

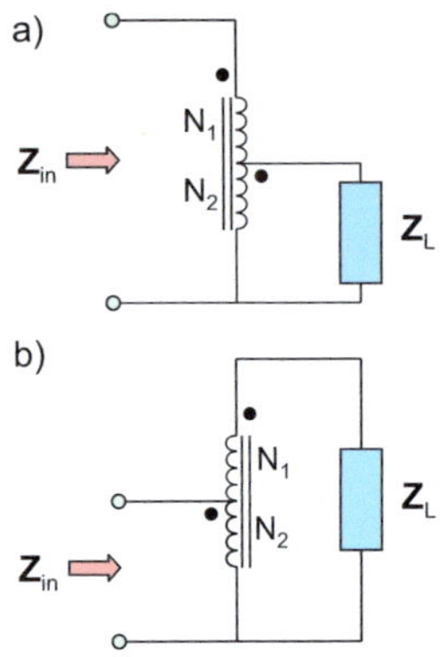

12.3.2 Multiwinding Transformer

12.3.3 Center-Tapped Transformer: Single-Ended to Differential Transformation

Problem 12.37. In the circuit shown in the following figure, $N_1 = 2N_2 = N_3$. How is the instantaneous power partitioned between the two secondary windings if both of them are terminated into the same load resistances? To answer this question, express both $p_2(t)$ and $p_3(t)$ in terms of $p_1(t)$.

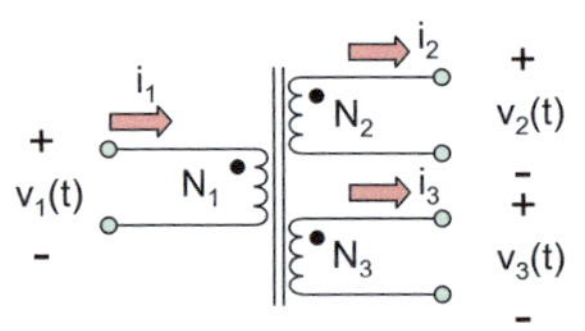

Problem 12.38. In the circuit of the previous problem, $N_1 = 2N_2 = 2N_3$. How is the instantaneous power partitioned between the two secondary windings if winding #2 is terminated into resistance R and winding #3 is terminated into resistance $3R$?

Problem 12.39. Determine the turns ratio for the center-tapped transformer in Fig. 12.13.

Problem 12.40. A household is using an ideal center-tapped distribution transformer shown in the following figure. All voltage values are the rms values. The resistive loads include a TV, a microwave, and a kitchen range. The powers for every individual load are shown in the figure. Determine:

A. Turns ratio, $N_1 : N_2$ of the transformer
B. rms value of input current i_1

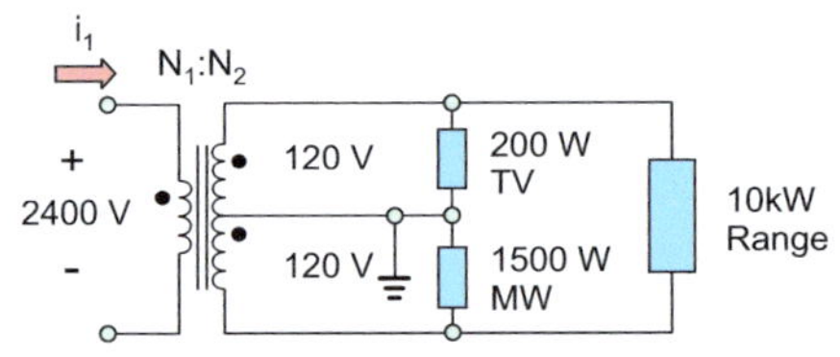

Problem 12.41. Determine phasor currents $\mathbf{I}_{R1}$, $\mathbf{I}_{R2}$, and $\mathbf{I}_{RO}$ for the center-tapped balanced transformer circuit shown in the following figure. You are given the source phasor voltage $\mathbf{V}_S = 10\angle 0^\circ$ [V], resistance values $R = 50\ \Omega$, and turns ratio $N_1 : N_2 = 1$.

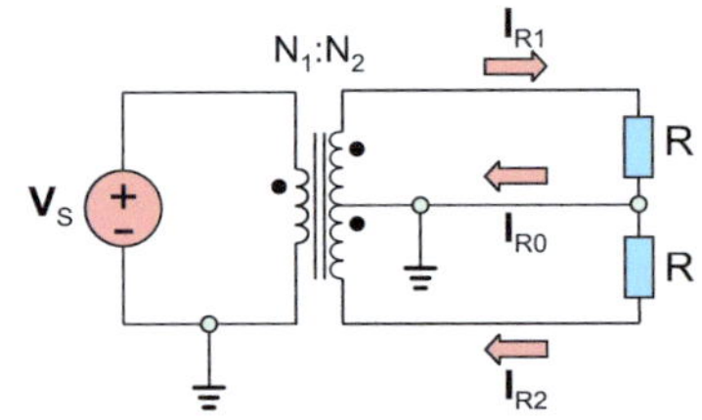

Problem 12.42. Solve the previous problem in a general form, i.e., express phasor currents $\mathbf{I}_{R1}$, $\mathbf{I}_{R2}$, and $\mathbf{I}_{RO}$ in the (generally unbalanced) circuit shown in the following figure in terms of given circuit parameters $\mathbf{V}_S$, $\mathbf{Z}_1$, $\mathbf{Z}_2$, and turns ratio $a = N_1 : N_2$.

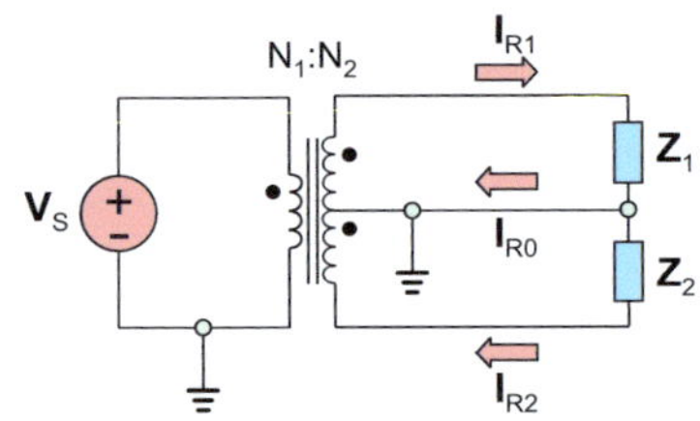

12.3.4 Current Transformer

Problem 12.43. Determine current $i_2(t)$ in an ideal current transformer shown in the figure given that $i_1(t) = 10\cos\omega t$ [A]. Count the number of turns.

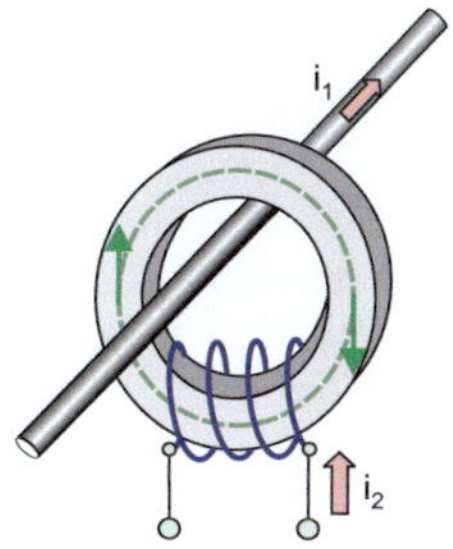

12.4 Real-Transformer Model

12.4.1 Model of a Nonideal Low-Frequency Transformer

12.4.2 Model Parameters and Their Extraction

12.4.3 Analysis of Nonideal Transformer Model

12.4.4 Voltage Regulation and Transformer Efficiency

Problem 12.44

A. Name all seven components of the low-frequency nonideal transformer model.

B. Draw the corresponding circuit diagram.

Problem 12.45. A practical power transformer is characterized by the following nameplate information: **20 kVA 2400:240 V 60 Hz.**

A. Determine rated-load phasor voltage $\mathbf{V}_L$ (show units and assign phase zero).

B. Determine rated-load phasor current $\mathbf{I}_L$ for power factor of 0.8 lagging (show units).

C. Determine rated-load active power P_L (show units).

Problem 12.46. Repeat the previous problem for the power transformer characterized by **2 kVA 230:115 V 50 Hz.**

Problem 12.47. A practical power transformer is characterized by the following information:

Element nameplate	**20 kVA 2400:240 V 60 Hz**
Magnetizing reactance $X_m = \omega L_m$, Ω	15,000
Core loss resistance, R_c, Ω	100,000
Primary leakage reactance $X_{l1} = \omega L_{l1}$, Ω	6.5
Secondary leakage reactance, $X_{l12} = \omega L_{l12}$, Ω	0.07
Primary ohmic resistance R_1, Ω	3.0
Secondary ohmic resistance R_2, Ω	0.03

Check whether or not equalities (12.38) for the well-designed transformer are satisfied.

Problem 12.48. In a well-designed transformer, the number of turns of both primary and secondary windings is usually quite large. Why is it so? Can we just design a 10:1 transformer with 10 and 1 turns, respectively?

Problem 12.49. The nonideal transformer model is shown in the following figure in frequency domain. Given the rated load with a power factor of 0.8 lagging, determine the source phasor voltage $\mathbf{V}_S$ and the source phasor current $\mathbf{I}_S$ for transformer #1 model from Table 12.2.

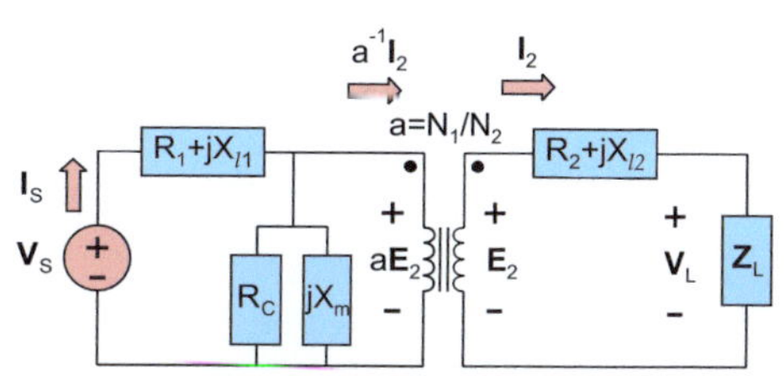

Problem 12.50. Repeat the previous problem for transformer #2 model from Table 12.2.

Problem 12.51. Repeat Problem 12.49 for transformer #3 model from Table 12.2.

Problem 12.52. The nonideal transformer model is shown in the following figure in frequency domain. Given the rated load with a power factor of 0.7 lagging, determine:

A. Percentage regulation
B. Percentage efficiency

for transformer #1 model from Table 12.2

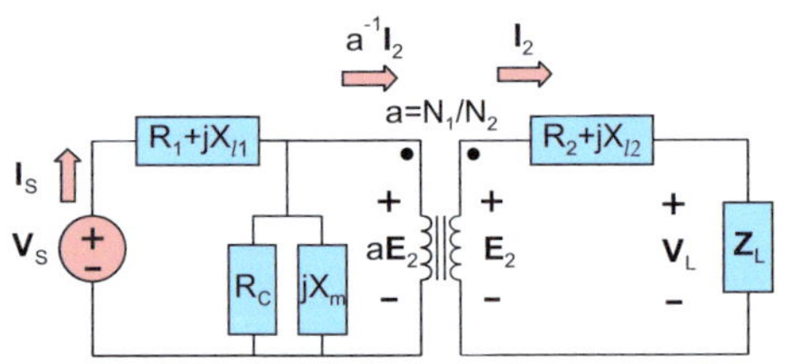

Problem 12.53. Repeat the previous problem for transformer #2 model from Table 12.2.

Problem 12.54. Repeat Problem 12.52 for transformer #3 model from Table 12.2.

12.5 Model of Coupled Inductors

12.5.1 Model of Two Coupled Inductors

12.5.2 Analysis of Circuits with Coupled Inductors

12.5.3 Coupling Coefficient

Problem 12.55. Solve a circuit with two coupled inductors in the figure below in frequency domain:

A. Determine the load phasor current and the source phasor current.
B. Determine phasor voltages $\mathbf{V}_1$ and $\mathbf{V}_2$

given that $\mathbf{V}_S = 10\angle 0^\circ$ [V] and $R_L = 10\ \Omega$.

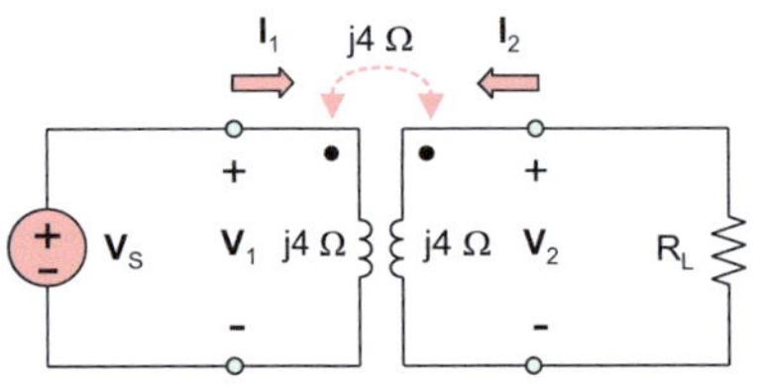

Problem 12.56. Solve the previous problem when the mutual inductance is exactly zero.

Problem 12.57. Solve Problem 12.55 when the mutual reactance is equal to 3 Ω.

Problem 12.58. In the circuit shown in the following figure, determine source phasor current $\mathbf{I}_S$ given that $R_L = 1\ \Omega$, $R_S = 2\ \Omega$, and $\mathbf{V}_S = 15\angle 0^\circ$ [V].

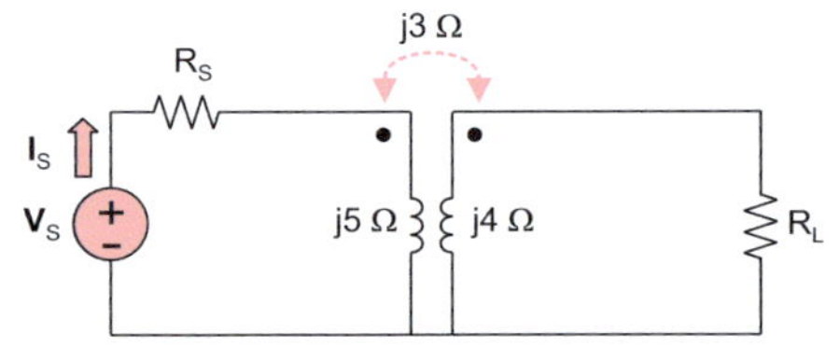

12.5.4 Application Example: Wireless Inductive Power Transfer

12.5.5 Application Example: Coupling of Nearby Magnetic Radiators

Problem 12.59. Two small coaxial ceramic-core coils with $r_1 = r_2 = 2.0$ cm and with $N_1 = N_2 = 100$ are separated by 1 m. What is the voltage signal induced in the second coil (RX) if the current in the first coil (TX) is given by $i_1 = 100\ \text{mA} \times \sin(\omega t)$? The operation frequency is 1 MHz.

Problem 12.60. Repeat the previous problem when the operation frequency changes to 10 MHz.

Problem 12.61. In the circuit shown in the following figure, find source phasor current $\mathbf{I}_S$ in time domain $i_S(t)$ given that $v_S(t) = 10\cos\omega t$ [V].

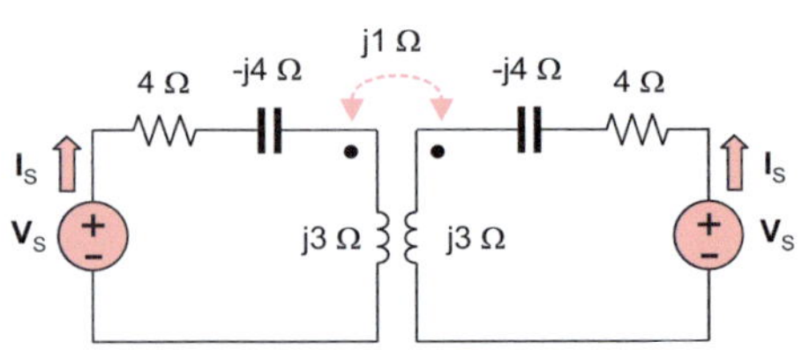

Index

S. N. Makarov et al., *Practical Electrical Engineering*, https://doi.org/10.1007/978-3-319-96692-2

K

L